W9-CRY-512

SENSORY SYSTEMS NEUROSCIENCE

This is Volume 25 in the

FISH PHYSIOLOGY series
Edited by Anthony P. Farrell and Colin J. Brauner
Honorary Editor: William S. Hoar and David J. Randall

A complete list of books in this series appears at the end of the volume

SENSORY SYSTEMS NEUROSCIENCE

Edited by

TOSHIAKI J. HARA
Department of Fisheries and Oceans
Freshwater Institute, Winnipeg
Manitoba, Canada
and
Department of Zoology
University of Manitoba
Winnipeg, Manitoba
Canada

BARBARA S. ZIELINSKI
Department of Biological Sciences
University of Windsor, Windsor
Ontario, Canada

AMSTERDAM • BOSTON • HEIDELBERG • LONDON
NEW YORK • OXFORD • PARIS • SAN DIEGO
SAN FRANCISCO • SINGAPORE • SYDNEY • TOKYO
Academic Press is an imprint of Elsevier

Front Cover Photograph: The image of the Sockeye Salmon is reproduced courtesy of Walker Weber/National Geographic Image Collection ©. The background for the cover is an immunofluorescence image of neuroepithelial cells and associated innervations of the zebrafish gills viewed by confocal microscopy (modified adaptation from Chapter 3, Fig. 6 (Jonz and Nurse, 2003) with permission).

Academic Press is an imprint of Elsevier
525 B Street, Suite 1900, San Diego, California 92101-4495, USA
84 Theobald's Road, London WC1X 8RR, UK

This book is printed on acid-free paper. ∞

For information on all Elsevier Academic Press publications
visit our Web site at www.books.elsevier.com

ISBN-13: 978-0-12-350449-4
ISBN-10: 0-12-350449-X

PRINTED IN THE UNITED STATES OF AMERICA
07 08 09 10 9 8 7 6 5 4 3 2 1

CONTENTS

CONTRIBUTORS ix

PREFACE xi

1. Olfaction
Barbara S. Zielinski and Toshiaki J. Hara

1. Introduction 1
2. Olfactory Repertoire 4
3. An Evolutionary Assessment of the Function of the Nasal Cavity 6
4. Olfactory Sensory Neurons 13
5. The Olfactory Bulb 22
6. Central Processing of Olfactory Signals 31
7. Concluding Remarks 32
References 33

2. Gustation
Toshiaki J. Hara

1. Introduction 45
2. Structural Organization 46
3. Functional Properties 63
4. Gustatory Behaviors 75
5. Conclusions and Prospects 85
References 87

3. Branchial Chemoreceptor Regulation of Cardiorespiratory Function
Kathleen M. Gilmour and Steve F. Perry

1. Introduction 97
2. Cardiorespiratory Responses 98
3. Chemoreceptors 110

4. Central Integration and Efferent Pathways 133
5. Conclusions and Future Directions 138
References 139

4. Nociception
Lynne U. Sneddon

1. Introduction 153
2. Neural Apparatus 154
3. Central Nervous System 160
4. Moleculer Markers of Nociception 166
5. Whole Animal Responses 170
6. Conclusions 173
References 174

5. Visual Sensitivity and Signal Processing in Teleosts
Lei Li and Hans Maaswinkel

1. Introduction 180
2. Characteristics of the Visual System 180
3. Absolute Visual Sensitivity 188
4. Circadian Regulation of Visual Sensitivity 195
5. Chemosensory Modulation of Visual Sensitivity 204
6. Inherited and Acquired Impairments of Visual Sensitivity 210
7. Contrast Visual Sensitivity 214
8. Spectral Visual Sensitivity 221
9. Conclusions 226
References 227

6. Molecular and Cellular Regulation of Pineal Organ Responses
Jack Falcón, Laurence Besseau, and Gilles Boeuf

1. Introduction 244
2. Functional Organization of the Pineal 246
3. The Fish Pineal Organ: A Light Sensor 253
4. The Fish Pineal Organ: A Melatonin Factory 263
5. Intracellular Regulation of Arylalkylamine *N*-Acetyltransferase 2 275
6. Photoperiodic Versus Circadian Control Melatonin Production 279
7. Nonphotic Regulation of Pineal Organ Output Signals 284
8. Conclusions and Perspectives 290
Abbreviations 293
References 293

7. Electroreception: Object Recognition in African Weakly Electric Fish

Gerhard von der Emde

1. Introduction 307
2. Fish Electrogenesis 308
3. Fish Electroreception 308
4. Passive Electrolocation 310
5. Production of Electric Signals 312
6. Active Electrolocation 314
References 330

8. Magnetoreception

Michael M. Walker, Carol E. Diebel, and Joseph L. Kirschvink

1. Introduction 337
2. Introduction to Magnetic Field Stimuli 339
3. How Can Magnetic Fields Be Detected? 344
4. Structure of Candidate Magnetite-Based Magnetoreceptors 346
5. Behavioral Responses to Magnetic Fields in the Laboratory 349
6. Neural Responses to Magnetic Fields in the Laboratory 355
7. Neuroanatomy 359
8. Use of the Magnetic Sense in Navigation 362
9. What is Known About the Navigational Abilities of Fish? 367
10. Concluding Remarks 371
References 372

9. Neural and Behavioral Mechanisms of Audition

Andrew H. Bass and Zhongmin Lu

1. Introduction 377
2. Behavioral Studies of Audition 378
3. Peripheral and Central Auditory Pathways 382
4. Neurophysiological Mechanisms of Audition 387
5. Auditory Lateral Line Integration 399
6. Vocal Modulation of Inner Ear and Lateral Line 399
7. Steroid Hormones and Seasonal Changes in Hearing 400
8. Future Directions 404
References 405

10. The Lateral Line System of Fish

Horst Bleckmann

1. Introduction 411
2. The Lateral Line Periphery 412
3. Central Physiology 430
4. Conclusions 441

References 443

11. Neuromodulatory Functions of Terminal Nerve-GnRH Neurons

Hideki Abe and Yoshitaka Oka

1. Introduction 455
2. Electrophysiological and Morphological Features of Single TN-GnRH Neurons Revealed by Intracellular Recording and Labeling 457
3. GnRH Release Demonstrated by RIA 462
4. Pacemaker Mechanism of TN-GnRH Neurons 463
5. Modulation of Pacemaker Frequencies of TN-GnRH Neurons by GnRH 464
6. Autocrine/Paracrine Control of TN-GnRH Neuron Pacemaker Frequencies by GnRH 466
7. Cellular Mechanisms of Modulation of Pacemaker Frequencies by GnRH 469
8. Multimodal Sensory Inputs to TN-GnRH System 478
9. Neuromodulatory Action of GnRH Peptides 485
10. Distribution of GnRH Receptors in the Brain 487
11. Nonsynaptic Release of GnRH 488
12. Modulation of Neural Functions by GnRH 489
13. Behavioral Functions of TN-GnRH System 491
14. Working Hypothesis 492

References 495

INDEX 505

OTHER VOLUMES IN THE SERIES 521

CONTRIBUTORS

The numbers in parentheses indicate the pages on which the authors' contributions begin.

HIDEKI ABE *(455), Department of Biological Sciences, Graduate School of Science, University of Tokyo, 7-3-1 Hongo, Bunkyo-ku, Tokyo, Japan*

ANDREW H. BASS *(377), Department of Neurobiology and Behavior, Cornell University, Ithaca, New York, USA*

LAURENCE BESSEAU *(243), Laboratoire Arago, UMR 7628, CNRS and University Pierre and Marie Curie, Banyuls sur Mer, France*

HORST BLECKMANN *(411), Institute of Zoology, University of Bonn, Poppelsdorfer Schloss, Bonn, Germany*

GILLES BOEUF *(243), Laboratoire Arago, UMR 7628, CNRS and University Pierre and Marie Curie, Banyuls sur Mer, France*

CAROL E. DIEBEL *(337), School of Biological Sciences, University of Auckland, Auckland, New Zealand*

JACK FALCÓN *(243), Laboratoire Arago, UMR 7628, CNRS and University Pierre and Marie Curie, Banyuls sur Mer, France*

KATHLEEN M. GILMOUR *(97), Department of Biology and Centre for Advanced Research in Environmental Genomics, University of Ottawa, Ottawa, Canada*

TOSHIAKI J. HARA *(1, 45), Department of Fisheries and Oceans, Freshwater Institute, Winnipeg, Manitoba, Canada; Department of Zoology, University of Manitoba, Winnipeg, Manitoba, Canada*

JOSEPH L. KIRSCHVINK *(337), Division of Geological and Planetary Sciences, California Institute of Technology, Geology Division, Pasadena, California, USA*

LEI LI *(179), Department of Biological Sciences, University of Notre Dame, Notre Dame, Indiana, USA*

Zhongmin Lu *(377), Department of Biology, University of Miami, Coral Gables, Florida, USA*

Hans Maaswinkel *(179), Department of Biological Sciences, University of Notre Dame, Notre Dame, Indiana, USA*

Yoshitaka Oka *(455), Department of Biological Sciences, Graduate School of Science, University of Tokyo, 7-3-1 Hongo, Bunkyo-ku, Tokyo, Japan*

Steve F. Perry *(97), Department of Biology and Centre for Advanced Research in Environmental Genomics, University of Ottawa, Ottawa, Canada*

Lynne U. Sneddon *(153), School of Biological Sciences, University of Liverpool, The BioScience Building, Liverpool, United Kingdom*

Gerhard von der Emde *(307), Institute of Zoology, University of Bonn, Endenicher Allee 11-13, Bonn, Germany*

Michael M. Walker *(337), School of Biological Sciences, University of Auckland, Auckland, New Zealand*

Barbara S. Zielinski *(1), Department of Biological Sciences, University of Windsor, Windsor, Ontario, Canada*

PREFACE

This volume examines the sensory systems and some aspects of nervous and endocrine integrations. The study of fish sensory systems has undergone revolutionary changes since Volume 5 of the Fish Physiology series focused specifically on sensory systems over 35 years ago. Since that time, numerous sensory adaptations have been revealed, yet these fish systems continue as valuable models for other vertebrates. The past several decades have also witnessed the technical progress in a variety of different experimental approaches, including anatomy, electrophysiology, biochemistry, molecular biology, imaging, and genetics. Each of these developments has contributed towards understanding the complex workings of the sensory systems. These advancements have also expanded the knowledge base for subsequent investigation and for innovative application of new technologies.

The sensory systems are the windows onto the world, outside and inside the body. They detect chemical and physical energy and relay thereof information to the brain. The present volume examines how fish sense chemical (Chapters 1–4 and 11), mechanical (Chapters 4, 7, 9, and 10), and electromagnetic (Chapters 5, 6 and 8) energy. All of these forms of energy activate respective receptor cells, either neurons or neuron-like cells, by specific mechanisms known collectively as sensory transduction. These generate receptor potentials – voltage changes across the cell membrane that lead to firing of action potentials. These action potentials travel down the axons to the brain where the process of sensation takes place. In this way, chemical, mechanical, or electromagnetic energy is converted to neural activities that communicate to the brain.

The volume extends beyond the traditional senses that are commonly studied. It includes the branchial chemoreceptors for regulating cardiorespiratory functions (Chapter 3), light transduction in the pineal for circadian rhythm control through melatonin production (Chapter 6), and extends to the gonadotropin-releasing hormone (GnRH) containing terminal nerve (TN) which plays a role in controlling motivational states (Chapter 11).

This volume is dedicated to the memory of William Hoar, honorary editor, and to Aubrey Gorbman, who introduced us to the field of fish physiology, and guided us to the path of integrative functioning of both nervous and endocrine systems. The idea for this volume was conceived by David Randall, honorary editor, who devoted three volumes of the Fish Physiology to the nervous system. In assembling this volume, we have been greatly helped and encouraged by the diligence and enthusiasm of the chapter authors. We wish to express our gratitude to the following people for their advice and suggestions: John Caprio, Peter Ekström, Robert Evans, Richard Fay, Anne Hansen, Dennis Higgs, John Janssen, Sönke Johnsen, Masashi Kawasaki, Sadao Kiyohara, James Rose, William Stell, Steve Yazulla, and Celeste Wirsig-Wiechmann. We would also like to thank series editors, Tony Farrell and Colin Brauner for their guidance and support, and Els Bosma, Andrew Richford and especially Kirsten Funk at Elsevier Academic for publishing advice and encouragement.

Toshiaki J. Hara
Barbara S. Zielinski

1

OLFACTION

BARBARA S. ZIELINSKI
TOSHIAKI J. HARA

1. Introduction
2. Olfactory Repertoire
3. An Evolutionary Assessment of the Function of the Nasal Cavity
4. Olfactory Sensory Neurons
   4.1. Morphology and Central Projections
   4.2. The Transduction of Olfactory Signals
   4.3. The Specificity of Odorant Detection: Odorant Receptors
   4.4. Odorant Responses
5. The Olfactory Bulb
   5.1. Neural Composition
   5.2. Information Flow
6. Central Processing of Olfactory Signals
7. Concluding Remarks

1. INTRODUCTION

The olfactory system contributes to a fish's success in sustaining life—to feeding, avoiding predation, spawning migration and reproductive activity, even parental care, and offspring–parent interactions. The goal of this chapter is to present a world of scent from the perspective of the cells and neural pathways that respond to odors and channel olfactory input to behavioral responses. There are likely diverse adaptations of the olfactory system in the 28,000 extant fish species (Nelson, 1994; Nelson *et al.*, 2004) that flourish in the many bodies of water on earth. Consequently, this is a comparative review, which strives to point out variation in the morphology and function of the olfactory system of fishes.

Studies of several model species, pond dwelling cyprinids (zebrafish, *Danio rerio*; goldfish, *Carassius auratus*; channel catfish, *Ictalurus punctatus*) and sporting fish (rainbow trout, *Onchorhynchus mykiss*) have advanced

Sensory Systems Neuroscience: Volume 25
FISH PHYSIOLOGY

DOI: 10.1016/S1546-5098(06)25001-5

understanding of molecular and neural mechanisms. The homologous expression of odorant receptors in transgenic zebrafish (Sato *et al.*, 2005) offers unequivocal means for finally linking odorant quality to cellular response and central pathways.

This chapter starts with an overall view of the olfactory system, followed by a phylogenetic assessment of the gross anatomy of the peripheral olfactory organ, a synopsis of olfactory receptors (ORs), of transduction events and the channeling of responses to brain centers. For earlier comprehensive reviews of the anatomy and physiology of the olfactory system in fish, the reader is directed to Hara (1982, 1992), Laberge and Hara (2001), Caprio and Finger (2003), Zeiske *et al.* (1992), Zielinski and Hara (2001); to Stacey and Sorensen (2002, 2005) for a survey of pheromone communication; and to Eisthen (1997) for an evolutionary perspective of the olfactory system of vertebrates.

The sensory systems used for vertebrate chemoreception include olfaction, gustation, as well as the common chemical sense. The olfactory system uniquely uses cranial nerve I (the olfactory nerve) for transmitting chemosensory signals from the nose to the olfactory bulb of the brain. Odorous molecules enter the nasal cavity and interact with receptor proteins on the apical dendritic surface of olfactory sensory neurons (OSNs) located in the olfactory epithelium. The odorants bind onto G-protein–coupled receptor proteins on the apical dendritic endings and stimulate the formation of second messenger signaling molecules for gating the flow of ions through membrane channels. Graded change in membrane potential—the receptor potential—ensues and action potentials are triggered in the axons. These OSN axons enter the olfactory nerve, then form synaptic contacts onto olfactory bulb mitral cells—output neurons projecting into the forebrain. In some fish, pedunculated olfactory bulbs are well separated from the telencephalon, and in others, sessile olfactory bulbs are adjacent to the telencephalon (Figure 1.1). There have been reports of extrabulbar projections from the olfactory epithelium, extending directly into the telencephalon (Szabo *et al.*, 1991; Becerra *et al.*, 1994; Anadon *et al.*, 1995; Hofmann and Meyer, 1995; Laberge and Hara, 2003). The olfactory nerve also contains fibers of the terminal nerve (cranial nerve 0, *nervus terminalis*), and fibers of the trigeminal nerve (cranial nerve V) enter the olfactory mucosa. A review by Oka (Chapter 11, this volume) presents information on the terminal nerve system.

In many tetrapods, there are two spatially distinct olfactory subsystems—the main olfactory system and the vomeronasal system. The vomeronasal system has been linked with the chemoreception of pheromones, chemicals produced by one member of a species and detected by another member in which it produces a physiological or behavioral response, with the requirement that the chemical communication be mutually beneficial to sender and receiver (Meredith, 2001).

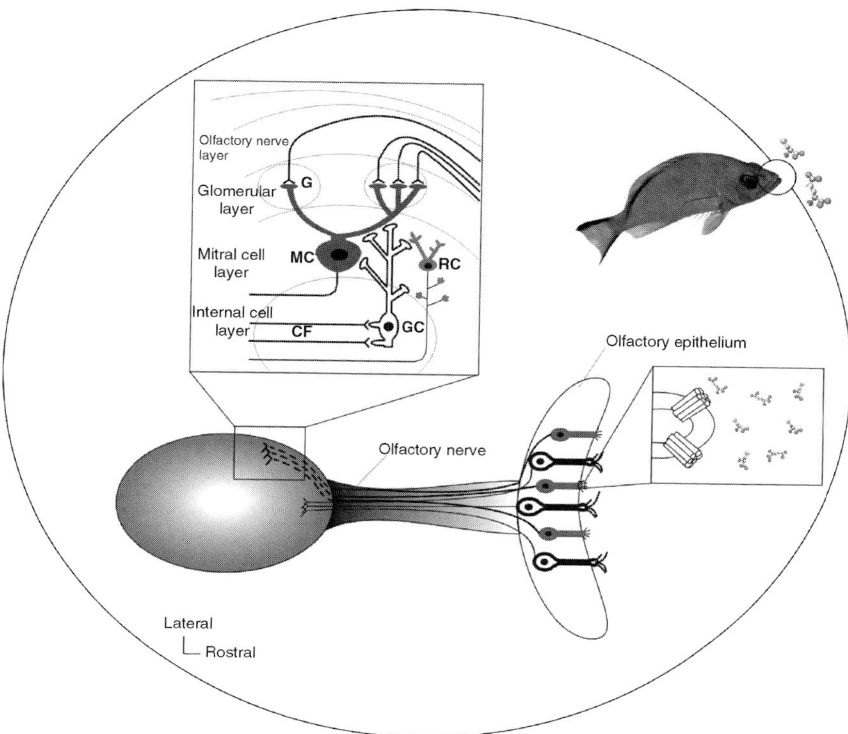

Fig. 1.1. A diagramatic representation of the primary olfactory pathway in teleost fish. Small soluble molecules enter the nasal cavity where receptor proteins on ciliated and microvillous olfactory sensory neurons bind these odorants. The axons of these sensory neurons extend through the olfactory nerve into spatially distinct regions of the olfactory bulb. In the olfactory bulb, the axons of the olfactory sensory neurons form synaptic contacts onto mitral cells (MC), output neurons of the olfactory bulb. The mitral cells as well as ruffed cells (RC), also output neurons, form synaptic contacts onto interneurons, granule cells (GC). Centrifugal fibers (CF) extend from central brain regions into the olfactory bulb. Not shown, juxtaglomerular cells.

In teleost fish, a single peripheral olfactory organ contains two well-defined morphological OSN subtypes, ciliated and microvillous. Tracing and genetic labeling (Hansen *et al.*, 2004; Sato *et al.*, 2005) studies have shown that the ciliated and microvillous OSN cell types innervate spatially distinct glomerular territories in the olfactory bulb and that mitral cells innervated by these OSNs axon terminals are chemotopically organized in channel catfish and project to specific pathways along the olfactory tract (summarized in Figure 1.1). A third cell type, the crypt cell, in the olfactory epithelium of some fish species, has been by retrogradely labeled from the olfactory bulb in channel catfish, and crucian carp (Hansen *et al.*, 2003; Hamdani and Døving, 2006).

2. OLFACTORY REPERTOIRE

Summated OSN generator potentials are measured from whole-field voltage transients recorded from the surface of the olfactory epithelium by a technique known as the electroolfactogram (EOG) (Ottoson, 1956; Scott and Scott-Johnson, 2002). In the underwater EOG system for fish (Silver *et al.*, 1976), a gelatin-filled glass capillary microelectrode is positioned on the olfactory epithelium, while water flows through the nasal cavity. A solenoid-driven switching device allows seemless alternating between water and water/odorant delivery without any mechanostimulation of nasal tissues (Sveinsson and Hara, 1990, 2000). The EOG technique has shown picomolar to micromolar response threshold for odor ligands, and larger responses with increased odor concentration. The EOG dose–response curve continues to rise with increase in odorant concentration, as high ligand concentrations stimulate irritation responses, and nonspecific cell responses. Consequently, when identifying odorous molecules, it is important to consider near threshold concentration values. These are often representative of natural environmental conditions and correlate to behavioral attraction or avoidance responses. Odor ligands are also identified by recording multiunit spikes from the surface of the olfactory epithelium (Caprio *et al.*, 1989; Kang and Caprio, 1991), compound action potentials from the olfactory nerve, and of OSN channel activity. Behavioral responses have also demonstrated odorant activity when paired with olfactory sensory deprivation negative controls. However, the EOG technique has predominated for examining fish olfactory sensitivity.

The odorous molecules are of relatively low molecular weight and encompass a wide range of compounds: L-amino acids, bile acids, steroids, prostaglandins, and nucleotides (reviewed by Hara, 1992). The amino acids are indicators of food. In cyprinids, the freshness of a food source is perceived through nucleotides, such as ATP. Bile acids and sex pheromones may convey the social information and endogenous status of animals to conspecifics or other animals (Sorensen and Caprio, 1998; Sorensen *et al.*, 2005a). Molecules contributing to alarm responses in cyprinids may contain nitrogen oxides (Brown *et al.*, 2003). Olfactory epithelial response thresholds are in the picomolar to nanomolar range for steroids and prostaglandins, and in the nanomolar to micromolar range for amino acids (Hara, 1992).

In teleosts, olfactory feeding cues include a large variety of L-amino acids. Some of these (cysteine, arginine, proline, and serine) stimulate gustatory as well as olfactory responses (Caprio, 1982). In lampreys, responsiveness to amino acids is limited to L-arginine and other basic amino acids. Micromolar concentrations of polyamines (spermine, cadaverine, putrescine) evoked olfactory epithelial responses, as well as feeding behavior in goldfish (Rolen *et al.*, 2002).

Bile compounds, cholesterol derivatives formed by the liver and released in feces, are potent odor ligands (Zhang *et al.*, 2001), evoke behavioral responses (Sola and Tosi, 1993) and may form part of reproductive pheromone signaling in teleosts (Miranda *et al.*, 2005). Petromyzonol sulfate, and its derivatives that are pheromones are released by larval sea lampreys during spawning migration of adults (Li *et al.*, 1995; Fine *et al.*, 2004; Sorensen *et al.*, 2005b). The sulfated bile alcohol, 3 keto-petromyzonol sulfate, released by spermiated male lampreys is attractive to ovulating female sea lampreys (Li *et al.*, 2002).

Cyprinids respond to nucleotides with an atypical reversed polarity EOG response (Rolen *et al.*, 2002). Salmonids may not have evolved ORs for nucleotides as this class of compounds has not been observed to elicit olfactory responses in these fish (Hara, personal communication).

Steroidal compounds and prostaglandins synthesized by gonadal tissues in some teleosts (Stacey and Sorensen, 2005) function as reproductive pheromones when released into the water. The EOG response threshold extends down to the picomolar range for some steroids. In the perciform teleost, the round goby, picomolar amounts of the androstenedione derivatives 11-keto testosterone, etiocholanolone, and etiocholanolone glucuronide, evoke EOG responses; whereas 0.1 micromolar (μM) L-alanine is needed to evoke an EOG response (Murphy *et al.*, 2001; Jasra *et al.*, 2006). Behavioral studies have suggested that major histocompatability complex peptide products released by reproductive phase of the three-spined stickleback function as mate selection pheromones (Milinski *et al.*, 2005).

Several EOG studies have shown that the sensitivity and magnitude of olfactory epithelial responses vary with gender, life cycle stage or seasonality. The EOG response threshold concentration to water previously occupied by reproductive stages, amino acids, putative pheromones steroids, or some prostaglandins, depends on maturity in cyprinids (Irvine and Sorensen, 1993; Cardwell *et al.*, 1995); salmonids (Moore and Scott, 1991; Bjerselius and Olsen, 1993); the round goby (Belanger *et al.*, 2004). In roach, gender is a factor for EOG sensitivity to pheromones (Lower *et al.*, 2004). In tilapia and round goby, both percid actinopterygii teleosts, EOG sensitivity and response magnitude were greater in reproductively mature stages than in nonreproductive stages (Belanger *et al.*, 2004; Miranda *et al.*, 2005). The reason for these differences in EOG response magnitude is undetermined and may lie in the perireceptor events, or OSN channels, receptors or OSNs themselves.

3. AN EVOLUTIONARY ASSESSMENT OF THE FUNCTION OF THE NASAL CAVITY

The nasal cavity provides an environment of odorant–receptor interaction to occur at a rate and sensitivity for appropriate behavioral responses to occur. There may be evolutionary pairing of the morphology of the nasal cavity and the development of the accessory nasal structures, the nasal sacs which assist in the flow of water over the olfactory epithelium, an event that has compared to "sniffing" (Nevitt, 1991). Agnathans (lampreys) are monorhinic; water enters and leaves the nasal cavity through a single nostril located on the dorsal surface of the head. All other fish have two nasal cavities, one on each side of the head, rostral to the eyes. Water enters the nasal cavity through the anterior nostril, flows over the olfactory epithelium, and leaves by the posterior nostril. In some teleosts, two accessory nasal sacs, the ethmoid and the lacrimal sac, a single nasal sac, or a fused nasal sac, assist in ventilating unidirectional waterflow through the nasal cavity (Parker, 1910; Pipping, 1926; Liermann, 1933; Døving *et al.*, 1977; Melinkat and Zeiske, 1979; Kux *et al.*, 1988; Nevitt, 1991; Belanger *et al.*, 2003). The occurrence of these nasal sacs has long been associated with the need to regulate this flow in sedentary bottom-dwelling fish (Burne, 1909; Kapoor and Ojha, 1972), yet nasal sacs are absent from many demersal fish (Table 1.1).

In most fish, the olfactory epithelium is located on multilamellar mucosal folds that form a flowerlike arrangement, termed the olfactory rosette. In some fish, these lamellar folds are numerous (up to 230-folds in the reef-dwelling perciform, the barred snapper, *Hoplopagrus guentheri*; Pfeiffer, 1964). In others, lamellar folding is minimal or even absent (Burne, 1909; Kleerekoper, 1969; Hara, 1975; Døving *et al.*, 1977; Yamamoto, 1982; Zeiske *et al.*, 1992; Hansen and Reutter, 2004). The multilamellar olfactory epithelium has been explained as increasing the surface area for odorant–receptor interaction. However, olfactory sensory activity is robust in fish with a relatively flat olfactory epithelium (Figures 1.2C and 1.4A, B, C), as seen from responses to diverse compounds in the round goby (Murphy *et al.*, 2001). Burne (1909) suggested that flattened olfactory epithelia are indicative of demersal (interacting with the substrate) or benthic (living at the bottom) habitat yet the olfactory chamber has a single fold in pelagic fish such as blennies and gars (Burne, 1909; Theisen *et al.*, 1980; Yamamoto, 1982). Possibly, nasal sac pumping activity (sniffing) (Nevitt *et al.*, 1991; Murphy *et al.*, 2001; Murphy and Stacey, 2002; Belanger *et al.*, 2006) enhances odorant sampling, so that investment into the development of elaborate lamellar folding is not required to fulfill olfactory sensory input for survival.

Table 1.1

The Gross Morphology of the Peripheral Olfactory Organ in Teleost Fishes, Showing the Occurrence of Olfactory Epithelial Folding

Taxon (some common names)	Olfactory epithelium–genus species	Number of nasal sacs–genus species	References	Habitat
Osteoglossiformes (knifefishes)	ML–*Mormyrus*	0–*Mormyrus*	Burne, 1909	Demersal
	–*Pantodon buchholzi*		Hansen and Finger, 2000	Pelagic
	–*Pollimyrus castelnaui*			Demersal
Anguilliformes (eels)	ML–*Anguilla anguilla*	0–*A. anguilla*	Burne, 1909; Teichmann, 1954	Demersal
	–*Conger vulgaris*	*C. vulgaris*	Yamamoto, 1982	Demersal
	–*A. japonica, C. myriaster*			
Clupeiformes (herrings)	ML–*Clupea harengus*	1–*C. harengus*	Burne, 1909	Demersal
	–*Clupanodon punctatus*		Yamamoto, 1982	Demersal
Cypriniformes (carps and minnows)	ML–*Carassius auratus*		Liermann, 1933	Demersal
	–*Cyprinus carpio*	0–*C. carpio*	Døving *et al.*, 1977; Yamamoto, 1982	Demersal/Pelagic
	–*Danio rerio*		Hansen and Zeiske, 1993	Pelagic
	–*Phoxinus phoxinus*		Teichmann, 1954; Døving *et al.*, 1977	Pelagic
	–*Tinca tinca*	0–*T. tinca*		
Siluriformes (catfishes)	ML–*Ictalurus punctatus*		Caprio and Raderman-Little, 1978	Demersal
	–*Siluris glanis*	1–*S. glanis**	Burne, 1909; Jakubowski and Kunysz, 1979	Demersal
	–*Plotosus lineatus*		Theisen *et al.*, 1991	Demersal
Esociformes (pikes)	ML–*Esox lucius*	0–*E. lucius*	Burne, 1909; Teichmann, 1954	Demersal
Salmoniformes (trouts and salmons)	ML–*Onchorynchus mykiss*		Teichmann, 1964	Pelagic
	–*Salmo trutta*	0–*S. trutta*, 1–*Salmo salar*	Døving *et al.*, 1977; Burne, 1909	Demersal
	–*Salvelinus alpinus*		Hanson and Finger, 2000	Demersal
	–*Coregonus clupeaformis*		Hara *et al.*, 1973	Demersal

(continued)

Table 1.1 (*continued*)

Taxon (some common names)	Olfactory epithelium–genus species	Number of nasal sacs–genus species	Reference	Habitat
Gadiformes (cods)	ML–*Melanogrammus aeglefinus*,[a] *Gadus morhua*		Burne, 1909	Demersal
	–*Lota lota*		Teichmann, 1954	Demersal
	–*Muraenolepis microps*	0–*M. microps*	Eastman and Lanoo, 2001	Demersal
Atheriniformes (silversides)	1L–*Cololabis saira*,		Yamamoto, 1982	Pelagic
	Hemiramphus sajori[b]		Yamamoto, 1982	
	ML–*Chilatherina sentaniensis*	1–*C. sentaniensis*		Demersal
	–*Glossolepis incisus*	–*G. incisus*	Zeiske *et al.*, 1979	Demersal
	–*Melanotaenia sentaniensis*	–*Melanotaenia* sp.	Yamamoto, 1982	Reef
	–*Iso hawaiiensis*			
Beloniformes (gars)	F–*Oryzias latipes*		Yamamoto, 1982	
	1L–*Belone vulgaris*	0–*B. vulgaris*	Burne, 1909; Theisen *et al.*, 1980	Demersal
	–*Cheilopogon agoo*		Yamamoto, 1982	
Cyprinodontiformes (killifishes, swordtails)	F–*Austrofundulus transilis*	1–*A. transilis*	Zeiske, 1974	Pelagic
	–*Cynolebias whitei*,[c]	–*C. whitei*,	Zeiske, 1974	
	Pterolebias peruensis[d]	*P. peruensis*	Yamamoto, 1982	
	–*Gambusia affinis*		Zeiske, 1974	Demersal
	F:1L–*Cynolebias nigripinnis*[e]	–*C. nigripinnis*		
	–*Rivulus cylindraceus*	–*R. cylindraceus*	Zeiske, 1974	Benthopelagic
	1L-3L–*Aplocheilus lineatus*, *Aplocheilus dayi*	–*A. lineatus*, *A. dayi*	Kux *et al.*, 1988	Pelagic
	ML–*Fundulus chrysotus*			Demersal
	Valencia hispanica			Demersal
	–*Procatopus similis*			Demersal
	–*Aplocheilichthys spilauchen*			Benthopelagic
	–*Xiphophorus helleri*			Demersal

Gasterosteiformes (seahorses, pipefishes, sticklebacks)	F–*Hippocampus coronatus*[f]		Yamamoto, 1982	Demersal
	–*Syngnathus*[f]		Liermann, 1933	Demersal
	1L–*Spinachia spinachia*	1–*S. spinachia*	Theisen, 1982	Demersal
	2L–*Gasterosteus aculeatus*		Wunder, 1957	Demersal
Synbranchiformes	F–*Fluta alba*[g]		Yamamoto, 1982	Demersal
Perciformes (perches, clingfishes, gobies, tunas, cichlids)	F–*Conidens laticephalus*[h]		Yamamoto, 1982	
	–*Omobranchus elegans*		Yamamoto, 1982	Demersal
	1L–*Acanthogobius flavimanus*			Pelagic
	–*Ammodytes personatus*		Belanger *et al.*, 2003	Pelagic
	–*Odonoamblyopus rubicundus*		Burne, 1909	Demersal
	–*Neogobius melanostomus*	2–*N. melanostomus*	Burne, 1909; Teichmann, 1954	
	ML–*Mugil chelo*[i]	1–*M. chelo*	Yamamoto, 1982	Benthopelagic
	Trachinus vipera[j]	1–*T. vipera*	Yamamoto, 1982	Demersal
	–*Perca fluviatilis*	2–*P. fluviatilis*	Yamamoto, 1982	Pelagic
	–*Leiognathus nuchalis*,		Fishelson, 1997	Pelagic
	Toxotes jaculator[k]		Hansen and Finger, 2000	Pelagic
	–*Sphyraena japonica*,		Figure, present study	Pelagic
	Thunnus thynnus			Pelagic
	–*Scomber japonicus*			Demersal
	–*Tilapia zillii*	2–*T. zillii*		Reef
	–*Archocentrus nigrofasciatus*			Reef
	–*Kuhlia sandvicensis*			
Pleuronectiformes (flounders)	ML–*Pleuronectes flesus*[l]	2–*P. flesus*	Burne, 1909	Demersal

(*continued*)

Table 1.1 (*continued*)

Taxon (some common names)	Olfactory epithelium–genus species	Number of nasal sacs–genus species	Reference	Habitat
Tetraodontiformes (puffers, boxfishes)	F–*Rudarius ercodes*,		Yamamoto, 1982	Reef
	Ostracion tuberculatus		Yamamoto, 1982	
	1L–*Navodon modestus*[m]		Yamamoto, 1982	
	ML–*Fugu niphobles*[n]			Demersal

[a]*Melanogrammus aeglefinus* was originally *Gadus aeglefinus* (Fishbase).
[b]*Hemiramphus sajori* is also known as *Hyporhamphus sajori* (Fishbase).
[c]*Cynolebias whitei* is the original name for this species, also currently called *Nematolebias whitei* (Fishbase).
[d]*Pterolebias peruensis* is the original name for this species, also currently called *Aphyolebias peruensis* (Fishbase).
[e]*Cynolebias nigripinnis* is the original name for this species, also currently called *Austrolebias nigripinnis* (Fishbase).
[f]*Hippocampus coronatus* and *Syngnathus* species are listed under the Syngnathiformes order in Fishbase.
[g]*Fluta alba* is also known as *Muraena alba* and *Monopterus albus* (Fishbase).
[h]*Conidens laticephalus* was formally listed under the Gobiesociformes order.
[i]*Mugil chelo* is also known as *Mugil labrosus* and more currently, *Chelon labrosus* (Fishbase).
[j]*Trachinus vipera* is the original name for this species, also currently called *Echiichthys vipera* (Fishbase).
[k]*Toxotes jaculator* was originally known as *Sciaena jaculatrix* and is also currently known as *Toxotes jaculatrix* (Fishbase).
[l]*Pleuronectes flesus* is the original name for this species, also currently called *Platichthys flesus* (Fishbase).
[m]*Navodon modestus* was originally known as *Monacanthus modestus* and is also currently known as *Thamnaconus modestus* (Fishbase).
[n]*Fugu niphobles* was originally known as *Spheroides niphobles* and is also currently known as *Takifugu niphobles* (Fishbase).

ML, Multilamellar, olfactory epithelium on a multilamellar structure in the olfactory chamber; F, flat–olfactory epithelium on a flat surface on the floor of the olfactory chamber; 1L, olfactory epithelium on a single lamellar fold on the floor of the olfactory chamber; F:1L, the floor of the olfactory chamber contains a small rostral fold and a small caudal fold and is otherwise flat; *, slight posterior extension.

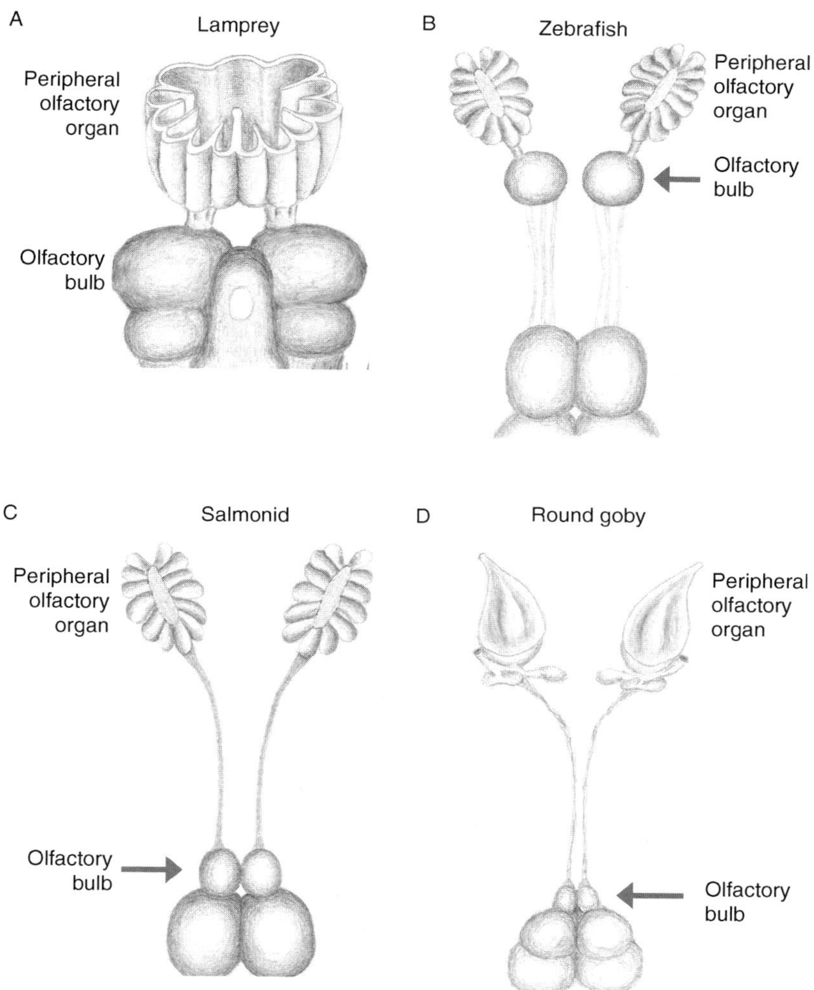

Fig. 1.2. A diagrammatic representation of the morphological arrangement of the peripheral olfactory organ, the olfactory nerve, and the olfactory bulb in four fish. (A) Lampreys are monorhinic with a single bilaterally symmetric multilamellar peripheral olfactory organ. Two olfactory nerves extend from the peripheral olfactory organ to the olfactory bulbs. (B) In cyprinids, such as zebrafish, the multilamellar peripheral olfactory organ takes on a rosette form. The short olfactory nerves connect the pedunculted olfactory bulb, and olfactory tracts lead to the telencephalon. (C) In salmonids, the olfactory rosette is connected by a long olfactory nerve to a sessile olfactory bulb, adjacent to the telencephalon. (D) In the round goby, an Acanthopterygian percid, the olfactory epithelium lines a relatively flat peripheral olfactory organ, with a single low longitudinal ridge. Nasal sacs are caudal to the olfactory epithelium. A long olfactory nerve connects to the sessile olfactory bulb, adjacent to the telencephalon.

Taxonomic survey of peripheral olfactory organ morphology reveals an evolutionary relationship between peripheral olfactory organ polymorphism and nasal sac development (Table 1.1, Figure 1.1). A multilamellar olfactory epithelium is present, and nasal sacs are largely absent from Agnathans, Chondrithyes, and in early evolutionary lines of teleosts: knifefishes (Osteoglossiformes), eels (Anguilliformes), and herrings (Clupeiformes); as well as carps and minnows (Cypriniformes), catfish (Siluriformes), pikes (Esociformes), salmons (Salmoniformes), and cods (Gadiformes). However, there are exceptions with single air sacs in Clupeiformes (*Clupea harengus*), Siluriformes (*Siluris glanis*), and Salmoniformes (*Salmo salar*).

In Acanthopterygii, teleosts with stiff fin spines that allow for locomotor modifications (Helfman *et al.*, 1997), the olfactory epithelium varies from flat, unilamellar to multilamellar and air sacs have evolved (Table 1.1, Figure 1.2). Although our survey includes only 50 species from these orders, it may be representative of the thousands of species grouped as the Acanthopterygii. This survey suggests that variability of olfactory epithelial folding and of the ethmoid and lacrimal nasal sacs evolved during the Cenozoic radiation of Acanthopterygii.

In Acanthoptery, accessory nasal sacs ventilates the olfactory epithelium. The nasal sac ventilation coincides with opercular movement (Johnson and Brown, 1962). In the round goby, *Neogobius melanostomus*, putative pheromones (steroidal compounds) enhance gill ventilation activity through olfactory sensory input (Murphy and Stacey, 2002; Belanger *et al.*, 2006), and with larger for gill ventilation increases in reproductive compared to nonreproductive stages (Belanger *et al.*, 2003).

In silversides (Atheriniformes), the olfactory chamber is unilamellar [*Cololabis saira*, *Hemiramphus sajori*; (Yamamoto, 1982)] or multilamellar [*Chilatherina sentaniensis*, *Glossolepis incisus*, *Melanotaenia (Nematocentris) maccullochi* (Zeiske *et al.*, 1979), *Iso hawaiiensis* (Yamamoto, 1982)]. A single nasal sac is present in *C. sentaniensis*, *G. incisus*, and *Melanotaenia* sp. In gars (Beloniformes), the olfactory chamber is flat in *Oryzias latipes* (Yamamoto, 1982) and unilamellar in *Belone vulgaris* (Burne, 1909; Theisen *et al.*, 1980) and in *Cheilopogon agoo* (Yamamoto, 1982). The demersal unilamellar *B. vulgaris* lacks any nasal sacs.

The killifish (Cyprinodontiformes) possess a single nasal sac (Table 1.1), and the olfactory chamber morphology is highly variable—from entirely flat (*Austrofundulus transilis*, *Pterolebias peruensis*), flat olfactory epithelium interspersed with small mounds or ridges with nonolfactory epithelium (*Cynolebias whitei*, *Cynolebias nigripinnis*, *Rivulus cylindracens*), with one to three lamellar folds (*Aplocheilus lineatus*, *Aplocheilus dayi*; Zeiske, 1974), or multilamellar (*Xiphophorus helleri*; Kux *et al.*, 1988).

In seahorses, pipefishes, and sticklebacks (Gasterosteiformes), the olfactory chamber is flat in *Hippocampus coronatus* (Yamamoto, 1982) and in *Syngnathus* (Liermann, 1933); unilamellar in *Spinachia spinachia* (Theisen, 1982); and has two lamellae in the three-spined stickleback, *Gasterosteus aculeatus* (Wunder, 1957). A single nasal sac is present in *S. spinachia* (Theisen, 1982).

The perciforms are an especially diverse order, first known from the early Cenozoic and now with 7800 species grouped into 150 families (Nelson, 1994). The morphology of the olfactory chamber varies in the 16 species included in this survey, yet all possess one or two nasal sacs (Table 1.1, Figure 1.2). The olfactory epithelium is flat [*Conidens laticephalus*, *Omobranchus elegans* (Yamamoto, 1982)]; unilamellar [*Acanthogobius flavimanu*, *Ammodytes personatus*, *Odonoamblyopus rubicundus* (Yamamoto, 1982), *N. melanostomus* (Belanger *et al.*, 2003)]; and multilamellar [*Mugil chelo*, *Trachinus vipera* (Burne, 1909), *Perca fluviatilis* (Burne, 1909; Teichmann, 1954), *Leiognathus nuchalis*, *Toxotes jaculator*, *Sphyraena japonica*, *Thunnus thynnus*, and *Scomber japonicus* (Yamamoto, 1982), *Tilapia zillii* (Fishelson, 1997), and *Archocentrus nigrofasciatus* (Hansen and Finger, 2000)].

This taxon-based survey shows that in species that diverged prior to the Cenozoic radiation (Carroll, 1988), the olfactory chamber is multilamellar and a single nasal sac has seldom evolved. In bony rayed fishes (Acanthopterygii), the olfactory chamber is polymorphic, ranging from flat to multilamellar, and paired often with the evolution of an ethmoid and lacrimal air sac. The reason for the variation in lamellar configuration in teleosts is unknown at this time. Correlation to habitat is not clear. Within the Acanthopterygii, nasal sacs are present in demersal as well as pelagic species with the exception of the demersal *B. vulgaris*, without nasal sacs. In the other fish species, a single nasal sac is present in demersal species (*C. harengus*, *Siluris glanis*, *Salmo trutta*).

4. OLFACTORY SENSORY NEURONS

During the past 20 years, physiological, morphological, immunochemical, and molecular studies have investigated linkage between OSN morphology with odorant transduction signaling, receptor expression, and systematic spatial representation of odorant responses (chemotopy).

4.1. Morphology and Central Projections

The OSNs are bipolar first-order neurons. The dendritic ending is immersed in the mucus of the nasal cavity, the cell body is within the olfactory epithelium, and the axon extends into the olfactory nerve, terminating in neuropil within the olfactory bulb at the rostral portion of the brain.

The OSN axon terminals in olfactory bulb neuropil—the olfactory glomeruli—represent functional units of information processing in the central olfactory system (Shepherd, 1994). An essential component of vertebrate OSN receptor expression was revealed in 1996 by Peter Mombaerts, when he showed that OSNs that expressed a particular receptor dispersed in the olfactory epithelium converged onto spatially distinct olfactory glomeruli.

Although most OSN axons terminate in the olfactory bulb, there is physiological and anatomical evidence for OSN fibers passing through the ventral region of olfactory bulb and terminating directly onto the telencephalon. In fish, the olfactory epithelium contains the OSNs as well as ciliated nonsensory cells, sustentacular cells, and basal cells in the olfactory epithelium (Zeiske *et al.*, 1992; Hansen and Zielinski, 2005). As in other vertebrates, the OSNs undergo continuous cell turnover during the life cycle, and basal cells provide the progenitor cell source for newly formed OSNs.

In teleost fishes, two morphological OSN forms, ciliated and microvillous, intermingle in the olfactory epithelium and project to specific regions of the olfactory bulb. A third OSN morph, the crypt cell, which is also ciliated, but with a distinct apical location, was discovered by Anne Hansen and her collaborators (2000). Lampreys have only ciliated OSNs (Thornhill, 1967; Vandenbossche *et al.*, 1995), and sharks have only microvillous OSNs. The ciliated lamprey OSNs display polymorphisms based on dendritic length and the position of the cell body within the olfactory epithelium (Laframboise *et al.*, 2006).

A ring of cilia surrounds the olfactory knob at the apical dendritic ending of the ciliated OSNs, and a tuft of microvilli extends from the apex of microvillous OSNs (Thommesen, 1983; Zielinski and Hara, 1988; Hara and Zielinski,1989). The dendrite of ciliated OSNs is thin ($\sim$1 μm), and the cell body is located in the lower third region (deep) of the olfactory epithelium. Subpopulations of ciliated OSNs with glutathione conjugation capabilities have been identified (Starcevic *et al.*, 1993; Starcevic and Zielinski, 1995; Yanagi *et al.*, 2004). The dendrite of microvillous OSNs is wider than in ciliated OSNs ($\sim$5 μm), and the cell body is in the center of the olfactory epithelium.

The olfactory epithelial identity of OSN extrabulbar fibers that project to the telencephalon has not been determined as these have been visualized through anterograde dye labeling of the olfactory epithelium, which stains the entire olfactory epithelium (Von Bartheld *et al.*, 1984; Von Bartheld, 2004). Functional studies in whitefish (*Coregonus clupeaformis*), which show neural responses in the ventral medial telencephalon, not the olfactory bulb, to nasal application of prostaglandin F (PGF), support the idea of OSN extrabulbar innervation. These extrabulbar ventral telencephalic territories overlap onto terminal nerve territories, which also extend along the olfactory nerve. Consequently, there has been concern that the extrabulbar fibers may in fact be terminal nerve fibers; however, the terminal nerve has a neuromodulatory rather than a chemosensory function.

In 1912, Sheldon observed spatially and morphologically distinct medial and lateral olfactory nerve bundles in carp. We know that the different OSN types (ciliated and microvillous) project to different glomerular territories in different fish. In cyprinids, ciliated OSNs project largely into medial olfactory bulbar glomeruli and microvillous to lateral regions (zebrafish, Sato *et al.*, 2005; channel catfish, Hansen *et al.*, 2003; Morita and Finger, 1998). Fluorescent expression of ciliated OSNs in transgenic zebrafish showed innervation of medial as well as dorsal, rostral, and sparse lateral olfactory bulb glomeruli (Table 1.2; Sato *et al.*, 2005). Ciliated OSNs were retrogradely filled following both medial and lateral, as well as rostroventral injections of the lipophilic carbocyanine dye, DiI (1,1′-dioctadecyl-3,3,3′3′-tetramethyl-indocarbocyanine perchlorate), in channel catfish (Hansen *et al.*, 2003) and caudomedial injections to the crucian carp (Hamdani and Døving, 2002). Microvillous OSNs, expressing the transient receptor potential channel C2 (TRPC2), projected to lateral and ventrocaudal glomeruli in zebrafish (Sato *et al.*, 2005). In channel catfish, dorsolateral as well as rostrodorsal and caudodorsal DiI injections labeled microvillous OSNs (Hansen *et al.*, 2003). These data show differences in the central projection territories for the ciliated and microvillous OSNs in these two cyprinid species, although the general medial/ciliated OSN and lateral/microvillous OSN pattern is conserved.

The oval-shaped crypt cells are dispersed throughout the apical region of the olfactory epithelium. In these cells, a single cilium and several microvilli extend from the surface that is submerged compared to the other cellular components of the olfactory epithelium. Unlike other cells of the olfactory epithelium, the crypt cells express the calcium binding protein S-100 (Germana *et al.*, 2004). Crypt cells occur in Ostariophysian and Actinopterygian fish (Belanger *et al.*, 2003; Hansen *et al.*, 2003), as in lampreys (Laframboise *et al.*, 2006), but are absent from sarcopterygians (Hansen and Finger, 2000). The injection of DiI into olfactory bulbar ventrocaudal and ventrocentral regions in the channel catfish labeled crypt cells in channel catfish (Hansen *et al.*, 2003). In crucian carp, axons from crypt cells type pass through, or terminate in this ventral region of the olfactory bulb (Hamdini and Døving, 2006).

Not surprisingly, there are functional links between OSN morphology and central projections at the level of OR expression, signal transduction, and odorant responses. These will be explored in the following sections.

4.2. The Transduction of Olfactory Signals

In addition to projecting into different spaces within the olfactory bulb, the transduction pathways differ in the microvillous and ciliated OSNs

(Cao *et al.*, 1998; Speca *et al.*, 1999; Belanger *et al.*, 2003; Hansen *et al.*, 2003, 2004; Pfister and Rodriguez, 2005). In both, odorant binding onto a G-protein–coupled membrane receptor protein gates channel activity through second messenger activity. The identity of the second messenger cascades differ between the ciliated and microvillous OSNs.

Odorant binding onto an OR promotes a conformational change in the intracellular domain that affects its interaction with the G-protein, the second protein in the signal transduction pathway. On activation by the receptor, the G-protein binds guanosine triphosphate (GTP) and the α-subunit of this heteromeric GTP-binding protein on the cytosolic side of the plasma membrane, and dissociates and stimulates the formation of cyclic adenosine monophosphate (cAMP) by adenylyl cyclase. In 1989, Jones and Reed discovered the stimulatory G-protein subunit, $G_{\alpha olf/s}$, in the plasma membrane of mammalian OSNs. Stimulatory G-protein α-subunits were later found in catfish OSNs (Abogadie *et al.*, 1995; DellaCorte *et al.*, 1996), and the $G_{\alpha olf/s}$ protein was localized to the cilia and plasma membrane of ciliated OSNs of the channel catfish (Hansen *et al.*, 2003) and goldfish (Hansen *et al.*, 2004), and in the perciform round goby (Figures 1.3 and 1.4D–F, Belanger *et al.*, 2003). Hansen *et al.* (2003) showed cAMP-mediated olfactory sensory transduction in response to amino acids and to bile acids in channel catfish.

A second class of receptors is coupled through a G-protein to a plasma membrane phospholipase C that is specific for the plasma membrane lipid phosphatidylinositol 4,5-biphosphate. This enzyme catalyzes the formation of two potent second messengers, diacylglyercol and inositol 1,4,5-triphosphate (IP_3). When a ligand binds its specific receptor in the plasma membrane, the receptor–ligand complex catalyzes GTP–GDP exchange on an associated G-protein, $G_{\alpha q}$, activating it exactly as the odorant activates $G_{\alpha olf/s}$. The activated $G_{\alpha q}$, in turn activates a specific membrane-bound phospholipase C, which catalyzes the production of the two second messengers by hydrolysis of phosphatidylinositol 4,5-bisphosphate in the plasma membrane. The stimulatory G-protein subunits $G_{\alpha 0}$ and $G_{\alpha q/11}$ activate the phosholipase C/IP_3-mediated sensory transduction cascade in the mammalian vomeronasal organ (Wekesa and Anholt, 1999). Evidence for an IP_3-mediated pathway for amino acid stimulation has been shown in salmon (Lo *et al.*, 1993) and zebrafish (Ma and Michel, 1998). Biochemical and physiological evidence for both cAMP and IP_3 transduction mechanisms in catfish was presented by Miyamoto *et al.* (1992) and Restrepo *et al.* (1993). There are several subpopulations of microvillous OSNs utilizing different G-protein subunits. In the round goby, microvillous OSNs and crypt cells were immunoreactive to $G_{\alpha 0}$ (Figure 3 in Belanger *et al.*, 2003), although this immunoreactivity was not consistent to all lots of the antisera tested

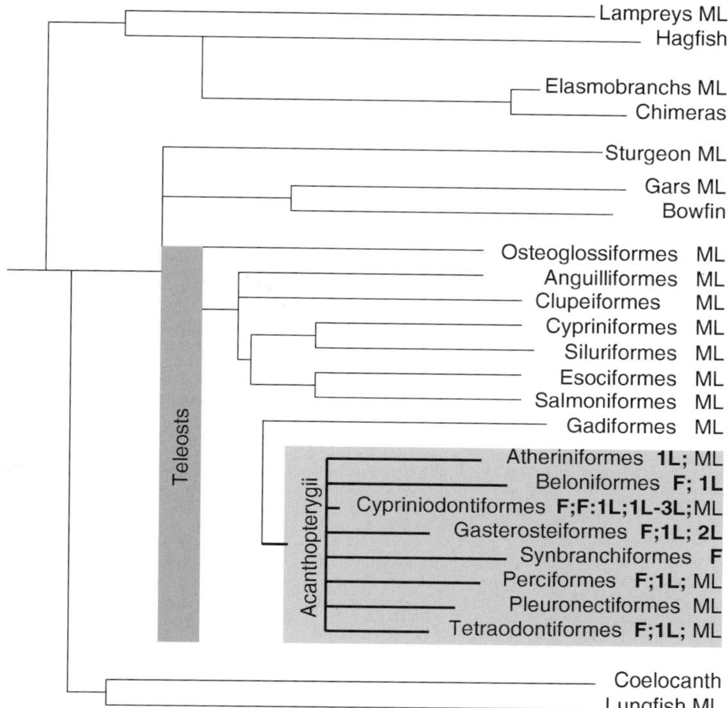

Fig. 1.3. Phyletic tree of the Teleostei, based on Nelson (1994) showing the occurrence of a multilamellar (ML) peripheral olfactory organ, compared to a flat olfactory chamber, onefold (1L), twofolds (2L), and threefolds (3L), summarized from Table 1.1. Phylogeny based on the Tree of Life project.

(unpublished results). In channel catfish, microvillous OSNs express $G_{\alpha q/11}$ and crypt cells express $G_{\alpha 0}$ (Hansen *et al.*, 2003). In contrast, in the goldfish, many microvillous OSNs express $G_{\alpha 0}$, very few express $G_{\alpha q/11}$ or the inhibitory G-protein subunit $G_{\alpha i\text{-}3}$; crypt cells express $G_{\alpha q/11}$ and the submerged cilia of these cells contain $G_{\alpha 0}$ (Hansen *et al.*, 2004).

In zebrafish, the channels directed by G-protein activity differ according to OSN subtype: the cyclic nucleotide–gated A2 subunit is expressed—together with OR-type odorant receptor molecules—in ciliated OSNs. The TRP2 channel, together with V2R-type OR molecules, are localized in microvillous OSNs. In mammals, TRP2 channels are in vomeronasal neurons and are likely gated by diacyl glycerol. In TRP2 knockouts, responses to urine and pheromone responses were abolished or strongly reduced.

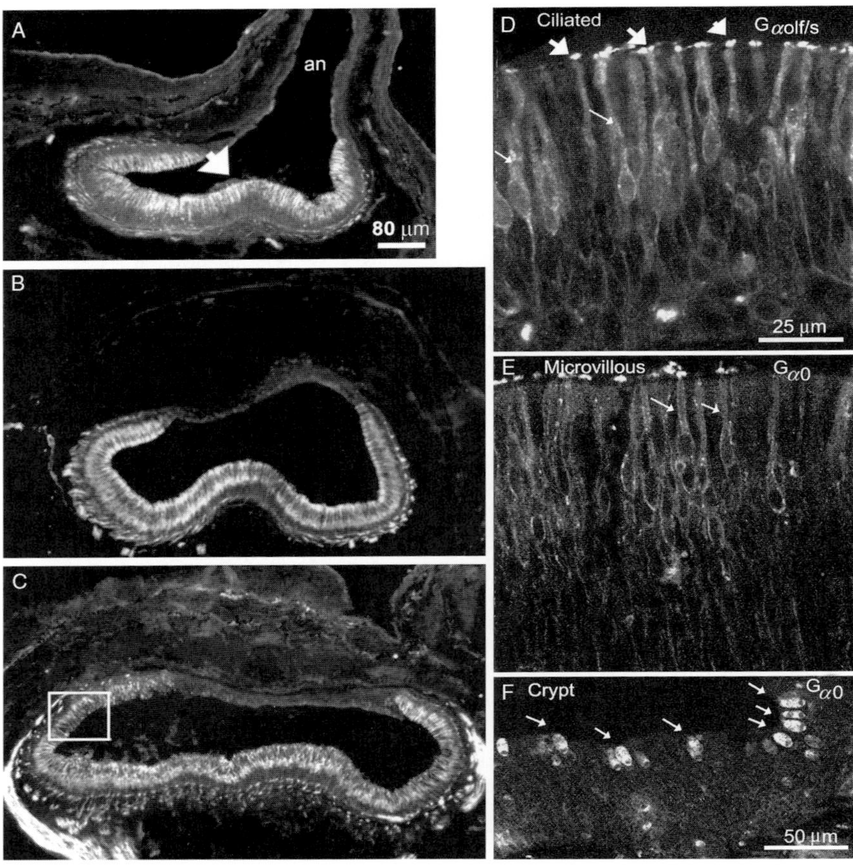

Fig. 1.4. (A–C) Cross-sectional views of the olfactory chamber in the round goby, *Neogobius melanostomus* show a single low longitudinal fold. Cryostat sections immunostained for $G_{\alpha olf/s}$ to show the distribution of ciliated olfactory sensory neurons in the olfactory epithelium. (D–F) The localization G proteins in the olfactory sensory neurons of the round goby, *Neogobius melanostomus.* A, B, and C are the same magnification; D and E are the same magnification. (A) At the rostral edge of the olfactory chamber, the olfactory epithelium is relatively flat with a slight elevation (arrowhead) in the center. The olfactory epithelium is close to the anterior nostril (an), and also lines the lateral and dorsal sides of the nasal cavity. (B) In the center of the olfactory chamber, a pronounced fold is present along the floor of the nasal cavity. Olfactory epithelium is absent from the roof of the olfactory chamber, but present along the ventral, lateral and mediodorsal surfaces, where $G_{\alpha olf/s}$ immunoreactivity of olfactory sensory neurons is strong. The $G_{\alpha olf/s}$ immunostaining extends into small nerve fascicles in the lamina propria beneath the olfactory epithelium. (C) The floor of the olfactory chamber is relatively flat at the caudal edge of the olfactory epithelium. Prominent large, $G_{\alpha olf/s}$ nerve fibers are seen at the lateral and medial edges of the olfactory chamber. (D) Ciliated olfactory sensory neurons were $G_{\alpha olf/s}$ immunoreactive. The ciliated surfaces were strongly immunoreactive (arrowheads).

Nitric oxide may also have a role during olfactory signal transduction of fish OSNs. Nitric oxide synthase is expressed robustly in the olfactory epithelium. In the sea lamprey, substrate blockade of this enzyme by arginine derivatives inhibited olfactory response to L-arginine (Zielinski *et al.*, 1996).

4.3. The Specificity of Odorant Detection: Odorant Receptors

Physiological cross-adaptation experiments (Sutterlin and Sutterlin, 1971; Hara, 1982; Thommesen, 1982; Caprio and Byrd, 1984; Ohno *et al.*, 1984; Michel and Debridge, 1997) and biochemical ligand-binding competition studies (Cagan and Zeiger, 1978) conducted over 25 years ago initiated the search for receptor proteins binding onto the many amino acids and bile acid fish odorants. This approach revealed relatively independent ORs for acidic, basic, and neutral amino acids (Brown and Hara, 1981; Hara, 1982, 1992, 2005; Bruch and Rulli, 1988) and for bile acids (Zhang and Hara, 1994). We know now that as in other vertebrates (Buck and Axel, 1991; Dulac and Axel, 1995) teleost OSNs express G-protein–coupled receptor proteins (Barth *et al.*, 1996; Byrd *et al.*, 1996; Freitag *et al.*, 1997). The approaches of EOG cross-adaptation, odorant mixture experiments and single-unit recording in the olfactory epithelium in brown and rainbow trout have shown that amino acids, a bile acid, and a PGF activate independent receptors. However, within amino acids, receptor types are only partially independent, with increasing concentration of an amino acid recruiting responding neurons (Laberge and Hara, 2004).

The ORs are seven transmembrane (serpentine) proteins; all are members of the rhodopsin G-protein–coupled receptor family. The OR genes form genomic clusters that are scattered on many chromosomes. Each gene is ~310 codons long on average, without any introns in their coding regions, and the size of the OR repertoire may be 100 genes in fish (Dugas and Ngai, 2001).

In mammals, the OR-type genes are found in the olfactory epithelium. V2R genes are in vomeronasal cells with cell bodies in the basal region of the vomeronasal epithelium, and the V1R genes are in vomeronasal cells with cell bodies in the apical part epithelium. The expression of V2R receptors may receive endocrine modulation as the mouse V2R receptor increases expression following testosterone treatment (Alekseyenko *et al.*, 2006).

Transgenic and *in situ* hybridization studies in fish have shown that OR-type genes are found in ciliated OSNs, and the V2R-families, intermediate

---

(E) Microvillous olfactory sensory neurons were $G_{\alpha 0}$ immunoreactive. The cell bodies of these cells were in the upper half of the olfactory epithelium, and relatively thick dendrites were prominent (arrows). (F) Crypt cells (arrows) were identified by their ovoid shape in the apical region of the olfactory epithelium. These cells were $G_{\alpha o}$ positive. (Adapted from Hansen and Zielinski, 2005 with kind permission of Springer Science and Business Media.)

and short OSNs, and presumptive microvillous OSNs (Cao *et al.*, 1998; Speca *et al.*, 1999; Hansen *et al.*, 2003, 2004; Sato *et al.*, 2005); and a single V1R may be in crypt or microvillous OSNs (Asano-Miyoshi *et al.*, 2000; Pfister and Rodriguez, 2005). The OR-type and also the V2R-type receptor genes have been identified in several species [zebrafish, *Danio rerio* (Barth *et al.*, 1996; Byrd *et al.*, 1996; Hashiguchi and Nishida, 2005; Sato *et al.*, 2005); catfish, *Ictalurus punctatus* (Ngai *et al.*, 1993); goldfish, *Carassius auratus* (Cao *et al.*, 1998); medaka, *Oryzias latipes* (Yasuoka *et al.*, 1999; Kondo *et al.*, 2002); pufferfish, *Fugu rubripes* (Asano-Miyoshi *et al.*, 2000); carp, *Cyprinus carpio*; Atlantic salmon, *Salmo salar* (Wickens *et al.*, 2001)]. The V1R-type receptors have been shown in zebrafish, medaka, telapia (*Oryzias latipes*) as well as in *Takifugu rubripes* and *Tetraodon nigroviridis* (Pfister and Rodriguez, 2005).

The OR gene categories class I and class II are expressed in mammals, whereas only class I groups, which may bind water soluble odor ligands, are present in teleosts. These class I OR genes may have evolved from approximately two ancestral genes common to most fish species (Irie-Kushiyama *et al.*, 2004). Niimura and Nei (2005) have shown that the OR gene family is more diverse in fish (zebrafish, *Danio rerio* and pufferfish, *Fugu rubripes*) than in mammals, with eight of nine ancestral OR gene groups present in fish, while only two of these nine groups are in mammals. Fish may have retained ancestral genes with functional properties for aquatic chemoreception or amplified OR genes. OR gene family groups present in birds or mammals are absent or rare in fish; these may have been lost from fish.

A single V1R gene is expressed in fish (medaka, *Danio rerio*, *Takifugu rubripes*, *Tetraodon nigroviridus*). The apical location of the V1R cells means that these may be crypt or microvillous OSNs, and the uniqueness of this receptor indicates the importance of the unknown ligand that interacts with this receptor (Pfister and Rodriguez, 2005).

The V2R gene family is subdivided into 12 subfamilies with 2 subfamilies containing most genes in zebrafish (Hashiguchi and Nishida, 2005). Broad specificity, rather than group-restricted binding has been shown by functional studies carried out on a single V2R receptor from goldfish, receptor 5.24. Researchers in John Ngai's lab (Luu *et al.*, 2004) showed detection of several amino acids at the 20–1000 μM range and to the basic amino acids arginine and lysine in the micromolar range. They used a molecular modeling strategy to identify a binding pocket for receptor 5.24's amino acid residues used for ligand binding and selectivity, followed by site-directed mutagenesis to confirm many aspects of the model through receptor activity assays in a heterogenous vector.

With the availability of transgenic zebrafish for the OR and V2R receptors, loss-of-function paradigms may be fruitful in examining odor ligand–receptor specificity, unless of course, there is redundancy in amino acid receptors or receptors for particular chemical structures. These techniques may also lead to an understanding of the regulation of ligand receptor expression (Feng *et al.*, 2005).

4.4. Odorant Responses

It now appears that OSN types for certain odorants/pheromones are distributed randomly over the olfactory epithelium and converge onto a rough chemotopic pattern in the olfactory bulb with slight variation in different fish species (Table 1.2). With substantial evidence pointing to specificity with respect to central projections, odor ligand transduction, and receptor expression for the microvillous, ciliated, and crypt OSNs, the idea of odorant responses from these cell types will now be addressed. The ciliated and microvillous OSNs converge onto spatially distinct glomerular territories, and the odors activating these cells are likewise represented spatially in the olfactory bulb (Table 1.2).

Many single-unit studies show responses by single units to several amino acids (MacLeod, 1976; Bodznick, 1978; Meredith, 1981; Kang and Caprio, 1995a,b,c). Whole-cell voltage-clamp by Sato and Suzuki (2001) showed that microvillous OSNs respond to amino acid odorants in rainbow trout, and OR studies by Speca *et al.* (1999) and Luu *et al.* (2004) revealed microvillous OSNs expressing the V2R arginine receptor in goldfish. Experimental data from live tissue imaging, electrophysiology, tract tracing, and homologous gene expression experiments show a partial correlation between odorants and OSN phenotype in cyprinids (Table 1.2). For example, in zebrafish and channel catfish, amino acid responses extended beyond territories innervated by microvillous OSNs. Olfactory bulb regions receiving synaptic input from microvillous OSNs as well as from other bulbar regions responded to amino acids. In zebrafish, amino acids stimulated rostrolateral and lateral bulbar regions (Friedrich and Korsching, 1998), innervated largely by microvillous OSNs but also sparsely by ciliated OSNs (Sato *et al.*, 2005). In channel catfish, amino acids stimulated dorsolateral (Nikonov and Caprio, 2001), receiving microvillous OSN axons (Hansen *et al.*, 2003), as well as ventrolateral regions (Nikonov and Caprio, 2001), innervated by ciliated OSNs (Hansen *et al.*, 2003). Taken together, these data support that microvillous OSNs are amino acid detectors, but suggest that ciliated OSNs may also respond to amino acids. Ontogenetic development of ciliated OSNs with amino acid and bile acid responses preceding microvillous OSN development in rainbow trout (Zielinski and Hara, 1988) supports this idea.

The linkage between ciliated OSNs and bulbar responses to bile acids is evident in zebrafish and channel catfish. In zebrafish, bile acid responses were seen from the rostromedial region of the olfactory bulb (Friedrich and Korsching, 1998), which contains glomeruli from ciliated OSNs (Sato *et al.*, 2005). In the channel catfish, bile acid responses were recorded from dorsomedial and ventramedial sites (Nikonov and Caprio, 2001), where DiI injections by Hansen *et al.* (2003) largely labeled ciliated OSNs but also some microvillous OSNs. Whole-cell clamp data support the idea that bile acids stimulate ciliated OSNs (Sato and Suzuki, 2001). In salmonids and cyprinids, both ciliated OSNs and microvillous OSNs respond to amino acids (Erickson and Caprio, 1984; Ma and Michel, 1998; Speca *et al.*, 1999; Nikonov and Caprio, 2001; Sato and Suzuki, 2001; Hansen *et al.*, 2003).

Nucleic acids, which evoke unconventional EOG wave forms (Rolen *et al.*, 2002), stimulate olfactory bulb responses from caudolateral recording sites in cyprinid species [zebrafish (Friedrich and Korsching, 1998); goldfish, channel catfish (Nikonov and Caprio, 2001)]. The caudolateral olfactory bulb receives OSN projections from microvillous OSNs in zebrafish (Sato *et al.*, 2005) and channel catfish (Hansen *et al.*, 2003).

In the crucian carp, *Carassius carassius*, alarm response to conspecific skin extract stimulated the posterior medial portion of the olfactory bulb, which receives input from ciliated OSNs (Hamdani and Døving, 2002, 2003).

In whitefish (*Coregonus clupeaformis*), PGF stimulated unidentified OSNs that project beyond the olfactory bulb to the ventral medial telencephalon (Laberge and Hara, 2003). In zebrafish, reproductive pheromone 17,20 βP stimulated the ventromedial region of the olfactory bulb (Freidrich and Korsching, 1998), an area which contains microvillous OSN innervation in this species. In the crucian carp, crypt cells project to ventral olfactory bulb, where steroid and prostaglandin responses take place (Lastein *et al.*, 2006). On the other hand, Schmachtenberg (2006) found crypt cells responses to relatively high concentrations (1 mM) of amino acids, and electrical coupling to sustentacular cells, in the Pacific jack mackerel (*Trachurus symmetricus*).

Much work needs to be done to gain a full understanding of how OSNs are able to respond to the diverse olfactory signaling molecules, and how these mechanisms have diverged (or converged) in the many different taxa. Continued studies using genetic (Higashijima *et al.*, 2003; Vitebsky *et al.*, 2005), physiological, and morphological probes of appropriate fish models offer promising avenues of progress toward this goal.

5. THE OLFACTORY BULB

The OSN synaptic inputs representing each odorant/pheromone are integrated in the olfactory bulb and are channeled toward the central locations for memory or for specific motor responses. While there is a general understanding of cells comprising the olfactory bulb, knowledge of specific neural mapping and interactions that drive this function is an ongoing task. Here, the contributions of chemotopy, lateral inhibition, and temporal factors are important factors.

5.1. Neural Composition

While there is considerable evidence for an OSN subpopulation extending directly into the ventral telencephalon, the olfactory bulb receives overwhelming olfactory sensory input. Centrifugal fibers from the brain also extend into the olfactory bulb (Figure 1.2). Mitral cells and ruffed cells are the output neurons of the olfactory bulb; granule cells and juxtaglomerular cells are interneurons that modulate neural activity within the olfactory bulb.

The olfactory bulb is generally divided into three or four concentric layers. The superficial layer, the olfactory nerve layer, contains OSN axons (Figure 1.5) and the innermost layer is the internal cell or the granule cell layer (Byrd and Brunjes, 1995). Between, either a single layer—the glomerular layer containing mitral cells—or separate glomerular and mitral cell layers are denoted.

The olfactory glomeruli are roughly spherical masses of neuropil containing the glutamatergic synapses between sensory neuron axons (Figure 1.5) and the dendrites of bulbar projection neurons and interneurons. The glomeruli are arranged in specific pattern in rainbow trout (Riddle and Oakley, 1992) zebrafish (Baier and Korsching, 1994) and in sea lamprey (Frontini *et al.*, 2003). There are seven main groups in each, but there are considerable differences between size and location of medial glomeruli between zebrafish. The other domains: dorsal cluster, anterior plexus, lateral chain, ventral ring, and ventral cluster are conserved in Agnatha [lamprey (Frontini *et al.*, 2003)], Ostariophysi [zebrafish (Baier and Korsching, 1994)], and in the Acanthopterygii [round goby (unpublished results)].

The mitral cells are distributed diffusely between the olfactory nerve layer and the internal cell layer. Allison (1953), Satou (1990), and Dryer and Graziadei (1994) described mitral cells with several main dendrites reaching out to branch in two or more glomeruli. Surveys of mitral cells by confocal microscopy in zebrafish (Fuller and Byrd, 2005a) and lamprey (personal observations) have included mitral cells with single dendrites. These mitral cells may be sampling neural activity from single glomeruli rather than from multiple glomeruli. The output axons of mitral cells form separate lateral and medial tracts in cyprinids, and some other teleosts.

The ruffed cells, with a lamellar ruffle located along the proximal axonal segment, are much less common than the mitral cells. The ruffed cells, unique to fish, are located in the glomerular layer and internal cell layer in zebrafish, and may be present in all fish species, but with slight differences with respect to the positioning of the ruff (Fuller and Byrd, 2005a). During odorant application, the ruffed cells and the mitral cells show contrasting interactions (Zippel *et al.*, 1999).

Edwards and Michel (2002) describe several neuronal phenotypes in the adult zebrafish olfactory bulb based on labeling with γ-aminobutyric acid (GABA), glutamate, and tyrosine hydroxylase. Mitral cells are labeled with glutamate, and juxtaglomerular cells are labeled with either tyrosine hydroxylase or with GABA and tyrosine hydroxylase.

The internal cell layer contains granule cells, interneurons intrinsic to the olfactory bulb. The granule cells do not have axons; the dendrites extend superficially into the mitral cell layer, where they form dendro-dendritic synapses onto mitral cells. Satou (1990) has observed short axon cells in the olfactory bulb of carp.

The olfactory bulb also contains centrifugal fibers extending from the telencephalic hemisphere and terminating mainly in the internal cell layer, probably onto granule cells. Most are ipsilateral, but also some extend from contralateral telencephalic hemisphere.

There are terminal nerve fibers located in the rostromedial part of the olfactory bulb in the transition between the caudoventral olfactory bulb and the ventrorostral telencephalon (Uchida *et al.*, 2005). These cells, called ganglion cells of the terminal nerve, are multipolar or bipolar neurons that project caudally to various areas.

5.2. Information Flow

Methods for observing neural activity in the olfactory bulb have included electroencephalographic (EEG) recording from neuronal populations, single-unit recordings (Kang and Caprio 1995b,c; Zippel *et al.*, 1999; Nikonov and Caprio, 2004), and imaging techniques with synthetic calcium or voltage indicators, or with transgenic animals (zebrafish) expressing fluorescent proteins. The EEG records slow rhythmic waves of about 3–15 Hz following stimulation of the olfactory epithelium, and similar waves have been recorded from the olfactory tract in teleosts with a pedunculated olfactory bulb. Some olfactory tract fibers discharge synchronously with the olfactory bulb. These olfactory tract waves are summated action potentials. There is a rhythmic discharge pattern of teleost secondary neurons and this may be due to membrane potential oscillation from inputs: tonic olfactory nerve, dendro-dendritic synaptic interactions between secondary neurons and granule cells (Satou and Ueda, 1978). The granule cell dendrites may contribute to the rhythm generating through dendro-dendritic synaptic interactions between mitral cells and granule cells.

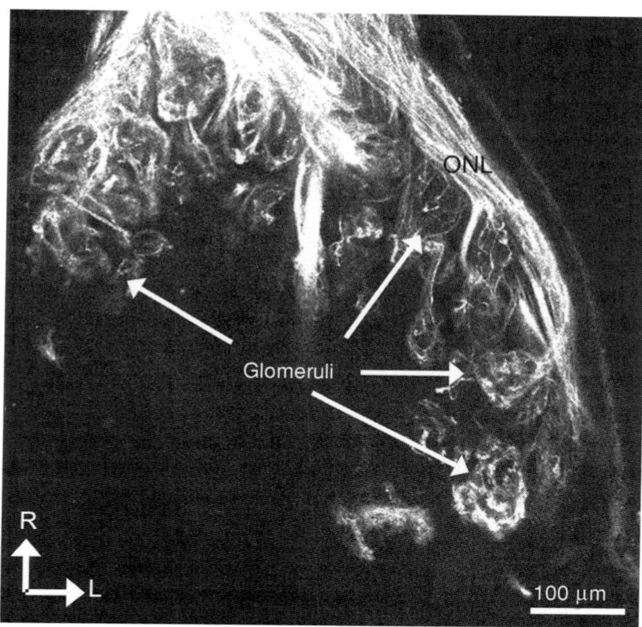

Fig. 1.5. Fluorescence image of a horizontal section from the olfactory bulb of a round goby *Neogobius melanostomn*, showing diI labeled olfactory sensory neuron axons. The diI was applied into the nasal cavity, *post mortem*, thus allowing for anterograde filling of olfactory sensory neurons. Labeled axons of olfactory sensory neurons can be seen in the outermost layer of the olfactory bulb, the olfactory nerve layer (ONL). The endings of these axons form spherical groupings in the olfactory glomeruli (arrows). R is rostral; L is lateral.

The odors activating the ciliated and microvillous OSNs converging onto spatially distinct glomerular territories are represented spatially in the olfactory bulb (Table 1.2). EEG topography has been shown in salmonids (Thommesen, 1978; Døving *et al.*, 1980; Hara and Zhang, 1996, 1998) and unit/few unit patterns in channel catfish (Nikonov and Caprio, 2001). Imaging in the olfactory bulb has been done by Friedrich and Korsching (1997, 1998) and Fuss and Korsching (2001). In the zebrafish, different odors show specific spatial activity in OSN afferents with some overlap of responses for odorants with similar structure in these afferent maps. In catfish, representation of odor quality is maintained in a specific pattern in the olfactory bulb and the brain (Nikonov *et al.*, 2005). Functional partitioning between lateral and medial olfactory bulbs has been observed with the lateral part responding to amino acids and the medial part responding to bile acids in salmonids (Thommesen, 1978; Døving *et al.*, 1980; Hara and Zhang, 1998; Laberge and Hara, 2003),

Table 1.2

A Summary of the Spatial Coordinates for Projection of Ciliated Olfactory Sensory Neurons into the Olfactory Bulb, and of Olfactory Bulb Responses to Odorant Classes

	OSN projections to the olfactory bulb		Olfactory bulb response		
Genus species Reference	Ciliated	Microvillous	Amino acids	Bile acids	Steroids, prostaglandins
Danio rerio Friedrich and Korsching, 1998; Sato *et al.*, 2005	Rostromedial Medial Dorsal Lateral (some)	Rostrolateral Lateral Lateroposterior Ventral	Rostrolateral Lateral	Rostromedial	Center ventral
*Ictalurus punctatus** Hansen *et al.*, 2003; Nikonov and Caprio, 2001	Medial Rostroventral Lateral	Dorsolateral Rostrodorsal Caudodorsal	Dorsolateral Ventrolateral	Dorsomedial Ventromedial	Unknown
Carassius carassius Hamdani *et al.*, 2001, 2002	Caudomedial Lateral	Lateral	Not tested	Not tested	Center ventral
Onchorynchus mykiss Hara and Zhang, 1998	Unknown	Unknown	Lateral^{+++} $\text{Caudolateral}^{+++}$ Anterior^{+}	Medial^{+} Lateral^{+++}	Undetected[†]

Salvelinus namaycush					
Hara and Zhang, 1998	Unknown	Unknown	Lateral^{+++} Caudolateral^{+++} Anterior^{+}	Medial^{+} Lateral^{++}	Undetected†
Salmo trutta					
Hara and Zhang, 1998	Unknown	Unknown	Lateral^{+++} Caudolateral^{+++} Anterior^{++} Medial^{+}	Medial^{+++} Lateral^{+++}	Undetected†
Coregonus clupeaformis					
Hara and Zhang, 1998; Laberge and Hara, 2003	Unknown	Unknown	Lateral^{+} Caudolateral^{+} Anterior^{+}	Central	Ventromedial telencephalon**

*In *Ictalurus punctatus*, DiI injections into two ventral areas, rostral to the central region, and caudal, loaded crypt cells (Hansen *et al.*, 2003).
**Responses to F-prostaglandins were recorded not from the olfactory bulb, but from the ventromedial telencephalon.
†When recorded electroencephalographically (EEG).
+Weak response.
++Strong response.
+++Very strong response.

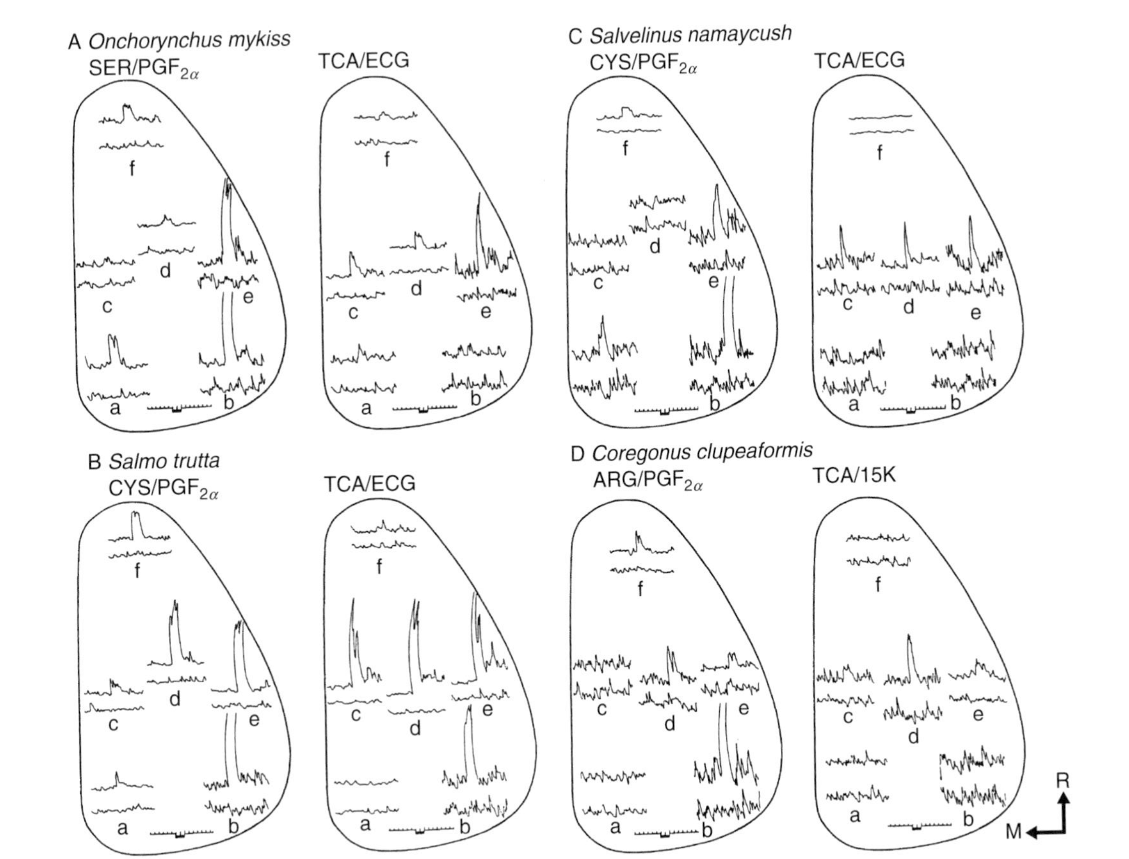

A *Onchorynchus mykiss*
SER/PGF$_{2\alpha}$
TCA/ECG
B *Salmo trutta*
CYS/PGF$_{2\alpha}$
TCA/ECG
C *Salvelinus namaycush*
CYS/PGF$_{2\alpha}$
TCA/ECG
D *Coregonus clupeaformis*
ARG/PGF$_{2\alpha}$
TCA/15K
a
b
c
d
e
f
R
M

zebrafish (Friedrich and Korsching, 1998), and catfish (Nikonov and Caprio, 2001). In salmonids, these chemotopic zones are not strictly adhered to as broad regions of the olfactory bulb respond to amino acid stimuli (Figure 1.6, Table 1.2). The EEG studies in salmonids showed overlapping chemotopic maps for unrelated odors (Hara and Zhang, 1998). In cyprinids, chemotopic zones extend to the output neurons, with the secondary olfactory neurons of the lateral olfactory tract (LOT) responsible for food odors, including amino acids (Von Rekowski and Zippel, 1993; Hamdani *et al.*, 2001a), while medial olfactory bulb and medial olfactory tract (MOT) receiving social or reproductive related stimulus (Demski and Dulka, 1984; Sorensen *et al.*, 1991; Hamdani *et al.*, 2000; Weltzien *et al.*, 2003). Retrograde dye-labeling studies in cyprinids have also shown convergence of microvillous, ciliated, or crypt OSNs onto fairly large olfactory bulb domains (Hansen *et al.*, 2004). In zebrafish, the microvillous and ciliated OSN projections were seen in adjacent glomerular territories (Sato *et al.*, 2005). These studies show that the olfactory bulb is not a clearly chemotopic structure with respect to OSN projections and odorant responses.

In four salmonid species, amino acid responses extended beyond lateral territories to rostral and medial regions of the olfactory bulb (Figure 1.6, Table 1.2; Zhang and Hara, 1994). Both the activation of different parts of the olfactory bulb and the presence of multiple glomeruli within certain response modules suggest a combinatorial odor code involving multiple odorant receptors and multiple glomeruli underlying the perception of even single, pure odorants. When considering spatial organization for odorant coding, the concentration of the stimulating odorant should be taken into consideration. In single-unit responses from rainbow trout, the neurons gained responsiveness to more amino acids when the stimulus concentration was increased (Laberge and Hara, 2004). This finding is corroborated by experiments with the goldfish 5.44 V2R receptor on olfactory sensory neurons, showing high affinity for arginine and low-affinity binding to other amino acids (Luu *et al.*, 2004). Alternatively, the two dendrites extending

---

Fig. 1.6. Spatial distributions of EEG responses to L-serine, $PGF_{2\alpha}$, taurocholic acid (TCA), and etiocholanolone-3α-ol-17-one glucuronide (ECG) onto the surface of the right olfactory bulb in (A) rainbow trout (*Onchorynchus mykiss*), (B) brown trout (*Salmo trutta*), (C) lake char (*Salvelinus namaycush*), and (D) lake whitefish (*Coregonus clupeaformis*). Recordings were made at six positions: (a) medioposterior, (b) lateroposterior, (c) mid-medial, (d) central, (e) mid-lateral, and (f) rostral. For illustration purposes, response to two chemicals are superimposed at each position; for example, L-serine/$PGF_{2\alpha}$ indicates that the upper tracings are for L-serine and the lower tracings for $PGF_{2\alpha}$ and TCA/ECG for TCA and ECG. Responses to L-serine at position (b) scaled off due to their large size, and a large portion of these responses were therefore cut off and on the bottom portion was shown. Stimulus duration of 10 sec is shown at the bottom. R, rostral; M, medial. (Adapted from Hara and Zhang, 1998.)

from mitral cells may be converging olfactory sensory input from separate glomeruli stimulated by the different amino acids (Allison, 1953; Fuller and Byrd, 2005b).

Diversity in chemotopy is seen from the different spatial odorant responses to candidate reproductive pheromones. Prostaglandins stimulate specific glomeruli in zebrafish and the ventral telencephalon in whitefish. In the goldfish olfactory bulb, responses to hormonal metabolites sex pheromones by extracellular single-unit recordings showed presumed mitral cells, the medial part of the olfactory bulb, responding by excitation or by inhibition (Fujita *et al.*, 1991). In all of these, the pheromone stimulated medial rather than lateral domains.

The afferent map may be preserved or modified in bulbar-processing steps beyond the synaptic activity between OSN afferents (Nikonov *et al.*, 2002). The chemotopic bulbar activation patterns and individual mitral cells are not specific to the individual amino acids. This effect can be explained by a combinatorial process for assessing odorant quality. For example, arginine selective units may be responding to amino acids with three methylene groups and a terminal amide or guanidium group. However, low-affinity responses for high odorant concentrations sometimes reported in the literature must be considered here. For example, alanine units responded to similar amino acids as well (Nikonov and Caprio, 2001). The presence of olfactory bulb neurons not specific to an odor class (Hanson and Sorensen, 2001; Masterman *et al.*, 2001) counters the combinatorial strategy for odor identification by the olfactory bulb.

Suppressive responses from mitral cells to odorants likely enhance contrast. For example, a mitral cell may be stimulated by arginine or suppressed by methione and alanine, and cells responding to a particular amino acid may be very far apart.

In nature, the likely scenario for the odor landscape is that of mixtures rather than single compounds. Behaviorally, fish are able to identify mixtures (Valentincic *et al.*, 2000). When odor mixtures were applied, single units that responded to one component also responded to the mixture, and the mixture interactions were weak in rainbow trout (Laberge and Hara, 2001, 2004) and zebrafish (Tabor *et al.*, 2004), although there is some evidence for mixture effects at the cellular level in catfish (Kang and Caprio, 1997). At some level, the brain discards the information about individual constituents when an odor mixture is presented, and individual components acquire a mixture-specific property, the naturally occurring circumstance. This information may reside more with the relative response intensity rather than the identity of active glomeruli (Tabor *et al.*, 2004). These mixture effects may arise from olfactory bulb interneurons as sometimes mitral cells that were excited by a single component were not responsive when that component was presented in a mixture (Tabor *et al.*, 2004). These may be due to mitral cell dendritic or circuit interactions through local inhibition. Mitral cells and the interneurons form a reciprocally connected network that processes sensory input and produces temporally patterned olfactory bulb output activity. This decorrelation of mitral cell outputs by interneuron activity was supported through spatiotemporal activity patterns across thousands of neurons through calcium imaging in zebrafish genetically encoded cell-type markers (Yaksi and Friedrich, 2006). There may also be temporal coding taking place here (Friedrich and Laurent, 2001) during the time course of a stimulus, and this may be important for identifying responses and channeling these responses for integration (Friedrich *et al.*, 2004) and motor output. In addition to these functions, the olfactory bulb may participate in memory formation (Satou *et al.*, 2005).

6. CENTRAL PROCESSING OF OLFACTORY SIGNALS

The output neurons from the olfactory bulb project through the olfactory tracts largely to telencephalic territories. In fish with pedunculated olfactory bulbs, anatomically distinct lateral and LOTs and MOTs convey behaviorally distinct information. These two distinct nerve bundles were observed by Sheldon (1912) in the carp from the medial and lateral parts of the olfactory bulb, respectively. This spatial separation may also be present in other species. In carp, the LOT is used for feeding behavior (Von Rekowski and Zippel, 1993; Hamdani *et al.*, 2001b) and the MOT is used for reproductive behavior and alarm behavior (Stacey and Kyle, 1983; Demski and Dulka, 1984; Sorensen *et al.*, 1991; Hamdani *et al.*, 2000; Weltzien *et al.*, 2003).

The MOT and LOT tract bundles contain fibers that originate from different portions of the olfactory bulb (Dubois-Dauphin *et al.*, 1980; Satou *et al.*, 1983; Satou, 1990). Each bundle connects to different brain areas, although some overlap between bundles seems to exist (Finger, 1975; Rooney *et al.*, 1992). The MOT axons project to the preoptic area in *Carassius auratus* (Oka *et al.*, 1982; Von Bartheld *et al.*, 1984) and the cod (*Gadus morhua*) (Rooney *et al.*, 1992). Axons from the ganglion cells of the terminal nerve run through the ventromedial part of the olfactory tract in the carp. Specific regions of the telencephalon and diencephalon receive LOT and MOT projections. The terminal fields are in the telencephalon, a terminal field in the hypothalamus.

In sturgeon, *Acipenser baeri*, secondary olfactory projections extend to the dorsomedial telencephalon, anterior commissure, dorsocentral telencephalon, preoptic area, thalamus, habenula, and hypothalamus (Heusa *et al.*, 2005). In cod, *Gadus morhua* L., secondary projections labeled the telencephalon, anterior commissural and preoptic areas, habenula, dorsal thalamus, and the diencephalons (Rooney *et al.*, 1992). The LOT projected to lateral, dorsal posterior regions of the dorsal telencephalon, and the ventral region of the telencephalon. The lateral MOT projected to the lateral and central regions of the dorsal telencephalon. The medial fibers of the MOT projected to the ventral telencepahlon and the posterior region of the dorsal telecephalon. In goldfish, secondary olfactory projections were traced to the ventral and dorsal lateral areas of the telencephalon, and the preoptic area and the posterior tubercule in the diencephalons. Fibers were seen crossing to the contralateral side in the anterior and habenular commissures also (Von Bartheld *et al.*, 1984) Telencephalic–olfactory connections have also been characterized in rainbow trout (Matz, 1995; Folgueira *et al.*, 2004a,b).

In summary, injection of tracers to the olfactory bulb in teleost fish results in labeling in discrete areas of the dorsal and ventral telencephalon and in specific regions of the diencephalon.

Two previous studies have documented the central neural connectivity of the olfactory bulb in lampreys, and spatial aspects are under investigation. In the silver lamprey (*Ichthyomyzon unicuspis*) and river lamprey (*Lampetra fluviatilis*), there were connections between the olfactory bulb and dorsal and lateral pallium, the septum and preoptic area (Northcutt and Puzdrowski, 1988; Polenova and Vesselkin, 1993). The pallium refers to broad areas in the medial, dorsal, and lateral telencephalon. In the river lamprey, the olfactory bulb connected to pallial areas, primordial hippocamp, and the dorsal thalamus. In these species, as well as in the sea lamprey (*Petromyzon marinus*; Chang, 2006; Ren *et al.*, 2006), there are projection differences between medial and lateral territories.

7. CONCLUDING REMARKS

The nasal cavity of fish enables the interaction of odorants to odor receptors and olfactory activity. There is diversity in the anatomy of the nasal cavity and the development of one or two accessory nasal sacs that assist in guiding the flow of odorants over the olfactory epithelium, particularly in the Acanthopterygii.

The ciliated and microvillous olfactory OSNs and crypt cells are intermixed in the olfactory epithelium but project to spatially defined glomerular territories in the olfactory bulb. There is also a population of OSNs projecting to extrabulbar targets. The transduction cascades using cAMP and IP_3 are found in ciliated and microvillous OSNs, respectively. There are few mechanistic studies of interactions between odor ligand binding to specific characterized receptors and consequent channel activity. With increased availability of transgenic animals expressing fluorescently expressed receptors in fish, such as the OR, V1R, and V2R types, we look forward to exciting progress toward understanding the interactions between receptors and specific chemical moieties to channel interactions. Correlative physiological studies, pairing *in vivo* electrophysiological activity with behavior, such as those for trout amino acid receptors by Hara and collaborators (Hara, 2005), will be particularly informative. Here, we suggest that researchers consider the odorants and behavioral responses of fish in their natural environment such as the amino acid feeding behavior observed by Hara.

The olfactory bulb is important for odorant recognition and projects olfactory sensory input into appropriate brain regions. However, the mechanism for channeling through the olfactory bulb is not entirely understood, and the specific role of interneurons in olfactory memory and odor recognition is also unknown. Clearly, we have a long way to go to understand fish olfaction. Progress since Linda Buck and Richard Axel's discovery of ORs 15 years ago has triggered an outburst of research activity of the olfactory system. We look forward to answers of these exciting questions in the future.

ACKNOWLEDGMENTS

Barbara S. Zielinski is grateful to assistance from Rachelle Belanger, Steven Chang, Alyson Laframboise, Yolanta Kita, Dina Kokh, Heather Macdonald, Dr. Xiang Ren, and Jessica Vaissica, and to NSERC and the Great Lakes Fishery Commission for research support.

REFERENCES

Abogadie, F. C., Bruch, R. C., and Farbman, A. I. (1995). G-protein subunits expressed in catfish olfactory receptor neurons. *Chem. Senses* **20,** 199–206.

Alekseyenko, O. V., Baum, M. J., and Cherry, J. A. (2006). Sex and gonadal steroid modulation of pheromone receptor gene expression in the mouse vomeronasal organ. *Neuroscience* **140,** 1349–1357.

Allison, A. C. (1953). The structure of the olfactory bulb and its relationship to the olfactory pathways in the rabbit and the rat. *J. Comp. Neurol.* **98,** 309–353.

Anadon, R., Manso, M. J., Rodriguez-Moldes, I., and Becerra, M. (1995). Neurons of the olfactory organ projecting to the caudal telencephalon and hypothalamus: A carbocyanine-dye labelling study in the brown trout (Teleostei). *Neurosci. Lett.* **191,** 157–160.

Asano-Miyoshi, M., Suda, T., Yasuoka, A., Osima, S., Yamashita, S., Aber, K., and Emori, Y. (2000). Random expression of main and vomeronasal olfactory receptor genes in immature and mature olfactory epithelia of *Fugu rubripes*. *J. Biochem.* **127,** 915–924.

Baier, H., and Korsching, S. (1994). Olfactory glomeruli in the zebrafish from an invariant pattern and are identifiable across animals. *J. Neurosci.* **14,** 219–230.

Barth, A. L., Justice, N. J., and Ngai, J. (1996). Asynchronous onset of odorant receptor expression in the developing zebrafish olfactory system. *Neuron* **16,** 23–34.

Becerra, M., Manso, M. J., Rodriguez-Moldes, I., and Anadon, R. (1994). Primary olfactory fibers project to the ventral telencephalon and preotic region in trout (*Salmo trutta*): A developmental immunocytochemical study. *J. Comp. Neurol.* **342,** 131–143.

Belanger, A. J., Arbuckle, W. J., Corkum, L. D., Gammon, D. B., Li, W., Scott, A. P., and Zielinski, B. S. (2004). Behavioural and electrophysiological responses by reproductive female *Neogobius melanostomus* to odours released by conspecific males. *J. Fish Biol.* **65,** 933–946.

Belanger, R. M., Smith, C. M., Corkum, L. D., and Zielinski, B. S. (2003). Morphology and histochemistry of the peripheral olfactory organ in the round goby, *Neogobius melanostomus* (Teleostei: Gobiidae). *J. Morphol.* **257,** 62–71.

Belanger, R. M., Corkum, L. D., Li, W., and Zielinski, B. S. (2006). Olfactory sensory input increases gill ventilation in male round gobies (*Neogobius melanostomus*) during exposure to steroids. *Comp. Biochem. Physiol. A Mol. Integr. Physiol.* **144,** 196–202.

Bjerselius, R., and Olsen, K. H. (1993). A study of the olfactory sensitivity of the crusian carp (*Carassius carassius*) and goldfish (*Carassius auratus*) to 17α,20β-dihydroxyprogesterone and prostaglandin F2α. *Chem. Senses* **18,** 427–436.

Bodznick, D. (1978). Characterization of olfactory bulb units of sockeye salmon with behaviourally relevant stimuli. *J. Comp. Physiol. A* **127,** 147–155.

Brown, G. E., Adrian, J. C., Jr., Naderi, N. T., Harvey, M. C., Harvey, M. C., and Kelly, J. M. (2003). Nitrogen oxides elicit antipredator responses in juvenile channel catfish, but not in convict cichlids or rainbow trout: Conservation of the ostariophysan alarm pheromone. *J. Chem. Ecol.* **29,** 1781–1796.

Brown, S. B., and Hara, T. J. (1981). Accumulation of chemostimulatory amino acids by a sedimentable fraction isolated from olfactory rosettes of rainbow trout (*Salmo gairdneri*). *Biochim. Biophys. Acta* **675,** 149–162.

Bruch, R. C., and Rulli, R. D. (1988). Ligand binding specificity of a neutral L-amino acid olfactory receptor. *Comp. Biochem. Physiol. B* **91,** 535–540.

Buck, L., and Axel, R. (1991). A novel multigene family may encode odorant receptors: A molecular basis for odor recognition. *Cell* **65,** 175–187.

Burne, R. H. (1909). The Anatomy of the olfactory organ of Teleostean fishes. *Proc. Zool. Soc. Lond.* **2,** 610–663.

Byrd, C. A., and Brunjes, P. C. (1995). Organization of the olfactory system in the adult zebrafish: Histological, immunohistochemical and quantitative analysis. *J. Comp. Neurol.* **358,** 247–259.

Byrd, C. A., Jones, J. T., Quattro, J. M., Rogers, M. E., Brunjes, P. C., and Vogt, R. G. (1996). Ontogeny of odorant receptor gene expression in zebrafish, *Danio rerio*. *J. Neurobiol.* **29,** 445–458.

Cagan, R. H., and Zeiger, W. N. (1978). Biochemical studies of olfaction: Binding specificity of radioactively labelled stimuli to an isolated olfactory preparation from rainbow trout (*Salmo gairdneri*). *Proc. Natl. Acad. Sci. USA* **75,** 4679–4683.

Cao, Y., Oh, B. C., and Stryer, L. (1998). Cloning and localization of two multigene receptor families in goldfish olfactory epithelium. *Proc. Natl. Acad. Sci. USA* **95,** 11987–11992.

Caprio, J. (1982). High sensitivity and specificity of olfactory and gustatory receptors of catfish to amino acids. *In* "Chemoreception in Fishes" (Hara, T. J., Ed.), pp. 109–134. Elsevier, Amsterdam.

Caprio, J., and Byrd, R. P. (1984). Electrophysiological evidence for acidic, basic and neutral amino acids olfactory receptor sites in the catfish. *J. Gen. Physiol.* **84,** 403–422.

Caprio, J., and Finger, T. E. (2003). The olfactory system in catfishes. *In* "Catfishes" (Kapoor, Arratia, Chardon, and Diogo, Eds.). Oxford & IBH Publishing Co., Pvt. Ltd., New Delhi.

Caprio, J., and Raderman-Little, R. (1978). Scanning electron microscopy of the channel catfish olfactory lamellae. *Tissue Cell Res.* **10,** 1–9.

Caprio, J., Dudek, J., and Robinson, J. J., II (1989). Electro-olfactogram and multiunit olfactory receptor responses to binary and trinary mixtures of amino acids in the channel catfish, *Ictalurus Puncatus. J. Gen. Physiol.* **93,** 245–262.

Cordwell, J. R., Stacey, N. E., Tan, E. S. P., McAdam, D. S. O., and Lang, S. L. C. (1995). Androgen increases olfactory receptor response to a vertebrate sex pheromone. *J. Comp. Physiol. A* **176,** 55–61.

Carroll, R. L. (1988). "Vertebrate Paleontology and Evolution." WH Freeman, New York.

Chang, S. (2006). The neural connectivity of olfactory bulb regions in ovulated female lampreys (*Petromyzon marinus*). M.Sc. thesis. University of Windsor.

DellaCorte, C., Restrepo, D., Menco, B. P., Andreini, I., and Kalionski, D. L. (1996). G alpha 9/ G alpha 11: Immunolocalization in the olfactory epithelium of the rat (*Rattus rattus*) and the channel catfish (*Ictalutus punctatus*). *Neuroscience* **74,** 261–273.

Demski, L. S., and Dulka, J. G. (1984). Functional-anatomical studies on sperm release evoked by electrical stimulation of the olfactory tract in goldfish. *Brain Res.* **291,** 241–247.

Døving, K. B., Dubois-Dauphin, M., Holley, A., and Jourdan, F. (1977). Functional anatomy of the olfactory organ of fish and the ciliary mechanism of water transport. *Acta Zool.* **58,** 245–255.

Døving, K. B., Selset, R., and Thommesen, G. (1980). Olfactory sensitivity to bile acids in salmonid fishes. *Acta Physiol. Scand.* **108,** 123–131.

Dryer, L., and Graziadei, P. P. (1994). Mitral cell dendrites: A comparative approach. *Anat. Embryo.* **189,** 91–106.

Dubois-Dauphin, M., Døving, K. B., and Holley, A. (1980). Topographical relation between the olfactory-bulb and the olfactory tract in tench (*Tinca tinca* L.). *Chem. Senses* **5,** 159–169.

Dugas, J. C., and Ngai, J. (2001). Analysis and characterization of an odorant receptor gene cluster in the zebrafish genome. *Genomics* **71,** 53–65.

Dulac, C., and Axel, R. (1995). A novel family of genes encoding putative pheromone receptors in mammals. *Cell* **83,** 195–206.

Eastman, J. T., and Lanoo, M. J. (2001). Anatomy and histology of the brain and sense organs of the Antarctic eel. *Muraenolepis microps. J. Morph.* **250,** 34–50.

Edwards, J. G., and Michel, W. C. (2002). Odor stimulated glutamatergic neurotransmission in the zebrafish olfactory bulb. *J. Comp. Neurol.* **454,** 294–309.

Eisthen, H. L. (1997). Evolution of vertebrate olfactory system. *Brain Behav. Evol.* **50,** 222–233.

Erickson, J. R., and Caprio, J. (1984). The spatial distribution of ciliated and microvillous olfactory receptor neurons in the channel catfish is not matched by a differential specificity to amino acid and bile salt stimuli. *Chem. Senses* **9,** 127–141.

Feng, B., Bulchand, S., Yaksi, E., Fredrich, R. W., and Jusuthasan, S. (2005). The recombination activation gene I (Rag I) is expressed in a subset of zebrafish olfactory neurons but is not essential for axon targeting or amino acid detection. *BMC Neurosci.* **6,** 1–12.

Fine, J. M., Vrieze, L. A., and Sorensen, P. W. (2004). Evidence that petromyzontid lampreys employ a common migratory pheromone that is partially comprised of bile acids. *J. Chem. Ecol.* **30,** 2091–2110.

Finger, T. E. (1975). The distribution of the olfactory tracts in the bullhead catfish,. *Ictalurus nebulosus. J. Comp. Neurol.* **161,** 125–142.

Fishelson, L. (1997). Comparative ontogenesis and cytomorphology of the nasal organs in some species of cichlid fish (Cichlidae, Teleoste). *J. Zool.* **243,** 281–294.

Folgueira, M., Anadon, R., and Yanez, J. (2004a). An experimental study of the connections of the telencephalon in rainbow trout (Onchorhynchus mykiss). I: Olfactory bulb in ventral area. *J. Comp. Neurol.* **480,** 180–203.

Folgueira, M., Anadon, R., and Yanez, J. (2004b). Experimental study of the telencephalon in the rainbow trout (Onchorhynchus mykiss). II: Dorsal are and preoptic region. *J. Comp. Neurol.* **480,** 204–233.

Freitag, J., Ludwig, G., and Breer, H. (1997). Evolution of the olfactory receptor gene family. *Chem. Senses* **22,** 682.

Friedrich, R., and Korsching, S. I. (1997). Combinatorial and chemotropic odorant coding in the zebrafish olfactory bulb visualized by optical imaging. *Neuron* **18,** 737–752.

Friedrich, R., and Korsching, S. I. (1998). Chemotopic, combinatorial, and noncombinatorial odorant representations in the olfactory bulb revealed using a voltage-sensitive axon tracer. *J. Neurosci.* **18,** 9977–9988.

Friedrich, R. W., and Laurent, G. (2001). Dynamic optimization of odor representation by slow temporal patterning of mitral cell activity. *Science* **291,** 889–894.

Friedrich, R. W., Habermann, C. J., and Laurent, G. (2004). Multiplexing using synchrony in the zebrafish olfactory bulb. *Nat. Neurosci.* **7,** 862–871.

Frontini, A., Zaidi, A. U., Hua, H., Wolak, T. P., Greer, C. A., Kafitz, K. W., Li, W., and Zielinski, B. S. (2003). Glomerular territorries in the olfactory bulb from the larval stage of the sea lamprey *Petromyzon marinus. J. Comp. Neurol.* **465,** 27–37.

Fujita, I., Sorensen, P. W., Stacey, N. E., and Hara, T. J. (1991). The olfactory system, not the terminal nerve, functions as the primary chemosensory pathway mediating responses to sex pheromones in male goldfish. *Brain Behav. Evol.* **38,** 313–321.

Fuller, C. L., and Byrd, C. A. (2005a). Characterization of three distinct types of output neurons in the adult zebrafish olfactory bulb. *Chem. Senses* **30,** A10.

Fuller, C. L., and Byrd, C. A. (2005b). Ruffed cells identified in the adult zebrafish olfactory bulb. *Neurosci. Lett.* **379,** 190–194.

Fuss, S. H., and Korsching, S. I. (2001). Odorant feature detection: Activity mapping of structure response relationship in the zebrafish olfactory bulb. *J. Neurosci.* **21,** 8396–8407.

Germana, A., Montalbano, G., Laura, R., Ciriaco, E., Del Valle, M. E., and Vega, J. A. (2004). S100 protein-like immunoreactivity in the crypt olfactory neurons of the adult zebrafish. *Neurosci. Lett.* **371,** 196–198.

Hamdani, E. H., and Døving, K. B. (2002). The alarm reaction in crucian carp in mediated by olfactory neurons with long dendrites. *Chem. Senses* **27,** 395–398.

Hamdani, E. H., and Døving, K. B. (2003). Sensitivity and selectivity of neurons in the medial region of the olfactory bulb to skin extract from cospesifics in crucian carp, *Carassius carassius. Chem. Senses* **28**(3), 181–189.

Hamdani, E. H., and Døving, K. B. (2006). Specific projection of the sensory crypt cells in the olfactory system in crusian carp, *Carassius carassius. Chem. Senses* **31,** 63–67.

Hamdani, E. H., Stabell, O. B., Alexander, G., and Døving, K. B. (2000). Alarm reaction in the crucian carp is mediated by the medial bundle of the medial olfactor. *Chem. Senses* **25,** 103–109.

Hamdani, E. H., Alexander, G., and Døving, K. B. (2001a). Projection of sensory neurons with microvilli to the lateral olfactory tract indicates their participation in feeding behavior in crucian carp. *Chem. Senses* **26,** 1139–1144.

Hamdani, E. H., Kasumyan, A., and Døving, K. B. (2001b). Is feeding behaviour in crucian carp mediated by the lateral olfactory tract? *Chem. Senses* **26,** 1133–1138.

Hansen, A., and Finger, T. E. (2000). Phyletic distribution of crypt-type olfactory receptor neurons in fishes. *Brain Behav. Evol.* **55,** 100–110.

Hansen, A., and Reutter, K. (2004). Chemosensory systems in fish: Structural, functional and ecological aspects. *In* "The Senses of Fish Adaptations for the Reception of Natural Stimuli" (Von der Emde, G., Mogdans, J., and Kapoor, B. G., Eds.), pp. 55–89. Narosa Publishing House, New Delhi.

Hansen, A., and Zeiske, E. (1993). Development of the olfactory organ in the zebrafish, *Brachydanio rerio*. *J. Comp. Neurol.* **333,** 289–300.

Hansen, A., and Zielinski, B. (2005). Diversity in the olfactory epithelium of bony fishes: Development, lamellar arrangement, sensory neuron cell types and transduction components. *J. Neurocytol.* **34,** 183–208.

Hansen, A., Rolen, S. H., Anderson, K. T., Morita, Y., Caprio, J., and Finger, T. E. (2003). Correlation between olfactory receptor cell type and function in the channel catfish. *J. Neurosci.* **23,** 9328–9339.

Hansen, A., Anderson, K. T., and Finger, T. E. (2004). Differential distribution of olfactory receptor neurons in goldfish: Structural and molecular correlates. *J. Comp. Neurol.* **477,** 347–359.

Hanson, L. R., and Sorensen, P. W. (2001). Single-unit recording demonstrates that pheromones are discriminated by a combination of labeled-line and population coding in the goldfish olfactory bulb. The Twenty-Third Annual Meeting of the Association for Chemoreception Sciences, April 25–29, Sarasota, Florida.

Hara, T. J. (1975). Olfaction in fish. *Prod. Neurobiol.* **5,** 271–335.

Hara, T. J. (Ed.) (1982). Structure-activity relationships of amino acids as olfactory stimuli. *In* "Chemoreception in Fishes," pp. 135–158. Elsevier, Amsterdam.

Hara, T. J. (Ed.) (1992). "Fish Chemoreception." Chapman and Hall, London.

Hara, T. J. (2005). Olfactory responses to amino acids in rainbow trout: Revisited. *In* "Fish Chemosenses" (Reutter, K., and Kapoor, B. G., Eds.). Science Publishers, Enfield, NH.

Hara, T. J., and Zhang, C. (1996). Spatial projections to the olfactory bulb of functionally distinct and randomly distributed primary neurons in salmonid fishes. *Neurosci. Res.* **26,** 65–74.

Hara, T. J., and Zhang, C. (1998). Topographic bulbar projections and dual neural pathways of the primary olfactory neurons in Salmonid fishes. *Neuroscience* **82,** 301–313.

Hara, T. J., and Zielinski, B. (1989). Structural and functional development of the olfactory organ in teleosts. *Trans. Am. Fish. Soc.* **118,** 183–194.

Hara, T. J., Law, Y. M., and Hobden, B. R. (1973). Comparison of the olfactory response to amino acids in rainbow trout, brook trout and whitefish. *Comp. Biochem. Physiol. A* **45,** 969–977.

Hashiguchi, Y., and Nishida, M. (2005). Evolution of vomeronasal-type odorant receptor genes in the zebrafish genome. *Gene* **362,** 19–28.

Helfman, G. S., Collette, B. B., and Facey, D. E. (1997). "The Diversity of Fishes." Blackwell Sciences, Malden, MA.

Higashijima, S., Masino, M. A., Mandel, G., and Fetcho, J. R. (2003). Imaging neuronal activity during zebrafish behavior with a genetically encoded calcium indicator. *J. Neurophysiol.* **90,** 3986–3997.

Hofmann, M. H., and Meyer, D. L. (1995). The extrabulbar olfactory pathway: Primary olfactory fibers bypassing the olfactory bulb in bony fishes? *Brain Behav. Evol.* **46,** 378–388.

Heusa, G., Anadon, R., and Yanez, J. (2005). Olfactory projections in a chondrostean fish: Acipenser baeri: An experimental study. *J. Comp. Neurol.* **426,** 145–158.

Irie-Kushiyama, S., Asano-Miyoshi, M., Suda, T., Abe, K., and Emori, Y. (2004). Identification of 24 hour genes and two pseudogenes coding for olfactory receptors in Japanese loach, classified into four subfamilies: A putative evolutionary process for fish olfactory receptor genes by comprehensive phylogenetic analysis. *Gene* **325,** 123–135.

Irvine, I. A. S., and Sorensen, P. W. (1993). Acute olfactory sensitivity of wild common carp, *Cyprinus caprio*, to goldfish sex pheromones is influenced by gonadal maturity. *Can. J. Zool.* **71,** 2199–2210.

Jakubowski, M., and Kunysz, E. (1979). Anatomy and morphometry of the olfactory organ of the wels, *Silurus glanis* L. (Siluridae, Pisces). *Zeitschrift für Mikroskopisch*-anatomische *Forschung* **93,** 728–735.

Jasra, S. K., Avci, Z., Corkum, L., Scott, A. P., Li, W., and Zielinski, B. (2006). Putative steroidal pheromones: Synthesis sites and olfactory epithelial responses in the Round Goby (*Neogobius melanostomus*). Association of Chemoreception Sciences 28th Annual Meeting. Abstract 290.

Johnson, H. E., and Brown, J. C. D. (1962). Olfactory apparatus in the black rockfish, Sebastodes melanops. *Copeia* **1962,** 838–840.

Jones, D. T., and Reed, R. R. (1989). Golf: An olfactory neuron specific-G protein involved in odorant signal transduction. *Science* **244,** 790–795.

Kang, J., and Caprio, J. (1991). Electro-olfactogram and multiunit olfactory receptor responses to complex mixtures of amino acids in the channel catfish, *Ictalurus punctatus. J. Gen. Physiol.* **98,** 699–721.

Kang, J., and Caprio, J. (1995a). *In vivo* responses of single olfactory receptor neurons in the channel catfish, *Ictalurus punctatus. J. Neurophysiol.* **73,** 172–177.

Kang, J., and Caprio, J. (1995b). Electrophysiological responses of single olfactory bulb neurons to amino acids in the channel catfish, *Ictalurus punctatus. J. Neurophysiol.* **74,** 1421–1434.

Kang, J., and Caprio, J. (1995c). Electrophysiological responses of single olfactory bulb neurons to binary mixtures of amino acids in the channel catfish, Ictalurus punctatus. *J. Neurophysiol.* **74,** 1435–1443.

Kang, J., and Caprio, J. (1997). *In vivo* responses of single olfactory receptor neurons of channel catfish to binary mixtures of amino acids. *J. Neurophysiol.* **77,** 1–8.

Kapoor, A. S., and Ojha, P. P. (1972). Studies on ventilation of the olfactory chambers of fishes with a critical reevaluation of the role of accessory nasal sacs. *Arch. Biol.* **83,** 167–178.

Kleerekoper, H. (1969). *In* "Olfaction in Fishes," pp. 1–222. Indiana University Press, Bloomington.

Kondo, R., Kaneko, S., Sun, H., Sakaizumi, M., and Chigusa, S. I. (2002). Diversification of olfactory receptor genes in the Japanese medaka fish, *Oryzias latipes. Gene* **282,** 113–120.

Kux, J., Zeiske, E., and Osawa, Y. (1988). Laser Doppler velocimetry measurement in the model flow of a fish olfactory organ. *Chem. Senses* **13,** 257–265.

Laberge, F., and Hara, T. J. (2001). Neurobiology of fish olfaction: A review. *Brain Res. Brain Res. Rev.* **36,** 46–59.

Laberge, F., and Hara, T. J. (2003). Non-oscilatory discharges of F-Prostoglandin responsive neuron population in the oflactory bulb-telencephalon transition area in Lake fish. *Neuroscience* **116,** 1089–1095.

Laberge, F., and Hara, T. J. (2004). Electrophysiological demonstration of independent olfactory receptor types and associated neuronal reponses in the trout olfactory bulb. *Comp. Biochem. Physiol. A Mol. Integr. Physiol.* **137,** 397–408.

Laframboise, A., Chang, S., Ren, X., Dubuc, R., and Zielinski, B. (2006). Sea Lamprey (*Petromyzon marinus*) olfactory sensory neurons display polymorphisms. Association of Chemoreception Sciences 28th Annual Meeting. Poster 54.

Lastein, S., Hamdani, E. H., and Døving, K. B. (2006). Gender distinction in neural discrimination of sex pheromones in the olfactory bulb of crucian carp, *Carassius carassius. Chem. Senses* **31,** 69–77.

Li, W., Sorensen, P. W., and Gallaher, D. D. (1995). The olfactory system of migratory adult sea lamprey (*Petromyzon marinus*) is specifically and acutely sensitive to unique bile acids released by conspecific larvae. *J. Gen. Physiol.* **105,** 569–587.

Li, W., Scott, A. P., Siefkes, M. J., Yan, H., Liu, Q., Yun, S. S., and Gage, D. A. (2002). Bile acid secreted by male sea lamprey that acts as a sex pheromone. *Science* **296,** 138–141.

Liermann, K. (1933). Über den Bau des Geruchsorgans der Teleostier. *Z. Anat. Entwicklungs* **100,** 1–39.

Lo, Y. H., Bradley, T. M., and Rhoads, D. E. (1993). Stimulation of Ca(2+)-regulated olfactory phospholipase C by amino acids. *Biochemistry* **32,** 12358–12362.

Lower, N., Scott, A. P., and Moore, A. (2004). Release of sex steroids into the water by roach. *J. Fish Biol.* **46,** 16.

Luu, P., Acher, F., Bertrand, H. O., Fan, J., and Ngai, J. (2004). Molecular determinants of ligand selectivity in a vertebrate odorant receptor. *J. Neurosci.* **24,** 10128–11137.

Ma, L., and Michel, W. C. (1998). Drugs affecting phospholipase C-mediated signa transduction block the olfactory cyclic nucleotide-gated current of adult zebrafish. *J. Neurophysiol.* **79,** 1183–1192.

MacLeod, N. K. (1976). Spontaneous activity of single neurons in the olfactory bulb of the rainbow trout (*Salmo gairdneri*) and its modulation by olfactory stimulation with amino acids. *Exp. Brain Res.* **25,** 267–278.

Masterman, R., Hanson, L. R., and Sorensen, P. W. (2001). The goldfish olfactory bulb encodes pheromone information using at least two spatial maps. The Twenty-Third Annual Meeting of the Association for Chemoreception Sciences, April 25–29, Sarasota, Florida.

Matz, S. P. (1995). Connections of the olfactory bulb in the Chinook Salmon, *Oncorhynchus tshawytscha. Brain Behav. Evol.* **46,** 108–120.

Melinkat, R., and Zieske, E. (1979). Functional morphology of ventilation if the olfactory organ in Bedotia geayi Pellegrin 1909 (*Teleostei athrinidae*). *Zool. Anz.* **203,** 354–368.

Meredith, M. (1981). The analysis of response similarity in single neurons of the goldfish olfactory bulb using amino-acids as ododr stimuli. *Chem. Senses* **6,** 277–293.

Meredith, M. (2001). Human vomeronasal organ function: A critical review of best and worst cases. *Chem. Senses* **26,** 433–445.

Michel, W. C., and Debridge, D. S. (1997). Evidence of distinct amino acid and bile salt receptors in the olfactory system of the zebrafish, *Danio rerio. Brain Res.* **764,** 179–187.

Milinski, M., Griffiths, S., Wegner, K. M., Reursch, T. B., Haas-Assenbaum, A., and Boehm, T. (2005). Mate choice decisions of stickleback females predictably modified by MHC peptide ligands. *Proc. Natl. Acad. Sci. USA* **102,** 4414–4418.

Miranda, A., Almeida, O. G., Hubbard, P. C., Barata, E. N., and Canario, A. V. (2005). Olfactory discrimination of female reproductive status by male tilapia (*Oreochromis mossambicus*). *J. Exp. Biol.* **208**(Pt. 11), 2037–2043.

Miyamoto, T., Restrepo, D., Cragoe, E. J., Jr., and Teeter, J. H. (1992). IP3- and camp–induced responses in isolated olfactory receptor neurons from the channel catfish. *J. Membr. Biol.* **127,** 173–183.

Moore, A., and Scott, A. P. (1991). Testosterone is a potent odorant in precocious male Atlantic salmon (*Salmo salar* L.) parr. *Phil. Trans. Roy. Soc. Lond. B* **332,** 241–244.

Mombaerts, P., Wang, F., Dulac, C., Chao, S. K., Names, S., Mendelsoh, M., Edmondson, J., and Axel, R. (1996). Visualizing an olfactory sensory map. *Cell* **87,** 675–686.

Morita, Y., and Finger, T. E. (1998). Differential projections of ciliated and microvillous olfactory receptor cells in the catfish, *Ictalurus punctatus. J. Comp. Neurol.* **398,** 539–550.

Murphy, C. A., and Stacey, N. E. (2002). Methyl-testosterone induces male-typical behavioural responses to putative steroidal pheromones in female round gobies (*Neogobius melanostomus*). *Horm. Behav.* **42,** 109–115.

Murphy, C. A., Stacey, N. E., and Corkum, L. D. (2001). Putative steroidal pheromones in the round goby, *Neogobius melanostomus*: Olfactory and behavioral responses. *J. Chem. Ecol.* **27,** 443–470.

Nelson, J. S. (1994). "Fishes of the World," 3rd edn., John Wiley and Sons, New York.

Nelson, J. S., Crossman, E. J., Espinosa-Perez, H., Findley, L. T., Gilbert, C. R., Lea, R. N., and Williams, J. D. (2004). "Common and Scientific Names of Fishes from the United States, Canada, and Mexico." American Fisheries Society, Special Publication 29, Bethesda, Maryland.

Nevitt, G. A. (1991). Do fish sniff? A new mechanism of olfactory sampling in pleuronectid flounders. *J. Exp. Biol.* **157,** 1–18.

Ngai, J., Dowling, M. M., Buck, L., Axel, R., and Chess, A. (1993). The family of genes encoding odorant receptors in the channel catfish. *Cell* **72,** 657–666.

Niimura, Y., and Nei, M. (2005). Evolutionary dynamics of olfactory receptor genes in fishes and tetrapods. *Proc. Natl. Acad. Sci. USA* **102,** 6039–6044.

Nikonov, A. A., and Caprio, J. (2001). Electrophysiological evidence for a chemotopy of biologically relevant odors in the olfactory bulb of the channel catfish. *J. Neurophysiol.* **86,** 1869–1876.

Nikonov, A. A., and Caprio, J. (2004). Odorant specificity of single olfactory bulb neurons to amino acids in the channel fish. *J. Neurophysiol.* **92,** 123–134.

Nikonov, A. A., Parker, G. M., and Caprio, J. (2002). Odorant–induced olfactory receptor neural oscillations and of olfactory bulbar responses in the channel fish. *J. Neurosci.* **22,** 2352–2362.

Nikonov, A. A., Finger, T. E., and Caprio, J. (2005). Beyond the olfactory bulb: An odotopic map in the forebrain. *Proc. Natl. Acad. Sci. USA* **102,** 18688–18693.

Northcutt, R. G., and Puzdrowski, R. L. (1988). Projections of the olfactory bulb and nervus terminalis in the silver lamprey. *Brain Behav. Evol.* **32,** 96–107.

Ohno, T., Yoshii, K., and Kurihara, K. (1984). Multiple receptor types for amino acids in the carp olfactory cells revealed by quantitative cross-adaptation method. *Brain Res.* **310,** 13–21.

Oka, Y., Ichikawa, M., and Ueda, K. (1982). Synaptic organization of the olfactory bulb and central projection of the olfactory tract. *In* "Chemoreception in Fishes" (Hara, T. J., Ed.), pp. 61–75. Elsevier, Amsterdam.

Ottoson, D. (1956). Analysis of the electrical activity of the olfactory epithelium. *Acta Physiol. Scand.* **35**(Suppl. 122), 1–83.

Parker, G. H. (1910). Olfactory reaction in fishes. *J. Exp. Zool.* **8,** 535–542.

Pfeiffer, W. (1964). The morphology of the olfactory organ in *Hoplopagrus guentheri* Gill. *Can. J. Zool.* **42,** 235–237.

Pfister, P., and Rodriguez, I. (2005). Olfactory expression of a single and highly variable V1R pheromone receptor-like gene in fish species. *Proc. Natl. Acad. Sci. USA* **102,** 5239–5494.

Pipping, M. (1926). Der Geruchssinn der Fische mit besonderer Berücksichtigung seiner Bedeutung für das Aufsuchen des Futters. *Societas Scientiarum Fennica. Commentiones Biologicae* **2,** 1–28.

Polenova, O. A., and Vesselkin, N. P. (1993). Olfactory and nonolfactory projections in the river lamprey (*Lampetra fluviatilis*) telencephalon. *J. Hirnforsch.* **34,** 261–279.

Ren, X., Chang, S., Auclair, F., Dubuc, R., and Zielinski, B. (2006). Spatially distinct sensory input to medial olfactory bulb glomeruli and output projections into habenula and ventral thalamus in the sea lamprey *Petromyzon marinus*. Association for Chemoreception Sciences, 28th Annual Meeting, Sarasota Florida April 26–30.

Restrepo, D., Boekhoff, I., and Breer, H. (1993). Rapid kinetic measurements of second messenger formation in olfactory cilia from channel catfish. *Am. J. Physiol.* **264,** C906–C911.

Riddle, D. R., and Oakley, B. (1992). Immunocytochemical identification of primary olfactory afferents in rainbow trout. *J. Comp. Neurol.* **324,** 575–589.

Rolen, S. H., Sprensen, P. W., Mattson, D., and Caprio, J. (2002). Polyamines as olfactory stimuli in the goldfish *Carassius auratus*. *J. Exp. Biol.* **206**(Pt. 10), 1683–1696.

Rooney, D., Døving, K. B., Ravaille-Veron, M., and Szabo, T. (1992). The central connections of the olfactory bulbs in cod, *Gadus morhua L. J. Hirnforsch.* **33,** 63–75.

Sato, K., and Suzuki, N. (2001). Whole-cell response characteristics of ciliated and microvillous olfactory receptor neurons to amino acids, pheromone candidates and urine in rainbow trout. *Chem. Senses* **26,** 1145–1156.

Sato, Y., Miyasaka, N., and Yoshihara, Y. (2005). Mutually exclusive glomerular innervation by two distinct types of olfactory sensory neurons revealed in transgenic zebrafish. *J. Neurosci.* **25,** 4889–4897.

Satou, M. (1990). Synaptic organization, local neuronal circuitry, and functional segregation of the teleost olfactory bulb. *Prog. Neurobiol.* **34,** 115–142.

Satou, M., and Ueda, K. (1978). Synchronized rhythmic discharges of the secondary olfactory neurons in carp. *Brain Res.* **158,** 313–329.

Satou, M., Fujita, I., Ichikawa, M., and Ueda, K. (1983). Field potential and intracellular potential studies of the olfactory bulb in the carp: Evidence for a fuctional separation of the olfactory bulb into lateral and medial subdivisions. *J. Comp. Physiol.* **152,** 319–333.

Satou, M., Anzai, S., and Huruno, M. (2005). Long-term potentiation and olfactory memory formation in the carp (*Cyprinus carpio* L.) olfactory bulb. *J. Comp. Physiol. A Neuroethol. Sens. Neural. Behav. Physiol.* **191,** 421–434.

Schmachtenberg, O. (2006). Histological and electrophysiological properties of crypt cells from the olfactory epithelium of the marine teleost. *Trachurus symmetrices. J. Comp. Neurol.* **495,** 113–121.

Scott, J. W., and Scott-Johnson, P. E. (2002). The electroolfactogram: A review of its history and uses. *Microsc. Res. Tech.* **58,** 152–160.

Sheldon, R. E. (1912). The olfactory tracts and centers in teleosts. *J. Comp. Neurol.* **22,** 177–339.

Sheperd, G. M. (1994). The extent of adaptation in bullfrog saccular hair cells. *J. Neurosci.* **14,** 6217–6229.

Silver, W. L., Caprio, J., Blackwell Joan, F., and Tucker, D. (1976). The underwater electro-olfactorgram: A tool for the study of the sense of smell of marine fishes. *Cell. Mol. Life Sci.* **32,** 1216–1217.

Sola, C., and Tosi, L. (1993). Bile salts and taurine as chemical stimuli for glass eels, *Anguilla anguilla*: A behavioural study. *Environ. Biol. Fishes* **37,** 197–204.

Sorensen, P. W., and Caprio, J. (1998). Chemoperception. *In* "The Physiology of Fishes" (Evans, D. H., Ed.), pp. 375–405. CRC Press, Boca Raton, FL.

Sorensen, P. W., Hara, T. J., and Stacey, N. E. (1991). Sex pheromones selectively stimulate the medial olfactory tracts of the male goldfish. *Brain Res.* **558,** 343–347.

Sorensen, P. W., Pinillos, M., and Scott, A. P. (2005a). Sexually mature male goldfish release large quantities of androstenedione into the water where it functions as a pheromone. *Gen. Comp. Endocrinol.* **140,** 164–175.

Sorensen, P. W., Fine, J. M., Dvornikons, V., Jeffrey, C. S., Shao, F., Wang, J., Vrieze, L. A., Anderson, K. R., and Hoye, T. R. (2005b). Mixture of new sulfated steroids functions as a migratory pheromone in the sea lamprey. *Nat. Chem. Biol.* **1,** 324–328.

Speca, D. J., Lin, D. M., Sorensen, P. W., Isacoff, E. Y., Ngai, J., and Dittman, A. H. (1999). Functional identification of a goldfish odorant receptor. *Neuron* **23,** 487–498.

Stacey, N. E., and Kyle, A. L. (1983). Effects of olfactory tract lesions on sexual and feeding behavior in the goldfish. *Physiol. Behav.* **30,** 621–628.

Stacey, N., and Sorensen, P. (2002). Hormonal pheromones in fish. *In* "Hormones, Brain and Behaviour" (Pfaff, D. W., Arnold, A. P., Etgen, A. M., Fahrbach, S. E., and Fahrbach, R. T., Eds.), Vol. 2, pp. 375–434. Elsevier Science, San Diego, CA.

Starcevic, S. L., and Zielinski, B. S. (1995). Immunohistochemical localization of glutathione *S*-transferase pi in rainbow trout olfactory receptor neurons. *Neurosci. Lett.* **183,** 175–178.

Starcevic, S. L., Muruganandam, A., Mutus, B., and Zielinski, B. S. (1993). Glutathione in the olfactory mucosa of rainbow trout (*Oncorhynchus mykiss*). *Chem. Senses* **18,** 57–65.

Stacey, N. E., and Sorensen, P. W. (2002). Fish hormonal pheromones. *In* "Hormones, Brain, and Behaviour" (Pfaff, D. W., Arnold, A. P., Etgen, A., Fahrbach, S., and Rubin, R., Eds.), Vol. 2, pp. 375–435. Academic Press, New York.

Stacey, N. E., and Sorensen, P. W. (2005). Reproductive pheromones. *In* "Behaviour and Physiology of Fish" Vol. 24, pp. 359–411.

Stacey, N., and Sorensen, P. (2005). Reproductive pheromones. *Behav. Physiol. Fish* **24,** 359–412.

Sutterlin, A. M., and Sutterlin, N. (1971). Electrical responses of the olfactory epithelium of Atlantic salmon (*Salmo salar*). *J. Fish. Res. Board Can.* **28,** 565–572.

Sveinsson, T., and Hara, T. J. (1990). Multiple olfactory receptors for amino acids in the Arctic char (*Salvelinus alpinus*) evidenced by cross-adaptation experiments. *Comp. Biochem. Physiol. A Mol. Intgr. Physiol.* **97,** 289–293.

Sveinsson, T., and Hara, T. J. (2000). Olfactory sensitivity and specificity of Arctic char, *Salvelinus alpinus*, to a putative male pheromone, prostaglandin f(2)alpha. *Physiol. Behav.* **69,** 301–307.

Szabo, T., Blahser, S., Denizot, J. P., and Ravaille-Veron, M. (1991). Extensive primary olfactory projections beyond the olfactory bulb in teleost fish. *Neurobiol.* **312,** 555–560.

Tabor, R., Yaksi, E., Weislogel, J. M., and Friedrich, R. W. (2004). Processing of odor mixtures in the zebrafish olfactory bulb. *J. Neurosci.* **24,** 6611–6620.

Teichmann, H. (1954). Vergleichende Untersuchungen an der Nase der Fische. *Z. Morphol. Oekol. Tiere* **43,** 171–212.

Teichmann, H. (1964). Experimente zur Nasenentwicklung der Regenbogenforelle (*Salmo irideus* W. Gibb.). *Wilhelm Roux Archiv fur Entwicklungs Mechanik der Organismen* **155,** 129–143.

Theisen, B. (1982). Functional morphology of the olfactory organ in *Spinachia spinachia* (L.) (Teleostei, Gasterosteidae). *Acta Zool.* **63,** 247–254.

Theisen, B., Breucker, H., Zeiske, E., and Melinkat, R. (1980). Structure and development of the olfactory organ in the garfish *Belone belone* (L.) (Teleostei, Atheriniformes). *Acta Zool.* **61,** 161–170.

Theisen, B., Zeiske, E., Silver, W. L., Marui, T., and Caprio, J. (1991). Morphological and physiological studies on the olfactory organ of the striped eel catfish, *Plotosus lineatus*. *Mar. Biol.* **110,** 127–135.

Thommesen, G. (1978). The spatial distribution of odour induced potentials in the olfactory bulb of char and trout (*Salmonidae*). *Acta Physiol. Scand.* **102,** 205–217.

Thommesen, G. (1982). Specificity and distribution of receptor cells in the olfactory mucosa of char (*Salmo alpinus* L.). *Acta Physiol. Scand.* **115,** 47–56.

Thommesen, G. (1983). Morphology, distribution, and specifity of olfactory receptor cells in salmonid fishes. *Acta Physiol. Scand.* **117,** 241–250.

Thornhill, R. A. (1967). The ultrastructure of the olfactory epithelium of the lamprey *Lampetra fluviatilis*. *J. Cell Sci.* **2,** 591–602.

Uchida, H., Ogawa, S., Harada, M., Matushita, M., Iwata, M., Sakuma, Y., and Parhar, I. S. (2005). The olfactory organ modulates gonadotropin-releasing hormone types and nest-building behavior in the tilapia *Oreochromis niloticus*. *J. Neurobiol.* **65,** 1–11.

Valentincic, T., Kralj, J., Stenovec, M., Koce, A., and Caprio, J. (2000). The behavioral detection of binary mixtures of amino acids and their individual components by catfish. *J. Exp. Biol.* **203**(Pt. 21), 3307–3317.

VanDenbossche, J., Seelye, J. G., and Zielinski, B. S. (1995). The morphology of the olfactory epithelium in larval, juvenile and upstream migrant stages of the sea lamprey, *Petromyzon marinus*. *Brain Behav. Evol.* **45,** 19–24.

Vitebsky, A., Reyes, R., Sanderson, M. J., Michel, W. C., and Whitlock, K. E. (2005). Isolation and characterization of the laure olfactory behavioral mutant in the zebrafish, *Danio rerio*. *Dev. Dyn.* **234,** 229–242.

Von Bartheld, C. S. (2004). The terminal nerve and its relation with extrabulbar "olfactory" projections: Lessons from lampreys and lungfishes. *Microsc. Res. Tech.* **65**(1–2), 13–24.

Von Bartheld, C. S., Meyer, D. L., Fiebig, Z., and Ebbesson, S. O. E. (1984). Central connections of the olfactory bulb in the goldfish, Carassius auratus. *Cell Tissue Res.* **238,** 475–487.

Von Rekowski, C., and Zippel, H. P. (1993). In goldfish the qualitative discriminative ability for odors rapidly returns after bilateral nerve axotomy and lateral olfactory tract transection. *Brain Res.* **618,** 338–340.

Wekesa, K. S., and Anholt, H. (1999). Differential expression of G protein in the mouse olfactory system. *Brain Res.* **837,** 117–126.

Weltzien, F. A., Hoglund, E., Hamdani, E. H., and Døving, K. B. (2003). Does the lateral bundle of the medial olfactory tract mediate reproductive behavior in male crucian carp? *Chem. Senses* **28,** 293–300.

Wickens, A., May, D., and Rand-Weaver, M. (2001). Molecular characterization of a putative Atlantic salmon (*Salmo salar*) odorant receptor. *Comp. Biochem. Physiol.* **129,** 653–660.

Wunder, W. (1957). Die Sinnesorgane der Fische. *Allg. Fisch. Ztg.* **82,** 1–24.

Yaksi, E., and Friedrich, R. W. (2006). Recontruction of firing rate changes across neuronal populations by temporally deconvolved Ca2+ imaging. *Nat. Methods* **3,** 377–383.

Yamamoto, M. (1982). Comparative morphology of the peripheral olfactory organ in teleosts. *In* "Chemoreception in Fishes" (Hara, T. J., Ed.), pp. 39–59. Amsterdam, Elsevier.

Yanagi, S., Kudo, H., Doi, Y., Yamauchi, K., and Ueda, H. (2004). Immunohistochemical demonstration of salmon olfactory glutathione *S*-transferase class pi (N24) in the olfactory system of lacustrine sockeye salmon during ontogenesis and cell proliferation. *Anat. Embryol.* **208,** 231–238.

Yasuoka, A., Endo, K., Asano-Miyoshi, M., Abe, K., and Emori, Y. (1999). Two subfamilies of olfactory receptor genes in medaka fish, *Oryzias latipes*: Genomic organization and differential expression in olfactory epithelium. *J. Biochem.* **126,** 866–873.

Zeiske, E. (1974). Morphologische und morphometrische Untersuchungen am Geruchsorgan oviparer Zahnkarpfen (Pisces). *Zeitschrift für Morphologie der Tiere* **77,** 19–50.

Zeiske, E., Breucker, H., and Melinkat, R. (1979). Gross morphology and fine structure of the olfactory organ of rainbow fish (Atheriniformes, Melanotaeniidae). *Acta Zool.* **60,** 173–186.

Zeiske, E., Theisen, B., and Breucker, H. (1992). Structure, development, and evolutionary aspects of the peripheral olfactory system. *In* "Fish Chemoreception" (Hara, T. J., Ed.), pp. 13–39. Chapman and Hall, London.

Zhang, C., and Hara, T. J. (1994). Multiplicity of salmonid olfactory receptors for bile acids as evidenced by cross-adaptation and ligand binding assay. *Chem. Senses* **19,** 579.

Zhang, C., Brown, S. B., and Hara, T. J. (2001). Biochemical and physiological evidence that bile acids produced and released by lake char (*Salvelinus namaycush)* function as chemical signals. *J. Comp. Physiol.* **171,** 161–171.

Zielinski, B. S., and Hara, T. J. (1988). Morphological and physiological development of olfactory receptor cells in rainbow trout (*Salmo gairdneri*) embryos. *J. Comp. Neurol.* **271,** 300–311.

Zielinski, B. S., and Hara, T. J. (2001). The neurobiology of fish olfaction. *In* "Sensory Biology of Jawed Fishes" (Kapoor, B. G., and Hara, T. J., Eds.), pp. 347–366. Science Publishers Inc., Enfield.

Zielinski, B. S., Osahan, J. K., Hara, T. J., Hosseini, M., and Wong, E. (1996). Nitric oxide synthase in the olfactory mucosa of the larval sea lamprey (*Petromyzon marinus*). *J. Comp. Neurol.* **365,** 18–26.

Zippel, H. P., Reschke, C., and Korff, V. (1999). Simultaneous recordings from two physiologically different types of relay neurons, mitral cells and ruffed cells, in the olfactory bulb of goldfish. *Cell Mol. Biol.* **45,** 327–337.

2

GUSTATION

TOSHIAKI J. HARA

1. Introduction
2. Structural Organization
   2.1. Taste Buds
   2.2. Central Gustatory Nuclei and Pathways
3. Functional Properties
   3.1. Responses to Chemical Stimuli
   3.2. Responses to Mechanical/Tactile Stimuli
4. Gustatory Behaviors
   4.1. Feeding Behavior
   4.2. Aversive Behavior
5. Conclusions and Prospects

1. INTRODUCTION

Vertebrates possess three major classes of chemosensory systems, the gustatory organs, the olfactory organs, and the common chemical senses. Gustation is defined as the chemical sense that is mediated by specific receptors of gustatory cells within taste buds. Olfactory responses are undisputedly mediated through specific protein receptors expressed in the receptor neuron membranes, and therefore are structurally and functionally in strong contrast with the gustatory and common chemical sense (Chapter 1, this volume). Collectively defined as the common chemical sense are all other chemical senses that are perhaps mediated by nonspecific nervous structures such as free nerve endings. All three sensory systems are innately stimulated by chemical substances, whether specific or nonspecific, and prone to mechanical stimulation.

Gustation mediates one of the most fundamental processes for the survival of individuals and species: feeding. The relation of gustation to food intake is manifested in all vertebrates including humans, but many aspects of the complex integration of chemosensory input with antecedent or concomitant

Sensory Systems Neuroscience: Volume 25
FISH PHYSIOLOGY

DOI: 10.1016/S1546-5098(06)25002-7

physiological activities remain mostly unclear. The evolutionary development of the chemosensory systems associated with feeding followed two different courses, one in invertebrates and one in vertebrates. In invertebrates, gustatory cells preferentially sensitive to food materials can be found almost anywhere on the body surface, for example, on antennae, tentacles, or legs. In vertebrates, on the other hand, the presence of specialized taste buds in the oral cavity has been taken as evidence of an ability to taste. Taste perception is responsible for basic food appraisal and bestows the organism with valuable discriminatory power. Nevertheless, whether invertebrates or vertebrates, the general principles on which they operate are essentially the same.

Fish, constituting slightly more than one-half of the total number of ~50,000 recognized living vertebrate species, are a pivotal group that has served as a vertebrate model for the study of the gustatory system. Taste buds were first described in fish in the early 1800s, and by the turn of the century the peripheral and central gustatory neural organization had been clearly illustrated. The year 2005 marked a centenary since the publication of Herrick's (1905) award-winning detailed description of the central gustatory pathways in the brains of bony fishes. Further elaboration has since been made using modern techniques focusing on topographic representations of the gustatory and motor roots in the fish brains (Finger, 1976; Kanwal and Caprio, 1987; Morita and Finger, 1987; Puzdrowski, 1987; Hayama and Caprio, 1989; Kiyohara *et al.*, 2002). Certain fish groups have evolved a vast system of gustatory receptors over virtually the entire body surface, along with an elaborately organized complex central neural organization. The fish gustatory receptor system, however, is unique, in that unlike those of the terrestrial, receptors are stimulated by dilute solutions, or "distance chemical receptors." With the exception just mentioned, however, there exists a broad line of similarity between both the peripheral and the central gustatory paths in all vertebrate types, and fishes with enormous hypertrophy of the gustatory centers may serve as a guide to point the way for researches on the gustatory pathways of higher vertebrates where the system is less accessible. Thus, this chapter is concerned with aspects of the fish gustatory system in perspective of those of all vertebrates, with an emphasis on new insights into the taste bud development, central neural pathways, and physiological as well as behavioral implications.

2. STRUCTURAL ORGANIZATION

2.1. Taste Buds

2.1.1. Distribution

The taste buds (also termed "terminal buds, end buds, or cutaneous buds") constitute the structural basis of the peripheral gustatory organ.

The term "gustatory system" is used to designate these organs, together with their nervous pathways toward and within the brain. The gustatory system is thus distinguished from those unspecialized sensory fiber systems serving various viscera without special sensory organs. In teleosts, taste buds are distributed in five subpopulations, to a varying degrees: (1) oral, (2) palatal and laryngeal, (3) branchial (gills), (4) cutaneous, and (5) barbels (Figure 2.1). Two major fish groups, cyprinids (~2700 species including carps, minnows, goldfish, and so on) and silurids (~2300 species including catfishes), have a vast array of taste buds over virtually the entire body surface. Yellow bullhead catfish (*Ictalurus natalis*), for example, have more than 175,000 taste buds on the entire body surface alone (Atema, 1971). Carp (*Cyprinus carpio*) and goldfish (*Carassius auratus*), on the other hand, have evolved a highly sophisticated food separation system on the roof of the mouth (palatal organ) that is studded with thousands of taste buds (Sibbing and Uribe, 1985). The taste bud distribution patterns reflect fishes' feeding habits, strategies, and habitats (Kiyohara *et al.*, 1980; Gomahr *et al.*, 1992; Fishelson and Delarea, 2004). In rainbow trout (*Oncorhynchus mykiss*), a visual feeder, the highest taste bud density of 30 per mm^2 is found in the palate, totaling 3000–4000 taste buds (Hara *et al.*, 1993). The number of taste buds in the oral cavity of amago salmon (*Oncorhynchus rhodurus*) increases slowly during the first 60 days posthatching, then increases sharply over the next 300 days, reaching a total number of more than 15,000, at which time smolt transformation normally takes place (Komada, 1993). One area that has been overlooked is the gill arches/rakers where the total number of taste buds is estimated at 12,500 in bullhead (Atema, 1971). Among marine species that have generally less taste buds on the gill arches/rakers, the sea catfish, *Plotosus lineatus*, has exceptionally large number of taste buds (Iwai, 1964). The development of branchial taste buds is thought to be an adaptation to freshwater habitats where a wider range of physical and chemical changes exist than in sea water. There is evidence suggesting the sea catfish is a recent invader from fresh water (Iwai, 1963).

Thus, the peripheral gustatory system is very well developed in fishes in general, and the arrangement of the subpopulations of taste buds is similar across teleosts, although the abundance and distribution of taste buds vary considerably among species. However, cyprinids and silurids are notable exceptions to the general condition among teleosts, as they evolved thousands of extra taste buds, which is a result of the independent evolution that occurred a number of times among teleosts (Northcutt, 2005).

2.1.2. Structure

Most taste buds are bulbiform, or flask shaped, vary in size (45–75 μm in height and 30–50 μm in width) depending on the thickness of the epithelial layer, and oriented perpendicular to the skin surface (Iwai, 1964;

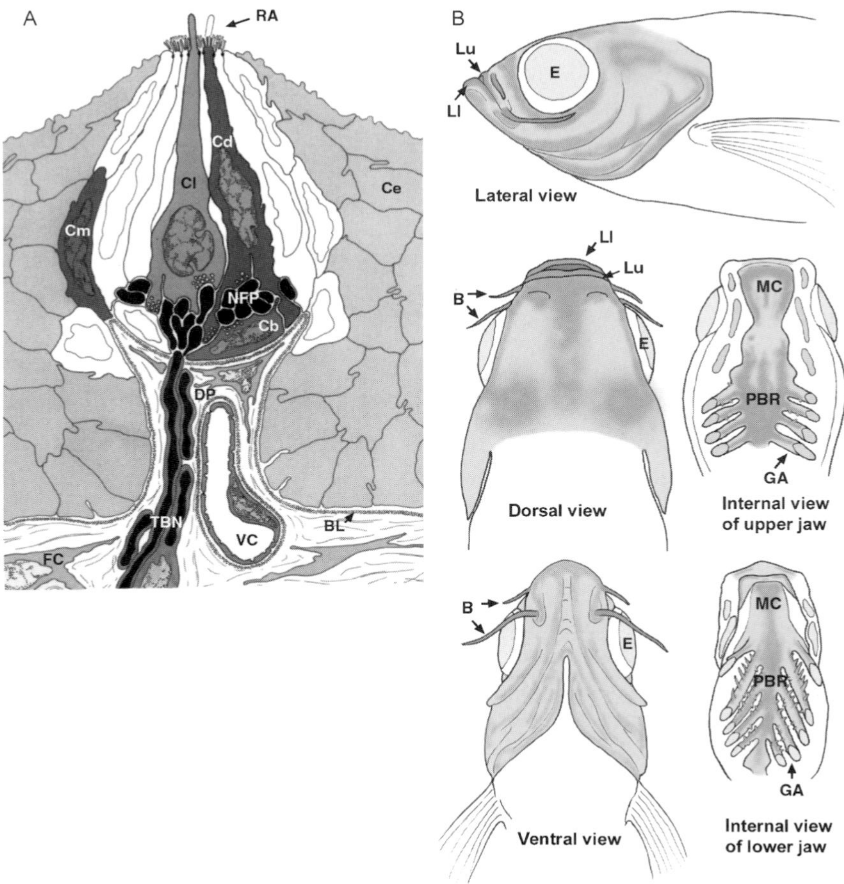

Fig. 2.1. (A) Schematic drawing of a taste bud (TB) typical for teleosts. Dark cells (Cd) with microvilli, light cells (Cl) with a single rodlike protrusion, and basal cells (Cb) constitute a pear- or onion-shaped TB, which sits on a dermal papilla (DP). Marginal cells (Cm), not belonging to the TB proper, form the interface between TB and the epithelial cells (Ce). The TB nerve (TBN) reaches into the TB to form the nerve fiber plexus (NFP). BL, basal lamina; RA, receptive area; and VC, capillary vessel. (B) Distribution of TBs in juvenile and adult zebrafish. High density of TBs is marked by darker shades of color. B, barbels; E, eye; GA, gill arches; Ll, lower lip; Lu, upper lip; MC, mouth cavity; and PBR, pharyngobranchial region. [Adapted from Braun and Northcutt (1995) and Hansen *et al.* (2002) with permission.] (See Color Insert.)

Kiyohara *et al.*, 1980) (Figure 2.2). Taste buds vary in shape considerably among species and location even in the same species. In channel catfish, for example, taste buds innervated by vagal nerves are taller and more slender than those innervated by facial nerves (Eram and Michel, 2005). The sensory area, or taste bud pore, is usually ~10 μm in diameter, but up to 20 μm in the goatfish, *Parupeneus* sp., barbels, and characterized by a rod-shaped protrusion or tightly packed microvilli, on the tips of the taste buds (Kiyohara *et al.*, 2002). A taste bud is buried in the epithelial layers or sits on a dermal papilla without being partitioned by membranes or any other tissues from the surrounding epithelial cells. At least three gustatory cell types constitute a fish taste bud: (1) those ending with a rod-shaped apical protrusion or cilia (0.5-μm thick and 1.5- to 3.0-μm long; rod cells), (2) those ending with microvilli (0.1–0.2 μm in diameter and 0.5–1.0 μm in length; microvillar cells), and (3) basal cells (Figure 2.1A). The rod and microvillar cells have traditionally been termed "light" and "dark" cells, respectively, based on the electron density, however, the light and electron microscopical appearance of gustatory cells

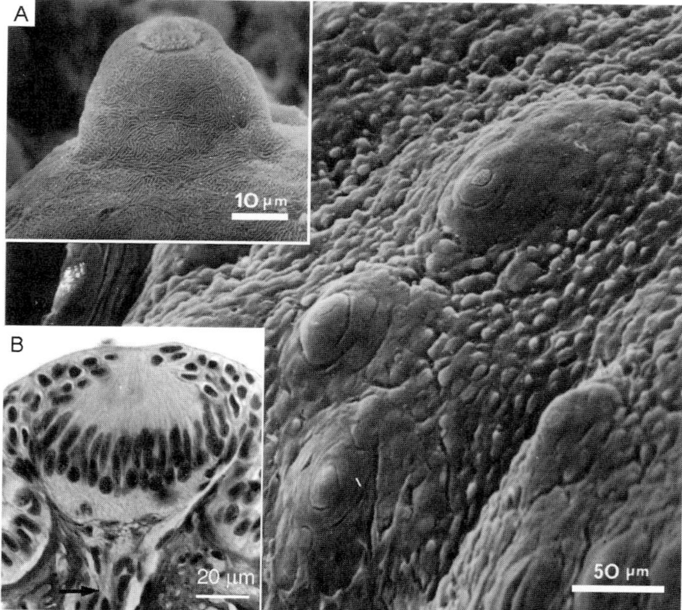

Fig. 2.2. Morphology and distribution of taste buds on the palate of rainbow trout *Oncorhynchus mykiss*. (A) Scanning electron micrograph of a single taste bud located on an elevated papilla. (B) Light micrograph of a longitudinal section through a taste bud on the palate. (Courtesy of R. E. Evans and B. Zielinski.)

has provoked considerable controversy with respect to cell classification (Reutter, 1978; Jakubowski and Whitear, 1990; Reutter and Witt, 1993; Figure 2.1A). The total number of cells in a taste bud varies considerably, for example, 5 in *Pomatoschistus*, 67 in *Corydoras*, and as many as 100 in *Ciliata mustela* (60 rod cells and 49 microvillar cells) (Crisp *et al.*, 1975; Jakubowski and Whitear, 1990). In the minnows, *Pseudorasbora parva*, the ratio of rod cells to microvillar cells within a taste bud varies considerably (0.9–2.7), with the highest those inside the lower lip (Kitoh *et al.*, 1987). Taste buds innervated by facial nerves contain significantly more cells than those innervated by vagal nerves (216.3 ± 34.6 vs. 135.0 ± 19.2) in the channel catfish (Eram and Michel, 2005). These two taste bud types of channel catfish show heterogeneous distribution patterns of immunoreactivities for metabolites, notably more γ-aminobutyric acid (GABA) positive cells (microvillar cells) present in the vagal innervated taste buds. An additional cell type, the fusiform cell, which is low in electron density and has a brush-like apical ending with several microvilli is present in zebrafish (*Danio rerio*) (Hansen *et al.*, 2002).

The rod cell, most likely a receptor cell, is cylindrical to spindle shaped. Its supranuclear cytoplasm contains numerous electron-dense tubules (0.4–0.6 μm), and the infranuclear portion is characterized by the presence of abundant vesicles of 50–70 nm in diameter. Tubules sometimes form a central and peripheral pattern, similar to that found in a variety of motile cilia. The number of rod cells within a taste bud varies considerably, for example, in the minnows (*P. parva*), 48–73% of the total cells (27–98) in a taste bud are rod cells (Kitoh *et al.*, 1987). The microvillar cell, often believed to be a supporting cell, could well be a receptor cell, also. Its cytoplasm is characterized by numerous fine filaments of ~50 Å in diameters, running through the entire cytoplasm either diffused or in bundles. Junctional complexes are usually present between microvillar cells and between the rod and microvillar cells (Hirata, 1966).

The basal cells, typically one to five per taste bud, are situated at the bottom of the taste bud in depressions of the basal lamina, perpendicular to its longitudinal axis. They attach to both receptor and supporting cells by desmosomes (Jakubowski and Whitear, 1990). The cytoplasm contains many lucent vesicles, 40–50 nm in diameter. The basal cells contain serotonin (5-HT) and possess synaptic connections with both cell types as well as with nerve fibers, suggesting that they might act as modulators of gustatory activity (Uga and Hama, 1967; Reutter, 1978; Toyoshima *et al.*, 1984). Unlike those of the mammalian taste buds, however, basal cells in fish are not the precursors of other cell types within the taste buds.

Generally, bundles of unmyelinated nerve fibers enter from below into the basal portion of taste buds passing by the side of the basal cells.

However, the number and pattern of bundles supplying a taste bud vary among species and location even within the same species. For example, a single nerve bundle in the goatfish barbels (Kiyohara *et al.*, 2002), two bundles in the sea catfish barbels (Sakata *et al.*, 2001), and more than two bundles in the pelvic fin of a gadid fish (Kotrschal *et al.*, 1993) enters the basal portion of the taste bud. On entering, they further branch out and make an intricate intragemmal nerve plexus, some components of which make synaptic contacts either with the rod cells or microvillar cells. The synaptic specialization such as those found in the central nervous system (CNS) is not always clear at these nerve-ending receptor junctions (Jakubowski and Whitear, 1990; Reutter and Witt, 1993).

2.1.3. Heterogeneity of Gustatory Cells Within a Taste Bud

Whether the observed heterogeneity of taste bud cells represents different receptor cell types with distinct functional properties is unknown. However, it is safe to say that at least the rod cells are receptor cells. Gustatory cells are extremely small, and generally not easily accessible for conventional single cell recording techniques, consequently little is known about the function of individual gustatory cells. It is thus unclear whether gustatory receptor cells express one or more specific receptor types or whether an individual taste bud is selective for a particular chemical type. In the simplest model, different gustatory stimuli could be encoded by different gustatory cells expressing different receptors, possibly with transduction pathways. Alternatively, gustatory receptor cells could be broadly tuned within or between gustatory stimuli. For example, an immunohistochemical study on the channel catfish gustatory epithelium shows that individual receptor cells preferentially express receptors for either L-alanine or L-arginine, two of the most potent stimulants, although a small percentage of cells may express lower levels of receptors for both amino acids (Finger *et al.*, 1996; see in the following section). This indicates that taste buds are not selectively tuned to specific chemical stimuli. Further, that the two probes used in this study label only a small portion of the receptor field of taste buds may indicate that other receptor types exist not recognized by these probes. In the goatfish (*Parupeneus trifasciatus*) barbels, by contrast, taste buds are innervated by a longitudinally running main nerve trunk (LNB), a branch of the facial nerve, running through the barbel's core (Kiyohara *et al.*, 2002). Each taste bud cluster, containing ~14 taste buds, is innervated by one of the two circumferential nerve branches (CNB), originating from the LNB containing estimated 90 fibers (Figure 2.3). All together, four taste bud clusters, containing a total of ~56 taste buds, and therefore all the gustatory receptor cells, are innervated by a single nerve fiber. Although how many receptor cells within a taste bud are innervated by a single nerve fiber are unknown, these receptor cells may function as on-off switches, responding only the presence of a

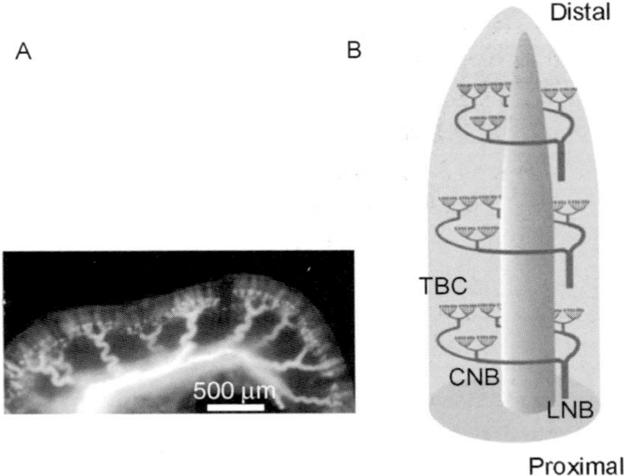

Fig. 2.3. Innervation of taste buds on the goatfish *P. trifasciatus* barbel. (A) Cross-section through a barbel showing each taste bud cluster (TBC) innervated by a single radial nerve bundle emanating from the circumferential nerve branches (CNB). Scale bar, 500 μm. (B) A longitudinal nerve branch (LNB) originating from the main nerve trunk sends a pair of CNB medially and laterally around the margins of the barbel to innervate four TBCs at each proximal-distal level. See text for details. [Adapted from Kiyohara *et al.* (2002) with permission.]

stimulus with no regard to quality. Furthermore, in the recurrent facial nerve of the channel catfish, the number of taste buds innervated by a single fiber increases vastly as fish grow, implying that each fiber innervates more taste buds in the larger fish (Finger *et al.*, 1991). In the rockling (*C. mustela*) barbels, each taste bud containing 50–70 receptor cells is innervated by 6–7 nerve bundles, each of which contains an average of 35 axons. So, the average taste bud is innervated by about 230 nerve axons (Crisp *et al.*, 1975).

In mammals, two families of receptors, T1Rs and T2Rs, mediating responses to sweet, amino acids, and bitter compounds, are expressed in nonoverlapping populations of receptor cells within individual taste buds of the tongue and palate epithelium in rats (Hoon *et al.*, 1999; Zhang *et al.*, 2003). This implies that sweet, amino acids, and bitter modalities are encoded separately by the activation of distinct cell types, which is consistent with that of channel catfish.

2.1.4. Development

Gustatory cells of vertebrates, unlike all other sensory receptor cells and neurons that originate from neurogenic ectoderms (neural tube, neural crest, or ectodermal placodes), are unique in that they originate from local epithelial tissue elements of primarily endoderms. Like neurons, gustatory receptor cells form synapses and are capable of generating receptor potentials, and

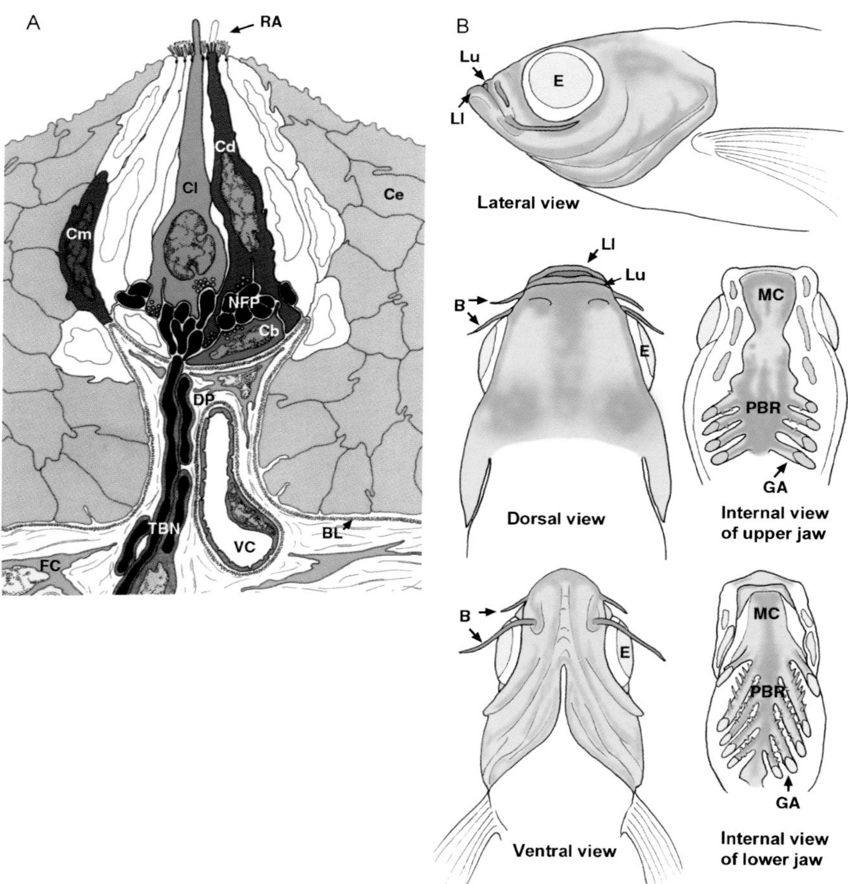

Chapter 2, Fig. 1.

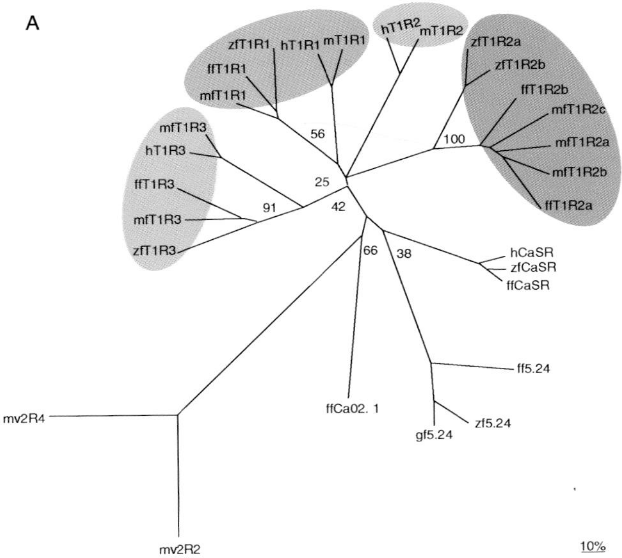

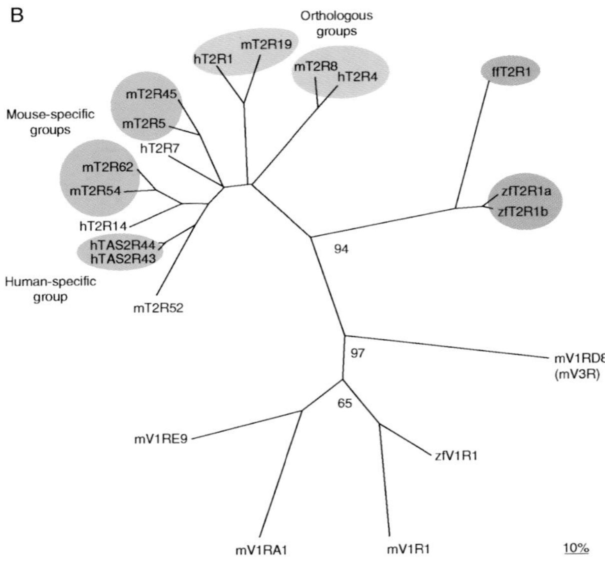

Chapter 2, Fig. 13.

like epithelial cells they have a limited life span and are regularly replaced (Raderman-Little, 1979; Kitoh *et al.*, 1987; Roper, 1993; Stone *et al.*, 1995).

Ontogenetically, taste buds generally develop later than their counterpart olfactory organ. Taste bud primordia appear first around the mouth and within the oropharynx regions a few days prior to hatching in zebrafish and channel catfish (Hansen *et al.*, 2002; Northcutt, 2005). The development of trunk taste buds lags far behind that of the head region. In zebrafish, the primordial cells accumulated beneath the single layer of epithelial cells begin to elongate from the basal lamina to the epithelial surface by 4 days post-fertilization. By 10 days postfertilization, the primordia begin to erupt as mature taste buds, at which time basal cells and nerve fibers are clearly seen and the dermal papillae begin to form beneath the basal cells. In rainbow trout the earliest taste buds are seen only in larvae 8 days posthatching (Twongo and MacCrimmon, 1977). Although details of the development of the innervation of taste buds are still unknown, the above-mentioned data suggest that taste buds arise from within the specified epithelium (early specification) rather than being induced (nerve induction) by peripheral nerves (Barlow and Northcutt, 1995; Northcutt, 2004).

One of the most characteristic features of the taste bud is that, while maintaining its entity, it is a constituent of the epithelium without partition from the surrounding epithelial cells. The average life span of taste bud cells on the channel catfish barbel is 12–42 days, depending on surrounding temperature (Raderman-Little, 1979). Rod cells appear to renew at a faster rate than the microvillar in minnows (Kitoh *et al.*, 1987). Unlike the olfactory epithelium, in which the basal cells function as progenitors for the receptor neurons, epithelial cells surrounding the taste buds divide and some of their daughter cells migrate into taste buds to form gustatory cells (Raderman-Little, 1979).

2.1.5. Comparison with Higher Vertebrate Taste Buds

In mammals, taste buds are restricted to the oropharyngeal cavity including larynx and the entrance of the esophagus. Taste buds are especially abundant on the dorsal surface of the tongue, where they are distributed on the three types of gustatory papillae: fungiform, foliate, and circumvallate papillae. The apical processes of gustatory cells project into a pit and open to the environment through a narrow pore. Generally four, sometimes five, cell types are recognized: type I, II, III, and IV cells. There seems to be a general consensus that type III cells, although comprising the smallest fraction 5–15% of the total gustatory cell populations, are receptor cells (Kinnamon, 1987; Reutter and Witt, 1993). Type IV cells, basal cells, are generally considered to be the stem cells of the other cell types (Roper, 1989). Synaptic contacts with nerve endings within mammalian gustatory cells are not all clear-cut either. A high-voltage electron microscopic study of mouse taste buds showed that fungiform gustatory cells, although fewer than those in the circumvallate papillae, have synapses onto nerve fibers, all of which are afferent, that is, from the gustatory cells to nerves (Kinnamon *et al.*, 1993). Synapses are characterized by clusters of 40–70 nm vesicles within the cell cytoplasm adjacent to the presynaptic membrane. Fluorescence-histochemical and immunohistochemical investigations in mouse, rabbit, and monkey show that serotonin (5-HT) is a transmitter, especially of type III cells (Takeda *et al.*, 1982; Fujimoto *et al.*, 1987). In fact, mouse taste buds release 5-HT when depolarized with KCl or stimulated with bitter, sweet, or acid taste substances (Huang *et al.*, 2005). However, what excites serotonergic gustatory cells is still not clear. They may not be primary gustatory receptor cells, but instead may modulate the function of receptor cells (Kim and Roper, 1995). A study with mice strongly suggests that adenosine 5′-triphosphate (ATP) serves as a key neurotransmitter linking taste buds to sensory nerve fibers (Finger *et al.*, 2005). These authors demonstrate that mice with genetically knocked-out ionotropic purinergic receptors (P2Xs) either reduce or eliminate gustatory nerve and behavioral responses to sweeteners, glutamate, and bitter substances, leaving the responsiveness to touch and temperature intact. Stimulation of taste buds *in vitro* evokes release of ATP. Furthermore, 5-HT receptor (5-HT_{3A}) knockout mice reduce the behavioral responsiveness to pain, whereas the gustatory behavior is identical to that of the wild-type controls for each taste quality tested. Thus, 5-HT does not act on neural receptors to transmit gustatory information from taste buds to the gustatory nerve.

Amphibians exhibit the most complex taste buds, termed taste disks, of any vertebrate. Although taste buds of urodeles are generally similar to those in fishes, taste buds of anurans, especially of frogs are flat and disk-like (up to 200 μm in diameter) located on large fungiform papillae (Reutter and Witt, 1993). Taste disks are encircled by either marginal or ciliated cells. Developmentally, these large and complex taste buds are formed just before metamorphosis of tadpoles. In the adult frog taste buds, eight different cell types lie within the three different sensory epithelial layers—basal, intermediate, and superficial strata. The intermediate stratum is composed mainly of microvillar, rod, and wing cells, whereas the basal stratum consists of two types of basal cells: (1) basal cells comparable to other vertebrates and (2) basal stem cells, believed to be responsible for generation of new receptor cells. Electrophysiologically, rod cells possess the voltage-gated Ca^{2+} channels and do classical synaptic transmission (Ca-influx) (Suwabe and Kitada, 2004). Two types of nerve fibers, thick and thin, enter the taste disk, and the former make contacts exclusively with ordinary basal cells, whereas the latter, after forming a nerve plexus, make complex synaptic contacts with various cell components.

2.1.6. Solitary Chemosensory Cells and Common Chemical Sense

In fishes, solitary chemosensory cells (SCCs), differentiated epithelial cells closely resembling gustatory receptor cells without organizing into discrete end organs, are scattered over the surface of the skin and inside the mouth, and believed to be innervated by either recurrent facial or spinal nerves (Kotrschal, 1991; Whitear, 1992; Finger, 1997b). In some cyprinids, average densities of SCCs of 2000–4000 per mm^2 are estimated, and total SCC numbers are higher than those for the entire taste bud cells. A highly specialized, and best studied, SCC system can be found in the anterior dorsal fin of rocklings, of which the epidermis contains ~5 million SCC cells that account for nearly 15% of the entire fin ray epidermis (Kotrschal, 1991; Kotrschal *et al.*, 1998). They are innervated by the dorsal recurrent facial nerve fibers. Each cell makes one to four synaptic contacts resembling those found in taste buds and several hundred cells synapse onto one single nerve fiber. In the course of evolution, secondary sensory cells capable of responding to environmental stimuli differentiate in the epithelia of primarily aquatic vertebrates and certain of these cells may have incorporated into discrete taste buds, while others failed to have a special association with surrounding epithelial cells. PHA-E lectin, specifically inhibit binding of L-arginine to the channel catfish taste buds, also label the apices of SCCs, suggesting that the two systems share similar carbohydrate moieties on membrane proteins (see in a later section; Finger *et al.*, 1996). The electrophysiological responsiveness of the rockling fin rays to mucoid has provoked speculation about their roles in possible interspecific interactions (Whitear and Kotrschal, 1988; Peters *et al.*, 1991; Kotrschal, 1995). Further studies on the functional properties of the SCC as well as the olfactory and gustatory systems will be required to unequivocally claim their primary chemosensory function.

The common chemical sense, mediated by nonspecific nervous structures such as free nerve endings, is widely distributed in all fish groups. Nerve fibers are believed to come to literally free endings within the epidermis without any specialized structure (Whitear, 1971). A specialized form of the common chemical sense is found in the pectoral fin rays of the sea robins (*Prionotus* sp.; Herrick, 1907; Scharrer *et al.*, 1947; Finger, 1982, 1997b). The fin rays lack taste buds, and are innervated solely by spinal nerve endings, and no known receptor exists. The spinal nerves arise from bipolar cells in the dorsal root ganglion, whose central processes terminate within the dorsal horn of the first few spinal segments, termed accessory spinal lobes. These lobes project rostrally to the lateral funicular nucleus situated at the spinomedullary junctions. This funicular nucleus relays the ascending spinal input to the contralateral midbrain and ventral diencephalon, where, while maintaining distinct channels within the CNS, signals from the olfactory, gustatory, and common chemical senses are to be coordinated (Kotrschal and Finger, 1996). The sea robin fin rays respond specifically to relatively low concentrations of compounds including the actual feeding stimuli (Silver and Finger, 1984). Because, sea robins, like the dorsal fin rays of rocklings, respond vigorously to tactile and proprioceptive stimuli, further studies on extensive feeding experiments involving both olfactory and gustatory stimuli will be required before assigning a novel function to this unique sensory system.

2.2. Central Gustatory Nuclei and Pathways

2.2.1. Primary Gustatory Nuclei

The gustatory receptor cells that have no axons are innervated by axons of three cranial nerves, facial (VII), glossopharyngeal (XI), and vagul (X) nerves, which detect and transmit gustatory signals to the primary gustatory nucleus (PGN). The PGN is a mammalian equivalent of the nucleus of the solitary tract (NST), regardless of which nerve they enter the brain. This is a ventrolateral, epibrachial placodal afferent system lies medial to the octavolateral nuclei and dorsal to the visceral motor nuclei of the rostral medulla (Figure 2.4A). Generally, all cutaneous taste buds on the body surface and rostral oral regions are innervated by the facial nerve, while taste buds located within the posterior oral cavity and pharyngobranchial region are innervated by the glossopharyngeal and vagal nerves. Similarly, in mammals, the chorda tympani nerve, a branch of the facial nerve, innervates taste buds on the fungiform papillae located on the anterior two-thirds of the tongue, the glossopharyngeal nerve the posterior third, and the superior laryngeal nerve innervates the taste buds on the esophagus and epiglottis. Although the details vary, the configuration of the gustatory neural connections is essentially similar in all fish species studied (catfishes, cyprinids, mullet, tilapia, trouts, and so on), with only minor variations. It is important to note that the gustatory system is the only vertebrate sensory system in which three cranial nerves carry all peripheral gustatory information. Somatotopic organization generally exists in the termination of axons in the PGN. The gustatory nerves from the different regions of the body thus enter the nucleus in approximately the same order as they are located in the body, that is, axons from more rostral regions of the mouth (or body) enter the more rostral portions of the nucleus (Figure 2.4A; Finger, 1987; Butler and Hodos, 1996). The primary gustatory column also receives descending projections from neurons of the lateral hypothalamus and tertiary gustatory nucleus, TGN (Section 2.2.2).

Specialization of the gustatory system has occurred independently in cyprinids and silurids. Goldfishes and carps, for example, have a highly

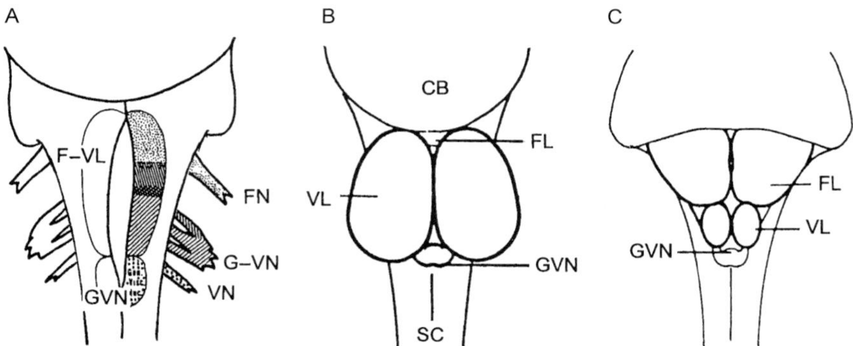

Fig. 2.4. Diagram showing the dorsal view of gustatory and general visceral nuclei in the medulla of unspecialized teleost (A), goldfish (B), and catfish (C). Facial (FN), glosso-pharyngeal–vagal (G–VN), and vagal nerves (VN) terminate in rostrocaudal sequence in gustatory lobes. CB, cerebellum; FL, facial lobe; F–VL, facial–vagal lobe; GVN, general visceral nucleus; and SC, spinal cord. [Adapted from Finger (1988) and Kanwal and Finger (1992) with permission.]

developed palatal organ, which is a muscular structure attached to the roof of the mouth and the surface of the gill arches. They are studded with numerous taste buds innervated by the palatine branches of the vagal nerve. With the aid of the palatal organ, these fish are able to separate nonpalatable food materials and spit them out (Sibbing *et al.*, 1986). The vagal nerve terminates in the gustatory nucleus as in other fishes, but the gustatory nuclei in these fishes have largely bulged out to form a prominent vagal lobe (Figure 2.4B). The internal structure of the vagal lobe is topographically organized, that is, the entire structural feature of the oropharyngeal cavity is mapped onto the vagal lobe so that, for example, the anterior end of the palatal organ is represented anteriorly in the lobe and progressing posteriorly (Morita and Finger, 1985a). In addition, gustatory nerves from each of the oropharyngeal structures terminate in specific layers of the lobe, that is, nerves innervating taste buds of the palatal organ project to layer 6, those innervating the gill arches project to layers 2 and 4, and so on (Morita and Finger, 1985b). In parallel with the sensory neuron inputs, the outputs of the motor neurons innervating musculatures of the palatal organ and gill arches, occupying the deepest layer of the lobe, are also organized in a topographic fashion (Morita and Finger, 1987). These motoneurons are not present in recognizable numbers in other teleosts correlated with the absence of the palatal organ. Thus, the function of a neural structure seems to be the primary determinant of its substructural organization and intrinsic neuronal connectivity.

In catfishes, on the other hand, the facial lobe is large and organized into six longitudinal columns or lobules extending rostrocaudally and arranged alongside each other (Figure 2.4C; Hayama and Caprio, 1989; Kiyohara and Caprio, 1996). Each lobule receives segregated input from discrete portions of the body surface. In channel catfish, for example, the three medial lobules receive input from the medial and lateral mandibular barbels and the maxillary barbel, respectively. The two ventral lobules receive inputs from the nasal barbel and pectoral fin and a large dorsolateral lobule from the face and flank. The entire extraoral surface of the body is thus mapped somatotopically onto the facial lobe in a well-defined manner, which seems to be a feature seen throughout the teleosts (Kanwal and Finger, 1992). Cell bodies of neurons are organized into small clusters or glomeruli, to which the incoming primary gustatory fibers terminate. A gustatory viscerotopy also exists in mammals (Travers *et al.*, 1986), and thus a topographic organization of primary gustatory input may be present in many, if not all, vertebrates. The facial lobe lacks a defined motor layer corresponding with that of the vagal lobe, instead, contains large neurons which give rise to motor root fibers responsible for more general body movements such as turning and seizing (Herrick, 1905).

Both the facial and vagal lobes have two main types of secondary neurons: (1) small intrinsic neurons, filling the interior of the lobe; and (2) larger specialized neurons superficially arranged over the lobes—the chief secondary gustatory neurons (Herrick, 1905). The intrinsic neurons function as interneurons within the lobes, while the chief gustatory neurons, directly receiving the peripheral gustatory terminals of the first order, give rise to the long paths of the secondary connection primarily to the superior secondary gustatory nucleus (SGN) in the isthmus. But in no case does a peripheral gustatory neuron connect directly with a peripheral motor neuron (Herrick, 1905).

2.2.2. Secondary and Tertiary Gustatory Nuclei

The major efferent projections of the primary gustatory center are ascending via two pathways (Figure 2.5; Finger, 1978; Morita and Finger, 1985b; Rink and Wullimann, 1998; Yoshimoto *et al.*, 1998; Folgueira *et al.*, 2003). Typically, the primary centers project to the superior SGN in the isthmus (or pons), to the tertiary gustatory relay nucleus in the caudal diencephalon (pTGN), and then to the telencephalic gustatory area (e.g., the caudal dDm). The SGN neurons also project to the hypothalamic inferior lobe (IL) and to a posterior thalamic nucleus. The caudal part of the diencephalon is called the preglomerular nuclear complex or the posterior thalamic nuclear complex in different species. Catfishes have two such preglomerular centers, the nucleus of the lateral thalamus, homologous to pTGN, and a more posteriorly located

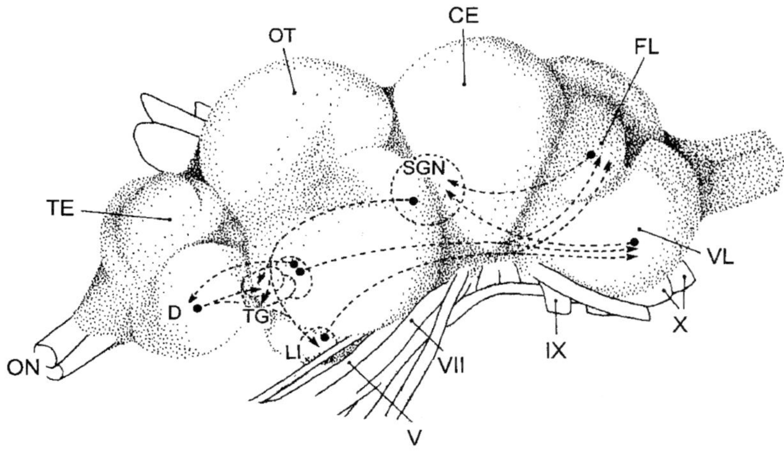

Fig. 2.5. Schematic diagram of the dorsolateral view of the brain of a cyprinid fish showing the entrance of trigeminal (V), facial (VII), glossopharyngeal (IX), and vagal (X) nerves, and the main central gustatory connections. CE, cerebellum; D, dorsal telencephalic area; Fl, facial lobe; LI, hypothalamic inferior lobe; ON, olfactory nerve; OT, optic tectum; SGN, secondary gustatory nucleus; TE, telencephalon; and TG, tertiary gustatory nucleus. [Adapted from Herrick (1905), Luiten (1975), and Rink and Wullimann (1998) with permission.]

nucleus lobobulbaris (Lamb and Caprio, 1993a). Electrophysiological responses of the channel catfish SGN neurons show that the precise topographic projections displayed in the facial and vagal lobes are not maintained at the SGN (Lamb and Caprio, 1992), suggesting that spatial aspects of gustatory maps may not be relevant in the encoding of gustatory information. Neurons in nuclei in the caudal inferior lobe exhibit differential responsiveness to amino acid and tactile stimuli, confirming previous anatomical identification as recipients of projections from the primary and secondary gustatory information. The distinct electrophysiological responsiveness of these nuclei, combined with their different connectivity patterns, suggest that the gustatory nuclei in the inferior lobe are involved in various different sensory processing mechanisms (Lamb and Caprio, 1993b). In the rainbow trout, neurons of the preoptic nucleus, highly immunoreactive to neuropeptides, project to the SGN, suggesting that preoptic SGN and/or preoptic primary gustatory column projections may be important neuroregulatory circuits in some teleosts (Folgueira *et al.*, 2003). The pTGN is, because of its ascending telencephalic connections, sometimes referred to as the thalamic gustatory nucleus and could be homologous to the subthalamus of mammals (Yoshimoto *et al.*, 1998). The IL is a large and highly differentiated brain part in teleosts comparable to other major brain parts such as cerebellum, optic tectum, or telecephalon, and is believed to represent a multisensory integration centre (Rink and Wullimann, 1998; Folgueira *et al.*, 2003). Thus, the ascending gustatory connection pattern is shared by all teleosts and seems to represent the ancestral patterns for teleosts (Wullimann, 1998). The SGN is the homologue of the parabrachial nucleus (PBN), that is, the pontine taste area of mammals. In humans, axons from the NST bypass the pontine PBN and terminate in the medial part of the ventrobasal thalamus, which in turn, project to two areas of primary gustatory cortex. The hypothalamic projection makes contact with the visceral and endocrine control cell groups in the hypothalamus.

2.2.3. Telencephalic Gustatory Areas

Telencephalic projections of the secondary and tertiary gustatory neurons have been identified in carp (Murakami *et al.*, 1986), channel catfish (Kanwal *et al.*, 1988; Lamb and Caprio, 1993a), goldfish (Rink and Wullimann, 1998), tilapia (Yoshimoto *et al.*, 1998), and rainbow trout (Folgueira *et al.*, 2003). Generally, gustatory nuclei in the telencephalic region include medial and central portions of area dorsalis, which receive projections from the pTGN. In tilapia, pTGN projects to the dorsal region of area dorsalis pars medialis (dMm), whereas in rainbow trout it projects to the medial region of the dorsal telencephalic area (Dm) and the ventral telencephalon. In addition, the tilapia SGN projects to area ventralis pars intermedia (Vi) and areas dorsalis pars posterior (Dp). Note that the Vi and the Dp of tilapia and Dm and the ventral telencephalon in salmonids receive direct projections from the olfactory bulb (Matz, 1995). In mammals, the SGN, or PBN, projects directly to the amygdala, which receives direct projections from the olfactory bulb. Phylogenetically, the Vi and Dp are considered as a part of the amygdala and the lateral pallium, respectively (Northcutt and Davis, 1983). As Herrick (1905) pointed out with regards to the major importance of the relation of gustation and olfaction, the secondary centre is the area olfactoria of the forebrain, whose commissure bears the same relation to the secondary tracts as do those of the secondary gustatory nuclei. The main tertiary tract passes, as before, to the inferior lobe, which is, in fishes, the central correlation station for all sensory impression (Figure 2.6). The olfactory and gustatory tertiary tracts end together throughout the inferior lobe and they have a common descending conduction path, the tractus lobobulbaris.

The studies described in earlier sections demonstrate that the ascending fiber connections of the central gustatory system in teleosts are essentially similar to those of mammals. The lemniscal, or direct, pathway from the primary gustatory centers through the SGN and pTGN to the dDm, for example, in tilapia, corresponds with the mammalian gustatory pathway

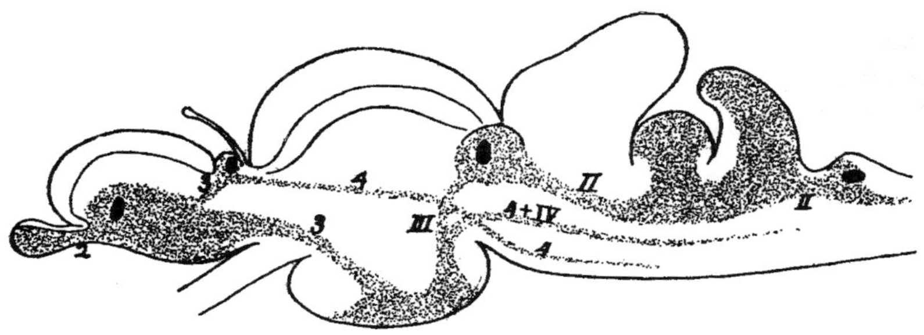

Fig. 2.6. Diagram showing the relations of the gustatory and olfactory centers in teleosts, as represented by the shaded areas. The black spots represent the position of commissures of secondary or tertiary fibers. Numbers 2, 3, and 4 represent the olfactory conduction paths of the second, third, and fourth orders, respectively, and II, III, and IV represent the gustatory conduction paths of the second, third, and fourth orders, respectively. [From Herrick (1905) with permission.]

from the solitary nucleus to gustatory neocortices. In parallel, the gustatory pathway from the primary gustatory centers through the SGN to the Vi and Dp corresponds with the pathway from the solitary nucleus through the pontine taste areas to the amygdala in mammals.

2.2.4. Descending Projections to Facial and Vagal Lobes

The inferior lobe and lobobulbar lobe develop extensive descending fiber projections to the primary gustatory centers in the facial (cyprinids) and vagal (catfishes) lobes as well as in the viscerosensory column (trout). The vagal lobe, as described, projects to motoneurons that innervate the oropharyngeal musculature, which controls swallowing or intraoral food handling in cyprinids. In contrast, the facial lobe of catfishes projects to somatosensory nuclei in the caudal medulla, which may provide a means to correlate gustatory and somatosensory inputs from a single locus on the body surface (Finger, 1987). Furthermore, the return path for all sensory activities of the inferior lobe is the tractus lobobulbaris passing into the oblongata, through which the descending neurons of the inferior lobe communicate with the peripheral motoneurons of the oblongata, and probably also of the spinal cord, thus providing for the most complex reflexes of which the fish is capable (Herrick, 1905).

2.2.5. Alliance with the Trigeminal Nerve

The complex gustatory system as has been described makes itself even more complex by the fact that the facial nerve is closely allied to the trigeminal nerve. In vertebrates, the trigeminal nerve generally carries the sensory nerves from the jaws and head as well as the motor neurons that control the jaw muscles. It is responsible for sensation including pain, temperature, touch, and proprioception (Chapter 4, this volume). The trigeminal system is best developed in animals with a prominent snout, especially on vibrissae (whiskers). The sensory part of the trigeminal is exclusively of tactile-carrying stimuli from the same region innervated by the facial nerve (Herrick, 1905; Luiten, 1975). Barbels of catfishes are heavily innervated by trigeminal nerves mixed with the facial, both of which project to the facial lobe by three branches, ophthalmic, maxillary, and mandibular. The trigeminal fibers terminate not only in the principal and spinal trigeminal nuclei but also throughout the facial lobe with the exception of the lateralmost lobule which contains a representation of taste buds innervated by the recurrent branch of the facial nerve (Kiyohara *et al.*, 1999). The trigeminal fibers of channel catfish are coarser than those of the facial which terminate within the same structural loci, and its input to the primary gustatory complex is restricted to those portions of the nucleus receiving sensory inputs from the face and barbels. Contrary to its central projections, virtually nothing is known about the peripheral termination of the trigeminal nerve in fishes, except for the observations that in sea catfish and goatfish some nerve fibers terminate inside taste buds perigemmally without ending on gustatory cells (Sakata *et al.*, 2001; Kiyohara *et al.*, 2002). It thus appears as if the trigeminal fibers have failed to locate their target cells, that is, they become orphaned gustatory nerves. In humans, the lingual nerve, which is one of the main trigeminal sensory roots innervating the anterior two-thirds of the tongue, terminates as free nerve endings in the epidermis, in and around taste buds, and in the papillary dermis. Some of the lingual nerve fibers completely surround taste buds and a few even terminate in them. However, lingual nerve fibers do not seem to form synapses with gustatory cells (Simon and Wang, 1993). The relationship between gustatory and tactile sensibility is thus well recognized but its implications are not fully understood. One implication is that there is probably no "pure gustatory area" at any level of the gustatory neuraxis.

Three remarkable specialized adaptations of the trigeminal nerve have been known in vertebrates: (1) infrared receptors in the sensory pits on the heads in snakes (Molenaar, 1991), (2) mechano- and electroreceptors on the prominent snout in an egg-laying mammal platypus (Scheich *et al.*, 1986), and (3) magnetoreception in birds (Semm and Beason, 1990; Lohmann and Johnsen, 2000) and fish (Walker *et al.*, 1997). In fish, some neurons of the superficial ophthalmic branch of the trigeminal nerve of rainbow trout are specialized to respond to magnetic stimuli, most likely detecting the electromagnetic alignment of magnetites present in the nasal cavity (Chapter 8, this volume).

3. FUNCTIONAL PROPERTIES

3.1. Responses to Chemical Stimuli

3.1.1. General

The first electrophysiological study was conducted by Hoagland (1933) who recorded gustatory responses of the facial nerves of bullhead catfish (*Ameiurus nebulosus*) innervating lips and barbels in response to chemical and tactile stimulation. This pioneering work demonstrated the great sensitivity to mechanical stimuli (touching and water current), whereas responses to chemical stimulation such as acetic acid, NaCl, and meat juice were small and barely detectable. Some three decades later, more thorough investigations of electrical responses of the gustatory receptors to various chemicals were undertaken by Tateda (1964), Konishi *et al.* (1966), and Bardach *et al.* (1967a) in sea catfish (*Plotosus anguillaris*), catfish (*Ameiurus melas*), and sea robin (*Prionotus carolinus*), respectively. However, it was not until gustatory responses of Atlantic salmon (*Salmo salar*) (Sutterlin and Sutterlin, 1970), channel catfish (Caprio, 1975), and puffer (*Fugu pardalis*) (Kiyohara *et al.*, 1975) to amino acids were demonstrated that a resurgence of research into fish gustatory physiology began (see also Chapter 4, Vol. 5 of this series; Marui and Caprio, 1992). Unlike in terrestrial animals, the fish gustatory receptors are a "distance chemical sense," stimulated by dilute aqueous solution of low-molecular-weight substances, and the majority of studies over the years focused on chemicals including amino acids, peptides, nucleotides, and organic acids. Of these, amino acids are the best studied gustatory stimuli for fish. Gustatory response spectra for amino acids vary considerably among fish species, contrasting with relatively consistent spectra of the olfactory (Hara, 2005; Chapter 1, this volume). A total of nearly 30 teleost species whose gustatory responses to amino acids have been examined systematically can be grouped roughly into two: (1) those that respond to a few amino acids and (2) those that respond to many amino acids (Hara and Zielinski, 1989; Hara, 1993, 1994). The former are best represented by salmonids (Hara *et al.*, 1993, 1999) and the latter by channel catfish (Caprio *et al.*, 1993).

3.1.2. Amino Acids

In most studies, electrical responses of the peripheral gustatory receptors are recorded from branches of the facial nerve innervating taste buds inside the upper (e.g., minnow, puffer, rainbow trout) or lower lips (e.g., carp), or maxillary barbels (e.g., channel catfish). Gustatory responses to amino acids are generally characterized by a fast adapting, rapidly returning to the baseline activity even with continuous stimulation, which contrasts with a slow-adapting olfactory response. Threshold concentrations for the more stimulatory amino acids range from 10^{-8} to 10^{-7} M for most species to 5×10^{-11} M for channel catfish, which are much below than those recorded for amphibians, birds, and mammals. However, inconsistencies in experimental conditions, especially those involved with perfusing water, could have had profound effects on the threshold determination. Generally, the L-isomer of an amino acid is more stimula ory than its D-enantiomer.

All species in Group 1 respond well to L-proline, L-alanine, and a few other related amino acids. Typical, and probably the prototype of the group, are the salmonids. In rainbow trout, for example, only L-proline (L-Pro), hydroxy-L-proline (L-Hpr), L-alanine (L-Ala), L-leucine (L-Leu), L-phenylalanine (L-Phe), L-α-aminoguanidinopropionic acid (Agp), and betaine (Bet) are stimulatory (Figure 2.7; Marui *et al.*, 1983a; Yamashita *et al.*, 2006). No significant differences exist in the response pattern to amino acids between the facial and glossopharyngeal nerves (Kohbara and Caprio, 2001). Cross-adaptation (desensitization) data indicate that these amino acids are detected by at least three independent gustatory receptor types: (1) Pro-receptor (L-Pro, L-Hpr, L-Ala), (2) Bet-receptor (Bet, Agp), and (3) Leu-receptor (L-Leu, L-Phe) (Figure 2.8; Yamashita *et al.*, 2006). It is emphasized that the concentration–response (C–R) relationship and cross-adaptation along with mixture experiments for determining receptor specificities is of potential value in assessing the specificity of a response to a particular receptor population. Although some strain differences exist among rainbow trout, these receptors are commonly distributed among salmonids, but many nonrainbow trout salmonids lack the Leu-receptors (Figure 2.9; Hara *et al.*, 1999). Arg, although responsive only under high pHs ($\geq$8.0), is inhibited by Agp, indicating that Agp, Bet, and Arg all share the same receptor type (Marui *et al.*, 1983a; Yamashita *et al.*, 2006). Pro, Ala, and Bet are also major stimulating amino acids for both Carp (*C. carpio*) and puffer, with a few more stimulatory amino acids than salmonids, may well belong to this group (Marui *et al.*, 1983b; Kiyohara and Hidaka, 1991). The most characteristic feature common to all species in Group 1 is that L-Pro is the most dominant stimulus of all amino acids tested. Also, note that Pro is the only naturally occurring amino (more accurately imino) acid that is not detected by the olfactory system. Thus, the gustatory and olfactory systems of these fishes detect distinct spectra of amino acids, consequently minimizing the redundancy (cf. Hara, 1994, 2005).

By contrast, channel catfish, representing Group 2 fishes, detect virtually all common amino acids through gustation, with the sensitivities higher (lower thresholds) than those of the olfactory counterpart. Recordings from the facial nerve innervating the maxillary barbels show that thresholds for L-Ala, the

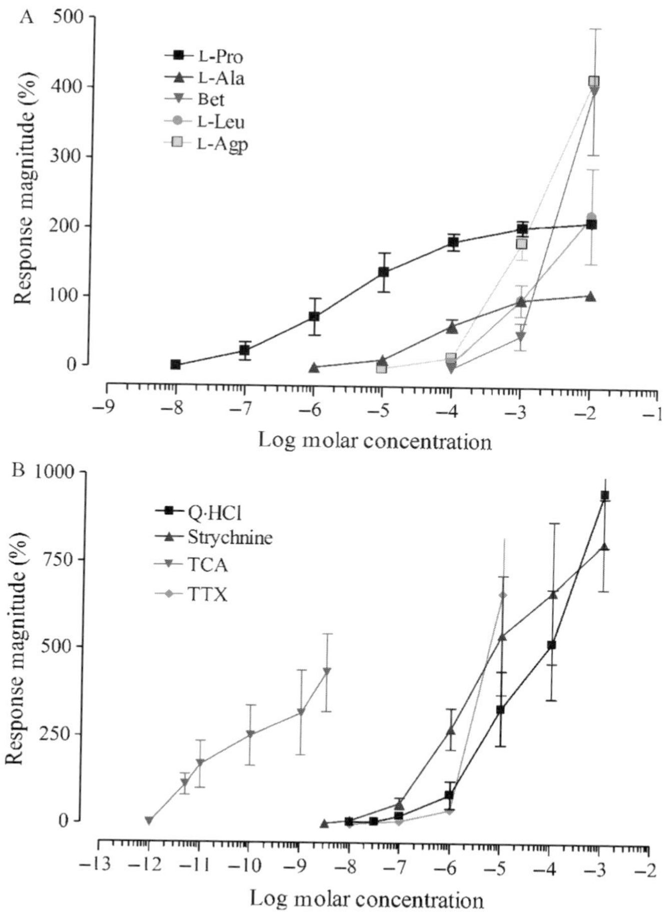

Fig. 2.7. Concentration–response (C–R) relationships of gustatory responses recorded from the maxillary branch of the palatine nerve innervating inside the upper jaw of rainbow trout. (A) L-Pro, L-proline; L-Ala, L-alanine; Bet, betaine; L-Leu, L-leucine; L-Agp, L-2-amino-3-guanidinopropionic acid. (B) Q·HCl, quinine hydrochloride; TCA, taurocholic acid; and TTX, tetrodotoxin. [Adapted from Yamashita *et al.* (2006) with permission.]

most effective gustatory stimulus for this species, average 5×10^{-11} M (Caprio, 1978). Cross-adaptation results indicate that all amino acids are detected mainly by L-Ala and L-Arg receptors, supplemented by D-Ala, L-Pro, D-Pro, D-Arg, D-His, and L-Lys receptors (Wegert and Caprio, 1991; Kohbara *et al.*, 1992; Caprio *et al.*, 1993). A cluster analysis of single facial nerve innervating the barbels further reveals two major clusters (Ala and Arg fibers) and two Arg

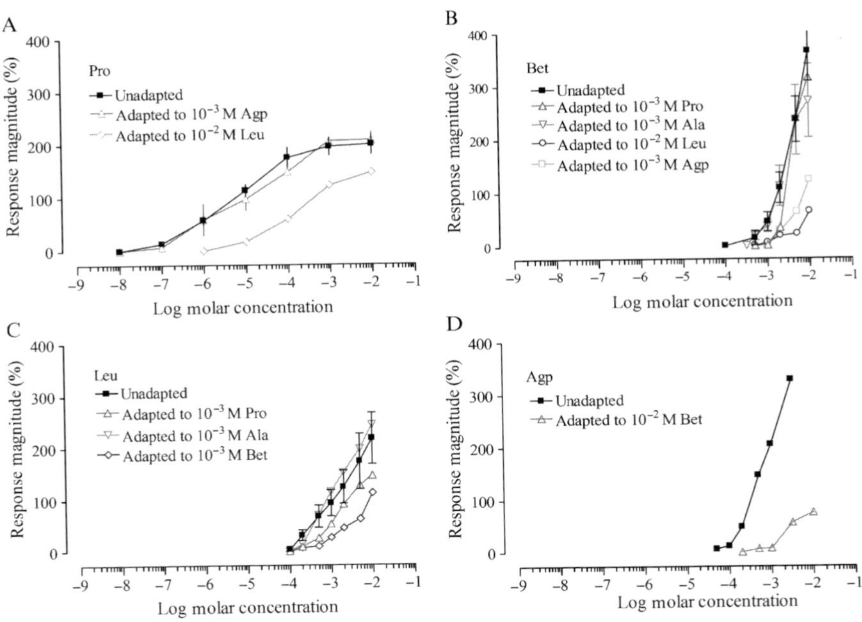

Fig. 2.8. Reciprocal cross-adaptation (desensitization) between amino acids and betaine to determine receptor specificities in rainbow trout. In each graph, Pro (A), Bet (B), Leu (C), or Agp (D) were tested over the concentrations (10^{-8}–10^{-2} M) before (unadapted) and during (adapted to) adaptation to the chemicals listed. In (A), the C–R curve for Pro was shifted parallel to the right by adapting to Leu, indicating competitive inhibition. [Adapted from Yamashita *et al.* (2006) with permission.]

subclusters (high-affinity Arg and low-affinity Pro fibers) responsive to amino acids (Kohbara *et al.*, 1992; Caprio *et al.*, 1993). Arg fibers respond to fewer amino acids, that is, narrowly tuned, while fibers comprising the Ala cluster respond to a number of other neutral L-amino acids and D-Ala. Thus, in channel catfish, the two major types of gustatory fibers, the Ala and Arg, transmit amino acid information to the central gustatory centers (Kohbara *et al.*, 1992; Caprio *et al.*, 1993). However, which amino acid is detected by which receptor type is not all clear. Also, cautious interpretation is required for single fiber recording, since to date whether single gustatory fibers innervate the same or different receptor cell types within single or multiple taste buds is not known. High sensitivity and specificity to Ala and Arg are maintained throughout the flank innervated by the recurrent branch of the facial nerve (VII) and oropharyngeal gustatory system innervated by glossopharyngeal (IX) and vagus (X) nerves in channel catfish (Davenport and Caprio, 1982;

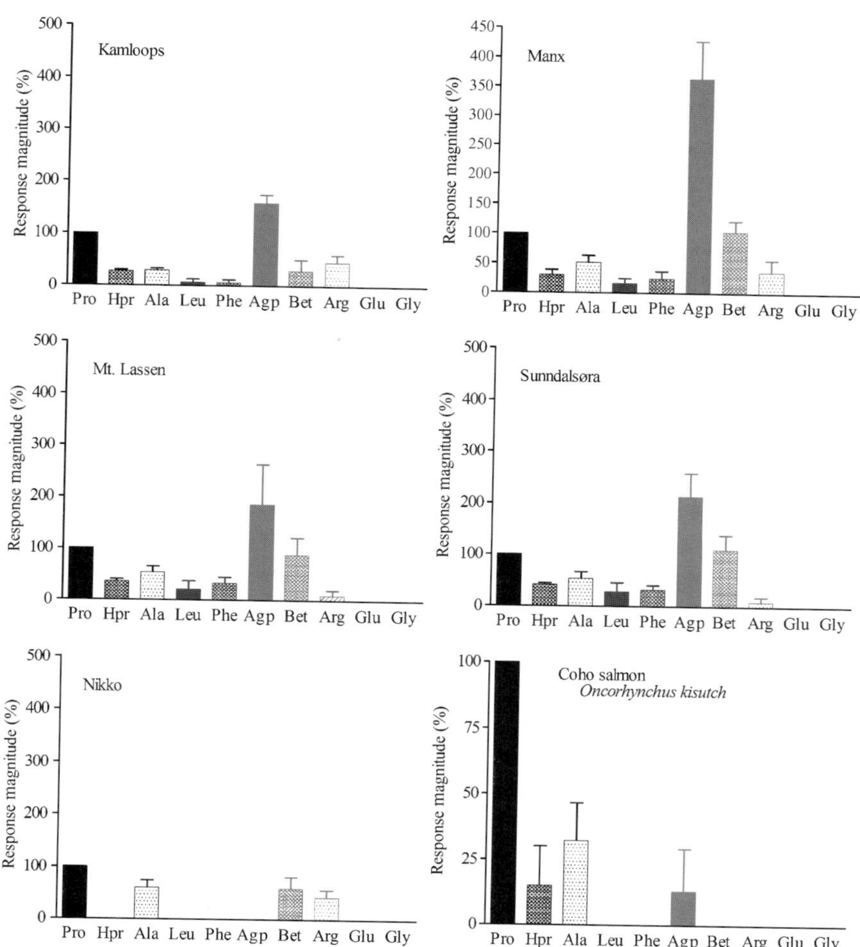

Fig. 2.9. Relative stimulatory effectiveness of amino acids tested on palatine nerve responses in six strains of rainbow trout. All chemicals are tested at 10^{-3} M, and the response magnitude is expressed as a percentage of the response to Pro. Pro, L-proline; Hpr, hydroxyl-L-proline; Ala, L-alanine; Leu, L-leucine; Phe, L-phenylalanine; Agp, L-2-amino-3-guanidinopropionic acid; Bet, betaine; Arg, L-arginine; Glu, L-glutamic acid; Gly, glycine. [Adapted from Hara *et al.* (1999) with permission.]

Kanwal and Caprio, 1983). The data further show that Pro is as stimulatory as Ala and Arg, especially for IX nerve, and that Pro cross-adapts Ala, suggesting they share the same receptor types (Kanwal and Caprio, 1983). It should be noted that: (1) responses to amino acids in channel catfish were recorded at pH $\geq$8.0, which may have been biased for basic amino acid (e.g., Arg) stimulation (Marui *et al.*, 1983a), (2) the cross-adaptation method used was not of true competition (or desensitization), and (3) single fiber recordings were made at or close to saturation concentrations ($\geq 10^{-4}$ M), where the receptor specificity is decreased or lost.

Fishes in which Ala and Arg are the two major gustatory amino acids include the Japanese eels, *Anguilla japonica* (Yoshii *et al.*, 1979), mullet, *Mugil cephalus* and red sea bream, *Chrysophyrys major* (Goh and Tamura, 1980b), topmouth minnow, *P. parva* (Kiyohara *et al.*, 1981), tigerfish, *Therapon oxyrhynchus* (Hidaka and Ishida, 1985), tilapia, *Tilapia zillii* (Johnsen *et al.*, 1990), and sea catfish, *Arius felis* (Michel *et al.*, 1993). Japanese eels and marine fishes are characterized by high sensitivities to Gly in addition to Ala and Arg (Hidaka and Ishida, 1985; Ishida and Hidaka, 1987). Not surprisingly, amino acid response spectra are different between mullet and red sea bream but not between two stocks of red sea bream (Goh and Tamura, 1980a), since gustatory receptors are perhaps genetically determined to respond to a set of amino acids. However, behavioral responses to those amino acids and mixtures within the spectrum may be learned (see in a later section).

In summary, although an accurate estimation of the number of effective amino acids and their detection thresholds for each fish species may be difficult to make due to inconsistencies in experimental conditions employed by different researchers, it is not unreasonable to conclude that Pro dominates in Group 1 fishes, minimizing the redundancy between the gustatory and olfactory systems. In Group 2 fishes although both Ala and Arg are dominant, considerable overlaps exist between the two systems.

3.1.3. Miscellaneous Chemicals

a. Bile Acids. Some bile acids, known as olfactory stimulants, also stimulate the salmonid gustatory receptors (Hara *et al.*, 1984; Zhang *et al.*, 2001; Yamashita *et al.*, 2006). In rainbow trout, taurocholic acid (TCA), one of the most potent bile acids tested, has threshold concentration at about 10^{-12} M, with the maximum response magnitude almost double that of Pro (Figure 2.7). Both taurine- and glycine-conjugated forms are generally more stimulatory than the free. Responses to bile acids are characterized by their slow adapting, tonic responses. Responses to all tested bile acids are inhibited by taurolithocholic acid (TLC), the most potent bile acid, suggesting that all bile acids share the same receptor type. To date, the only other species whose gustatory receptors have shown similar sensitivities and specificities to bile acids are three char species (*Salvelinus alpinus*, *S. fontinalis*, and *S. namaycush*) (Hara *et al.*, 1993). It is noteworthy that so far the gustatory sensitivity to bile acids is restricted to salmonids. The gustatory systems of carp, sea catfish, and the African chichlid, *Tilapisa nilotica* are not sensitive to bile acids up to 10^{-4} M (Marui and Caprio, 1992).

b. Quinine, Strychnine, and Tetrodotoxin. Quinine hydrochloride (Q·HCl) and strychnine, extremely bitter to humans, stimulate the gustatory receptors of salmonids. Rainbow trout detect Q·HCl and strychnine at threshold concentrations of 10^{-8}–10^{-7} M, with extremely high response magnitudes reaching nearly four times that of Pro (Figure 2.7; Yamashita *et al.*, 2006). Both chemicals are moderately stimulatory for three char species, with thresholds ranging between 10^{-7} and 10^{-6} M (Hara *et al.*, 1993). Although the two chemicals are very different in molecular conformation, adaptation to strychnine completely suppresses gustatory responses to Q·HCl in rainbow trout, possibly sharing the same receptors. In channel catfish, both glossopharyngeal and vagal nerves are also extremely sensitive to Q·HCl, with thresholds ranging between 10^{-9} and 10^{-6} M (Kanwal and Caprio, 1983). Some fish respond specifically to Q·HCl but others are sensitive to both chemicals with more responsiveness to Q·HCl than Ala. At concentration 10^{-3} M, Q·HCl suppresses activities of bimodal fibers responsive to both chemical and tactile stimulation, but not tactile-specific fibers (Ogawa *et al.*, 1997). Q·HCl also suppresses responses to the bitter substances, denatonium benzoate and caffeine, neither of which suppresses amino acid responses. These authors argue that the Q·HCl inhibition is caused by blocking voltage-gated calcium channels in the gustatory cells. In mammals, molecular biological studies show that the T2R receptor family encodes bitter receptors (Chandrashekar *et al.*, 2000). T2Rs are coexpressed in one cell population, indicating that cells with T2Rs recognize a wide range of structurally diverse bitter compounds by virtue of having a large number of receptors. Some bitter substances with amphophilic properties including Q·HCl also interact directly with G-proteins bypassing the receptors. A genetic tracing study further demonstrated that bitter sensitive neurons project to the restricted gustatory regions in the brainstem, thalamus, and cortex distinct from those of sweet sensitive neurons (Sugita and Shiba, 2005).

Puffer toxin [tetrodotoxin (TTX)] and a paralytic shellfish toxin [saxitoxin (STX)] are potent gustatory stimuli for salmonids (Yamamori *et al.*, 1988; Hara *et al.*, 1993; Kitada and Hara, 1994). In rainbow trout, TTX is detected at 2×10^{-7} M, and at 10^{-5} M it evokes a response magnitude four times that of 10^{-3} M Pro (Figure 2.7). STX is even more stimulatory (thresholds, 10^{-8} M) than TTX, and its C–R curve is almost identical to that of Pro (Yamamori *et al.*, 1988). Arctic char, by contrast, respond to TTX and STX with thresholds at 10^{-8} and 10^{-7} M, respectively, but the response magnitudes never exceed that of Pro. These two chemicals are vastly different structurally, do not cross-adapt, and are therefore most unlikely to share the same receptor. TTX suppresses Q·HCl, and vice versa, partially sharing the same receptors (Yamashita *et al.*, 2006). TTX and STX are known as two of the most potent neurotoxins, with a lethal toxicity to humans 300 times that of cyanide. Both exert essentially the same toxic action by blocking the voltage-sensitive sodium channels in neural tissues including the gustatory receptors in higher vertebrates. In fish, however, the treatment of the gustatory receptors with TTX does not affect amino acid induced responses, suggesting that TTX-sensitive sodium channels may not be involved in gustatory transduction in fishes (Yamamori *et al.*, 1988; Kitada and Hara, 1994).

c. CO_2 and pH. Gustatory receptors of some teleosts have long been known to be sensitive to CO_2 (Konishi *et al.*, 1969; Hidaka, 1970; Yoshii *et al.*, 1980; Yamashita *et al.*, 1989). In pH-controlled decarbonated natural water, rainbow trout detect CO_2 with a threshold 4×10^{-5} M, which is slightly higher than the CO_2 level found in natural water (1.8×10^{-5} M) equilibrated with air (Yamashita *et al.*, 1989). The threshold for H^+ is approximately 4×10^{-5} M. Responses to CO_2 are independent of pH, and little affected by adaptation to amino acids or bile acids. None of the single gustatory fibers responsive to CO_2 respond to other stimuli, except Q·HCl. These data clearly indicate that the rainbow trout gustatory receptors are capable of distinguishing between CO_2 and H^+, and that CO_2 receptors are distinct from those that detect other gustatory stimuli. CO_2 responsive fibers are also characterized by higher amplitudes of action potentials compared with those for amino acids. C–R curves for both CO_2 and H^+ responses recorded from a whole palatal nerve bundle shows two discrete components; gradual increases from the thresholds up to about 2×10^{-3} M and pH 3, respectively, followed by sharp increments with further increase in concentration (Yamashita *et al.*, 1989). The former perhaps represent responses mediated by respective receptors and transmitted through the palatine nerve, whereas the latter responses due possibly to stimulation of nonreceptor mechanisms such as involving trigeminal nerves. The C–R curve of the palatal gustatory receptors of the Japanese eels to CO_2, recorded from the facial nerves that lack the trigeminal nerve, exhibits a single component supporting the above-mentioned contention (Yoshii *et al.*, 1980). In the carp, the majority of the sensory fibers of the facial root, as described earlier, terminate on the facial lobe, a small fraction of the fibers of this nerve, however, terminate on the glossopharyngeal and vagal lobes (Luiten, 1975). These areas process not only gustatory information but also signals from other sources, for example, proprioception from the gills (Sutterlin and Saunders, 1969).

In mammals, although taste buds are predominantly distributed in the oral cavity, especially on the tongue and palate, they are also found on the epithelia of the pharynx and larynx. These taste buds are located at the entrance of the respiratory tract, suggesting that they may play a role in respiratory modulation rather than food appreciation (Nishijima *et al.*, 2004). The larynx is innervated by the superior laryngeal nerve, a branch of the vagus nerve containing CO_2-sensitive fibers. Both the intraepithelial free nerve endings and the taste buds have been identified as CO_2 receptors in rats (Nishijima *et al.*, 2004). Immunohistochemical studies further show that type III taste bud cells that have synapses with the nerve may be involved in the CO_2 detection (Nishijima and Atoji, 2004).

d. Other Chemicals. A number of carboxylic acids are shown to be stimulatory for the gustatory systems of Atlantic salmon (Sutterlin and Sutterlin, 1970), Japanese eels (Yoshii *et al.*, 1979), and carp (Marui and Caprio, 1992). Threshold concentrations are generally higher than those for amino acids, and C–R curves are steeper without saturating up to 10^{-2} M in all species. In both salmon and eels, response magnitudes increase with increasing carbon chain length, whereas in carp responses peak with compounds containing three to five carbon chains. Amino acids and carboxylic acids do not cross-adapt each other.

3.1.4. Receptor and Transduction Mechanisms

a. Amino Acid Receptors. Receptor-binding studies using a partially prepared membrane preparation of taste bud-rich epithelium from barbels of channel catfish have shown that two separate binding sites for L-Ala and L-Arg exist (Krueger and Cagan, 1976; Kalinoski *et al.*, 1989). The former, having the apparent dissociation constant (K_d) near 1.5×10^{-6} M, competes with other neutral amino acids, while the latter, with K_d of 1.8×10^{-9} M and 1.3×10^{-6} M, competes with several basic amino acids. Neither competes with each other, suggesting that they interact with separate receptor sites. These binding parameters have been claimed to be consistent with electrophysiological cross-adaptation data (Caprio *et al.*, 1993). However, the data showing that D-Ala and D-Arg inhibit L-Ala and L-Arg binding, respectively, in the same fashion as their L-enantiomers indicate the total lack of receptor specificity, that is, they are full agonists for the same receptors. One serious critique about the direct reversible ligand binding, however, is that even superbly high-affinity systems can often give inadequate binding. A calculation shows that, in the case of the channel catfish barbel experiment, "somehow one must increase the receptors in the bag by a factor of about 2000-fold if the extra counts are to be as large as the background" (O'Brien, 1979).

These two major classes of binding sites exhibit differential sensitivities to lectins (Kalinoski *et al.*, 1992). The L-Arg receptor is further shown to be a ligand-gated ion channel receptor that activates an ion channel of low selectivity, which in turn allows influx of Na and Ca present in the mucus covering the gustatory epithelium (Grosvenor *et al.*, 2004). An odorant receptor preferentially tuned to recognize Arg and closely related basic amino acids is also present in the goldfish microvillar receptor neurons. The receptor is a member of a multigene family of G-protein-coupled receptors, sharing sequence similarities with the calcium sensing, metabotrophic glutamate, and V2R class of vomeronasal receptors (Speca *et al.*, 1999). The receptor exhibits a single binding site with a K_d of 12.1×10^{-10} M, an order of magnitude lower than that of the channel catfish gustatory receptors.

Two families of gustatory receptors are known in mammals: T1R and T2R. The T1R family, consisting of three genes T1R1, T1R2, and T1R3, belong to the G-protein-coupled receptors, along with metabotrophic Glu and GABA receptors, and are selectively expressed in subsets of gustatory receptor cells (Li *et al.*, 2002; Nelson *et al.*, 2002; Scott, 2004). T1R1 and T1R3 combine to function as a broadly tuned L-amino acid sensor specifically responding to most of the 20 natural amino acids. By contrast, the T2R family containing approximately 30 genes is bitter receptors, detecting diverse organic molecules originating from plants including caffeine, nicotine, quinine, and strychnine. Unlike T1Rs, most T2Rs are coexpressed in one cell population that does not contain T1Rs, suggesting that each gustatory receptor cell may be capable of recognizing multiple bitter substances. Thus, the discovery and functional characterization of the cells and receptors for bitter, sweet, and amino acids taste in mammals provide a view that receptors for these taste substances are expressed in distinct populations of gustatory cells that operate independently of each other to trigger stereotypic behavioral responses (Zhao *et al.*, 2003). Analyses of genomic sequences encoding proteins homologous to mammalian T1Rs and T2Rs identified two families of candidate fish gustatory receptors T1Rs and T2Rs in puffer fish (*Fugu rubripes*), medaka fish (*Oryzias latipes*), and zebrafish (Ishimaru *et al.*, 2005). These receptors show high (60–70%) similarities to those of mammals; T1R genes consisting of three types are expressed in different subsets of gustatory cells, whereas T2Rs consisting of two to three types show low degree of similarity to those of mammals and expressed in different cells from those for T1R genes (Figure 2.13). These results suggest that vertebrates commonly possess two types of distinct gustatory receptors and signaling pathways. High degrees of the T1Rs identity between fish and mammals also suggest that the divergence in these gustatory receptor genes occurred prior to that of fish and mammals (Figure 2.13). However, the implication that the two families of gustatory receptors identified based on genomic databases have definite pairing with those natural ligands characterized earlier physiologically and behaviorally should be verified by genetic engineering (Zhao *et al.*, 2003).

b. Transduction Mechanisms. Two possible transduction sequences for gustation have been suggested (Brand *et al.*, 1991; Brand and Bruch, 1992). The binding of stimulus such as L-Ala, L-Arg, and L-Pro to receptors activates G-proteins, which in turn, activate second messenger producing enzyme phospholipase C (PLC) (Yasuoka *et al.* 2004). PLC metabolizes phosphatidylinositol biphosphate (PIP2) into two second messengers: inositol 1,4,5-triphosphate (IP_3) and diacylglycerol (DAG). IP_3 releases Ca^{2+}, while DAG activates protein kinase C (PKC). An increase in Ca^{2+} activity would trigger neurotansmitter release. In contrast, binding of L-Arg and L-Pro to their respective receptors are directly coupled to the activation of nonselective cation channels. Experiments with lipid bilayers into which purified gustatory membrane vesicles derived from channel catfish gustatory epithelium have been incorporated further indicate that L-Arg and L-Pro increase membrane conductance by directly activating different classes of cation channels (Kumazawa *et al.*, 1998).

Stimulation of gustatory receptor cells release neurotransmitters onto afferent nerves that innervate taste buds. In analogy to neurotransmitter release in other excitable cells, it is expected that neurotransmitter release in gustatory cells is dependent on an increase in intracellular Ca^{2+}—$[Ca^{2+}]$. In isolated channel catfish gustatory cells, some respond to L-Arg with an increase in $[Ca^{2+}]$, while others elicit decrease in $[Ca^{2+}]$ (Zviman *et al.*, 1996). These heterogeneous responses of the $[Ca^{2+}]$ to L-Arg is consistent with depolarization or hyperpolarization of gustatory cells, suggesting that excitatory and inhibitory responses to a single gustatory stimulus can potentially play a role in increasing the sensitivity of the gustatory system.

3.2. Responses to Mechanical/Tactile Stimuli

Behavioral studies indicate that gustation and mechanoreception are closely involved in feeding in many fishes. Catfish, for example, find food not by vision but by touch; at the moment food touches a barbel or the fish's body the fish turns toward it with reflex speed and snaps it up (Herrick, 1904; Olmsted, 1918; Biedenbach, 1973). The need for direct contact with food suggests that relevant receptors are mechanosensitive and that central mechanosensory pathways are involved in activating the feeding behavior. However, as we have seen, the barbels are richly supplied with taste buds exclusively supplied by the facial nerve and paralleled by the free nerve-ending plexus via the trigeminal nerve. A question then arises whether the mechanosensitivity originates from the taste bud proper (gustatory nerves) or surrounding nervous components such as trigeminal nerves.

In whole bundle or multifiber recordings, as all fish chemosensory electrophysiologists experience, whether gustatory or olfactory, it is essential to eliminate responses caused by mechanical stimuli due to changing water flow that masks the chemical responses (e.g., Hara *et al.*, 1973; Sveinsson and Hara, 2000). In his pioneering study, Hoagland (1933) not only found high sensitivities of the catfish facial nerve complex but also noticed the striking differences in the impulse amplitudes, that is, low amplitudes due to chemical stimuli and high amplitudes to the mechanical stimuli (touching or water currents). He suggested that the specificity of impulses for the two sense modalities is correlated with cell sizes of origin of the axons: large cells in the Gasserian ganglion supplying tactile (trigeminal) nerve endings and small cells of origin in the geniculate ganglion sending axons to taste buds. However, since the finding that the carp glossopharyngeal nerve, consisting primarily of gustatory activity, responded to both chemical and tactile stimuli (Konishi and Zotterman, 1961), a number of studies has ensued identifying the origin of the bimodal nature of the gustatory nerve activities in catfishes (Biedenbach, 1973; Davenport and Caprio, 1982; Kanwal and Caprio, 1983; Ogawa *et al.*, 1997), carp (Marui and Funakoshi, 1979), puffer (Kiyohara *et al.*, 1985), and rainbow trout (Yamashita *et al.*, 1989). These studies show that the majority of fibers recorded are highly sensitive to mechanical stimuli (touching, sliding, brushing, and water current). Although most fibers respond to either chemical or tactile stimuli, small fractions of fibers of the facial nerve innervating maxillary barbels and recurrent facial nerve innervating the flank of the channel catfish contain bimodal fibers responding to both gustatory (amino acids) and tactile stimulation. Electrophysiological recordings from secondary neurons both in the facial and vagal lobes of carp and catfishes show that gustatory and tactile inputs are overlapped and loosely organized in a somatotopic manner (Marui and Caprio, 1982; Kanwal and Caprio, 1988; Marui *et al.*, 1988).

Thus, the VIIth, IXth, and Xth gustatory nerves all comprise chemical as well as mechanosensitive fibers. The next logical question is whether all the mechanical sensitivity is derived from taste buds. Some authors suggest that basal cells, like Merkel cells of higher vertebrates, might function as mechanoreceptors (Reutter and Witt, 1993). It is, however, unlikely that such gentle mechanical stimuli as light brushing and changes in water flow directly affect the activity of basal cells sitting deep at the bottom of taste buds without interfering with neighboring gustatory cells. One of the morphological features shared by both the gustatory and olfactory, let alone the lateral-line system is that all have developed ciliated cells of one type or another, directly exposed to the aqueous environment. Ciliated olfactory receptor neurons and lateral-line hair cells are characterized by their cilia having a $9 + 2$ microtubule configuration. Evidence exists that suggest a remnant of microtubule arrangement in rod cells of the sea catfish taste buds (Kitoh *et al.*, 1987). It can be speculated that the mechanosensitivity of the gustatory nerve is in part attributable to rod cells, leaving the chemosensitivity to other receptor cells. The relationship between gustatory and tactile sensibility is now not only between the trigeminal nerve but also between receptors within the taste bud.

4. GUSTATORY BEHAVIORS

4.1. Feeding Behavior

4.1.1. Gustatory, Tactile, or Olfactory

Feeding behavior of fish is complex and diverse among species. Typically, three phases of food search can be differentiated: (1) an initial period of arousal or excitement, wherein the fish is alerted to the presence of the stimulus, (2) a subsequent search or exploratory phase to locate the source, and (3) a consummation phase, in which the fish seeks to ingest the potential food (Atema, 1971; Jones, 1992). However, as will be discussed later, they are in reality a continuum without necessarily distinct transition. Various sensory systems contribute to the process, and their role and significance differ at different phases and in different fish species. It has long been a subject of investigation by many whether feeding behavior in fishes is mediated by gustatory, tactile, or olfactory stimuli, or in combination. From his extensive observations on several species of barbeled fishes, Herrick (1904, 1905) concluded that they, especially catfish, are able to precisely localize food by a combination of tactile and gustatory stimulation, unless very strong gustatory stimuli were applied to a small area of the percipient organ, a "local sign." He also emphasized that "memory" plays a part in gustatory and tactile discrimination, and that, as mentioned earlier, vision and olfaction assume greater importance in feeding. Both Parker (1910, 1912) and Olmsted (1918) also claim that responses to food substances are due to stimulation of the olfactory organ, and that blinded or barbel-less fish find food as readily as normal fish. The barbels are of use in finding food only by coming into direct contact with it. On the contrary, catfishes can find distant chemical clues by means of gustation alone; olfactory deprivation does not impair their searching ability (Bardach *et al.*, 1967b). When all barbels, or barbels as well as flank taste buds were made inoperative, these catfish made a one-sided looping movement. Sea catfish detect food even when the olfactory and gustatory capabilities, especially barbels are all removed, and are able to discriminate between bags containing worms and gravel (Satô, 1937b). Mullets seem to rely exclusively on chin barbels for feeding. Japanese goatfish, *Upeneoides bendsasi*, for example, are unable to recognize a packet of lugworm meat, or to find food hidden under mud when the barbels are removed (Satô, 1937a). Elimination of olfaction has no effects. The contents of the digestive cavity of freshly caught goatfish often contain annelids, their natural food, indicating that they recognize prey worms in mud using their barbels in natural habitat. Similarly, a preliminary experiment shows that in the Hawaiian goatfish, *Parupeneus porphyreus*, barbels are capable of mediating all phases of food (crab extract) localization including the initial arousal to the presence of food organisms somewhere in the general vicinity (Holland, 1978). Anosmic animals show no differences in feeding arousal or food search from normal animals, whereas barbel-less animals are much more refractory and do not display food search characteristics.

Selective ablation of the entire sensory area of the facial and vagal lobes has further characterized distinct functional roles played by the two systems in catfish (Atema, 1971); the former is involved in localization and pickup of food (in combination with tactile inputs), whereas the latter in the control of swallowing. The entire observation is based only on a limited number of fish (two facial-ablated TL_1 and TL_2, and two vagal-ablated TL_3 and TL_4), and, as the author noted, no explanation is given how the feeding behavior has been restored in the facial or vagal lobe ablated fishes after several weeks of operation. It is also questionable whether the facial and vagal, especially the latter, ablation is truly selective, since, as will be seen later, increasingly more evidence has accumulated indicating that the IX and X gustatory nerves innervating O_2 and CO_2 receptors on the gill arches control ventilatory activity.

4.1.2. Search for Feeding Stimulants

Identifying the active ingredients in food has attracted the interests of many investigators, purely from scientific curiosity or for the potential for practical application (Jones, 1992). Some of the more extensive studies include:

1. Exploratory and feeding responses of the Japanese eel, *A. japonica*, to extract of the short-necked clam, *Tapes japonica* (Hashimoto *et al.*, 1968; Konosu *et al.*, 1968). The extracts contain several amino acids, mostly Gly, Ala, Glu, Asp, and Arg, with traces of Ser and Thr. A synthetic amino acid mixture evoked feeding responses, but not as strong as the original clam extract. When tested individually, only Arg, Ala, and Gly were effective at thresholds 5×10^{-8}, 1×10^{-7}, and 2×10^{-7} mmoles, respectively. Importantly, although types of diets had no effects on the observed feeding responses of the eels whether they had been fed either on beef liver, worms, *Tubifex*, or frozen saury meat.
2. Bet in extracts of the pink shrimp, *Penaeus duorarum*, was the major single compound to induce feeding behavior (strikes of perforated rubber bulb containing stimulants) in the pigfish, *Orthopristis chrysopterus*, and the pinfish, *Lagodon rhomboides* (Carr, 1982). In pigfish, an artificial mixture containing Bet plus 19 amino acids was fully as stimulatory as the extract itself, with Bet accounting for ca. 39% of the total potency, whereas in pinfish, only Bet had significant activity

when tested individually, yet it accounted for less than 10% of the activity of the mixture (Bet plus 20 amino acids).

3. Gly and Ala have been shown to be the most active amino acids for winter flounder, *Pseudopleuronectes americanus*, mummichog, *Fundulus heteroclitus*, and Atlantic silverside, *Menidia menidia* in an attraction study in natural habitat (Sutterlin, 1975). A number of other amino acids were also effective attractants for these species. Although there is some overlap, different species preferred different compounds.
4. Puffer, fed on short-necked clam meat, consumed starch pellets if they contained clam extracts. The amino acids (Ala, Glu, Gly, Pro, and Ser) that stimulated lip taste buds electrophysiologically were generally effective in stimulating feeding response, but the concentration of the amino acids necessary to release the behavior was greater by two to three orders of magnitude than that required to elicit gustatory neural activity (Hidaka, 1982). Q·HCl was rejected when offered alone at 10^{-3} M and strongly suppressed the responses if combined with clam extracts. In another marine species, bottom feeding cod, *Godus morhua*, of the isolated single components in shrimp extracts tested, Gly was the most potent, followed by Ala, Pro, and Arg, in inducing food search behavior (Ellingsen and Døving, 1986).

The studies described earlier, examining different aspects of feeding behavior elicited by natural food and/or food extract in diverse fish species, have shown conclusively that amino acids acting singly or in combination stimulate feeding behavior. Each species selectively respond to a specific mixture of compounds, but synthetic mixtures of amino acids never attain the effectiveness of the original extracts of natural foods. It is important to note that gustatory receptors, and the olfactory for that matter, are genetically fixed, by which their response capacity is determined. However, subsequent behavioral responses to chemical compounds may be learned by experience, consequently the correlation does not always exist between electrophysiological and behavioral data. An observation on juvenile sockeye salmon, *Oncorhynchus nerka*, is very suggestive; the salmon respond only to extracts of those foods constituting their current diet, but they change their extract preference concomitantly after gradual weaning over to a new diet (McBride *et al.*, 1962). A complex sensory experience, or flavor, is built and probably stored in a temporary memory.

4.1.3. Feeding Behavior Triggered by Single Amino Acids

The mechanisms by which feeding behavior are triggered by amino acids, singly or in combination, will be examined in two fish groups: (1) one in which gustation is solely involved throughout feeding (e.g., channel catfish), and (2) another in which olfaction is primarily used, complemented by gustation (e.g., salmonids, goldfish).

Channel catfish elicit the entire sequence of feeding behavioral patterns in response to single amino acids (Caprio *et al.*, 1993; Valentinčič and Caprio, 1994b). L-Arg, L-Ala, and L-Pro ($>10^{-6}$–10^{-4} M) induce consummatory behavioral patterns such as turning, pumping of water across the gill arches, and biting-snapping. L-Arg is most effective in eliciting biting-snapping behavior at thresholds 3×10^{-7} M. The behavioral pattern elicited by both L-Ala and L-Arg is surprisingly similar, despite considerable differences in the biochemical (receptor binding) and biophysical (transduction) characteristics between the two. Also, note that the olfactory receptors of this species are equally sensitive to these two chemicals. The appetitive and consummatory feeding behavior are identical in intact and anosmic channel catfish, indicating that the entire sequence of feeding behavior is mediated by the gustatory system alone. This is contradictory to Parker's (1910) suggestion that "the olfactory apparatus of the catfish is serviceable in sensing food at a distance much beyond that at which the organs of taste are capable of acting; in other words, catfishes truly scent their food." Furthermore, when conditioned to L-Pro, channel catfish responded to the conditioned stimulus at 1/30 the gustatory threshold concentration (3×10^{-6} M) determined electrophysiologically (Valentinčič and Caprio, 1994a). Therefore, these authors claim that olfaction rather than gustation is involved in the conditioned responses. Information on which amino acids induce what aspects of appetitive and consummatory feeding behavioral patterns at exactly what concentrations is still missing. In addition, which subpopulations of the taste bud system (e.g., oral cavity, barbels, flank, gills, and so on) are involved in what specific feeding patterns have yet to be determined.

Naïve rainbow trout, lake char *Salvelinus namaycush*, lake whitefish *Coregonus clupeaformis*, and goldfish respond to 10^{-6} M Cys, the most potent olfactory stimulating amino acid determined electrophysiologically, by increasing locomotor activity, an initial arousal behavior, in exactly the same fashion as to food extracts. The enhanced locomotor activity is followed by distinct species-specific search behavioral patterns: (1) bottom searching in rainbow trout, (2) surfacing/jumping in lake char, (3) exploratory behavior against the trough window in lake char, and (4) gravel pecking in goldfish (Figure 2.10; Hara, 2006). Pro as well as Ala and Ser at 10^{-6} M are all effective, to varying degrees, in eliciting feeding behavior. However, effects of Arg and Glu are different in different species, suppressing locomotor activity in rainbow trout, but triggering feeding behavior in others (Figures 2.11 and 2.12). Thus, in rainbow trout if a mixture of Cys and Arg is presented, the elevated swimming activity by Cys is suppressed (Figure 2.11). Goldfish display a typical gravel pecking searching behavior,

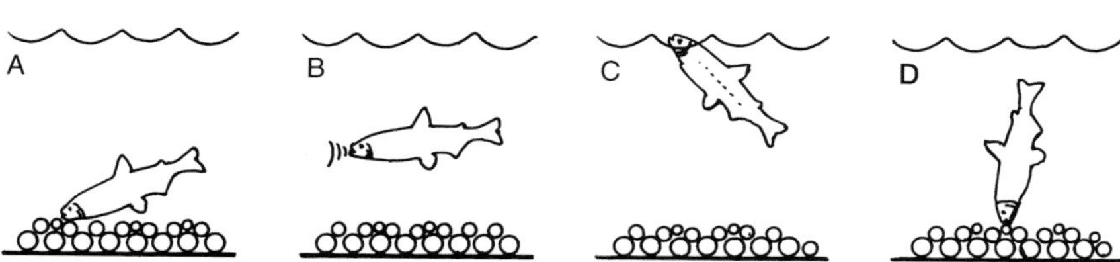

Fig. 2.10. Diagrams showing various feeding behavioral patterns elicited by food extract and chemical stimulation of fishes. On detection of stimuli, fish increase locomotor activity (appetitive behavior) followed by species-specific search behavioral patterns: (A) bottom searching in rainbow trout, (B) exploratory/escape behavior against the trough window in lake whitefish, (C) surfacing/jumping in lake char, and (D) gravel pecking in goldfish. [From Hara (2005) with permission.]

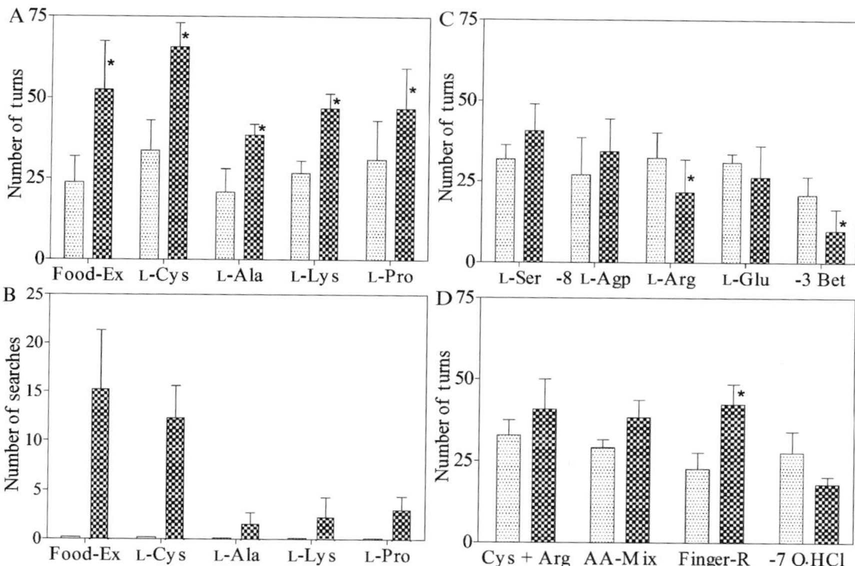

Fig. 2.11. Locomotor activity (A, C, and D) and search behavior (B) of rainbow trout elicited by food extract (Food-Ex) and chemical stimuli. Light bars, control; dark bars, stimulus. With the exception of food extract and finger rinse (Finger-R), all chemicals are tested at 10^{-6} M, unless otherwise mentioned. L-Cys, L-Cysteine; L-Lys, L-lysine; L-Ser, L-serine; L-Arg, L-arginine; L-Glu, L-glutamic acid; AA-Mix, amino acid mixture containing all tested amino acids at 10^{-6} M. [Adapted from Hara (2006) with permission.]

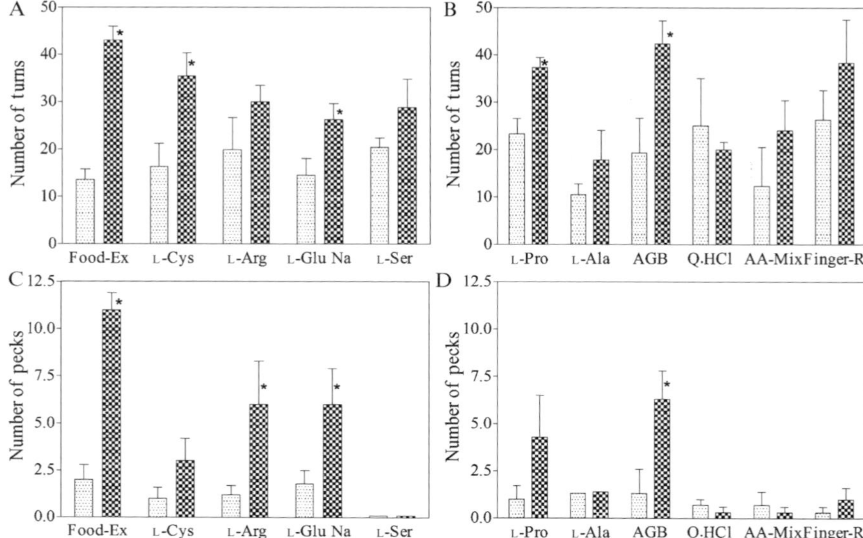

Fig. 2.12. Locomotor activity (A and B) and bottom pecking behavior (C and D) of goldfish elicited by Food-Ex and chemical stimuli. Light bars, control; dark bars, stimulus. L-GluNa, monosodium L-glutamate; AGB, 1-amino-4-guanidinobutane (agmatine). [Adapted from Hara (2006) with permission.]

positioning their body almost perpendicular to the trough bottom, accompanied by increased locomotor activity (Figure 2.10). They continue to display a series of pecking motions, lasting several seconds at a time, while circling the entire length of the trough. During one particular series of pecking motions, goldfish repeatedly peck gravel several times. Arg, Glu, and Pro are as effective as food extracts in inducing both locomotor and pecking activities, but Cys, the most potent amino acid in stimulating locomotor activity, does not trigger the pecking behavior (Figure 2.13). The behavioral responses in these fishes are in close accordance with the electrophysiological threshold and specificity. These results clearly indicate that single amino acids initiate appetitive feeding behavior (arousal and searching) primarily through olfaction, and interchageably and/or complemented by gustation in naive fishes. It is thus plausible that in the absence of visual cues the olfactory system plays a major role in triggering feeding behavior in these species. The gustatory system also plays a role, in cases where there were no olfactory cues available. The results from rainbow trout are generally consistent with the earlier observation by Valentinčič and Caprio (1997) using the Ljubliana strain rainbow trout. However, their

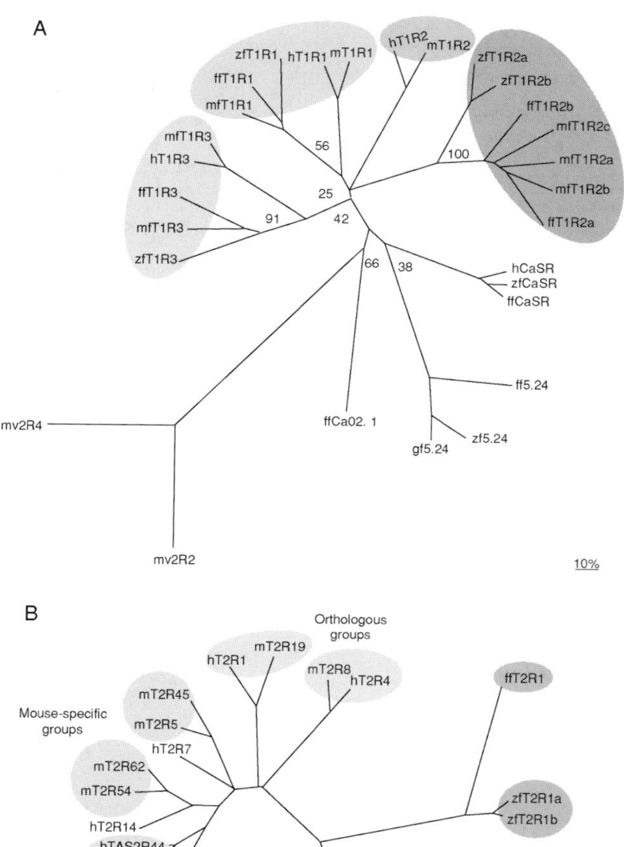

Fig. 2.13. Phylogenetic trees showing fish and mammalian T1Rs and T2Rs and related receptors. (A) Fish T1Rs are categorized into three groups: a group including mammalian T1Ra (blue), a group including mammalian T1R3s (green), and a group specific for fish T1R (red). zf5.24 and ff5.24 genes are orthologs of goldfish olfactory receptor 5.24 gene; zfCaSR and

postulation that the olfactory system detects potent gustatory stimuli such as Pro and provides the afferent input for arousal and the release of all feeding activity patterns, since amino acids that are highly potent physiological gustatory stimuli do not release either feeding behavior or reflex biting/snapping actions in adults anosmic rainbow trout, cannot be warranted since no evidence exists indicating that Pro is detected by olfaction in rainbow trout, and most likely many other species.

Brown trout, *Salmo trutta*, and Atlantic salmon, *S. salar*, fry elicit feeding behavior including snapping, darting, and increased swimming in response to Arg, Ala, Gly, or Pro at a threshold 5×10^{-8} M (Mearns, 1989). Feeding behavior in red sea bream is also activated by single amino acids by either olfaction or gustation (Goh and Tamura, 1980b). In a study testing the palatability, rainbow trout swallowed cotton pellets flavored with shrimp extract, whereas they consistently rejected pellets soaked in distilled water alone (Jones, 1989). Of 23 pure chemicals tested, the trout responded positively only to Pro, Leu, Phe, Glu, Arg, and Try. Pro, the most effective compound was active at 10^{-4} M. There is fairly good agreement between these behavioral results and the electrophysiological data on the palatal nerve sensitivity, suggesting that the same amino acids not only mediate the initial phase of appetitive feeding behavior but also palatability, and possibly the final consummatory phase.

Of chemicals closely related to amino acids, agmatine, 1-amino-4-guanidinobutane (AGB), or decarboxylated arginine, elicits strong feeding behavior in goldfish, most likely through olfaction since it evokes electro-olfactogram (EOG) responses with a threshold at 10^{-9} M in goldfish, lake whitefish, and walleye, but not in rainbow trout (Figure 2.12; Hara and Zhang, 1997). AGB, a nonvolatile amine, is converted from Arg by the decarboxylation of bacterial enzyme, most notably in decomposing invertebrate muscle tissues (Yamanaka *et al.*, 1986). As decomposition progresses, AGB along with amino acids are further converted to form polyamines such as putrescine, cadavarine, and spermine, which stimulate the goldfish olfactory system with thresholds 10^{-8}–10^{-7} M, and elicit feeding behavior (Rolen *et al.*, 2003). Omnivorous cyprinids and ictalurid catfishes are not only predators of live invertebrates but also they are scavengers of dead bait.

---

ffCaSR genes are orthologs of human calcium-sensing receptor hCaSR genes; and ffCa02.1 is a Fugu receptor distantly related to the mouse vomeronasal receptors, mV2R2 and mV2R4. (B) Fish T2Rs (red) constitute fish-specific groups similar to mouse- or human-specific groups (blue); and mammalian groups (green). ff, puffer fish; gf, goldfish; h, human; m, medaka fish; zf, zebrafish. [Adapted from Ishimaru *et al.* (2005) with permission.] (See Color Insert.)

4.2. Aversive Behavior

Some organic molecules including Q·HCl, strychnine, and caffeine stimulate the fish gustatory receptors at extremely low concentrations and cause repulsive behavior (cf. Figures 2.8 and 2.12). These substances are extremely bitter to humans which warn us not to ingest these potentially harmful compounds. The fish model will help understand how bitter receptor mechanisms were evolved in higher vertebrates.

Q·HCl induce avoidance behavior at 10^{-6} M in lake whitefish and suppresses locomotor activity in rainbow trout and goldfish (cf. Figure 2.12; Jones and Hara, 1985; Hara, 2005). The aversion and suppression by low concentrations of Q·HCl are most likely through its specific receptor mechanism, and not by nonspecific effects on voltage-gated ion channels in gustatory cells shown in channel catfish that requires much higher concentrations (Ogawa, *et al.*, 1997). Further, in goldfish gelatin pellets flavored with Q·HCl or caffeine at concentrations 10^{-5} and 10^{-3} M and above, respectively, are rejected (Lamb and Finger, 1995). Pellets containing a mixture of Q·HCl and food or Q·HCl and either Ala or Pro are frequently ingested, but rejected after extended sorting (rinsing and backwashing) behavior. Also, if food pellets soaked in Q·HCl solution are dropped into an aquarium, trout, and goldfish catch them, most likely by vision, the former ingest them immediately, but the latter spit them out after a brief mastication (Hara, 2005). These results indicate that fish species with a well-developed palatal organ such as goldfish use chemical cues to drive the intraoral sorting and rejection/ingestion behaviors to spit out foods containing unpalatable materials, whereas species with underdeveloped or without the palatal organ are able to avoid them prior to intake. Whichever mechanism is involved in the behavior, it suggests that the central recognition and discrimination rather than local reflexes of high levels of recognition are involved in the processes.

Puffer toxin, TTX, distributed in aquatic organisms much wider than originally thought, is one of the most potent neurotoxins known, with its lethal toxicity to humans 300 times that of cyanide. Puffers, when stimulated by handling or electrical shock, secrete TTX into the surrounding water from the unique gland-like structure in the skin containing high concentrations of TTX, usually accompanied by swelling of the abdomen (Kodama *et al.*, 1985; Saito *et al.*, 1985). Rainbow trout and several other marine species reject toxic puffer livers and food pellets containing TTX (Yamamori *et al.*, 1980). A paralytic shellfish toxin, STX, believed to be only one of 10–12 closely related naturally occurring toxins produced by dinofalgellates, also has pharmacological actions similar to TTX. The sensitive, specific gustatory receptor system for the toxins may provide dual functions providing adaptive significance both to producing and receiving organisms.

4.2.1. CO_2/H^+-Driven Behavior—Ventilatory Control

Fish, if given a choice, avoid ambient water with unfavorably high levels of CO_2 and H^+. The main directive factor in combined pH/CO_2 gradients is CO_2, particularly within the pH range 7.4–5.5 (Höglund, 1961). Roach (*Leuciscus rutilus*), minnows (*Phoxinus phoxinus*), and Atlantic salmon detect and avoid CO_2 separately from the accompanying pH. Arctic char, *S. alpinus*, avoid CO_2 plume at plume/ambient gradients greater than 50 μmol/liter but is attracted at lower concentrations (Jones *et al.*, 1985). The response to acidified natural water closely follows the C–R curves for CO_2 but not H^+, indicating that fish behaviorally discriminate between H^+ and CO_2, relying principally on the latter for orientation. The avoidance reactions to CO_2 have long been suspected to be attributable to chemoreceptors in the gill region, because ablation of the olfactory organs and the sectioning of the nerve innervating the lateral-line organs do not alter the reactions.

The avoidance of CO_2 is most likely a behavioral strategy meant primarily to prevent respiratory distress. If, however, fish are trapped in environmental acidosis, hypercapnia, and arterial acidosis, they do react by reducing, and not increasing, ventilatory activity (Konishi *et al.*, 1969). CO_2 is "painful" in other animals as it causes a peripheral acidosis that excites niciceptors (Chapter 4, this volume). There is a growing body of experimental evidence suggesting that a significant CO_2/H^+-driven ventilatory action exists in fishes (Burleson and Smatresk, 2000; Gilmour, 2001; Perry and Gilmour, 2002). In some species, bilateral denervation of the gills abolish the cardiorespiratory effects of hypercapnia, indicating that CO_2/H^+ chemoreceptors are exclusively located in the branchial structures. The first gill arch seems to be a particularly important site of branchial location of chemoreception in rainbow trout (Chapter 3, this volume).

In mammals including humans, although taste buds are predominantly located in the oral cavity, especially on the tongue and palate, substantial numbers of taste buds are also found on the epiglottal and laryngeal surface. Because food or fluid would seldom contact them during normal ingestion, the functions of the laryngeal and epiglottal taste buds may be quite different from those of the lingual. Some consider that they may play a role in respiratory modulation rather than food appreciation (Nishijima *et al.*, 2004). The superior laryngeal nerves detect CO_2 in the larynx (Wang *et al.*, 2001; Nishijima and Atoji, 2004). Because the ancestral fish from which both modern fish and land vertebrates evolved had lungs that enabled them to live in stagnant, poorly aerated water for long periods of time when necessary, they may participate in a series of reflexes that function to modulate respiration or to block the airway.

5. CONCLUSIONS AND PROSPECTS

In fishes, the gustatory receptors, taste buds, are located in five different subpopulations, all of which are innervated by either of the three cranial nerves, that is, facial (VIIth), glossopharyngeal (IXth), and vagal (Xth). The gustatory system is the only vertebrate sensory system in which three cranial nerves carry all peripheral gustatory information to the CNS, suggesting the diversity of roles they play. The gustatory system is very well developed in fishes in general. Two suborders especially have become highly specialized for use of the gustatory sense: cyprinids and silurids. They have, along with expansive peripheral receptors, developed elaborate central structures, probably the most complex neural organization found in any CNS. Intra-and extraoropharyngeal distributions of taste buds provide fishes with detailed chemical and likely tactile environments as well as the means to detect food items and reject inedible or noxious substances. These processes are accomplished through extraordinary central gustatory structures: complex facial and vagal lobes with topographic organization, of which neurons project to the SGN, which in turn, projects to the telencephalon and several diencephalic nuclei including the inferior hypothalamic lobe. The former pathway corresponds to the mammalian gustatory pathway from the solitary nucleus to the gustatory neocortices, whereas the latter to the pathway through the amygdala. The inferior lobe, as Herrick emphasized, is the "central correlation station for all sensory impression." The functional significance of the telencephalic gustatory centres of fishes is, however, at present unknown. In mammals, gustatory neurons in the orbitofrontal cortex are believed to be involved with learning and associative processes, and these issues should be explored in fish gustation studies as well. Also, keep in mind that the hippocampus is a possible repository of memory, the capacity to store and recall information, which is a necessary component of learning.

Electrophysiologically, research has centred on facial gustatory sensitivity to amino acids for over three decades. Proline dominates one group led by salmonids, while alanine and arginine dominate the other led by channel catfish; in the former all naturally occurring amino acids are detected via at least three receptor types, Pro-, Leu-, and Bet-receptors, whereas in the latter by Ala- and Arg-receptors. These are supplemented by several other receptors, all of which are transmitted by two main fiber clusters: alanine and arginine. The amino acid response spectra are complementary with those of the olfactory counterparts for the former group, and overlapping for the latter. Fish gustatory receptors are generally highly sensitive to bitter (to humans) substances including marine toxin, but the extreme sensitivity to bile acids appears restricted to salmonids, suggesting a specific function for this fish group. There is growing evidence suggesting that the CO_2 sensitivity of (gustatory) receptors on gill arches might be involved in ventilatory/respiratory regulation in fishes.

Behaviorally, feeding is triggered by single amino acids through gustation in catfishes, whereas in salmonids and others primarily through olfaction, complemented by gustation. Feeding, one of the two most fundamental processes for the survival of individuals and species, thus is not operated exclusively by a single sensory system. In the absence of visual cues, feeding is initiated primarily by olfaction, complemented by gustation in many fish species. Fishes are capable of avoiding noxious substances prior to intake of food into the mouth. Bottom feeders such as cyprinids suck up mouthfuls of sediments, extract the edible, and eject the remainder with the aid of the specially developed palatal organ. Functions of the barbel and cutaneous gustatory receptors which catfishes developed have yet to be characterized. Regardless of the methods, feeding requires mechanisms for differentiating between edible and inedible and between more favored and less favored. The ability to identify amino acids and related chemicals is particularly important, as it provides the means to seek out food sources, while the perception of unfavorable substances is important for its protective value, enabling the avoidance of harmful and potentially deadly environmental toxins.

With a ground plan laid by Herrick a century ago, a great deal of effort has been expended to work out the details of the facial/vagal lobe systems in cyprinids and silurids, which greatly enhanced our understanding of the way gustatory information is processed in the CNS. Future research should be focused on: (1) identifying gustatory receptor cells that have a high percentage of tactile sensitivity, whether it be the light and dark, or rod and microvillar cells; (2) characterizing homo- or heterogeneity of receptor cells within a taste bud and synaptic contacts with innervating gustatory nerves; (3) analyzing functional gustatory representations in the brain, now that the central pathways in fishes are much clearer; (4) exploring potential functional roles the branchial and other gustatory receptor subpopulations might play such as a sensor for the peripheral ventilatory regulatory centre; and (5) although several candidate gustatory receptor genes have been identified, no particular receptor has been definitely paired with any ligand. Fish with genetically ablated specific T1R or T2R subunits would help define the relationships between receptors and gustatory physiology and behavior.

ACKNOWLEDGMENTS

I wish to thank my collaborators and students for their encouragement, ideas, and assistance. I also thank R. Evans, S. Kiyohara, and L. Sneddon for constructive comments on this chapter. Research in the author's laboratory was supported in part by Natural Science and Engineering Research Council of Canada.

REFERENCES

Atema, J. (1971). Structures and functions of the sense of taste in the catfish (*Ictalurus natalis*). *Brain Behav. Evol.* **4,** 273–294.

Bardach, J. E., Fujiya, M., and Holl, A. (1967a). Investigation of external chemoreceptors of fishes. *In* "Olfaction and Taste II" (Hayashi, T., Ed.), pp. 647–665. Pergamon Press, Oxford.

Bardach, J. E., Todd, J. H., and Crickmer, R. (1967b). Orientation by taste in fish of the genus *Ictalurus*. *Science* **155,** 1276–1278.

Barlow, L. A., and Northcutt, R. G. (1995). Embryonic origin of amphibian taste buds. *Dev. Biol.* **169,** 273–285.

Biedenbach, M. A. (1973). Functional properties and projection areas of cutaneous receptors in catfish. *J. Comp. Physiol.* **84,** 227–250.

Brand, J. G., and Bruch, R. C. (1992). Molecular mechanisms of chemosensory transduction: Gustation and olfaction. *In* "Fish Chemoreception" (Hara, R. C., Ed.), pp. 126–149. Chapman & Hall, London.

Brand, J. G., Teeter, J. H., Kumazawa, T., Huque, T., and Bayley, D. L. (1991). Transduction mechanisms for the taste of amino acids. *Physiol. Behav.* **49,** 899–904.

Braun, C. B., and Northcutt, R. G. (1995). Distribution and innervation of the cutaneous chemosensory systems in zebrafish (*Danio rerio*). *Soc. Neurosci.* **21,** 691.

Burleson, M. L., and Smatresk, N. J. (2000). Branchial chemoreceptors mediate ventilatory responses to hypercapnic acidosis in channel catfish. *Comp. Biochem. Physiol.* **125A,** 403–414.

Butler, A. B., and Hodos, W. (1996). "Comparative Vertebrate Neuroanatomy: Evolution and Adaptation." John Wiley & Sons, New York.

Caprio, J. (1975). High sensitivity of catfish taste receptors to amino acids. *Comp. Biochem. Physiol.* **52A,** 247–251.

Caprio, J. (1978). Olfaction and taste in the channel catfish: An electrophysiological study of the responses to amino acids and derivatives. *J. Comp. Physiol.* **123,** 357–371.

Caprio, J., Brand, J. G., Teeter, J. H., Valentincic, T., Kalinoski, D. L., Kohbara, J., Kumazawa, T., and Wegert, S. (1993). The taste system of the channel catfish: From biophysics to behavior. *Trends Neurosci.* **16,** 192–197.

Carr, W. E. S. (1982). Chemical stimulation of feeding behavior. *In* "Chemoreception in Fishes" (Hara, T. J., Ed.), pp. 259–273. Elsevier, Amsterdam.

Chandrashekar, J., Mueller, K. L., Hoon, M. A., Adler, E., Feng, L., Guo, W., Zuker, C. S., and Ryba, N. J. (2000). T2Rs function as bitter taste receptors. *Cell* **100,** 703–711.

Crisp, M., Lowe, G. A., and Laverack, M. S. (1975). On the ultrastructure and permeability of taste buds of the marine teleosts *Ciliata mustela*. *Tissue Cell* **7,** 191–202.

Davenport, C. J., and Caprio, J. (1982). Taste and tactile recordings from the ramus recurrens facialis innervating flank taste buds in the catfish. *J. Comp. Physiol.* **147,** 217–229.

Ellingsen, O. F., and Døving, K. B. (1986). Chemical fractionation of shrimp extracts inducing bottom food search behavior in cod (*Gadus morhua* L.). *J. Chem. Ecol.* **12,** 155–168.

Eram, M., and Michel, W. C. (2005). Morphological and biochemical heterogeneity in facial and vagal nerve innervated taste buds of the channel catfish, *Ictalurus punctatus*. *J. Comp. Neurol.* **486,** 132–144.

Finger, T. E. (1976). Gustatory pathways in the bullhead catfish. I. Connections of the anterior ganglion. *J. Comp. Neurol.* **165,** 513–526.

Finger, T. E. (1978). Gustatory pathways in the bullhead catfish. II. Facial lobe connections. *J. Comp. Neurol.* **180,** 691–706.

Finger, T. E. (1982). Somatotopy in the representation of the pectral fin and free rays in the spinal cord of the sea robin, *Prionotus carolinus. Biol. Bull.* **163,** 154–161.

Finger, T. E. (1987). Gustatory nuclei and pathways in the central nervous system. *In* "Neurobiology of Taste and Smell" (Finger, T. E., and Silver, W. L., Eds.), pp. 285–310. John Wiley & Sons, New York.

Finger, T. E. (1988). Organization of chemosensory systems within the brain of bony fishes. *In* "Sensory Biology of Aquatic Animals" (Atema, J., Fay, R. R., Popper, A. N., and Tavolga, W. N., Eds.), pp. 339–363. Springer-Verlag, New York.

Finger, T. E. (1997a). Gustatory nuclei and pathways in the central nervous system. *In* "Neurobiology of Taste and Smell" (Finger, T. E., and Silver, W. L., Eds.), pp. 331–353. John Wiley & Sons, New York.

Finger, T. E. (1997b). Evolution of taste and solitary chemosensory cell system. *Brain Behav. Evol.* **50,** 234–243.

Finger, T. E., Drake, S. K., Kotrschal, K., Womble, M., and Dockstader, K. C. (1991). Postlarval growth of the peripheral gustatory system in the channel catfish, *Ictalurus punctatus. J. Comp. Neurol.* **314,** 55–66.

Finger, T. E., Bryant, B. P., Kalinoski, D. L., Teeter, J. H., Bottger, B., Grosvenor, W., Cagan, R. H., and Brand, J. G. (1996). Differential localization of putative amino acid receptors in taste buds of the channel catfish, *Ictalurus punctatus. J. Comp. Neurol.* **373,** 129–138.

Finger, T. E., Danilova, V., Barrows, J., Bartel, D. L., Vigers, A. J., Stone, L., Hellekant, G., and Kinnamon, S. C. (2005). ATP signaling is crucial for communication from taste buds to gustatory nerves. *Science* **310,** 1495–1499.

Fishelson, L., and Delarea, Y. (2004). Taste buds on the lips and mouth of some blenniid and gobiid fishes: Comparative distribution and morphology. *J. Fish Biol.* **65,** 651–665.

Folgueira, M., Anadón, R., and Yañez, J. (2003). Experimental study of the connections of the gustatory system in the rainbow trout, *Oncorhunchus mykiss. J. Comp. Neurol.* **465,** 604–619.

Fujimoto, S., Ueda, H., and Kagawa, H. (1987). Immunocytochemistry on the localization of 5-hydroxytryptamine in monkey and rabbit taste buds. *Acta Anat.* **128,** 80–83.

Gilmour, K. M. (2001). The CO_2/pH ventilatory drive in fish. *Comp. Biochem. Physiol.* **130A,** 219–240.

Goh, Y., and Tamura, T. (1980a). Olfactory and gustatory responses to amino acids in two marine teleosts: Red sea bream and mullet. *Comp. Biochem. Physiol.* **66C,** 217–224.

Goh, Y., and Tamura, T. (1980b). Effect of amino acids on the feeding behaviour in red sea bream. *Comp. Biochem. Physiol.* **66C,** 225–229.

Gomahr, A., Palzenberger, M., and Kotrschal, K. (1992). Density and distribution of external taste buds in cyprinids. *Env. Biol. Fish.* **33,** 125–134.

Grosvenor, W., Kaulin, Y., Spielman, A. I., Bayley, D. L., Kalinoski, D. L., Teeter, J. H., and Brand, J. G. (2004). Biochemical enrichment and biophysical characterization of a taste receptor for L-arginine from the catfish, *Ictalurus punctatus. BMC Neurosci.* **5**(25), 1–19.

Hansen, A., Reutter, K., and Zeiske, E. (2002). Taste bud development in the zebrafish, *Danio rerio. Dev. Dyn.* **223,** 483–496.

Hara, T. J. (1993). Chemoreception. *In* "The Physiology of Fishes" (Evans, D. H., Ed.), pp. 191–218. CRC Press, Boca Raton.

Hara, T. J. (1994). The diversity of chemical stimulation in fish olfaction and gestation. *Rev. Fish Biol. Fish.* **4,** 1–35.

Hara, T. J. (2005). Olfactory responses to amino acids in rainbow trout: Revisited. *In* "Fish Chemosenses" (Reutter, K., and Kapoor, B. G., Eds.), pp. 31–64. Science Publishers, Enfield.

Hara, T. J. (2006). Feeding behaviour in some teleosts is triggered by single amino acids primarily through olfaction. *J. Fish Biol.* **68,** 810–825.

Hara, T. J., and Zhang, C. (1997). Olfactory responses to guanidine compounds in telost fishes. Presented at the International Symposium. Olfaction and Taste XII and AChemS XIX, San Diego, CA, 7–12 July, 1997.

Hara, T. J., and Zielinski, B. (1989). Structural and functional development of the olfactory organ in teleosts. *Trans. Am. Fish. Soc.* **118,** 183–194.

Hara, T. J., Law, Y. M. C., and van der Veen, E. (1973). A stimulatory apparatus for studying the olfactory activity in fishes. *J. Fish. Res. Board Can.* **30,** 283–285.

Hara, T. J., Macdonald, S., Evans, R. E., Marui, T., and Arai, S. (1984). Morpholine, bile acids and skin mucus as possible chemical cues in salmonid homing: Electrophysiological re-evaluation. *In* "Mechanisms of Migration in Fishes" (McCleave, J. D., Arnold, G. P., Dodson, J. J., and Neill, W. H., Eds.), pp. 363–378. Plenum, New York.

Hara, T. J., Sveinsson, T., Evans, R. E., and Klaprat, D. A. (1993). Morphological and functional characteristics of the olfactory and gustatory organs of three *Salvelinus* species. *Can. J. Zool.* **71,** 414–423.

Hara, T. J., Carolsfeld, J., and Kitamura, S. (1999). The variability of the gustatory sensibility in salmonids, with special reference to strain differences in rainbow trout, *Oncorhynchus mykiss*. *Can. J. Fish. Aquat. Sci.* **56,** 13–24.

Hashimoto, Y., Konosu, S., Fusetani, N., and Nose, T. (1968). Attractants for eels in the extracts of short-necked clam: I. Survey of constituents eliciting feeding behavior by the omission test. *Bull. Jpn. Soc. Sci. Fish.* **34,** 78–83.

Hayama, T., and Caprio, J. (1989). Lobule structure and somatotopic organization of the medullary facial lobe in the channel catfish *Ictalurus punctatus*. *J. Comp. Neurol.* **285,** 9–17.

Herrick, C. J. (1904). The organ and sense of taste in fishes. *Bull. US Comm.* **22,** 237–272.

Herrick, C. J. (1905). The central gustatory paths in the brains of bony fishes. *J. Comp. Neurol. Psychol.* **15,** 375–456.

Herrick, C. J. (1907). The tactile centers in the spinal cord and brain of the sea robin, *Prionotus corolinus* L. *J. Comp. Neurol.* **17,** 307–327.

Hidaka, I. (1970). The effect of carbon dioxide on the carp palatal chemoreceptors. *Bull. Jpn. Soc. Sci. Fish.* **36,** 1034–1039.

Hidaka, I. (1982). Taste receptor stimulation and feeding behavior in the puffer. *In* "Chemoreception in Fishes" (Hara, T. J., Ed.), pp. 243–257. Elsevier, Amsterdam.

Hidaka, I., and Ishida, Y. (1985). Gustatory response in the shimaisaki (tigerfish) *Therapon oxyrhynchus*. *Bull. Jpn. Soc. Sci. Fish.* **51,** 387–391.

Hirata, Y. (1966). Fine structure of the terminal buds on the barbells of some fishes. *Arch. Histol. Jpn.* **26,** 507–523.

Hoagland, H. (1933). Specific nerve impulses from gustatory and tactile receptors in catfish. *J. Gen. Physiol.* **16,** 685–693.

Höglund, L. B. (1961). The reactions of fish in concentration gradients. *Rep. Inst. Freshwater Res. Drottningholm* **43,** 1–147.

Holland, K. (1978). Chemosensory orientation to food by a Hawaian goatfish (*Parupeneus porphyreus*, Mullidae). *J. Chem. Ecol.* **4,** 173–186.

Hoon, M. A., Adler, E., Lindemeier, J., Battey, J.f., Ryba, N. J. P., and Zuker, C. S. (1999). Putative mammalian taste receptors: A class of taste-specific GPCRs with distinct topographic selectivity. *Cell* **96,** 541–551.

Huang, Y.-J., Murayama, Y., Lu, K.-S., Pereira, E., Plonsky, I., Baur, J. E., Wu, D., and Roper, S. D. (2005). Mouse taste buds use serotonin as a neurotransmitter. *J. Neurosci.* **26,** 843–847.

Ishida, Y., and Hidaka, I. (1987). Gustatory response profiles for amino acid, glycinebetaine, and nucleotides in several marine teleosts. *Nippon Suisan Gakkaishi* **53,** 1391–1398.

Ishimaru, Y., Okada, S., Naito, H., Nagai, T., Yasuoka, A., Matsumoto, I., and Abe, K. (2005). Two families of candidate taste receptors in fishes. *Mech. Dev.* **122,** 1310–1321.

Iwai, T. (1963). Taste buds on the gill rakers and gill arches of the sea catfish, *Plotosus anguillaris* (Lacépède). *Copeia* **2,** 271–274.

Iwai, T. (1964). A comparative study of the taste buds in gill rakes and gill arches of teleostean fishes. *Bull. Misaki Mar. Biol. Inst., Kyoto Univ.* **7,** 19–34.

Jakubowski, M., and Whitear, M. (1990). Comparative morphology and cytology of taste buds. *Z. Mikrosk. Anat. Forsch.* **104,** 529–560.

Johnsen, P. B., Zhou, H., and Adams, M. A. (1990). Gustatory sensitivity of the herbivore *Tilapia zillii* to amino acids. *J. Fish Biol.* **36,** 587–593.

Jones, K. A. (1989). The palatability of amino acids and related compounds to rainbow trout, *Salmo gairdneri* Richardson. *J. Fish Biol.* **34,** 149–160.

Jones, K. A. (1992). Food search behaviour in fish and the use of chemical lures in commercial and sports fishing. *In* "Fish Chemoreception" (Hara, T. J., Ed.), pp. 288–320. Chapman & Hall, London.

Jones, K. A., and Hara, T. J. (1985). Behavioural responses of fishes to chemical cues: Results from a new bioassay. *J. Fish Biol.* **27,** 495–504.

Jones, K. A., Hara, T. J., and Scherer, E. (1985). Locomotor response by Arctic char (*Salvelinus alpinus*) to gradients of H^+ and CO_2. *Physiol. Zool.* **58,** 413–420.

Kalinoski, D. L., Bryant, B. P., Shaulsky, G., Brand, J. G., and Harpaz, S. (1989). Specific L-arginine taste receptor sites in the catfish, *Ictalurus punctatus*: Biochemical and neurophysiological characterization. *Brain Res.* **488,** 163–173.

Kalinoski, D. L., Johnson, L. C., Bryant, B. P., and Brand, J. G. (1992). Selective interactions of lectins with amino acid taste receptors in channel catfish. *Chem. Senses* **17,** 381–390.

Kanwal, J. S., and Caprio, J. (1983). An electrophysiological investigation of the oro-pharyngeal (IX-X) taste system in the channel catfish, *Ictalurus punctatus. J. Comp. Physiol. A* **150,** 345–357.

Kanwal, J. S., and Caprio, J. (1987). Central projections of the glossopharyngeal and vagal nerves in the channel catfish, *Ictalurus punctatus*: Clues to differential processing of visceral inputs. *J. Comp. Neurol.* **264,** 216–230.

Kanwal, J. S., and Caprio, J. (1988). Overlapping taste and tactile maps of the oropharynx in the vagal lobe of the channel catfish, *Ictalurus punctatus. J. Neurobiol.* **19,** 211–222.

Kanwal, J. S., and Finger, T. E. (1992). Central representation and projections of gustatory systems. *In* "Fish Chemoreception" (Hara, T. J., Ed.), pp. 79–102. Chapman & Hall, London.

Kanwal, J. S., Finger, T. E., and Caprio, J. (1988). Forebrain connections of the gustatory system in ictalurid catfishes. *J. Comp. Neurol.* **278,** 353–376.

Kim, D. J., and Roper, S. D. (1995). Localization of serotonin in taste buds: A comparative study in four vertebrates. *J. Comp. Neurol.* **353,** 364–370.

Kinnamon, J. C. (1987). Organization and innervation of taste buds. *In* "Neurobiology of Taste and Smell" (Finger, T. E., and Silver, W. L., Eds.), pp. 277–353. John Wiley & Sons, New York.

Kinnamon, J. C., Henzler, D. M., and Royer, S. M. (1993). HVEM ultrastructural analysis of fungiform taste buds, cell types, and associated synapses. *Microsc. Res. Tech.* **26,** 142–156.

Kitada, Y., and Hara, T. J. (1994). Effects of diluted natural water and altered ionic environments on gustatory responses in rainbow trout (*Oncorhynchus mykiss*). *J. Exp. Biol.* **186,** 173–186.

Kitoh, J., Kiyohara, S., and Yamashita, S. (1987). Fine structure of taste buds in the minnow. *Nippon Suisan Gakkaishi* **53,** 1943–1950.

Kiyohara, S., and Caprio, J. (1996). Somatotopic organization of the facial lobe of the sea catfish *Arius felis* studied by transganglionic transport of horseradish peroxidase. *J. Comp. Neurol.* **368,** 121–135.

Kiyohara, S., and Hidaka, I. (1991). Receptor sites for alanine, proline, and betaine in the palatal taste system of the puffer, *Fugu pardalis. J. Comp. Physiol. A* **169,** 523–530.

Kiyohara, S., Hidaka, I., and Tamura, T. (1975). The anterior cranial gustatory pathway in fish. *Experientia* **31,** 1051–1053.

Kiyohara, S., Yamashita, S., and Kitoh, J. (1980). Distribution of taste buds on the lips and inside the mouth in the minnow, *Pseudorasbora parva. Physiol. Behav.* **24,** 1143–1147.

Kiyohara, S., Yamashita, S., and Harada, S. (1981). High sensitivity of minnow gustatory receptors to amino acids. *Physiol. Behav.* **26,** 1103–1108.

Kiyohara, S., Hidaka, I., Kitoh, J., and Yamashita, S. (1985). Mechanical sensitivity of the facial nerve fibers innervating the anterior palate of the puffer, *Fugu pardalis*, and their central projections in the primary taste center. *J. Comp. Physiol. A* **157,** 705–716.

Kiyohara, S., Yamashita, S., Lamb, C. F., and Finger, T. E. (1999). Distribution of trigeminal fibers in the primary facial gustatory center of channel catfish, *Ictalurus punctatus. Brain Res.* **841,** 93–100.

Kiyohara, S., Sakata, Y., Yoshitomi, T., and Tsukahara, J. (2002). The 'goatee' of goatfish: Innervation of taste buds in the barbells and their representation in the brain. *Proc. R. Soc. Lond. B* **269,** 1773–1780.

Kodama, M., Ogata, T., and Satô, S. (1985). External secretion of tetrodotoxin from puffer fishes stimulated by electric shock. *Mar. Biol.* **87,** 199–202.

Kohbara, J., and Caprio, J. (2001). Taste responses of the facial and glossopharyngeal nerves to amino acids in the rainbow trout. *J. Fish Biol.* **58,** 1062–1072.

Kohbara, J., Michel, W., and Caprio, J. (1992). Responses of single facial taste fibers in the channel catfish, *Ictalurus punctatus*, to amino acids. *J. Neurophysiol.* **68,** 1012–1026.

Komada, N. (1993). Distribution of taste buds in the oropharyngeal cavity of fry and fingerling amago salmon, *Oncorhynchus rhodurus. Jpn. J. Ichthyol.* **40,** 110–116.

Konishi, J., and Zotterman, Y. (1961). Taste function in the carp. An electrophysiological study on gustatory fibers. *Acta Physiol. Scand.* **52,** 150–161.

Konishi, J., Uchida, M., and Mori, Y. (1966). Gustatory fibers in the sea catfish. *Jpn. J. Physiol.* **16,** 194–204.

Konishi, J., Hidaka, I., Toyota, M., and Matsuda, H. (1969). High sensitivity of the palatal chemoreceptors of the carp to carbon dioxide. *Jpn. J. Physiol.* **19,** 327–341.

Konosu, S., Fusetani, N., Nose, T., and Hashimoto, Y. (1968). Attractants for eels in the extracts of short-necked clam: II. Survey of constituents eliciting feeding behavior by fractionation of the extracts. *Bull. Jpn. Soc. Sci. Fish.* **34,** 84–87.

Kotrschal, K. (1991). Solitary chemosensory cells: Taste, common chemical sense or what? *Rev. Fish Biol. Fish.* **1,** 3–22.

Kotrschal, K. (1995). Ecomorphology of solitary chemosensory cell systems in fish: A review. *Env. Biol. Fish.* **44,** 143–155.

Kotrschal, K., and Finger, T. E. (1996). Secondary connections of the dorsal and ventral facial lobes in a teleosts fish, the rockling (*Ciliata motella*). *J. Comp. Neurol.* **370,** 415–426.

Kotrschal, K., Royer, S., and Kinnamon, J. C. (1998). High-voltage electron microscopy and 3-D reconstruction of solitary chemosensory cells in the anterior dorsal fin of the gadid fish *Ciliata mustela* (Teleostei). *J. Struct. Biol.* **124,** 59–69.

Kotrschal, K., Whitear, M., and Finger, T. E. (1993). Spinal and facial innervation of the skin in the gadid fish *Ciliaa mustela* (Teleostei). *J. Comp. Neurol.* **331,** 407–417.

Krueger, J.M, and Cagan, R. H. (1976). Biochemical studies of taste sensation. *J. Biol. Chem.* **251,** 88–97.

Kumazawa, T., Brand, J. G., and Teeter, J. H. (1998). Amino acid-activated channels in the catfish taste system. *Biophys. J.* **75,** 2757–2766.

Lamb, C. F., and Caprio, J. (1992). Convergence of oral and extraoral information in the superior secondary gustatory nucleus of the channel catfish. *Brain Res.* **588,** 201–211.

Lamb, C. F., and Caprio, J. (1993a). Diencephalic gustatory connections in the channel catfish. *J. Comp. Neurol.* **337,** 400–418.

Lamb, C. F., and Caprio, J. (1993b). Taste and tactile responsiveness of neurons in the posterior diencephalon of the channel catfish. *J. Comp. Neurol.* **337,** 419–430.

Lamb, C. F., and Finger, T. E. (1995). Gustatory control of feeding behavior in goldfish. *Physiol. Behav.* **57,** 483–488.

Li, X., Staszewski, L., Xu, H., Durick, K., Zoller, M., and Adler, E. (2002). Human receptors for sweet and umami taste. *Proc. Natl. Acad. Sci. USA* **99,** 4692–4696.

Lohmann, K. J., and Johnsen, S. (2000). The neurobiology of magnetoreception in vertebrate animals. *Trends Neurosci.* **23,** 153–159.

Luiten, P. G. M. (1975). The central projections of the trigeminal, facila and anterior lateral line nerves in the carp (*Cyprinus carpio* L.). *J. Comp. Neurol.* **160,** 399–417.

Marui, T., and Caprio, J. (1982). Electrophysiological evidence for the topographical arrangement for taste and tactile neurons in the facial lobe of the channel catfish. *Brain Res.* **231,** 185–190.

Marui, T., and Caprio, J. (1992). Teleost gustation. *In* "Fish Chemoreception" (Hara, T. J., Ed.), pp. 171–198. Chapman & Hall, London.

Marui, T., and Funakoshi, M. (1979). Tactile input to the facial lobe of the carp, *Cyprinus carpio* L. *Brain Res.* **177,** 479–488.

Marui, T., Evans, R. E., Zielinski, B., and Hara, T. J. (1983a). Gustatory responses of the rainbow trout (*Salmo gairdneri*) palate to amino acids and derivatives. *J. Comp. Physiol.* **153,** 423–433.

Marui, T., Harada, S., and Kasahara, Y. (1983b). Gustatory specificity for amino acids in the facial taste system of the carp, *Cyprinus carpio* L. *J. Comp. Physiol. A* **153,** 299–308.

Marui, T., Caprio, J., Kiyohara, S., and Kasahara, Y. (1988). Topographical organization of taste and tactile neurons in the facial lobe of the sea catfish, *Protosus lineatus*. *Brain Res.* **446,** 178–182.

Matz, S. P. (1995). Connections of the olfactory bulb in the Chinook salmon (*Oncorhynchus tshawytscha*). *Brain Behav. Evol.* **46,** 108–120.

McBride, J. R., Idler, D. R., Jonas, R. E. E., and Tomlinson, N. (1962). Olfactory perception in juvenile salmon. I. Observations on response of juvenile sockeye to extracts of foods. *J. Fish. Res. Board. Can.* **19,** 327–334.

Mearns, K. J. (1989). Behavioural responses of salmonid fry to low amino acid concentrations. *J. Fish Biol.* **34,** 223–232.

Michel, W. C., Kohbara, J., and Caprio, J. (1993). Amino acid receptor sites in the facial taste system of the sea catfish *Arius felis*. *J. Comp. Physiol. A* **172,** 129–138.

Molenaar, G. J. (1991). Anatomy and physiology of infrared sensitivity in snakes. *In* "Biology of the Reptilia: Neurology C. Sensorimotor Integration" (Gans, C., and Ulinski, P. S., Eds.), Vol. 17, pp. 367–453. Chicago University Press, Chicago.

Morita, Y., and Finger, T. E. (1985a). Reflex connections of the facial and vagal gustatory systems in the brainstem of the bullhead catfish, *Ictalurus nebulosus*. *J. Comp. Neurol.* **231,** 547–558.

Morita, Y., and Finger, T. E. (1985b). Topographic ands laminar organization of the vagal gustatory system in the goldfish, *Carassius auratus*. *J. Comp. Neurol.* **238,** 187–201.

Morita, Y., and Finger, T. E. (1987). Topographic representation of the sensory and motor roots of the vegus nerve in the medulla of goldfish, *Carassius auratus*. *J. Comp. Neurol.* **264,** 231–249.

Murakami, T., Fukuoka, T., and Ito, H. (1986). Telencephalic ascending acousticolateral system in a teleosts (*Sebastiscus marmoratus*), with special reference to the fiber connections of the nucleus preglomerulosus. *J. Comp. Neurol.* **247,** 383–397.

Nelson, G., Chandrashekar, J., Hoon, M. A., Feng, L., Zhao, G., Ryba, N. J. P., and Zuker, C. S. (2002). An amino-acid taste receptor. *Nature* **416,** 199–202.

Nishijima, K., and Atoji, Y. (2004). Taste buds and nerve fibers in the rat larynx: An ultrastructural and immunohistochemical study. *Arch. Histol. Cytol.* **67,** 195–209.

Nishijima, K., Tsubone, H., and Atoji, Y. (2004). Contribution of free nerve endings in the laryngeal epithelium to CO_2 reception in rats. *Auton. Neurosci.* **110,** 81–88.

Northcutt, R. G. (2004). Taste buds: Development and evolution. *Brain Behav. Evol.* **64,** 198–206.

Northcutt, R. G. (2005). Taste bud development in the channel catfish. *J. Comp. Neurol.* **482,** 1–16.

Northcutt, R. G., and Davis, R. E. (1983). Telencephalic organization in ray-finned fishes. *In* "Fish Neurobiology: Higher Brain Areas and Functions" (Davis, R. E., and Northcutt, R. G., Eds.), Vol. 2, pp. 202–236. University of Michiga, Ann Arbor.

O'Brien, R. D. (1979). Problems and approaches in noncatalytic biochemistry. *In* "The Receptors. A Comprehensive Treatise: General Principles and Procedures" (O'Brien, R. D., Ed.), Vol. 1, pp. 311–335. Plenum Press, New York.

Ogawa, K., Marui, T., and Caprio, J. (1997). Quinine suppression of single facial taste fiber responses in the channel catfish. *Brain Res.* **769,** 263–272.

Olmsted, J. M. D. (1918). Experiments on the nature of the sense of smell in the common catfish, *Amiurus nebulosus* (Lesueur). *Am. J. Physiol.* **46,** 443–458.

Parker, G. H. (1910). Olfactory reactions in fishes. *J. Exp. Zool.* **8,** 535–542.

Parker, G. H. (1912). The relations of smell, taste, and the common chemical sense in vertebrates. *Proc. Acad. Nat. Sci. Philadelphia* **15,** 221–234.

Perry, S. F., and Gilmour, K. M. (2002). Sensing and transfer of respiratory gases at the fish gill. *J. Exp. Biol.* **293,** 249–263.

Peters, R. C., Kotrschal, K., and Krautgartner, W.-D. (1991). Solitary chemoreceptor cells of *Ciliata mustela* (Gadidae, Teleostei) are tuned to mucoid stimuli. *Chem. Senses* **16,** 31–42.

Puzdrowski, R. L. (1987). The peripheral distribution and central projections of the sensory rami of the facial nerve in goldfish, *Carassius auratus*. *J. Comp. Neurol.* **259,** 382–392.

Raderman-Little, R. (1979). The effect of temperature on the turnover of taste bud cells in catfish. *Cell. Tissue Kinet.* **12,** 269–280.

Reutter, K. (1978). Taste organ in the bullhead (Teleostei). *Adv. Anat. Embryol. Cell Biol.* **55,** 1–98.

Reutter, K., and Witt, M. (1993). Morphology of vertebrate taste organs and their nerve supply. *In* "Mechanisms of Taste Transduction" (Simon, S. A., and Roper, S. D., Eds.), pp. 29–82. CRC Press, Boca Raton.

Rink, E., and Wullimann, M. F. (1998). Some forebrain connections of the gustatory system in the goldfish *Carassius auratus* visualized by separate DiI application to the hypothalamic inferior lobe and the torus lateralis. *J. Comp. Neurol.* **394,** 152–170.

Rolen, S. H., Sorensen, P. W., Mattson, D., and Caprio, J. (2003). Polyamines as olfactory stimuli in the goldfish *Carassius auratus*. *J. Exp. Biol.* **206,** 1683–1696.

Roper, S. D. (1989). The cell biology of vertebrate taste receptors. *Annu. Rev. Neurosci.* **12,** 329–353.

Roper, S. D. (1993). Synaptic interactions in taste buds. *In* "Mechanisms of Taste Transduction" (Simon, S. A., and Roper, S. D., Eds.), pp. 275–293. CRC Press, Boca Raton.

Saito, T., Noguchi, T., Harada, T., Murata, O., and Hashimoto, K. (1985). Tetrodotoxin as a biological defense agent for puffers. *Bull. Jpn. Soc. Sci. Fish.* **51,** 1175–1180.

Sakata, Y., Tsukahara, J., and Kiyohara, S. (2001). Distribution of nerve fibers in the barbels of sea catfish *Plotosus lineatus*. *Fish. Sci.* **67,** 1136–1144.

Satô, M. (1937a). Further studies on the barbells of a Japanese goatfish, *Upeneoides bensasi* (Temminck & Schlegel). *Sci. Rep. Tohoku Imp. Univ. Ser. 4* **11,** 297–302.

Satô, M. (1937b). On the barbells of a Japanese sea catfish, *Plotosus anguillaris* (Lacépède). *Sci. Rep. Tohoku Imp. Univ. Ser. 4* **11,** 323–332.

Scharrer, E., Smith, S. W., and Palay, S. L. (1947). Chemical sense and taste in the fishes, *Prionotus* and *Trichogaster*. *J. Comp. Neurol.* **86,** 183–193.

Scheich, H., Langner, G., Tidemann, C., Cole, R. B., and Guppy, A. (1986). Electroreception and electrolocation in the platypus. *Nature (Lond.)* **319,** 401–402.

Scott, K. (2004). The sweet and the bitter of mammalian taste. *Curr. Opin. Neurobiol.* **14,** 423–427.

Semm, P., and Beason, R. C. (1990). Responses to small magnetic variations by the trigeminal system of the bobolink. *Brain Res. Bull.* **25,** 735–740.

Sibbing, F. A., and Uribe, R. (1985). Regional specializations in the oro-pharyngeal wall and food processing in the carp (*Cyprinus carpio* L.). *Neth. J. Zool.* **35,** 377–422.

Sibbing, F. A., Osse, J. W. M., and Terlouw, A. (1986). Food handling in the carp (*Cyprinus carpio* L.), its movement patterns, mechanisms and limitations. *J. Zool. (Lond.) Ser. A* **210,** 161–203.

Silver, W. L., and Finger, T. E. (1984). Electrophysiological examination of a non-olfactory, non-gustatory chemosense in the searobin, *Prionotus carolinus*. *J. Comp. Physiol. A* **154,** 167–174.

Simon, S. A., and Wang, Y. (1993). Chemical responses of lingual nerves and lingual epithelia. *In* "Mechanisms of Taste Transduction" (Simon, S. A., and Roper, S. D., Eds.), pp. 225–252. CRC Press, Boca Raton.

Speca, D. J., Lin, D. M., Sorensen, P. W., Isacoff, E. Y., Ngai, J., and Dittman, A. H. (1999). Functional identification of a goldfish odorant receptor. *Neuron* **23,** 487–498.

Stone, L. M., Finger, T. E., Tam, P. P. L., and Tan, S.-S. (1995). Taste receptor cells arise from local epithelium, not neurogenic ectoderm. *Proc. Natl. Acad. Sci. USA* **92,** 1916–1920.

Sugita, M., and Shiba, Y. (2005). Genetic tracing shows segregation of taste neuronal circuitries for bitter and sweet. *Science* **309,** 781–785.

Sutterlin, A. M. (1975). Chemical attraction of some marine fish in their natural habitat. *J. Fish. Res. Board Can.* **32,** 729–738.

Sutterlin, A. M., and Saunders, R. L. (1969). Proprioceptors in the gills of teleosts. *Can. J. Zool.* **47,** 1209–1212.

Sutterlin, A. M., and Sutterlin, N. (1970). Taste responses in Atlantic salmon (*Salmo salar*) parr. *J. Fish. Res. Board Can.* **27,** 1927–1942.

Suwabe, T., and Kitada, Y. (2004). Voltage-gated inward currents of morphologically identified cells of the frog taste disc. *Chem. Senses* **29,** 61–73.

Sveinsson, T., and Hara, T. J. (2000). Olfactory sensitivity and specificity of Arctic char, *Salvelinus alpinus*, to a putative male pheromone, prostaglandin $F_{2\alpha}$ *Physiol. Behav.* **69,** 301–307.

Takeda, M., Shishido, Y., Kitao, K., and Suzuki, Y. (1982). Monoamines of taste buds in the fungiform and foliate papillae of the mouse. *Arch. Histol. Jpn.* **45,** 239–246.

Tateda, H. (1964). The taste response of the isolated barbells of the catfish. *Comp. Biochem. Physiol.* **11,** 367–378.

Toyoshima, K., Nada, O., and Shimamura, A. (1984). Fine structure of monoamine containing basal cells in the taste buds of the barbells of three species of teleosts. *Cell Tissue Res.* **235,** 479–484.

Travers, S. P., Pfaffmann, C., and Norgren, R. (1986). Convergence of lingual ans palatal gustatory neural activity in the nucleus of the solitary tract. *Brain Res.* **365,** 305–320.

Twongo, T. K., and MacCrimmon, H. R. (1977). Histogenesis of the oropharynfeal and oesophageal mucosa as related to early feeding in rainbow trout, *Salmo gairdneri* Richardson. *Can. J. Zool.* **55,** 116–125.

Uga, S., and Hama, K. (1967). Electron microscopic studies on the synaptic region of the taste organ of carps and frogs. *J. Electron Microsc.* **16,** 269–276.

Valentinčič, T., and Caprio, J. (1994a). Chemical and visual control of feeding and escape behaviors in the channel catfish *Ictalurus punctatus*. *Physiol. Behav.* **55,** 845–855.

Valentinčič, T., and Caprio, J. (1994b). Consummatory feeding behavior to amino acids in intact and anosmic channel catfish *Ictalurus punctatus*. *Physiol. Behav.* **55,** 857–863.

Valentinčič, T., and Caprio, J. (1997). Visual and chemical release of feeding behavior in adult rainbow trout. *Chem. Senses* **22,** 375–382.

Walker, M. M., Diebel, C. E., Haugh, C. V., Pankhurst, P. M., and Montgomery, J. C. (1997). Structure and function of the vertebrate magnetic sense. *Nature* **390,** 371–376.

Wang, Z. H., Bradford, A., and O'Regan, R. G. (2001). Effects of carbonic anhydrase inhibition on the responsiveness of laryngeal receptors in cats to CO_2. *Exp. Physiol.* **86,** 641–649.

Wegert, S., and Caprio, J. (1991). Receptor sites for amino acids in the facial taste system of the channel catfish. *J. Comp. Physiol. A* **168,** 201–211.

Whitear, M. (1971). The free nerve endings in fish epidermis. *J. Zool., Lond.* **163,** 231–236.

Whitear, M. (1992). Solitary chemosensory cell. *In* "Fish Chemoreception" (Hara, T. J., Ed.), pp. 103–125. Chapman & Hall, London.

Whitear, M., and Kotrschal, K. (1988). The chemosensory anterior dorsal fins in rocklings (*Gaidropsarus* and *Ciliatus*, Teleostei, Gadidae): Activity, fine structure and innervation. *J. Zool.* **216,** 339–366.

Wullimann, M. F. (1998). The central nervous system. *In* "The Physiology of Fishes" (Evans, D. H., Ed.), pp. 245–282. CRC Press, Boca Raton.

Yamamori, K., Nakamura, M., and Kamiya, H. (1980). Behavioral responses of fishes to pufferfish toxin, tetrodotoxin. Presented at the Annual Meeting of the Japanese Society of Fisheries Science, Tokyo, 3–5 April, 1980.

Yamamori, K., Nakamura, M., Matsui, T., and Hara, T. J. (1988). Gustatory responses to tetrodotoxin and saxitoxin in fish: A possible mechanism for avoiding marine toxins. *Can. J. Fish. Aquat. Sci.* **45,** 2182–2186.

Yamanaka, H., Shimakura, K., Shiomi, K., and Kikuchi, T. (1986). Changes in non-volatile amine contents of the meats of sadine and saury pike during storage. *Bull. Jpn. Soc. Sci. Fish.* **52,** 127–130.

Yamashita, S., Evans, R. E., and Hara, T. J. (1989). Specificity of the gustatory chemoreceptors for CO_2 and H^+ in rainbow trout (*Oncorhynchus mykiss*). *Can. J. Fish. Aquat. Sci.* **46,** 1730–1734.

Yamashita, S., Yamada, T., and Hara, T. J. (2006). Gustatory responses to feeding- and non-feeding-stimulant chemicals, with special emphasis on amino acids, in rainbow trout. *J. Fish Biol.* **68,** 783–800.

Yasuoka, A., Aihara, Y., Matsumoto, I., and Abe, K. (2004). Phospholipase C-beta 2 as a mammalian taste signaling marker is expressed in the multiple gustatory tissues of medaka fish, *Oryzias latipes*. *Mech. Dev.* **121,** 985–989.

Yoshii, K., Kamo, N., Kurihara, K., and Kobatake, Y. (1979). Gustatory responses of eel palatine receptors to amino acids and carboxylic acids. *J. Gen. Physiol.* **74,** 301–317.

Yoshii, K., Kashiwayanagi, M., Kurihara, K., and Kobatake, Y. (1980). High sensitivity of the eel palatine receptors to carbon dioxide. *Comp. Biochem. Physiol.* **66A,** 327–330.

Yoshimoto, M., Albert, J. S., Sawai, N., Shimizu, M., Yamamoto, N., and Ito, H. (1998). Telencephalic ascending gustatory system in a cichlid fish, *Oreochromis* (*Tilapia*) *niloticus*. *J. Comp. Neurol.* **392,** 209–226.

Zhang, C., Brown, S. B., and Hara, T. J. (2001). Biochemical and physiological evidence that bile acids produced and released by lake char (*Salvelinus namaycush*), function as chemical signals. *J. Comp. Physiol. B* **171,** 161–171.

Zhang, Y., Hoon, M. A., Chandrashekar, J., Mueller, K. L., Cook, B., Wu, D., Zuker, C. S., and Ryba, N. J. P. (2003). Coding of sweet, bitter, and umami tastes: Different receptor cells sharing similar signaling pathways. *Cell* **112,** 293–301.

Zhao, G. Q., Zhang, Y., Hoon, M. A., Chandrashekar, J., Erienbach, I., Ryba, N. J. P., and Zuker, C. S. (2003). The receptors for mammalian sweet and umami taste. *Cell* **115,** 255–266.

Zviman, M. M., Restrepo, D., and Teeter, J. H. (1996). Single taste stimuli elicit either increase or decrease in intracellular calcium in isolated catfish taste cells. *J. Membr. Biol.* **149,** 81–88.

3

BRANCHIAL CHEMORECEPTOR REGULATION OF CARDIORESPIRATORY FUNCTION

KATHLEEN M. GILMOUR
STEVE F. PERRY

1. Introduction
2. Cardiorespiratory Responses
   2.1. Cardiovascular Responses Linked to Activation of Chemoreceptors
   2.2. Ventilatory Responses Linked to Chemoreceptor Activation
   2.3. Endocrine Responses Mediated by Chemoreceptor Activation
3. Chemoreceptors
   3.1. Chemoreceptor Location and Orientation
   3.2. Morphology of (Presumptive) Chemosensory Cells
   3.3. Chemotransduction Mechanisms
4. Central Integration and Efferent Pathways
5. Conclusions and Future Directions

1. INTRODUCTION

The cardiorespiratory system in fish, as in other vertebrates, functions to meet the metabolic requirements of the cells for O_2 and to remove the CO_2 produced by cellular metabolism. Oxygen delivery and CO_2 removal must be matched to cellular requirements in the face of not only variable metabolic demand but also fluctuating environmental O_2 and/or CO_2 tensions—aquatic environments are subject to significant spatial and/or temporal variation in environmental gas levels (Nikinmaa and Rees, 2005), far more so than is the case for most terrestrial environments. It is therefore essential that fish be able to sense and respond to changes in both environmental gas levels and metabolic O_2 demands. Branchial chemoreceptors represent a key component of the array of sophisticated sensory and control mechanisms that is used to adjust cardiorespiratory function in fish to meet such environmental and metabolic demands.

Sensory Systems Neuroscience: Volume 25
FISH PHYSIOLOGY

DOI: 10.1016/S1546-5098(06)25003-9

Chemoreceptive reflexes involve stimulus detection and afferent transmission of the sensory information to central locations for integration and processing, followed by appropriate adjustment of cardiovascular and ventilatory effectors in response to efferent motor output. Despite the large body of work documenting cardiorespiratory responses to various environmental and metabolic challenges (see reviews by Jones and Randall, 1978; Johansen, 1982; Randall, 1982, 1990; Randall and Daxboeck, 1984; Wood and Perry, 1985; Shelton *et al.*, 1986; Fritsche and Nilsson, 1993; Gilmour, 2001; Perry and Gilmour, 2002), our understanding of the sensory mechanisms and afferent pathways underlying these responses remains woefully incomplete. Specifically, little is known of the location, and particularly, function of peripheral chemoreceptors, even though these receptors are of critical importance in sensing the conditions that necessitate cardiorespiratory adjustments. Because it is only relatively recently that CO_2 and/or pH have been identified as specific regulators of cardiorespiratory function in fish (see reviews by Milsom, 1995a,b, 1998, 2002; Gilmour, 2001; Perry and Gilmour, 2002), much of what is known about peripheral chemoreceptors in fish is based on studies of O_2 chemoreflexes (see reviews by Smatresk, 1990; Burleson *et al.*, 1992; Fritsche and Nilsson, 1993; Burleson, 1995; Milsom *et al.*, 1998, 1999; Burleson and Milsom, 2003).

The objective of the present chapter, then, is to provide a synthesis of current knowledge on O_2- and CO_2/pH-based chemoreflexive regulation of cardiorespiratory function in fish. First, cardiovascular and ventilatory responses will be summarized, as it is generally on the basis of eliciting or eliminating such responses that chemoreceptor involvement is inferred. The location, orientation, and specificity of O_2- and CO_2/pH-sensitive chemoreceptors will then be discussed together with the cellular basis of O_2 and CO_2/pH sensing. Central processing and efferent information transmission will be summarized briefly, and outstanding unanswered questions and potential future research directions will form the focus of the final section. Because the emphasis of this chapter is on branchial chemoreceptors, the responses of water-breathing fish to water-borne stimuli (e.g., aquatic hypoxia or hypercapnia) will form the body of the material discussed. However, the responses of bimodally breathing fish to stimulation of branchial chemoreceptors will be considered where possible.

2. CARDIORESPIRATORY RESPONSES

The sensing by chemoreceptors of changes in the gas composition of the external or internal environment promotes an array of physiological and behavioral responses. These include adjustments of cardiovascular function, ventilation, locomotion, and a suite of responses linked to changes in circulating hormone levels. While many of these chemoreceptor-driven responses are functionally beneficial, the physiological significance of others, in particular some of the cardiovascular adjustments, is less obvious. Thus, while describing the various responses mediated by chemoreceptor activation, in this chapter attempts will also be made to ascribe, where possible, physiological benefit to the various responses. Another goal is to dispel the widely held view that the cardiorespiratory adjustments that accompany chemoreceptor stimulation in selected model species, such as rainbow trout, are representative of fish in general. Instead, the data provided demonstrate the profound interspecific variation in the responses of fish species to lowered ambient O_2 (hypoxia) and elevated CO_2 (hypercapnia).

2.1. Cardiovascular Responses Linked to Activation of Chemoreceptors

The cardiac responses to environmental perturbation have been discussed in numerous comprehensive reviews (Randall, 1982; Butler and Metcalfe, 1988; Perry and Wood, 1989; Burleson *et al.*, 1992; Bushnell and Jones, 1992; Taylor, 1992; Fritsche and Nilsson, 1993; Olson, 1998; Burleson, 1995; Perry and Gilmour, 1999, 2002; Milsom *et al.*, 1999). Arguably, the most widely studied cardiac reflex response is a reduction in cardiac frequency (f_H; bradycardia) that accompanies both hypoxia (Randall and Shelton, 1963; Randall and Smith, 1967) and hypercapnia (Kent and Peirce, 1978). The origin of the bradycardia is an increase in parasympathetic activity and subsequent stimulation of cardiac muscarinic receptors (Wood and Shelton, 1980; Taylor, 1992). Although often described as a "typical" response, it is clear that hypoxic bradycardia does not occur in all species that have been examined (Table 3.1). Of the 29 species for which data were readily available in the literature (Table 3.1), 11 either show no change in f_H or show a slight increase. Moreover, there are several examples of certain species exhibiting different cardiovascular responses to hypoxia depending on the particular study (e.g., see data for *Oncorhynchus mykiss*, *Gadus morhua*, and *Myoxocephalus scorpius* in Table 3.1). It is noteworthy that bradycardia does not appear to accompany aquatic hypoxia in any genus of lungfish (*Protopterus*, *Lepidosiren*, *Neoceratodus*). It has been argued that the absence of cardiorespiratory responses associated with aquatic hypoxia may indicate that external O_2 chemoreceptors are absent in lungfish and thus evolved only in the actinopterygian lineage (Perry *et al.*, 2005). Similarly, hypoxic bradycardia has not been observed in hagfish; this may reflect the absence of parasympathetic innervation of the hagfish heart (Nilsson, 1983). The interspecific variation in the cardiac response to hypercapnia is at least as large as for hypoxia; four out of nine species for which data are available (Table 3.2, Figure 3.1) do not exhibit hypercapnic bradycardia. Some of the

Table 3.1
Cardiovascular Responses to Acute Aquatic Hypoxia in a Variety of Species

Species	Δf_H	ΔV_b	ΔP_{art}	References
Oncorhynchus mykiss[a]	↓	NC	↑	Holeton and Randall, 1967; Marvin and Burton, 1973; Wood and Shelton, 1980
	↓	↓	*↑	Sundin and Nilsson, 1997
	↓	↓	NC	Perry *et al.*, 1999
	↓	NC	↓	Sandblom and Axelsson, 2005
Gadus morhua[b]	↓	NC	↑	Fritsche and Nilsson, 1989, 1990
	NC	↑	↑	Axelsson and Fritsche, 1991
Myoxocephalus scorpius[c]	↓	↓	↑	Turesson and Sundin, 2003; MacCormack and Driedzic, 2004
	↓	↓	NC	MacCormack and Driedzic, 2004
Colossoma macropomum[d]	↓	–	NC	Fritsche, 1990; Sundin *et al.*, 2000
Anguilla anguilla[e]	↓	↓	↓	Peyraud–Waitzenegger and Soulier, 1989
Katsuwonus pelamis[f]	↓	↓	NC	Bushnell and Brill, 1992
Thunnus albacares[g]	↓	↓	NC	Bushnell and Brill, 1992
Hoplias malabaricus[h]	↓	–	↓	Sundin *et al.*, 1999
Zoarces viviparus[i]	↓	–	↓	Fritsche, 1990
Hoplerythrinus unitaeniatus[j]	↓	–	↑	Oliveira *et al.*, 2004
Scyliorhinus canicula[k]	↓	NC	↓	Butler and Taylor, 1971; Short *et al.*, 1979; Taylor and Barrett, 1985
Hemiscyllium ocellatum[l]	↓	–	↓	Stenslokken *et al.*, 2004
Acipenser naccarii[m]	↓	–	NC	McKenzie *et al.*, 1995
Acipenser baeri[n]	↓	–	↑	Maxime *et al.*, 1995
Ciliata mustela[o]	NC	–	↑	Fritsche, 1990
Neoceratodus forsteri[p]	NC	↑p/↓s	NC	Fritsche *et al.*, 1993
Pagothenia bernacchii[q]	↑	NC	↑	Axelsson *et al.*, 1992
Protopterus dolloi[r]	NC	–	NC	Perry *et al.*, 2005
Lepidosiren paradoxa[s]	NC	–	–	Sanchez *et al.*, 2001a
Gobius cobitis[t]	NC	–	–	Berschick *et al.*, 1987
Hemitripterus americanus[u]	NC	–	↑	Saunders and Sutterlin, 1971
Tinca tinca[v]	↓	–	–	Randall and Shelton, 1963
Lepomis macrohirus[w]	↓	–	–	Marvin and Burton, 1973
Ameiurus nebulosus[x]	↓	–	–	Marvin and Burton, 1973
Leiopotherapon unicolor[y]	↓	–	–	Gehrke and Fielder, 1988
Liposarcus pardalis[z]	NC	–	–	MacCormack *et al.*, 2003
Glyptoperichthys gibbiceps[a′]	NC	–	–	MacCormack *et al.*, 2003
Epatretus cirrhatus[b′]	NC	↑	↑	Forster *et al.*, 1992
Myxine glutinosa[c′]	NC	NC	↑	Axelsson *et al.*, 1990

intraspecific variation in the cardiac responses to hypoxia or hypercapnia might reflect differing thresholds for stimulating chemoreceptors. For example, some teleost species are quick to develop a hypoxia-induced bradycardia [e.g., tench, *Tinca tinca* (Randall and Shelton, 1963); bluegill, *Lepomis macrochirus* (Marvin and Burton, 1973); spangled perch, *Leiopotherapon unicolor* (Gehrke and Fielder, 1988)]. However, other species require severely hypoxic conditions before responding with a significant bradycardia, [e.g., brown bullhead, *Ameiurus nebulosus* (Marvin and Burton, 1973) and sea raven, *Hemitripterus americanus* (Saunders and Sutterlin, 1971)].

Table 3.2
Cardiovascular Responses to Acute Aquatic Hypercapnia in a Variety of Species

Species	Δf_H	ΔV_b	ΔP_{art}	References
Oncorhynchus mykiss[a]	↓	NC	↑	Perry *et al.*, 1999; McKendry and Perry, 2001
Salmo salar[b]	↓	↑	↑	Perry and McKendry, 2001
Citahrychthus sordidus[c]	↓	–	NC	S. F. Perry and J. E. McKendry; unpublished data
Colossoma macropomum[d]	↓	–	↑	Sundin *et al.*, 2000
	↓	NC	NC	Gilmour *et al.*, 2005
Anguilla rostrata[e]	NC	NC	NC	S. F. Perry and J. E. McKendry; unpublished data
Hoplias malabaricus[f]	↓	–	↓	Reid *et al.*, 2000
Ameiurus nebulosus[g]	NC	NC	NC	S. F. Perry and J. E. McKendry; unpublished data
Squalus acanthias[h]	↓	↓	↓	McKendry *et al.*, 2001; Perry and McKendry, 2001
Acipenser transmontanus[i]	↑	↑	↑	Crocker *et al.*, 2000
Tinca tinca[j]	↑	–	–	Randall and Shelton, 1963

[a]Rainbow trout; [b]Atlantic salmon; [c]Pacific sandab; [d]tambaqui; [e]American eel; [f]traira; [g]brown bullhead; [h]spiny dogfish; [i]white sturgeon; [j]tench.

f_H, cardiac frequency; V_b, cardiac output; P_{art}, arterial blood pressure; NC, no change; –, not measured.

---

[a]Rainbow trout; [b]Atlantic cod; [c]shorthorn sculpin; [d]tambaqui; [e]European eel; [f]skipjack tuna; [g]yellowfin tuna; [h]traira; [i]eel pout; [j]jeju; [k]lesser spotted dogfish; [l]epaulette shark, [m]Adriatic sturgeon; [n]Siberian sturgeon; [o]five-bearded rockling; [p]Australian lungfish; [q]emerald rockcod; [r]slender African lungfish; [s]South American lungfish; [t]giant goby; [u]sea raven; [v]tench; [w]bluegill; [x]brown bullhead (also referred to as *Ictalurus nebulosus*); [y]spangled perch; [z]long-fin armored catfish (proper scientific name may be *Liposarcus multiradiatus*); [a′]leopard pleco; [b′]Pacific hagfish; [c′]Atlantic hagfish.

f_H, cardiac frequency; V_b, cardiac output or blood flow (p = pulmonary, s = systemic); P_{art}, arterial blood pressure (dorsal aortic or ventral aortic); NC, no change; –, not measured; *Indicates an increase in only ventral aortic pressure.

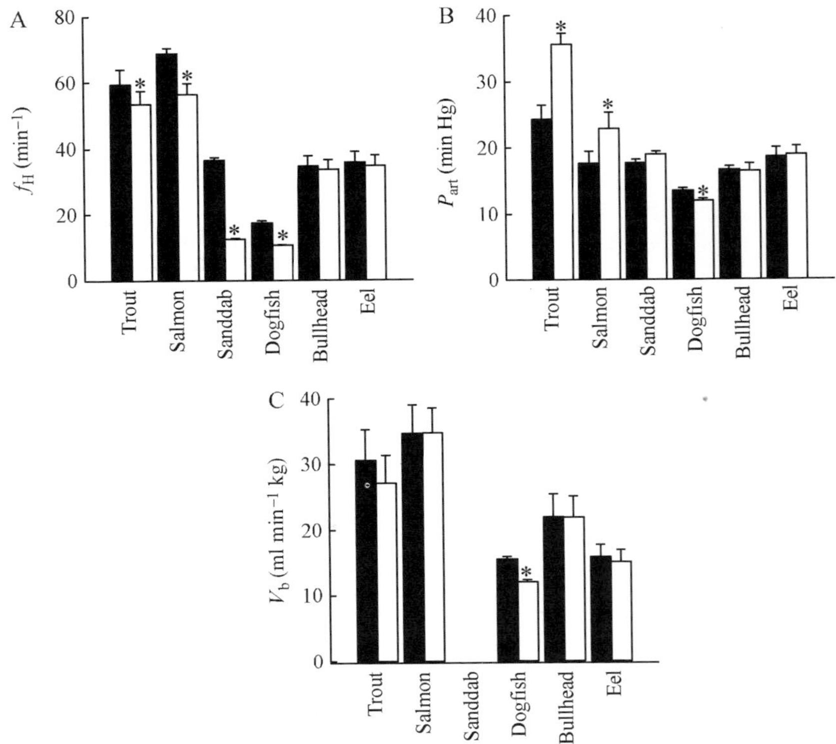

Fig. 3.1. The effects of hypercapnia (P_WCO_2 = 6.1 ± 0.4 mm Hg) on (A) heart rate (f_H), (B) arterial blood pressure (P_{art}), and (C) cardiac output ($\dot{V}_b$) in two marine teleosts (Atlantic salmon, $N = 7$; Pacific sanddab, $N = 7$), three freshwater teleosts (rainbow trout, $N = 9$; brown bullhead, $N = 8$; American eel, $N = 8$), and one marine elasmobranch (spiny dogfish, $N = 9$). *Denotes a statistically significant difference between normocapnia (filled bars) and the maximal response during hypercapnia (unfilled bars) for each species ($P < 0.05$; paired Student's t-test). The data for rainbow trout were reported previously in McKendry and Perry (2001).

In several species, including rainbow trout (*O. mykiss*), Atlantic cod (*G. morhua*), and dogfish (*Scyliorhinus canicula*), cardiac stroke volume is increased during hypoxic bradycardia to maintain (or even increase) cardiac output (however, see exceptions in Table 3.1). The precise mechanisms underlying the rise in stroke volume during hypoxia are not well established but might involve the stimulatory effects of circulating catecholamines, increased activity of sympathetic nerve fibres, increased venous pressure (see later), as well as the increased filling time associated with the bradycardia (Starling's law of the heart). Although relatively few measurements have been made (Table 3.1), it is clear that maintenance of cardiac output during hypoxia is *not* a generalized response in fish. A decrease in cardiac output would appear to be a more common response among the species that have been examined (Table 3.1). These two different strategies of cardiac output management during hypoxia may have important implications for branchial gas transfer because blood residence time in the gill will be increased only in those fish experiencing a fall in cardiac output (see later). Even fewer measurements have been made of cardiac output during hypercapnia (Table 3.2), but from the available data, it is obvious that, as with hypoxia, no single response predominates.

Another cardiovascular response that has been reported to accompany both hypoxia and hypercapnia is an elevation of arterial blood pressure that is caused by an increase in systemic vascular resistance. In those species that have been examined in detail, the increase in resistance is a consequence of the stimulation of vascular smooth muscle α-adrenergic receptors owing to the combined actions of increased sympathetic discharge and elevated levels of circulating catecholamines. As previously discussed for other chemoreceptor-driven cardiovascular responses, the hypertension accompanying hypoxia or hypercapnia in rainbow trout and a few other selected species (Tables 3.1 and 3.2; Figure 3.1) is simply one example of several possible responses associated with these environmental conditions. For instance, some species exhibit a marked reduction in blood pressure while in others arterial pressure remains constant. Because relatively few studies have simultaneously measured blood pressure and cardiac output (so as to calculate systemic resistance), it is often difficult to determine whether blood pressure changes reflect adjustments of blood flow and/or systemic resistance. As reported for hypoxic yellowfin tuna [*Thunnus albacares*; (Bushnell and Brill, 1992)], arterial blood pressure might remain constant, despite an increase in systemic resistance, owing to decreased cardiac output.

2.1.1. Is There any Physiological Benefit of Bradycardia or Hypertension During Hypoxia or Hypercapnia?

The functional role(s) of bradycardia and elevated blood pressure, responses that are linked to chemoreceptor activation, has been the subject of considerable debate for decades. Arguably, the most significant issue relates to whether bradycardia or elevated blood pressure can enhance branchial gas transfer during hypoxia or hypercapnia. Based on theoretical considerations (Randall and Daxboeck, 1984) and limited experimental data (Davie and Daxboeck, 1982), it has been suggested that bradycardia, even when cardiac output was kept constant, could augment gas transfer because of lamellar recruitment or increased gas permeability, both a result of increased pulsatility. Similarly, an elevation of mean ventral aortic pressure or a rise in systolic pressure may elicit an increase in the functional surface area of the gill by the recruitment of previously unperfused distal lamellae and by altering blood flow patterns within individual lamellae (Booth, 1979; Farrell *et al.*, 1979, 1980; Soivio and Tuurala, 1981). In addition, an elevation of ventral aortic pressure would be expected to distend the intralamellar vascular sheet leading to reduced blood-to-water diffusion distances (Farrell *et al.*, 1980). Theoretically, the net consequence of such changes, if occurring during hypoxia or hypercapnia, would be an increase in the diffusing capacity of the gill.

Despite the potential for chemoreceptor-driven cardiovascular responses to enhance branchial gas exchange, only three studies have directly examined the consequences of these cardiovascular changes to gas transfer and only two species have been assessed [dogfish, *S. canicula*; (Short *et al.*, 1979; Taylor and Barrett, 1985) and rainbow trout, *O. mykiss* (Perry and Desforges, 2006)]. In these studies, hypoxic bradycardia was prevented either by bilateral vagotomy (Short *et al.*, 1979) or by pharmacological blockade of cardiac muscarinic receptors using atropine (Taylor and Barrett, 1985; Perry and Desforges, 2006). Because evidence for a beneficial role of the hypoxic or hypercapnic bradycardia was provided only in one of these studies (Taylor and Barrett, 1985), it is difficult to draw any general conclusions except that there are few data supporting the view that bradycardia aids gas transfer, despite prevailing theory. Although there appears to be no obvious benefit of bradycardia on blood gases in trout and the results for dogfish are equivocal, these findings do not exclude an important function in other species. Unlike trout and dogfish that tend to maintain V_b during hypoxic bradycardia, there are several species (Table 3.1) that experience a fall in V_b. The increase in blood transit time through the lamellae, associated with the fall in V_b, might be expected to increase the efficiency of both CO_2 and O_2 transfers. Branchial O_2 transfer, while normally perfusion limited, becomes diffusion limited during hypoxia (Julio *et al.*, 2000) and thus an increase in blood transit time would be expected to increase PaO_2 during hypoxia. Because CO_2 transfer, at least across the trout gill, behaves as a diffusion-limited system at rest (Desforges *et al.*, 2002), increases in transit time would be expected to decrease $PaCO_2$ during normoxia or hypoxia.

Even fewer studies have evaluated the possibility that hypertension during hypoxia or hypercapnia augments gas transfer. The available data, however, suggest that in both Atlantic cod (Kinkead *et al.*, 1991) and rainbow trout (Perry and Desforges, 2006), there is no obvious advantage associated with hypertension in modulating gas transfer across the gill. Because elevated levels of CO_2 in the circulation of rainbow trout cause direct vasodilation of systemic blood vessels and a fall in vascular resistance (McKendry and Perry, 2001), it is possible that the reflex chemoreceptor-mediated peripheral vasoconstriction serves to counterbalance the dilatory effects of CO_2, thereby preventing a fall in arterial blood pressure during external hypercapnia.

2.1.2. Other Cardiovascular Adjustments Potentially Mediated by Chemoreceptors

Although less well studied, there is compelling evidence that branchial vascular resistance increases during hypoxia owing to vasoconstriction (Holeton and Randall, 1967; Pettersson and Johansen, 1982; Sundin, 1995; Perry *et al.*, 1999) at least in rainbow trout and Atlantic cod. The consequence of such branchial vasoconstriction is believed to be an elevation of perfusion pressure that could promote lamellar recruitment to increase the effective surface area for gas transfer. Changes in branchial vascular resistance have not been observed in trout exposed to hypercapnia (Perry *et al.*, 1999). Another response that has been observed in trout and epaulette shark [*Hemiscyllium ocellatum* (Stenslokken *et al.*, 2004)] during hypoxia is a redistribution of blood flow from the arterio-arterial circulation to the arterio-venous circulation of the gill (Sundin and Nilsson, 1997; see review by Olson, 2002). Increasing the percentage of blood flowing to the arteriovenous circulation may promote the return of oxygenated blood to the heart.

Although intuitively, it might be expected that shunting blood away from the visceral circulation during hypoxia would be beneficial, surprisingly little attention has been paid to the redistribution of systemic blood flow during hypoxia in fish. However, based on available data from a single species [Atlantic cod (Axelsson and Fritsche, 1991)], it would appear that blood flow to the gut is severely curtailed during hypoxia owing to increased vascular resistance of the celiac and mesenteric arteries.

Effects on the venous circulation constitute another aspect of chemoreceptor-mediated cardiovascular control that has largely been ignored. Two studies, however, have reported an increase in central venous pressure during acute hypoxia in rainbow trout (Perry *et al.*, 1999; Sandblom and Axelsson, 2005). The physiological significance, if any, of an increase in venous pressure during hypoxia is unclear although it may serve to aid cardiac filling and thus help to maintain cardiac output during hypoxic bradycardia (Sandblom and Axelsson, 2005).

2.2. Ventilatory Responses Linked to Chemoreceptor Activation

Several reviews have focused, in whole or in part, on the chemoreceptor-mediated control of breathing in fishes (Shelton *et al.*, 1986; Perry and Wood, 1989; Fritsche and Nilsson, 1993; Smatresk, 1994; Milsom, 1995b, 2002; Gilmour, 2001; Perry and Gilmour, 2002). The majority of studies examining chemoreceptor-mediated control of breathing has concentrated on the effects of aquatic hypoxia, although there has been increasing emphasis on the ventilatory responses to aquatic hypercapnia.

2.2.1. Hypoxia

All water-breathing fishes that have been examined respond to aquatic hypoxia with an increase in gill ventilation volume, and to hyperoxia with a decrease in ventilation volume. Hypoxic hyperventilation is accomplished usually by large changes in opercular stroke volume and comparatively smaller changes in frequency (f_V) (see Table 1 in Gilmour, 2001). Aquatic hypoxia elicits an increase in aerial respiration in many bimodal (water and air) breathers including gar, *Lepisosteus*, and bowfin, *Amia calva* (reviewed by Smatresk, 1994), and the facultative air-breathing Australian lungfish, *Neoceratodus forsteri* (Fritsche *et al.*, 1993; Kind *et al.*, 2002). The situation appears to be more complex for the obligate air-breathing lungfish; while the South American lungfish, *Lepidosiren paradoxa*, does not exhibit an increase in pulmonary ventilation with aquatic hypoxia (Sanchez *et al.*, 2001a), mixed results have been obtained from studies on African lungfish, *Protopterus*. For example, increased pulmonary ventilation during aquatic hypoxia was observed in *P. annectens* (Babiker, 1979) but conflicting results were obtained from two studies examining the hypoxic responses of *P. aethiopicus*; pulmonary hyperventilation was observed in one study (Jesse *et al.*, 1967), while breathing remained constant in another (Johansen and Lenfant, 1968). Finally, lung ventilation was unaffected by aquatic hypoxia in *P. dolloi* (Perry *et al.*, 2005). The situation is much clearer for aerial hypoxia, in that all species of lungfish respond to a decrease in inspired PO_2 with increased pulmonary ventilation. The hyperventilatory response to aerial hypoxia presumably is mediated by internal chemoreceptors associated with the pulmonary airways or the vasculature. The chemoreceptors responding to fluctuations in aerial PO_2 in *P. dolloi* remain fully functional in aestivating animals. Figure 3.2 presents representative recordings of buccal PO_2 in two specimens of aestivating lungfish exposed to hypoxia (Figure 3.2A) or hyperoxia (Figure 3.2B). In these recordings, the abrupt spikes in PO_2 represent breathing episodes lasting ~3–10 sec (S. F. Perry, R. Euverman, Y. K. Ip, S. F. Chew, T. Wang, and K. M. Gilmour, unpublished data); breathing frequency is increased during hypoxia and decreased during hyperoxia (Figure 3.2).

2.2.2. Hypercapnia

Although the most common ventilatory effect of environmental hypercapnia in water breathers is an increase in ventilation amplitude (V_{AMP}) and frequency (f_V), there is marked interspecific diversity (see Table 2 in Gilmour, 2001). To illustrate this variability, Figure 3.3 depicts the ventilatory responses of five different species to environmental hypercapnia ($P_WCO_2 = 6.1 \pm 0.4$ mm Hg; S. F. Perry and J. McKendry, unpublished data). This figure reveals a variety of responses to hypercapnia ranging from

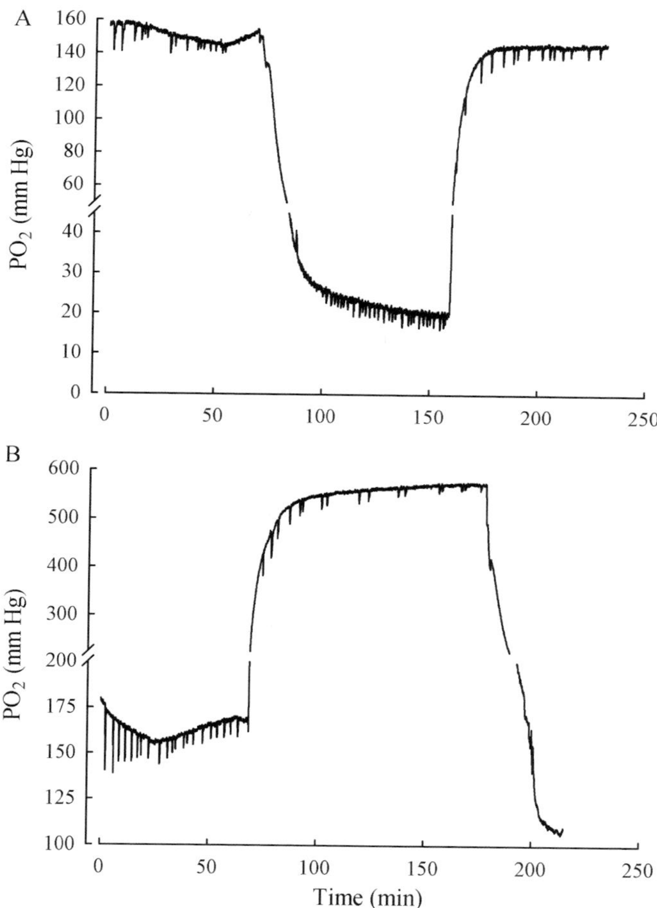

Fig. 3.2. Representative recordings of buccal PO_2 in two aestivating African lungfish (*Protopterus dolloi*) exposed to (A) aerial hypoxia or (B) aerial hyperoxia. The spikes in the PO_2 traces represent distinct breathing episodes lasting ~3–10 sec. Note the increase in breathing frequency associated with hypoxia and the decrease associated with hyperoxia.

increases in both V_{AMP} and f_V (Atlantic salmon, Pacific sanddab, and spiny dogfish), to increases in f_V only (brown bullhead), to no change in either variable (American eel). Another response pattern that has been reported for several species is an exclusive increase in V_{AMP} [spotted dogfish (Randall *et al.*, 1976); rainbow trout (Smith and Jones, 1982; Perry and Gilmour, 1996); common carp *Cyprinus carpio* (Soncini and Glass, 2000), and zebrafish *Danio rerio* (Vulesevic *et al.*, 2006)]. Previous studies on rainbow trout

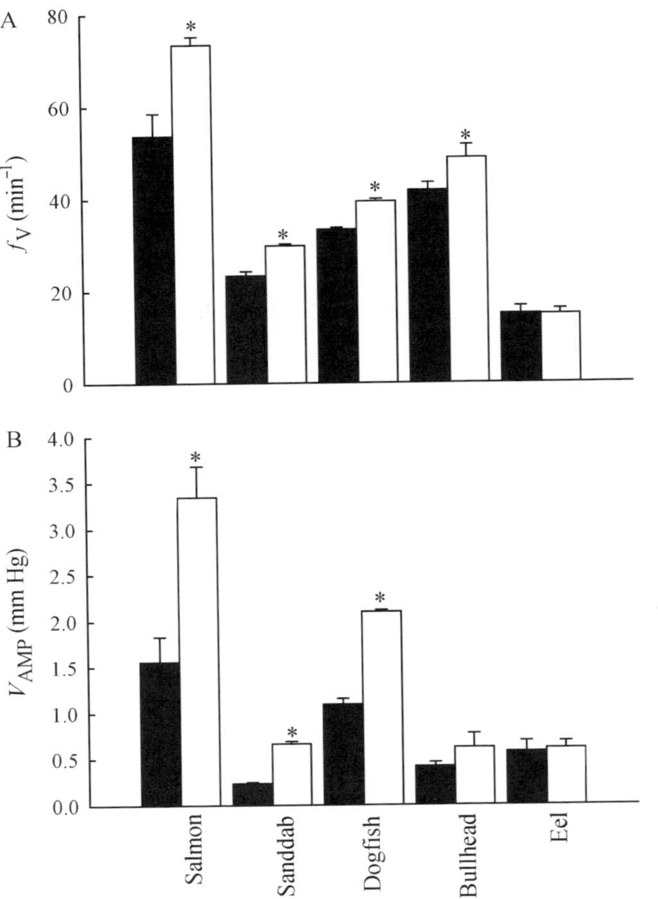

Fig. 3.3. The effects of hypercapnia ($P_WCO_2 = 6.1 \pm 0.4$ mm Hg) on (A) ventilation frequency (f_V) and (B) ventilation amplitude (V_{AMP}) in two marine teleosts (Atlantic salmon, $N = 11$; Pacific sanddab, $N = 10$), two freshwater teleosts (brown bullhead, $N = 8$; American eel, $N = 8$), and a marine elasmobranch (spiny dogfish, $N = 8$). *Denotes a statistically significant difference between normocapnia (filled bars) and the maximal response during hypercapnia (unfilled bars) for each species ($P < 0.05$; paired Student's t-test).

(Smith and Jones, 1982; Thomas and Le Ruz, 1982; Thomas, 1983; Gilmour and Perry, 1994; Perry and Gilmour, 1996; Perry *et al.*, 1999) have demonstrated that an increase in V_{AMP} is the most consistent response to hypercapnia. The magnitude of the increases (12–317%), however, varies enormously, due at least in part to differing P_WCO_2 levels used in the various studies.

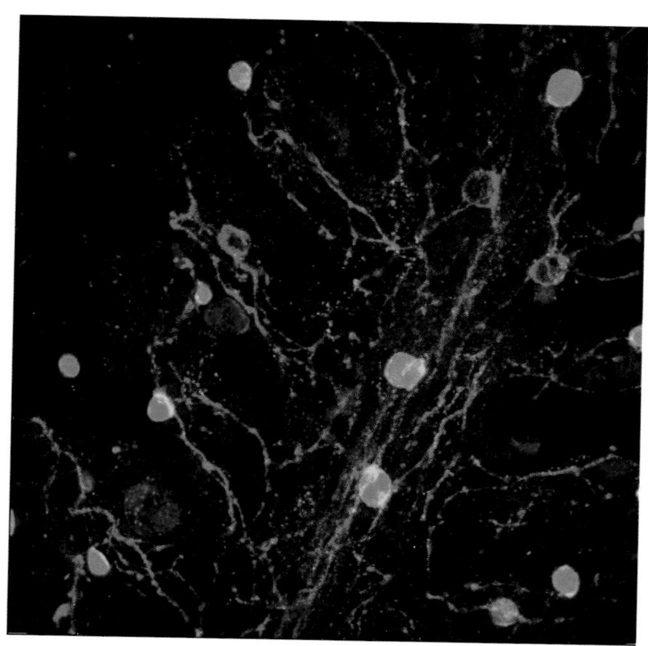

Chapter 3, Fig. 6.

Only half of the aforementioned studies observed a significant increase in f_V and the extent of these changes was minor (17–43%) in comparison to the changes in V_{AMP}. However, as illustrated in Figure 3.3, a change to V_{AMP} is not always the predominant ventilatory response to environmental hypercapnia. For example, Sundin *et al.* (2000) observed a 45% increase in f_V in the tambaqui during elevated P_WCO_2 with no significant change in V_{AMP}. A similar pattern was observed in channel catfish (*Ictalurus punctatus*) exposed to hyperoxic hypercapnia (Burleson and Smatresk, 2000).

The functional significance of hyperventilation during aquatic hypercapnia has been the subject of considerable speculation because hyperventilation during hypercapnia cannot prevent the development of respiratory acidosis or the accompanying reduction in blood O_2 content (in those fish displaying Root effects). Nevertheless, hyperventilation during hypercapnia is likely to be physiologically significant for several reasons (Gilmour, 2001). First, while respiratory acidosis is unavoidable during exposure to hypercapnic water, there is no doubt that hyperventilation will serve to minimize the extent of the acidosis by lowering arterial PCO_2. Owing to the log–linear relationship between $PaCO_2$ and pH, even small reductions in $PaCO_2$ can cause relatively large increases in blood pH. Second, the elevation of arterial PO_2 caused by hyperventilation, while unable to protect against Root effects, may nevertheless enhance O_2 loading at the gill that otherwise would be compromised by Bohr effects. This strategy would be particularly effective in fish normally displaying high-affinity hemoglobins and low PaO_2 values. Third, because hypercapnia in the natural environment normally occurs during periods of hypoxia, it is possible that the hyperventilatory responses evoked by slight elevations in CO_2 have evolved as an "early warning system" to prepare for ensuing severe hypoxia.

2.2.3. Other Ventilatory Responses

A strategy used by several species of hypoxia tolerant tropical species to cope with hypoxic water is aquatic surface respiration (Val, 1995). In this presumed reflex response, fish ventilate their gills by skimming O_2-enriched water from the air–water interface. In species such as the tambaqui (*Colossoma macropomum*), skimming of the surface water is aided by the swelling of the lower lip. Although lip swelling and aquatic surface respiration in tambaqui are presumed to arise from chemoreceptor stimulation, total gill denervation was without effect on the frequency of surface respiration during hypoxia and only marginally influenced the extent of lip swelling (Sundin *et al.*, 2000). Thus, it has been suggested that there are extrabranchial receptors linked to these responses in tambaqui. Lip swelling and aquatic surface respiration may be cued specifically by hypoxic stimuli because tambaqui exposed to hypercapnia failed to exhibit either response over a 6-h period (Florindo *et al.*, 2004).

2.3. Endocrine Responses Mediated by Chemoreceptor Activation

Exposure of fish to hypoxia elicits an increase in the circulating levels of the stress hormones: cortisol and catecholamines (adrenaline and noradrenaline) [see Tables 2.3 and 2.4 in Perry and Gilmour (1999)]. While the rise in catecholamine and cortisol levels is presumably associated with activation of chemoreceptors, a definitive link has only been recently established for catecholamines (Reid and Perry, 2003). The study of Reid and Perry (2003) demonstrated the presence of both external (water monitoring) and internal (blood oriented) branchial O_2 chemoreceptors linked to catecholamine secretion in rainbow trout. Whereas the external receptors appear to be confined to the first gill arch, the distribution of the internal receptors is more widespread. Because extirpation of the first gill arch (and hence removal of external O_2 chemoreceptors linked to catecholamine release) did not prevent the mobilization of catecholamines during hypoxia, it would appear that stimulation of the internally oriented chemoreceptors is the predominant mechanism initiating catecholamine release during hypoxia.

Catecholamine release into the circulation also occurs during aquatic hypercapnia (Reid *et al.*, 1998). It has largely been assumed that the stimulatory effects of hypercapnia on catecholamine secretion are indirect and reflect an impairment of blood O_2 transport related to respiratory acidosis (Perry *et al.*, 1989). However, there is now evidence that aquatic hypercapnia can also directly induce catecholamine secretion by activating a broadly distributed population of externally oriented branchial CO_2 chemoreceptors (Perry and Reid, 2002).

3. CHEMORECEPTORS

The initial step in the reflexes leading to the cardiorespiratory responses described earlier is detection of changes in blood (as a proxy of changes in metabolic demand) and/or environmental gas tensions by chemoreceptors. In tetrapod vertebrates, O_2-, CO_2-, and pH-sensitive peripheral chemoreceptors and central chemoreceptors that monitor CO_2/pH are involved in initiating cardiorespiratory reflexes (see reviews by González *et al.*, 1992, 1994; Milsom, 1995a,b, 1998; Nattie, 1999; Taylor *et al.*, 1999; Burleson and Milsom, 2003; Milsom *et al.*, 2004). Changes in CO_2/pH constitute the more important stimulus, in keeping with the O_2-rich nature of air as the ventilated medium, and at least within mammals, central chemoreceptors predominate over the peripheral CO_2/pH-sensitive receptors that monitor blood, and to a lesser extent, airways, in terms of driving cardiorespiratory responses (Milsom, 1998, 2002; Nattie, 1999; Taylor *et al.*, 1999; Milsom *et al.*, 2004; Putnam *et al.*, 2004). O_2-mediated chemoreflexes are dominated by the receptors of the carotid body or its homologues in, respectively, mammals and nonmammalian tetrapods, and function primarily to adjust cardiorespiratory function in the event of significant hypoxemia (González *et al.*, 1992; Milsom, 1998; Nattie, 1999; Taylor *et al.*, 1999). Although the exact anatomical sites of O_2- and particularly CO_2/pH-sensitive cells in fish remain uncertain, abundant evidence points to the gills as the main location for chemoreceptors involved in cardiorespiratory regulation. The gill arches of fish share a common embryonic origin with the carotid and aortic arches of tetrapods such that the branchial distribution of chemoreceptors linked to cardiorespiratory reflexes in fish is phylogenetically consistent with the distribution of peripheral receptors that monitor blood O_2 and CO_2/pH in tetrapods (Milsom, 1998, 2002; Taylor *et al.*, 1999; Burleson and Milsom, 2003). Unlike the situation in tetrapods, however, the branchial chemoreceptors of fish are of overwhelming importance in initiating cardiovascular and ventilatory adjustments to changes in both O_2 and CO_2/pH. In some species, the failure of total branchial denervation to eliminate ventilatory and cardiovascular responses to hypoxic and/or hypercapnic stimuli implicates the involvement of extrabranchial chemoreceptors, generally those associated with the orobranchial cavity, in cardiorespiratory chemoreflexes (Hughes and Shelton, 1962; Saunders and Sutterlin, 1971; Butler *et al.*, 1977; Burleson and Smatresk, 1990a; McKenzie *et al.*, 1991b; Sundin *et al.*, 1999, 2000; Reid *et al.*, 2000; Milsom *et al.*, 2002). With the possible exception of air-breathing fish, however, there is little experimental support for a role for central CO_2 chemosensitivity in fish equivalent to that found in tetrapod vertebrates (see later) (see reviews by Smatresk, 1994; Milsom, 1995a, 2002; Gilmour, 2001; Remmers *et al.*, 2001).

3.1. Chemoreceptor Location and Orientation

3.1.1. Oxygen

The majority of studies has focused on identifying the anatomical location of O_2-sensitive chemoreceptors, a focus that is not surprising in view of the primary O_2-keyed ventilatory drive in water breathers (see reviews by Burleson *et al.*, 1992; Fritsche and Nilsson, 1993; Burleson, 1995; Milsom *et al.*, 1999; Gilmour, 2001; Perry and Gilmour, 2002; Burleson and Milsom, 2003). These studies have relied largely on indirect approaches to localize O_2-sensitive chemoreceptors, including the evocation of hypoxic cardiorespiratory responses—particularly bradycardia and hyperventilation since these are most readily measured—by selective application of hypoxic (Saunders and Sutterlin, 1971; Daxboeck and Holeton, 1978; Smith and Jones, 1978) or pharmacological stimuli, such as the potent O_2-chemoreceptor stimulant cyanide (Eclancher and Dejours, 1975; Smatresk, 1986; Smatresk *et al.*, 1986; Burleson and Smatresk, 1990b; McKenzie *et al.*, 1995), or the elimination of such responses by the transection of branchial nerves (Hughes and Shelton, 1962; Butler *et al.*, 1977; Smith and Jones, 1978; Smith and Davie, 1984; Fritsche and Nilsson, 1989; Burleson and Smatresk, 1990a; McKenzie *et al.*, 1991b; Hedrick and Jones, 1999; Sundin *et al.*, 1999, 2000) or gill extirpation (Reid and Perry, 2003). Evidence for the presence of at least two functional populations of O_2 chemoreceptors in the gills has been provided by such studies. One population responds preferentially to stimuli delivered in the water flow across the gills, whereas the second is activated by the arrival of stimuli via the circulation. It is not yet clear whether these functional populations are represented by distinct cells or clusters of cells, as it is conceivable that a receptor cell located appropriately within the multilayered gill epithelium could be capable of sensing both water and blood O_2 tensions (Perry and Gilmour, 2002). Additionally, subpopulations may exist, particularly within the internally oriented chemoreceptor population where separate chemosensitive sites on the afferent and efferent side of the gill circulation could provide information on, respectively, venous O_2 tensions, which are likely to reflect metabolic demands, and arterial O_2 tensions, which are more strongly influenced by external O_2 availability (Perry and Gilmour, 2002).

Functionally, however, the two O_2 chemoreceptor populations can be distinguished through the selective application of stimuli in the ventilatory water versus the circulatory system. Thus, individually, either administration of cyanide into the inspired water to activate externally oriented chemoreceptors or injection of cyanide into the circulation to stimulate internally oriented receptors is sufficient stimulus to trigger bradycardia and/or hyperventilation in a variety of species (Smatresk, 1986; Burleson and Smatresk, 1990b; McKenzie *et al.*, 1995; Sundin *et al.*, 1999, 2000; Reid and Perry, 2003) as well as catecholamine mobilization in rainbow trout (Reid and Perry, 2003). Similarly, in experiments in which the O_2 tensions of water flowing over the two sets of gills were regulated independently, bradycardia was induced by exposure of one set of gills to hypoxic water even though arterial O_2 tension was maintained above normoxic values by irrigation of the contralateral gills with hyperoxic water (Smith and Jones, 1978). Hypoxemia in the absence of environmental hypoxia also elicits cardiorespiratory adjustments. Hyperventilation occurs when blood O_2 content is reduced by exposure to carbon monoxide (Holeton, 1971; Soncini and Glass, 2000) or the induction of anemia (Wood *et al.*, 1979; Smith and Jones, 1982; although, not always, see Cameron and Davis, 1970), but is accompanied by increases in heart rate, stroke volume, and consequently, cardiac output (Holeton, 1971; Wood *et al.*, 1979). The absence of bradycardia, coupled with the fact that blood O_2 tension is maintained or elevated (Holeton, 1971;

Wood *et al.*, 1979; Smith and Jones, 1982), raises questions both about the nature of the proximate stimulus, partial pressure or concentration, for O_2-sensitive chemoreceptor activation, and the association of specific physiological responses with particular receptor populations. With respect to the latter point, detailed studies carried out on nonteleost Actinopterygian species, such as sturgeon and bowfin (McKenzie *et al.*, 1991b, 1995) as well as the teleosts channel catfish and rainbow trout (Burleson and Milsom, 1995b; Burleson and Smatresk, 1990b), suggested that externally oriented O_2-sensitive receptors mediated both ventilation and cardiovascular chemoreflexes, but that only ventilatory responses were elicited by activation of internally oriented receptors (Burleson *et al.*, 1992; Burleson, 1995; Milsom *et al.*, 1999; Burleson and Milsom, 2003). However, differences among studies on individual species [e.g., contrasting effects on heart rate of cyanide injection into the circulation were obtained in three different studies on rainbow trout (Eclancher and Dejours, 1975; Burleson and Milsom, 1995b; Reid and Perry, 2003)] and expansion of the number of species for which data are available (Sundin *et al.*, 1999, 2000) cast doubt on the broad applicability of such generalizations—a diversity of response patterns is apparent (Table 3.3).

Table 3.3

The Association of Bradycardia and Hyperventilation with O_2-Sensitive Chemoreceptors Monitoring Blood or Water, on the Basis of NaCN Injection into the Circulation or Inspired Water to Stimulate Internally or Externally Oriented Receptors, Respectively

Species	$\downarrow f_H$	$\uparrow f_V$	$\uparrow V_{AMP}$	References
Oncorhynchus mykiss[a]	E	E/I	E/I	Burleson and Milsom, 1995b
	I	I	NA	Eclancher and Dejours, 1975
	E/I	NA	E/I	Reid and Perry, 2003
Ictalurus punctatus[b]	E	E/I	E/I	Burleson and Smatresk, 1990b
Hoplias malabaricus[c]	E/I	E/I	E/I	Sundin *et al.*, 1999
Colossoma macropomum[d]	E/I	E/I	E/I	Sundin *et al.*, 2000
Acipenser naccarii[e]	E	E/I	E/I	McKenzie *et al.*, 1995
Lepisosteus osseus[f]	I	NS	I	Smatresk *et al.*, 1986
*	NS	I	I	Smatresk, 1986
Amia calva[g]	E	I	E/I	McKenzie *et al.*, 1991b

[a]Rainbow trout; [b]channel catfish; [c]traira; [d]tambaqui; [e]Adriatic sturgeon; [f]longnose gar; [g]bowfin.

$\downarrow f_H$, bradycardia; $\uparrow f_V$, increased ventilation frequency; $\uparrow V_{AMP}$, increased ventilation amplitude.

E, I, and E/I indicate that response was elicited by external application of cyanide, internal application, or both external and internal, whereas NS indicates that neither the response to internal cyanide nor that to external cyanide was significant. NA indicates that the variable in question was not monitored.

*External NaCN not tested.

Attempts to localize to different gill arches O_2 chemoreceptors linked to specific physiological responses using nerve transection approaches further illustrate the wealth of species diversity (Table 3.4). In fish, branches of cranial nerves VII (facial), IX (glossopharyngeal), and X (vagus) carry sen-sory information from the gills (see reviews by Sundin and Nilsson, 2002; Burleson and Milsom, 2003). The pseudobranch or spiracle, where present, is innervated by the facial and glossopharyngeal nerves, and the glossopharyngeal, together with cranial nerve V (trigeminal), also provide afferent innervation to the orobranchial cavity. All gill arches receive innervation from the vagus nerve, while the first gill arch receives additional innervation from the glossopharyngeal. On the basis of the summary presented in Table 3.4, bilateral section of the vagus and glossopharyngeal nerves is, in *most* species, sufficient to abolish or significantly attenuate *most* cardiorespiratory responses to hypoxic stimuli. Moreover, certain species, including trout, coho salmon, cod, and traira exhibit a distinct dependence of the hypoxic bradycardia on the first gill arch. Again, however, it is the diversity of patterns present that is most striking (Table 3.4), particularly in contrast to the highly structured situation that exists in mammals, where cardiorespiratory (particularly ventilatory) responses to hypoxia are dominated by the blood-sensing O_2-sensitive chemoreceptors of the carotid body (González *et al.*, 1992; Marshall, 1994; Milsom, 1998; Taylor *et al.*, 1999), chemoreceptors which are the phylogenetic homologues of the O_2-sensitive receptors on the first gill arch of fish (Burleson and Milsom, 2003). The phylogenetic and functional significance of fish–tetrapod differences is discussed by Burleson and Milsom (2003), but a satisfactory explanation for the interspecific variability in patterns of O_2-sensitive chemoreceptor control remains elusive.

Direct evidence for the existence of O_2-sensitive chemoreceptors in fish gills has been obtained from afferent nerve recordings of fibres that responded to hypoxic stimuli or to the application of NaCN with an increase in discharge frequency. Multiunit recordings of activity in a branch of the glossopharyngeal were collected for isolated, perfused pseudobranch preparations from rainbow trout (Laurent, 1967; Laurent and Rouseau, 1969, 1972; see also Laurent and Dunel-Erb, 1984). Perfusion with hypoxic saline or inclusion of NaCN in the perfusate produced rapid but low-amplitude increases in activity suggestive of the presence of O_2-sensitive cells. Notably, though, there is little evidence from other approaches that the pseudobranch contributes to cardiorespiratory chemoreflexes (Randall and Jones, 1973; Smith and Jones, 1978; Perry and Reid, 2002; Reid and Perry, 2003; but see Hedrick and Jones, 1999). Isolated, perfused first gill arch preparations from tuna and rainbow trout were used to generate more detailed descriptions of single unit afferent activity in the vagus or glossopharyngeal nerves, respectively (Milsom and Brill, 1986; Burleson and Milsom, 1990, 1993, 1995a).

Table 3.4

The Association of Bradycardia and Hyperventilation with Internally and Externally Oriented O_2-Sensitive Branchial and Extrabranchial Chemoreceptors

<table>
<tr><th rowspan="2">Species</th><th colspan="2">$\downarrow f_H$</th><th colspan="2">$\uparrow f_V$</th><th colspan="2">$\uparrow V_{AMP}$</th><th rowspan="2">References</th></tr>
<tr><th>Internal</th><th>External</th><th>Internal</th><th>External</th><th>Internal</th><th>External</th></tr>
<tr><td>Oncorhynchus mykiss[a]</td><td colspan="2">G1</td><td colspan="2">NA</td><td colspan="2">NA</td><td>Smith and Jones, 1978</td></tr>
<tr><td></td><td colspan="2">G1–4, E</td><td colspan="2">NA</td><td colspan="2">G1–4, E</td><td>Reid and Perry, 2003</td></tr>
<tr><td></td><td>G1</td><td>G1</td><td>NA</td><td>NA</td><td>G1</td><td>G2–4, E</td><td></td></tr>
<tr><td>Oncorhynchus kisutch[b]</td><td colspan="2">G1</td><td colspan="2">NA</td><td colspan="2">NA</td><td>Smith and Davie, 1984</td></tr>
<tr><td>Ictalurus punctatus[c]</td><td colspan="2">E</td><td colspan="2">G1–4</td><td colspan="2">G1–4</td><td>Burleson and Smatresk, 1990a</td></tr>
<tr><td></td><td>NA</td><td>G1–4</td><td>NA</td><td>G1–4</td><td>NA</td><td>G1–4</td><td></td></tr>
<tr><td>Gadus morhua[d]</td><td colspan="2">G1</td><td colspan="2">NA</td><td colspan="2">NA</td><td>Fritsche and Nilsson, 1989</td></tr>
<tr><td>Tinca tinca (N = 1)[e]</td><td colspan="2">NA</td><td colspan="2">G1–4</td><td colspan="2">E</td><td>Hughes and Shelton, 1962</td></tr>
<tr><td>Hoplias malabaricus[f]</td><td colspan="2">G1</td><td colspan="2">G1–4</td><td colspan="2">G1–4, E</td><td>Sundin et al., 1999</td></tr>
<tr><td></td><td>G1</td><td>G1–4</td><td>G1</td><td>G1–4</td><td>G1–4</td><td>G1–4</td><td></td></tr>
<tr><td>Colossoma macropomum[g]</td><td colspan="2">G1–4, E</td><td colspan="2">G1–4</td><td colspan="2">E</td><td>Sundin et al., 2000</td></tr>
<tr><td></td><td>G1–4, E</td><td>G1–4</td><td>G1–4</td><td>G1–4, E</td><td>E</td><td>G1–4, E</td><td></td></tr>
<tr><td>Hemitripterus americanus[h]</td><td colspan="2">NS</td><td colspan="2">G1–4, E</td><td colspan="2">G1–4, E</td><td>Saunders and Sutterlin, 1971</td></tr>
<tr><td>Amia calva[i]</td><td>NS</td><td>G1–4</td><td>G1–4, E</td><td>NS</td><td>G1–4, E</td><td>G1–4, E</td><td>McKenzie et al., 1991b</td></tr>
<tr><td>Scyliorhinus canicula[j]</td><td colspan="2">O, S, G1–4</td><td colspan="2">NA</td><td colspan="2">NA</td><td>Butler et al., 1977</td></tr>
</table>

[a]Rainbow trout; [b]coho salmon; [c]channel catfish; [d]Atlantic cod; [e]tench; [f]traira; [g]tambaqui; [h]sea raven; [i]bowfin; [j]dogfish.

$\downarrow f_H$, bradycardia; $\uparrow f_V$, increased ventilation frequency; $\uparrow V_{AMP}$, increased ventilation amplitude.

G1 indicates that the response is largely or solely mediated by chemoreceptors located on the first gill arch; G1–4 or G2–4 indicates that the response is mediated by chemoreceptors located on one or more of the four gill arches; E, O, and S refer to the involvement of extrabranchial chemoreceptors, orobranchial chemoreceptors, and receptors located in the spiracle, respectively. NA indicates that the variable was not monitored; NS indicates that the variable in question does not change in response to hypoxic stimuli. Differentiation between internally and externally oriented chemoreceptors was possible in studies that utilized cyanide injection into the vasculature and inspired water, respectively; where responses to exposure to hypoxia were examined, such distinctions could not be made.

Independent manipulation of the perfusion fluid and the solution bathing the gill arch revealed in both cases the presence of fibres that responded preferentially to one or both of external versus internal stimuli, a finding that is in agreement with the results of *in vivo* studies of cardiorespiratory responses to stimuli presented in the water versus the blood. As with the *in vivo* work, species differences were apparent with O_2-sensitive fibres in tuna being distributed between those that responded to both internal and external hypoxia (65% of those tested), and those that responded only to internal hypoxia (35%; Milsom and Brill, 1986), whereas O_2-sensitive fibres in trout were approximately equally distributed among those that responded only to internal stimuli (37%), only to external stimuli (37%), or to both types of stimuli (26%; Burleson and Milsom, 1993).

Recently, cellular approaches have been used to identify O_2-sensitive cells within the gill epithelium (Mercer *et al.*, 2000; Jonz *et al.*, 2003, 2004). Whole-cell patch-clamp recordings have provided evidence for the presence in neuroepithelial cells derived from zebrafish gill tissue of an O_2-sensitive K^+ channel that closes in response to hypoxia, leading to membrane depolarization (Jonz *et al.*, 2004). O_2-sensitive K^+ channels were first described in the O_2-sensing glomus cells of the mammalian carotid body and have been found in all hypoxia-sensitive neurosecretory cells examined to date, where they are thought to form a core element of the O_2-sensing mechanism (González *et al.*, 1994; Peers and Buckler, 1995; López-Barneo *et al.*, 2001, 2003; Nurse, 2005). Thus, the detection of an O_2-sensitive K^+ channel in zebrafish neuroepithelial cells strongly suggests that these cells function as branchial O_2 chemoreceptors. In mammalian glomus cells, activation of the O_2-sensing mechanism (see later) results in the release of neurotransmitters, in proportion to the intensity of the hypoxic stimulus, and these in turn influence discharge frequency in the associated afferent neuron (González *et al.*, 1992, 1994; Peers and Buckler, 1995; Eyzaguirre and Abudara, 1999; Prabhakar, 2000). A similar series of events presumably occurs in fish O_2 chemoreceptors and will need to be described to definitively support the hypothesis that fish gill neuroepithelial cells (or at least a population of neuroepithelial cells) act as branchial O_2 chemoreceptors. The identification of an O_2-sensitive K^+ channel in zebrafish neuroepithelial cells (Jonz *et al.*, 2004) constitutes an important step in this regard. In addition, the extensive network of innervation to the neuroepithelial cells by neurons of extrinsic origin (i.e., cell bodies not located in the gill) (Dunel-Erb *et al.*, 1982; Bailly *et al.*, 1992; Sundin *et al.*, 1998b; Jonz and Nurse, 2003), the presence in neuroepithelial cells of numerous dense-cored (secretory) vesicles containing a variety of neurochemicals (Dunel-Erb *et al.*, 1982; Bailly *et al.*, 1992; Zaccone *et al.*, 1992, 1995, 1996; Goniakowska-Witalinska *et al.*, 1995; Sundin *et al.*, 1998b; Jonz and Nurse, 2003), and the preliminary report, suggestive of hypoxia-induced exocytosis, of decreased vesicle number and vesicle degranulation on exposure to hypoxia (Dunel-Erb *et al.*, 1982), hold promise in terms of the critical demonstration of hypoxia-induced neurotransmitter release and its impact on sensory neuron activity.

3.1.2. CO_2 AND/OR pH

Fewer studies have examined the location and orientation of CO_2/pH-sensitive chemoreceptors in fish than is the case for O_2 chemoreceptors. However, the available evidence suggests that, at least in some respects, patterns of CO_2 chemoreception in fish are somewhat less diverse than are those for O_2 chemoreceptors. Although the cardiorespiratory responses evoked by exposure to hypercapnia are highly variable (see in an earlier section), the chemoreceptors that initiate cardiorespiratory responses to hypercapnic stimuli seem, in the species investigated to date, to be dominated by externally oriented branchial receptors that are principally activated by changes in water CO_2 tension rather than by accompanying changes in water pH. Preliminary results suggesting the presence of CO_2-sensitive fibres were obtained for an isolated, perfused pseudobranch preparation from rainbow trout. Increases in discharge frequency during perfusion with solutions of elevated CO_2 were observed in multiunit recordings of afferent activity (Laurent, 1967; Laurent and Rouzeau, 1972; see also Laurent and Dunel-Erb, 1984). Apart from these pilot data, however, neither afferent nerve recordings nor cellular characterization have been applied to the identification of branchial CO_2-sensitive chemoreceptors, and therefore evidence for the location, orientation, and stimulus specificity of CO_2-sensitive chemoreceptors in fish relies almost entirely on evidence provided by indirect approaches (i.e., evocation of responses by CO_2 or pH stimuli and their elimination by nerve sectioning or gill extirpation).

The branchial location of CO_2-sensitive chemoreceptors is supported by the elimination or marked attenuation of hypercapnia-induced cardiorespiratory responses by total branchial denervation or gill extirpation in the five species on which such experiments have been carried out, channel catfish (Burleson and Smatresk, 2000), tambaqui (Sundin *et al.*, 2000; Florindo *et al.*, 2004), traira (Reid *et al.*, 2000), dogfish (McKendry *et al.*, 2001), and rainbow trout (Perry and Reid, 2002). Attempts to link specific physiological responses to CO_2-sensitive chemoreceptors in particular locations have been made for three species and suggest that the absence of global trends that characterizes the distribution of O_2-linked cardiorespiratory chemoreceptors will apply equally well to CO_2 (Table 3.5). For example, selective branchial denervation or gill extirpation revealed a distinct dependence of the hypercarbic bradycardia on the first gill arch in trout and tambaqui, but not traira, whereas the ventilation amplitude response to hypercapnia was also initiated by chemoreceptors confined primarily to the

Table 3.5

A Comparison Among Three Teleost Fish of the Association of Specific Cardiorespiratory Responses Evoked by Hypercapnia with CO_2 Chemoreceptors in Different Locations

Cardiorespiratory response	Rainbow trout (*Onochynchus mykiss*)	Tambaqui (*Colossoma macropomum*)	Traira (*Hoplias malabacricus*)
Bradycardia	G1	G1	G2–4
Blood pressure	G2–4 and/or E	E	G1
$\uparrow f_V$	NS	G2–4	G2–4
$\uparrow V_{AMP}$	G1	NS	G2–4, E
Reference	Perry and Reid, 2002	Sundin *et al.*, 2000	Reid *et al.*, 2000

$\uparrow f_V$, increased ventilation frequency; $\uparrow V_{AMP}$, increased ventilation amplitude.

G1 indicates that the response is largely or solely mediated by chemoreceptors located on the first gill arch; G2–4 indicates that the response is mediated by chemoreceptors located on one or more of the remaining gill arches; E refers to the involvement of extrabranchial chemoreceptors; NS indicates that the variable in question does not change in response to hypercapnia. In trout, dorsal aortic blood pressure was monitored; in tambaqui and traira, ventral aortic blood pressure was monitored.

first gill arch in trout, but not tambaqui or traira (Reid *et al.*, 2000; Sundin *et al.*, 2000; Perry and Reid, 2002). As with O_2-sensitive chemoreceptors, the basis of the interspecific variability in the distribution of CO_2-sensitive chemoreceptors and the cardiorespiratory responses they initiate is unclear.

The apparently exclusively external orientation of branchial CO_2-sensitive chemoreceptors contrasts with the initiation of O_2 chemoreflexes by changes in water and/or blood O_2 levels but is supported by several lines of evidence. The injection of CO_2-enriched water into the inspired water stream to preferentially stimulate externally oriented receptors elicits CO_2 tension-dependent (e.g., Figure 3.4; see also Gilmour *et al.*, 2005) cardiorespiratory responses that are typical of hypercapnia, including bradycardia, increased systemic resistance, changes in blood pressure, and hyperventilation, depending on the species examined (Perry and McKendry, 2001; Perry and Reid, 2002; Gilmour *et al.*, 2005). These CO_2-specific responses are not obtained when a bolus of CO_2-enriched saline is injected into the venous circulation to preferentially stimulate internally oriented receptors; minor cardiovascular adjustments are often observed in such experiments but can be attributed to volume loading because they are also detected following the injection of air-equilibrated saline (Perry and McKendry, 2001; Perry and Reid, 2002; Gilmour *et al.*, 2005). In addition to the divergent responses to the presentation of CO_2 stimuli in water versus blood, cardiorespiratory variables are resistant to internal hypercapnia in the absence of elevated water CO_2 tension. Use of the carbonic anhydrase inhibitor acetazolamide to inhibit red blood cell carbonic anhydrase

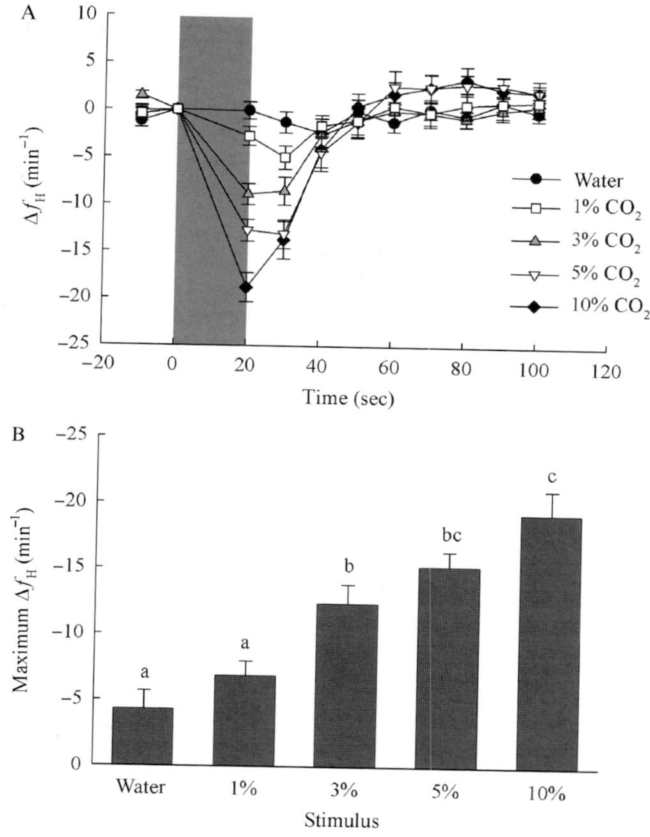

Fig. 3.4. Heart rate responses to the injection of a bolus of air-equilibrated water or CO_2-enriched water (four different levels; nominally 1%, 3%, 5%, and 10% CO_2) into the inspired water stream are presented for tambaqui *Colossoma macropomum* to illustrate the CO_2-dependent nature of changes in cardiorespiratory variables. (A) The data are presented as time courses in which mean heart rate data (Δf_H; $N = 16–17$) were compiled for 10 sec intervals and are plotted for the 10 sec preceding and 80 sec following bolus injection; the shaded area represents the 20 sec period of injection. To allow comparisons across different injections, heart rate data for individual fish were normalized (prior to calculating means) by subtracting from each value the heart rate at time = 0 sec. The mean maximum change in heart rate for injections of air-equilibrated water and water equilibrated with 1%, 3%, 5%, or 10% CO_2 in air are presented in (B). Groups that do not share a letter are significantly different from one another (one-way repeated measures analysis of variance, $P < 0.001$). Values are means ± 1 S.E.M. Data recalculated and redrawn from Gilmour *et al.* (2005).

and promote CO_2 retention (Henry and Heming, 1998; Tufts and Perry, 1998) produces a respiratory acidosis (Hoffert and Fromm, 1973). However, despite a marked internal hypercapnia in which arterial CO_2 tension may increase two- to threefold, ventilatory and cardiovascular responses typical of hypercapnic exposure are not observed unless the acetazolamide-treated fish is exposed to external hypercapnia (McKendry and Perry, 2001; Gilmour *et al.*, 2005; see also Table 3 in Gilmour, 2001). Moreover, under these conditions cardiorespiratory responses track the water CO_2 tension much more closely than that of blood (Figure 3.5; Gilmour *et al.*, 2005). Similar results were obtained for cardiovascular parameters in experiments in which trout were exposed to hyperoxia to elevate internal CO_2 levels in the absence of external hypercapnia (McKendry and Perry, 2001), in this case by lowering

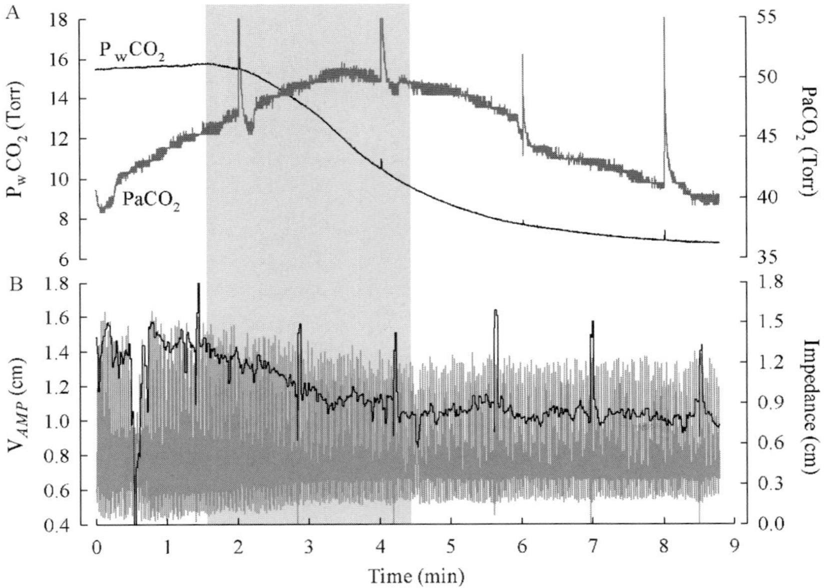

Fig. 3.5. Representative data acquisition traces collected for ventilation parameters, as well as blood and water CO_2 tension for tambaqui (*Colossoma macropomum*) illustrating the lack of correspondence between changes in arterial PCO_2 ($PaCO_2$, A) and ventilation (B) and the tracking of changes in water PCO_2 (P_WCO_2, A) by ventilation. Ventilation is presented as the raw trace from an impedance converter linked to small metal plates sutured to the opercula (gray, B) overlain by the calculated ventilation amplitude (V_{AMP}) trace (black line, B). The gray shaded area emphasizes a time during which blood and water PCO_2 values are moving in opposite directions; ventilation tracks P_WCO_2. [Reproduced from Gilmour *et al.* (2005) with permission from the Company of Biologists.]

the O_2-linked ventilatory drive and hence promoting CO_2 retention (Wood and Jackson, 1980).

These externally oriented hypercapnia-sensitive chemoreceptors appear to respond preferentially to CO_2 rather than to accompanying changes in water pH. Whereas the injection of CO_2-enriched water into the buccal cavity produces cardiorespiratory responses characteristic of hypercapnia, responses to parallel injections of CO_2-free water acidified to the pH of the CO_2-enriched water are at best modest and typically elicited by injection of the most acidic water (Reid *et al.*, 2000; Sundin *et al.*, 2000; Perry and McKendry, 2001; Gilmour *et al.*, 2005). Similarly, fish exposed to acidic water in which PCO_2 remains constant exhibit little or no hyperventilation (Neville, 1979; Thomas and Le Ruz, 1982). More specific localization of branchial CO_2-sensitive chemoreceptors awaits detailed afferent nerve recordings and/or the application of cellular approaches to the identification of CO_2-sensitive cells.

Although the results of experiments designed to directly distinguish between externally and internally oriented CO_2 chemoreceptors establish that changes in water rather than blood CO_2 levels are the key stimuli for CO_2-linked cardiorespiratory chemoreflexes, the existence of internally oriented CO_2- and/or pH-sensitive receptors involved specifically in the regulation of ventilation was suggested by earlier studies (reviewed by Gilmour, 2001). In particular, Wood and Munger (1994) reported a close correspondence between arterial acid–base status and ventilation following exhaustive exercise in rainbow trout. Treatment of trout with carbonic anhydrase to enhance CO_2 excretion attenuated both the extent of the postexercise respiratory acidosis and the normal postexercise hyperventilation with ventilation being more closely correlated to arterial pH than to arterial PCO_2 (Wood and Munger, 1994). Similarly, changes in ventilation in elasmobranch fish exposed to hyperoxia or hypercapnia were closely correlated with changes in arterial pH and to a lesser extent or not at all to changes in arterial PCO_2 (Heisler, 1988; Heisler *et al.*, 1988; Graham *et al.*, 1990; Wood *et al.*, 1990). These data argue for the existence of a ventilatory drive linked to arterial pH. Bolus intra-arterial injections of acid or base, on the other hand, have yielded conflicting results, including no significant effects on ventilation (McKenzie *et al.*, 1991a; Gilmour and Perry, 1996; Reid *et al.*, 2000; Sundin *et al.*, 2000), increases in ventilation amplitude but not frequency (Aota *et al.*, 1990; Gilmour and Perry, 1996), and increases in both amplitude and frequency (but dominated by amplitude changes; Janssen and Randall, 1975; McKenzie *et al.*, 1993). Fewer studies have assessed cardiovascular responses to internal acid–base status. Heart rate and blood pressure were unaffected by intra-arterial acid injections in tambaqui (Sundin *et al.*, 2000), but bradycardia was observed in traira (Reid *et al.*, 2000) and hypertension in bowfin (McKenzie *et al.*, 1991a). In many of these studies, however, blood O_2 status was not monitored, and therefore activation of blood-oriented O_2-chemoreceptor reflexes owing to acidosis-induced decreases in blood O_2 content cannot be ruled out, at least in the teleost fish.

Thus, the role of arterial acid–base status in regulating cardiorespiratory function remains uncertain as do the location and stimulus specificity of any chemoreceptors involved in mediating such responses. By analogy with the peripheral and central CO_2-sensitive chemoreceptive sites in tetrapods (González *et al.*, 1992, 1994; Milsom, 1995a,b, 1998; Nattie, 1999; Taylor *et al.*, 1999; Burleson and Milsom, 2003; Milsom *et al.*, 2004), the two obvious locations for internally oriented CO_2 chemoreceptors in fish are the gills and the brain. Ventilatory responses to hypercapnic exposure in skate were better correlated with arterial pH than with brain intracellular pH or cerebrospinal fluid pH (Wood *et al.*, 1990). However, injections of acid- or CO_2-enriched saline into prebranchial blood were largely without effect on cardiorespiratory function in intact fish used in gill denervation or extirpation studies (Reid *et al.*, 2000; Sundin *et al.*, 2000; Perry and Reid, 2002), making it impossible to implicate internally oriented branchial chemoreceptors. Definitive experimental evidence that demonstrates a role for central CO_2-sensitive chemoreceptors in the regulation of cardiorespiratory function in fish has been equally difficult to obtain. In preliminary experiments on tench, changes in the amplitude and/or frequency of ventilation following the injection of acidic solutions into the posterior region of the medulla provided evidence for the presence of central CO_2/pH-chemosensitive areas (Hughes and Shelton, 1962). Similarly, early work on an isolated lamprey brain preparation indicated that the frequency of respiratory motor output (fictive breathing) was influenced by the pH of the superfusion solution (Rovainen, 1977). However, neither ventilation (including air breathing in bowfin) nor cardiovascular variables were affected by manipulation of the pH or CO_2 tension of solutions used to perfuse the cerebral ventricles or to superfuse the brainstem of, respectively, bowfin and tambaqui (Hedrick *et al.*, 1991; Milsom *et al.*, 2002). Apart from the data for bowfin, studies of bimodal breathers have yielded more support for the presence of central chemosensitivity in fish. Fictive air-breathing frequency was increased when isolated brainstem preparations of gar and *Betta splendens* were superfused with hypercarbic solutions (Wilson *et al.*, 2000; Harris *et al.*, 2001; see also Gilmour, 2001; Remmers *et al.*, 2001), and pulmonary ventilation in lungfish (*L. paradoxa*) increased significantly as the pH of a solution perfusing the IVth cerebral ventricle was lowered (Sanchez *et al.*, 2001b). Taken together, these data suggest that fish that are strictly water breathers lack central CO_2 chemosensitivity, but that CO_2/pH-sensitive chemoreceptors may be present in the brains of species that are bimodal breathers, where they contribute exclusively to the regulation of air breathing. Given the developmental shift in central CO_2 chemosensitivity that occurs in tadpoles as a result of the transition from water breathing (where central CO_2 chemoreceptors stimulate fictive gill ventilation, at least in later stage tadpoles) to air breathing (where fictive lung but not gill ventilation is increased by superfusion of the brainstem preparation with hypercapnic solutions) during metamorphosis (reviewed by Remmers *et al.*, 2001), insight might be gained from an examination of teleost fish species, such as *Arapaima gigasi*, that undergo a similar developmental transition from aquatic to aerial respiration (Brauner *et al.*, 2004).

3.1.3. Do the Same Chemoreceptors Sense O_2 and CO_2?

The glomus cells of the mammalian carotid body sense changes in blood O_2 and CO_2 tensions and pH (González *et al.*, 1992, 1994; Peers and Buckler, 1995; Dasso *et al.*, 2000), as do chemoreceptive cells in carotid homologues in other tetrapods (Smatresk, 1990; Hempleman *et al.*, 1992; Kusakabe, 2002). Similarly, aortic body or aortic arch chemoreceptors in mammals and amphibians, respectively, which are also homologues of fish branchial chemoreceptors (Burleson and Milsom, 2003), respond to both hypoxic and hypercapnic stimuli (Lahiri *et al.*, 1981; Van Vliet and West, 1992; West and Van Vliet, 1993). It would therefore be expected that fish branchial chemoreceptors would detect changes in both O_2 and CO_2, but such a conclusion awaits characterization of chemoreceptor sensitivity to hypoxic and hypercapnic stimuli at the cellular and/or afferent neuron levels. Moreover, the dual sensitivity of tetrapod chemoreceptors to O_2 and CO_2 does not preclude the possibility that additional populations of chemoreceptors responsive to a single stimulus modality are present in fish, particularly given that fish chemoreceptors respond to either water- or blood-borne stimuli, whereas the tetrapod homologues monitor only blood.

Supporting the concept of dual sensitivity to O_2 and CO_2 are the similar although not necessarily identical distributions, in the species for which direct comparisons are possible (Table 3.6), of chemoreceptors involved in mediating cardiorespiratory responses to environmental hypoxia versus hypercapnia. A striking difference, however, is the activation of cardiorespiratory responses by changes in blood O_2 tension (Table 3.3) but not blood CO_2 tension (McKendry and Perry, 2001; Perry and McKendry, 2001; Perry and Reid, 2002; Gilmour *et al.*, 2005), an observation that argues for the existence of at least two branchial chemoreceptor populations, one that senses changes in both O_2 and CO_2 and one that is sensitive only to hypoxic stimuli. In addition, the lack of an additive effect of hypoxic and hypercapnic stimuli on cardiorespiratory chemoreflexes in tambaqui is inconsistent with carotid chemoreceptor-mediated O_2–CO_2 stimulus interaction effects on ventilation

Table 3.6

A Comparison of the Distribution of Chemoreceptors Involved in Mediating Selected Cardiorespiratory Responses to Hypercapnia Versus Hypoxia in Several Teleost Species

Species	Variable	Hypoxia (external NaCN injection)	Hypercapnia (CO_2-enriched water injection)
Rainbow trout (*Onochynchus mykiss*)	Bradycardia	>G1[a], G1[b] (G1[a])	G1[c] (G1[c])
	$\uparrow V_{AMP}$	>G1[a] (>G1[a])	G1[c] (G1[c])
	Catecholamine mobilization	>G1[a] (G1[a])	>G1[c] (>G1[c])
Channel catfish (*Ictalurus punctatus*)	Bradycardia	E[d]	NS[e]
Tambaqui (*Colossoma macropomum*)	Bradycardia	G4, E[f]	G1[f]
	$\uparrow f_V$	G4[f]	G4[f]
	$\uparrow V_{AMP}$	E[f]	NS[f]
Traira (*Hoplias malabaricus*)	Bradycardia	G1[g]	G4[h]
	$\uparrow f_V$	G4[g]	G4[h]
	$\uparrow V_{AMP}$	G4, E[g]	G4, E[h]

[a]Reid and Perry, 2003; [b]Smith and Jones, 1978; [c]Perry and Reid, 2002; [d]Burleson and Smatresk, 1990a; [e]Burleson and Smatresk, 2000; [f]Sundin *et al.*, 2000; [g]Sundin *et al.*, 1999; [h]Reid *et al.*, 2000.

$\uparrow f_V$, increased ventilation frequency; $\uparrow V_{AMP}$, increased ventilation amplitude.

G1 indicates that denervation of the first gill arch eliminated the response; >G1 indicates that denervation of the first gill arch alone (the only condition tested in studies where this symbol is reported) was insufficient to eliminate the response; G4 indicates that total branchial denervation was required to eliminate the response; E indicates the involvement of extrabranchial receptors; NS indicates that the response in intact animals was not significant. Ventilation data for channel catfish are not reported because the only conditions tested were intact and total branchial denervation, which was sufficient to abolish all ventilatory responses to both hypoxia and hypercapnia; it was not possible to look for different chemoreceptor distributions with this all-or-none type of experimental design. A comparison of responses to the injection of hypoxic (NaCN) or hypercabic (CO_2-enriched water) stimuli into the inspired water was possible for rainbow trout only; the data are included in parentheses.

in mammals, potentially implying that separate chemoreceptors are responsible for O_2- and CO_2-induced chemoreflexes (Reid *et al.*, 2005). Reid *et al.* (2005) found that while existing respiratory disturbances modulated cardiorespiratory responses to further respiratory challenges in tambaqui, the effects were less than additive, that is, aquatic hypercapnia was without effect or attenuated chemorespiratory responses to NaCN, and aquatic hypoxia blunted CO_2-evoked ventilation and heart rate responses. By contrast, ventilatory (Lahiri and DeLaney, 1975a) and neural (Fitzgerald and Parks, 1971; Lahiri and DeLaney, 1975a,b; Fitzgerald and Dehghani, 1982; Roy *et al.*, 2000) responses to hypercapnic stimuli in mammals are potentiated by hypoxia, and vice versa, an effect that is thought to arise at least in part through interactive effects of O_2 and CO_2 stimuli within the glomus cell (Dasso *et al.*, 2000; Roy *et al.*, 2000).

3.2. Morphology of (Presumptive) Chemosensory Cells

The neuroepithelial cells of the gill epithelium have long been implicated as branchial O_2 chemoreceptors (see reviews by Dunel-Erb *et al.*, 1982; Bailly *et al.*, 1992; Zaccone *et al.*, 1992; Sundin *et al.*, 1998b; Jonz and Nurse, 2003; Jonz *et al.*, 2003, 2004; Saltys *et al.*, 2006) owing to the common embryological origin (the neural crest) and morphological traits that they share with chemoreceptor cells in other vertebrates, including mammalian carotid body glomus cells and neuroepithelial body cells, as well as their favorable location for the monitoring of either or both aquatic and blood gas levels (see reviews by Laurent, 1984; Burleson *et al.*, 1992; Zaccone *et al.*, 1994, 1995, 1997, 1999; Burleson, 1995; Burleson and Milsom, 2003). Neuroepithelial cells have been described in the gills of all fish species examined to date, where they are found scattered along the length of the filamental epithelium on all filaments (Figure 3.6), but particularly in the distal regions of the filament and on the leading edge (i.e., facing the incoming ventilatory flow of water) (Dunel-Erb *et al.*, 1982; Bailly *et al.*, 1992; Goniakowska-Witalinska *et al.*, 1995; Zaccone *et al.*, 1996, 1997; Sundin *et al.*, 1998a,b; Jonz and Nurse, 2003; Saltys *et al.*, 2006). Within the filamental epithelium, neuroepithelial cells may be located in the outer epithelial layers, having direct contact with the ventilatory water flow (open-type cells), or on the basal lamina, separated from the ventilatory flow by one or more cell layers (closed-type cells) but in closer proximity to the vascular compartment (Dunel-Erb *et al.*, 1982; Bailly *et al.*, 1992; Zaccone *et al.*, 1992; Goniakowska-Witalinska *et al.*, 1995). Neuroepithelial cells have also been described in the lamellar epithelium of zebrafish and bowfin (Goniakowska-Witalinska *et al.*, 1995; Jonz and Nurse, 2003; Saltys *et al.*, 2006). In zebrafish, the lamellar neuroepithelial cells are of the open type, are smaller than the filamental neuroepithelial cells, and are present

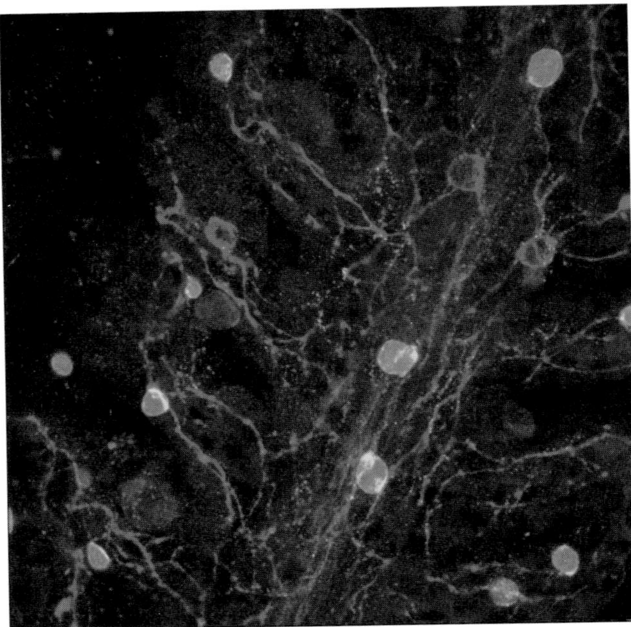

Fig. 3.6. An immunofluorescent image of a portion of a filament and several lamellae from a zebrafish (*Danio rerio*) gill prepared as a whole mount and viewed by confocal microscopy. Neuroepithelial cells were identified by the presence of positive immunoreactivity against serotonin (green) and/or the synaptic vesicle protein SV2 (red); colocalization of serotonin and SV2 is indicated by a yellow-orange signal. Innervation of the neuroepithelial cells was investigated using an antiserum against a zebrafish neuron-specific surface antigen (zn-12; red)—all neuroepithelial cells, regardless of the presence of serotonin immunoreactivity or location (filament or lamella), received innervation from a plexus of zn-12 immunoreactive nerve fibres. [Reproduced from Jonz and Nurse (2003) with permission from Wiley-Liss.] (See Color Insert.)

primarily in the proximal lamellae of all gill arches. Regardless of location, however, neuroepithelial cells receive a rich and often complex innervation (Dunel-Erb *et al.*, 1982; Bailly *et al.*, 1992; Sundin *et al.*, 1998b; Jonz and Nurse, 2003; Saltys *et al.*, 2006).

Morphologically, neuroepithelial cells exhibit a well-developed Golgi apparatus and endoplasmic reticulum but are identified by the presence of numerous dense-cored vesicles that may be congregated near the basal lamina and/or nerve profiles (Dunel-Erb *et al.*, 1982; Bailly *et al.*, 1992; Goniakowska-Witalinska *et al.*, 1995). Zebrafish, goldfish, and trout neuroepithelial cells are immunopositive for SV2, a synaptic vesicle protein that is found in neuronal and endocrine cells, supporting the premise that the dense-cored vesicles are secretory (Jonz and Nurse, 2003; Saltys *et al.*, 2006). Histochemical analysis (specifically the Falck-Hillarp method, the emission of fluorescence following formaldehyde treatment) indicated that branchial neuroepithelial cells contain significant concentrations of monoamines, particularly serotonin (Dunel-Erb *et al.*, 1982). The presence of serotonin was confirmed by immunohistochemical analysis (Bailly *et al.*, 1992; Zaccone *et al.*, 1992, 1997; Goniakowska-Witalinska *et al.*, 1995; Sundin *et al.*, 1998a,b; Jonz and Nurse, 2003; Saltys *et al.*, 2006), although non-serotonin-containing neuroepithelial cells have also been identified in the zebrafish and goldfish gills on the basis of SV2 immunoreactivity (Jonz and Nurse, 2003; Saltys *et al.*, 2006). Even among serotonin-containing neuroepithelial cells, it is likely that different subtypes occur. For example, Jonz and Nurse (2003) described size differences between filamental and lamellar neuroepithelial cells as well as a subpopulation of lamellar neuroepithelial cells that was immunopositive for a purine receptor. Moreover, neuroepithelial cells exhibit variable immunoreactivity for other neurochemicals, including leu-5-enkephalin, met-5-enkephalin, and endothelin (Zaccone *et al.*, 1992, 1996; Goniakowska-Witalinska *et al.*, 1995). The neurochemical content of fish neuroepithelial cells certainly warrants further investigation given the wide range and interspecific variability of biogenic amines, neuropeptides, and neurotransmitters that have been localized to mammalian chemoreceptor cells (González *et al.*, 1994; Nurse, 2005). From the point of view of CO_2 chemosensitivity, branchial neuroepithelial cells should also be assessed for the presence of carbonic anhydrase, an enzyme that plays a key role in the mammalian CO_2 chemotransduction mechanism and is present in glomus cells (Nurse, 1990; Iturriaga *et al.*, 1991; Iturriaga, 1993; González *et al.*, 1994; Peers and Buckler, 1995; Lahiri and Forster, 2003; Zhang and Nurse, 2004).

Additional evidence that branchial neuroepithelial cells serve as O_2 chemoreceptors has been generated by studies of neuroepithelial cell plasticity and development in zebrafish. Acclimatization of adult zebrafish to hypoxic water was accompanied by increases in neuroepithelial cell size and number (Jonz *et al.*, 2004), changes that are similar to those observed in glomus cells of the carotid body in mammals exposed to chronic hypoxia and that are thought, through mechanisms that remain unclear, to contribute to the enhanced sensitivity of the carotid chemoreceptors to hypoxia under these conditions (Wang and Bisgard, 2002). In addition, a substantial enhancement of the hyperventilatory response to hypoxia in developing zebrafish, together with the appearance of quinidine sensitivity, were coincident with the emergence of consistent innervation of filamental neuroepithelial cells (Jonz and Nurse, 2005). Quinidine is a specific blocker of the O_2-sensitive K^+ channel and therefore mimics the effects of hypoxia (Jonz *et al.*, 2004), so the finding that quinidine sensitivity of the hyperventilatory response develops at the time that neuroepithelial cells become consistently innervated suggests a causal relationship and implicates the neuroepithelial cells as the sensory site in a hypoxia-activated ventilatory chemoreflex (Jonz and Nurse, 2005).

The existence of subpopulations of neuroepithelial cells varying in location, innervation pattern, neurochemical content, and response to hypoxia [only 60% of neuroepithelial cells examined by Jonz *et al.* (2004) responded to hypoxia] lends support to the hypothesis that multiple populations of chemoreceptors showing distinct patterns of stimulus sensitivity (O_2 vs. CO_2), orientation (blood vs. water), and evoked response (ventilation vs. cardiovascular variables) are present in the gills of fish.

3.2.1. Insight from the Development of Chemoreceptors, Their Plasticity, and Consequences on Cardiorespiratory Responses

Owing to short generation times, transparent embryos and larvae, and easy access to large number of externally fertilized eggs, the zebrafish (*D. rerio*) is emerging as a powerful model organism to study the ontogeny and plasticity of chemoreceptor-mediated cardiorespiratory responses (Pelster, 2002). Although hyperventilatory or behavioral responses to hypoxia can be observed as early as 2 days postfertilization (dpf) in zebrafish, it is not until the neuroepithelial cells of the gill filament are fully innervated at 7 dpf that maximal ventilatory responses to hypoxia are elicited (Jonz and Nurse, 2005). The zebrafish cardiac M_2 muscarinic receptor is able to initiate bradycardia in response to cholinergic agonists as early as 3 dpf (Hsieh and Liao, 2002), well before the full maturation of the branchial neuroepithelial cells. Thus, if dependent on fully functional neuroepithelial cells, hypoxic bradycardia may only occur several days after the maturation of the cardiac M_2 muscarinic receptor. Peripheral vasoconstriction, often observed during hypoxia or hypercapnia, can be elicited by α-adrenoreceptor agonists at 8 dpf (Bagatto, 2005). Thus, maturation of the α-adrenoreceptor appears to coincide closely with the maturation of a functional neuroepithelial cell. The rate at which cardiovascular control mechanisms develop can be influenced by environmental factors including water oxygenation status, temperature, and current velocity (Bagatto, 2005). For example, the development of adrenergic tachycardia and peripheral vasoconstriction are accelerated by hypoxia. It is not clear, however, whether the development of the branchial chemoreceptors controlling these functions is affected in a similar manner; this area clearly warrants further investigation.

The developmental plasticity of respiratory control in the zebrafish was investigated in a study (Vulesevic and Perry, 2006) by exposing fish to hypoxia, hyperoxia, or hypercapnia during the first 7 dpf, and then assessing normal breathing patterns as well as acute responses to ventilatory stimuli in the adult fish. The results clearly demonstrated that chemoreceptor-mediated responses in adult fish could be profoundly affected by the rearing environment. For example, the respiratory responses of fish reared in hyperoxic water to acute hypoxia, hypercapnia, or external cyanide were blunted (hypoxia, cyanide) or eliminated (hypercapnia). Future studies should attempt to link the plasticity of these ventilatory responses to changes in chemoreceptor presence and/or function.

Adult fish also are capable of exhibiting chemoreceptor plasticity that can influence cardiorespiratory responses. For example, exposure of zebrafish for 28 days to hyperoxic water ($P_WO_2 = 350$ mm Hg) caused a blunting of the ventilatory responses to acute hypoxia or hypercapnia (Vulesevic *et al.*, 2006) that was associated with a significant reduction in the density of gill filament neuroepithelial cells. Although long-term (60 day) exposure of zebrafish to hypoxia ($P_WO_2 = 35$ mm Hg) caused hypertrophy of gill filament neuroepithelial cells in zebrafish (Jonz *et al.*, 2004), the ventilatory response of the fish to acute hypoxia (at least after 28 days) was actually blunted (Vulesevic *et al.*, 2006). These findings are in marked contrast to the results of Burleson *et al.* (2002), who demonstrated that prior exposure of channel catfish to moderate hypoxia for 7 days markedly increased the ventilatory response to acute severe hypoxia.

Although relatively few investigations of chemoreceptor and chemoreflex development and plasticity have been carried out to date, the approaches show promise of being powerful tools both for linking specific cardiorespiratory responses to chemoreceptor function, and potentially, for gaining insight into the extent of and reasons for intra- and interspecific variability in chemoreceptor-mediated cardiorespiratory reflexes.

3.3. Chemotransduction Mechanisms

While not without controversy and unresolved questions, hypoxia chemotransduction in mammalian glomus is widely thought to involve the inhibition of an O_2-sensitive K^+ channel leading to membrane depolarization that, in turn, promotes Ca^{2+} entry via voltage-gated Ca^{2+} channels. Finally, the rise in intracellular Ca^{2+} concentration triggers neurosecretion with the released neurotransmitters modulating action potential frequency in adjacent afferent neurons (González *et al.*, 1992, 1994; Peers and Buckler, 1995; Prabhakar, 2000; López-Barneo *et al.*, 2003; Nurse, 2005). A number of different hypoxia-sensitive K^+ channels exist and are variable in occurrence among species, chemoreceptor sites (i.e., carotid body, aortic body, lung neuroepithelial bodies), and even within individual chemosensory cells (Prabhakar, 2000; Patel and Honore, 2001; López-Barneo *et al.*, 2003).

An outstanding question includes the molecular identity of the O_2 sensor and the signaling pathway linking this sensor to closure of K^+ channels with O_2-sensitive channel subunits, the mitochondrial electron transport chain, and a membrane-associated heme protein (e.g., NADPH oxidase, hemoxygenase-2) being favored hypotheses (Prabhakar, 2000; López-Barneo *et al.*, 2001, 2003; Patel and Honore, 2001; Kummer and Yamamoto, 2002; Williams *et al.*, 2004; Wyatt and Buckler, 2004). The inhibition of a K^+ channel is also thought to form the basis of hypercapnia and acid chemotransduction in mammalian chemoreceptor cells. In this case, K^+ channels, which may be different from those involved in O_2 sensing, close in response to the lowering of intracellular pH, either by the (extracellular) acid stimulus itself or as a result of the intracellular conversion of CO_2 (during hypercapnia) to H^+ catalyzed by carbonic anhydrase (Iturriaga, 1993; González *et al.*, 1994; Peers and Buckler, 1995; Lahiri and Forster, 2003; Putnam *et al.*, 2004; Nurse, 2005). Molecular CO_2 may also have direct effects on membrane potential (Summers *et al.*, 2002; Putnam *et al.*, 2004). As with the transduction of hypoxia stimuli, the resultant membrane depolarization causes Ca^{2+}-dependent neurosecretion. The similarity of the hypoxia and hypercapnia/acid chemotransduction mechanisms, coupled with the fact that both sensory modalities are present within an individual cells, provides a mechanism for the potentiation of the response to hypercapnic stimuli by hypoxia and vice versa—the increase in Ca^{2+} current and intracellular Ca^{2+} concentration is significantly greater when both stimuli are applied simultaneously than in the presence of either stimulus individually (Dasso *et al.*, 2000; Summers *et al.*, 2002; see also Lahiri and Forster, 2003; Putnam *et al.*, 2004). The nature of the neurotransmitter or neurotransmitters responsible for converting the glomus cell receptor potential generated by hypoxic or hypercapnic/acidic stimuli into an excitatory postsynaptic response remains unclear, although corelease of ATP and acetylcholine seems likely, at least for the rat carotid body (Fitzgerald *et al.*, 1997; Zhang *et al.*, 2000; Zhang and Nurse, 2004; reviewed by Prabhakar, 2000; Iturriaga and Alcayaga, 2004; Spyer *et al.*, 2004; Nurse, 2005).

The above-mentioned models of hypoxia and hypercapnia/acid stimuli chemotransduction by mammalian glomus cells provide a road map for determining the signal transduction mechanisms present in fish branchial chemoreceptors. As yet, however, only a few steps have been taken along this path. The evidence (Figure 3.7), reported by Jonz and colleagues (Jonz *et al.*, 2004), supporting the presence of an O_2-sensitive K^+ channel in zebrafish gill neuroepithelial cells is arguably the most significant advance to date. Zebrafish neuroepithelial cells expressed a K^+ current under voltage-clamp conditions that decreased in a reversible, O_2-dependent fashion during hypoxia and was sensitive to quinidine, a blocker of mammalian

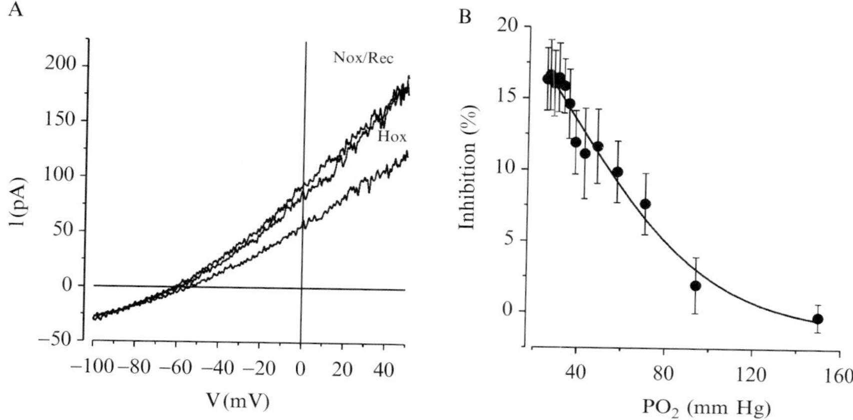

Fig. 3.7. Whole-cell voltage-clamp recording of an O_2-sensitive current in isolated neuroepithelial cells of the zebrafish gill. (A) A whole-cell recording from an O_2-sensitive neuroepithelial cell showing the current–voltage (I–V) relationship during exposure to normoxia (Nox, 150 Torr), hypoxia (Hox, 25 Torr), and after recovery in normoxia (Rec) is depicted. Currents were evoked by changing the voltage from −100 to +50 mV, following a ramp protocol, from a holding potential of −60 mV. Calculation of the O_2-sensitive current (i.e., the difference between normoxic and hypoxic currents) indicated that it reversed near E_K and thus is carried predominantly by K^+ ions (I_K). The relationship between PO_2 and the extent of inhibition of the O_2-sensitive I_K by hypoxia is illustrated in (B). At a PO_2 of 25 Torr, the O_2-sensitive I_K was reduced by 16.6%. Values are means ± standard error of the mean for eight cells. [Reproduced from Jonz *et al.* (2004) with permission from Blackwell.]

O_2-sensitive background or leak K^+ channels (Buckler *et al.*, 2000). In current-clamp recordings, hypoxic stimuli produced membrane depolarization that was associated with a decrease in conductance and that was blocked by quinidine. Taken together, the data suggested the presence of an O_2-sensitive background K^+ channel in zebrafish branchial neuroepithalial cells, and hence a mechanism of O_2 sensing similar to that found in mammalian O_2-sensitive chemoreceptors (Jonz *et al.*, 2004).

Progress has also been made in identifying the neurochemicals that may be involved in translating chemoreceptor stimulation into changes in afferent neuron activity. Acetylcholine and nicotine strongly stimulated discharge frequency in afferent nerve recordings of an isolated, perfused first gill arch preparation from rainbow trout (Figure 3.8), producing responses similar to those evoked by hypoxia or NaCN and suggesting that acetylcholine may act, via nicotinic receptors, as a neurotransmitter in the chemotransduction mechanism of fish chemoreceptors (Burleson and Milsom, 1995a), as in mammals (González *et al.*, 1994; Prabhakar, 2000; Iturriaga and Alcayaga, 2004;

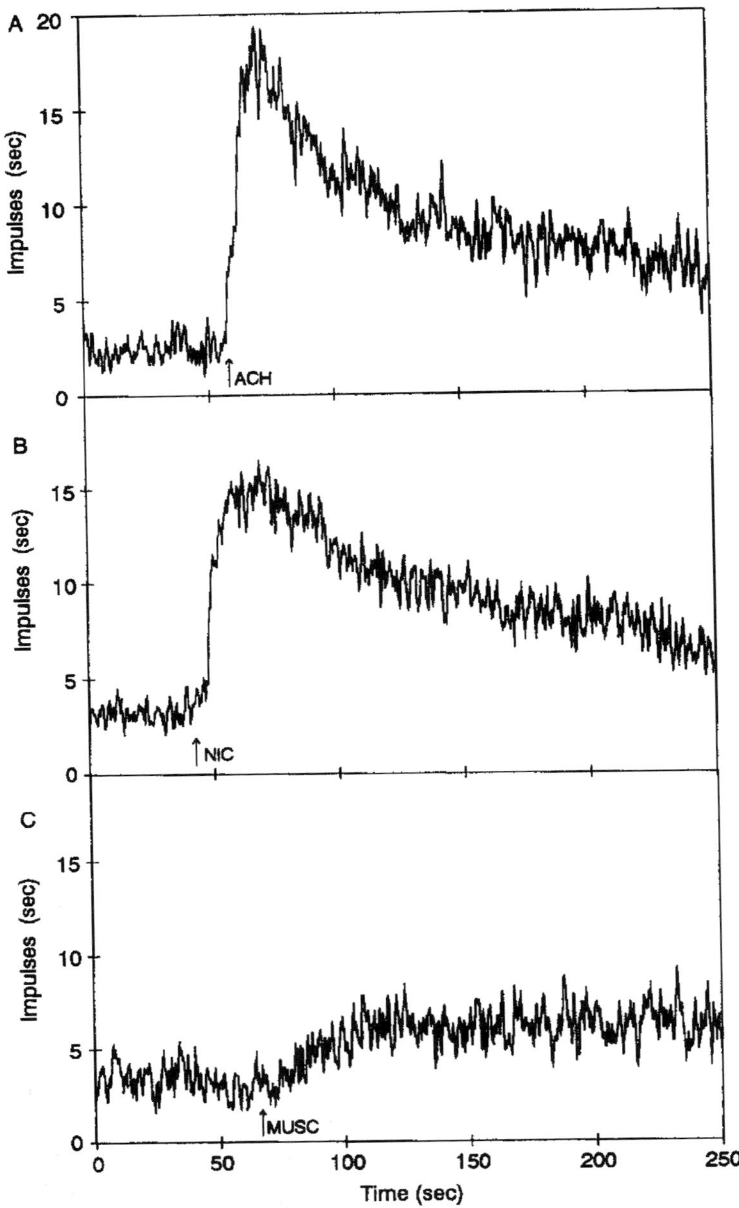

Fig. 3.8. Afferent nerve recordings from the glossopharyngeal nerve of an isolated, saline-perfused first gill arch preparation from rainbow trout. Mean discharge frequencies are presented prior to and following the injection into the perfusate (at the arrow) of a 0.1 mL bolus of (A) acetylcholine (ACH; $N = 8$), (B) nicotine (NIC; $N = 12$), or (C) muscarine (MUSC; $N = 7$). [Reproduced from Burleson and Milsom, (1995a) with permission from Elsevier.]

Nurse, 2005). Neuromodulatory roles for dopamine and serotonin in fish chemoreceptor stimulus transduction were postulated on the basis of the modest, transient responses in afferent nerve activity they elicited (Burleson and Milsom, 1995a). Again, this situation resembles that in mammalian glomus cells, which contain a diversity of neuroactive ligands that exert neuromodulatory effects by acting either on the glomus cell itself or at the sensory nerve endings (González *et al.*, 1994; Peers and Buckler, 1995; Eyzaguirre and Abudara, 1999; Prabhakar, 2000; Iturriaga and Alcayaga, 2004; Nurse, 2005). For example, dopamine is a particularly abundant biogenic amine in mammalian glomus cells that appears to play both inhibitory and excitatory neuromodulatory roles through interactions with, respectively, dopaminergic receptors on the glomus cell and serotonergic receptors on the postsynaptic sensory neuron (González *et al.*, 1994; Iturriaga and Alcayaga, 2004; Nurse, 2005).

In summary, it is apparent that although our knowledge of stimulus transduction mechanisms in mammalian chemoreceptors is far from complete, the characterization of chemotransduction pathways in fish lags far behind (Figure 3.9). Clearly, this area is one on which attention must be focused. Given the apparent conservation of stimulus transduction mechanisms across vertebrates, characterization of the chemotransduction pathways of fish may well provide insight into those of mammalian systems, helping to resolve the current uncertainty and often conflicting views.

4. CENTRAL INTEGRATION AND EFFERENT PATHWAYS

Sensory information from branchial chemoreceptors is transmitted to the brain in the afferent fibres of the glossopharyngeal and vagus cranial nerves (see in an earlier section). These afferent fibres project in a rostrocaudally ordered sequence to a visceral sensory area within the medulla that comprises a pair of longitudinal zones placed dorsolaterally on either side of the IVth ventricle with approximately one-third of the column length extending posteriorly beyond the caudal apex of the IVth ventricle (obex) (Nieuwenhuys and Pouwels, 1983; Taylor, 1992; Meek and Nieuwenhuys, 1998; Taylor *et al.*, 1999; Sundin *et al.*, 2003b). Little is known of the central pathways through which information from peripheral chemoreceptors is integrated to elicit cardiorespiratory chemoreflexes nor have the neurotransmitters and receptors involved been described in any detail. Some evidence suggests that, as in mammals (Machado *et al.*, 1997; Talman, 1997; Sapru, 2002), the excitatory amino acid glutamate may act as a neurotransmitter at the afferent terminals, exerting effects via ionotropic (ligand-gated ion channel) glutamate receptors including both *N*-methyl-D-aspartate (NMDA) and

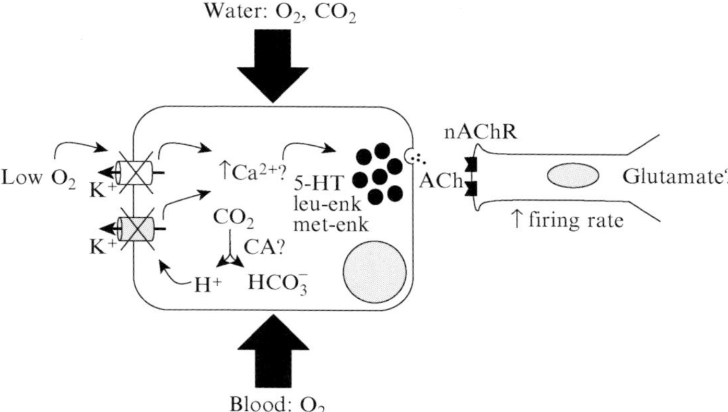

Fig. 3.9. A model of O_2- and CO_2-stimulated excitation of branchial chemoreceptors leading to neurotransmitter release and a consequent increase in the firing rate within an associated afferent nerve fibre. The model for O_2 activation is supported in part by the work of Jonz *et al.* (2004). The model for CO_2 activation is completely speculative, but based on experimentally supported models of CO_2 activation of mammalian carotid glomus cells (Peers and Buckler, 1995; Lahiri and Forster, 2003; Putnam *et al.*, 2004). The presence of serotonin (5-HT), leu-5-enkaphalin and met-5-enkaphalin is supported by experimental data (see text). The suggestion of acetylcholine (ACh) as the primary excitatory neurotransmitter, through its actions on postsynaptic nicotinic cholinergic receptors (nAChR) is supported by the data of Burleson and Milsom (1995a). The possible role of glutamate in neurotransmission at the central projections of afferent neurons is suggested by recent work on shorthorn sculpin and channel catfish (Sundin *et al.*, 2003a,b; Turesson and Sundin, 2003).

non-NMDA receptors. Glutamate immunoreactivity was detected in afferent fibres of the vagal nerve trunk as well as the vagal visceral sensory area within the medulla in shorthorn sculpin (*M. scorpius*; Turesson and Sundin, 2003) but not channel catfish (Sundin *et al.*, 2003a). Similarly, microinjections of glutamate into the vagal sensory area elicited cardiorespiratory responses reminiscent of those elicited by branchial chemoreceptor activation in shorthorn sculpin (Sundin *et al.*, 2003b) but were without effect in channel catfish (Sundin *et al.*, 2003a); Sundin *et al.* (2003b) were also able to use microinjections of glutamate to map different combinations of cardiorespiratory responses to discrete locations within the visceral sensory area, suggesting the existence of distinct reflexogenic zones. Although these findings suggest that glutamate serves as an afferent neurotransmitter in shorthorn sculpin but not channel catfish, NMDA receptor immunoreactivity was detected in the visceral sensory area of both species (Sundin *et al.*, 2003a; Turesson and Sundin, 2003). In addition, microinjection into the visceral sensory area of the broad spectrum glutamate receptor antagonist kynurenic acid eliminated ventilatory responses to hypoxia in channel catfish (Sundin *et al.*, 2003a), while systemic administration of the NMDA receptor-specific antagonist MK-801 blocked some but not all cardiorespiratory responses to hypoxia and NaCN exposure in shorthorn sculpin (Turesson and Sundin, 2003). Taken together, these data provide persuasive evidence for the involvement of glutamatergic pathways in the central processing of sensory input from the branchial chemoreceptors, but clearly, additional studies are required to better define mechanisms of central integration of branchial chemosensory information.

The visceral sensory area to which the chemoreceptor afferent fibres project is in close proximity to the motor nuclei of the trigeminal, facial, glossopharyngeal, and vagal cranial nerves that innervate and drive the respiratory muscles; the sensory and motor zones are separated by the sulcus limitans of His (Nieuwenhuys and Pouwels, 1983; Taylor, 1992; Meek and Nieuwenhuys, 1998; Taylor *et al.*, 1999; Sundin *et al.*, 2003b). Whereas the major ventilatory muscles of the operculum and jaw (forming the "respiratory pump") are innervated primarily by the facial and trigeminal cranial nerves, vagal and glossopharyngeal motor output to the gills may be more involved in the adjustment of filament position and the regulation of the branchial vasculature (Taylor, 1992; Burleson and Smith, 2001; Sundin and Nilsson, 2002). The central pattern generator responsible for generating the rhythmic motor output that initiates the muscle contractions of ventilation is also located in the medulla, probably within the reticular formation, which has both efferent and afferent linkages with the motor zone, as well as higher brain areas (Nieuwenhuys and Pouwels, 1983; Shelton *et al.*, 1986; Taylor, 1992; Meek and Nieuwenhuys, 1998; Taylor *et al.*, 1999). Thus, afferent input from the branchial chemoreceptors and other peripheral sensory receptors (e.g., mechanoreceptors, nociceptors, proprioceptors) must be integrated with the basic output of the central pattern generator and input from higher brain centers to ultimately determine the pattern, frequency, and amplitude of ventilatory movements, effects which are implemented via control of the efferent output from the interconnected motor nuclei of the trigeminal, facial, glossopharyngeal, and vagal cranial nerves (Shelton *et al.*, 1986; Roberts and Ballintijn, 1988; Burleson *et al.*, 1992; Taylor, 1992; Taylor *et al.*, 1999; Burleson and Smith, 2001; Sundin and Nilsson, 2002). The specific pathways, neurotransmitters, and receptors involved remain to be determined (Figure 3.10).

Vagal motor output also has an inhibitory effect on cardiac function that is mediated by the action of acetylcholine released from efferent terminals on muscarinic cholinergic receptors (Taylor, 1992; Taylor *et al.*, 1999). The reflex bradycardia associated with stimulation of branchial chemoreceptors during exposure to hypoxic stimuli (hypoxia or NaCN) or hypercapnia is

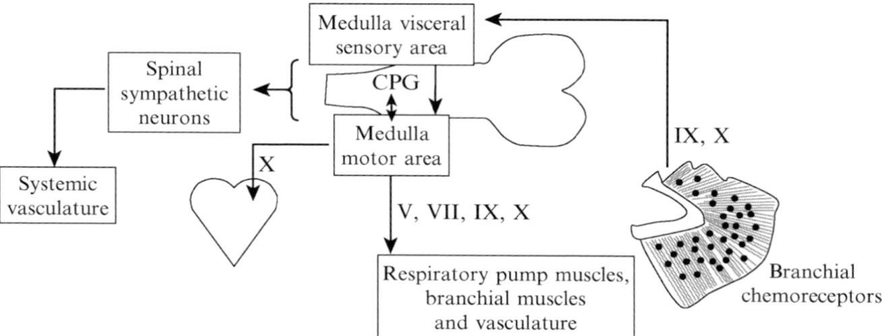

Fig. 3.10. A schematic of branchial chemoreceptor reflex regulation of cardiovascular function in fish. Afferent information from branchial chemoreceptors is transmitted via the glossopharyngeal (IX) and vagal (X) cranial nerves, which project to the visceral sensory area of the medulla. This afferent input is combined with input from the respiratory central pattern generator (CPG) and higher brain centers to influence the activity of motor nuclei for the trigeminal (V), facial (VII), glossopharyngeal and vagal cranial nerves, which regulate the skeletal muscles of the respiratory pump (primarily V and VII; to control ventilation frequency and amplitude) as well as filament position and the branchial vasculature (primarily IX and X; to control branchial resistance). Vagal stimulation of the heart has a cardioinhibitory effect. Central output may also adjust systemic resistance via activation of the spinal sympathetic neurons to the systemic vasculature.

implemented largely or wholly via this efferent vagal pathway (Figure 3.10) since it can be markedly attenuated or eliminated by administration of the muscarinic antagonist atropine (or cardiac vagotomy; Taylor *et al.*, 1977) in both elasmobranch (Butler and Taylor, 1971; Taylor *et al.*, 1977; Kent and Peirce, 1978; Taylor and Barrett, 1985; McKendry *et al.*, 2001) and teleost species (Wood and Shelton, 1980; Peyraud-Waitzenegger and Soulier, 1989; Fritsche and Nilsson, 1990; McKenzie *et al.*, 1995; Perry and Reid, 2002). In the one species, epaulette shark (*H. ocellatum*), in which atropine treatment was without effect on the hypoxic bradycardia, the atypically slow onset of the decrease in heart rate during hypoxia suggested the involvement of a mechanism other than a nervous chemoreflex (Stenslokken *et al.*, 2004). Somewhat unexpectedly, atropine treatment attenuated ventilatory responses to stimulation of branchial chemoreceptors in dogfish (McKendry *et al.*, 2001) and sturgeon (McKenzie *et al.*, 1995). This effect may have been mediated at the afferent limb of the ventilatory reflex, where a likely neuromodulatory role for muscarinic cholinergic receptors in the transduction mechanism of O_2 chemoreception, similar to that in mammals (Iturriaga and Alcayaga, 2004; Nurse, 2005), may be inferred from the observation that muscarine elicited a slowly appearing, moderate increase in afferent discharge frequency in an isolated, perfused first gill arch preparation from rainbow trout (Figure 3.8), but the response of this preparation to hypoxia was not blocked by atropine (Burleson and Milsom, 1995a).

The efferent mechanisms responsible for the activation of vascular resistance and pressor responses to hypoxia and hypercapnia, like the responses themselves (see earlier), appear to vary with both stimulus and species, thereby defying both neat categorization and broad generalizations. In addition, the mechanisms underlying these responses have received less attention than is the case for heart rate. A final difficulty is that blood pressure in fish is regulated through a complex interplay of nervous, hormonal, and local mechanisms that, in many cases, have yet to be fully characterized (Bushnell *et al.*, 1992; Nilsson and Holmgren, 1992; Nilsson, 1994; Olson, 1998; Farrell and Olson, 2000). That having been said, a few mechanisms underlying adjustments of blood pressure and vascular resistance during hypoxia or hypercapnia have been identified (Figure 3.10). Where systemic resistance is increased by hypoxia or hypercapnia in teleost fish (see in an earlier section), often with a resultant rise in dorsal aortic blood pressure, activation of vasoconstrictory α-adrenoreceptors, primarily by increased sympathetic output to the systemic vasculature but with a potential role for circulating catecholamines, appears to be responsible (Wood and Shelton, 1980; Fritsche and Nilsson, 1990; Axelsson and Fritsche, 1991). Thus, treatment with the α-adrenoreceptor antagonist yohimbine prevented systemic resistance from rising during hypercapnia (Perry *et al.*, 1999) or hypoxia (Wood and Shelton, 1980) in rainbow trout. Similarly, inhibition of neuronal noradrenaline release using bretylium in cod (*G. morhua*) significantly reduced the extent of the hypoxia-induced hypertension (Fritsche and Nilsson, 1990; Axelsson and Fritsche, 1991). Elasmobranch fish, in contrast, often exhibit a hypercapnia- or hypoxia-induced hypotension (see in an earlier section) that is not the result of reduced systemic resistance, but rather may be secondary to chemoreflex-induced reductions in heart rate and blood flow because elimination of the hypercapnic or hypoxic bradycardia with atropine also prevented or reduced the extent of the fall in blood pressure (Butler and Taylor, 1971; McKendry *et al.*, 2001). The hypoxic hypotension in dogfish was also eliminated, without relieving the hypoxic bradycardia, by inhibiting nitric oxide synthesis or by scavenging nitric oxide, suggesting that hypoxia-stimulated production of the potent vasodilator nitric oxide may contribute to the hypoxic hypotension (Swenson *et al.*, 2005), perhaps, given the effect of atropine treatment, through a cholinergic mechanism. Finally, in some species (see in an earlier section) hypoxia or hypercapnia causes increases in branchial vascular resistance that are atropine sensitive and point to a vagally mediated constriction of the efferent branchial vasculature, probably the sphincter at the base of the efferent filamental artery (Kent and Peirce, 1978; Sundin and Nilsson, 1997; Stenslokken *et al.*, 2004), although roles for other vasoactive substances cannot be ruled out (Sundin and Nilsson, 1996, 1997; Stenslokken *et al.*, 2004).

5. CONCLUSIONS AND FUTURE DIRECTIONS

Branchial chemoreceptors play a vital role in initiating the reflex responses that adjust cardiorespiratory function to meet the requirements of the fish for O_2 uptake and CO_2 excretion under fluctuating conditions of physiological demand and environment O_2 and CO_2 tensions. Although significant progress has been made in recent years in understanding branchial chemoreceptor regulation of cardiorespiratory function in fish, notably with respect to the identification of CO_2-specific cardiorespiratory reflexes and characterization of the cellular basis of O_2 sensing, our knowledge remains far from complete. A number of questions merit concentrated research effort.

A substantial body of data exists on the ventilatory and cardiovascular responses of different fish species to hypoxia, and to a lesser extent, hypercapnia. In a limited number of cases, attempts have been made to attribute specific cardiorespiratory responses to the activation of particular populations of chemoreceptors. To date, however, unifying patterns have largely failed to emerge from what is at present a very diverse collection of data. The diversity of responses undoubtedly reflects, at least in part, the variety of experimental protocols that have been employed—the severity of the hypoxic or hypercapnic stimulus, the rapidity of exposure to the stimulus, and any prior history of exposure all are likely to affect the response of an individual fish, while phylogeny and environmental adaptation may contribute to interspecific variation. Although it is unreasonable to expect uniformity given the variability among fish species in general, coming to grips with the current diversity will aid in understanding the functional and phylogenetic significance of branchial chemoreceptor regulation of ventilation and cardiovascular variables.

This objective will also be facilitated by the identification of O_2- and CO_2-sensitive chemoreceptor cells within the gills, and by the characterization of the location and orientation of populations of these cells. Do the same cells sense O_2 and CO_2? Do the same cells monitor water and blood? Is stimulation of the same cells linked to activation of ventilation and cardiovascular responses? Answers to these questions will not only aid in discerning underlying patterns of cardiorespiratory control by branchial chemoreceptors but may also be of use in describing the phylogenetic transition from water-breathing vertebrates possessing a multitude of widely dispersed chemoreceptors that monitor both ambient and internal environments, to the concentration of peripheral chemoreceptors in a single, dominant site that monitors only blood in birds and mammals (Milsom, 1998, 2002; Burleson and Milsom, 2003). Moreover, once the chemoreceptor cells have been identified, morphometric effects of factors, such as development and acclimation to particular environments, can be investigated as possible contributors to the plasticity of chemoreceptor-evoked responses observed during ontogeny (Jonz and Nurse, 2005) or as a result of exposure to conditions of altered gas tensions.

Finally, information on the mechanisms of chemotransduction underlying the excitation of fish chemoreceptor cells by hypoxic and hypercapnic/acidic stimuli and the neurotransmitters involved in communicating this excitation to sensory neurons and the central nervous system is limited at best. Elucidating these mechanisms is of importance in evaluating the effectiveness of specific stimuli as well as the potential impact of other factors on stimulus effectiveness, and again may provide insight into the phylogeny of O_2 and CO_2/pH chemoreception in vertebrates.

ACKNOWLEDGMENTS

Original research of the authors reported above was supported by NSERC of Canada Discovery and Research Tools and Instruments grants. Thanks are extended to W. K. Milsom and S. G. Reid for reading the chapter, and to M. G. Jonz, C. A. Nurse, M. L. Burleson, and W. K. Milsom for permission to reproduce previously published figures.

REFERENCES

Aota, S., Holmgren, K. D., Gallaugher, P., and Randall, D. J. (1990). A possible role for catecholamines in the ventilatory responses associated with internal acidosis or external hypoxia in rainbow trout *Oncorhynchus mykiss*. *J. Exp. Biol.* **151,** 57–70.

Axelsson, M., and Fritsche, R. (1991). Effects of exercise, hypoxia and feeding on the gastrointestinal blood flow in the Atlantic cod *Gadus morhua*. *J. Exp. Biol.* **158,** 181–198.

Axelsson, M., Farrell, A. P., and Nilsson, S. (1990). Effects of hypoxia and drugs on the cardiovascular dynamics of the Atlantic hagfish *Myxine glutinosa*. *J. Exp. Biol.* **151,** 297–316.

Axelsson, M., Davison, W., Forster, M. E., and Farrell, A. P. (1992). Cardiovascular responses of the red-blooded Antarctic fishes *Pagothenia bernacchii* and *P. borchgrevinki*. *J. Exp. Biol.* **167,** 179–201.

Babiker, M. M. (1979). Respiratory behaviour, oxygen consumption and relative dependence on aerial respiration in the African lungfish (*Protopterus annectens*, Owen) and an air-breathing teleost (*Clarias lazera*, C.). *Hydrobiologia* **65,** 177–187.

Bagatto, B. (2005). Ontogeny of cardiovascular control in zebrafish (*Danio rerio*): Effects of developmental environment. *Comp. Biochem. Physiol. A* **141,** 391–400.

Bailly, Y., Dunel-Erb, S., and Laurent, P. (1992). The neuroepithelial cells of the fish gill filament: Indolamine-immunocytochemistry and innervation. *Anat. Rec.* **233,** 143–161.

Berschick, P., Bridges, C. R., and Grieshaber, M. K. (1987). The influence of hyperoxia, hypoxia and temperature on the respiratory physiology of the intertidal rockpool fish *Gobius cobitis* Pallas. *J. Exp. Biol.* **130,** 369–387.

Booth, J. H. (1979). The effects of oxygen supply, epinephrine, and acetylcholine on the distribution of blood flow in trout gills. *J. Exp. Biol.* **83,** 31–39.

Brauner, C. J., Matey, V., Wilson, J. M., Bernier, N. J., and Val, A. L. (2004). Transition in organ function during the evolution of air-breathing; insights from *Arapaima gigas*, an obligate air-breathing teleost from the Amazon. *J. Exp. Biol.* **207,** 1433–1438.

Buckler, K. J., Williams, B. A., and Honore, E. (2000). An oxygen-, acid- and anaesthetic-sensitive TASK-like background potassium channel in rat arterial chemoreceptor cells. *J. Physiol.* **525,** 135–142.

Burleson, M. L. (1995). Oxygen availability: Sensory systems. *In* "Biochemistry and Molecular Biology of Fishes, 5. Environmental and Ecological Biochemistry" (Hochachka, P. W., and Mommsen, T. P., Eds.), pp. 1–18. Elsevier, Amsterdam.

Burleson, M. L., and Milsom, W. K. (1990). Propranolol inhibits O_2-sensitive chemoreceptor activity in trout gills. *Am. J. Physiol.* **258,** R1089–R1091.

Burleson, M. L., and Milsom, W. K. (1993). Sensory receptors in the first gill arch of rainbow trout. *Respir. Physiol.* **93,** 97–110.

Burleson, M. L., and Milsom, W. K. (1995a). Cardio-ventilatory control in rainbow trout: I. Pharmacology of branchial, oxygen-sensitive chemoreceptors. *Respir. Physiol.* **100,** 231–238.

Burleson, M. L., and Milsom, W. K. (1995b). Cardio-ventilatory control in rainbow trout: II. Reflex effects of exogenous neurochemicals. *Respir. Physiol.* **101,** 289–299.

Burleson, M. L., and Milsom, W. K. (2003). Comparative aspects of O_2 chemoreception. Anatomy, physiology, and environmental applications. *In* "Oxygen Sensing. Responses and Adaptation to Hypoxia" (Lahiri, S., Semenza, G. L., and Prabhakar, N. R., Eds.), pp. 685–707. Marcel Dekker, New York.

Burleson, M. L., and Smatresk, N. J. (1990a). Effects of sectioning cranial nerves IX and X on cardiovascular and ventilatory reflex responses to hypoxia and NaCN in channel catfish. *J. Exp. Biol.* **154,** 407–420.

Burleson, M. L., and Smatresk, N. J. (1990b). Evidence for two oxygen-sensitive chemoreceptor loci in channel catfish, *Ictalurus punctatus*. *Physiol. Zool.* **63,** 208–221.

Burleson, M. L., and Smatresk, N. J. (2000). Branchial chemoreceptors mediate ventilatory responses to hypercapnic acidosis in channel catfish. *Comp. Biochem. Physiol. A* **125,** 403–414.

Burleson, M. L., and Smith, R. L. (2001). Central nervous control of gill filament muscles in channel catfish. *Respir. Physiol.* **126,** 103–112.

Burleson, M. L., Smatresk, N. J., and Milsom, W. K. (1992). Afferent inputs associated with cardioventilatory control in fish. *In* "Fish Physiology, Volume XIIB The Cardiovascular System" (Hoar, W. S., Randall, D. J., and Farrell, A. P., Eds.), pp. 389–426. Academic Press, San Diego.

Burleson, M. L., Carlton, A. L., and Silva, P. E. (2002). Cardioventilatory effects of acclimatization to aquatic hypoxia in channel catfish. *Respir. Physiol. Neurobiol.* **131,** 223–232.

Bushnell, P. G., and Brill, R. W. (1992). Oxygen transport and cardiovascular responses in skipjack tuna (*Katsuwonus pelamis*) and yellowfin tuna (*Thunnus albacares*) exposed to acute hypoxia. *J. Comp. Physiol. B* **162,** 131–143.

Bushnell, P. G., Jones, D. R., and Farrell, A. P. (1992). The arterial system. *In* "Fish Physiology, Volume XIIB The Cardiovascular System" (Hoar, W. S., Randall, D. J., and Farrell, A. P., Eds.), pp. 89–139. Academic Press, New York.

Butler, P. J., and Metcalfe, J. D. (1988). Cardiovascular and respiratory systems. *In* "Physiology of Elasmobranch Fishes" (Shuttleworth, T. J., Ed.), pp. 1–47. Springer-Verlag, Berlin.

Butler, P. J., and Taylor, E. W. (1971). Response of the dogfish (*Syciorhinus canicula* L.) to slowly induced and rapidly induced hypoxia. *Comp. Biochem. Physiol. A* **39,** 307–323.

Butler, P. J., Taylor, E. W., and Short, S. (1977). The effect of sectioning cranial nerves V, VII, IX and X on the cardiac response of the dogfish *Scyliorhinus canicula* to environmental hypoxia. *J. Exp. Biol.* **69,** 233–245.

Cameron, J. N., and Davis, J. C. (1970). Gas exchange in rainbow trout (*Salmo gairdneri*) with varying blood oxygen capacity. *J. Fish. Res. Bd. Canada* **27,** 1069–1085.

Crocker, C. E., Farrell, A. P., Gamperl, A. K., and Cech, J. J., Jr. (2000). Cardiorespiratory responses of white sturgeon to environmental hypercapnia. *Am. J. Physiol.* **279,** R617–R628.

Dasso, L. L. T., Buckler, K. J., and Vaughan-Jones, R. D. (2000). Interactions between hypoxia and hypercapnic acidosis on calcium signaling in carotid body type I cells. *Am. J. Physiol.* **279,** L36–L42.

Davie, P. S., and Daxboeck, C. (1982). Effect of pulse pressure on fluid exchange between blood and tissues in trout gills. *Can. J. Zool.* **60,** 1000–1006.

Daxboeck, C., and Holeton, G. F. (1978). Oxygen receptors in the rainbow trout, *Salmo gairdneri*. *Can. J. Zool.* **56,** 1254–1259.

Desforges, P. R., Harman, S. S., Gilmour, K. M., and Perry, S. F. (2002). The sensitivity of CO_2 excretion to changes in blood flow in rainbow trout is determined by carbonic anhydrase availability. *Am. J. Physiol.* **282,** R501–R508.

Dunel-Erb, S., Bailly, Y., and Laurent, P. (1982). Neuroepithelial cells in fish gill primary lamellae. *J. Appl. Physiol.* **53,** 1342–1353.

Eclancher, B., and Dejours, P. (1975). Contrôle de la respiration chez les poissons téléostéens: Existence de chémorécepteurs physiologiquement analogues aux chémorécepteurs des vertébrés supérieurs. *C. R. Acad. Sc. Paris Série D* **280,** 451–453.

Eyzaguirre, C., and Abudara, V. (1999). Carotid body glomus cells: Chemical secretion and transmission (modulation?) across cell-nerve ending junctions. *Respir. Physiol.* **115,** 135–149.

Farrell, A. P., and Olson, K. R. (2000). Cardiac natriuretic peptides: A physiological lineage of cardioprotective hormones? *Physiol. Biochem. Zool.* **73,** 1–11.

Farrell, A. P., Daxboeck, C., and Randall, D. J. (1979). The effect of input pressure and flow on the pattern and resistance to flow in the isolated perfused gill of a teleost fish. *J. Comp. Physiol.* **133,** 233–240.

Farrell, A. P., Sobin, S. S., Randall, D. J., and Crosby, S. (1980). Intralamellar blood flow patterns in fish gills. *Am. J. Physiol.* **239,** R428–R436.

Fitzgerald, R. S., and Dehghani, G. A. (1982). Neural responses of the cat carotid and aortic bodies to hypercapnia and hypoxia. *J. Appl. Physiol.* **52,** 596–601.

Fitzgerald, R. S., and Parks, D. C. (1971). Effect of hypoxia on carotid chemoreceptor response to carbon dioxide in cats. *Respir. Physiol.* **12,** 218–229.

Fitzgerald, R. S., Shirahata, M., and Ide, T. (1997). Further cholinergic aspects of carotid body chemotransduction of hypoxia in cats. *J. Appl. Physiol.* **82,** 819–827.

Florindo, L. H., Reid, S. G., Kalinin, A. L., Milsom, W. K., and Rantin, F. T. (2004). Cardiorespiratory reflexes and aquatic surface respiration in the neotropical fish tambaqui (*Colossoma macropomum*): Acute responses to hypercarbia. *J. Comp. Physiol. B* **174,** 319–328.

Forster, M. E., Davison, W., Axelsson, M., and Farrell, A. P. (1992). Cardiovascular responses to hypoxia in the hagfish, *Eptatretus cirrhatus*. *Resp. Physiol.* **88,** 373–386.

Fritsche, R. (1990). Effects of hypoxia on blood pressure and heart rate in three marine teleosts. *Fish Physiol. Biochem.* **8,** 85–92.

Fritsche, R., and Nilsson, S. (1989). Cardiovascular responses to hypoxia in the Atlantic cod, *Gadus morhua*. *Exp. Biol.* **48,** 153–160.

Fritsche, R., and Nilsson, S. (1990). Autonomic nervous control of blood pressure and heart rate during hypoxia in the cod, *Gadus morhua*. *J. Comp. Physiol. B* **160,** 287–292.

Fritsche, R., and Nilsson, S. (1993). Cardiovascular and ventilatory control during hypoxia. *In* "Fish Ecophysiology" (Rankin, J. C., and Jensen, F. B., Eds.), pp. 180–206. Chapman & Hall, London.

Fritsche, R., Axelsson, M., Franklin, C. E., Grigg, G. G., Holmgren, S., and Nilsson, S. (1993). Respiratory and cardiovascular responses to hypoxia in the australian lungfish. *Respir. Physiol.* **94,** 173–187.

Gehrke, P. C., and Fielder, D. R. (1988). Effects of temperature and dissolved oxygen on heart rate, ventilation rate and oxygen consumption of spangled perch, *Leiopotherapon unicolor* (Gunther 1859), (Percoidei, Teraponidae). *J. Comp. Physiol. B* **157,** 771–782.

Gilmour, K. M. (2001). The CO_2/pH ventilatory drive in fish. *Comp. Biochem. Physiol. A* **130,** 219–240.

Gilmour, K. M., and Perry, S. F. (1994). The effects of hypoxia, hyperoxia or hypercapnia on the acid-base disequilibrium in the arterial blood of rainbow trout, *Oncorhynchus mykis*. *J. Exp. Biol.* **192,** 269–284.

Gilmour, K. M., and Perry, S. F. (1996). Effects of metabolic acid-base disturbances and elevated catecholamines on the acid-base disequilibrium in the arterial blood of rainbow trout. *J. Exp. Zool.* **274,** 281–290.

Gilmour, K. M., Milsom, W. K., Rantin, F. T., Reid, S. G., and Perry, S. F. (2005). Cardiorespiratory responses to hypercarbia in tambaqui (*Colossoma macropomum*): Chemoreceptor orientation and specificity. *J. Exp. Biol.* **208,** 1095–1107.

Goniakowska-Witalinska, L., Zaccone, G., Fasulo, S., Mauceri, A., Licata, A., and Youson, J. H. (1995). Neuroendocrine cells in the gills of the bowfin *Amia calva*. An ultrastructural and immunocytochemical study. *Folia Histochem. Cytobiol.* **33,** 171–177.

González, C., Almaraz, L., Obeso, A., and Rigual, R. (1992). Oxygen and acid chemoreception in the carotid body chemoreceptors. *Trends Neurosci.* **15,** 146–153.

González, C., Almaraz, L., Obeso, A., and Rigual, R. (1994). Carotid body chemoreceptors: From natural stimuli to sensory discharges. *Physiol. Rev.* **74,** 829–898.

Graham, M. S., Turner, J. D., and Wood, C. M. (1990). Control of ventilation in the hypercapnic skate *Raja ocellata*: I. Blood and extradural fluid. *Respir. Physiol.* **80,** 259–277.

Harris, M. B., Wilson, R. J. A., Perry, S. F., and Remmers, J. E. (2001). Central respiratory pattern generation and CO_2/H^+ sensitivity: New insights from old fish. *FASEB J.* **15,** A151.

Hedrick, M. S., and Jones, D. R. (1999). Control of gill ventilation and air-breathing in the bowfin *Amia calva*. *J. Exp. Biol.* **202,** 87–94.

Hedrick, M. S., Burleson, M. L., Jones, D. R., and Milsom, W. K. (1991). An examination of central chemosensitivity in an air-breathing fish (*Amia calva*). *J. Exp. Biol.* **155,** 165–174.

Heisler, N. (1988). Acid-base regulation. *In* "Physiology of Elasmobranch Fishes" (Shuttleworth, T. J., Ed.), pp. 215–251. Springer-Verlag, Berlin.

Heisler, N., Toews, D. P., and Holeton, G. F. (1988). Regulation of ventilation and acid-base status in the elasmobranch *Scyliorhinus stellaris* during hyperoxia-induced hypercapnia. *Respir. Physiol.* **71,** 227–246.

Hempleman, S. C., Powell, F. L., and Prisk, G. K. (1992). Avian arterial chemoreceptor resopnses to steps of CO_2 and O_2. *Respir. Physiol.* **90,** 325–340.

Henry, R. P., and Heming, T. A. (1998). Carbonic anhydrase and respiratory gas exchange. *In* "Fish Respiration" (Perry, S. F., and Tufts, B. L., Eds.), pp. 75–111. Academic Press, San Diego.

Hoffert, J. R., and Fromm, P. O. (1973). Effect of acetazolamide on some hematological parameters and ocular oxygen concentration in rainbow trout. *Comp. Biochem. Physiol.* **45A,** 371–378.

Holeton, G. F. (1971). Oxygen uptake and transport by the rainbow trout during exposure to carbon monoxide. *J. Exp. Biol.* **54,** 239–254.

Holeton, G. F., and Randall, D. J. (1967). Changes in blood pressure in the rainbow trout during hypoxia. *J. Exp. Biol.* **46,** 297–305.

Hsieh, D. J., and Liao, C. F. (2002). Zebrafish M2 muscarinic acetylcholine receptor: Cloning, pharmacological characterization, expression patterns and roles in embryonic bradycardia. *Br. J. Pharmacol.* **137,** 782–792.

Hughes, B., and Shelton, G. (1962). Respiratory mechanisms and their nervous control in fish. *Adv. Comp. Physiol. Biochem.* **1,** 275–364.

Iturriaga, R. (1993). Carotid body chemoreception: The importance of CO_2-HCO_3^- and carbonic anhydrase. *Biol. Res.* **26,** 319–329.

Iturriaga, R., and Alcayaga, J. (2004). Neurotransmission in the carotid body: Transmitters and modulators between glomus cells and petrosal ganglion nerve terminals. *Brain Res. Rev.* **47,** 46–53.

Iturriaga, R., Lahiri, S., and Mokashi, A. (1991). Carbonic anhydrase and chemoreception in the cat carotid body. *Am. J. Physiol.* **261,** C565–C573.

Janssen, R. G., and Randall, D. J. (1975). The effects of changes in pH and PCO_2 in blood and water on breathing in rainbow trout, *Salmo gairdneri*. *Respir. Physiol.* **25,** 235–245.

Jesse, M. J., Shub, C., and Fishman, A. P. (1967). Lung and gill ventilation of the African lungfish. *Respir. Physiol.* **3,** 267–287.

Johansen, K. (1982). Respiratory gas exchange of vertebrate gills. *In* "Gills" (Houlihan, D. F., Rankin, J. C., and Shuttleworth, T. J., Eds.), pp. 99–128. Cambridge University Press, Cambridge.

Johansen, K., and Lenfant, C. (1968). Respiration in the African lungfish, *Protopterus aetipoicus*. II. Control of breathing. *J. Exp. Biol.* **49,** 453–468.

Jones, D. R., and Randall, D. J. (1978). The respiratory and circulatory systems during exercise. *In* "Fish Physiology" (Hoar, W. S., and Randall, D. J., Eds.), pp. 425–501. Academic Press, London.

Jonz, M. G., and Nurse, C. A. (2003). Neuroepithelial cells and associated innervation of the zebrafish gill: A confocal immunofluorescence study. *J. Comp. Neur.* **461,** 1–17.

Jonz, M. G., and Nurse, C. A. (2005). Development of oxygen sensing in the gills of zebrafish. *J. Exp. Biol.* **208,** 1537–1549.

Jonz, M. G., Fearon, I. M., and Nurse, C. A. (2003). Potential oxygen sensing pathways in the zebrafish gill. *In* "Chemoreception: From Cellular Signalling to Functional Plasticity, Adv. Exp. Med. Biol" (Pequignot, J. M., Gonzalez, C., Nurse, C. A., Prabhakar, N. R., and Dalmaz, Y., Eds.), Vol. 536, pp. 217–223. Kluwer Academic/Plenum Publishers, New York.

Jonz, M. G., Fearon, I. M., and Nurse, C. A. (2004). Neuroepithelial oxygen chemoreceptors of the zebrafish gill. *J. Physiol.* **560,** 737–752.

Julio, A. E., Desforges, P., and Perry, S. F. (2000). Apparent diffusion limitations for carbon dioxide excretion in rainbow trout (*Oncorhynchus mykiss*) are relieved by intravascular injections of carbonic anhydrase. *Respir. Physiol.* **121,** 53–64.

Kent, B., and Peirce, E. C. I. (1978). Cardiovascular responses to changes in blood gases in dogfish shark, *Squalus acanthias*. *Comp. Biochem. Physiol. C* **60,** 37–44.

Kind, P. K., Grigg, G. C., and Booth, D. T. (2002). Physiological responses to prolonged aquatic hypoxia in the Queensland lungfish *Neoceratodus forsteri*. *Respir. Physiol. Neurobiol.* **132,** 179–190.

Kinkead, R., Fritsche, R., Perry, S. F., and Nilsson, S. (1991). The role of circulating catecholamines in the ventilatory and hypertensive responses to hypoxia in the Atlantic cod (*Gadus morhua*). *Physiol. Zool.* **64,** 1087–1109.

Kummer, W., and Yamamoto, Y. (2002). Cellular distribution of oxygen sensor candidates—oxidases, cytochromes, K^+-channels—in the carotid body. *Microsc. Res. Tech.* **59,** 234–242.

Kusakabe, T. (2002). Carotid labyrinth of amphibians. *Microsc. Res. Tech.* **59,** 207–226.

Lahiri, S., and DeLaney, R. G. (1975a). Relationship between carotid chemoreceptor activity and ventilation in the cat. *Respir. Physiol.* **24,** 267–286.

Lahiri, S., and DeLaney, R. G. (1975b). Stimulus interaction in the respones of carotid body chemoreceptor single afferent fibers. *Respir. Physiol.* **24,** 249–266.

Lahiri, S., and Forster, R. E. (2003). CO_2/H^+ sensing: Peripheral and central chemoreception. *Int. J. Biochem. Cell Biol.* **35,** 1413–1435.

Lahiri, S., Mokashi, A., Mulligan, E., and Nishino, T. (1981). Comparison of aortic and carotid chemoreceptor responses to hypercapnia and hypoxia. *J. Appl. Physiol.* **51,** 55–61.

Laurent, P. (1967). La pseudobranchie des téléostéens: Preuves électrophysiologiques de ses fonctions chémoréceptrice et barorécepetrice. *C. R. Acad. Sc. Paris Série D* **264,** 1879–1882.

Laurent, P. (1984). Gill internal morphology. *In* "Fish Physiology, Volume X Gills A" (Hoar, W. S., and Randall, D. J., Eds.), pp. 73–183. Academic Press, London.

Laurent, P., and Dunel-Erb, S. (1984). The pseudobranch: Morphology and function. *In* "Fish Physiology, Volume X Gills B Ion and Water Transfer" (Hoar, W. S., and Randall, D. J., Eds.), pp. 285–323. Academic Press, San Diego.

Laurent, P., and Rouseau, J. D. (1969). Action de la PO_2 sur l'activité nerveuse afférente provenant de la pseudobranchie d'un Téléostéen. *J. Physiol., Paris* **61**(Suppl. 1), 145.

Laurent, P., and Rouzeau, J.-D. (1972). Afferent neural activity from pseudobranch of teleosts. Effects of PO_2, pH, osmotic pressure and Na^+ ions. *Respir. Physiol.* **14,** 307–331.

López-Barneo, J., Pardal, R., and Ortega-Sáenz, P. (2001). Cellular mechanisms of oxygen sensing. *Annu. Rev. Physiol.* **63,** 259–287.

López-Barneo, J., del Toro, R., Levitsky, K. L., Chiara, M. D., and Ortega–Sáenz, P. (2003). Regulation of oxygen sensing by ion channels. *J. Appl. Physiol.* **96,** 1187–1195.

MacCormack, T. J., and Driedzic, W. R. (2004). Cardiorespiratory and tissue adenosine responses to hypoxia and reoxygenation in the short-horned sculpin *Myoxocephalus scorpius*. *J. Exp. Biol.* **207,** 4157–4164.

MacCormack, T. J., McKinley, R. S., Roubach, R., Almeida-Val, V. M. F., Val, A. L., and Driedzic, W. R. (2003). Changes in ventilation, metabolism, and behaviour, but not bradycardia, contribute to hypoxia survival in two species of Amazonian armoured catfish. *Can. J. Zool.* **81,** 272–280.

Machado, B. H., Mauad, H., Chianca, D. A., Haibara, A. S., and Colombari, E. (1997). Autonomic processing of the cardiovascular reflexes in the nucleus tractus solitarii. *Braz. J. Med. Biol. Res.* **30,** 533–543.

Marshall, J. M. (1994). Peripheral chemoreceptors and cardiovascular regulation. *Physiol. Rev.* **74,** 543–594.

Marvin, D. E., and Burton, D. T. (1973). Cardiac and respiratory responses of rainbow trout, bluegills and brown bullhead catfish during rapid hypoxia and recovery under normoxic conditions. *Comp. Biochem. Physiol.* **46A,** 755–765.

Maxime, V., Nonnotte, G., Peyraud, C., Williot, P., and Truchot, J. P. (1995). Circulatory and respiratory effects of an hypoxic stress in the siberian sturgeon. *Resp. Physiol.* **100,** 203–212.

McKendry, J. E., and Perry, S. F. (2001). Cardiovascular effects of hypercarbia in rainbow trout (*Oncorhynchus mykiss*): A role for externally oriented chemoreceptors. *J. Exp. Biol.* **204,** 115–125.

McKendry, J. E., Milsom, W. K., and Perry, S. F. (2001). Branchial CO_2 receptors and cardiorespiratory adjustments during hypercarbia in Pacific spiny dogfish (*Squalus acanthias*). *J. Exp. Biol.* **204,** 1519–1527.

McKenzie, D. J., Aota, S., and Randall, D. J. (1991a). Ventilatory and cardiovascular responses to blood pH, plasma PCO_2, blood O_2 content and catecholamines in an air-breathing fish, the bowfin (*Amia calva*). *Physiol. Zool.* **64,** 432–450.

McKenzie, D. J., Burleson, M. L., and Randall, D. J. (1991b). The effects of branchial denervation and pseudobranch ablation on cardioventilatory control in an air-breathing fish. *J. Exp. Biol.* **161,** 347–365.

Mckenzie, D. J., Taylor, E. W., Bronzi, P., and Bolis, C. L. (1995). Aspects of cardioventilatory control in the adriatic sturgeon (*Acipenser naccarii*). *Respir. Physiol.* **100,** 45–53.

Meek, J., and Nieuwenhuys, R. (1998). Holosteans and teleosts. *In* "The Central Nervous System of Vertebrates" (Nieuwenhuys, R., ten Donkelaar, H. J., and Nicholson, C., Eds.), Vol. 2, pp. 759–937. Springer, Berlin.

Mercer, S. E., Wilk-Blaszczak, M. A., and Burleson, M. L. (2000). A comparative model for the investigation of the electrophysiological basis of oxygen sensing. *FASEB J.* **14,** A45.

Milsom, W. K. (1995a). Regulation of respiration in lower vertebrates: Role of CO_2/pH chemoreceptors. *In* "Advances in Comparative and Environmental Physiology, Volume 21 Mechanisms of Systemic Regulation: Acid–Base Regulation, Ion Transfer and Metabolism" (Heisler, N., Ed.), pp. 61–104. Springer, Berlin.

Milsom, W. K. (1995b). The role of CO_2/pH chemoreceptors in ventilatory control. *Braz. J. Med. Biol. Res.* **28,** 1147–1160.

Milsom, W. K. (1998). Phylogeny of respiratory chemoreceptor function in vertebrates. *Zoology* **101,** 316–332.

Milsom, W. K. (2002). Phylogeny of CO_2/H^+ chemoreception in vertebrates. *Respir. Physiol. Neurobiol.* **131,** 29–41.

Milsom, W. K., and Brill, R. W. (1986). Oxygen sensitive afferent information arising from the first gill arch of yellowfin tuna. *Respir. Physiol.* **66,** 193–203.

Milsom, W. K., Sundin, L., Reid, S. G., Kalinin, A. L., and Rantin, F. T. (1999). Chemoreceptor control of cardiovascular reflexes. *In* "Biology of Tropical Fishes" (Val, A. L., and Almeida-Val, V. M. F., Eds.), pp. 363–374. INPA, Manaus, Brazil.

Milsom, W. K., Reid, S. G., Rantin, F. T., and Sundin, L. (2002). Extrabranchial chemoreceptors involved in respiratory reflexes in the neotropical fish *Colossoma macropomum* (the tambaqui). *J. Exp. Biol.* **205,** 1765–1774.

Milsom, W. K., Abe, A. S., Andrade, D. V., and Tattersall, G. J. (2004). Evolutionary trends in airway CO_2/H^+ chemoreception. *Respir. Physiol. Neurobiol.* **144,** 191–202.

Nattie, E. (1999). CO_2, brainstem chemoreceptors and breathing. *Prog. Neurobiol.* **59,** 299–331.

Neville, C. M. (1979). Ventilatory responses of rainbow trout (*Salmo gairdneri*) to increased H^+ ion concentration in blood and water. *Comp. Biochem. Physiol.* **63A,** 373–376.

Nieuwenhuys, R., and Pouwels, E. (1983). The brain stem of actinopterygian fishes. *In* "Fish Neurobiology, Volume 1 Brain Stem and Sense Organs" (Northcutt, R. G., and Davis, R. E., Eds.), pp. 25–87. The University of Michigan Press, Ann Arbor.

Nikinmaa, M., and Rees, B. B. (2005). Oxygen-dependent gene expression in fishes. *Am. J. Physiol.* **288,** R1079–R1090.

Nilsson, S. (1983). Autonomic nerve function in the vertebrates. "Zoophysiology," Vol. 13, p. 253. Springer-Verlag, Berlin/Heidelberg, New York.

Nilsson, S. (1994). Evidence for adrenergic nervous control of blood pressure in teleost fish. *Physiol. Zool.* **67,** 1347–1359.

Nilsson, S., and Holmgren, S. (1992). Cardiovascular control by purines, 5-hydroxytryptamine, and neuropeptides. *In* "Fish Physiology, Volume XIIB The Cardiovascular System" (Hoar, W. S., Randall, D. J., and Farrell, A. P., Eds.), pp. 301–341. Academic Press, San Diego.

Nurse, C. A. (1990). Carbonic anhydrase and neuronal enzymes in cultured glomus cells of the carotid body of the rat. *Cell Tissue Res.* **261,** 65–71.

Nurse, C. A. (2005). Neurotransmission and neuromodulation in the chemosensory carotid body. *Auton. Neurosci.* **120,** 1–9.

Oliveira, R. D., Lopes, J. M., Sanches, J. R., Kalinin, A. L., Glass, M. L., and Rantin, F. T. (2004). Cardiorespiratory responses of the facultative air-breathing fish jeju, *Hoplerythrinus unitaeniatus* (Teleostei, Erythrinidae), exposed to graded ambient hypoxia. *Comp. Biochem. Physiol. A* **139,** 479–485.

Olson, K. R. (1998). The cardiovascular system. *In* "The Physiology of Fishes" (Evans, D. H., Ed.), pp. 129–154. CRC Press, Boca Raton, Florida.

Olson, K. R. (2002). Vascular anatomy of the fish gill. *J. Exp. Zool.* **293,** 214–231.

Patel, A. J., and Honore, E. (2001). Molecular physiology of oxygen-sensitive potassium channels. *Eur. Respir. J.* **18,** 221–227.

Peers, C., and Buckler, K. J. (1995). Transduction of chemostimuli by the type I carotid body cell. *J. Membr. Biol.* **144,** 1–9.

Pelster, B. (2002). Developmental plasticity in the cardiovascular system of fish, with special reference to the zebrafish. *Comp. Biochem. Physiol. A* **133,** 547–553.

Perry, S. F., and Desforges, P. R. (2006). Does bradycardia or hypertension enhance gas transfer in rainbow trout (*Oncorhynchus mykiss*)? *Comp. Biochem. Physiol. A* **144,** 163–172.

Perry, S. F., and Gilmour, K. M. (1996). Consequences of catecholamine release on ventilation and blood oxygen transport during hypoxia and hypercapnia in an elasmobranch (*Squalus acanthias*) and a teleost (*Oncorhynchys mykiss*). *J. Exp. Biol.* **199,** 2105–2118.

Perry, S. F., and Gilmour, K. M. (1999). Respiratory and cardiovascular systems. *In* "Stress Physiology" (Balm, P. H. M., Ed.), pp. 52–107. Sheffield Academic Press, Sheffield.

Perry, S. F., and Gilmour, K. M. (2002). Sensing and transfer of respiratory gases at the fish gill. *J. Exp. Zool.* **293,** 249–263.

Perry, S. F., and McKendry, J. E. (2001). The relative roles of external and internal CO_2 *versus* H^+ in eliciting the cardiorespiratory responses of *Salmo salar* and *Squalus acanthias* to hpercarbia. *J. Exp. Biol.* **204,** 3963–3971.

Perry, S. F., and Reid, S. G. (2002). Cardiorespiratory adjustments during hypercarbia in rainbow trout (*Oncorhynchus mykiss*) are initiated by external CO_2 receptors on the first gill arch. *J. Exp. Biol.* **205,** 3357–3365.

Perry, S. F., and Wood, C. M. (1989). Control and coordination of gas transfer in fishes. *Can. J. Zool.* **67,** 2961–2970.

Perry, S. F., Kinkead, R., Gallaugher, P., and Randall, D. J. (1989). Evidence that hypoxemia promotes catecolamine release during hypercapnic acidosis in rainbow trout (*Salmo gairdneri*). *Respir. Physiol.* **77,** 351–364.

Perry, S. F., Fritsche, R., Hoagland, T., Duff, D. W., and Olson, K. R. (1999). The control of blood pressure during external hypercapnia in the rainbow trout (*Oncorhynchus mykiss*). *J. Exp. Biol.* **202,** 2177–2190.

Perry, S. F., Gilmour, K. M., Vulesevic, B., McNeill, B., Chew, S. F., and Ip, Y. K. (2005). Circulating catecholamines in hypoxic lungfish (*Protopterus dolloi*): A comparison of aquatic and aerial hypoxia. *Physiol. Biochem. Zool.* **78,** 325–334.

Pettersson, K., and Johansen, K. (1982). Hypoxic vasoconstriction and the effects of adrenaline on gas exchange efficiency in fish gills. *J. Exp. Biol.* **97,** 263–272.

Peyraud-Waitzenegger, M., and Soulier, P. (1989). Ventilatory and circulatory adjustments in the European eel (*Anguilla anguilla* L.) exposed to short term hypoxia. *Exp. Biol.* **48,** 107–122.

Prabhakar, N. R. (2000). Oxygen sensing by the carotid body chemoreceptors. *J. Appl. Physiol.* **88,** 2287–2295.

Putnam, R. W., Filosa, J. A., and Ritucci, N. A. (2004). Cellular mechanisms involved in CO_2 and acid signaling in chemosensitive neurons. *Am. J. Physiol.* **287,** C1493–C1526.

Randall, D. J. (1982). The control of respiration and circulation in fish during exercise and hypoxia. *J. Exp. Biol.* **100,** 275–288.

Randall, D. J. (1990). Control and co-ordination of gas exchange in water breathers. *In* "Advances in Comparative and Environmental Physiology" (Boutilier, R. G., Ed.), pp. 253–278. Springer-Verlag, Berlin.

Randall, D. J., and Daxboeck, C. (1984). Oxygen and carbon dioxide transfer across fish gills. *In* "Fish Physiology, Volume X Gills A" (Hoar, W. S., and Randall, D. J., Eds.), pp. 263–314. Academic Press, London.

Randall, D. J., and Jones, D. R. (1973). The effect of deafferentation of the pseudobranch on the respiratory response to hypoxia and hyperoxia in the trout (*Salmo gairdneri*). *Respir. Physiol.* **17,** 291–301.

Randall, D. J., and Shelton, G. (1963). The effects of changes in environmental gas concentrations on the breathing and heart rate of a teleost fish. *Comp. Biochem. Physiol.* **9,** 229–239.

Randall, D. J., and Smith, J. C. (1967). The regulation of cardiac activity in fish in a hypoxic environment. *Physiol. Zool.* **40,** 104–113.

Randall, D. J., Heisler, N., and Drees, F. (1976). Ventilatory response to hypercapnia in the larger spotted dogfish *Scyliorhinus stellaris*. *Am. J. Physiol.* **230,** 590–594.

Reid, S. G., and Perry, S. F. (2003). Peripheral O_2 chemoreceptors mediate humoral catecholamine secretion from fish chromaffin cells. *Am. J. Physiol.* **284,** R990–R999.

Reid, S. G., Bernier, N., and Perry, S. F. (1998). The adrenergic stress response in fish: Control of catecholamine storage and release. *Comp. Biochem. Physiol. A* **120,** 1–27.

Reid, S. G., Sundin, L., Kalinin, A. L., Rantin, F. T., and Milsom, W. K. (2000). Cardiovascular and respiratory reflexes in the tropical fish, traira (*Hoplias malabaricus*): CO_2/pH chemoresponses. *Respir. Physiol.* **120,** 47–59.

Reid, S. G., Perry, S. F., Gilmour, K. M., Milsom, W. K., and Rantin, F. T. (2005). Reciprocal modulation of O_2 and CO_2 cardiorespiratory chemoreflexes in the tambaqui. *Respir. Physiol. Neurobiol.* **146,** 175–194.

Remmers, J. E., Torgerson, C., Harris, M. B., Perry, S. F., Vasilakos, K., and Wilson, R. J. A. (2001). Evolution of central respiratory chemoreception: A new twist on an old story. *Respir. Physiol.* **129,** 211–217.

Roberts, B. L., and Ballintijn, C. M. (1988). Sensory interaction with central "generators" during respiration in the dogfish. *J. Comp. Physiol. A* **162,** 695–704.

Rovainen, C. M. (1977). Neural control of ventilation in the lamprey. *Fed. Proc.* **36,** 2386–2389.

Roy, A., Rozanov, C., Mokashi, A., and Lahiri, S. (2000). PO_2-PCO_2 stimulus interaction in $[Ca^{2+}]_i$ and CSN activity in the adult rat carotid body. *Respir. Physiol.* **122,** 15–26.

Saltys, H. A., Jonz, M. G., and Nurse, C. A. (2006). Comparative study of gill neuroepithelial cells and their innervation in teleosts and *Xenopus* tadpoles. *Cell Tissue Res.* **323,** 1–10.

Sanchez, A., Soncini, R., Wang, T., Koldkjaer, P., Taylor, E. W., and Glass, M. L. (2001a). The differential cardio-respiratory responses to ambient hypoxia and systemic hypoxaemia in the South American lungfish, Lepidosiren paradoxa. *Comp. Biochem. Physiol. A* **130,** 677–687.

Sanchez, A. P., Hoffmann, A., Rantin, F. T., and Glass, M. L. (2001b). Relationship between cerebro-spinal fluid pH and pulmonary ventilation of the South American lungfish, *Lepidosiren paradoxa* (Fitz.). *J. Exp. Zool.* **290,** 421–425.

Sandblom, E., and Axelsson, M. (2005). Effects of hypoxia on the venous circulation in rainbow trout (*Oncorhynchus mykiss*). *Comp. Biochem. Physiol. A* **140,** 233–239.

Sapru, H. N. (2002). Glutamate circuits in selected medullo-spinal areas regulating cardiovascular function. *Clin. Exp. Pharmacol. Physiol.* **29,** 491–496.

Saunders, R. L., and Sutterlin, A. M. (1971). Cardiac and respiratory responses to hypoxia in the sea raven, *Hemitripterus americanus*, and an investigation of possible control mechanisms. *J. Fish. Res. Bd. Canada* **28,** 491–503.

Shelton, G., Jones, D. R., and Milsom, W. K. (1986). Control of breathing in ectothermic vertebrates. *In* "Handbook of Physiology, Section 3. The Respiratory System, Volume 2 Control of Breathing" (Cherniak, N. S., and Widdicombe, J. G., Eds.), pp. 857–909. American Physiological Society, Bethesda.

Short, S., Taylor, E. W., and Butler, P. J. (1979). The effectiveness of oxygen transfer during normoxia and hypoxia in the dogfish (*Scyliohinus canicula* L.) before and after cardiac vagotomy. *J. Comp. Physiol. B* **132,** 289–295.

Smatresk, N. J. (1986). Ventilatory and cardiac reflex responses to hypoxia and NaCN in *Lepisosteus osseus*, an air-breathing fish. *Physiol. Zool.* **59,** 385–397.

Smatresk, N. J. (1990). Chemoreceptor modulation of endogenous respiratory rhythms in vertebrates. *Am. J. Physiol.* **259,** R887–R897.

Smatresk, N. J. (1994). Respiratory control in the transition from water to air breathing in vertebrates. *Amer. Zool.* **34,** 264–279.

Smatresk, N. J., Burleson, M. L., and Azizi, S. Q. (1986). Chemoreflexive responses to hypoxia and NaCN in longnose gar: Evidence for two chemoreceptor loci. *Am. J. Physiol.* **251,** R116–R125.

Smith, F. M., and Davie, P. S. (1984). Effects of sectioning cranial nerves IX and X on the cardiac response to hypoxia in the coho salmon, *Oncorhynchus kisutch. Can. J. Zool.* **62,** 766–768.

Smith, F. M., and Jones, D. R. (1978). Localization of receptors causing hypoxic bradycardia in trout (*Salmo gairdneri*). *Can. J. Zool.* **56,** 1260–1265.

Smith, F. M., and Jones, D. R. (1982). The effect of changes in blood oxygen carrying capacity on ventilation volume in the rainbow trout (*Salmo gairdneri*). *J. Exp. Biol.* **97,** 325–334.

Soivio, A., and Tuurala, H. (1981). Structural and circulatory responses to hypoxia in the secondary lamellae of *Salmo gairdneri* gills at two temperatures. *J. Comp. Physiol.* **145,** 37–43.

Soncini, R., and Glass, M. L. (2000). Oxygen and acid–base status related drives to gill ventilation in carp. *J. Fish Biol.* **56,** 528–541.

Spyer, K. M., Dale, N., and Gourine, A. V. (2004). ATP is a key mediator of central and peripheral chemosensory transduction. *Exp. Physiol.* **89,** 53–59.

Stenslokken, K. O., Sundin, L., Renshaw, G. M., and Nilsson, G. E. (2004). Adenosinergic and cholinergic control mechanisms during hypoxia in the epaulette shark (*Hemiscyllium ocellatum*), with emphasis on branchial circulation. *J. Exp. Biol.* **207,** 4451–4461.

Summers, B. A., Overholt, J. L., and Prabhakar, N. R. (2002). CO_2 and pH independently modulate L-type Ca^{2+} current in rabbit carotid body glomus cells. *J. Neurophysiol.* **88,** 604–612.

Sundin, L., and Nilsson, G. E. (1996). Branchial and systemic roles of adenosine receptors in rainbow trout: An in vivo microscopy study. *Am. J. Physiol.* **271,** R661–R669.

Sundin, L., and Nilsson, G. E. (1997). Neurochemical mechanisms behind gill microcirculatory responses to hypoxia in trout: *In vivo* microscopy study. *Am. J. Physiol.* **272,** R576–R585.

Sundin, L., and Nilsson, S. (2002). Branchial innervation. *J. Exp. Zool.* **293,** 232–248.

Sundin, L., Davison, W., Forster, M. E., and Axelsson, M. (1998a). A role of 5-HT_2 receptors in the gill vasculature of the Antarctic fish *Pagothenia borchgrevinki. J. Exp. Biol.* **201,** 2129–2138.

Sundin, L., Holmgren, S., and Nilsson, S. (1998b). The oxygen receptor of the teleost gill? *Acta Zool.* **79,** 207–214.

Sundin, L., Reid, S. G., Kalinin, A. L., Rantin, F. T., and Milsom, W. K. (1999). Cardiovascular and respiratory reflexes: The tropical fish, traira (*Hoplia malabaricus*) O_2 chemoresponses. *Respir. Physiol.* **116,** 181–199.

Sundin, L., Reid, S. G., Rantin, F. T., and Milsom, W. K. (2000). Branchial receptors and cardiorespiratory reflexes in the neotropical fish, Tambaqui (*Colossoma macropomum*). *J. Exp. Biol.* **203,** 1225–1239.

Sundin, L., Turesson, J., and Burleson, M. L. (2003a). Identification of central mechanisms vital for breathing in in the channel catfish, *Ictalurus punctatus. Respir. Physiol. Neurobiol.* **138,** 77–86.

Sundin, L., Turesson, J., and Taylor, E. W. (2003b). Evidence for glutamatergic mechanisms in the vagal sensory pathway initiating cardiorespiratory reflexes in the shorthorn sculpin *Myoxocephalus scorpius. J. Exp. Biol.* **206,** 867–876.

Sundin, L. I. (1995). Responses of the branchial circulation to hypoxia in the Atlantic cod, *gadus morhua. Am. J. Physiol.* **37,** R771–R778.

Swenson, K. E., Eveland, R. L., Gladwin, M. T., and Swenson, E. R. (2005). Nitric oxide (NO) in normal and hypoxic vascular regulation of the spiny dogfish, *Squalus acanthias. J. Exp. Zool.* **303A,** 154–160.

Talman, W. T. (1997). Glutamatergic transmission in the nucleus tractus solitarii: From server to peripherals in the cardiovascular information superhighway. *Braz. J. Med. Biol. Res.* **30,** 1–7.

Taylor, E. W. (1992). Nervous control of the heart and cardiorespiratory interactins. *In* "Fish Physiology, Volume XII B The Cardiovascular System" (Hoar, W. S., Randall, D. J., and Farrell, A. P., Eds.), pp. 343–387. Academic Press, San Diego.

Taylor, E. W., and Barrett, D. J. (1985). Evidence of a respiratory role for the hypoxic bradycardia in the dogfish *Scyliohinus canicula* L. *Comp. Biochem. Physiol. A* **80,** 99–102.

Taylor, E. W., Short, S., and Butler, P. J. (1977). The role of the cardiac vagus in the response of the dogfish *Scyliorhinus canicula* to hypoxia. *J. Exp. Biol.* **70,** 57–75.

Taylor, E. W., Jordan, D., and Coote, J. H. (1999). Central control of the cardiovascular and respiratory systems and their interactions in vertebrates. *Physiol. Rev.* **79,** 855–916.

Thomas, S. (1983). Changes in blood acid-base balance in trout (*Salmo gairdneri* Richardson) following exposure to combined hypoxia and hypercapnia. *J. Comp. Physiol. B* **152,** 53–57.

Thomas, S., and Le Ruz, H. (1982). A continuous study of rapid changes in blood acid-base status of trout during variations of water PCO_2. *J. Comp. Physiol.* **148,** 123–130.

Tufts, B. L., and Perry, S. F. (1998). Carbon dioxide transport and excretion. *In* "Fish Respiration, Volume 17 Fish Physiology" (Perry, S. F., and Tufts, B. L., Eds.), pp. 229–281. Academic Press, San Diego.

Turesson, J., and Sundin, L. (2003). *N*-methyl-D-aspartate receptors mediate chemoreflexes in the shorthorn sculpin *Myoxocephalus scorpius. J. Exp. Biol.* **206,** 1251–1259.

Val, A. L. (1995). Oxygen-transfer in fish-morphological and molecular adjustments. *Braz. J. Med. Biol. Res.* **28,** 1119–1127.

Van Vliet, B. N., and West, N. H. (1992). Functional characteristics of arterial chemoreceptors in an amphibian (*Bufo marinus*). *Respir. Physiol.* **88,** 113–127.

Vulesevic, B., and Perry, S. F. (2006). Developmental plasticity of ventilatory control in zebrafish, *Danio rerio. Respir. Physiol. Neurobiol.* (in press).

Vulesevic, B., McNeill, B., and Perry, S. F. (2006). Chemoreceptor plasticity and respiratory acclimatization in the zebrafish, *Danio rerio. J. Exp. Biol.* **209,** 1261–1273.

Wang, Z.-Y., and Bisgard, G. E. (2002). Chronic hypoxia-induced morphological and neurochemical changes in the carotid body. *Microscopy Res. Tech.* **59,** 168–177.

West, N. H., and Van Vliet, B. N. (1993). Sensory mechanisms regulating the cardiovascular and respiratory systems. *In* "Environmental Physiology of the Amphibians" (Feder, M. E., and Burggren, W. W., Eds.), pp. 151–182. The University of Chicago Press, Chicago.

Williams, S. E. J., Wootton, P., Mason, H. S., Bould, J., Iles, D. E., Riccardi, D., Peers, C., and Kemp, P. J. (2004). Hemoxygenase-2 is an oxygen sensor for a calcium-sensitive potassium channel. *Science* **306,** 2093–2097.

Wilson, R. J. A., Harris, M. B., Remmers, J. E., and Perry, S. F. (2000). Evolution of air-breathing and central CO_2/H^+ respiratory chemosensitivity: New insights from an old fish? *J. Exp. Biol.* **203,** 3505–3512.

Wood, C. M., and Jackson, E. B. (1980). Blood acid-base regulation during environmental hyperoxia in the rainbow trout (*Salmo gairdneri*). *Respir. Physiol.* **42,** 351–372.

Wood, C. M., and Munger, R. S. (1994). Carbonic anhydrase injection provides evidence for the role of blood acid-base status in stimulating ventilation after exhaustive exercise in rainbow trout. *J. Exp. Biol.* **194,** 225–253.

Wood, C. M., and Perry, S. F. (1985). Respiratory, circulatory, and metabolic adjustments to exercise in fish. *In* "Circulation, Respiration, and Metabolism" (Gilles, R., Ed.), pp. 2–22. Springer-Verlag, Berlin.

Wood, C. M., and Shelton, G. (1980). The reflex control of heart rate and cardiac output in the rainbow trout: Interactive influences of hypoxia, haemorrhage, and systemic vasomotor tone. *J. Exp. Biol.* **87,** 271–284.

Wood, C. M., McMahon, B. R., and McDonald, D. G. (1979). Respiratory, ventilatory, and cardiovascular responses to experimental anaemia in the starry flounder, *Platichthys stellatus. J. Exp. Biol.* **82,** 139–162.

Wood, C. M., Turner, J. D., Munger, R. S., and Graham, M. S. (1990). Control of ventilation in the hypercapnic skate *Raja ocellata*: II. cerebrospinal fluid and intracellular pH in the brain and other tissues. *Respir. Physiol.* **80,** 279–298.

Wyatt, C. N., and Buckler, K. J. (2004). The effect of mitochondrial inhibitors on membrane currents in isolated neonatal rat carotid body type I cells. *J. Physiol.* **556,** 175–191.

Zaccone, G., Lauweryns, J. M., Fasulo, S., Tagliafierro, G., Ainis, L., and Licata, A. (1992). Immunocytochemical localization of serotonin and neuropeptides in the neuroendocrine paraneurons of teleost and lungfish gills. *Acta Zool.* **73,** 177–183.

Zaccone, G., Fasulo, S., and Ainis, L. (1994). Distribution patterns of the paraneuronal endocrine cells inthe skin, gills and the airways of fishes as determined by immunohistochemical and histological methods. *Histochem. J.* **26,** 609–629.

Zaccone, G., Fasulo, S., and Ainis, L. (1995). Neuroendocrine epithelial cell system in respiratory organs of air-breathing and teleost fishes. *Int. Rev. Cytol.* **157,** 277–314.

Zaccone, G., Mauceri, A., Fasulo, S., Ainis, L., Lo Cascio, P., and Ricca, M. B. (1996). Localization of immunoreactive endothelin in the neuroendocrine cells of fish gill. *Neuropeptides* **30,** 53–57.

Zaccone, G., Fasulo, S., Ainis, L., and Licata, A. (1997). Paraneurons in the gills and airways of fishes. *Microsc. Res. Tech.* **37,** 4–12.

Zaccone, G., Mauceri, A., Ainis, L., Fasulo, S., and Licata, A. (1999). Paraneurons in the skin and gills of fishes. *In* "Ichthyology Recent Research Advances" (Saksena, D. N., Ed.), pp. 417–447. Science Publishers Inc., Enfield, NH.

Zhang, M., and Nurse, C. A. (2004). CO_2/pH chemosensory signaling in co-cultures of rat carotid body receptors and petrosal neurons: Role of ATP and ACh. *J. Neurophysiol.* **92,** 3433–3445.

Zhang, M., Zhong, H., Vollmer, C., and Nurse, C. A. (2000). Co-release of ATP and ACh mediates hypoxic signalling at rat carotid body chemoreceptors. *J. Physiol.* **525,** 143–158.

4

NOCICEPTION

LYNNE U. SNEDDON

1. Introduction
2. Neural Apparatus
   2.1. Nociceptor Anatomy
   2.2. Nociceptor Electrophysiology
3. Central Nervous System
   3.1. Brain Structure
   3.2. Pathways to the Brain
4. Moleculer Markers of Nociception
   4.1. GABA
   4.2. Substance P and the Preprotachykinins
   4.3. NMDA
   4.4. Opioids, Endogenous Opioids, and Enkephalins
   4.5. Global Gene Expression
5. Whole Animal Responses
   5.1. Avoidance Learning
   5.2. *In Vivo* Observations
6. Conclusions

1. INTRODUCTION

Nociception is the detection of a noxious stimulus and is usually accompanied by a reflex withdrawal response away from that stimulus immediately on detection. Noxious stimuli are those that can or potentially could cause tissue damage so stimuli, such as high mechanical pressure, extremes of temperature and chemicals, for example, acids, venoms, prostaglandins, and so on, excite nociceptive nerve fibers. This important somatosensory system acts as an alarm system to alert an individual to damage that may be detrimental to survival. In humans this can give rise to the experience of pain where damage does occur, although pain is not just a sensory event but incorporates a psychological component. The International Association for the Study of Pain (IASP) has defined nociception as reflex activity induced

Sensory Systems Neuroscience: Volume 25
FISH PHYSIOLOGY

DOI: 10.1016/S1546-5098(06)25004-0

in the nociceptor and nociceptive pathways by a noxious stimulus but clearly states that pain is an unpleasant sensory and emotional experience associated with actual or potential tissue damage, or described in terms of such damage. The IASP adds that "the inability to communicate verbally does not negate the possibility that an individual is experiencing pain and is in need of appropriate pain-relieving treatment." Although the IASP agreed on this note with regards to neonates and children, we must also apply it to animals since if we have never been an animal then how do we know if it does not consciously experiences a potentially painful event. In comparison to other sensory systems described in this volume, nociception is relatively underexplored in fish. However, studies have examined the properties of peripheral receptors that preferentially detect noxious stimuli and nociceptors have been identified in fish. Research on neuroanatomy has identified similar central nervous system pathways in fish that are essential for nociception to occur in higher vertebrates. Fish also possess many of the substances associated with nociception such as substance P, *N*-methyl-D-aspartate (NMDA), opioids, and endogenous opioids. Studies have shown that fish are capable of learning to avoid noxious stimuli and that morphine blocks learning. Finally, adverse behavioral and physiological responses have been observed in fish enduring a noxious event that are ameliorated by the administration of painkillers or analgesics. This chapter shall review the nociceptive system in fish describing the neural apparatus and pathways leading to higher brain areas, the potential involvement of brain areas during a noxious event, and compare the properties of the fish system to that of higher vertebrates (amphibians, birds, and mammals). Research on whole animal responses to noxious and potentially painful stimuli shall also be discussed.

2. NEURAL APPARATUS

2.1. Nociceptor Anatomy

Generally nociceptors are free nerve endings in the periphery, internal organs, and viscera and are usually of two fiber or nerve types, small myelinated A-delta fibers and smaller unmyelinated C fibers (Lynn, 1994). In mammalian systems, C fibers can range in diameter from 0.2 to 3.0 μm whereas the larger A-delta fibers range from 2 to 14 μm. The teleost fish skin has numerous free nerve endings of unknown function over the whole body (Whitear, 1983), yet few studies have looked anatomically for the presence of A-delta or C fibers in this group. One such study on the rainbow trout, *Oncorhynchus mykiss*, examined the fifth cranial nerve, the trigeminal nerve, which conveys nociceptive information from oral and facial areas in higher vertebrates, and found both fiber types in all of the three main branches of the trigeminal nerve (Sneddon, 2002). The trigeminal leaves the brain stem directly fusing into a ganglion where the nerve cell bodies are located and splits into three branches: the ophthalmic branch that conveys information from the snout, the maxillary branch that conveys information from the upper jaw, and the mandibular which travels to the lower jaw. The fibers in *O. mykiss* matched the size range found in higher vertebrates (Figure 4.1). C fibers were found in discrete bundles with a diameter that ranged between 0.1 and 1.2 μm and comprised of 4% of total fiber type (3.8% ophthalmic; 4.1% maxillary; 4.1% mandibular). A-delta fibers are composed of 33% of total fiber type and had diameters in the range of 0.4 to 5.8 μm. The trigeminal branches were also composed of A-alpha fibers (9%) that are usually associated with the innervation of muscle and A-beta fibers (53%) that convey somatosensory information in higher vertebrates. Although studies on only one species, the rainbow trout, has been published by the Sneddon laboratory has similar data from common carp, *Cyprinus carpio*; goldfish, *Carassius auratus*; and zebrafish, *Danio rerio* (Ashley *et al.*, unpublished observations; Reilly *et al.*, unpublished observations).

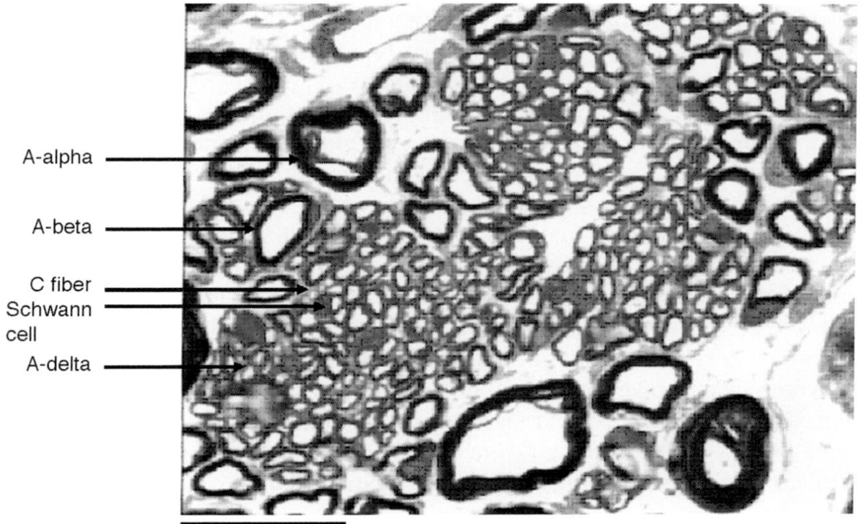

Fig. 4.1. Section of the maxillary branch of the trigeminal nerve of the rainbow trout showing the presence of A-delta and C fibers (Sneddon, 2002). Scale bar: 1000× magnification and the scale bar is 2 μm. (Reprinted from *Neuroscience Letters*, 319, Sneddon (© 2002), with permission from Elsevier.)

Studies have reported that C fibers comprise 50–65% of total fiber type in higher vertebrates including amphibians, birds, and mammals (Lynn, 1994), yet *O. mykiss* had only 4% C fibers within the trigeminal. A-delta fibers are normally outnumbered from four to one by C fibers in mammals (Lynn, 1994) but A-delta fibers comprise 25% of fiber type in the fish. This reduced number of C fibers in fish relative to the high number found in mammals may be explained by the advance onto land in evolution. Terrestrial vertebrates may be at greater risk of injury due to gravitational forces, noxious gases, and extremes of temperature whereas in the aquatic environment, buoyancy counteracts gravity, chemicals may be diluted, and there are generally no great fluctuations in temperature as are seen on land. So perhaps this teleost fish has not devoted as much neural wiring to a nociceptive system as the terrestrial vertebrates have, who have a much more comprehensive system to deal with the increased risk of damage. This hypothesis remains to be tested (Sneddon, 2004).

Most other studies on the anatomy of peripheral nerves have examined elasmobranch fish and have discovered a lack of C fibers although A-delta fibers are present (e.g., stingray, *Dasyatis sabina*; spotted eagle ray, *Aetobatus narinari*; cownose ray, *Rhinoptera bonasus*; longtailed ray, *Himantura* sp.; Coggeshall *et al.*, 1978; Leonard, 1985; Snow *et al.*, 1996). However, in the shovelnose ray, *Rhinobatus battilium*, and the black-tip shark, *Carcharhinus melanopterus*, unmyelinated fibers that could be C fibers were found but the researchers suggest that myelination was incomplete in the small specimens used (Snow *et al.*, 1993). Leonard (1985) also states that only the sensory nerves that enter the dorsal root ganglion of the spinal cord contains fewer than 1% unmyelinated fibers but more are found in the cranial nerves and whether the trigeminal nerve was included in the cranial nerves is not divulged in the article. When thinking about the life history of elasmobranch species, it seems intuitive that they may not possess a classical nociceptive system since their courtship and copulation involves the male biting the female and many females have copulatory scarring (Kajiura *et al.*, 2000). It would seem adaptive in this instance for the female to not perceive the mechanical and damaging properties of the copulatory bite, as noxious otherwise copulation may not occur. However, we simply do not know if the motivational drive to reproduce overrides any noxious experience as elasmobranchs show aversion responses to electric fields and noxious chemicals used as shark repellents (Gilbert and Gilbert, 1973; Sisneros and Nelson, 2000).

2.2. Nociceptor Electrophysiology

Nociceptors are generally characterized by their slowly adapting response to mechanical stimulation where a prolonged response to mechanical stimuli is seen with a reduction in the firing rate of the nerve over the duration of the stimulation, their response to noxious heat (>40 °C) and they may also respond to noxious chemicals (Lynn, 1994). The conduction velocity of the smaller C fibers in mammals is ~0.3–1.2 m/sec, whereas A-delta conduct at a speed of 5–30 m/sec. In the lamprey, *Petromyzon marinus*, the oldest living predecessor of the fishes, receptive fields were identified that gave a slowly adapting response to mechanical stimulation, responded to noxious heat and to damaging stimuli (Matthews and Wickelgren, 1978). This study has not been followed up in this species but studies have failed to find nociceptors in elasmobranchs (Coggeshall *et al.*, 1978; Leonard, 1985). Investigations have mostly been conducted on the rainbow trout, and have shown the presence of nociceptors on the head of the fish that were innervated by the trigeminal nerve (Sneddon, 2003a). The majority of these receptors innervated by the trigeminal nerve were slowly adapting mechanoreceptors as well as 22 receptive fields from a total of 58 responsive to noxious heat (over 40 °C, Figure 4.2; Sneddon, 2003a; Sneddon *et al.*, 2003a). Out of the 22 receptive fields, 18 also responded to application of a noxious chemical, 1% acetic acid, and could be classified as polymodal nociceptors whereas the remaining 4 did not respond to acid and could be classified as mechanothermal nociceptors (Sneddon, 2003a).

Out of the 22 nociceptors, only 1 was a C fiber with a conduction velocity of 0.67 m/sec whereas the others were A-delta fibers that conducted in the range of 1.6–8.5 m/sec. Therefore, the majority of polymodal nociceptors were A-delta fibers in the trout. However, in the higher vertebrate skin, C fibers usually act as polymodal nociceptors (Lang *et al.*, 1990). In mammals, polymodal A-delta fibers are usually found in oral mucosa (Toda *et al.*, 1997), skeletal muscle (Kumazawa and Mizumura, 1977) and visceral organs (Kumazawa and Mizumura, 1980; Haupt *et al.*, 1983). The nociceptors of these areas are mainly polymodal possibly due to the various aqueous and hard substances that they come into contact with that provide a mixture of mechanical, chemical, and thermal stimulation. Fish inhabit an aqueous world so this may be why the majority of nociceptors on the skin are polymodal (Sneddon, 2003a).

Electrophysiological studies have to date failed to find slowly adapting mechanoreceptors, a property of nociceptors, or receptors that responded to temperature increases in the stingray, *D. sabina* (Coggeshall *et al.*, 1978; Snow *et al.*, 1996). This is in contrast to studies on the rainbow trout. Therefore, studies in elasmobranchs have failed to find nociceptors yet one teleost fish appears to possess unmyelinated fibers and nociceptors. The lamprey, *P. marinus*, has only unmyelinated fibers and electrophysiological recordings did find slowly adapting receptors that responded to noxious heat and were, therefore, possibly nociceptive (Matthews and Wickelgren, 1978).

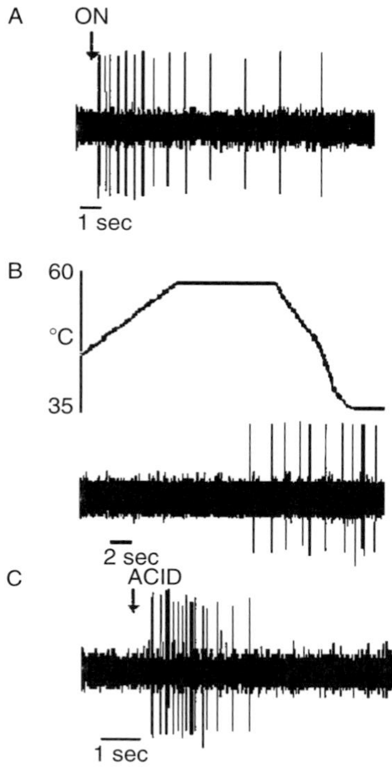

Fig. 4.2. A polymodal nociceptor innervated by the trigeminal nerve of the rainbow trout responding to mechanical (A), thermal (B), and chemical stimulation (C; 1% acetic acid). The receptor is slowly adapting to mechanical stimulation (A; ON indicates application of stimulus), has a thermal threshold of 58 °C (B), and responds to application of a drop of acetic acid onto the receptive field (C) (Sneddon *et al.*, 2003a). (Adapted from *Proceedings of the Royal Society of London B*, 270, Sneddon *et al.* (© 2003) with permission from the Royal Society of London.)

This may represent an evolutionary divergence between the teleost and elasmobranch groups where some elasmobranchs have lost unmyelinated fibers and teleosts have retained these. Future studies should expand the neuroanatomical study of different groups of agnathans, elasmobranchs, and teleost fish to determine the evolutionary anatomical relationships of these lower vertebrate groups.

Many nociceptor properties are common to *O. mykiss* and higher vertebrates. For example, the diameter of the receptive fields ranged from 1.6 to 9 mm, which is a similar diameter found in birds (Gentle and Tilston, 2000) and mammals (Toda *et al.*, 1997); the fish nociceptors have large, broad action potentials with slow depolarization as also seen in mammals (Gallego, 1983; López deArmentia *et al.*, 2000); and finally the conduction velocity of the fish nociceptors are within the mammalian A-delta and C fiber range (Lynn, 1994).

There are some interesting differences between fish and higher animals. Mechanical thresholds are much lower in the fish nociceptors with some being stimulated below 0.1 g (Sneddon, 2003a). A minimum pressure of 0.6 g is required to stimulate mammalian nociceptors (Lynn, 1994), yet the fish nociceptors appear to be much more sensitive with this low thresholds only seen in corneal nociceptors of mammalian models (Belmonte and Gallar, 1996). When comparing the electrophysiological properties of the fish nociceptors to those found on the snake head (Liang and Terashima, 1993) and those found on the mouse cornea (López deArmentia *et al.*, 2000), most of the properties are strikingly similar (Table 4.1). Characteristics such as rate of firing, maximum rate of depolarization, and after hyperpolarization amplitude duration are slower but this is probably due to the temperature difference between the cold-blooded fish at 18 °C and the mammal at 37 °C. This temperature difference may mean that the fish physiological characteristics are two to four times slower. Since the teleost fish shares common properties with both a reptile and a mammal, this suggests that these characteristics may have evolved in a predecessor of the teleost group and, therefore, the electrophysiological properties of nociceptors in the agnathans warrant further investigation to determine their detailed electrophysiological properties.

The current study in the Sneddon laboratory has explored nociceptor physiology in more detail and we have found that it is possible to hypersensitize the rainbow trout nociceptors (Ashley *et al.*, unpublished observations).

Table 4.1

Electrophysiological Properties of A-Delta Nociceptors in a Fish (Sneddon, 2003a), a Snake (Liang and Terashima, 1993), and a Mouse (López deArmentia *et al.*, 2000)

	A-delta nociceptors		
Electrophysiological properties	Fish	Snake	Mouse
Conduction Velocity (m/sec)	0.7–5.5	3.8	0.7–5.7
AP amplitude (mV)	10–90	91	70–89
AP Duration (msec)	0.8–2.4	2.4	0.7–2.8
AHP Amplitude (mV)	1.8–5.5	11.9	6–12
dV/dt_{max} (V/sec)	63–226	182	115–291

Mean values are shown for conduction velocity, action potential (AP) amplitude and duration, after hyperpolarization (AHP) amplitude, and the maximum rate of depolarization (dV/dt_{max}). (Data reprinted from *Brain Research*, 972, Sneddon, © 2003, and *Neuroscience*, 101, López de Armentia *et al.*, © 2000, with permission from Elsevier and from the Journal of Comparative Neurology, 328, Liang & Terashima, © 1993, with permission from Wiley Interscience.)

This is a phenomenon similar to hyperalgesia seen in humans and mammals where damage to the nociceptor results in previously nonnoxious innocuous stimuli such as a light touch becoming painful after damage. This presents the possibility of a prolonged nociceptive experience in the fish species with nociceptors with sensitivity around a damaged area. The Sneddon laboratory has also tested ~50 nociceptors to examine if they have cold sensitivity since low temperatures below 4 °C excite human nociceptors. To date, we have not identified cold nociceptors which may not be surprising since many fish species live in temperatures below 4 °C (Ashley *et al.*, unpublished observations). This also has important welfare implications since many fish are slaughtered using live chilling or are placed on ice (Skjervold *et al.*, 2001), however, suffocation in air does elicit a stress response (Sharpe *et al.*, 1998).

3. CENTRAL NERVOUS SYSTEM

3.1. Brain Structure

The fishes have the necessary brain areas for nociceptive processing to occur (Figure 4.3; e.g., pons, medulla, and thalamus), however, an area of debate is the cortex located in the telencephalon. The most highly evolved vertebrates, humans, and primates have the most developed cortex with the evolution of the neocortex. As we descend the evolutionary tree of vertebrates (Figure 4.4), the cortex becomes less differentiated but fishes do possess a rudimentary cortex area. Reviews of the telencephalic structure in lampreys (Nieuwenhuys, 1977) and hagfishes (Bone, 1963) indicate that the telencephalon consists of olfactory bulbs, cerebral hemispheres, and a telencephalon medium. The cerebral hemispheres are composed of a roof called the pallium and floor called the subpallium. The striatum runs over the subpallium that extends into the telencephalon medium. In the elasmobranchs, the cerebral hemispheres are characterized by large well-defined cell groups with a well-developed thalamic input where the pallial center receives substantial ascending sensory inputs from the visual, trigeminal, and auditory systems (Cohen *et al.*, 1973; Platt *et al.*, 1974; Bullock and Corwin, 1979; Northcutt, 1981). The telencephalon of chimeras (holocephalons) or ratfishes differ from those in elasmobranchs in that the olfactory bulb arise from the rostral pole of the telencephalon rather than far laterally as in most elasmobranch species. The chimeras also lack pallial formations which characterize the telencephalon of all elasmobranchs. In teleosts, this rudimentary cortex or the cerebral hemisphere is better developed with the hemispheric zones possessing complex projections to the diencephalon and midbrain (Northcutt, 1981). Tracing studies in the zebrafish, *D. rerio*,

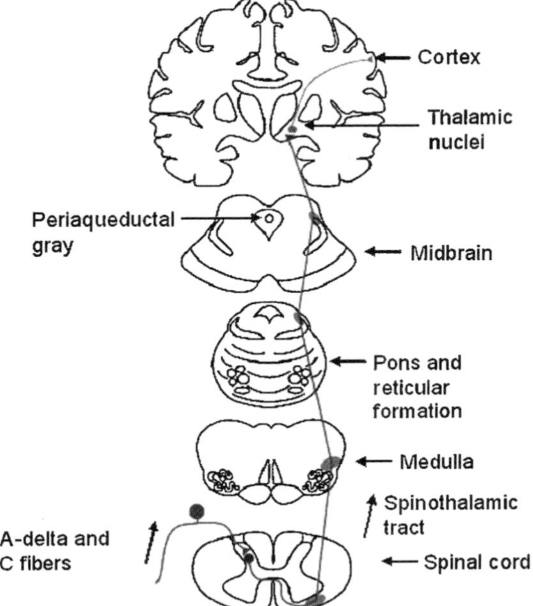

Fig. 4.3. A diagrammatic representation of the spinothalamic tract in humans. Nociceptive information in mammals is conducted via C and A-delta fibers from the periphery of the body to the spinal cord via the dorsal horn and terminates in the substantia gelatinosa. The information then crosses the spinal cord to the spinothalamic tract and ascends through the pons, medulla, and thalamus to the cortex.

identified afferent nuclei to the pallium: the olfactory bulb, dorsal entopeduncular, parvocellular preoptic and suprachiasmatic nuclei, anterior, dorsal, and central posterior dorsal thalamic, rostrolateral nuclei, periventricular nucleus of the posterior tuberculum, posterior tuberal nucleus, tuberal hypothalamic nuclei, dorsal tegmental nucleus, superior reticular nucleus, locus coeruleus, and the superior raphe nucleus (Figure 4.5; Rink and Wulliman, 2004). Efferent projections terminate in other telencephalic areas, habenula periventricular pretectum, paracommisural nucleus, posterior dorsal thalamus, preoptic region, midline posterior tuberculum, tuberal hypothalamus, and interpeduncular nucleus. The thalamus is essential for pain processing in humans and ascending information is conveyed to cortex where descending information is relayed backz to the thalamus. These connections between the cortical regions and the thalamus also exist in *D. rerio* (Rink and Wulliman, 2004) but have yet to be studied with regard to nociceptive processing.

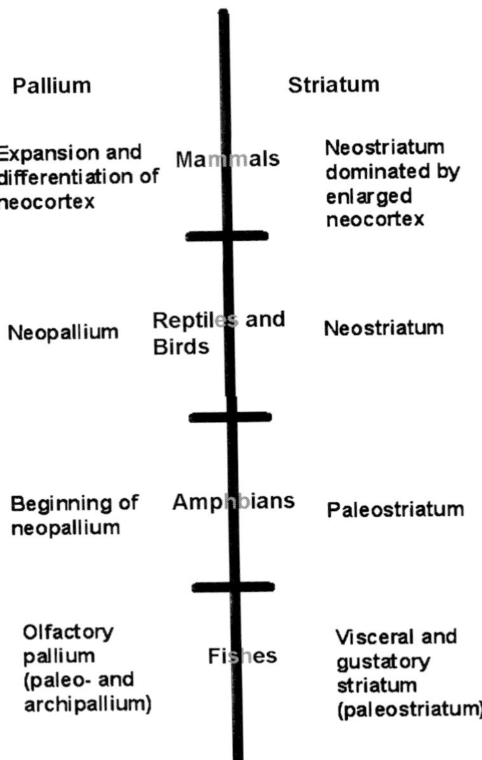

Fig. 4.4. A hypothetical linear evolutionary diagram showing the cortical development from fishes to mammals (Northcutt, 1981). (Adapted and reprinted, with permission, from the *Annual Review of Neuroscience*, 4, © 1981 by Annual Reviews *www.annualreviews.org*; Northcutt, 1981.)

Research into the intricacies of brain areas and their function has been explored using lesion studies. Lesions of the lateral and medial pallia have resulted in deficits in learning and memory with regards to electric shock training, which is a potentially painful, noxious stimulus (Portavella *et al.*, 2004). The medial pallium lesions resulted in a deficit in emotional tasks and as such are thought to have the same function as the mammalian amygdala, whereas the lateral pallium resulted in impaired performance in spatial learning and memory retrieval and has a similar function to the mammalian hippocampus. The pallium is part of the fish cortex and although when given an electric shock, the goldfish did perform and escape response, they failed to learn to avoid that response as perhaps the aversive experience could not be perceived since part of the cortex was lesioned (Portavella, personal communication).

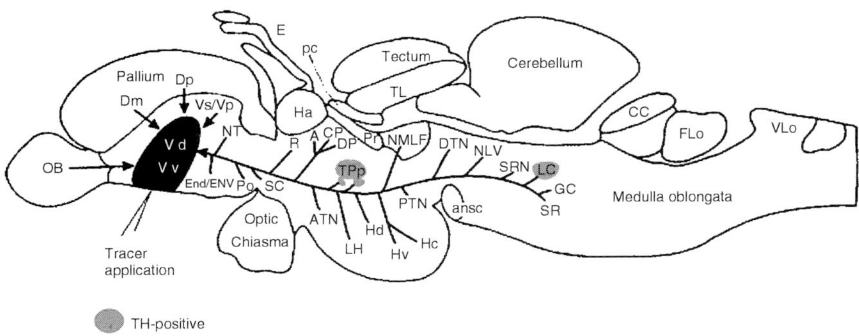

Fig. 4.5. Lateral view of zebrafish brain showing all inputs to subpallium (Vd/Vv). Black area: tracer application site; gray shaded areas: neurons double labeled for tyrosine hydroxylase and retrograde neuronal tracing from ventral telencephalic area (after Rink and Wullimann, 2004). A, anterior thalamic nucleus; ansc, ansulate commissure; ATN, anterior tuberal nucleus; CC, cerebellar crest; CP, central posterior thalamic nucleus; Dm, medial zone of dorsal telencephalic area; Dp, posterior zone of dorsal telencephalic area; DP, dorsal posterior thalamic nucleus; DTN, dorsal tegmental nucleus; E, epiphysis; ENd, dorsal part of entopeduncular nucleus; ENv, ventral part of entopeduncular nucleus; FLo, facial lobe; GC, central gray; Ha, habenula; Hc, caudal zone of periventricular hypothalamus; Hd, dorsal zone of periventricular hypothalamus; Hv, ventral zone of periventricular hypothalamus; LC, locus coeruleus; LH, lateral hypothalamic nucleus; NLV, nucleus lateralis valvulae; NMLF, nucleus of the medial longitudinal fascicle; NT, nucleus taeniae; OB, olfactory bulb; pc, posterior commissure; Po, preoptic region; Pr, periventricular pretectum; PTN, posterior tuberal nucleus; R, nucleus rostrolateralis; SC, suprachiasmatic nucleus; SR, superior raphe; SRN, superior reticular nucleus; TL, longitudinal torus; TPp, periventricular nucleus of posterior tuberculum; Vd, dorsal nucleus of ventral telencephalic area; VLo, vagal lobe; Vs, supracommissural nucleus of the ventral telencephalic area; Vv, ventral nucleus of ventral telencephalic area. (Reprinted from *Brain Research*, 1011, © 2004, with permission from Elsevier.)

A study by Dunlop and Laming (2005) demonstrated that there was neural activity in the cortical area as well as the diencephalon and rhombencephalon (Figure 4.6). Nociceptive and nonnociceptive stimuli were given to rainbow trout, *O. mykiss*, and goldfish, *C. auratus*, which elicited different responses from the same brain areas demonstrating that the higher brain is active while the fish are stimulated peripherally. In the goldfish, most of the noxious information was conveyed to the forebrain via C fibers but A-delta fibers predominated in the rainbow trout (Figure 4.6) which confirms previous studies (Sneddon, 2002, 2003a). This illustrates that there are important species differences which are also seen in higher vertebrates (Benson and Thurmon, 1987). The Sneddon laboratory has embarked on a functional magnetic resonance imaging study to determine if the brain is active during noxious stimulation. Using common carp, *C. carpio*, it was demonstrated that during noxious stimulation there was significant activity in the brain,

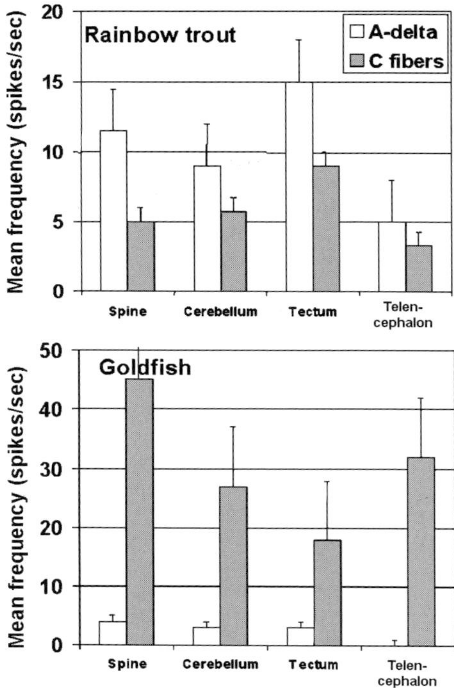

Fig. 4.6. The electrophysiological activity recorded in the rostral spinal cord (Spine), hindbrain (Cerebellum), midbrain (Tectum), and forebrain (Telencephalon) of rainbow trout (top) and goldfish (bottom) during noxious stimulation. (Adapted from *Journal of Pain*, 9, Dunlop & Laming © 2005, with permission from Elsevier.)

with most activity in the forebrain followed by midbrain and hindbrain (Sneddon *et al.*, unpublished observations in the Van der Linden Laboratory). This study confirms that noxious stimulation is not confined to the spinal cord and brain stem and does involve higher brain centers. Study in this field is incomplete but should be directed at understanding the function of these areas in nociceptive processing.

3.2. Pathways to the Brain

The major tracts involved in nociceptive processing and relaying information to the brain are the trigeminal tract conveying information from the head, and the spinothalamic tract conveying information from the rest of the body. Both tracts have been extensively studied in lower vertebrate groups. In the aganathan, the Pacific hagfish, *Eptatretus stouti*, horseradish peroxidase tracing demonstrated that the trigeminal nerve descended through the medulla to the rostral spinal cord as it does in higher vertebrates (Ronan, 1988). The trigeminal tract also ascended into the mesencephalic tectum, although distinct mesencephalic and descending nuclei of the trigeminal system are yet to be found in the agnathans. This was also the case in the lamprey, *P. marinus* (Northcutt, 1979). However, studies in the teleost fish, carp, *C. carpio* (Luiten, 1975) and sturgeon, *Acipenser oxyrhynchus* (New and Northcutt, 1984) have shown that there are distinct nuclei namely the mesencephalic nucleus and descending nucleus that are also found in higher vertebrates. The appearance of the two trigeminal nuclei in the teleosts but not in the predecessors of the fishes suggests that these nuclei evolved after the agnathans and these nuclei are present in an elasmobranch, the dogfish, *Scyliorhinus canicula*, (Rodríguez-Moldes *et al.*, 1993; Anadón *et al.*, 2000), and thus must have evolved between the agnathans and the emergence of the fishes. In the carp, *C. carpio*, the trigeminal clearly projects to the thalamus as it does in other vertebrates, therefore, the trigeminal tract exists in the fishes.

The spinal cord of elasmobranchs has been considerably studied, showing that the gray matter of the spinal cord can be divided into seven laminae (Iwahori *et al.*, 1998). The first lamina is cell dense and occupies the major part of the dorsal horn and relates to lamina 1 and the substantia gelatinosa of the spinal cord of mammals, birds, and reptiles where nociception is mainly processed. Ascending fibers in the spinal cord of the nurse shark, *Ginglymostoma cirratum*, reach the reticular formation and also project to the dorsal motor nucleus of the vagus, the medulla oblongata, the central brain stem, the cerebellar cortex, the cerebellar nucleus, the nucleus intercollicularis, the mesencephalic tectum, and the thalamus (Ebbesson and Hodde, 1981). Within the reticular formation of two elasmobranch species, the thornback guitarfish (*Platyrhinoidis triseriata*) and the horn shark (*Heterodontus francisci*), there are 19 reticular nuclei with spinal projections indicating that the elasmobranch reticular formation is complexly organized into many of the same nuclei as found in higher vertebrate groups (Cruce *et al.*, 1999). Therefore, the elasmobranch groups have the same basic components of ascending spinal projections (Ebbesson and Hodde, 1981). This has also been confirmed in studies on teleost fish [e.g., sea robin, *Prionotus carolinus* (Finger, 2000); and channel catfish, *Ictalurus punctatus* (Goehler and Finger, 1996)]. In *P. carolinus*, ascending projections were seen from spinal sensory nerves to the lateral funicular complex, the medial lemniscus of the medulla where some nerves terminate ventrolateral medullary reticular formation, torus semicircularus, the ventral optic tectum, the ventromedial nucleus of the thalamus, and the lateral subnucleus of the nucleus preglomerulosus of the thalamus (Finger, 2000). A prominent spinocerebellar system in this teleost was also reported. Finger (2000) stated that the similarities in connectivity between *P. carolinus* and other vertebrates are striking. Therefore, fish do possess a spinothalamic and trigeminal tract that is comparable to higher vertebrates but studies on how nociceptive information is conveyed by these tracts to higher brain areas are lacking in fish.

4. MOLECULER MARKERS OF NOCICEPTION

In higher vertebrates, a powerful battery of techniques have yielded important insights into the mechanisms by which noxious, potentially painful stimuli are transduced and processed (Lewin *et al.*, 2004). This has identified a number of key molecular players that are involved in the molecular and cellular mechanisms underlying central nervous system processing of painful stimuli in humans. Unfortunately, these molecules have not been investigated with regards to nociception in fish but many studies have examined the presence and distribution of these molecules. Some of these molecules are gamma-aminobutyric acid (GABA), NMDA, preprotachykinins, substance P, calcitonin gene related peptide (CGRP), opioids, endogenous opioids, and enkephalins. Therefore, the following part of this chapter shall review the existence of these molecules in the fish nervous system. For example, the distribution of immunoreactivity to serotonin, substance P, somatostatin, CGRP, neuropeptide Y, and bombesinin in a variety of elasmobranch species is strikingly similar to mammals (Ebbesson and Hodde, 1981; Cameron *et al.*, 1990). Although crucial in nociception, these proteins are also involved in other functions.

4.1. GABA

With respect to nociception in mammalian models, GABA is involved in allodynia, where damaged tissue becomes hypersensitized and previously innocuous stimuli which would not stimulate the nociceptors of the area before damage are now extremely painful (Lewin *et al.*, 2004). In development, GABA neurons appear 51 h posthatch in the stickleback, *Gasterosteus aculeatus* (Ekstrom and Ohlin, 1995). These GABA neurons appear in the ventral telencephalon or subpallium caudal to the optic recess, in the ventral diencephalon, and in the spinal cord. As development proceeds, GABA neurons appear in the rostral telencephalon and ventral rhombencephalon, hypothalamus and pretectum, preoptic region, thalamus, and throughout the spinal cord by 96 h posthatch. Studies have shown that GABA is responsive to energetic stress with a decrease during exhaustive exercise in the teleost fish, the red drum, *Sciaenops ocellatus*, with an increase in the brain postexercise (Ortiz and Lutz, 1995) and this may serve a protective role to prevent neuronal damage in the energy-compromised state. This role may be confirmed by an increase in GABA during anoxia in the crucian carp, *Carassius carassius* (Hylland and Nilsson, 1999).

4.2. Substance P and the Preprotachykinins

Substance P is released by nociceptors when damaged by noxious stimuli and causes blood cells to dilate and also stimulates mast cells to produce histamine causing an inflammatory response. Substance P and preprotachykinin family are involved in the mechanisms of cell damage in mammalian models. The vasodilator action of the tachykinin family including substance P is mediated primarily through neurokinin (NK1) receptors (Waugh *et al.*, 1993). Both mammalian substance P and dogfish (*S. canicula*) substance P have a vasodilatory action on the rat but not on unrestrained dogfish, suggesting that NK1 receptors do not innervate the cardiovascular system of this species. However, intestinal cells of the rainbow trout, *O. mykiss*, do degranulate in response to substance P application and capsaicin, which is a noxious stimulus, also has this effect (Powell *et al.*, 1993). The role of substance P in nociception in fish has not been explored but many studies have examined the distribution of this tachykinin. When examining the telencephalon of the Pacific hagfish, *E. stouti*, substance P distribution was dissimilar compared with other vertebrates (Wicht and Northcutt, 1994). Whereas in the dogfish, *S. canicula*, substance P-like immunoreactive cell bodies and fibers are widely dispersed throughout the brain (Rodríguez Moldes *et al.*, 1993). In the telencephalon, substance P is found in the olfactory bulbs, pallium, and subpallium. In the diencephalon, substance P is found in the preoptic recess organ, organon vasculosum hypothalami, and hypothalamus. In the rhombencephalon, substance P activity is found in the isthmal region, the trigeminal descending root, the visceral sensory area, commissural nucleus, visceromotor column, and in high density of the substantia gelatinosa of the rostral spinal cord. In the gymnotiform fish, the brown ghost fish, *Apteronotus leptorhynchus*, a similar distribution was seen but labeling was also seen in the funicular nucleus which has a nociceptive function in other vertebrates (Weld and Maler, 1992). Substance P labeling has also been seen in the cranial and spinal sensory ganglia in the filefish, *Stephanolepsis cirrhifer* (Funakoshi *et al.*, 2000). Therefore, substance P appears to be found in the areas where nociceptive processing occurs in higher vertebrates.

4.3. NMDA

NMDA receptors have been shown to be involved in the sensitization of pain in the central nervous system of mammals and contribute to chronic, pathological pain in humans (Lewin *et al.*, 2004). NMDA receptor 1 has been identified in the teleosts, *A. leptorynchus*, and the zebrafish, *D. rerio* (Bottai *et al.*, 1997; Edwards and Michel, 2003). NMDA receptor 1 was found to be highly enriched in the telencephalon especially in the pallial region and hypothalamus with lower amounts in the brain stem and cerebellum (Bottai *et al.*, 1997). Nothing is known about NMDA and nociception in fish.

4.4. Opioids, Endogenous Opioids, and Enkephalins

Opioid receptors, endogenous opioids, and enkephalins are one of the requirements to determine whether nociception can occur in an animal (Bateson, 1991). The nervous systems of cyclostomes (lampreys and hagfish), holosteans (bowfins), teleosts (bony fish), and dipnoans (lungfish) produce endogenous opioid neuromodulators (Dores and Joss, 1988; Dores *et al.*, 1989; McDonald and Dores, 1991). The classic opioid gene family comprises three neuropeptide precursors encoded on the following genes: proopiomelanocortin (end products: endorphin, melanocyte-stimulating hormone-associated peptides), proenkephalin (end products: met-enkephalin, leu-enkephalin), and prodynorphin (end products: dynorphins and neoendorphin). All of these have been found in fish (McDonald and Dores, 1991). Opioids elicit antinociception or analgesia through three distinct types of receptors designated as mu, kappa, and delta in mammals (Newman *et al.*, 2000) and these three receptors have been identified in the zebrafish, *D. rerio* (Stevens, 2004), as well as several other species including the Pacific hagfish (Li *et al.*, 1996). In higher vertebrates, opioid receptors and substances are located particularly in the regions involved in the processing of nociceptive and pain information, for example, the spinal cord, the raphe nucleus, the reticular formation, the periaqueductal gray, and the thalamus (Simantov *et al.*, 1977). In nonmammalian vertebrates, enkephalin-like immunoreactivity has been demonstrated in birds (Reiner *et al.*, 1982), reptiles (Reiner, 1987), and amphibians (Merchenthaler *et al.*, 1987). Opioid receptors have been found in fish (Buatti and Pasternak, 1981) as well as enkephalin-like substances in various brain areas of goldfish, *C. auratus* (Finger, 1981; Schulman *et al.*, 1981); catfish, *Clarias batrachus* (Finger, 1981); African lungfish, *Protopterus annectens* (Reiner and Northcutt, 1987), and rainbow trout, *O. mykiss* (Vecino *et al.*, 1991). Within the brain of fish, enkephalins show a similar distribution pattern to that seen in higher vertebrates (Vecino *et al.*, 1992) and are found in the telencephalon, nucleus ventromedialis of the thalamus, nucleus lateralis tuberis, nucleus recessus lateralis, and nucleaus recessus posterioris. Enkephalin-like activity was also seen in the mesencephalic tegmentum, medial torus semicircularis, and the cerebellum. The highest density of enkephalin-like stained fibers were found in the telencephalon in the area ventralis telencephali, the mesencephalic tegmentum, and the dorsal horn of the spinal cord of *O. mykiss* (Vecino *et al.*, 1992). Enkephalin-like activity shows a similar pattern in *Salmo salar* (Vecino *et al.*, 1995). In the spinal cord, enkephalin-like immunoreactivity is most dense in the superficial portion of lamina A, which is thought to be similar to the substantia gelatinosa of mammals (Snow *et al.*, 1996). In elasmobranchs, both met-enkephalin and leu-enkephalin have been identified but dynorphin-related peptides appear to be absent (Vallarino *et al.*, 1994). These two enkephalins are found in distinct nuclei of the dogfish, *S. canicula*, brain, and in particular the pallium of the telencephalon. The distribution is similar to that found in tetrapods.

The family of RFamide-related peptides are believed to be endogenous opioid agonists and are involved in nociception (Kanetoh *et al.*, 2003). In the catfish *C. batrachus*, there is a close relationship with FMRF-related peptides and the central opioid system with morphine causing transport or release of the FMRFamide peptides (Khan *et al.*, 1998). The actions of morphine can also be blocked by the antagonists naloxone and MIF-1 (Ehrensing *et al.*, 1982). Therefore, a comparable distribution of opioid receptors and substances is found in both elasmobranchs and teleosts with the fundamental physiological mechanisms similar to that seen in higher animals.

4.5. Global Gene Expression

With the advent of genome-wide screening tools, it is now possible to examine global gene expression using microarrays or gene chips in response to a noxious stimulus (Reilly *et al.*, 2004). This has been proved to be a useful approach in mammalian studies in the search for the molecular basis of nociception (Mogil *et al.*, 2000; Reilly *et al.*, 2004). In the Sneddon laboratory, common carp, *C. carpio*, and rainbow trout, *O. mykiss*, were either given injections of noxious stimuli or saline as a control and then forebrain, midbrain, and hindbrain were sampled at subsequent intervals. The majority of gene expression changes during nociception occurred in the forebrain where the cortex is situated (975 genes), followed by the midbrain (777 genes) with least expression changes occurring in the midbrain (485 genes; all genes were significantly different from the control samples, Reilly and Sneddon, unpublished data). All of the key players identified above were found to be significantly expressed in the fish brain and so are evolutionarily conserved. These results together with the electrophysiological data from Dunlop and Laming (2005) demonstrate that higher brain areas are active at the molecular and cellular level during a noxious experience and, therefore, this is not confined to reflex activity in the spinal cord and brain stem.

5. WHOLE ANIMAL RESPONSES

5.1. Avoidance Learning

One of the criteria for nociception is that the animal must be able to learn to avoid a noxious stimulus. Ehrensing *et al.* (1982) conditioned goldfish to avoid electric shock using negative conditioning as a test paradigm. When morphine was administered the fish failed to learn and a high voltage of electric shock was needed to elicit a response, whereas when the anatagonists MIF-1 and naloxone were also administered, avoidance was generated at a much lower voltage (Figure 4.7). A few other studies have shown that teleost fish are capable of associating a stimulus with a noxious experience and learning to avoid it subsequently (Beukema, 1970a,b). These involved common carp, *C. carpio*, and pike, *Esox lucius*, avoiding hooks in angling trials.

5.2. *In Vivo* Observations

To try and gauge whether an animal is perceiving pain rather than a simple nociceptive reflex, behavioral observations have been made on fish experiencing a noxious, potentially painful event. If the animal's normal behavior is adversely affected and these responses are not simple reflexes then there may be a psychological component to the experience that may be negative. This is a very subjective method but the primary method is also available for assessing pain in animals (Bateson, 1991). When rainbow trout, *O. mykiss*, were given subcutaneous injections of acetic acid and bee venom (algesics) to the lips, they showed an enhanced respiration rate for ~3 h as

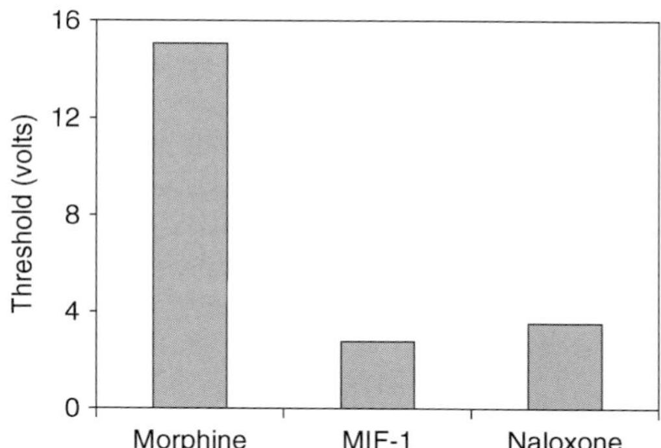

Fig. 4.7. The voltage of electric shock needed for an avoidance response in goldfish administered with morphine or the opioid antagonists MIF-1 and naloxone. (Adapted from *Pharmacology, Biochemistry and Behavior*, 17, Ehrensing *et al.*, © 1982, with permission from Elsevier.)

well as performance of anomalous behaviors and during this period they did not feed (Sneddon *et al.*, 2003a). Handled controls and saline-injected fish did not perform these anomalous behaviors and did not show such a great increase in respiration rate and began feeding around 80 min after treatment. The noxiously stimulated fish only resumed feeding once the behavioral and physiological affects of the bee venom and acetic acid had subsided (~180 min). Further testing showed that noxiously stimulated trout did not show an appropriate fear response to a fear-causing stimulus and it was suggested that the noxious experience dominated attention and the fish could not divert attention to the fear stimulus (Sneddon *et al.*, 2003b). This could be interpreted as the noxious experience being the imperative in this test paradigm and many clinical studies have shown that human also do not perform on other tasks when in pain (Kuhajda *et al.*, 2002). Therefore, the studies in the rainbow trout demonstrate that the negative effects of a noxious experience are complicated in nature, suggesting higher processing is involved and thus there is the potential for pain perception in this teleost fish. Little work has been conducted on other species but common carp, *C. carpio*, and zebrafish, *D. rerio*, do show similar anomalous behaviors while enduring acetic acid injection that trout were given. (Reilly *et al.*, unpublished observations).

Very little is known about analgesia in fish since it has been found that they possess nociceptors (Sneddon, 2002, 2003a). The few studies that are available address the effects of morphine, an opioid analgesic. The *in vivo* responses to subcutaneous acetic acid injection were quantified in the rainbow trout, *O. mykiss*. These adverse behavioral responses included the rubbing of the affected area and rocking from side to side on either pectoral fin which was not seen in handled controls or saline-injected fish (Sneddon, 2003b; Sneddon *et al.*, 2003a). When morphine was administered to fish injected with acid, there was a dramatic reduction in this rubbing behavior (Figure 4.8A Sneddon, 2003b) as well as rocking behavior. Additionally, the enhanced respiration rate seen in acid-injected fish was also ameliorated by morphine affecting behavioral output (Figure 4.8B). Therefore, morphine appears to reduce nociceptive reponses in this teleost fish. This has not been repeated in any of the other fish groups.

These indirect measures of the response to a potentially painful event demonstrate the complicated nature of the response and combined with the neurophysiology of the central and peripheral nervous system; we have conclusive evidence that fish are capable of nociception. Since the behavior is not limited to a simple reflex response and the higher brain centers are active, and the fish studied show suspension of normal behavior, one could conclude that the experience is an adverse one. Therefore, there is the potential for pain perception. Hopefully, this chapter has shown that fish

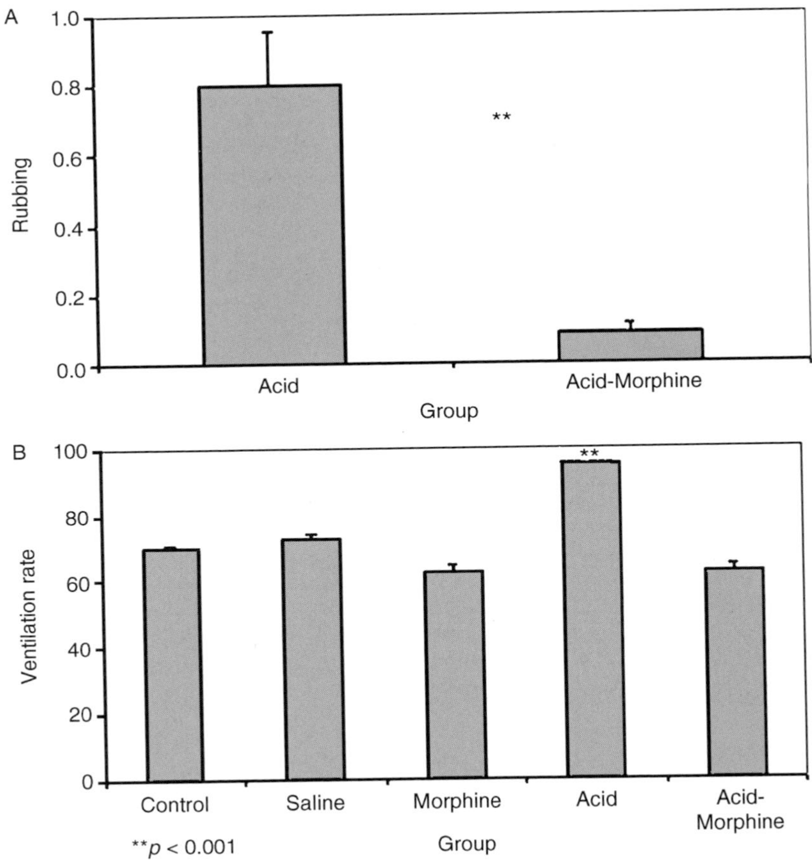

Fig. 4.8. (A) The mean frequency of rubbing performed by rainbow trout injected subcutaneously with 0.1% acetic acid (Acid) and fish also injected with the acid but administered with morphine (Acid-Morphine). (B) The mean respiration rate (rate of ventilation of gills) of handled controls (Control), saline injected fish (Saline), control fish administered with morphine (Morphine), acid injected fish (Acid), as well as acid injected fish that were treated with morphine (Acid-Morphine). (Reprinted from *Applied Animal Behaviour Science*, 83, Sneddon, © 2003, with permission from Elsevier.)

may be capable of sensory pain but unless one can get into the animal mind and know how an animal is "feeling" it is impossible to know whether the fish is suffering. The author allows the readers to make up their own mind after presenting the evidence. Since we cannot truly know what a fish experiences, it may be best to give them the benefit of doubt and treat them humanely as if they are capable of pain perception.

6. CONCLUSIONS

Agnathans and teleosts appear to possess the neural apparatus to detect noxious stimuli and studies have confirmed the possession of nociceptors. These have yet to be identified in elasmobranches, although this group does show aversion to noxious stimuli. The physiological and anatomical properties of these nociceptors are similar to those found in higher vertebrates. The fish groups do possess the brain areas and pathways leading to higher brain centers that are necessary for nociception to occur in other vertebrate models. Many of the key molecules involved in nociception in mammalian models are present in all fish groups and generally distributed in similar areas of the brain and spinal cord. The fish species tested have learned to avoid a noxious stimulation and in two teleost species profound physiological and behavioral changes are apparent while the fish endure a noxious event. Work on the fish groups is in its infancy compared with other animals and by studying a variety of species we can gain an insight into the evolution of this important sensory system. However, much remains to be done and by characterizing nociceptors in more detail we shall gain an insight into the comparative aspects of nociception. Studies should not only be restricted to anatomy and electrophysiology but should also assess the involvement of the brain, changes at the molecular level, and also investigating the behavioral responses to noxious stimulation to include the use of analgesics. Another key point to address is species differences since there are behavioral and physiological differences between common carp, goldfish, zebrafish, and rainbow trout when subjected to a noxious event (Dunlop and Laming, 2005; Reilly *et al.*, unpublished observations). Fish groups have an amazing and diverse array of life histories, physiological tolerances, and will encounter varied environmental conditions. For example, species living in polar regions will not encounter such high temperatures as those living in tropical regions; fish living in the deep sea will be subject to greater pressure than fish with a coastal shelf distribution; and fish may encounter a variety of chemicals in industrial and urban areas which are not found in pristine waters. Therefore, it is vital that a wide range of species are tested to truly understand their capacity for nociception and pain.

ACKNOWLEDGMENTS

I am grateful to Toshiaki Hara for inviting me to write this chapter and for an NERC fellowship, BBSRC research grant, and financial support from the Wellcome Trust and the Royal Society of London.

REFERENCES

Anadón, R., Molist, P., Rodríguez-Moldes, I., López, J. M., Quintela, I., Cerviño, M. C., Barja, P., and González, A. (2000). Distribution of choline acetyltransferase immunoreactivity in the brain of an elasmobranch, the lesser spotted dogfish (*Scyliorhinus canicula*). *J. Comp. Neurol.* **420,** 139–170.

Ashley, P. A., McCrohan, C. R., and Sneddon, L. U. Unpublished observations.

Bateson, P. (1991). Assessment of pain in animals. *Anim. Behav.* **42,** 827–839.

Belmonte, C., and Gallar, J. (1996). Corneal nociceptors. *In* "Neurobiology of Nociceptors" (Belmonte, C., and Cervero, F., Eds.), pp. 146–183. Oxford University Press, Oxford.

Benson, G. J., and Thurmon, J. C. (1987). Species differences as a consideration in the alleviation of animal pain and distress. *J. Am. Vet. Med. Assoc.* **191,** 1227–1230.

Beukema, J. J. (1970a). Angling experiments with carp (*Cyprinus carpio* L.) II. Decreased catchability through one trial learning. *Neth. J. Zool.* **19,** 81–92.

Beukema, J. J. (1970b). Acquired hook avoidance in the pike *Esox lucius* L. fished with artificial and natural baits. *J. Fish Biol.* **2,** 155–160.

Bone, Q. (1963). The central nervous system. *In* "The Biology of Myxine" (Bordal, A., and Fange, R., Eds.), pp. 50–91. Universitetsforlaget, Oslo.

Bottai, D., Dunn, R. J., Ellis, W., and Maler, L. (1997). N-methyl-D-aspartate receptor 1 mRNA distribution in the central nervous system of the weakly electric fish *Apteronotus leptorhynchus*. *J. Comp. Neurol.* **389,** 65–80.

Buatti, M. C., and Pasternak, G. W. (1981). Multiple opioid receptors: Phylogenetic differences. *Brain Res.* **218,** 400–405.

Bullock, T. H., and Corwin, J. T. (1979). Acoustic evoked activity in the brain of sharks. *J. Comp. Physiol.* **129,** 23–34.

Cameron, A. A., Plenderleith, M. B., and Snow, P. J. (1990). Organisation of the spinal cord in four species of elasmobranch fish: Cytoarchitecture and distribution of serotonin and selected peptides. *J. Comp. Neurol.* **297,** 201–218.

Coggeshall, R. E., Leonard, R. B., Applebaum, M. L., and Willis, W. D. (1978). Organisation of peripheral nerves of the Atlantic stingray, *Dasyatis sabina*. *J. Neurophysiol.* **41,** 97–107.

Cohen, D. H., Duff, T. A., and Ebbesson, S. O. E. (1973). Electrophysiological identification of a visual area in shark telencephalon. *Science* **182,** 492–494.

Cruce, W. L. R., Stuesse, S. L., and Northcutt, R. G. (1999). Brainstem neurons with descending projections to the spinal cord of two elasmobranch fishes: Thornback guitarfish, *Platyrhinoidis triseriata*, and horn shark, *Heterodontus francisci*. *J. Comp. Neurol.* **403,** 534–560.

Dores, R. M., and Joss, J. M. P. (1988). Immunological evidence for multiple forms of alpha-melanotrophin in the pars intermedia of the Australian lungfish *Neoceratodus forsteri*. *Gen. Comp. Endocrinol.* **71,** 468–474.

Dores, R. M., McDonald, L. K., and Crim, J. W. (1989). Flexible body dynamics of the goldfish: Implications for reticulospinal command mechanisms. *J. Neurosci.* **8,** 2758–2768.

Dunlop, R., and Laming, P. (2005). Mechanoreceptive and nociceptive responses in the central nervous system of goldfish (*Carassius auratus*) and trout (*Oncorhynchus mykiss*). *J. Pain* **9,** 561–568.

Ebbesson, S. O. E., and Hodde, K. C. (1981). Ascending spinal systems in the nurse shark, *Ginglymostoma cirratum*. *Cell Tissue Res.* **216,** 313–331.

Edwards, J. G., and Michel, W. C. (2003). Pharmacological characterisation of ionotropic glutamate receptors in the zebrafish olfactory bulb. *Neurosci.* **122,** 1037–1047.

Ehrensing, R. H., Michell, G. F., and Kastin, A. J. (1982). Similar antagonism of morphine analgesia by MIF-1 and naxolone in *Carassius auratus*. *Pharm. Biochem. Behav.* **17,** 757–761.

Ekstrom, P., and Ohlin, L. M. (1995). Ontogeny of GABA-immunoreactive neurons in the central nervous system of in a teleost, *Gasterosteus aculeatus*. *J. Chem. Neuroanat.* **9,** 271– 288.

Finger, T. E. (1981). Fish that taste with their feet: Spinal sensory pathways in the sea robin, *Prionotus carolinus*. *Biol. Bull.* **161,** 154–161.

Finger, T. E. (2000). Ascending spinal systems in the fish, *Prionotus carolinus*. *J. Comp. Neurol.* **422,** 106–122.

Funakoshi, K., Kadota, T., Atobe, Y., Nakano, M., Goris, R. C., and Kishida, R. (2000). Differential distribution of nerve terminals immunoreactive for substance P and cholecystokinin in the sympathetic preganglionic cell column of the filefish *Stephanolepsis cirrhifer*. *J. Comp. Neurol.* **428,** 174–189.

Gallego, R. (1983). The ionic basis of action potentials in petrosal ganglion cells of the cat. *J. Physiol.* **342,** 591–602.

Gentle, M. J., and Tilston, V. L. (2000). Nociceptors in the legs of poultry: Implications for potential pain in pre-slaughter shackling. *Anim. Welf.* **9,** 227–236.

Gilbert, P. W., and Gilbert, C. (1973). Sharks and shark deterrents. *Underwater J.* **5,** 69–79.

Goehler, L. E., and Finger, T. E. (1996). Visceral afferent and efferent columns in the spinal cord of the teleost, *Ictalurus punctatus*. *J. Comp. Neurol.* **371,** 437–447.

Haupt, P., Janig, W., and Kohler, W. (1983). Response pattern of visceral afferent fibers, supplying the colon, upon chemical and mechanical stimuli. *Eur. J. Physiol.* **398,** 41–47.

Hylland, P., and Nilsson, G. E. (1999). Extracellular levels of amino acids neurotransmitters during anoxia and forced energy deficiency in crucian carp brain. *Brain Res.* **823,** 49–58.

Iwahori, N., Kawawaki, T., and Baba, J. (1998). Neuronal organisation of the spinal cord in the red stingray (*Dasyatids akajei: Chondrichthyes*). *J. Brain Res.* **39,** 103–116.

Kajiura, S. M., Sebastian, A. P., and Tricas, T. C. (2000). Dermal bite wounds as indicators of reproductive seasonality and behaviour in the Atlantic stingray, *Dasyatis sabina*. *Env. Biol. Fishes* **58,** 23–31.

Kanetoh, T., Sugikawa, T., Sasaki, I., Muneoka, Y., Minakata, H., Takabatake, I., and Fujimoto, M. (2003). Identification of a novel frog RFamide and its effect on the latency of the tail-flick response of the newt. *Comp. Biochem. Physiol. C* **134,** 259–266.

Khan, F. A., Jain, M. R., Saha, S. G., and Subhedar, N. (1998). FMRFamide-like immunoreactivity in the olfactory system responds to morphine treatment in the teleost *Clarias batrachus*: Involvement of opioid receptors. *Gen. Comp. Endocrinol.* **110,** 79–87.

Kuhajda, M. C., Thorn, B. E., Klinger, M. R., and Rubin, N. J. (2002). The effect of headache pain on attention (encoding) and memory (recognition). *Pain* **97,** 213–221.

Kumazawa, T., and Mizumura, K. (1977). Thin-fibre receptors responding to mechanical, chemical and thermal stimulation in skeletal muscle of the dog. *J. Physiol.* **273,** 179–194.

Kumazawa, T., and Mizumura, K. (1980). Chemical responses of polymodal receptors of the scrotal contents of dogs. *J. Physiol.* **299,** 219–230.

Lang, E., Novak, A., Reeh, P. W., and Handwerker, H. O. (1990). Chemosensitivity of fine afferents from the rat skin *in vitro*. *J. Neurophysiol.* **63,** 887–901.

Leonard, R. B. (1985). Primary afferent receptive field properties and neurotransmitter candidates in a vertebrate lacking unmyelinated fibres. *Prog. Clin. Res.* **176,** 135–145.

Lewin, G. R., Lu, Y., and Park, T. J. (2004). A plethora of pain molecules. *Curr. Opin. Neurobiol.* **14,** 443–449.

Li, X., Keith, D. E., Jr., and Evans, C. J. (1996). Mu opioid receptor-like sequences are present throughout vertebrate evolution. *J. Mol. Evol.* **43,** 179–184.

Liang, Y., and Terashima, S. (1993). Physiological properties and morphological characteristics of cutaneous and mucosal mechanical nociceptive neurons with A-δ peripheral axons in the trigeminal ganglia of crotaline snakes. *J. Comp. Neurol.* **328,** 88–102.

López deArmentia, M., Cabanes, C., and Belmonte, C. (2000). Electrophysiological properties of identified trigeminal ganglion neurons innervating the cornea of the mouse. *Neurosci.* **101,** 1109–1115.

Luiten, P. G. M. (1975). The central projections of the trigeminal, facial, and anterior lateral line nerves in the carp (*Cyprinus carpio* L.). *J. Comp. Neurol.* **160,** 399–418.

Lynn, B. (1994). The fibre composition of cutaneous nerves and the classification and response properties of cutaneous afferents, with particular reference to nociception. *Pain Rev.* **1,** 172–183.

Matthews, G., and Wickelgren, W. O. (1978). Trigeminal sensory neurons of the sea lamprey. *J. Comp. Physiol. A* **123,** 329–333.

McDonald, L. K., and Dores, R. M. (1991). Detection of met-enkephalin in the CNS of teleosts *Anguilla rostrata* and *Oncorhynchus kisutch*. *Peptides* **12,** 541–547.

Merchenthaler, I., Maderdrut, J. L., Lazar, G., Gulyas, J., and Pertrusz, P. (1987). Immunocytochemical analysis of proenkephalin-derived peptides in the amphibian hypothalamus and optic tectum. *Brain Res.* **416,** 219–227.

Mogil, J. S., Yu, L., and Basbaum, A. I. (2002). Pain genes?: Natural variation and transgenic mutants. *Ann. Rev. Neurosci.* **23,** 777–811.

New, J. G., and Northcutt, R. G. (1984). Primary projections of the trigeminal nerve in two species of sturgeon: *Acipenser oxyrynchus* and *Scaphirhynchus platorynchus*. *J. Morphol.* **182,** 125–136.

Newman, L. C., Wallace, D. R., and Stevens, C. W. (2000). Selective opioid agonist and antagonist competition for [H3]-naloxone binding in amphibian spinal cord. *Brain Res.* **884,** 184–191.

Nieuwenhuys, R. (1977). The brain of the lamprey in a comparative perspective. *Ann. NY Acad. Sci.* **299,** 97–145.

Northcutt, R. G. (1979). Experimental determination of the primary trigeminal projections in lampreys. *Brain Res.* **163,** 323–327.

Northcutt, R. G. (1981). Evolution of the telencephalon in nonmammals. *Ann. Rev. Neurosci.* **4,** 301–350.

Ortiz, M., and Lutz, P. L. (1995). Brain neurotransmitter changes associated with exercise and stress in a teleost fish (*Sciaenops ocellatus*). *J. Fish Biol.* **46,** 551–562.

Platt, C. J., Bullock, T. H., Czěh, G., Kovačević, N., Konjević, D., and Gojković, M. (1974). Comparison of electroreceptor, mechanoreceptor and optic evoked potentials in the brains of some rays and sharks. *J. Comp. Physiol.* **9,** 323–355.

Portavella, M., Torres, B., Salas, C., and Papini, M. R. (2004). Lesions of the medial pallium, but not of the lateral pallium, disrupt spaced-trial avoidance learning in goldfish (*Carassius auratus*). *Neurosci. Lett.* **362,** 75–78.

Powell, M. D., Wright, G. M., and Burka, J. F. (1993). Morphological and distributional changes in the eosinophilic granule cell (EGC) population of the rainbow trout (*Oncoryhnchus mykiss*) intestine following systemic administration of capsaicin and substance P. *J. Exp. Zool.* **266,** 19–30.

Reilly, S. C., Cossins, A. R., Quinn, J. P., and Sneddon, L. U. (2004). Discovering genes: The use of microarrays and laser capture microdissection in pain research. *Brain Res. Rev.* **46,** 225–233.

Reilly, S. C., Cossins, A. R., Quinn, J. P., and Sneddon, L. U. Unpublished observations.

Reiner, A. (1987). The distribution of proenkephalin-derived peptides in the central nervous system of turtles. *J. Comp. Neurol.* **259,** 65–91.

Reiner, A., and Northcutt, R. G. (1987). An immunohistochemistry study of the telencephalon of the African lungfish, *Protopterus annectens*. *J. Comp. Neurol.* **256,** 463–481.

Reiner, A., Karten, H. J., and Brecha, N. C. (1982). Enkephalin-mediated basal ganglia influences over the optic tectum: Immunohistochemistry of the tectum and lateral spiriform nucleus in pigeon. *J. Comp. Neurol.* **208,** 37–53.

Rink, E., and Wulliman, M. F. (2004). Connections of the ventral telencephalon (subpallium) in the zebrafish (*Danio rerio*). *Brain Res.* **1011,** 206–220.

Rodríguez-Moldes, I., Manso, M. J., Becerra, M., Molist, P., and Anadón, R. (1993). Distribution of substance P-like immunoreactivity in the brain of the elasmobranch *Scyliorhinus canicula. J. Comp. Neurol.* **335,** 228–244.

Ronan, M. (1988). The sensory trigeminal tract of Pacific hagfish. Primary afferent projections and neurons of the tract nucleus. *Brain Behav. Ecol.* **32,** 169–180.

Schulman, J. A., Finger, T. E., Brecha, N. C., and Karten, H. J. (1981). Enkephalin immunoreactivity in Golgi cells and mossy fibres of mammalian, avian and teleost cerebellum. *Neuroscience* **6,** 2407–2416.

Sharpe, C. S., Thompson, D. A., Blankenship, H. L., and Schreck, C. B. (1998). Effects of routine handling and tagging procedures on physiological stress responses in juvenile chinook salmon. *Prog. Fish Cult.* **60,** 81–87.

Simantov, R., Kuhar, M. J., Uhl, G. R., and Snyder, S. H. (1977). Opioid peptide enkephalin: Immunohistochemical mapping in rat central nervous system. *Proc. Natl. Acad. Sci. USA* **74,** 2167–2171.

Sisneros, J. A., and Nelson, D. R. (2000). Surfactants as chemical shark repellents: Past, present and future. *Environ. Biol. Fishes* **60,** 117–129.

Skjervold, P. O., Fjaera, S. O., Østby, P. B., and Einen, O. (2001). Live-chilling and crowding stress before slaughter of Atlantic salmon (*Salmo salar*). *Aquaculture* **192,** 265–280.

Sneddon, L. U. (2002). Anatomical and electrophysiological analysis of the trigeminal nerve in a teleost fish, *Oncorhynchus mykiss. Neurosci. Lett.* **319,** 167–171.

Sneddon, L. U. (2003a). Trigeminal somatosensory innervation of the head of the rainbow trout with particular reference to nociception. *Brain Res.* **972,** 44–52.

Sneddon, L. U. (2003b). The evidence for pain perception in fish: The use of morphine as an analgesic. *App. Anim. Behav. Sci.* **83,** 153–162.

Sneddon, L. U. (2004). Evolution of nociception in vertebrates: Comparative analysis of lower vertebrates. *Brain Res. Rev.* **46,** 123–130.

Sneddon, L. U., Braithwaite, V. A., and Gentle, M. J. (2003a). Do fish have nociceptors: Evidence for the evolution of a vertebrate sensory system. *Proc. Roy. Soc. Lond. B* **270,** 1115–1122.

Sneddon, L. U., Braithwaite, V. A., and Gentle, M. J. (2003b). Novel object test: Examining pain and fear in the rainbow trout. *J. Pain* **4,** 431–440.

Sneddon, L. U., Verhoye, M., and Van der Linden, A. Unpublished observations.

Snow, P. J., Plenderleith, M. B., and Wright, L. L. (1993). Quantitative study of primary sensory neurone populations of three species of elasmobranch fish. *J. Comp. Neurol.* **334,** 97–103.

Snow, P. J., Renshaw, G. M. C., and Hamlin, K. E. (1996). Localization of enkephalin immunoreactivity in the spinal cord of the long-tailed ray *Himantura fai. J. Comp. Neurol.* **367,** 264–273.

Stevens, C. W. (2004). Opioid research in amphibians: An alternative pain model yielding insights on the evolution of opioid receptors. *Brain Res. Rev.* **46,** 204–215.

Toda, K., Ishii, N., and Nakamura, Y. (1997). Characteristics of mucosal nocieptors in the rat oral cavity: An *in vitro* study. *Neurosci. Lett.* **228,** 95–98.

Vallarino, M., Bucharles, C., Facchinetti, F., and Vaudry, H. (1994). Immunocytochemical evidence for the presence of Met-enkephalin and Leu-enkephalin in distinct neurons in the brain of the elasmobranch fish *Scyliorhynus canicula. J. Comp. Neurol.* **347,** 585–597.

Vecino, E., Ekstrom, P., and Sharma, S. C. (1991). Enkephalin-like immunoreactive cells in the mesencephalic tegmentum project to the optic tectum in *Salmo gairdneri* and *Salmo salar*. *Cell Tissue Res.* **264,** 133–137.

Vecino, E., Piňuela, C., Arévalo, R., Lara, J., Alonso, J. R., and Aijón, J. (1992). Distribution of enkephalin like immunoreactivity in the central nervous system of the rainbow trout: An immunocytochemical study. *J. Anat.* **180,** 435–453.

Vecino, E., Perez, M. T. R., and Ekstrom, P. (1995). Localisation of enkephalinergic neurons in the central nervous system of the salmon (*Salmo salar*) by *in situ* hybridisation and immunocytochemistry. *J. Chem. Neuroanat.* **9,** 81–97.

Waugh, D., Wang, Y. X., Hazon, N., Balment, R. J., and Conlon, J. M. (1993). Primary structures and biological activities of substance P related peptides from the brain of dogfish, *Scyliorhinus canicula*. *Eur. J. Biochem.* **214,** 469–474.

Weld, M. M., and Maler, L. (1992). Substance-P-like immunoreactivity in the brain of the gymnotiform fish *Apteronotus leptorynchus*: Presence of sex differences. *J. Chem. Neuroanat.* **5,** 107–129.

Whitear, M. (1983). The question of free nerve endings in the epidermis of lower vertebrates. *Acta Biol. Hung.* **34,** 303–319.

Wicht, H., and Northcutt, R. G. (1994). An immunohistochemical study of the telencephalon and the diencephalons in a myxinoid jawless fish, the pacific hagfish, *Eptatretus stouti*. *Brain Behav. Evol.* **43,** 140–161.

5

VISUAL SENSITIVITY AND SIGNAL PROCESSING IN TELEOSTS

LEI LI
HANS MAASWINKEL

1. Introduction
2. Characteristics of the Visual System
   2.1. Structure of the Eye
   2.2. Brain Areas Involved in Vision
   2.3. Summary
3. Absolute Visual Sensitivity
   3.1. How to Measure Visual Sensitivity?
   3.2. Dark Adaptation
   3.3. Summary
4. Circadian Regulation of Visual Sensitivity
   4.1. Circadian Modulation of Rod and Cone Sensitivity
   4.2. Circadian Modulation of Rod–Cone Dominance
   4.3. Circadian Regulation of Dopamine and Melatonin Release
   4.4. Circadian Regulation of Opsin mRNA Expression
   4.5. Summary
5. Chemosensory Modulation of Visual Sensitivity
   5.1. The Terminal Nerve
   5.2. Olfactory Stimulation Affects Visual Sensitivity via the TN Projection to the Retina
   5.3. Mechanisms by Which the TN Modulates Visual Sensitivity
   5.4. Significance of the Retinal Projection of the TN
   5.5. Summary
6. Inherited and Acquired Impairments of Visual Sensitivity
   6.1. Night Blindness Mutations
   6.2. Acquired Impairments
   6.3. Summary
7. Contrast Visual Sensitivity
   7.1. Concepts of Contrast Sensitivity
   7.2. Development and Dysfunction of Contrast Sensitivity
   7.3. Summary

Sensory Systems Neuroscience: Volume 25
FISH PHYSIOLOGY

DOI: 10.1016/S1546-5098(06)25005-2

8. Spectral Visual Sensitivity
 8.1. Spectral Coding
 8.2. Tuning of Spectral Sensitivity
 8.3. Summary
9. Conclusions

1. INTRODUCTION

Fish have adapted to a wide range of habitats, including estuaries, deep sea, lakes, and caves. Accordingly, the characteristics of the visual system vary considerably among species. This chapter will mainly report the research carried out in a small number of species belonging to the cyprinids, such as goldfish (*Carassius auratus*) and zebrafish (*Danio rerio*). Some of the topics in this chapter focus on research that is related to our own attempts in the field of ophthalmic research.

As a first approach, visual sensitivity can be analyzed along three dimensions: absolute, contrast, and spectral sensitivity. Our own research is mainly concerned with absolute sensitivity, and to a lesser extent, contrast and spectral sensitivity. Accordingly, in this chapter we primarily focus on absolute sensitivity and describe the two other forms of sensitivity in a more concise form. After describing the characteristics of the visual system in teleost fish (Section 2), we present current research about absolute visual sensitivity (Section 3). In Section 4, we describe circadian modulation of absolute sensitivity and the roles of neurotransmitters, such as dopamine and melatonin. Chemosensory modulation of absolute sensitivity is the subject of Section 5. Subsequently, we describe some of the factors that may impair absolute sensitivity (Section 6). In Section 7, we review research concerning contrast sensitivity, which also plays a role in motion detection. Finally, in the last section (Section 8) we discuss some of the relevant studies in the field of spectral sensitivity in fish.

2. CHARACTERISTICS OF THE VISUAL SYSTEM

The visual system in teleosts is highly developed and shows a wide variety of structures, depending on the natural habitat of the species. Deep sea and cave fish, for instance, are often blind or deprived of color vision. Many diurnal fish living in mid- or shallow water possess rods and several types of cones, which might include cones sensitive to ultraviolet light. A special adaptation of the visual system to the environment is found in some species. For example, the four-eyed fish (*Anableps anableps*) live in estuaries and usually swim with the upper half of the eyes above the water line. They possess two separate visual mechanisms suited for either aerial or aquatic vision. In the following section, we characterize the basic structures of the eye and the brain areas involved in processing visual information, especially as described in cyprinids. We will also point out some of the typical differences to the homologous anatomical structures in mammals, especially in humans.

2.1. Structure of the Eye

The general organization of the teleost eye is similar to the organization of the mammalian eye. However, typical differences exist concerning optics, distribution of photoreceptors, range of photopigments, and the role of stem cells in the adult retina.

2.1.1. Optics

Because the cornea of the fish eye is in immediate contact with the surrounding water, its refractive power is weak. To compensate for this, the fish lens has a much higher refractive power than the mammalian lens (Sivak, 1990). In teleosts, the lens is usually spherical and very dense. Consequently, focusing (accommodation) can not be performed, as in mammals, by changing its shape. Instead, the lens moves toward or away from the retina, or moves in the plane of the pupil (Fernald and Wright, 1985). One consequence of this is that different parts of the retina can simultaneously focus on different distances. In some species the density of rods and cones is more or less constant across the retina, that is, the retina lacks specialized areas for acute vision that are characterized by increased density of cones, cone bipolar cells, and retinal ganglion cells (RGC), such as central areas, foveae, or visual streaks. Central areas are roughly circular. In some vertebrates, the inner retina of this area is thinner than normal, resulting in a depression. In this case, it is called fovea. Most primates, many birds and reptiles possess foveae (sometimes more than one per retina). Other vertebrates possess horizontally elongated retinal areas of increased cone density, called visual streaks. Visual streaks are found in turtles and rabbits, among others. Some animals, such as dogs, cats, and cheetahs, possess both a visual streak and a central area. Whether a species possesses a central area, a fovea, visual streak, or no obviously specialized area at all seems to be an adaptation to the natural habitat (Rodieck, 1998). Among teleosts, some species possess foveae, for example, pipefish (*Corythoichtyes paxtoni*), sea basses (Serranidae), blennies (Blenniidae), and several deep-sea fishes (Walls, 1942; Collin and Collin, 1999; Warrant, 2000). However, the majority of teleost species do not have clearly delineated foveae. Well-studied examples are zebrafish, goldfish, green sunfish (*Lepomis cyanellus*), the cichlid *Haplochromis burtonii*, and butterflyfish (*Chaetodon rainfordi*) (Mednick *et al.*, 1988; Fernald, 1990; Cameron and

Easter, 1993; Fritsches and Marshall, 2002; Cameron and Carney, 2004). But even in some of these species, a somewhat higher density of RGCs and cones has been found in parts of the retina, usually the temporal pole (Mednick *et al.*, 1988; Cameron and Easter, 1993; Zygar *et al.*, 1999).

Generally, adaptation to high light intensities can occur at photoreceptor (Perlman and Normann, 1998), network (Green *et al.*, 1975), or mechanical levels (Burnside, 2001). In mammals, for example, adaptation to varying light intensities at the mechanical level is achieved by changing the size of the pupil. This mechanism of light adaptation has a limited range ($\sim$1.2 log units in humans), but has the advantage of being fast. However, according to Perlman and Normann (1998), pupils mainly contribute to augmenting depth of field under high-intensity light conditions. In many teleosts the iris cannot contract. In some teleosts, amphibians, and birds, mechanical adaptation to changes in ambient light intensity is achieved by retinomotor movements (Burnside, 2001). During light adaptation, for example, the cones contract and the rods elongate so that their outer segments are surrounded by the fingerlike processes of retinal pigment epithelium (RPE) cells. Melanosomes migrate into these processes thereby shielding rods from overexposure to high-intensity light (Vihtelic *et al.*, 2006). It is generally assumed that in the light-adapted stage, melanosomes reduce stray light, which would degrade the image quality by decreasing the signal-to-noise ratio in cones (Marc, 1999). During dark adaptation, these steps are reversed.

2.1.2. Photopigments

The teleost retina normally possesses rods and up to four different spectral cone types, namely, long wavelength-sensitive (L), middle wavelength-sensitive (M), short wavelength-sensitive (S), and ultraviolet (UV)-sensitive cones. Analogous to human color vision, L, M, and S cones are sometimes referred as "red," "green," and "blue" cones, respectively. This terminology is not accurate. For example, the absorption maxima (λ_{max}) of L, M, S, and UV cones in goldfish are 623, 537, 447, and 356, respectively (Palacios *et al.*, 1998), whereas in zebrafish they are 564, 478, 407, and 360, respectively (Cameron, 2002). Thus, whereas L-cones in goldfish are most sensitive to "red" light, in zebrafish they are most sensitive to "orange" light. Overall, in zebrafish, the λ_{max} of all cone opsins are shifted to shorter wavelengths in comparison to cone opsins in goldfish, a phenomenon referred as "blue shift" of spectral sensitivity (Levine and MacNichol, 1979). Moreover, L-cone opsin gene in zebrafish is less abundantly expressed than other opsin genes (Chinen *et al.*, 2003), which might further accentuate the blue shift of spectral sensitivity.

UV cones have been described for some fish species. They might play a role in the detection of plankton (Bowmaker and Kunz, 1987). In some species, UV cones are only found in the larval stage. Other species possess UV cones during some stages of their adult lives, but not during other stages. The rainbow trout (*Oncorhynchus mykiss*), for example, loses its UV cones at the time of seaward migration and regenerates them again when returning to freshwater spawning grounds (Hawryshyn, 1998). Many teleost species such as guppy (*Poecilia reticulata*), zebrafish, damselfish (*Dascyllus* sp.), and roach (*Rutilus rutilus*) retain their UV cones throughout adulthood (Branchek and Bremiller, 1984; Douglas, 1986; Cameron, 2002; Hawryshyn *et al.*, 2003). In zebrafish, UV cones seem to contribute to the dark-adapted spectral sensitivity curve (Saszik and Bilotta, 1999). This is in contrast to many other cyprinids such as goldfish (Nussdorf and Powers, 1988) and giant danio (*Danio aequipinnatus*) (van Roessel *et al.*, 1997) where L-cones have been suggested to have such a role (Yang *et al.*, 1990).

As in other vertebrates, photopigments in teleosts (Bowmaker, 1995) are composed of an opsin and a chromophore. Opsins are synthesized by photoreceptor cells, whereas chromophores are converted from vitamin A in the RPE. Some fish possess polymorphic opsin genes. For example, zebrafish have four different M-cone opsin genes and two L-cone opsin genes (LWS-1 and LWS-2), each with a different λ_{max} (Chinen *et al.*, 2003). One of each type is dominantly expressed. Not much is known about the determinants of the expression patterns. A study by Nawrocki *et al.* (1985) found that in young zebrafish larvae, the λ_{max} of L-cones is shorter than in older larvae and adult fish. According to Chinen *et al.* (2003), Nawrocki's finding suggests that the retina switches from LWS-2 expression to LWS-1 expression during maturation.

In fish, as in amphibians and reptiles, two families of vitamin A and two types of chromophores exist: 11-*cis*-retinal, derived from vitamin A1, and 11-*cis*-3-dehydroretinal, derived from vitamin A2 (Loew and Dartnall, 1976; Bowmaker, 1995). 11-*cis*-retinal based photopigments are called rhodopsins and 11-*cis*-3-dehydroretinal based photopigments are called porphyropsins. In the literature, some terminological confusion occurs because the term "rhodopsin" is sometimes exclusively applied to rod rhodopsin or even to rod opsin. Throughout this chapter, we use the term "rhodopsin" as defined earlier, that is, designing the class of photopigments that contains the chromophore 11-*cis*-retinal (Rodieck, 1998). λ_{max} of porphyropsins are usually higher than λ_{max} of homologous rhodopsins. In photoreceptors, a mix of both classes of pigments might be found. The proportion of these pigments in some species depends on the season (Loew and Dartnall, 1976), stage of life (especially in migrating species) (Beatty, 1966), and water temperature (Tsin and Beatty, 1979). Other proposed factors are environmental lighting conditions, hormones, and age (Kusmic and Gualtieri, 2000).

2.1.3. Arrangement of Photoreceptors

The retinas of many teleosts do not possess regions of increased photoreceptor densities (see earlier), cones are more or less homogeneously distributed over the retina. A characteristic of many teleost retinas is the regular pattern of the cone arrangement. In goldfish and rainbow trout, for example, S-cones are surrounded by four pairs of L-M double cones and UV-cones are located in the corners (Marc and Sperling, 1976; Allison *et al.*, 2003). In zebrafish, rows of double (L-M) cones alternate with rows consisting of alternating S and UV cones (Robinson *et al.*, 1993). Some researchers likened these regular mosaic-like patterns to the ommatidia of insects. Many explanations for the functions of these patterns have been offered. Waterman and Forward (1970) suggested that they allow the detection of polarized light, which might play a role in navigation. Fernald (1990) suggests that such an arrangement "allows detailed chromatic patterns to be resolved equally well over the entire retina." This is in striking contrast to the human retina, where the fovea is characterized by a high density of M- and L-cones, but no S-cones and rods in its center. The peripheral retina is characterized by a high density of rods and a low density of all cone types (Rodieck, 1998). According to Roorda *et al.* (2001), the three cone types are arranged in a "random mosaic" pattern: cones of any given type (e.g., L-cones) occur in patches, however, these patches are random in size and shape and randomly distributed over the retina.

2.1.4. Significance of Stem, Progenitor, and Precursor Cells in the Adult Retina

In teleosts, the eyes grow throughout life (Powers and Raymond, 1990; Otteson and Hitchcock, 2003). Accordingly, the retina also grows continuously. This occurs by cell proliferation in the circumferential germinal zone (CGZ) and by "stretching" of the entire retina. The CGZ is the peripheral region of the retina, where progenitor cells divide and differentiate to neuronal cells (such as cone photoreceptor cells and horizontal cells) and glia cells (such as Müller cells). However, rod precursor cells are not present in the CGZ; they are only found in the outer nuclei layer (ONL) of the central retina. They are derived from progenitor cells in the inner nuclei layer (INL), which have migrated to the ONL (Johns, 1982; Julian *et al.*, 1998; Otteson *et al.*, 2001) using the processes of Müller cells as scaffolds. As a result of retinal stretching during growth, the overall density of cones and other cell types decreases. However, in the central retina, the density of rods remains more or less constant, because new rods are continuously added, and thus the rod–cone ratio increases during the lives of the fish. Furthermore, because of the continuous growth of the retina, the number of cones per unit visual angle increases (Fernald, 1990), thus, visual acuity increases (slightly) with this growth.

The teleost retina can regenerate after damage (Powers and Raymond, 1990; Otteson and Hitchcock, 2003). This has been examined after various types of retinal injuries caused by prolonged intensive light exposure (Vihtelic and Hyde, 2000; Vihtelic *et al.*, 2006), neurotoxic agents (Negishi *et al.*, 1991; Mensinger and Powers, 1999; Stenkamp *et al.*, 2001), and surgical (Yurco and Cameron, 2005) or laser-induced lesions (Wu *et al.*, 2001). The cells involved in retinal regeneration are progenitor cells located in the INL of the central retina (Raymond *et al.*, 1988; Wu *et al.*, 2001; Otteson and Hitchcock, 2003). The progenitors in the INL are locally associated with Müller cells (Vihtelic and Hyde, 2000; Yurco and Cameron, 2005; Vihtelic *et al.*, 2006). They seem to migrate along the processes of Müller cells, which has also been described during normal development (Raymond and Rivlin, 1987). Some studies suggest that in teleosts, Müller cells can transdifferentiate into other cell types. Transdifferentiation consists of two steps: the cells dedifferentiate to become neuronal progenitor cells again, which then can differentiate to other cell types. This notion is partially based on the observation that stressed Müller cells can enter the cell cycle and migrate from the INL to the ONL (Wu *et al.*, 2001). Other researchers point out that there is no direct evidence for transdifferentiation of Müller cells into other cell types (Otteson and Hitchcock, 2003).

Regeneration of the retina occurs within weeks. However, the original arrangement of the cone mosaic is not fully restored after ouabain lesion (in goldfish; Stenkamp *et al.*, 2001) and after intense light treatment (in zebrafish; Vihtelic and Hyde, 2000). Cones of a given spectral type might occur in clumps and the cone arrangement is quasirandom. It is not clear whether this affects visual functions. Mensinger and Powers (1999) report that in goldfish about 7 months after ouabain-induced lesion the electroretinogram (ERG) b-wave amplitude was 50% of that in the control group. However, the dorsal light reflex (DLR) was fully restored. Generally, fish tilt around their longitudinal axis to maintain perceived down-dwelling illumination equal in both eyes. Normally, the tilting angle should be zero; however, in fish with retinal lesions in one eye, it will deviate from zero. When vision in the lesioned eye restores, the tilting angle decreases. Experiments to assess impairments and restoration of more complex visual functions (e.g., pattern recognition) have not been performed in any regeneration study in fish.

Hitchcock and Cirenza (1994) studied the synaptic organization of goldfish retinas 16–20 weeks after performing surgical excisions of small patches. Using light and electron microscopy, they found that overall the regenerated retina was very similar to the normal retina. The density of neurons and synapses were also similar. The depth profile of the synapses within the IPL differed only slightly between groups: in the regenerated retinas the synapses of bipolar cell terminals were slightly displaced in sclerad direction, although the ultrastructure of the synapses looked normal. However, Marc *et al.* (2003) have shown that remodeling of neuronal circuitries occurs during retinal degeneration in mammals (where regeneration is minimal or absent). Theoretical considerations reveal that corrupted networks are the result of alterations of specific connectivities rather than of loss of number of connections. Thus, apparent restoration of morphological integrity, as assessed by histological methods, does not warrant functional recovery.

Regeneration of the injured retina also occurs in other vertebrates. Amphibians possess a ciliary marginal zone (CMZ, the equivalent of the teleost CGZ) that is active throughout the life of the animals (Reh and Levine, 1998). After retinal damage, RPE cells can transdifferentiate into other retinal cell types (Mitashov, 1997). In birds, Müller cells can transdifferentiate into other retinal cell types (Fischer and Reh, 2003a,b). However, this potential is lost a few months after hatching. In rats and humans, pigmented cells from the ciliary margin (PCM cells) can transdifferentiate *in vitro* into retinal progenitor cells (Ahmad *et al.*, 2000; Tropepe *et al.*, 2000; Tsonis and Del Rio-Tsonis, 2004) which can further differentiate into photoreceptors or glia cells. This seems not to occur *in vivo*. Several studies investigate the possibility of regenerating damaged retinas in mammals by transplanting stem or progenitor cells from brain, spinal cord or retina, or bone marrow stem cells (Lu *et al.*, 2002; Tomita *et al.*, 2002; Dong *et al.*, 2003; Qiu *et al.*, 2005). To restore normal retinal functions, the cells have to differentiate into retinal neurons, form synapses, and integrate into the retinal network. Thus far, the results are promising, but a full integration of the transplanted and differentiated cells has not yet been achieved. The study of the processes involved in regeneration of the injured teleost and amphibian retina has guided this research in the past and will probably do so in the future.

2.2. Brain Areas Involved in Vision

Much is known about the visual brain areas and their functions in mammals. The situation in fish is different. Not many studies deal with the functional aspects of those brain areas in fish, partly due to technical difficulties (Vanegas and Ito, 1983). Lesion and electrophysiological studies are often difficult to perform because the majority of the visual nuclei are small. They are sometimes less then 1 mm in diameter and may consist of less than 100 neurons. By far, the largest visual nucleus in most fish species is the optic tectum (OT), with the exception of weakly electrical fish (e.g., Mormyridae), where the torus semicircularis (TS, a multimodal brain area) is much larger. Here, we concisely review the functional aspects of the OT and TS (both belonging to the mesencephalon) as well as the functions of some other nuclei. Reviews of anatomical and electrophysiological properties of the visual brain areas in fish can be found in Douglas and Djamgoz (1990), Northcutt and Davis (1983), and Vanegas and Ito (1983).

The OT is homologous to the superior colliculus in mammals. RGCs project to the superficial layer of the OT in an ordered manner, forming a retinotopic map. Tectal neurons have been described in zebrafish that respond to stimuli moving in specific directions in the visual field (Sajovic and Levinthal, 1982a,b). Many cells respond preferentially to stimuli that move in the horizontal plane, especially when they move from the temporal to the nasal field. In goldfish, they respond to movements in certain velocity ranges (Wartzok and Marks, 1973). Like the superior colliculus in mammals (Dean *et al.*, 1989) and the OT in amphibians (Ewert, 1984), the OT in fish plays a role in both orientation and escape responses, which are probably important for prey pursuit and predator avoidance, respectively (Herrero *et al.*, 1998). This was discovered by electrical stimulation of the OT (Akert, 1949; Herrero *et al.*, 1998) and in ablation studies (Yager *et al.*, 1977). Both the superior colliculus in mammals (Krauzlis *et al.*, 2004) and the OT in teleosts have important roles in generating saccades (targeted eye movements) as part of the orienting response (Luque *et al.*, 2005). The type of neuronal response elicited by electrical stimulation depends on the exact OT location of the electrode and the stimulus parameters (e.g., intensity, frequency). In the natural context, the response might depend on brightness and contrast, location in the visual field, and movement direction and velocity of the visual stimulus.

The TS is homologous to the inferior colliculus in mammals. It is usually the second largest visual brain area after the OT, except in weakly electrical fish which use electrical signals for electrolocation (e.g., navigation, prey detection) and electrocommunication (e.g., social aggression, courtship behavior) (Moller, 1995; von der Emde, 1998). In these species, the size relationship between these two nuclei is reversed. The TS is a multimodal area that receives input from the visual (via the OT), acoustic, and lateral line systems, and from electroreceptor organs. Many neurons respond to stimulation of different modalities. Visual information processed by the TS does not seem to code precise information about wavelength, shape, contrast, position in space, direction, or velocity of movement. Some researchers suggest that one of the functions of the TS is to detect sudden changes in the visual field and to elicit an orientation response (Schellart, 1990). The TS projects to the reticular formation, which has a role in the arousal state of the brain and the orienting response.

Other visual nuclei in the diencephalon and telencephalon have been described (Schellart, 1990). As in other vertebrates, the diencephalic accessory optic system (AOS) and pretectum in teleosts (Northcutt and Butler, 1976;

Wullimann *et al.*, 1996) have been suggested to have important roles in visual self-motion detection and in optokinetic nystagmus, that is, eye/head movements performed to stabilize the retinal image of the global environment (Simpson, 1984; Klar and Hoffmann, 2002; Giolli *et al.*, 2006). Both areas are important for gaze control (Rodieck, 1998). Some of the telencephalic areas receiving visual input seem to be crucial for spatial orientation and visual memory formation (Lopez *et al.*, 2000; Portavella *et al.*, 2002; Rodriguez *et al.*, 2002; Saito and Watanabe, 2004). It has been suggested that the dorsolateral telencephalon is a functional homolog to the mammalian hippocampus (which has a role in spatial mapping and probably also contextual and/or configural learning) and the dorsomedial telencephalon is a functional homolog to the mammalian amygdala (which has a role in emotional learning) (Portavella *et al.*, 2002).

2.3. Summary

The overall organization of the teleost eye is similar to that of the mammalian eye. However, some characteristic features in many fish species are not common in mammals. These include the presence of a spherical lens, absence of a movable iris, visual focusing by movement of the lens, retinomotor movement of photoreceptors and RPE elements for the purpose of light adaptation, absence of a central area characterized by increased photoreceptor density, presence of two chromophores and several subtypes of opsin genes, regular mosaic-like pattern of cone arrangement, and stem, progenitor, and precursor cells that support lifelong growth of the retina and retinal regeneration after injuries.

The best-studied visual brain areas in teleosts are nuclei belonging to the mesencephalon (OT, TS). The OT is homologous to the mammalian superior colliculus and is often seen as the most important visual brain area for "seeing" in teleosts. The TS is homologous to the mammalian inferior colliculus and is a multisensory area engaged in the orientation response. Diencephalic areas such as the AOS and pretectum play probably important roles in gaze control. The functions of the telencephalon in visual processing have not been well studied. The dorsolateral and dorsomedial telencephalon might be important for complex learning processes involving visual information.

3. ABSOLUTE VISUAL SENSITIVITY

Absolute visual sensitivity is defined by the minimal light intensity that can be detected by the visual system after full dark adaptation. The exact meaning of "detection" depends on the visual process under investigation.

For example, "visual sensitivity" can refer to the minimal light intensity required to modulate photoreceptor cell processes, to evoke bipolar cell or RGC threshold responses, to generate visually evoked responses in the OT, or to elicit behavioral responses. It becomes clear that the study of visual sensitivity requires a multilevel approach.

3.1. How to Measure Visual Sensitivity?

The two most common methods for determining absolute visual sensitivity are behavioral and electrophysiological tests. Two electrophysiological methods for *in vivo* recording are frequently used: ERG and RGC recordings. They determine visual sensitivities at different levels of the visual information-processing network.

3.1.1. Behavioral Tests

There are two principal categories of behavioral tests to determine visual sensitivity: tests based on learned or conditioned behavior and tests based on innate or unconditioned behavior (Douglas and Hawryshyn, 1990). Conditioned behavioral tests require often time-consuming training of the subjects. On the positive side, the conditioned animal is motivationally primed (by reward, punishment, or an emotional state) to perform the requested task. Unconditioned tests have the advantage that training is not necessary. However, the lack of a conditioned stimulus will often give the animal a choice between different equally "desirable" behaviors. Thus, a greater number of test subjects or repeated testing of the same individuals is needed to obtain reliable results. The decision, which type of test to use, depends also on several other considerations. For example, when testing absolute visual sensitivity of night blindness mutants, the measurements should be independent of possible learning deficits which might be caused by the mutation of a gene whose expression is not restricted to the retina (Section 6). In some cases, conditioned and unconditioned tests of visual functions (e.g., acuity) might elicit different results because different central circuits are involved in unconditioned and conditioned visually guided behaviors such as the optokinetic response (OKR) and spatial discrimination (Douglas *et al.*, 2005; see also Section 7.1.4). Thus, the choice between conditioned and unconditioned tests will depend on a number of theoretical and practical considerations.

Powers and Easter (1978) studied absolute visual sensitivity in goldfish, using a conditioned test. Dark-adapted goldfish were fixed in a restraining box. Their heart or respiration rates were recorded by electrocardiography or thermistors. The fish were trained several times per day. They received 10–20 conditioning trials per session. A trial consisted of presenting the fish with a spot of light that was well above threshold level. The presentation of the stimulus was followed by an electrical shock. Fish were trained until a robust response was elicited, which consisted of a decrease in heart or respiration rate in anticipation of the shock. Once the fish were successfully conditioned, the threshold light intensity that elicited the conditioned response was determined. This was done by reducing the light intensity stepwise until the fish responded to only 50% of stimulus presentations.

Li and Dowling (1997) introduced the behavioral escape response paradigm to measure absolute visual threshold in zebrafish. This test can be performed without training the fish. It is based on the natural escape response in zebrafish, which can be easily elicited by confronting them with a dark silhouette. The apparatus is depicted in Figure 5.1. A transparent container containing a fish is inserted in the center of a rotating drum. The inside of the drum is covered with white paper that is marked with a black bar. A column in the center of the container prevents the fish from swimming through the center. Normally fish swim along the wall of the container but when they encounter the black segment they show an escape response. They either reverse their swimming direction 180° or veer off to the center and resume swimming close to the wall of the container after they pass the black bar. Under well-lit conditions, zebrafish show a very robust, nonhabituating escape response.

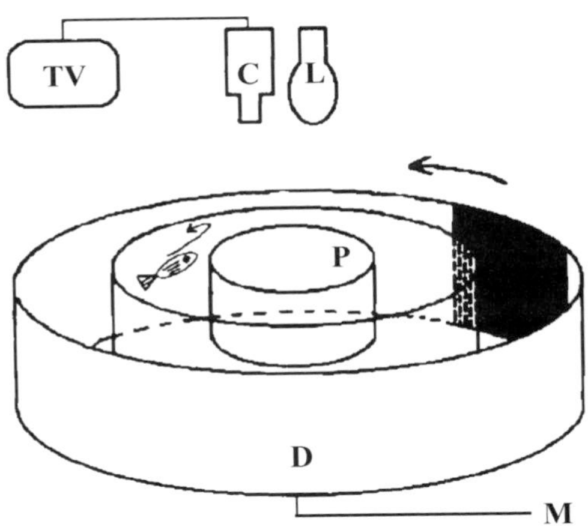

Fig. 5.1. Schematic presentation of the escape response assay. In the center of the container, a pole (P) prevents the fish from swimming through the center. A drum (D), which is marked on the inside with a black segment on a white background, is driven by a motor (M). The black bar serves as the visual stimulus that normally elicits an escape response by the fish. A light source (L) is suspended above the apparatus. The light intensity can be adjusted by adding or removing neutral density filters. An infrared camera (C) is connected to a TV monitor and allows observation of the behavior of the fish. [Adapted from Li and Dowling (1997) with permission.]

Using this setup, absolute visual sensitivity can be determined. After 30 min of dark adaptation, the fish are tested for the escape response to determine the minimal light intensity required to detect the black segment. The intensity of the light source suspended over the apparatus can be adjusted by adding or removing neutral density filters. At the start of the test, the light intensity is set at a dim level. If no escape response occurs, the light intensity is increased (in steps of 0.5 log units) until the fish respond in at least 5 of 10 encounters. This light intensity is noted as the visual threshold.

3.1.2. Electrophysiological Recordings

ERG is used to determine outer retinal sensitivity (Dowling, 1987). The fish is anesthetized and immobilized, and then placed on its side on a wet sponge and most of its body is covered with a wet paper towel. To keep the fish oxygenized, a slow stream of aerated system water is directed into its mouth via a tube. A beam of white light (generated by a halogen lamp) is directed via a mirror system to the eye onto which the glass electrode is placed.

The ERG shows a complex waveform in response to light flashes: a corneal negative a-wave, followed by corneal positive b- and d-waves (Figure 5.2). The a-wave is usually regarded as originating in photoreceptor cells, that is, rods and cones. The b-wave is generated by ON-bipolar cells in conjunction with Müller cells (Brown, 1968; Newman, 1980; Dowling, 1987; Stockton and Slaughter, 1989) and perhaps also with amacrine cells and RGCs (Dong

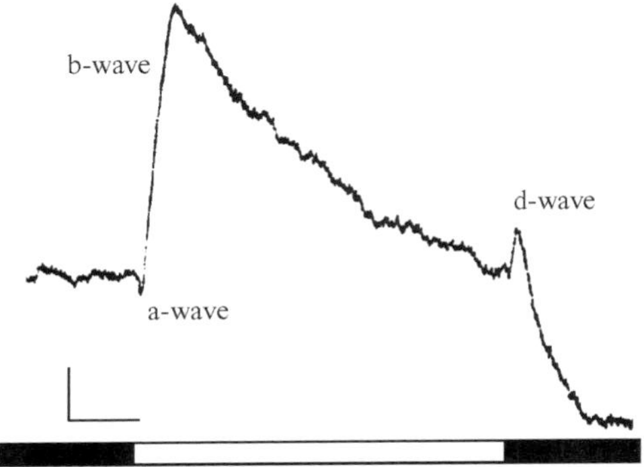

Fig. 5.2. Schematic presentation of the ERG that was elicited by a short light flash. The a-wave is generated by photoreceptor cells. It is partially masked by the early onset of the b-wave. The b-wave is generated by ON-bipolar cells, possibly in conjunction with Müller cells and other inner retinal cells. The d-wave is generated by OFF-bipolar cells. During the course of dark adaptation, the amplitude of the b-wave increases. The amplitude of the d-wave first increases and then decreases. The white segment in the bar at the bottom of the graph represents the light flash. [Adapted from Ren and Li (2004a) with permission.]

and Hare, 2000; Awatramani *et al.*, 2001). The d-wave is generated by OFF-bipolar cells (Stockton and Slaughter, 1989; Wong *et al.*, 2000). The a-wave in fish is partially masked by the following b-wave, which commences before the a-wave reaches its peak. To study the full a-wave in isolation, the b-wave can be pharmacologically suppressed by DL-2-amino-4-phosphonobutyric acid (APB) (DeMarco and Powers, 1989).

To determine absolute (rod) sensitivity, the fish are dark adapted for 30 min and the threshold light required to evoke threshold ERG is measured. The recording starts at a low light intensity, which is incrementally increased in steps of 0.5 log units. A computer-controlled shutter regulates the duration of the light stimulus. The lowest light intensity that generates a b-wave with an amplitude at or above a certain threshold value, the determination of which depend among others on the electrical noise, is noted as the ERG threshold.

To record cone functions selectively, flicker ERG (fERG) can be used. It is widely applied in primate vision research. In this setup, the stimulus consists of a train of high-frequency stimuli to which rods do not respond (Kondo and Sieving, 2001; Alexander *et al.*, 2003). fERG has only scarcely been used in fish (e.g., Mora-Ferrer and Behrend, 2004). For example, Branchek (1984) applied fERG to measure cone function in larval zebrafish. Cone system functions may also be evaluated by standard ERG, for example, after light adapting the fish with a background illumination (to saturate rod functions; see Hughes *et al.*, 1998), or by applying long-wavelength stimuli to which some cones (e.g., L-cones) respond more strongly than rods (Li and Dowling, 1998).

By recording RGC activities, overall retinal sensitivity can be assessed. RGCs generate both light-induced and spontaneous action potentials (APs). In response to a brief light flash, AP frequency either increases or decreases. This occurs either during stimulus onset (ON response), stimulus offset (OFF response), or during both stimulus onset and offset (ON-OFF response). The light intensity that elicits a 20–30% change (either increase or decrease) in spike frequency is often regarded as the RGC threshold response. To determine RGC sensitivity, fish are prepared in a similar way as for ERG recording. The eyes are slightly pulled out of the sockets to expose the optic nerve. RGC responses are recorded by a wire electrode inserted into the optic nerve.

In RGC recordings, the light stimulus can consist of diffuse whole field illumination or of light spots or annuli. These latter stimuli are applied to investigate the center-surround organization of receptive fields of RGCs. When using whole field illumination, stimulation of both the ON-center and the OFF-surround (or vice versa) might inhibit AP frequency changes that would have been obtained by stimulating the ON-center alone.

Nevertheless, whole field stimulation is suitable for quick assessment of overall retinal sensitivities.

In our studies, we apply both behavioral and electrophysiological tests. The behavioral test is fast and noninvasive. Thus, after testing we can select fish for breeding, follow their visual thresholds over a period of time (weeks or months), or perform other experiments with the same individuals. Electrophysiological recording in small animals such as zebrafish, on the other hand, is usually terminal. ERG and RGC recording often require that the fish be anesthetized and immobilized over a longer period. After this, it is usually not possible to revive the fish. In our studies, only the data of those fish that show a clearly visible heartbeat at the end of the experiment are used.

Behavioral and electrophysiological testing are different in what they measure: in the former we investigate function-specific pathways (within retina or brain), whereas in the latter we determine general properties of elements in the retina. For example, with behavioral testing we can determine spatial frequency properties of stimuli that are either important for gaze stabilization (OKR) or for visual discrimination task (Douglas *et al.*, 2005; Section 7). With electrophysiological methods, we can determine receptor field properties of retinal neurons irrespective of the functional pathways.

3.2. Dark Adaptation

After the transition from light to dark, visual sensitivity switches from cone dominance to rod dominance. During early dark adaptation, retinal sensitivity is governed by the cone system. During late dark adaptation, rods become dominant. This transition is accompanied by a shift of spectral sensitivity from longer to shorter wavelengths. This latter phenomenon is called the Purkinje shift (Dowling, 1987; Anstis, 2002).

Li and Dowling (1998) measured the course of dark adaptation of rod and cone systems in zebrafish using the behavioral escape response paradigm. This was done by measuring the light threshold that elicits escape responses at different times of dark adaptation after light adaptation. Threshold decreases during dark adaptation. This process consists of two phases. During early dark adaptation, the threshold decreases very rapidly. It reaches a plateau at 6 min, which represents the threshold for cones. At 8 min, the threshold decreases again, although at a slower rate than during early dark adaptation, until it reaches the next plateau at about 20 min. This is believed to represent the absolute threshold of rods (Figure 5.3).

During dark adaptation, relative rod and cone contributions to retinal sensitivity can be evaluated by using phototopically matched lights of specific wavelengths. In zebrafish, for example, sensitivity of rods and M-cones peaks

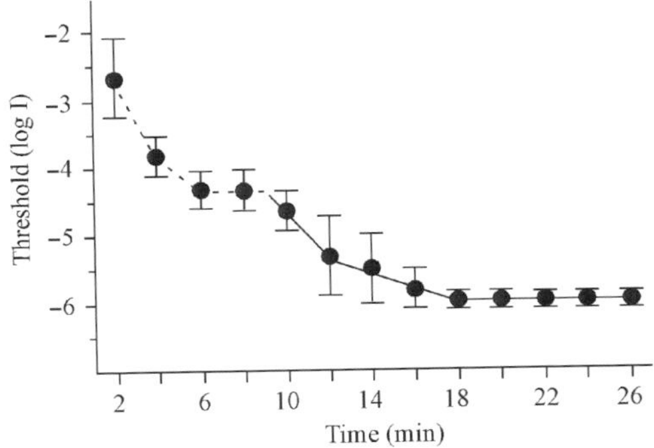

Fig. 5.3. Dark adaptation curve of retinal sensitivity in zebrafish as determined by the behavioral escape paradigm. During the first 6 min of dark adaptation, the threshold decreases quickly and levels off at 8 min. After 8 min, the threshold decreases again, albeit more slowly than during the first phase. It reaches its lowest level at about 20 min of dark adaptation. During early dark adaptation, retinal sensitivity is governed by cones, whereas it is governed by rods during late dark adaptation. [Adapted from Li and Dowling (1997) with permission.]

around 500 nm, whereas sensitivity of L-cones peaks around 570 nm. Rods are highly sensitive to 500-nm light but not to 625-nm light. Thus, 625-nm light can be used to determine the contribution of cones to visual sensitivity. During early dark adaptation (first 6 min), the threshold as determined under 625-nm light is similar to the threshold as determined under 500-nm light (Li and Dowling, 1998; Ren and Li, 2004a). This suggests that during this period, retinal sensitivity is regulated by the cone system. During late dark adaptation (after 8 min), the Purkinje shift occurs: the threshold for 625-nm light remains unchanged, whereas the threshold for 500-nm light decreases. This suggests that after the Purkinje shift, retinal sensitivity is dominated by rods.

3.3. Summary

Absolute visual sensitivity is defined by the lowest light intensity that elicits a visual response. Common methods to determine the visual threshold are behavioral and electrophysiological tests. Conditioned behavioral tests require training of the fish, whereas unconditioned behavioral tests exploit innate behavioral responses. ERG determines the sensitivity of the outer retina (photoreceptors, bipolar cells), and RGC recording determines the overall sensitivity of the retina. Depending on the signal parameters (intensity, wavelength, frequency, background illumination), these assays can be used to selectively measure signals in either the rod or cone pathway. Behavioral and electrophysiological tests differ in that the former are suited to investigate selected functional parameters, whereas the latter are suited to study basic properties of retinal neurons within the network.

During dark adaptation, relative contributions of rods and cones to retinal sensitivity change. During early dark adaptation (e.g., the first 6 min in zebrafish), the sensitivity is cone dominant. Thereafter, the sensitivity becomes rod dominant. The absolute visual (rod) threshold is reached after about 20 min of dark adaptation. The transition of cone-to-rod dominance causes the Purkinje shift of retinal sensitivity.

4. CIRCADIAN REGULATION OF VISUAL SENSITIVITY

Fish, like many other animals, show a cyclical pattern of activity that varies with the light-dark (LD) cycle. Depending on the species, fish may be most active during the day or during the night. Zebrafish, for example, are about two times more active (measured by locomotive activity) during the day than during the night (Hurd *et al.*, 1998). This effect persists in constant lighting conditions, suggesting that it is regulated by a circadian clock. Other behaviors, such as feeding, sleeping, mating, and spawning seem also to be under circadian control (Zhdanova and Reebs, 2006). Experiments have shown that in many fish species, including zebrafish (Li and Dowling, 1998) and goldfish (Bassi and Powers, 1987; Iigo and Tabata, 1996), visual system functions (such as absolute visual sensitivity) are regulated by a circadian clock. The circadian rhythm of visual sensitivity can be light entrained. For example, when the animals are kept for a period of time under a reversed LD cycle (lights on during the night, lights off during the day), the circadian rhythm of visual sensitivity reverses.

Circadian experiments are carried out while the animals are kept in constant darkness (DD condition) or constant light (LL condition), as opposed to the normal alternating LD condition. "Subjective day" is the period of DD or LL when it is daytime according to the LD condition. Similarly, "subjective night" is the period when it is nighttime according to the LD condition.

4.1. Circadian Modulation of Rod and Cone Sensitivity

In zebrafish the threshold of both rods and cones, as measured by the behavioral escape test, depends on the time of day (Li and Dowling, 1998). Under LD (lights on: 8 A.M.–10 P.M.), rods and cones had the highest threshold (lowest sensitivity) in the early morning (4 A.M.) and the lowest threshold in the late afternoon (6 P.M.). Between day and night, the threshold of cones fluctuated by about 1.4 log units, while the threshold of rods varied by about 2.2 log units. When zebrafish were kept under DD, the fluctuation of rod and cone thresholds followed the same pattern as under the LD condition, demonstrating that this fluctuation is regulated by a circadian clock. Similar results were found when the fish were kept under LL.

Under DD, the circadian rhythm persisted for several days and then gradually damped out. Under LL, it disappeared after two to three days. These observations were obtained by behavioral testing and were confirmed by ERG recording. The pattern of circadian rod and cone threshold variations can be light entrained. For example, when zebrafish were exposed to a 1-h light pulse, 4 h before normal light onset, the circadian rhythm was phase-shifted: in fish exposed to the light pulse, maximal visual sensitivity was reached ~4 h earlier than in fish that were not exposed to the light pulse.

The fluctuations of the absolute threshold between day and night as determined by ERG (~1 log unit) were smaller than the fluctuations as determined by the behavioral test (~2 log units). This suggests that either independent circadian oscillators are located in both the outer retina and the inner retina or beyond, or the gain of the photoreceptor-to-brain transmission is a nonlinear function of the state of dark adaptation. We will see in a later section that several neurotransmitters, such as dopamine, which play important roles in the circadian modulation of retinal signal transmission, modulate the retinal pathway at different levels (photoreceptors, horizontal cells, bipolar cells, RGCs). Thus, these neurotransmitters might alter the gain of neuronal elements.

4.2. Circadian Modulation of Rod–Cone Dominance

During dark adaptation, visual sensitivity switches from cone dominance to rod dominance. Experimental data suggest that the shift between rod and cone dominance also follows a circadian rhythm. The neurotransmitter dopamine has been found to have a crucial role in those transitions.

4.2.1. Evidence of Rod–Cone Dominance Transitions in the Outer Retina

The fish retina possesses both rod and cone horizontal cells (Negishi *et al.*, 1997; Connaughton *et al.*, 2004) that are specific for processing rod and cone signals, respectively. Mangel (2001) and colleagues have shown that the response of L (luminosity)-type cone horizontal cells (which react to stimuli of all wavelength with hyperpolarization; Section 8) to light stimuli varies in a circadian fashion. During the normal light period, their responses to flashes of white light have higher amplitudes than during the dark period.

This pattern of response persists under DD and is reversed when the eyes are previously exposed to reversed LD. Furthermore, Mangel and Wang (1996) demonstrated that during the subjective day, the spectral sensitivity of L-type cone horizontal cells is similar to that of L-cones, which present the major input to those horizontal cells (Section 8). During the subjective night, it resembles the spectral sensitivity of rods and rod horizontal cells. Since cone horizontal cells do not receive direct synaptic input from rods, Mangel (2001) suggests that rod input to cone horizontal cells at night might be achieved by gap-junctions between rods and cones and subsequent relaying of rod signals from cones to horizontal cells. This type of electrical coupling has been described in many vertebrates (Raviola and Gilula, 1973), including fish (Witkovsky *et al.*, 1974). These findings demonstrate that during the night, L-type cone horizontal cells are rod-dominant, whereas during the day they are cone-dominant.

Ren and Li (2004a,b) examined the circadian control of rod–cone dominance in zebrafish by recording ERG. The b-wave shows a Purkinje shift during dark adaptation, which is caused by a transition from cone to rod dominance. The d-wave is cone dominant at all times. During early dark adaptation, the thresholds of both b- and d-waves decrease. During late dark adaptation, only the threshold of the b-wave decreases, while the threshold of the d-wave increases. In frogs, the increase of d-wave threshold is likely caused by rod suppression of cone-driven activity (Dong *et al.*, 1988). The duration of dark adaptation needed for this dominance shift to occur follows a circadian rhythm. In the morning, when the fish is least sensitive to light, the rod–cone dominance transition (as measured by the ratio of the b- and d-amplitude) is slower than in the late afternoon, when the fish is most sensitive to light. One interpretation of these findings is that the contribution of rods to visual sensitivity is less dominant in the morning than in the afternoon. Circadian rod–cone dominance shifts have also been described in other vertebrates, such as birds (Manglapus *et al.*, 1998).

4.2.2. Effects of Dopamine

Dopamine plays a role in modulating the circadian rhythm of rod–cone dominance. In fish, as in many other animals, retinal dopamine release is higher during the subjective day than during the subjective night (Wirz-Justice *et al.*, 1984; McCormack and Burnside, 1993; Adachi *et al.*, 1999; Ribelayga *et al.*, 2002; Zawilska *et al.*, 2003). In fish retinas, dopaminergic interplexiform cells (DA-IPCs) are the only class of cells that synthesize dopamine (Witkovsky and Dearry, 1992). Witkovsky *et al.* (1988) have shown that in *Xenopus* exogenous dopamine increases cone input to cone horizontal cells during the night (when under normal circumstances the rod input is dominant) and that destruction of DA-IPCs by 6-hydroxydopamine (6-OHDA) increases rod input to cone horizontal cells during the day (when under normal circumstances the cone input is dominant). Mangel and Wang (1996) also found that dopamine increases the effectiveness of cone input in cone horizontal cells during the night. Changes in input can be caused by changes in synaptic transmission or by changes of gain in postsynaptic neurons. The authors suggest that dopamine decreases the conductance of rod–cone gap junctions via D_2 receptors.

Li and Dowling (2000a) studied the effect of dopamine depletion on visual sensitivity and on circadian regulation of retinal sensitivity in zebrafish. After destroying the DA-IPCs with 6-OHDA, the circadian rhythms of behavioral sensitivity diminished. Sensitivity remained at a low level for all times of the day and night. The rod system did not seem to contribute to behavioral sensitivity, whereas the cone system seemed to function normally. However, studying the ERG, the authors found that both the rod and cones contribute to outer retinal sensitivity. This is in agreement with the previous finding by Yazulla *et al.* (1996) that dopamine depletion in goldfish did not affect normal transition from rod to cone dominance. By recording RGC spikes, Li and Dowling (2000a) demonstrated that after dopamine depletion, rod signal transmission in the inner retina was blocked. This latter data suggest that dopamine plays a role in the circadian regulation of rod–cone sensitivity by modulating inner retinal circuitries. Gabriel *et al.* (2001) report that in rats, protein kinase C (PKC) activity in rod bipolar cells and some amacrine cells follows a circadian rhythm and can be light entrained. Depolarization of rod bipolar cells increased PKC transport to the synaptic terminals (Vaquero *et al.*, 1996). Whether dopamine has a role in the circadian expression of PKC or whether PKC expression has a role in rod–cone dominance shift is not known.

4.3. Circadian Regulation of Dopamine and Melatonin Release

The circadian clock function has traditionally been associated with the neurotransmitters dopamine and melatonin. Here we review some major findings concerning the connection between circadian regulation of visual functions and the release of dopamine and melatonin, as well as the interactions between these two neurotransmitters.

4.3.1. Assays to Determine Retinal Dopamine or Melatonin Release

The release of retinal dopamine and melatonin can be determined in several ways. Some earlier studies did not directly measure neurotransmitter release. Instead, they measured retinal functions that are presumably affected by those neurotransmitters, such as retinomotor movement or coupling of horizontal cells. From this, the amount of dopamine or melatonin release was inferred

(Dearry and Burnside, 1986; Dong and McReynolds, 1992). The retinal content of dopamine and melatonin can be determined after homogenizing the retina (Ribelayga *et al.*, 2002). Since this method provides the total retinal content of the neurotransmitter, the metabolites must also be quantified to estimate the amount of the released neurotransmitter. A direct method to determine the release of either neurotransmitter is *in vivo* microdialysis (Adachi *et al.*, 1999; Puppala *et al.*, 2004). Dopamine and melatonin accumulate in the vitreous humor. To determine whether the brain controls retinal release of dopamine and melatonin or whether the retina possesses an autonomous circadian clock that regulates the release of these neurotransmitters, eyecups are isolated and the release of dopamine or melatonin is determined independently of central regulation (Cahill, 1996; Ribelayga *et al.*, 2004).

4.3.2. Retinal Dopamine and Melatonin Release Follow a Circadian Rhythm

In vivo microdialysis studies with pigeons (Adachi *et al.*, 1999) have shown that in the retina, both dopamine and melatonin release follow a circadian rhythm but the two are out of phase. Similarly, studies in fish have shown that in the retina, the release of both dopamine (Ribelayga *et al.*, 2002) and melatonin (Iigo *et al.*, 1997) follows a circadian rhythm. Dopamine peaks during the subjective day, while melatonin peaks during the subjective night. Both neurotransmitters play a role in regulating (in opposite directions) some circadian retinal processes, for example, retinomotor movements (Pierce and Besharse, 1985; Dearry and Burnside, 1986), disk shedding (Pierce and Besharse, 1986; White and Fisher, 1989; Burnside, 2001), and photoreceptor sensitivity (Ko *et al.*, 2003; Wiechmann *et al.*, 2003).

Are the circadian rhythms of retinal dopamine and melatonin release regulated by a retinal clock or by a master clock located in the brain? In mammals, the hypothalamic suprachiasmatic nucleus is often seen as a "master" clock, setting the rhythms of "slave clocks" in the periphery, such as the retina (Tamai *et al.*, 2003). In fish, a central circadian oscillator has not been proposed (Ekström and Meissl, 1997), although circadian release of melatonin by the pineal gland has often been regarded as being central to other biological rhythms, such as metabolic rhythms (Zachmann *et al.*, 1992). Several studies have shown that in cultured retinas or eyecup preparations from both mammals (Tosini and Menaker, 1996) and fish (Cahill, 1996; Iigo *et al.*, 1997; Ribelayga *et al.*, 2004), the release of dopamine and melatonin is controlled by a circadian rhythm (Tosini and Fukuhara, 2003; Green and Besharse, 2004). This suggests that the retina harbors circadian oscillators. However, the endogenous circadian rhythms of melatonin release in isolated eyecups damp out very quickly (Cahill, 1996). It is not clear if this is due to a loss of normal physiological functions of the preparation or to the fact that the retinal circadian pacemakers need to be synchronized by a central pacemaker.

Dopamine and melatonin modulate each other's release (Cahill and Besharse, 1991; Behrens *et al.*, 2000; Ribelayga *et al.*, 2004). Ribelayga *et al.* (2004), for example, demonstrated that in goldfish dopamine release remains low at all times when melatonin is present in high concentrations and high at all times when the melatonin receptor antagonist luzindole is present. This suggests that dopamine release is under control of melatonin. They also demonstrated that melatonin decreases cone input to cone horizontal cells. Thus, it has the opposite effect of dopamine (see earlier). When administering luzindole during the night, the cone input to the cone horizontal cells increases, that is, they switch to a day state. This effect is absent when dopamine receptors are pharmacologically blocked. Ribelayga *et al.* (2004) concluded from these experiments that the effect of melatonin on horizontal cells is mediated by dopamine release from DA-IPCs. The fact that melatonin is released by photoreceptor cells and is controlled by a circadian rhythm (Cahill and Besharse, 1993) raises the question of whether the circadian rhythm of dopamine release is secondary to that of melatonin release. Many researchers are reluctant to make such a statement (Green and Besharse, 2004). Some studies have shown that dopamine inhibits melatonin release by photoreceptor cells (Cahill and Besharse, 1991). A possibility is that both photoreceptor cells and DA-IPCs have independent clocks, which can synchronize each other by means of melatonin or dopamine (the different models are discussed in Green and Besharse, 2004).

In mice, some early circadian genes are found to express in dopaminergic amacrine cells, suggesting the possibility of a melantonin-independent circadian regulation of dopamine release (Witkovsky *et al.*, 2003; Gustincich *et al.*, 2004). In chicks, research (Megaw *et al.*, 2006; Morgan and Boelen, 1996) suggests that circadian regulation has only a small effect on dopamine release compared to the effect of light. The circadian effect seems to be mediated by melatonin, which suggests that the clock is located in photoreceptor cells but not in dopaminergic amacrine cells. However, light can via the dopaminergic pathway reset the clock. In *Xenopus*, several clock genes are expressed in both photoreceptors and inner retinal neurons (Zhu *et al.*, 2000; Zhuang *et al.*, 2000). To date, not enough evidence is available to determine the origin of retinal circadian rhythms in teleosts.

4.3.3. Light Entrains the Circadian Rhythm of Dopamine and Melatonin Release

The circadian rhythm of dopamine and melatonin release can be light entrained. In fact, many studies have shown that both dopamine and melatonin release are under direct photic control (Kirsch and Wagner, 1989;

Iigo *et al.*, 1997; Ribelayga and Mangel, 2003; Tosini and Fukuhara, 2003; Puppala *et al.*, 2004). The relationship between dopamine and melatonin release and light is complex. It depends on factors such as time of day (subjective day or night), duration of light or dark adaptation, and characteristics of the light stimuli (e.g., continuous light vs. flickering light). For example, Weiler *et al.* (1997) have shown that in the white perch (*Morone americana*), retinal dopamine release increased after prolonged darkness, but only during the subjective night or when exposing the dark-adapted fish (at any time of the day) to flickering light stimuli. Kirsch and Wagner (1989) found that continuous light increased dopamine release but flickering light elicited a stronger response. Hamasaki *et al.* (1986), using very different stimulus parameters, found that flickering light decreased dopamine release. Puppala *et al.* (2004) found that in zebrafish, dopamine release was higher in light-adapted fish than dark-adapted fish. They also studied the effect of transitions from light to dark or dark to light on dopamine release. The light stimulus consisted of either continuous light or flickering light (1 Hz). The transition from dark to flickering light resulted in increased dopamine release, whereas the transition from flickering light to dark decreased dopamine release. When using continuous light instead of flickering light neither effect was seen. Both transitions involving flickering light were stronger later during the day (Figure 5.4).

Melatonin release is also under photic control. Iigo *et al.* (1997) showed that in goldfish eyecup preparations, melatonin release was higher during the night than during the day (circadian effect). However, when these eyecups were exposed to light during the night, melatonin release was strongly reduced (photic effect). In some salmonids (e.g., rainbow trout), melatonin release is under photic but not under circadian control (Iigo *et al.*, 1997).

4.4. Circadian Regulation of Opsin mRNA Expression

Both the circadian clock and light determine the level of opsin mRNA expression. For example, Korenbrot and Fernald (1989) have shown that in a cichlid (*Haplochromis burtoni*), rod opsin mRNA is high during the day and low during the night. When the fish were kept under DD, the same results were found. This demonstrates the involvement of a circadian rhythm. When the eye was stimulated by light during the night, rod opsin mRNA rose to daytime levels. Thus, rod opsin mRNA levels are also under photic control.

Li *et al.* (2005) measured L-cone opsin mRNA expression in zebrafish under different lighting conditions. During a 24-h period under LD, L-cone opsin mRNA expression was low in the early morning and high in the late afternoon. The expression followed the same pattern of fluctuation as behavioral visual

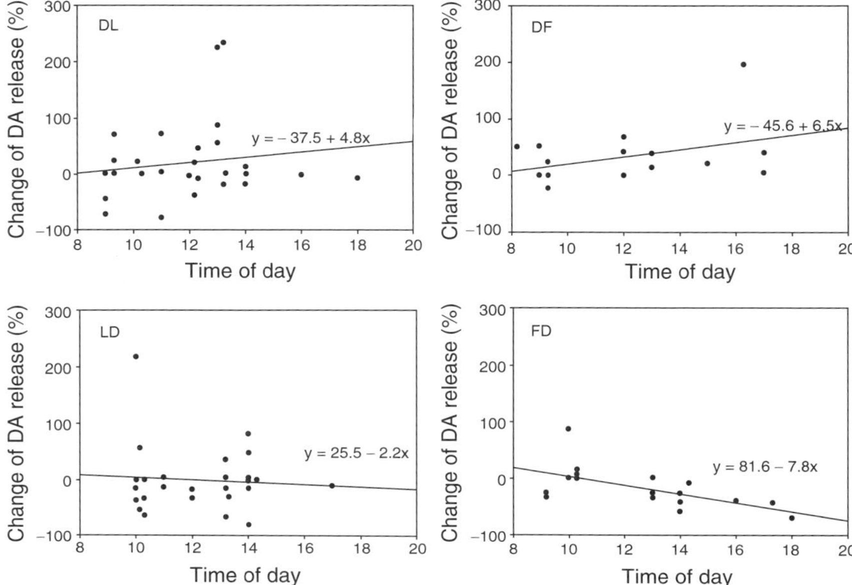

Fig. 5.4. Changes in vitreal dopamine accumulation are evoked by transitions between lighting conditions. Dopamine release was determined by *in vivo* microdialysis and HPLC. The transitions were dark to continuous light (DL), dark to flickering light (DF), continuous light to dark (LD), and flickering light to dark (FD). Each dot represents the percentage changes of dopamine release for an individual fish. These data are plotted against the time of day. The decrease in dopamine release in the FD condition was clearly dependent on the time of day ($p < 0.01$). [Adapted from Puppala *et al.* (2004) with permission.]

sensitivity. During a 24-h period under DD, the expression of L-cone opsin mRNA followed a similar pattern as under LD. These rhythmical variations, however, damped out faster than the circadian behavioral variations (which were still apparent on day 5): after 24 h of DD, L-cone opsin mRNA expression lost its periodicity and stabilized at a level between the normal circadian highest and lowest values. Under LL, L-cone opsin mRNA expression also lost its circadian pattern after 24 h and stabilized at a very low level, close to what is normally seen during the subjective morning (Figure 5.5). Thus, while many visual processes (such as behavioral sensitivity, opsin mRNA expression, disk shedding, and retinomotor movement) may correlate during the circadian cycle, this does not mean that a causal relationship between them exists. They might damp out at different rates or become desynchronized under certain circumstances (e.g., after dopamine depletion).

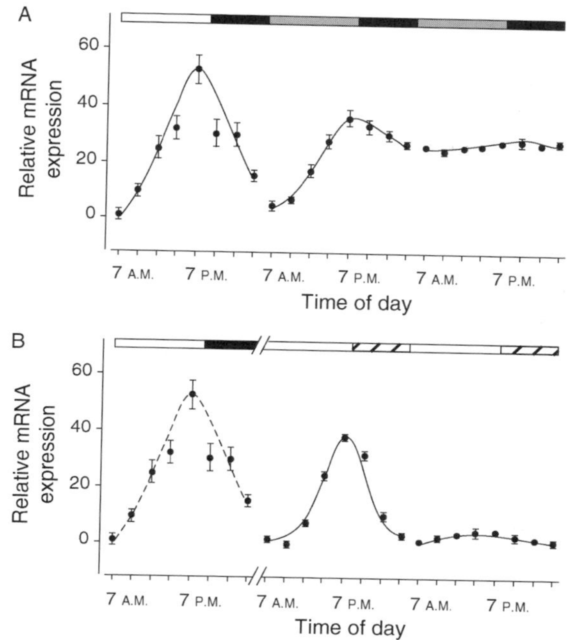

Fig. 5.5. Circadian expression of zebrafish L-cone opsin mRNA in LD and subsequently in DD or LL. In LD (first 24 h) the expression was low during the early morning and high during the late afternoon. In DD (A), the fluctuation of mRNA expression persisted for 24 h, then damped out. In LL (B), the rhythm damped out after 24 h, and thereafter remained at a low level. Relative opsin mRNA expression was determined by dividing the expression at each time point by the expression at 7 A.M. Horizontal bars at the top of the panels indicate lighting conditions. White bars, light during subjective day; black bars, dark during subjective night; gray bars, dark during subjective day; hatched bars, light during subjective night. Data represent mean ± S.E.M. [Adapted from Li *et al.* (2005) with permission.]

Studying the role of dopamine in L-cone opsin mRNA expression, Li *et al.* (2005) found that during the early morning, when opsin mRNA expression is normally low, the administration of dopamine increased it. During the late afternoon, when L-cone opsin mRNA expression is already high, dopamine did not have any effect. By using selective agonists for dopamine D_1 and D_2 receptors, they could show that the effect of dopamine on L-cone mRNA expression was mediated by D_1 receptors. It is not known whether photoreceptors possess D_1 receptors. Wagner and Behrens (1993) did not detect them on photoreceptors in rainbow trout. Mora-Ferrer *et al.* (1999) believed that some of the immunolabeling against D_1 receptors detected in the outer retina of goldfish might be due to labeling of cone axons, but could not demonstrate this with certainty. It is possible that dopamine regulates opsin expression via D_1 receptors located on horizontal or on RPE cells. Glutamate release of cones is modulated by horizontal cells that convey information from the cones located in the surrounding of their receptive fields (Section 8). Dopamine regulates the connectivity between horizontal cells (i.e., determine the size of the surrounds) and possibly the connectivity between horizontal cells and cone pedicles (Kirsch *et al.*, 1991). Photoreceptors also interact closely with RPE cells, which possess D_1 receptors. RPE cells provide the chromophores for rhodopsins and porhyropsins and are thus important for the visual cycle. Future research has to clarify if such interactions between photoreceptor and horizontal or RPE cells can play a role in opsin mRNA expression.

4.5. Summary

Visual sensitivity is regulated by a circadian clock. Rhythmical 24-h variations of visual sensitivity are maintained under DD and LL. The circadian rhythm of visual sensitivity can be determined by behavioral testing, ERG or RGC recordings. The modulation depth of the circadian rhythm is different when ERG and behavioral measurements are compared. This might be due to the existence of different circadian oscillators or due to different gains in elements of the visual pathway.

Rod–cone dominance shifts are also regulated by a circadian clock. For example, the spectral sensitivity curve of L-type cone horizontal cells coincides with the L-cone photoreceptor spectral sensitivity curve during the subjective day, whereas it coincides with the rod spectral sensitivity curve during the subjective night. Dopamine seems to play a role in circadian control of rod–cone dominance shift. The release of retinal dopamine and melatonin also follows a circadian rhythm; dopamine concentration peaks when melatonin concentration is lowest and vice versa. Both dopamine and melatonin modulate many retinal functions. It is a matter of debate whether independent circadian clocks are involved.

5. CHEMOSENSORY MODULATION OF VISUAL SENSITIVITY

5.1. The Terminal Nerve

In most vertebrates, including humans, the terminal nerve (TN) has been described (Demski, 1993; Wirsig-Wiechmann *et al.*, 2002; Behrens and Wagner, 2004). In many species, it is associated with the olfactory or the vomeronasal system. TN cell bodies receive input from sensory organs or brain areas involved in sensory functions, such as olfaction, vision, and somatosensory perception (Yamamoto and Ito, 2000), and project diffusely to many areas of the central nervous system, such as the olfactory epithelium, amygdala, septum, OT, and retina (Schwanzel-Fukuda and Silverman, 1980; Stell *et al.*, 1984; Jennes, 1987; Oka, 1992; Wirsig-Wiechmann and Wiechmann, 2001). Some of the TN neurons contain gonadotropin releasing hormone (GnRH), suggesting that the TN is involved in sexual and/or reproductive behavior (Demski and Northcutt, 1983; Stell *et al.*, 1987; Wirsig-Wiechmann *et al.*, 2002). The TN also contains FRamide-like peptide. It should be noted that localization of the neurons (olfactory bulb, ventral telencephalon) and afferent and efferent connectivities vary among species. It is not always clear which cells belong to the TN (Szabo *et al.*, 1991; for definition criteria of TN see Wirsig-Wiechmann *et al.*, 2002).

In goldfish, the TN has been found to project axons to the retina (Stell *et al.*, 1984). This has been confirmed for most fish species studied thus far (Demski, 1993). Only in a few other vertebrate species, such as vole (Wirsig-Wiechmann and Wiechmann, 2002), amphibians such as *Rana pipiens* (Wirsig-Wiechmann and Basinger, 1988), *Rana catesbeiana*, and *Xenopus laevis* (Uchiyama *et al.*, 1988), and possibly the rhesus macaque (Witkin, 1987), retinal TN projections have been reported. TN projections in the retina of many fish species contain both GnRH and RFamide-like peptides (Stell *et al.*, 1984; Ball *et al.*, 1989) and seem to make direct contact with DA-IPCs (Zucker and Dowling, 1987) and GABAergic and glycinergic amacrine cells (Ball *et al.*, 1989). The question inevitably arises as what is the function of these projections. Does olfactory or other sensory input to the TN affect visual system functions? Is there any indication that RFamide or GnRH modulates retinal sensitivity?

5.2. Olfactory Stimulation Affects Visual Sensitivity via the TN Projection to the Retina

Davis *et al.* (1988) investigated whether destruction of the TN affects visual sensitivity in goldfish. For this purpose, they destroyed the olfactory bulbs and the telencephalon. In a conditioned test, they measured the breathing suppression response evoked by a spot of "red" light that was conditioned to an electric shock. No effect of the lesion on the threshold light intensity was detected. Note that the authors neither tested the effect of sensory stimulation of the TN nor tested the rod threshold, which would have required white light or light with a wavelength around 500 nm. In a very different approach, Weiss and Meyer (1988) investigated whether olfactory stimulation could alter visual sensitivity. They administered either saline or water containing dissolved food extracts on the nostrils of the angelfish (*Pterophyllum scalare*) and recorded the ERG response to light flashes. They found that the b-wave amplitude in the food-extract group was significantly greater than that of the saline group. The authors suggested that the TN might have a critical role in chemosensory modulation of visual functions.

Maaswinkel and Li (2003b) performed a series of experiments to investigate whether olfactory stimulation (with amino acids) could modulate visual sensitivity in zebrafish. Amino acids are an important class of olfactory stimuli in fish (Caprio and Byrd, 1984). Valentinčič and Caprio (1997) have shown that amino acids can elicit feeding-related behaviors in fish. Glomeruli in the olfactory bulb of zebrafish encode information about different classes of olfactory stimuli, including amino acids (Friedrich and Korsching, 1997, 1998). Maaswinkel and Li (2003b) compared absolute visual sensitivity in zebrafish before and after adding amino acids (L-arginine, L-alanine, L-methionine, and L-aspartic acid; each is representative for one of four functional classes of amino acids) (Friedrich and Korsching, 1997, 1998) to the water by using the behavioral escape response paradigm. All four amino acids reduced the visual threshold in dark-adapted fish when applied at a higher concentration (Figure 5.6). Among them, the clearest effect was seen after administering 10^{-3} M L-methionine, which reduced the behavioral threshold by ~0.45 log units. The effect of methionine on the behavioral visual threshold was only seen in the early morning (when visual sensitivity is at a low level), not in the late afternoon (when visual sensitivity is already at its highest level). Maaswinkel and Li (2003b) also tested the effect of methionine on the ERG b-wave threshold. They found that the application of methionine to the contralateral nostril decreased the threshold by ~0.4 log units. This was also found only in the early morning, not in the afternoon.

Maaswinkel and Li (2003b) tested the effect of methionine on visual sensitivity after bilateral bulbectomy. They found that after the surgery, methionine lost its potential to modulate visual sensitivity. This demonstrates that amino acids exert their effects on vision via the olfactory system. Since the TN projects to many areas of the brain, including the OT, it might be that the effect of amino acids on vision is mediated by modulation of the function of visual nuclei rather than the function of the retina. Moreover, the effect of olfactory stimulation could also be mediated by olfactory pathways other than the TN (Behrens and Wagner, 2004). To distinguish between these possibilities, Maaswinkel and Li (2003b) investigated the effects of manipulating the retinal dopaminergic system, one of the targets of the TN, on olfactory modulation of visual sensitivity. They found that either intraocular injection of the dopamine D_2 receptor antagonist sulpiride or dopamine-depletion of the DA-IPCs with 6-OHDA blocked the olfactory modulation of sensitivity by methionine. These results combined with the findings of

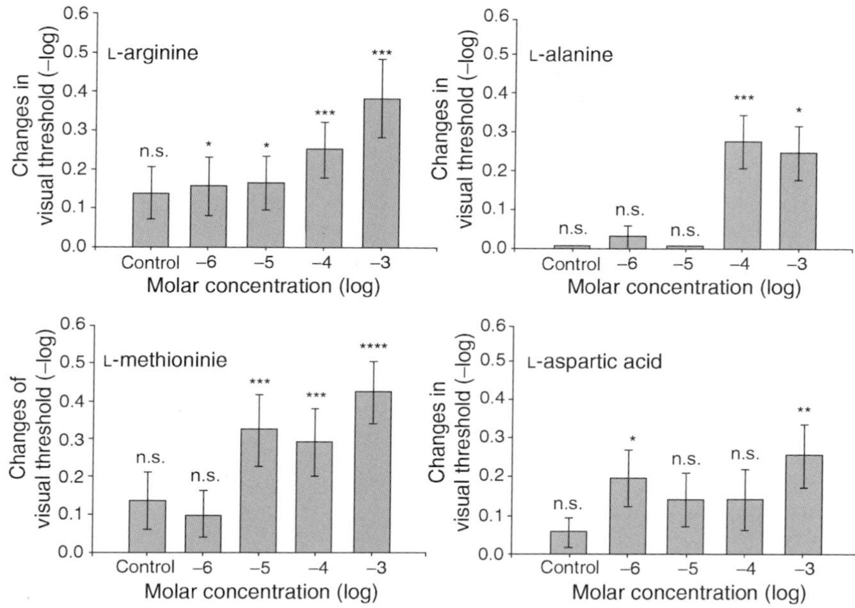

Fig. 5.6. Changes in behavioral threshold in zebrafish as determined by the behavioral escape paradigm after adding amino acids to the water. Four amino acids were tested: L-arginine, L-alanine, L-methionine, and L-aspartic acid. The effect of L-methionine was the strongest. Data present mean ± S.E.M; n.s., not significant; *, $p < 0.05$; **, $p < 0.01$; ***, $p < 0.005$; ****, $p < 0.001$. [Adapted from Maaswinkel and Li (2003b) with permission.]

earlier studies (see the following section), which show that TN transmitter GnRH and the transmitter homologue FMRFamide directly affect retinal functions (Walker and Stell, 1986; Umino and Dowling, 1991; Behrens *et al.*, 1993), suggest that the modulatory effect of olfactory stimulation on visual functions takes place in the retina and that the TN is the neuronal connection through which olfactory stimulation can exert its influence on visual sensitivity.

5.3. Mechanisms by Which the TN Modulates Visual Sensitivity

Walker and Stell (1986) found that GnRH increased spontaneous spike activity of RGCs in goldfish. The action of FMRFamide, the molluscan variant of RFamide, was less clear. Umino and Dowling (1991) have demonstrated that the effect of GnRH on cone horizontal cells of the white perch retina was very similar to the effect of dopamine. This included a decrease in size of receptive fields and depolarization of membrane potentials.

This effect of GnRH was eliminated by the depletion of DA-IPCs (using 6-OHDA) and by applying dopamine receptor antagonists. Umino and Dowling (1991) therefore suggested that GnRH promotes dopamine release. FMRFamide alone had no effect, but when administered together with GnRH, it inhibited the effect of the latter. Behrens *et al.* (1993) showed that in the blue acara (*Aequidens pulcher*), the formation of horizontal cell spinules (which are protrusions of the horizontal cells that contact the cone pedicles) increased when the eyecup preparations were exposed to GnRH. Kyle *et al.* (1995) and Fischer and Stell (1997) reported that FMRFamide used in the earlier studies does not have the exact structure as the RFamide found in goldfish. This raises questions about the validity of the findings from previous studies using FMRFamide.

In a study, Huang *et al.* (2005) found that olfactory stimulation with L-methionine increased RGC firing and decreased the threshold in response to visual stimuli. This effect was blocked by the dopamine D_1 receptor antagonist SCH23390, but not by the D_2 antagonist sulpiride. Furthermore, dopamine or the D_1 agonist SKF 38393 alone had the opposite effect on RGC activity as had olfactory stimulation with methionine, that is, they decreased cell firing and increased the threshold. By using *in vivo* microdialysis, the authors further found that olfactory stimulation decreased vitreal dopamine content. Patch-clamping isolated RGCs revealed that dopamine and SKF 38393 decreased calcium currents. Quinpirole, a D_2 receptor agonist, had no effect. These findings suggest that olfactory stimulation with amino acids decreases retinal dopamine release by the DA-IPCs. This effect might be mediated by the TN. The decrease of dopamine concentrations might directly affect RGC activity via the regulation of calcium channels. However, we have to be aware that the TN could affect RGC activities by many other pathways. For instance, RGCs possess GnRH receptors (Grens *et al.*, 2005). Thus, TN could directly modulate RGC activity. Furthermore, the role of TN projections to GABAergic and glycinergic amacrine cells (Stell *et al.*, 1987; Fischer and Stell, 1997) for visual system functions has thus far not been investigated. This pathway could also indirectly modulate RGC spiking. Finally, modulation of dopamine release by DA-IPCs can affect RGC activity via many pathways. Dopamine modulates elements of the retinal signaling pathway at many levels. Nonetheless, Huang *et al.* (2005) have shown (patch clamping of isolated cells) that dopamine can directly affect RGC responsiveness (see also Hayashida and Ishida, 2004).

5.4. Significance of the Retinal Projection of the TN

The behavioral significance of TN projection remains to be elucidated. It might play a role in visually guided behaviors, such as feeding or mating. Food extracts administered to the nostrils (Weiss and Meyer, 1988) increase the ERG b-wave amplitude. Amino acids administered in the same way have a similar effect and also increase the behavioral visual threshold (Maaswinkel and Li, 2003b) and decrease RGC threshold (Huang *et al.*, 2005). In fish, amino acids represent a category of olfactory (Hara, 1992) stimuli that elicit feeding-related behaviors (Valentinčič and Caprio, 1997; Lindsay and Vogt, 2004). Further studies are needed to determine whether olfactory stimulation can modulate visually guided foraging or prey catching by increasing visual sensitivity.

The TN has also been implicated in sexual and reproductive functions (Demski and Northcutt, 1983; Stell *et al.*, 1987; Wirsig-Wiechmann *et al.*, 2002). For example, lesion of GnRH-containing TN cells in the dwarf gourami (*Colisa lalia*) impairs male nest building (Yamamoto *et al.*, 1997). Furthermore, GnRH released by the TN in the olfactory and vomeronasal mucosa is hypothesized to modulate chemosensory responsiveness to pheromones (Wirsig-Wiechmann, 2001). Does the retinal projection of the TN have any role in mating or reproductive behavior? Ball *et al.* (1989) have shown that in goldfish GnRH content of the retina is lower in the spring than in the fall, when goldfish are sexually inactive. Walker and Stell (1986) found that GnRH applied to isolated retinas increased spontaneous RGC activity. In the spring, this effect was greater than in the fall. Whether sex pheromones can modulate TN activity is unknown. According to Fujita *et al.* (1991), they stimulate the olfactory nerve but not the TN. It remains to be determined whether stimulation of the olfactory epithelium [in fish, receptors of the vomeronasal system, which are stimulated by phermones, are located in the olfactory epithelium (Hansen *et al.*, 2004)] with sex pheromones affects visual functions and retinal physiology.

5.5. Summary

Research suggests that signal transmission in the retina is modulated by the TN. Amino acids or food extracts administered to the nostrils increase visual sensitivity as determined by ERG. Amino acids also increase behavioral visual sensitivity and decrease the threshold response of RGCs to light stimuli. There is ample evidence for the involvement of the dopaminergic system in the olfactory modulation of visual functions. Olfactory stimulation reduces retinal dopamine release as determined by microdialysis. The behavioral significance of the TN projection to the retina is not yet clear. One hypothesis states that the TN mediates mating or reproductive related behaviors. The effect of pheromones (applied to the nostrils) on visual sensitivity has not been investigated. Some researchers suggest that the retinal projection of the TN might play a role in foraging or prey catching.

One question for future research will be whether the modulation of visual functions by the TN only increases the absolute sensitivity or also fine-tunes specific visual parameters, such as spatiotemporal frequency characteristics of movement detection, contrast or spectral sensitivity. Given that the TN also projects to visual brain areas, such as the optic tectum, modulation of visual information processing might occur at different levels.

6. INHERITED AND ACQUIRED IMPAIRMENTS OF VISUAL SENSITIVITY

6.1. Night Blindness Mutations

Absolute visual sensitivity can be impaired by genetic diseases (Farrar *et al.*, 2002). These diseases are often, but not necessarily, characterized by photoreceptor cell degeneration. Several animal models are explored to understand the genetics of night blindness (Chader, 2002). Mutations of zebrafish genes that are involved in the development or functions of the visual system have become the subject of intensive research. Several night blindness mutations have been isolated (Li, 2001a,b).

To generate mutants with visual system defects, male zebrafish are treated repeatedly with a low dose of *N*-ethyl-*N*-nitrosourea, a chemical mutagen (Solnica-Krezel *et al.*, 1994). Mutagenized fish are then crossed with wild-type females. The F1 generation is screened for dominant mutations that cause visual defects. For the detection of dominant night blindness mutations, Li and Dowling (1997) developed the behavioral escape response paradigm (Section 3) to isolate mutants that show elevated visual thresholds. Subsequently, ERG recordings and histological analysis are performed to characterize the mutation. In some cases, RGC recordings (Li and Dowling, 2000b) and patch clamp studies (Yu and Li, 2003) can provide further information about the underlying mechanism of the disease.

Li and Dowling (1997) described the first dominant mutation (*nba*) in zebrafish that causes late-onset retinal degeneration. The *nba* heterozygous fish was characterized by a slow decline of visual sensitivity with age, as determined by behavioral and ERG testing. At 2–3 months of age, the mutants were undistinguishable from wild-type fish. Thereafter, their visual threshold increased slowly, and at 9 months of age it was 2.5 log units higher than in wild types. Over large parts of the retina, rods had completely disappeared. However, patches of intact rods remained (Figure 5.7). They might represent residual rods that survived the degeneration process. Alternatively, they might be newly differentiated rods that replaced the degenerated rods. Homozygous *nba* larvae were not viable. They died at 5 day postfertilization (dpf). At 2 dpf, the first signs of retinal degeneration were obvious, and at 4 dpf the retina was virtually destroyed. Thus, not only

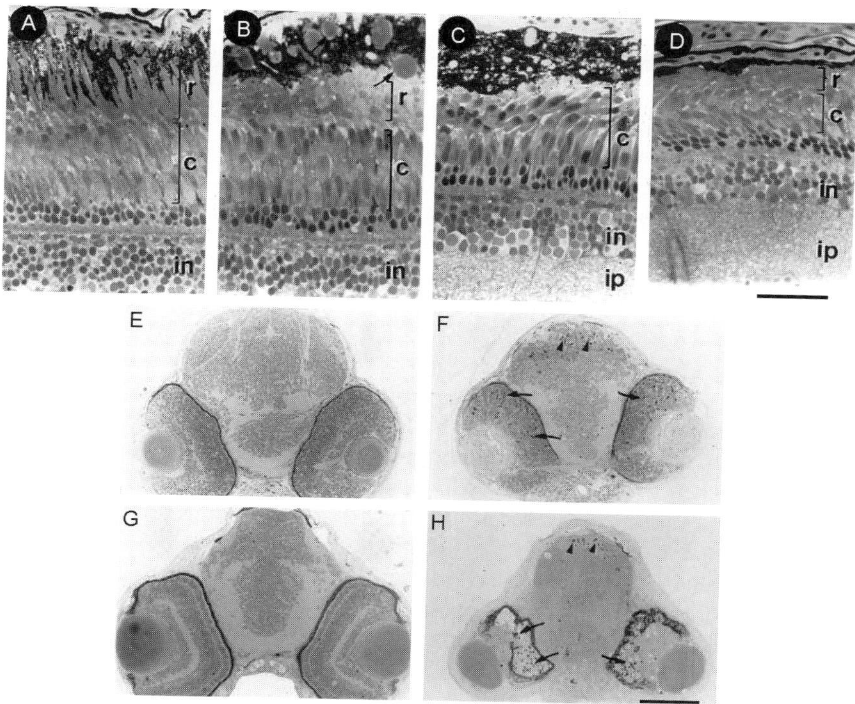

Fig. 5.7. Histological sections of the retina and the brain in wild-type and *nba* mutants. (A–D) Histological sections of a 13-month-old wild-type (A) and *nba* zebrafish (B–D). In the mutants, the degeneration is not uniformly across the retina. In some areas, rod outer segment (r) degeneration is obvious. The cone outer segments (c) are also sometimes affected (D). The arrowhead (B) shows lipid droplets that were sometimes observed in the RPE of mutants. (E and F) Transverse sections through the head and eyes of 2.5-day-old wild-type (E) and homozygous *nba* larvae (F). Apoptotic cells are seen throughout the tectum (arrowheads) and the retina (arrows). (G and H) Transverse sections through the head and eyes of 3.5-day-old wild-type (G) and homozygous *nba* larvae (H). By this time, most retinal cells were degenerated (arrows) in the mutants. Scale bars = 80 μm (B–D) and 100 μm (E–H). [Adapted from Li and Dowling (1997) with permission.]

photoreceptors but also other retinal cell types were affected. Furthermore, cell death was observed throughout the brain. Clearly, *nba* is not a photoreceptor cell-specific or even retina-specific mutation. Three other night blindness mutations characterized by extraretinal homozygous phenotypes are *nbb* (Li and Dowling, 2000b) (see later), *nbc* (Maaswinkel *et al.*, 2003b), and *nbd* (Maaswinkel *et al.*, 2003a).

Three dominant night blindness mutations (*nbe, nbf,* and *nbg*) without obvious extraretinal phenotypes have been reported (Maaswinkel *et al.*, 2005).

In *nbe*, despite severe rod degeneration, ERG sensitivities are only slightly reduced and the reduction of behavioral visual sensitivity is moderate. This suggests that compensation for the loss of photoreceptor cells may take place at the postreceptoral level. For example, signal transmission in bipolar cells might increase. It is also possible that other retinal neurons in the signaling network, such as RGCs and amacrine cells, contribute to the functional compensation of photoreceptor cell losses. Functional changes of postphotoreceptoral cells have been previously described in some other model systems for retinitis pigmentosa, for example, in the P23H transgenic rat (bearing a mutation of the rod opsin gene) and in the rd1 mouse (Aleman *et al.*, 2001; Strettoi *et al.*, 2002). Another zebrafish mutant, *nbg*, is characterized by inverse correlations between ERG parameters and behavioral threshold or photoreceptor cell degeneration. For example, the ERG b-wave threshold decreases with increasing behavioral visual threshold. This might indicate that considerable remodeling of the retinal circuitry (Marc *et al.*, 2003) has taken place.

Night blindness may also be found in mutants that show no signs of photoreceptor cell degeneration, for example, in *nbb* zebrafish (Li and Dowling, 2000b). In the behavioral test, *nbb* fish were characterized by elevated visual thresholds after prolonged (2 h) darkness. This effect could be reversed by bright light exposure. No abnormalities in the ERG were detected. Thus, the outer retina seems to function normally in *nbb*. However, inner retinal sensitivity in *nbb* decreased, that is, the threshold light levels required for modulating RGC spike frequencies were elevated. Immunolabeling and histology revealed degeneration of DA-IPCs and the retinal TN fiber network. To investigate whether these degenerations were responsible for the behavioral deficit, Li and Dowling (2000a,b) tested visual sensitivity of wild-type zebrafish with excised olfactory epithelium, after bulbectomy, or after dopamine depletion. They found that the visual deficits in these fish mimicked those described in *nbb* fish. This suggests that the degeneration of TN is at the core of the visual dysfunction in *nbb*.

6.2. Acquired Impairments

In teleosts, as in other vertebrates, visual sensitivity can be impaired as a result of environmental factors. These include exposure to neurotoxins, psychopharmacological agents, and abnormal lighting conditions.

Bilotta *et al.* (2002) studied the effect of ethanol exposure on young zebrafish larvae. They tested visual sensitivity a few days after terminating the treatment. The main finding was that exposure to ethanol at a very young age increased the ERG b-wave threshold in dark-adapted larvae. In addition, their eyes were smaller than the eyes of the control fish. In ethanol-exposed fish, visual acuity was lower than in control fish as assessed by the optomotor response paradigm (Section 7). Prenatal exposure to alcohol in humans also results in a number of retinal pathologies, including optic nerve hypoplasia and an increase in RGC death (Stromland and Sundelin, 1996).

Cocaine (another substance of abuse in humans) has been shown to reduce visual sensitivity in fish. It reduces dopamine reuptake in the synaptic cleft (Cooper *et al.*, 2002). In the behavioral escape task, cocaine dissolved in water increased the visual threshold of zebrafish by ~1 log unit (Darland and Dowling, 2001). ERG recordings confirmed that this effect was not simply due to altered motivation or swimming behavior. In this study, only acute cocaine effects were assessed. Whether chronic cocaine treatment results in dopamine receptor desensitization in the retina has not been studied. Desensitization occurs when the receptors become less responsive to the agonist after chronic exposure to it (Hammer *et al.*, 1997; Gainetdinov *et al.*, 2004). Reduction in ERG b-wave amplitude was also found in human drug addicts, who abuse cocaine, heroine, or cannabis (Gonzalez Perez *et al.*, 1995).

Abnormal light conditions during rearing of zebrafish larvae can reduce absolute visual sensitivity. Saszik and Bilotta (2001) reared zebrafish larvae during 1–6 dpf under DD. After this, the larvae were kept under normal LD. When tested between 6 and 9 dpf, the DD-reared larvae showed elevated visual dark-adapted ERG a- and b-wave thresholds, compared to LD-reared larvae. However, when tested between 11 and 13 dpf, the differences between both groups had disappeared. Thus, unlike the ERG threshold elevation seen after ethanol exposure (Bilotta *et al.*, 2002), the elevation of visual thresholds in larvae induced by DD rearing is reversible.

6.3. Summary

Inherited night blindness is characterized by increased scotopic threshold, that is, decreased absolute visual sensitivity. Several animal models exist to study the genetic basis and progress of the disease and possible interventions. Because random point mutation can be easily induced, zebrafish have become a model system to study the genetic basis of night blindness. Several dominant night blindness mutants have been isolated. In some of these, the mutation is not retina specific as demonstrated by the occurrence of extraretinal defects. In other mutants, only retinal defects have been detected. Similar to mammalian models of retinitis pigmentosa, in the zebrafish models, different involvements of postreceptoral elements and probable retinal remodeling have been described.

Environmental factors can also result in reduced absolute visual sensitivity of the dark-adapted retina. In zebrafish, exposure to ethanol, cocaine, or abnormal lighting conditions has been shown to permanently or transiently increase the visual threshold. Similar alterations of visual function have been described in humans.

7. CONTRAST VISUAL SENSITIVITY

The meaning of "contrast sensitivity" depends on the parameter of interest. It can refer to intensity, spectral, spatial, or temporal contrast. In fish, spatial contrast is usually determined by testing the discrimination between two different grating patterns or between one grating pattern and a homogenous field (Bilotta and Powers, 1991). Typically, the grating consists of contrasting bars (black and white or of different spectral characteristics). The grating pattern can be described by spatial frequency, that is, the number of repetitions of the pattern (cycles) per degree of visual angle. Intensity and/or spectral contrast of the bars are important variables of spatial contrast sensitivity. Mean luminance is also an important parameter. Temporal contrast is traditionally studied by presenting the subject with a sinusoidally flickering light stimulus (Bilotta *et al.*, 1998). The study of movement detection is closely related to the study of contrast sensitivity. In many studies, a moving grating pattern is used. Parameters of this pattern for the characterization of motion detection are, among others, spatial frequency, temporal frequency, and angular velocity (the velocity of the stimulus expressed in degrees of displacement in the visual field per second). The pattern used for motion detection can also consist of non-Fourier stimuli (Orger *et al.*, 2000) (see in a later section).

7.1. Concepts of Contrast Sensitivity

Traditionally, conditioned tests are used to investigate contrast sensitivity. Two frequently used unconditioned paradigms to measure contrast sensitivity are the OKR and the optomotor response (OMR) paradigms. In both conditioned and unconditioned tests, the stimulus consists typically of a striped pattern: either black and white bars or gradual changes between black and white, as modulated by a sinusoidal function. The unconditioned tests do not require training of the animals and allow for screening of a large number of fish in a short amount of time. Thus, they are ideally suited for the screening of visual mutants. However, we have to be aware that OKR and possibly OMR are components of the gaze control system, thus contrast measurements obtained by these tests might differ from those obtained by conditioned paradigms.

7.1.1. Contrast Sensitivity as Determined by Conditioned Response

Northmore and Dvorak (1979) studied contrast sensitivity as a function of spatial frequency in goldfish. During training, the fish received an electrical shock when a low-frequency high-contrast grating pattern was presented on an oscilloscope. Suppression of respiration was the conditioned response. After training, the fish were tested with grating patterns of different spatial frequencies and contrasts to determine the spatial contrast sensitivity curve. Using a similar procedure, Bilotta and Powers (1991) were also able to determine spatial contrast as a function of mean luminance and temporal frequency in goldfish.

7.1.2. Contrast Sensitivity Determined by OKR

The OKR is based on the optokinetic nystagmus, that is, the eye/head movements performed to stabilize retinal images (Moeller *et al.*, 2004). When a repeating stimulus pattern slowly moves across the visual field, the eyes track it. After a given angular rotation (which might be determined by anatomical constraints), the eyes move quickly back to their starting positions to "catch" the next pattern (this quick movement is called saccade) and restart the tracking movement. The OKR is easy to measure in zebrafish larvae (Clark, 1981; Brockerhoff *et al.*, 1995). The larvae are placed in a petri dish containing methylcellulose, which prevents body movements but allows eye movements. The petri dish is placed in the center of a drum. In the simplest setting, paper bearing a black and white grating pattern with given contrast and dimensions (which determine the spatial frequency) is attached to the inside of the drum, which can rotate at different speeds. Alternatively, the grating pattern is projected onto the drum and its parameters (spatial frequency, contrast, velocity) are computer controlled. The rotation of the eyes is recorded. Head movements as part of the gaze control system are usually not observed or studied in fish, however, they play an important role in mammals where they are sometimes used as the main parameter of OKR (Prusky *et al.*, 2004).

A common parameter used to express OKR is the gain, that is, the angular velocity of the eye movement divided by the angular velocity of the stimulus, which depends on the rotation speed of the drum and the apparent distance (taking the refractive index of the media into account) between the eye and the stimulus pattern. Rinner *et al.* (2005) have developed a computer program and paradigm to analyze the eye movements as a function of spatial and temporal frequency and contrast. They have shown that in zebrafish larvae the gain is dependent on angular velocity, spatial frequency, and contrast of the moving grating pattern.

7.1.3. Contrast Sensitivity as Determined by OMR

The OMR can be performed on both larval and adult fish by using a setup similar to the apparatus used for OKR testing. The petri dish is replaced by a circular container, allowing the fish to swim around freely. Normally, the fish swim along the wall of the container in the same direction as the moving grating pattern. By varying the parameters of the stimulus pattern, contrast sensitivity can be studied. Maaswinkel and Li (2003a) found that the temporal frequency curve has the characteristic of a low-pass filter for low spatial frequencies; for high spatial frequencies, it has the characteristic of a band-pass filter. Some zebrafish swim in the opposite direction for some spatiotemporal frequencies. A possible explanation for this is that, as in humans, the so-called wagon wheel effect exists also in fish. The wagon wheel effect is an apparent motion: the pattern is perceived as moving in the opposite direction. This effect possibly depends on the limited resolution of the retina (determined by the interphotoreceptor distance) and/or the sample frequency (Coletta *et al.*, 1990; Purves *et al.*, 1996). Similarly, the very high temporal frequencies (~300 cycle/sec) and angular velocities (over 1000°/sec), to which zebrafish responded, might also be explained by apparent motion. Neurons in the OT that respond to such high frequencies and velocities have not been described. The perceived temporal frequencies and velocities might be lower than the real ones.

Orger *et al.* (2000) analyzed the perception of Fourier and non-Fourier motion by 7-day-old larval zebrafish, using the OMR paradigm. Fourier motion detection is based on the detection of motion energy (e.g., moving patches of light intensities), whereas non-Fourier motion detection is based on other principles, which might involve pattern detection or coherence of stimuli. Orger *et al.* (2000) replaced the grating pattern (a typical example of Fourier motion stimuli) with a non-Fourier pattern and found that the larvae followed the movement of this pattern, thus demonstrating that they can detect non-Fourier motion. This is interesting because it is generally believed that in humans and other primates, non-Fourier motion detection is mediated by the cortex (Baker, 1999). However, zebrafish larvae do not possess a cortex. Orger *et al.* (2000) concluded that in fish, non-Fourier motion information is extracted at an earlier level of the visual pathway. Roeser and Baier (2003) demonstrated that after laser ablation of the OT in larval zebrafish, neither Fourier nor non-Fourier motion detection was impaired.

7.1.4. Processing of Contrast Sensitivity in Brain Areas

Do the three methods described (conditioned response, OKR, OMR) to determine spatial contrast sensitivity elicit comparable results? To our knowledge, no direct comparisons between the results obtained by these methods have been performed in teleosts.

Douglas *et al.* (2005) have shown that in rats, spatial contrast sensitivity as determined by OKR (measuring head tracking) was lower than the sensitivity obtained by a contrast discrimination task (a conditioned response task). They also found that the performance in the OKR task was not impaired by lesion of the striate cortex V1, whereas in the discrimination task the same lesion significantly reduced contrast sensitivity (Prusky and Douglas, 2004). They suggest that the difference between contrast sensitivities, as determined by both methods, is due to the involvement of different brain areas. The striate cortex (for which there is no homologous area in teleosts) is essential for the discrimination task. The authors further suggest that subcortical areas, such as the AOS, might be essential for eye or head tracking movements performed to stabilize retinal images as determined by OKR.

In teleosts (Northcutt and Butler, 1976; Wullimann *et al.*, 1996), as in some other vertebrates, the AOS receives projections from RGCs that possess large receptive fields and thus might be ideally suited for detecting optic flow. Simpson (1984) suggested that it probably has a role in the perception of self-movement. Ilg *et al.* (1993) have shown that lesion of the pretectal area, which is closely connected to the AOS (Simpson *et al.*, 1988), impairs OKR in rhesus monkeys. Klar and Hoffmann (2002) have found visual direction-selective neurons in the pretectum of the rainbow trout. AOS and pretectum are presumably part of the gaze control system (Rodieck, 1998; Giolli *et al.*, 2006). However, whether the AOS or the pretectum is a substrate for the OKR in teleosts has not been experimentally established.

In teleosts, the OT is often seen as the prime area for central visual information processing, although the current knowledge does not allow us to exclude an involvement of the telencephalon herein. Maaswinkel and Li (2003a) argued that the OT might be an essential substrate for OMR. The reasoning behind this was based on previous studies (Meyer *et al.*, 1970; Herrero *et al.*, 1998) that demonstrated that the OT has a role in approach and escape responses. Also, Springer *et al.* (1977) found that OMR (but not OKR) in goldfish was absent after bilateral OT ablation. However, Roeser and Baier (2003) found that in zebrafish larvae, laser ablation of the OT did not impair OMR.

7.1.5. Dual Channel Hypothesis of Contrast Sensitivity

Neumeyer (2003) studied visual acuity as a function of wavelength in goldfish using a conditioned behavioral test. Fish were rewarded for choosing a window showing a homogenous visual field over a window showing a grating pattern. Applying monochromatic illumination of different wavelengths, the authors found that the action spectrum of spatial resolution was similar to the photopic spectral sensitivity curve of the retina. In Neumeyer's interpretation, this means that the "color" channel (which receives input from all cone types) is used for the detection of spatial contrasts. Therefore, L-cones are not preferentially used. This is surprising because L-cones are the most abundant cone type in the goldfish retina (Marc and Sperling, 1976), thus they might seem to be the best choice for high spatial resolution. In another study, Schaerer and Neumeyer (1996) demonstrated that large field motion perception in goldfish is "color-blind" because L-cones mainly or exclusively contribute to OMR. In fact, M-cones seem to have an inhibitory effect. Anstis *et al.* (1998) and Krauss and Neumeyer (2003) obtained similar results for guppies and zebrafish, respectively. The color blindness of motion detection has also been shown for some other species, such as honeybees (Kaiser and Liske, 1974). From this and other experiments, Neumeyer (2003) concluded that spectral and spatial contrasts are processed by a color channel, using visual information in the entire spectral domain. Luminance, motion (as assessed by the OMR), and flicker (as assessed by the flicker fusion frequency test), on the other hand, are processed by a color-blind "brightness" channel mainly using L-cones.

In contrast to these findings, Orger and Baier (2005) found that in zebrafish larvae both L- and M-cones contribute to motion detection in the OMR. They suggest that both cone types contribute to a luminosity channel. However, the contributions of both cone types to OMR depend differently on the spatial frequency. One difference between this study and the study by Krauss and Neumeyer (2003) is that the latter is based on sensitivity measurements, whereas the former determines the contribution of both cone types at illuminations well above threshold level. Another difference is that the former study uses larvae, whereas the latter uses adult fish. Orger and Baier (2005) point out that during maturation, sensitivity of L-cones increases more dramatically than M-cones (Saszik *et al.,* 1999), thus in adult zebrafish the spectral sensitivity might be shifted to longer wavelengths. However, Hughes *et al.* (1998) have demonstrated by recording ERGs that in adult zebrafish, the retina is more sensitive to 480 nm than to 570 nm light. This is supported by the findings by Chinen *et al.* (2003) that L-cone opsin is less abundantly expressed than M- or S-cone opsin. Despite these differences, Orger and Baier (2005) agree with Krauss and Neumeyer (2003) that in zebrafish motion detection is color-blind. Wagner and Kröger (2005) concluded from their behavioral studies of the blue acara, reared under different monochromatic light conditions (Section 8), that OMR was driven by more than one spectral cone type.

7.1.6. Roles of Dopamine and Other Neurotransmitters in Contrast Sensitivity

Dopamine plays an important role in contrast sensitivity. Mora-Ferrer and Neumeyer (1996) found that dopamine D_1 receptor antagonists impaired spectral discrimination but not motion perception in goldfish. Mora-Ferrer and

Gangluff (2000) demonstrated that dopamine D_2 receptor antagonists impaired motion detection and reduced the flicker fusion frequency (FFF), a measure of temporal resolution (Mora-Ferrer and Gangluff, 2002). In an ERG study, Mora-Ferrer and Behrend (2004) found that both D_1 and D_2 receptor antagonists changed single flash responses, but only the latter changed the temporal transfer properties as determined by fERG. The upper limit frequency was reduced, which is consistent with the reduction of the FFF. Thus, the color and brightness channels of contrast sensitivity (Neumeyer, 2003) seem to be modulated by different dopamine receptor types. Dopamine has also been shown to have a role in human contrast sensitivity. In Parkinson's disease (which involves a reduction of central dopamine release), the visual contrast sensitivity curve is distorted but can be corrected by L-Dopa, a dopamine precursor often prescribed for the treatment of this disease. Apomorphine, a nonspecific dopamine receptor agonist, affects motion perception in humans (Masson *et al.*, 1993).

Mora-Ferrer *et al.* (2005) studied the effects of receptor antagonists of acetylcholine (ACh), GABA, dopamine, and glutamate (injected into the eyes) on the OMR in goldfish. They found that the nicotinic ACh-receptor blocker d-tubocurarine, the GABAa-receptor blocker bicuculline, and the GABAa/ c-receptor blocker picrotoxine eliminated the OMR completely for all rotation speeds of the drum bearing the stimulus pattern. The muscarinic ACh-receptor blocker atropine had no effect on OMR. Dopamine D_2-receptor blocker sulpiride and glutamate mGluR6-receptor blocker APB diminished the OMR only for high rotation speeds. The authors suggest that d-tubocurarine, bicuculline, and picrotoxine affect the functioning of direction-selective amacrine cells or ganglion cells, whereas sulpiride and APB reduce the upper-limit temporal frequency.

7.2. Development and Dysfunction of Contrast Sensitivity

Easter and Nicola (1996) have shown that in zebrafish the detection of a spatial pattern, as tested in the OKR test, is present at 73 h postfertilization (hpf), that is, only a few hours later than the first signs of the reaction to light (68 hpf). Before 73 hpf, the eye of zebrafish larvae is hyperopic (farsighted), but thereafter it becomes emmetropic (normal-sighted). The increase in the refractive index of the lens is responsible for this adaptation. Furthermore, the authors showed that the appearance of emmetropia did not depend on visual experience. Larvae reared in constant dark until 5 dpf showed the same tracking movements of the eyes (optokinetic nystagmus) in the OKR test as larvae reared under normal cyclic lighting conditions. This is in contrast to many other vertebrates, such as birds and mammals, in which the development from hyperopic to emmetropic vision depends on visual experience (Wallman and Winawer, 2004).

By testing OMR with grating patterns of different spatial frequencies, Bilotta (2000) found that visual acuity of zebrafish larvae increases with age. To investigate the role of lighting conditions, the author reared larvae under different schedules (LD, DD, LL) during the first 6 dpf. Thereafter, all the groups were kept under normal cycling light (LD). When tested at 12–14 dpf, the DD-reared larvae showed only slightly reduced visual acuity when compared to the LD-reared larvae. The LL-reared larvae, on the other hand, were strongly impaired. Larvae that were reared under LD, but missed one single dark cycle, also displayed decreased visual acuity. Bilotta (2000) concluded that LD periodicity is more important for the development of normal acuity than light intensity, especially since increasing the light intensity (from 300 to 1000 lux) in the LL-reared group had no additional effects. At 19–21 dpf, the distinction between groups had virtually disappeared. This suggests that the early rearing conditions only delayed the development of normal visual acuity. The effect of LL is not irreversible and the system is regulated back to normal when LD cycling is restored.

7.3. Summary

Contrast sensitivity is a complex notion. It can refer to intensity or spectral, spatial, or temporal contrast. Some tests to determine spatial contrast sensitivity are presented: conditioned response, OKR, and OMR. The visual brain areas possibly involved in contrast sensitivity are under investigation. Some studies suggest that the OT is essential to perform the OMR task, while other studies did not find such a connection. In teleosts, whether the AOS and/or pretectum are the substrates for OKR is still a matter of debate. In mammals, different assays to determine spatial acuity elicit different spatial contrast sensitivity curves. In teleosts, no studies have been performed to compare spatial acuity as determined by those tests.

The dual channel hypothesis proposes that a color channel is responsible for processing spectral and spatial information and a brightness channel is responsible for processing information about luminance, temporal modulation and motion. Some studies suggest that dopamine D_1 receptors have a role in the color channel and D_2 receptors in the brightness channel. Some other neurotransmitters, such as glutamate, Ach, and GABA, have also been shown to be involved in signal transmission in the brightness channel.

The development of the OKR in zebrafish larvae does not depend on the light conditions during early rearing. Visual acuity as determined by the OMR, on the other hand, is decreased in larvae reared under constant light. This effect, however, is transient and can be reversed when normal LD cycling is restored.

8. SPECTRAL VISUAL SENSITIVITY

Spectral visual sensitivity is the relative sensitivity of the visual system to stimuli of different wavelengths. Spectral sensitivity depends on the characteristics and distribution of photopigments, and chromatic processing at the photoreceptor and postreceptoral levels. One subject of major interest is tuning of spectral sensitivity. It can be studied in the context of the evolutionary adaptation, developmental adaptation under normal or abnormal rearing conditions, and adaptation to transient environmental conditions.

8.1. Spectral Coding

Here, we will concisely review the retinal network responsible for spectral opponency and further processing of wavelength information in teleosts. The different opsins and chromophores have been discussed in Section 2. In fish and some other lower vertebrates (e.g., turtles), the role of horizontal cells in spectral opponency has been thoroughly investigated. This is in contrast to mammals, where the circuitries responsible for spectral opponency are less understood (Dacey *et al.*, 1996). Excellent reviews about spectral information processing in fishes are available (Kamermans and Spekreijse, 1995; Toyoda and Shimbo, 1999; Twig *et al.*, 2003).

8.1.1. Horizontal Cells

According to Stell *et al.* (1975), the goldfish retina possesses one luminosity or L-type (H1) and two chromaticity or C-type (H2 and H3) cone horizontal cells. H1 cells are monophasic, that is, they respond to all wavelengths with hyperpolarization. They nearly exclusively receive input from L-cones. H2 cells mainly receive input from M-cones. However, inhibitory synaptic feedback from H1 cells to M-cones is responsible for depolarization of H2 cells by stimulation of L-cones. H2 cells are biphasic; they respond to light stimuli of lower wavelength with hyperpolarization and to stimuli of higher wavelengths (mediated by L-cones) with depolarization and some delay. The reversal (or null) wavelength is the wavelength at which the polarity of the cell reverses. H3 cells are triphasic. They mainly receive input from S-cones, which in turn receive inhibitory feedback from H2 cells. Triphasic horizontal cells are characterized by two response reversal (or null) wavelengths. Similar horizontal cell types have been found in other fish species, such as carp (*Carassius carassius*) and siberian sturgeon (*Acipenser baeri*) (Toyoda and Fujimoto, 1983; Govardoskii *et al.*, 1991). The model of spectral opponency described is known as the "cascade model." This model is currently accepted by most researchers, although some modifications have been made. According to

Kamermans and Spekreijse (1995), it is likely that all cone horizontal cell types receive input from all or most spectral cone types, but inputs are weighted differentially. These researchers also propose that biphasic horizontal cells provide feedback on L-cones. The differences between the two models (Stell, 1975; Kamermans and Spekreijse, 1995) are often difficult to evaluate. The picture is complicated by possible direct interactions between cones of different spectral types, either via telodendrites or photoreceptor coupling, which might exist between the members of double cones (Marchiafava *et al.*, 1985). The role of UV cones in spectral opponency has been insufficiently investigated.

One major discussion in spectral opponency research concerns the mechanism of the inhibitory feedback from horizontal cells to cones, an essential step in the cascade model (Kamermans and Spekreijse, 1999). Originally, GABA was suggested to play an important role in this (Marc *et al.*, 1978), however, many researchers agree that the mechanism involves calcium channels (Verweij *et al.*, 1996). Variations in extracellular electrical currents generated by cone and horizontal cell membrane conductance variations might be responsible for variations in the conductance of calcium channels in cone pedicles (Trifonov *et al.*, 1974; Kamermans *et al.*, 2001b). According to Kamermans *et al.* (2001a), hemichannels in horizontal cell dendrites that are located just opposite of voltage-gated calcium channels near the glutamate release sites in cone pedicles play an important role in locally restricted modulation of membrane potentials and thus of glutamate release.

8.1.2. Bipolar Cells

The receptive fields of spectrally coded bipolar cells are characterized by center and surrounding areas with different spectral characteristics. The receptive field of single-opponent bipolar cells has a monophasic ON-center and a monophasic OFF-surround or vice versa. However, most bipolar cells have double-opponent receptive fields (Kaneko and Tachibana, 1981; Shimbo *et al.*, 2000). They have biphasic centers (e.g., $M^+ L^-$) and opponent biphasic surrounds (e.g., $M^- L^+$). Shimbo *et al.* (2000) have described bipolar cells in carp with more complex receptive fields, for example, cells possessing triphasic centers ($S^- M^+ L^-$) and biphasic surrounds ($S^+ M^+ L^-$). Since the center can be multiphasic, it should be mentioned that classification of bipolar cells into ON and OFF cells is based on the depolarizing or hyperpolarizing response, respectively, of the cell to stimulation of the center with "white" light. Spectral-opponent bipolar cells can be either mixed or cone bipolar cells. Mixed bipolar cells receive input from both rods and cones (Scholes and Morris, 1973; Sherry and Yazulla, 1993). Rod-only bipolar cells, which are common in mammals, have to our knowledge not been described in those teleosts that possess cones (Connaughton *et al.*, 2004).

It is assumed (Shimbo *et al.*, 2000) that the center responses are not mediated by horizontal cells. This is because the small light spot required to stimulate the center only generates a small response in horizontal cells, which is too small to stimulate bipolar cells. Furthermore, in double-opponent cells, the response of the center to the L-component has a shorter latency than the response to the M-component. The cascade model of horizontal cell information processing (see earlier) predicts exactly the opposite; that is, a delay of L-signals because they are mediated via L-type horizontal cells. However, horizontal cells are perhaps responsible for the surround responses of bipolar cells (Shimbo *et al.*, 2000). A very low concentration of cobalt that blocks horizontal cell-to-cone feedback also reduces the surround response of RGCs in turtles (Vigh and Witkovsky, 1999). This suggests that horizontal input to bipolar cells is at least in part mediated by cones. However, Yang and Wu (1991) found in tiger salamander that blocking the cone-to-ON bipolar cell input by APB eliminates the center response, but seems to affect only partly the surround response. According to that study, horizontal cells contribute for one-third to the bipolar cell surround responses via a noncone pathway. The authors suggest that horizontal cells modulate bipolar cells directly by GABA. However, Vigh and Witkovsky (1999) argue that bipolar cell dendrites are insensitive to GABA and suggest that the noncone pathway possibly involves amacrine cells.

8.1.3. Amacrine, RGCs, and Visual Brain Areas

In the teleost retina, spectral opponency has also been described in amacrine cells (Kaneko, 1970; Djamgoz *et al.*, 1990). Amacrine cells receive input from bipolar cells, but unlike horizontal cells, their input and output connectivities are not exactly mapped. Not much is known about the role of amacrine cells in spectral coding. Spectral-coding RGCs in fish have been studied (Wagner *et al.*, 1960; Daw, 1968). In goldfish, about half of the RGCs studied are ML double-opponent cells. According to Sakai and colleagues (Sakai *et al.*, 1997a), spectral information in amacrine cells is not only coded by the polarity but also by the dynamics of the waveform. Similarly, spectral signals in RGCs are not only coded by increase or decrease of the spike frequency but also by the dynamics of the spike train (Sakai *et al.,* 1997b).

Spectral opponent cells have also been found in the OT. Most spectral-opponent cells in the rainbow trout OT (McDonald and Hawryshyn, 1999) exhibit a triphasic response pattern (S^+ M^- L^+). Some units are characterized by a biphasic response. The action spectrum of units in the TS in goldfish seems to be more or less flat (Gibbs and Northmore, 1998), thus these units seem not to be spectrally coded.

8.2. Tuning of Spectral Sensitivity

"Spectral tuning" of photopigments is defined by λ_{max} and the half-bandwidth of their spectral absorption curves. One subject of considerable interest is to understand spectral tuning of photopigments in context of evolutionary adaptation (Boughman, 2001). However, this research is part of a broader attempt to understand adaptive determinants of chromatic processing. This does not only involve spectral tuning of photopigments but also expression of different opsin genes and chromophores, distribution of cone types across the retina, and chromatic processing at photoreceptor and postreceptoral levels. Thus, by using the term "tuning of spectral sensitivity," we emphasize the characteristics and plasticity of chromatic processing at any level.

Carleton and Kocher (2001) have studied the molecular basis of tuning of spectral sensitivity of closely related cichlid fish species in Lake Malawi that occupy different ecological niches and display different sexual dimorphisms. They found that the differences between DNA sequences in orthologous genes in different species are small. Instead, tuning is achieved by differential expression of the opsin genes. An interesting question is whether gene expressions can be modified within a species, either during normal development or as an adaptation to environmental conditions. Nawrocki *et al.* (1985) found that L-cone opsin in young zebrafish larvae has a lower λ_{max} than in older larvae and adults. Chinen *et al.* (2003) suggest that this might be explained by switching the expression from the L-cone opsin gene LWS-2 to LWS-1 during development. In the Pacific pink salmon (*Oncorhynchus gorbuscha*), single cones of newly hatched fish express UV opsin. Later, the same cones express blue opsin (Cheng and Flamarique, 2004). Opsin genes are polymorphic in some fish species. For example, four different genes of M-opsin have been described in zebrafish, each with a different λ_{max} (467, 476, 488, and 505 nm) (Chinen *et al.*, 2003). It is not known whether these genes are differentially expressed under different environmental conditions.

One mechanism of tuning of spectral sensitivity is modifying the proportions of rhodopsin and porphyropsin. The two chromophores 11-*cis*-retinal and 11-*cis*-3-dehydroretinal are the basis for the distinction between rhodopsins and porphyropsins (Section 2). λ_{max} of porphyropsins are higher than λ_{max} of the homologous rhodopsins. These differences become greater with longer wavelengths. They range from a few nanometers for short wavelengths to more than 60 nm for long wavelengths. Under some conditions, the fish might exclusively possess rhodopsins, whereas under other conditions, it may exclusively possess porphyropsins or a mixture of both (Bowmaker, 1995). The switch between chromophores has usually been associated with seasonal variations and migration. Temperature-dependent variations of the rhodopsin/porhyropsin ratio have been described in goldfish (Tsin and Beatty, 1979).

More surprisingly, fish that normally spend their entire lives under relatively constant environmental conditions can also switch from possessing only rhodopsins to possessing a mixture of rhodopsins and phorphyropsins. This has been described in zebrafish (Saszik and Bilotta, 1999) after being transferred from warmer to cooler waters. They determined dark-adapted spectral sensitivity curve of the ERG b-wave. In zebrafish maintained in the high-temperature condition (28–30°C) the curve was shifted to shorter wavelengths as compared to the curve obtained from zebrafish that were maintained in the low-temperature condition (22–25°C). In the high-temperature condition, the spectral sensitivity curve fitted the rod rhodopsin absorption curve. In the cold temperature condition, it had values between the rod rhodopsin and rod porphyropsin sensitivity curves. Therefore, the authors suggested that in the warm temperature condition, the rod photopigment consists of rod rhodopsin, whereas in the cold temperature condition, it is a mixture of rod rhodopsin and rod porphyropsin. However, this was not confirmed by Allison *et al.* (2004). They determined the λ_{max} of rods and cones of zebrafish using microspectrophotometry. Water temperature did not alter the λ_{max}. The reason for the discrepancy between these two studies is not known. Allison *et al.* (2004) showed that thyroid hormone treatment changed λ_{max} of rods and cones to longer wavelengths, which suggests a shift from rhodopsins to porphyropsins.

Wagner and Kröger (2005) have studied tuning of spectral sensitivity in blue acara that were reared for at least one year under specified lighting conditions, starting from an early larval stage (14–28 dpf). Various lighting conditions were applied: dim white light, bright white light, or monochromatic light of the wavelengths: 450, 452, 485, 513, 534, 590, and 623 nm. The parameters analyzed were: absorbance, ratio of cone types, and lengths of cone outer segments; connectivities, spinule numbers, and spectral responses of horizontal cells; and OMR to stripes generated by either white or monochromatic lightsource.

Microspectrophotometric measurements revealed that in none of the groups was the expression of photopigments reduced. Furthermore, as in control fish, the visual pigments consisted exclusively of rhodopsins; no porphyropsins were detected. In the 485-nm group, 65% of the S-cones were eliminated, which was probably due to apoptosis. The length of the outer segments of M- and L-cones was increased.

Adaptation also occurred at the postreceptoral level. The blue acara possesses two types of cone horizontal cells, namely monophasic L-type and biphasic C-type horizontal cells. In the bright white light rearing condition, the connectivity between C-type horizontal cells and S-cones was reduced (as determined by the number of contacts), whereas in the 450/452-nm conditions, the connectivity between S-cones and L-type horizontal cells was slightly increased. In the 485-nm group, the number of mature spinules was significantly reduced. Spinules are small fingerlike evaginations of horizontal cell dendrites that seem to make contact with cone pedicles. Their function is still a matter of debate. The reduction of spinule numbers was transient; after keeping the 485-nm group in white light for 10 h, the number of spinules returned to control level. In the 450/452-nm group, the reversal wavelength of C-type horizontal cells shifted to a shorter wavelength. This might play a role in spectral constancy (see later).

Results obtained with the OMR test (Kröger *et al.*, 2003) demonstrated that fish reared under short- and middle-wavelength illumination were less sensitive to long-wavelength illumination than fish reared under long-wavelength or white light.

In fish and turtles, transient shifts of the reversal wavelengths of multiphasic horizontal cells have been found under monochromatic or high-intensity background illumination (Gottesman and Burkhardt, 1987; Kamermans and Spekreijse, 1995). The receptor field size of horizontal cells also changes with both light intensity and wavelength (Twig *et al.*, 2003). These adaptations to background illumination might be at the basis of spectral constancy that has been described in several fish species, such as goldfish (Neumeyer *et al.*, 2002) and carp (Dimentman *et al.*, 1972).

8.3. Summary

In teleosts, horizontal and bipolar cells play important roles in spectral coding. This especially has been well studied in horizontal cells. The cascade model proposed by Stell *et al.* more than 30 years ago is on the basis of our understanding of how spectral opponency is generated via cone-horizontal cell interactions. Spectral opponency in the surround of bipolar cells, but not within the center of bipolar cell receptive fields, is most possibly mediated by horizontal cells. Spectral coding was also demonstrated in amacrine cells, RGCs, and in the OT, but it has not been reported in the TS.

Tuning of spectral sensitivity in teleosts can depend on endogenous factors (e.g., stage of life) or environmental factors (e.g., temperature, intensity, and spectral characteristic of the ambient illumination, rearing light conditions). The retinal alterations needed to achieve tuning of spectral sensitivity can occur at the photoreceptor level (e.g., expression of opsins, chromophores, elimination of cone types) or at the postreceptoral level (e.g., connectivities and spectral properties of horizontal cells).

9. CONCLUSIONS

In this chapter, we reviewed the major lines of current research of visual sensitivity in fish. Visual sensitivity is regulated at different levels of the retinal network. Sensitivity of photoreceptor cells, for example, might be determined by the availability of opsins and the nature of the chromophore, as well as the functional state of transduction cascades. Second- and third-order neurons, that is, horizontal, bipolar, amacrine, and ganglion cells, can further determine the sensitivity of the system. The functioning of these different elements depends on neurotransmitters, such as melatonin and dopamine. The release of both transmitters is under circadian as well as photic control. Furthermore, dopamine release might be modulated by chemosensory stimulation via the TN. Signal processing in visual nuclei probably also contributes to visual sensitivity.

Thus, when assessing absolute, contrast, or spectral visual sensitivity, we have to distinguish the level of processing. Visual sensitivities, as determined by photon catch, ERG or RGC recordings, or behavioral escape responses, OMR, OKR, are different parameters that do not always correlate.

ACKNOWLEDGMENTS

We thank Dr. Tom Vihtelic for comments and Meghan Riley and Aprell Carr for proofreading the chapter. Some of the work of the authors was supported in part by NIH grants R01 EY13147 and EY13680 and by funding from the Center for Zebrafish Research at the University of Notre Dame.

REFERENCES

Adachi, A., Suzuki, Y., Nogi, T., and Ebihara, S. (1999). The relationship between ocular melatonin and dopamine rhythms in the pigeon: Effects of melatonin inhibition on dopamine release. *Brain Res.* **815,** 435–440.

Ahmad, I., Tang, L., and Pham, H. (2000). Identification of neural progenitors in the adult mammalian eye. *Biochem. Biophys. Res. Commun.* **270,** 517–521.

Akert, K. (1949). Der visuelle Greifreflex. *Helv. Physiol. Pharmacol. Acta* **7,** 112–134.

Aleman, T. S., LaVail, M. M., Montemayor, R., Ying, G., Maguire, M. M., Laties, A. M., Jacobson, S. G., and Cideciyan, A. V. (2001). Augmented rod bipolar cell function in partial receptor loss: An ERG study in P23H rhodopsin transgenic and aging normal rats. *Vision Res.* **41,** 2779–2797.

Alexander, K. R., Barnes, C. S., and Fishman, G. A. (2003). ON-pathway dysfunction and timing properties of the flicker ERG in carriers of X-linked retinitis pigmentosa. *Invest. Ophthalmol. Vis. Sci.* **44,** 4017–4025.

Allison, W. T., Dann, S. G., Helvik, J. V., Bradley, C., Moyer, H. D., and Hawryshyn, C. W. (2003). Ontogeny of ultraviolet-sensitive cones in the retina of rainbow trout (*Oncorhynchus mykiss*). *J. Comp. Neurol.* **461,** 294–306.

Allison, W. T., Haimberger, T. J., Hawryshyn, C. W., and Temple, S. E. (2004). Visual pigment composition in zebrafish: Evidence for a rhodopsin-porphyropsin interchange system. *Vis. Neurosci.* **21,** 945–952.

Anstis, S. (2002). The Purkinje rod-cone shift as a function of luminance and retinal eccentricity. *Vision Res.* **42,** 2485–2491.

Anstis, S., Hutahajan, P., and Cavanagh, P. (1998). Optomotor test for wavelength sensitivity in guppyfish (*Poecilia reticulata*). *Vision Res.* **38,** 45–53.

Awatramani, G., Wang, J., and Slaughter, M. M. (2001). Amacrine and ganglion cell contributions to the electroretinogram in amphibian retina. *Vis. Neurosci.* **18,** 147–156.

Baker, C. L., Jr. (1999). Central neural mechanisms for detecting second-order motion. *Curr. Opin. Neurobiol.* **9,** 461–466.

Ball, A. K., Stell, W. K., and Tutto, D. A. (1989). Efferent projections to the goldfish retina. *In* "NATO ASI Series: Neurobiology of the Inner Retina" (Weiler, R., and Osborne, N. N., Eds.), Vol. 31, pp. 103–116. Springer, Berlin.

Bassi, C. J., and Powers, M. K. (1987). Circadian rhythm in goldfish visual sensitivity. *Invest. Ophthalmol. Vis. Sci.* **28,** 1811–1815.

Beatty, D. D. (1966). A study of the succession of visual pigments in Pacific salmon (*Oncorhynchus*). *Can. J. Zool.* **44,** 429–455.

Behrens, U., and Wagner, H. J. (2004). Terminal nerve and vision. *Microsc. Res. Tech.* **65,** 25–32.

Behrens, U. D., Douglas, R. H., and Wagner, H. J. (1993). Gonadotropin-releasing hormone, a neuropeptide of efferent projections to the teleost retina induces light-adaptive spinule formation on horizontal cell dendrites in dark-adapted preparations kept *in vitro*. *Neurosci. Lett.* **164,** 59–62.

Behrens, U. D., Douglas, R. H., Sugden, D., Davies, D. J., and Wagner, H. J. (2000). Effect of melatonin agonists and antagonists on horizontal cell spinule formation and dopamine release in a fish retina. *Cell Tissue Res.* **299,** 299–306.

Bilotta, J. (2000). Effects of abnormal lighting on the development of zebrafish visual behavior. *Behav. Brain Res.* **116,** 81–87.

Bilotta, J., and Powers, M. K. (1991). Spatial contrast sensitivity of goldfish: Mean luminance, temporal frequency and a new psychophysical technique. *Vision Res.* **31,** 577–585.

Bilotta, J., Lynd, F. M., and Powers, M. K. (1998). Effects of mean luminance on goldfish temporal contrast sensitivity. *Vision Res.* **38,** 55–59.

Bilotta, J., Saszik, S., Givin, C. M., Hardesty, H. R., and Sutherland, S. E. (2002). Effects of embryonic exposure to ethanol on zebrafish visual function. *Neurotoxicol. Teratol.* **24,** 759–766.

Boughman, J. W. (2001). Divergent sexual selection enhances reproductive isolation of sticklebacks. *Nature* **411,** 944–948.

Bowmaker, J. K. (1995). The visual pigments in fish. *Prog. Retin. Eye Res.* **15,** 1–31.

Bowmaker, J. K., and Kunz, Y. W. (1987). Ultraviolet receptors, tetrachromatic colour vision and retinal mosaics in the brown trout (*Salmo trutta*): Age-dependent changes. *Vision Res.* **27,** 2101–2108.

Branchek, T. (1984). The development of photoreceptors in the zebrafish, *Brachydanio rerio*. II. Function. *J. Comp. Neurol.* **224,** 116–122.

Branchek, T., and Bremiller, M. (1984). The development of photoreceptors in the zebrafish, *Brachydanio rerio*. I. Strurture. *J. Comp. Neurol.* **224,** 107–115.

Brockerhoff, S. E., Hurley, J. B., Janssen-Bienhold, U., Neuhauss, S. C. F., Driever, W., and Dowling, J. E. (1995). A behavioral screen for isolating zebrafish mutants with visual system defects. *Proc. Natl. Acad. Sci. USA* **92,** 10545–10549.

Brown, K. T. (1968). The electroretinogram: Its components and their origins. *Vision Res.* **8,** 633–677.

Burnside, B. (2001). Light and circadian regulation of retinomotor movement. *Prog. Brain Res.* **131,** 477–485.

Cahill, G. M. (1996). Circadian regulation of melatonin production in cultured zebrafish pineal and retina. *Brain Res.* **708,** 177–181.

Cahill, G. M., and Besharse, J. C. (1991). Resetting the circadian clock in cultured *Xenopus* eyecups: Regulation of retinal melatonin rhythms by light and D2 dopamine receptors. *J. Neurosci.* **11,** 2959–2971.

Cahill, G. M., and Besharse, J. C. (1993). Circadian clock functions localized in *Xenopus* retinal photoreceptors. *Neuron* **10,** 573–577.

Cameron, D. A. (2002). Mapping absorbance spectra, cone fractions, and neuronal mechanisms to photopic spectral sensitivity in zebrafish. *Vis. Neurosci.* **19,** 365–372.

Cameron, D. A., and Carney, L. H. (2004). Cellular patterns in the inner retina of adult zebrafish: Quantitative analyses and a computational model of their formation. *J. Comp. Neurol.* **471,** 11–25.

Cameron, D. A., and Easter, S. S., Jr. (1993). The cone photoreceptor mosaic of the green sunfish, *Lepomis cyanellus*. *Vis. Neurosci.* **10,** 375–384.

Caprio, J., and Byrd, R. J. (1984). Electrophysiological evidence for acidic, basic, and neutral amino acid olfactory receptor sites in the catfish. *J. Gen. Physiol.* **84,** 403–422.

Carleton, K. L., and Kocher, T. D. (2001). Cone opsin genes of African cichlid fishes: Tuning spectral sensitivity by differential gene expression. *Mol. Biol. Evol.* **18,** 1540–1550.

Chader, G. J. (2002). Animal models in research on retinal degenerations: Past progress and future hope. *Vision Res.* **42,** 393–399.

Cheng, C. L., and Flamarique, I. N. (2004). New mechanism for modulating colour vision. *Nature* **428,** 279.

Chinen, A., Hamaoka, T., Yamada, Y., and Kawamura, S. (2003). Gene duplication and spectral diversification of cone visual pigments of zebrafish. *Genetics* **163,** 663–675.

Clark, D. T. (1981). Visual responses in developing zebrafish (*Brachydanio rerio*). University of Oregon, PhD Thesis.

Coletta, N. J., Williams, D. R., and Tiana, C. L. (1990). Consequences of spatial sampling for human perception. *Vision Res.* **30,** 1631–1648.

Collin, S. P., and Collin, H. B. (1999). The foveal photoreceptor mosaic in the pipefish, *Corythoichthyes paxtoni* (Syngnathidae, Teleostei). *Histol. Histopathol.* **14,** 369–382.

Connaughton, V. P., Graham, D., and Nelson, R. (2004). Identification and morphological classification of horizontal, bipolar and amacrine cells within the zebrafish retina. *J. Comp. Neurol.* **477,** 371–385.

Cooper, J. R., Bloom, F. E., and Roth, R. E. (2002). "The Biochemical Basis of Neuropharmacology," 8th edn., Oxford University Press, Oxford.

Dacey, D. M., Lee, B. B., Stafford, D. K., Pokorny, J., and Smith, V. C. (1996). Horizontal cells of the primate retina: Cone specificity without spectral opponency. *Science* **271,** 656–659.

Darland, T., and Dowling, J. E. (2001). Behavioral screening for cocaine sensitivity in mutagenized zebrafish. *Proc. Natl. Acad. Sci. USA* **98,** 11691–11696.

Davis, R. E., Kyle, A., and Klinger, P. D. (1988). Nervus terminalis innervation of the goldfish retina and behavioral visual sensitivity. *Neurosci. Lett.* **91,** 126–130.

Daw, N. W. (1968). Colour-coded ganglion cells in the goldfish retina: Extension of their receptive fields by means of new stimuli. *J. Physiol.* **197,** 567–592.

Dean, P., Redgrave, P., and Westby, G. W. (1989). Event or emergency? Two response systems in the mammalian superior colliculus. *Trends Neurosci.* **12,** 137–147.

Dearry, A., and Burnside, B. (1986). Dopaminergic regulation of cone retinomotor movement in isolated teleost retinas: I. Induction of cone contraction is mediated by D2 receptors. *J. Neurochem.* **46,** 1006–1021.

DeMarco, P. J., Jr., and Powers, M. K. (1989). Sensitivity of ERG components from dark-adapted goldfish retinas treated with APB. *Brain Res.* **482,** 317–323.

Demski, L. S. (1993). Terminal nerve complex. *Acta Anat.* (*Basel.*) **148,** 81–95.

Demski, L. S., and Northcutt, R. G. (1983). The terminal nerve: A new chemosensory system in vertebrates? *Science* **220,** 435–437.

Dimentman, A. M., Karas, A. Y., Maximov, V. V., and Orlov, O. Y. (1972). Constancy of object color perception in the carp (*Cyprinius carpio*). *Pavlov J. High. Nerv. Act.* **22,** 772–779. (In Russian; cited in Neumeyer *et al.*, 2002.).

Djamgoz, M. B., Spadavecchia, L., Usai, C., and Vallerga, S. (1990). Variability of light-evoked response pattern and morphological characterization of amacrine cells in goldfish. *J. Comp. Neurol.* **301,** 171–190.

Dong, C. J., and Hare, W. A. (2000). Contribution to the kinetics and amplitude of the electroretinogram b-wave by third-order retinal neurons in the rabbit retina. *Vision Res.* **40,** 579–589.

Dong, C. J., and McReynolds, J. S. (1992). Comparison of the effects of flickering and steady light on dopamine release and horizontal cell coupling in the mudpuppy retina. *J. Neurophysiol.* **67,** 364–372.

Dong, C. J., Qian, H. H., McReynolds, J. S., Yang, X. L., and Liu, Y. M. (1988). Suppression of cone-driven responses by rods in the isolated frog retina. *Vis. Neurosci.* **1,** 331–338.

Dong, X., Pilido, J. S., Qu, T., and Sugaya, K. (2003). Differentiation of human neural stem cells into retinal cells. *Neuroreport* **14,** 143–146.

Douglas, R. H. (1986). Photopic spectral sensitivity of a teleost fish, the roach (*Rutilus rutilus*), with special reference to its ultraviolet sensitivity. *J. Comp. Physiol. A* **159,** 415–421.

Douglas, R. H., and Djamgoz, M. (1990). "The Visual System of Fish." Chapman and Hall, London.

Douglas, R. H., and Hawryshyn, C. W. (1990). Behavioural studies of fish vision: An analysis of visual capabilities. *In* "The Visual System of Fish" (Douglas, R. H., and Djamgoz, M., Eds.), pp. 373–418. Chapman and Hall, London.

Douglas, R. M., Alam, N. M., Silver, B. D., McGill, T. J., Tschetter, W. W., and Prusky, G. T. (2005). Independent visual threshold measurements in the two eyes of freely moving rats and mice using a virtually-reality optokinetic system. *Vis. Neurosci.* **22,** 677–684.

Dowling, J. E. (1987). "The Retina: An Approachable Part of the Brain." Harvard University Press, Cambridge, MA.

Easter, S. S., and Nicola, G. N., Jr. (1996). The development of vision in the zebrafish (*Danio rerio*). *Dev. Biol.* **180,** 646–663.

Ekström, P., and Meissl, H. (1997). The pineal organ of teleost fishes. *Rev. Fish Biol. Fish.* **7,** 199–284.

Ewert, J. P. (1984). Tectal mechanisms that underlie prey-catching and avoidance behavior in toads. *In* "Comparative Neurology of the Optic Tectum" (Vanegas, H., Ed.), pp. 247–416. Plenum, New York.

Farrar, G. J., Kenna, P. F., and Humphries, P. (2002). On the genetics of retinitis pigmentosa and on mutation-independent approaches to therapeutic intervention. *EMBO J.* **21,** 857–864.

Fernald, R. D. (1990). *Haplochromis burtoni*: A case study. *In* "The Visual System of Fish" (Douglas, R. H., and Djamgoz, M., Eds.), pp. 443–463. Chapman and Hall, London.

Fernald, R. D., and Wright, S. E. (1985). Growth of the visual system in the African cichlid fish, *Haplochromis burtoni*: Accommodation. *Vision Res.* **25,** 163–170.

Fischer, A. J., and Reh, T. A. (2003a). Growth factors induce neurogenesis in the ciliary body. *Dev. Biol.* **259,** 225–240.

Fischer, A. J., and Reh, T. A. (2003b). Potential of Muller glia to become neurogenic retinal progenitor cells. *Glia* **43,** 70–76.

Fischer, A. J., and Stell, W. K. (1997). Light-modulated release of RFamide-like neuropeptides from nervus terminalis axon terminals in the retina of goldfish. *Neuroscience* **77,** 585–597.

Friedrich, R. W., and Korching, S. I. (1997). Combinatorial and chemotopic odorant coding in the zebrafish olfactory bulb visualized by optical imaging. *Neuron* **18,** 737–752.

Friedrich, R. W., and Korsching, S. I. (1998). Chemotopic, combinatorial, and noncombinatorial odorant representations in the olfactory bulb revealed using a voltage-sensitive axon tracer. *J. Neurosci.* **18,** 9977–9988.

Fritsches, K. A., and Marshall, N. J. (2002). Independent and conjugate eye movements during optokinesis in teleost fish. *J. Exp. Biol.* **205,** 1241–1252.

Fujita, I., Sorensen, P. W., Stacey, N. E., and Hara, T. J. (1991). The olfactory system, not the terminal nerve, functions as the primary chemosensory pathway mediating responses to sex pheromones in male goldfish. *Brain Behav. Evol.* **38,** 313–321.

Gabriel, R., Lesauter, J., Silver, R., Garcia-España, A., and Witkovsky, P. (2001). Diurnal and circadian variation of protein kinase C immunoreactivity in the rat retina. *J. Comp. Neurol.* **439,** 140–150.

Gainetdinov, R. R., Premont, R. T., Bohn, L. M., Lefkowitz, R. J., and Caron, M. G. (2004). Desensitization of G protein-coupled receptors and neuronal functions. *Annu. Rev. Neurosci.* **27,** 107–144.

Gibbs, M. A., and Northmore, D. P. (1998). Spectral sensitivity in the goldfish torus longitudinalis. *Vis. Neurosci.* **15,** 859–865.

Giolli, R. A., Blanks, R. H. I., and Lui, F. (2006). The accessory optic system: Basic organization with an update on connectivity, neurochemistry, and function. *Prog. Brain Res.* **151,** 407–440.

Gonzalez Perez, J., Parafita Mato, M., Segade Garcia, A., and Diaz Rey, A. (1995). Intraocular motility, electrophysiological tests and visual fields in drug addicts. *Opthalmol. Physiol. Opt.* **15,** 493–498.

Gottesman, J., and Burkhardt, D. A. (1987). Response properties of C-type horizontal cells in the retina of the bowfin. *Vision Res.* **27,** 170–189.

Govardoskii, V. I., Byzov, A. L., Zueva, L. V., Polisczuk, N. A., and Baburina, E. A. (1991). Spectral characteristics of photoreceptors and horizontal cells in the retina of the Siberian sturgeon *Acipenser baeri* Brandt. *Vision Res.* **31,** 2047–2056.

Green, C. B., and Besharse, J. C. (2004). Retinal circadian clocks and control of retinal physiology. *J. Biol. Rhythms* **19,** 91–102.

Green, D. G., Dowling, J. E., Siegel, I. M., and Ripps, H. (1975). Retinal mechanisms of visual adaptation in the skate. *J. Gen. Physiol.* **65,** 483–502.

Grens, K. E., Greenwood, A. K., and Fernald, R. D. (2005). Two visual processing pathways are targeted by gonadotropin-releasing hormone in the retina. *Brain Behav. Evol.* **66,** 1–9.

Gustincich, S., Contini, M., Gariboldi, M., Puopolo, M., Kadota, K., Bono, H., LeMieux, J., Walsh, P., Carninci, P., Hayashizaki, Y., Okazaki, Y., and Raviola, E. (2004). Gene discovery in genetically labeled single dopaminergic neurons of the retina. *Proc. Natl. Acad. Sci. USA* **101,** 5069–5074.

Hamasaki, D. I., Trattler, W. B., and Hajek, A. S. (1986). Light ON depresses and light OFF enhances the release of dopamine from the cat's retina. *Neurosci. Lett.* **68,** 112–116.

Hammer, R. P., Jr., Egilmez, Y., and Emmett-Oglesby, M. W. (1997). Neural mechanisms of tolerance to the effects of cocaine. *Behav. Brain Res.* **84,** 225–239.

Hansen, A., Anderson, K. T., and Finger, T. E. (2004). Differential distribution of olfactory receptor neurons in goldfish: Structural and molecular correlates. *J. Comp. Neurol.* **477,** 347–359.

Hara, T. J. (1992). Mechanisms of olfaction. *In* "Fish Chemoreception" (Hara, T. J., Ed.), pp. 150–170. Chapman and Hall, London.

Hawryshyn, C. W. (1998). Vision. *In* "The Physiology of Fishes" (Evans, D. H., Ed.), 2nd edn., pp. 345–374. CRC Press, Boca Raton.

Hawryshyn, C. W., Moyer, H. D., Allison, W. T., Haimberger, T. J., and McFarland, W. N. (2003). Multidimensional polarization sensitivity in damselfishes. *J. Comp. Physiol. A* **189,** 213–220.

Hayashida, Y., and Ishida, A. T. (2004). Dopamine receptor activation can reduce voltage-gated Na^+ current by modulating both entry into and recovery from inactivation. *J. Neurophysiol.* **92,** 3134–3141.

Herrero, L., Rodriguez, F., Salas, C., and Torres, B. (1998). Tail and eye movements evoked by electrical microstimulation of the optic tectum in goldfish. *Exp. Brain Res.* **120,** 291–305.

Hitchcock, P. F., and Cirenza, P. (1994). Synaptic organization of regenerated retina in the goldfish. *J. Comp. Neurol.* **343,** 609–616.

Huang, L., Maaswinkel, H., and Li, L. (2005). Olfactoretinal centrifugal input modulates zebrafish retinal ganglion cell activity: A possible role for dopamine-mediated Ca^{2+} signaling pathways. *J. Physiol.* **569,** 939–948.

Hughes, A., Saszik, S., Bilotta, J., Demarco, P. J., Jr., and Patterson, W. F., II (1998). Cone contributions to the photopic spectral sensitivity of the zebrafish ERG. *Vis. Neurosci.* **15,** 1029–1037.

Hurd, M. W., Debruyne, J., Straume, M., and Cahill, G. M. (1998). Circadian rhythms of locomotor activity in zebrafish. *Physiol. Behav.* **65,** 465–472.

Iigo, M., and Tabata, M. (1996). Circadian rhythms of locomotor activity in the goldfish *Carassius auratus*. *Physiol. Behav.* **60,** 775–781.

Iigo, M., Hara, M., Ohtani-Kaneko, R., Hirata, K., Tabata, M., and Aida, K. (1997). Photic and circadian regulations of melatonin rhythms in fishes. *Biol. Signals* **6,** 225–232.

Ilg, U. J., Bremmer, F., and Hoffmann, K. P. (1993). Optokinetic and pursuit system: A case report. *Behav. Brain Res.* **57,** 21–29.

Jennes, L. (1987). Sites of origin of gonadotropin releasing hormone containing projections to the amygdala and the interpeduncular nucleus. *Brain Res.* **404,** 339–344.

Johns, P. R. (1982). Formation of photoreceptors in larval and adult goldfish. *J. Neurosci.* **2,** 178–198.

Julian, D., Ennis, K., and Korenbrot, J. L. (1998). Birth and fate of proliferative cells in the inner nuclear layer of the mature fish retina. *J. Comp. Neurol.* **394,** 271–282.

Kaiser, W., and Liske, E. (1974). Die optomotorische Reaktion von fixiert fliegenden Bienen bei Reizung mit Spektrallichtern. *J. Comp. Physiol.* **89,** 391–408.

Kamermans, M., and Spekreijse, H. (1995). Spectral behavior of cone-driven horizontal cells in teleost retina. *Prog. Retin. Eye Res.* **14,** 313–360.

Kamermans, M., and Spekreijse, H. (1999). The feedback pathway from horizontal cells to cones. A mini review with a look ahead. *Vision Res.* **39,** 2449–2468.

Kamermans, M., Fahrenfort, I., Schultz, K., Janssen-Bienhold, U., Sjoerdsma, T., and Weiler, R. (2001a). Hemichannel-mediated inhibition in the outer retina. *Science* **292,** 1178–1180.

Kamermans, M., Kraaij, D., and Spekreijse, H. (2001b). The dynamic characteristics of the feedback signal from horizontal cells to cones in the goldfish retina. *J. Physiol.* **534,** 489–500.

Kaneko, A. (1970). Physiological and morphological identification of horizontal, bipolar and amacrine cells in goldfish retina. *J. Physiol.* **207,** 623–633.

Kaneko, A., and Tachibana, M. (1981). Retinal bipolar cells with double colour-opponent receptive fields. *Nature* **293,** 220–222.

Kirsch, M., and Wagner, H. J. (1989). Release pattern of endogenous dopamine in teleost retinae during light adaptation and pharmacological stimulation. *Vision Res.* **29,** 147–154.

Kirsch, M., Wagner, H. J., and Djamgoz, M. B. (1991). Dopamine and plasticity of horizontal cell function in the teleost retina: Regulation of a spectral mechanism through D1-receptors. *Vision Res.* **31,** 401–412.

Klar, M., and Hoffmann, K. P. (2002). Visual direction-selective neurons in the pretectum of the rainbow trout. *Brain Res. Bull.* **57,** 431–433.

Ko, G. Y., Ko, M. L., and Dryer, S. E. (2003). Circadian phase-dependent modulation of cGMP-gated channels of cone photoreceptors by dopamine and D2 agonist. *J. Neurosci.* **23,** 3145–3153.

Kondo, M., and Sieving, P. A. (2001). Primate photopic sine-wave flicker ERG: Vector modeling analysis of component origins using glutamate analogs. *Invest. Ophthalmol. Vis. Sci.* **42,** 305–312.

Korenbrot, J. I., and Fernald, R. D. (1989). Circadian rhythm and light regulate opsin mRNA in rod photoreceptors. *Nature* **337,** 454–457.

Krauss, A., and Neumeyer, C. (2003). Wavelength dependence of the optomotor response in zebrafish (*Danio rerio*). *Vision Res.* **43,** 1273–1282.

Krauzlis, R. J., Liston, D., and Carello, C. D. (2004). Target selection and the superior colliculus: Goals, choices and hypotheses. *Vision Res.* **44,** 1445–1451.

Kröger, R. H. H., Knoblauch, B., and Wagner, H. J. (2003). Rearing in different photic and spectral environments changes the optomotor response to chromatic stimuli in the cichlid fish *Aequidens pulcher*. *J. Exp. Biol.* **206,** 1643–1648.

Kusmic, C., and Gualtieri, P. (2000). Morphology and spectral sensitivities of retinal and extraretinal photoreceptors in freshwater teleosts. *Micron* **31,** 183–200.

Kyle, A. L., Luo, B., Magnus, T. H., and Stell, W. K. (1995). Substance P-, F8Famide-, and A18Famide-like immunoreactivity in the nervus terminalis and retina of goldfish, *Carassius auratus*. *Cell Tissue Res.* **280,** 605–615.

Levine, J. S., and MacNichol, E. F., Jr. (1979). Visual pigments in teleost fishes: Effects of habitat, microhabitat, and behavior on visual system evolution. *Sens. Process.* **3,** 95–131.

Li, L. (2001a). Zebrafish mutants: Behavioral genetic studies of visual system defects. *Dev. Dyn.* **221,** 365–372.

Li, L. (2001b). Genetic and epigenetic analysis of visual system functions of zebrafish. *Prog. Brain Res.* **131,** 555–563.

Li, L., and Dowling, J. E. (1997). A dominant form of inherited retinal degeneration caused by a non-photoreceptor cell-specific mutation. *Proc. Natl. Acad. Sci. USA* **94,** 11645–11650.

Li, L., and Dowling, J. E. (1998). Zebrafish visual sensitivity is regulated by a circadian clock. *Vis. Neurosci.* **15,** 851–857.

Li, L., and Dowling, J. E. (2000a). Effect of dopamine depletion on visual sensitivity of zebrafish. *J. Neurosci.* **20,** 1893–1903.

Li, L., and Dowling, J. E. (2000b). Disruption of the olfactoretinal centrifugal pathway may relate to the visual system defect in *night blindness b* mutant zebrafish. *J. Neurosci.* **20,** 1883–1892.

Li, P., Temple, S., Gao, Y., Haimberger, T. J., Hawryshyn, C. W., and Li, L. (2005). Circadian rhythms of behavioral cone sensitivity and long wavelength opsin mRNA expression: A correlation study in zebrafish. *J. Exp. Biol.* **208,** 497–504.

Lindsay, S. M., and Vogt, R. G. (2004). Behavioral responses of newly hatched zebrafish (*Danio rerio*) to amino acid chemostimulants. *Chem. Senses* **29,** 93–100.

Loew, E. R., and Dartnall, H. J. (1976). Vitamin A1/A2-based visual pigment mixtures in cones of the rudd. *Vision Res.* **16,** 891–896.

Lopez, J. C., Bingman, V. P., Rodriguez, F., Gomez, Y., and Salas, C. (2000). Dissociation of place and cue learning by telencephalic abalation in goldfish. *Behav. Neurosci.* **114,** 687–699.

Lu, B., Kwan, T., Kurimoto, Y., Shatos, M., Lund, R. D., and Young, M. J. (2002). Transplantation of EGF-responsive neurospheres from GFP transgenic mice into the eyes of rd mice. *Brain Res.* **943,** 292–300.

Luque, M. A., Pérez-Pérez, M. P., Herrero, L., and Torres, B. (2005). Involvement of the optic tectum and mesencephalic reticular formation in the generation of saccadic eye movements in goldfish. *Brain Res. Rev.* **49,** 388–397.

Maaswinkel, H., and Li, L. (2003a). Spatio-temporal frequency characteristics of the optomotor response in zebrafish. *Vision Res.* **43,** 21–30.

Maaswinkel, H., and Li, L. (2003b). Olfactory input increases visual sensitivity in zebrafish: A possible function for the terminal nerve and dopaminergic interplexiform cell. *J. Exp. Biol.* **206,** 2201–2209.

Maaswinkel, H., Mason, B., and Li, L. (2003a). ENU-induced late-onset night blindness associated with rod photoreceptor cell degeneration in zebrafish. *Mech. Ageing Dev.* **124,** 1065–1071.

Maaswinkel, H., Ren, J. Q., and Li, L. (2003b). Slow-progressing photoreceptor cell degeneration in night blindness c mutant zebrafish. *J. Neurocytol.* **32,** 1107–1116.

Maaswinkel, H., Riesbeck, L. E., Riley, M. E., Carr, A. L., Mullin, J. P., Nakamoto, A. T., and Li, L. (2005). Behavioral screening for nightblindness mutants in zebrafish reveals three new loci that cause dominant photoreceptor cell degeneration. *Mech. Ageing Dev.* **126,** 1079–1089.

Mangel, S. C. (2001). Circadian clock regulation of neuronal light responses in the vertebrate retina. *Prog. Brain Res.* **131,** 505–518.

Mangel, S. C., and Wang, Y. (1996). Circadian clock regulation of rod and cone pathways in the vertebrate retina. *Invest. Ophthalmol. Vis. Sci. Suppl.* **37,** S17.

Manglapus, M. K., Uchiyama, H., Buelow, N. F., and Barlow, R. B. (1998). Circadian rhythms of rod-cone dominance in the Japanese quail retina. *J. Neurosci.* **18,** 4775–4784.

Marc, R. E. (1999). The structure of the vertebrate retina. *In* "The Retinal Basis of Vision" (Toyoda, J. I., Murakami, M., Kaneko, A., and Saito, T., Eds.), pp. 3–19. Elsevier, Amsterdam.

Marc, R. E., and Sperling, H. G. (1976). The chromatic organization of the goldfish cone mosaic. *Vision Res.* **16,** 1211–1224.

Marc, R. E., Stell, W. K., Bok, D., and Lam, D. M. K. (1978). GABA-ergic pathways in the goldfish retina. *J. Comp. Neurol.* **182,** 221–246.

Marc, R. E., Jones, B. W., Watt, C. B., and Strettoi, E. (2003). Neural remodeling in retinal degeneration. *Prog. Retin. Eye Res.* **22,** 607–655.

Marchiafava, P. L., Strettoi, E., and Alpigiani, V. (1985). Intracellular recording from single and double cone cells isolated from the fish retina (*Tinca tinca*). *Exp. Biol.* **44,** 173–180.

Masson, G., Mestre, D., and Blin, O. (1993). Dopaminergic modulation of visual sensitivity in man. *Fundam. Clin. Pharmacol.* **7,** 449–463.

McCormack, C. A., and Burnside, B. (1993). Light and circadian modulation of teleost retinal tyrosine hydroxylase activity. *Invest. Ophthalmol. Vis. Sci.* **34,** 1853–1860.

McDonald, C. G., and Hawryshyn, C. W. (1999). Latencies and discharge patterns of color-opponent neurons in the rainbow trout optic tectum. *Vision Res.* **39,** 2795–2799.

Mednick, A. S., Berk, M. F., and Springer, A. D. (1988). Asymmetric distribution of cells in the inner nuclear and cone mosaic layers of the goldfish retina. *Neurosci. Lett.* **94,** 241–246.

Megaw, P. L., Boelen, M. G., Morgan, I. G., and Boelen, M. K. (2006). Diurnal patterns of dopamine release in chicken retina. *Neurochem. Int.* **48,** 17–23.

Mensinger, A. F., and Powers, M. K. (1999). Visual function in regenerating teleost retina following cytotoxic lesioning. *Vis. Neurosci.* **16,** 241–251.

Meyer, D. L., Schott, D., and Schaeffer, K. P. (1970). Reizversuche im Tectum opticum freischwimmender Kabeljaue bzw. Dorsche (*Gadus morrhua L.*). *Plügers Archiv* **314,** 240–252.

Mitashov, V. I. (1997). Retinal regeneration in amphibians. *Int. J. Dev. Biol.* **41,** 893–905.

Moeller, G. U., Kayser, C., Knecht, F., and Koenig, P. (2004). Interactions between eye movement systems in cats and humans. *Exp. Brain Res.* **157,** 215–224.

Moller, P. (1995). "Electric Fishes: History and Behavior." Chapman and Hall, London.

Mora-Ferrer, C., and Behrend, K. (2004). Dopaminergic modulation of photopic temporal transfer properties in goldfish retina investigated with the ERG. *Vision Res.* **44,** 2067–2081.

Mora-Ferrer, C., and Gangluff, V. (2000). D2-dopamine receptor blockade impairs motion detection in goldfish. *Vis. Neurosci.* **17,** 177–186.

Mora-Ferrer, C., and Gangluff, V. (2002). D2-dopamine receptor blockade modulates temporal resolution in goldfish. *Vis. Neurosci.* **19,** 807–815.

Mora-Ferrer, C., and Neumeyer, C. (1996). Reduction of red-green discrimination by dopamine D1 receptor antagonists and retinal dopamine depletion. *Vision Res.* **36,** 4035–4044.

Mora-Ferrer, C., Yazulla, S., Studholme, K. M., and Haak-Frendscho, M. (1999). Dopamine D1-receptor immunolocalization in goldfish retina. *J. Comp. Neurol.* **411,** 705–714.

Mora-Ferrer, C., Hausselt, S., Schmidt-Hoffmann, R., Ebisch, B., Schick, S., Wollenberg, K., Schneider, C., Teege, P., and Jürgens, K. (2005). Pharmacological properties of motion vision in goldfish measured with the optomotor response. *Brain Res.* **1058,** 17–29.

Morgan, I. G., and Boelen, M. K. (1996). A retinal dark-light switch: A review of the evidence. *Vis. Neurosci.* **13,** 399–409.

Nawrocki, L., Bremiller, R., Streisinger, G., and Kaplan, M. (1985). Larval and adult visual pigments of the zebrafish, *Brachydanio rerio*. *Vision Res.* **25,** 1560–1576.

Negishi, K., Sugawara, K., Shinagawa, S., Teransihi, T., Kuo, C. H., and Takasaki, Y. (1991). Induction of immunoreactive proliferating cell nuclear antigen (PCNA) in goldfish retina following intravitreal injection of tunicamycin. *Dev. Brain Res.* **63,** 71–83.

Negishi, K., Salas, R., and Laufer, M. (1997). Origins of horizontal cell spectral responses in the retina of marine teleosts (*Centropomus* and *Mugil* sp.). *J. Neurosci. Res.* **47,** 68–76.

Neumeyer, C. (2003). Wavelength dependence of visual acuity in goldfish. *J. Comp. Physiol. A* **189,** 811–821.

Neumeyer, C., Dörr, S., Fritsch, J., and Kardelky, C. (2002). Colour constancy in goldfish and man: Influence of surround size and lightness. *Perception* **31,** 171–187.

Newman, E. A. (1980). Current source-density analysis of the b-wave of frog retina. *J. Neurophysiol.* **43,** 1355–1366.

Northcutt, R. G., and Butler, A. B. (1976). Retinofugal pathways in the lingnose gar *Lepsisosteus osseus* (*Linnaeus*). *J. Comp. Neurol.* **166,** 1–15.

Northcutt, R. G., and Davis, R. E. (1983)."Fish Neurobiology," Vol. 2. The University of Michigan Press, Ann Arbor.

Northmore, D. P. M., and Dvorak, C. A. (1979). Contrast sensitivity and acuity of the goldfish. *Vision Res.* **19,** 255–261.

Nussdorf, J. D., and Powers, M. K. (1988). Spectral sensitivity of the electroretinogram b-wave in dark-adapted goldfish. *Vis. Neurosci.* **1,** 159–168.

Oka, Y. (1992). Gonadotropin-releasing hormone (GnRH) cells of the terminal nerve as a model neuromodulator system. *Neurosci. Lett.* **142,** 119–122.

Orger, M. B., and Baier, H. (2005). Channeling of red and green cone inputs to the zebrafish optomotor response. *Vis. Neurosci.* **22,** 275–281.

Orger, M. B., Smear, M. C., Anstis, S. M., and Baier, H. (2000). Pereception of Fourier and non-Fourier motion by larval zebrafish. *Nat. Neurosci.* **3,** 1128–1133.

Otteson, D. C., and Hitchcock, P. F. (2003). Stem cells in the teleost retina: Persistent neurogenesis and injury-induced regeneration. *Vision Res.* **43,** 927–936.

Otteson, D. C., D'Costa, A. R., and Hitchcock, P. F. (2001). Putative stem cells and the lineage of rod photoreceptors in the mature retina of goldfish. *Dev. Biol.* **232,** 62–76.

Palacios, A. G., Varela, F. J., Srivastava, R., and Goldsmith, T. H. (1998). Spectral sensitivity of cones in goldfish, *Carassius auratus*. *Vision Res.* **38,** 2135–2146.

Perlman, I., and Normann, R. A. (1998). Light adaptation and sensitivity controlling mechanisms in vertebrate photoreceptors. *Prog. Retin. Eye Res.* **17,** 523–563.

Pierce, M. E., and Besharse, J. C. (1985). Circadian modulation of retinomotor movements I. Interaction of melatonin and dopamine in the control of cone length. *J. Gen. Physiol.* **86,** 671–689.

Pierce, M. E., and Besharse, J. C. (1986). Melatonin and dopamine interactions in the regulation of rhythmic photoreceptor metabolism. *In* "Pineal and Retinal Relationships" (O'Brien, P. J., and Klein, D. C., Eds.), pp. 219–237. Academic Press, New York.

Portavella, M., Vargas, J. P., Torres, B., and Salas, C. (2002). The effects of telencephalic lesions on spatial, temporal, and emotional learning in goldfish. *Brain Res. Bull.* **57,** 397–399.

Powers, M., and Easter, S. S., Jr. (1978). Absolute visual sensitivity of the goldfish. *Vision Res.* **18,** 1137–1147.

Powers, M. K., and Raymond, P. A. (1990). Development of the visual system. *In* "The Visual System of Fish" (Douglas, R. H., and Djamgoz, M., Eds.), pp. 419–442. Chapman and Hall, London.

Prusky, G. T., and Douglas, R. M. (2004). Characterization of mouse cortical spatial vision. *Vision Res.* **44,** 3411–3418.

Prusky, G. T., Alam, N. M., Beekman, S., and Douglas, R. M. (2004). Rapid quantification of adult and developing mouse spatial vision using a virtual optomotor system. *Invest. Ophthalmol. Vis. Sci.* **45,** 4611–4616.

Puppala, D., Maaswinkel, H., Mason, B., Legan, S. J., and Li, L. (2004). An *in vivo* microdialysis study of light/dark-modulation of vitreal dopamine release in zebrafish. *J. Neurocytol.* **33,** 193–201.

Purves, D., Paydarfar, J. A., and Andrews, T. J. (1996). The wagon wheel illusion in movies and reality. *Proc. Natl. Acad. Sci. USA* **93,** 3693–3697.

Qiu, G., Seiler, M. J., Mui, C., Arai, S., Aramant, R. B., de Juan, E., Jr., and Sadda, S. V. (2005). Photoreceptors differentiation and integration of retinal progenitor cells transplanted into transgenic rats. *Exp. Eye Res.* **80,** 515–525.

Raviola, E., and Gilula, N. B. (1973). Gap junctions between photoreceptor cells in the vertebrate retina. *Proc. Natl. Acad. Sci. USA* **70,** 1677–1681.

Raymond, P. A., and Rivlin, P. K. (1987). Germinal cells in the goldfish retina that produce rod photoreceptors. *Dev. Biol.* **122,** 120–138.

Raymond, P. A., Reifler, M. J., and Rivlin, P. K. (1988). Regeneration of goldfish retina: Rod precursors are a likely source of regenerated cells. *J. Neurobiol.* **19,** 431–463.

Reh, T. A., and Levine, E. M. (1998). Multipotential stem cells and progenitors in vertebrate retina. *J. Neurobiol.* **36,** 206–220.

Ren, J. Q., and Li, L. (2004a). Rod and cone signaling transmission in the retina of zebrafish: An ERG study. *Int. J. Neurosci.* **114,** 259–270.

Ren, J. Q., and Li, L. (2004b). A circadian clock regulates the process of ERG b- and d-wave dominance transition in dark-adapted zebrafish. *Vision Res.* **44,** 2147–2152.

Ribelayga, C., and Mangel, S. C. (2003). Absence of circadian clock regulation of horizontal cell gap junctional coupling reveals two dopamine systems in the goldfish retina. *J. Comp. Neurology* **467,** 243–253.

Ribelayga, C., Wang, Y., and Mangel, S. C. (2002). Dopamine mediates circadian clock regulation of rod and cone input to fish retinal horizontal cells. *J. Physiol.* **544,** 801–816.

Ribelayga, C., Wang, Y., and Mangel, S. C. (2004). A circadian clock in the fish regulates dopamine release via activation of melatonin receptors. *J. Physiol.* **554,** 467–482.

Rinner, O., Rick, J. M., and Neuhauss, S. C. F. (2005). Contrast sensitivity, spatial and temporal tuning of the larval zebrafish optokinetic response. *Invest. Ophthalmol. Vis. Sci.* **46,** 137–142.

Robinson, J., Schmitt, E. A., Hárosi, F. I., Reece, R. J., and Dowling, J. E. (1993). Zebrafish ultraviolet visual pigment: Absorption spectrum, sequence, and localization. *Proc. Natl. Acad. Sci. USA* **90,** 6009–6012.

Rodieck, R. W. (1998). "The First Steps in Seeing." Sinauer Associates, Sunderland, MA.

Rodriguez, F., Lopez, J. C., Vargas, J. P., Broglio, C., Gomez, Y., and Salas, C. (2002). Spatial memory and hippocampal pallium through vertebrate evolution: Insights from reptiles and teleost fish. *Brain Res. Bull.* **57,** 499–503.

Roeser, T., and Baier, H. (2003). Visuomotor behaviors in larval zebrafish after GFP-guided laser ablation of the optic tectum. *J. Neurosci.* **23,** 3726–3734.

Roorda, A., Metha, A. B., Lennie, P., and Williams, D. R. (2001). Packing arrangement of the three cone classes in primate retina. *Vision Res.* **41,** 1291–1306.

Saito, K., and Watanabe, S. (2004). Spatial learning deficits after the development of dorsomedial telecephalon lesions in goldfish. *Neuroreport* **15,** 2695–2699.

Sajovic, P., and Levinthal, C. (1982a). Visual cells of zebrafish optic tectum: Mapping with small spots. *Neuroscience* **7,** 2407–2426.

Sajovic, P., and Levinthal, C. (1982b). Visual response properties of zebrafish tectal cells. *Neuroscience* **7,** 2427–2440.

Sakai, H. M., Machuca, H., and Naka, K. I. (1997a). Processing of color- and noncolor-coded signals in the gourami retina. II. Amacrine cells. *J. Neurophysiol.* **78,** 2018–2033.

Sakai, H. M., Machuca, H., Korenberg, M. J., and Naka, K. I. (1997b). Processing of color- and noncolor-coded signals in the gourami retina. III. Ganglion cells. *J. Neurophysiol.* **78,** 2034–2047.

Saszik, S., and Bilotta, J. (1999). The effects of temperature on the dark-adapted spectral sensitivity function of the adult zebrafish. *Vision Res.* **39,** 1051–1058.

Saszik, S., and Bilotta, J. (2001). Constant dark-rearing effects on visual adaptation of the zebrafish ERG. *Int. J. Dev. Neurosci.* **19,** 611–619.

Saszik, S., Bilotta, J., and Givin, C. M. (1999). ERG assessment of zebrafish retinal development. *Vis. Neurosci.* **16,** 881–888.

Schaerer, S., and Neumeyer, C. (1996). Motion detection in goldfish investigated with the optomotor response is "color blind." *Vision Res.* **36,** 4025–4034.

Schellart, N. A. M. (1990). The visual pathways and central non-tectal processing. *In* "The Visual System of Fish" (Douglas, R. H., and Djamgoz, M., Eds.), pp. 345–372. Chapman and Hall, London.

Scholes, J., and Morris, J. (1973). Receptor-bipolar connectivity patterns in fish retina. *Nature* **241,** 52–54.

Schwanzel-Fukuda, M., and Silverman, A. J. (1980). The nervus terminalis of the guinea pig: A new luteinizing hormone-releasing hormone (LHRH) neuronal system. *J. Comp. Neurol.* **191,** 213–225.

Sherry, D. M., and Yazulla, S. (1993). Goldfish bipolar cells and axon terminal patterns: A Golgi study. *J. Comp. Neurol.* **329,** 188–200.

Shimbo, K., Toyoda, J. I., Kondo, H., and Kujiraoka, T. (2000). Color-opponent responses of small and giant bipolar cells in the carp retina. *Vis. Neurosci.* **17,** 609–621.

Simpson, J. I. (1984). The accessory optic system. *Annu. Rev. Neurosci.* **7,** 13–41.

Simpson, J. I., Giolli, R. A., and Blanks, R. H. (1988). The pretectal nuclear complex and the accessory optic system. *Rev. Oculomot. Res.* **2,** 335–364.

Sivak, J. G. (1990). Optical variability of the fish lens. *In* "The Visual System of Fish" (Douglas, R. H., and Djamgoz, M., Eds.), pp. 63–80. Chapman and Hall, London.

Solnica-Krezel, L., Schier, A. F., and Driever, W. (1994). Efficient recovery of ENU-induced mutations from the zebrafish germline. *Genetics* **136,** 1401–1420.

Springer, A. D., Easter, S. S., Jr., and Agranoff, B. W. (1977). The role of the optic tectum in various visually mediated behaviors in goldfish. *Brain Res.* **128,** 393–404.

Stell, W. K. (1975). Horizontal cell axons and axon terminals in goldfish retina. *J. Comp. Neurol.* **159,** 503–520.

Stell, W. K., Lightfoot, D. O., Wheeler, T. G., and Leper, H. F. (1975). Goldfish retina: Functional polarization of cone horizontal cell dendrites and synapses. *Science* **190,** 989–990.

Stell, W. K., Walker, S. E., Chohan, K. S., and Ball, A. K. (1984). The goldfish nervus terminalis: A luteinizing hormone-releasing hormone and molluscan cardioexcitatory peptide immunoreactive olfactoretinal pathway. *Proc. Natl. Acad. Sci. USA* **81,** 940–944.

Stell, W. K., Walker, S. E., and Ball, A. K. (1987). Functional-anatomical studies on the terminal nerve projection to the retina of bony fishes. *Ann. NY Acad. Sci.* **519,** 80–96.

Stenkamp, D. L., Powers, M. K., Carney, L. H., and Cameron, D. A. (2001). Evidence for two distinct mechanisms and cellular pattern formation in regenerated goldfish retinas. *J. Comp. Neurol.* **431,** 363–381.

Stockton, R. A., and Slaughter, M. M. (1989). B-wave of the electroretinogram. A reflection of ON bipolar cell activity. *J. Gen. Physiol.* **93,** 101–122.

Strettoi, E., Porciatti, V., Falsini, B., Pignatelli, V., and Rossi, C. (2002). Morphological and functional abnormalities in the inner retina of the rd/rd mouse. *J. Neurosci.* **22,** 5492–5504.

Stromland, K., and Sundelin, K. (1996). Paediatric and ophthalmologic observations in offspring of alcohol abusing mothers. *Acta Paediatr.* **85,** 1463–1468.

Szabo, T., Blahser, S., Denizot, J. P., and Ravaille-Veron, M. (1991). The olfactoretinalis system = terminal nerve? *Neuroreport* **2,** 73–76.

Tamai, T. K., Vardhanabhuti, V., Arthur, S., Foulkes, N. S., and Whitmore, D. (2003). Flies and fish: Birds of a feather. *J. Neurendocrinol.* **15,** 344–349.

Tomita, M., Adachi, Y., Yamada, H., Takahashi, K., Kiuchi, K., Oyaizu, H., Ikebukuro, K., Kaneda, H., Matsumura, M., and Ikehara, S. (2002). Bone marrow-derived stem cells can differentiate into retinal cells in injured rat retina. *Stem Cells* **20,** 279–283.

Tosini, G., and Fukuhara, C. (2003). Photic and circadian regulation of retinal melatonin in mammals. *J. Neuroendocrinol.* **15,** 364–369.

Tosini, G., and Menaker, M. (1996). Circadian rhythms in cultured mammalian retina. *Science* **272,** 419–421.

Toyoda, J., and Fujimoto, M. (1983). Analyses of neural mechanisms mediating the effect of horizontal cell polarization. *Vision Res.* **23,** 1143–1150.

Toyoda, J. I., and Shimbo, K. (1999). Color processing in lower vertebrates. *In* "The Retinal Basis of Vision" (Toyoda, J. I., Murakami, M., Kaneko, A., and Saito, T., Eds.), pp. 199–213. Elsevier, Amsterdam.

Trifonov, J. A., Byzov, A. L., and Chailahian, L. M. (1974). Electrical properties of subsynaptic and nonsynaptic membranes of horizontal cells in fish retina. *Vision Res.* **14,** 229–241.

Tropepe, V., Coles, B. L., Chiasson, B. J., Horsford, D. J., Elia, A. J., McInnes, R. R., and van der Kooy, D. (2000). Retinal stem cells in the adult mammalian eye. *Science* **287,** 2032–2036.

Tsin, A. T., and Beatty, D. D. (1979). Scotopic visual pigment composition in the retinas and vitamins A in the pigment epithelium of the goldfish. *Exp. Eye Res.* **29,** 15–26.

Tsonis, P. A., and Del Rio-Tsonis, K. (2004). Lens and retina regeneration: Transdifferentiation, stem cells and clinical applications. *Exp. Ey Res.* **78,** 161–172.

Twig, G., Levy, H., and Perlman, I. (2003). Color opponency in horizontal cells of the vertebrate retina. *Prog. Retin. Eye Res.* **22,** 31–68.

Uchiyama, H., Reh, T. A., and Stell, W. K. (1988). Immunocytochemical and morphological evidence for a retinopetal projection in anuran amphibians. *J. Comp. Neurol.* **274,** 48–59.

Umino, O., and Dowling, J. E. (1991). Dopamine release from interplexiform cells in the retina: Effects of GnRH, FMRFamide, bicuculline, and enkephalin on horizontal cell activity. *J. Neurosci.* **11,** 3034–3046.

Valentinčič, T., and Caprio, J. (1997). Visual and chemical release of feeding behavior in adult rainbow trout. *Chem. Senses* **22,** 375–382.

van Roessel, P., Palacios, A. G., and Goldsmith, T. H. (1997). Activity of long-wave-length cones under scotopic conditions in the cyprinid fish *Danio aequipinnatus. J. Comp. Physiol. A* **181,** 493–500.

Vanegas, H., and Ito, H. (1983). Morphological aspects of the teleostean visual system: A review. *Brain Res.* **287,** 117–137.

Vaquero, C. F., Velasco, A., and de la Villa, P. (1996). Protein kinase C localization in the synaptic terminal of rod bipolar cells. *Neuroreport* **7,** 2176–2180.

Verweij, J., Kamermans, M., and Spekreijse, H. (1996). Horizontal cells feed back to cones by shifting the cone calcium-current activation range. *Vision Res.* **36,** 3943–3953.

Vigh, J., and Witkovsky, P. (1999). Sub-millimolar cobalt selectively inhibits the receptive field surround of retinal neurons. *Vis. Neurosci.* **16,** 159–168.

Vihtelic, T. S., and Hyde, D. R. (2000). Light-induced rod and cone cell death and regeneration in the adult albino zebrafish (*Danio rerio*) retina. *J. Neurobiol.* **44,** 289–307.

Vihtelic, T. S., Soverly, J. E., Kassen, S. C., and Hyde, D. R. (2006). Retinal regional differences in photoreceptor cell death and regeneration in light-lesioned albino zebrafish. *Exp. Eye Res.* **83,** 558–575.

von der Emde, G. (1998). Electroreception. *In* "The Physiology of Fishes" (Evans, D. H., Ed.), 2nd edn., pp. 313–343. CRC Press, Boca Raton.

Wagner, H. G., MacNichol, E. F., Jr., and Wolbarsht, M. L. (1960). The response properties of single ganglion cells in the goldfish retina. *J. Gen. Physiol.* **43,** 45–62.

Wagner, H. J., and Behrens, U. D. (1993). Microanatomy of the dopaminergic system in the rainbow trout retina. *Vision Res.* **33,** 1345–1358.

Wagner, H. J., and Kröger, R. H. H. (2005). Adaptive plasticity during the development of colour vision. *Prog. Retin. Eye Res.* **24,** 521–536.

Walker, S. E., and Stell, W. K. (1986). Gonadotropin-releasing hormone (GnRH), molluscan cardioexcitatory peptide (FMRFamide), enkephalin and related neuropetides affect goldfish retinal ganglion cell activity. *Brain Res.* **384,** 262–273.

Wallman, J., and Winawer, J. (2004). Homeostasis of eye growth and the question of myopia. *Neuron* **43,** 447–458.

Walls, G. L. (1942). "The Vertebrate Eye and its Adaptive Radiation." Hafner, New York.

Warrant, E. (2000). The eyes of deep-sea fishes and the changing nature of visual scenes with depth. *Philos. Trans. R. Soc. Lond. B Biol. Sci.* **355,** 1155–1159.

Wartzok, D., and Marks, W. B. (1973). Directionally selective visual units recorded in optic tectum of the goldfish. *J. Neurophysiol.* **36,** 588–604.

Waterman, T. H., and Forward, R. B., Jr. (1970). Field evidence for polarized light sensibility in the fish *Zenarchopterus. Nature* **288,** 85–87.

Weiler, R., Baldridge, W. H., Mangel, S. C., and Dowling, J. E. (1997). Modulation of endogenous dopamine release in the fish retina by light and prolonged darkness. *Vis. Neurosci.* **14,** 351–356.

Weiss, O., and Meyer, D. L. (1988). Odor stimuli modulate retinal excitability in fish. *Neurosci. Lett.* **93,** 209–213.

White, M. P., and Fisher, L. J. (1989). Effects of exogenous melatonin on circadian disc shedding in the albino rat retina. *Vision Res.* **29,** 167–179.

Wiechmann, A. F., Vrieze, M. J., Dighe, R., and Hu, Y. (2003). Direct modulation of rod photoreceptor responsiveness through a Mel_{1C} melatonin receptor in transgenic *Xenopus laevis* retina. *Invest. Ophthalmol. Vis. Sci.* **44,** 4522–4531.

Wirsig-Wiechmann, C. R. (2001). Function of gonadotropin-releasing hormone in olfaction. *Keio J. Med.* **50,** 81–85.

Wirsig-Wiechmann, C. R., and Basinger, S. F. (1988). FMRFamide-immunoreactive retinopetal fibers in the frog, Rana pipiens: Demonstration by lesion and immunocytochemical techniques. *Brain Res.* **449,** 116–134.

Wirsig-Wiechmann, C. R., and Wiechmann, A. F. (2001). The prairie vole vomeronasal organ is a target for gonadotropin-releasing hormone. *Chem. Senses* **26,** 193–202.

Wirsig-Wiechmann, C. R., and Wiechmann, A. F. (2002). Vole retina is a target for gonadotropin-releasing hormone. *Brain Res.* **950,** 210–217.

Wirsig-Wiechmann, C. R., Wiechmann, A. F., and Eisthen, H. L. (2002). What defines the nervus terminalis? Neurochemical, developmental, and anatomical criteria. *Prog. Brain Res.* **141,** 45–58.

Wirz-Justice, A., Da Prada, M., and Reme, C. (1984). Circadian rhythm in rat retinal dopamine. *Neurosci. Lett.* **45,** 21–25.

Witkin, J. W. (1987). Nervus terminalis, olfactory nerve, and optic nerve representation of luteinizing hormone-releasing hormone in primates. *Ann. NY Acad. Sci.* **519,** 174–183.

Witkovsky, P., and Dearry, A. (1992). Functional roles of dopamine in the vertebrate retina. *Prog. Retinal Res.* **11,** 247–292.

Witkovsky, P., Shakib, M., and Ripps, H. (1974). Interreceptoral junctions in the teleost retina. *Invest. Ophthalmol.* **13,** 996–1009.

Witkovsky, P., Stone, S., and Besharse, J. C. (1988). Dopamine modifies the balance of rod and cone inputs to horizontal cells of the *Xenopus* retina. *Brain Res.* **449,** 332–336.

Witkovsky, P., Veisenberger, E., LeSauter, J., Yan, L., Johnson, M., Zhang, D. Q., McMahon, D., and Silver, R. (2003). Cellular location and circadian rhythm of expression of the biological clock gene *Period 1* in the mouse retina. *J. Neurosci.* **23,** 7670–7676.

Wong, K. Y., Cohen, E. D., Adolph, A. R., and Dowling, J. E. (2000). The d-wave of the zebrafish ERG derives mainly from OFF-bipolar cell activity. *Invest. Ophthalmol. Vis. Sci.* **41,** S623.

Wu, D. M., Schneiderman, T., Burgett, J., Gokhale, P., Barthel, L., and Raymond, P. A. (2001). Cones regenerate from retinal stem cells sequestred in the inner nuclear layer of adult goldfish retina. *Invest. Ophthalmol. Vis. Sci.* **42,** 2115–2124.

Wullimann, M. F., Rupp, B., and Reichert, H. (1996). "Neuroanatomy of the Zebrafish Brain. A Topographic Atlas." Birhäuser, Basel.

Yager, D., Sharma, S. C., and Grover, B. G. (1977). Visual function in goldfish with unilateral and bilateral tectal ablation. *Brain Res.* **137,** 267–275.

Yamamoto, N., and Ito, H. (2000). Afferent sources to the ganglion of the terminal nerve in teleosts. *J. Comp. Neurol.* **428,** 355–375.

Yamamoto, N., Oka, Y., and Kawashima, S. (1997). Lesion of gonadotropin-releasing hormone-immunoreactive terminal nerve cells: Effects on the reproductive behavior of male dwarf gouramis. *Neuroendocrinology* **65,** 403–412.

Yang, X. L., and Wu, S. M. (1991). Feedforward lateral inhibition in retinal bipolar cells: Input-output relation of the horizontal cell-depolarizing bipolar cell synapse. *Proc. Natl. Acad. Sci. USA* **88,** 3310–3313.

Yang, X. L., Fan, T. X., and Li, J. D. (1990). Electroretinographic b-wave merely reflects the activity of the rod system in the dark-adapted carp retina. *Vision Res.* **30,** 993–999.

Yazulla, S., Lin, Z. S., and Studholme, K. M. (1996). Dopaminergic control of light-adaptive synaptic plasticity and role in goldfish visual behavior. *Vision Res.* **36,** 4045–4057.

Yu, C. J., and Li, L. (2003). The effect of dopamine on the ON bipolar cell outward K^+ current is lost in mutant zebrafish. *Invest. Ophthalmol. Vis. Sci.* **44,** S4140.

Yurco, P., and Cameron, D. A. (2005). Responses of Muller glia to retinal injury in adult zebrafish. *Vision Res.* **45,** 991–1002.

Zachmann, A., Ali, M. A., and Falcón, J. (1992). Melatonin and its effects in fishes: An overview. *In* "Rhythms in Fishes" (Ali, M. A., Ed.), pp. 149–165. Plenum, New York.

Zawilska, J. B., Bednarek, A., Berezinska, M., and Nowak, J. Z. (2003). Rhythmic changes in metabolism of dopamine in the chick retina: The importance of light versus biological clock. *J. Neurochem.* **84,** 717–724.

Zhdanova, I. V., and Reebs, S. G. (2006). Circadian rhythms in fish. *In* "Behaviour and Physiology of Fish" (Sloman, K. A., Wilson, R. W., and Balshine, S., Eds.), pp. 197–238. Elsevier Academic Press, Amsterdam.

Zhu, H., LaRue, S., Whiteley, A., Steeves, T. D. L., Takahashi, J. S., and Green, C. B. (2000). The *Xenopus Clock* gene is constitutively expressed in retinal photoreceptors. *Brain Res. Mol. Brain Res.* **75,** 303–308.

Zhuang, M., Wang, Y., Steenhard, B. M., and Besharse, J. C. (2000). Differential regulation of two period genes in the *Xenopus* eye. *Brain Res. Mol. Brain Res.* **82,** 52–64.

Zucker, C. L., and Dowling, J. E. (1987). Centrifugal fibers synapse on dopaminergic interplexiform cells in teleost retina. *Nature* **300,** 166–168.

Zygar, C. A., Lee, M. J., and Fernald, R. D. (1999). Nasotemporal asymmetry during teleost retinal growth: Preserving an area of specialization. *J. Neurobiol.* **41,** 435–442.

6

MOLECULAR AND CELLULAR REGULATION OF PINEAL ORGAN RESPONSES

JACK FALCÓN
LAURENCE BESSEAU
GILLES BOEUF

1. Introduction
2. Functional Organization of the Pineal
   2.1. Anatomy
   2.2. The Cell Types of the Pineal
   2.3. Conclusions
3. The Fish Pineal Organ: A Light Sensor
   3.1. Components of the Phototransduction Cascade
   3.2. Electrical Responses
   3.3. Release of a Neurotransmitter
   3.4. Conclusions
4. The Fish Pineal Organ: A Melatonin Factory
   4.1. Melatonin is Synthesized Within the Photoreceptor Cells
   4.2. The LD Cycle Synchronizes a Rhythm in Melatonin Production
   4.3. Conclusions
5. Intracellular Regulation of Arylalkylamine *N*-acetyltransferase 2
   5.1. Regulation of AANAT2 Protein
   5.2. Regulation of *AANAT2* Gene
   5.3. Conclusions
6. Photoperiodic Versus Circadian Control Melatonin Production
   6.1. Clock or No Clock
   6.2. The Circadian Clock and the Melatonin Output
   6.3. Conclusions
7. Nonphotic Regulation of Pineal Organ Output Signals
   7.1. External Factors
   7.2. Internal Factors
   7.3. Conclusions
8. Conclusions and Perspectives

Sensory Systems Neuroscience: Volume 25
FISH PHYSIOLOGY

DOI: 10.1016/S1546-5098(06)25006-4

The pineal gland of fish is an evagination of the roof of the diencephalon, which locates in a window below the skull. The pineal epithelium is organized like a retina that would contain cone-like photoreceptor cells, connecting to second-order neurons. Supporting "interstitial cells" would be the homologous of the retinal Müller cells. The fish pineal gland is a nonvisual photosensitive organ that transduces the light information and elaborates nervous and neurohormonal messages in response to changes in illumination—the nervous message is an excitatory neurotransmitter and the hormonal message is melatonin. The first one acts on the second-order neurons, which transmit immediately the information to brain centers via the ganglion cells. The second one is produced and released rhythmically into the cerebrospinal fluid and blood. Thus, a daily pattern with high levels during night and low levels during day provides the organism periodic information that allows synchronization of daily and annual physiological functions and behaviors. In most species, the rhythmic production of melatonin involves a circadian pacemaker located within the photoreceptor cells, which operate as a full cellular circadian system. Thus, light synchronizes the circadian clocks, which in turn drive the rhythmic patterns of melatonin secretion. Other external (mainly temperature) and internal (neurotransmitters, hormones) factors may modulate this pattern. This chapter focuses on the structural and functional properties of the fish pineal organ, and on the molecular mechanisms, which contribute to the synthesis of rhythmic output signals, including melatonin, the timekeeping molecule of the organism.

1. INTRODUCTION

Organisms have adapted to the cyclic variations of the external environment so that virtually all their biochemical, physiological, and behavioral events are rhythmic. The 24 h light/dark (LD) cycle is the major synchronizer of the daily rhythms. In most cases, endogenous clocks that maintain a cyclic activity with a period that approximates 24 h drive these rhythms. The so-called circadian[1] clock allows the right event to occur at the right time. This is a major improvement if one considers the myriad of events that cycle on a 24 h basis. In order to function properly, a biological clock needs an output system, which will provide a rhythmic signal to targeted areas and govern overt rhythms; it also needs an input pathway for synchronization. The pacemaker, together with its input and output pathways, make

[1]Circa: approximately; dian: day.

a circadian system. Many fish tissues possess circadian pacemakers (Whitmore *et al.*, 1998). However, the pineal organ of Teleost occupies a key position in the hierarchy of fish circadian clocks, as revealed by observing the effects of pinealectomy, for example, on free running activity rhythms (Underwood, 1989).[2] It was suggested that melatonin—known as the pineal hormone since 1958 (Lerner *et al.*, 1958)—"*... may be the mechanism by which the pineal communicates with the rest of the circadian system*" (Underwood, 1989).

The first studies dealing with teleost pineal organ emphasized its photoreceptive characteristics, and it soon appeared that it was involved in the control of functions and behaviors displaying daily rhythmicity (Zachmann *et al.*, 1992a; Ekström and Meissl, 1997; Falcón, 1999; Boeuf and Falcón, 2002). Thus, it was shown that pineal removal affected daily variations of locomotors activity and shoaling behavior, vertical migration and thermal preference, sleeplike state, skin pigmentation, or demand feeding. In some of these experiments, the effects could be reversed by melatonin administration. Similar studies allowed demonstrating that pineal melatonin affects biochemical parameters such as plasma ion levels, plasma and hepatic glucose levels, total lipid content, and hypothalamic monoamine content. Finally, indication was provided that the pineal organ and melatonin modulated neuroendocrine functions including growth and reproduction. However, these early studies were based on *in vivo* experiments, and they often led to conflicting results. The available data differed with gender, photoperiod, temperature, and reproductive stage. Molecular and pharmacological studies provided the first evidence that the pituitary is a target for melatonin, which modulates the release of pituitary hormones (Falcón *et al.*, 2003a). The effects are complex because they depend on the concentration and the season.

In brief, it is today well admitted that the pineal organ is a key component in the circadian organization of fish, and that it mediates the synchronizing effects of photoperiod on crucial physiological functions and behaviors. Melatonin appears as one neurohormonal messenger mediating these effects. However, a clear-cut picture of its modes of action is still missing. This awaits a more precise identification of its molecular and cellular targets, and elucidation of its mode of action. It is also of crucial importance to understand how are produced melatonin and other output signals from the pineal, in response to the daily changes in illumination. Our knowledge on this matter is reviewed below.

[2]Pinealectomy can cause either marked changes in the free running period, or splitting of the activity rhythm or arrhythmicity.

2. FUNCTIONAL ORGANIZATION OF THE PINEAL

2.1. Anatomy

The anatomical organization of the fish pineal organ has been described in many fish species (Ekström and Meissl, 1997; Falcón, 1999). The pineal organ is generally organized in two parts, an end vesicle connected to the dorsal epithalamus by a stalk (Figure 6.1). In adult fish, the end vesicle is located in a "window" below the skull—the skin covering the area is less pigmented, and the skull is thinner and may even form a kind of pit where the end vesicle enters (Figure 6.1C and D); this favors light transmission. The end vesicle may be as large as to cover the whole cerebral hemispheres and olfactory bulbs. Generally, the lumen of the organ communicates with the third ventricle, and is thus filled with cerebrospinal fluid (Figures 6.1D and E and 6.2A). In some cases the pineal appears either as a more or less

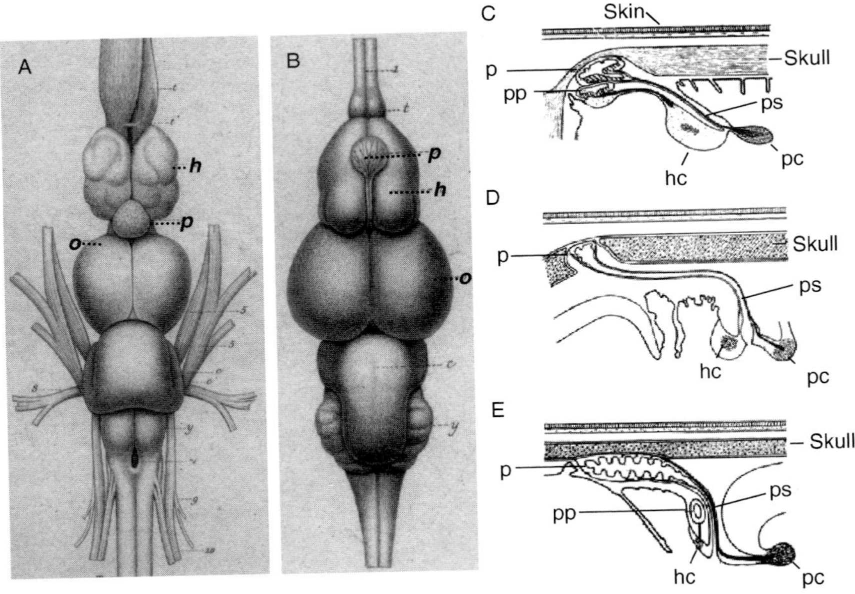

Fig. 6.1. Gross anatomy of the pineal organ. Dorsal view of the brains of *Conger vulgaris* (A) and *Mullus surmulletus* (B) showing the location of the pineal organ (p); from Baudelot (1883). Saggital sections through the epithalamic area of lampreys (C), chondrichtian (D), and teleostean (E) fish; from Studnicka (1905). h, cerebral hemisphere; hc, habenular commissure; o, optic tectum; ps, pineal stalk and nerve; pc, posterior commissure; pp, parapineal organ.

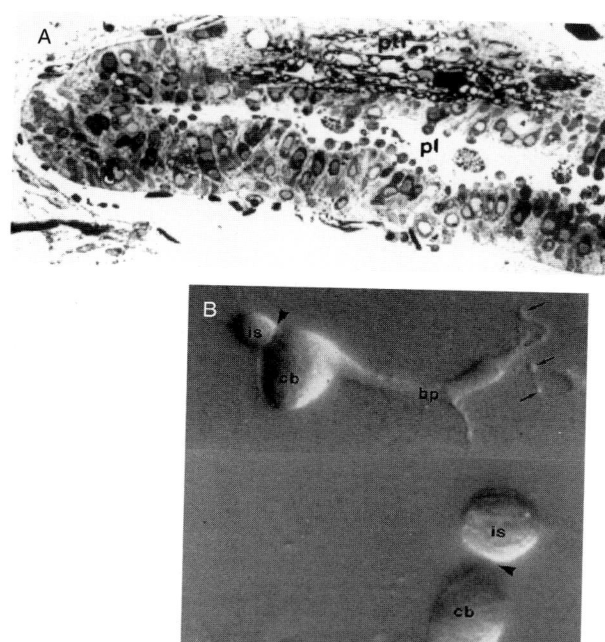

Fig. 6.2. Epithelium and photoreceptor cells from fish pineal organ. (A) Frontal section through the stalk of the pike pineal organ (Falcón, 1979a). The cells are radially organized around the pineal lumen (pl). The dark cells correspond to the photoreceptors. Their apical part protrudes into the lumen. On the dorsal part of the stalk is the pineal track (ptr). Capillaries (ca) surround the stalk. (B) Dissociated photoreceptor cells from the pineal organ of a trout (Bégay *et al.*, 1992). The cells display a segmental organization. In the apical part is the inner segment (is), divided into two by a neck (arrow head); the "is" protudes into the pineal lumen (pl). Below, the cell body (cb) sends one basal process (bp), which divides into several synaptic pedicles containing synaptic ribbons (sb).

folliculated vesicle or as a compact gland (Omura and Oguri, 1969). This prefigures the evolutionary trend observed in vertebrates, where all the intermediate stages are seen between the pineal vesicle, as seen in most fish, and the compact and more or less internalized gland as observed in snakes and mammals (Collin and Oksche, 1981; Collin *et al.*, 1989).

The pineal organ of fish is surrounded by fenestrated blood vessels that do not enter the epithelium (Figure 6.2). It lacks a blood–brain barrier, so that the epithelium is exposed to the haemal environment in its basal part, and to the cerebrospinal fluid in its apical part (Falcón, 1979a; Omura *et al.*, 1985; Ekström and Meissl, 1997). The pseudostratified epithelium may show more or less pronounced infoldings. Cytological and electron microscopic studies allowed distinguishing at least three cell types in the pineal parenchyma: photoreceptor cells, neurons, and glial (interstitial) cells (Figures 6.2B and 6.3A).

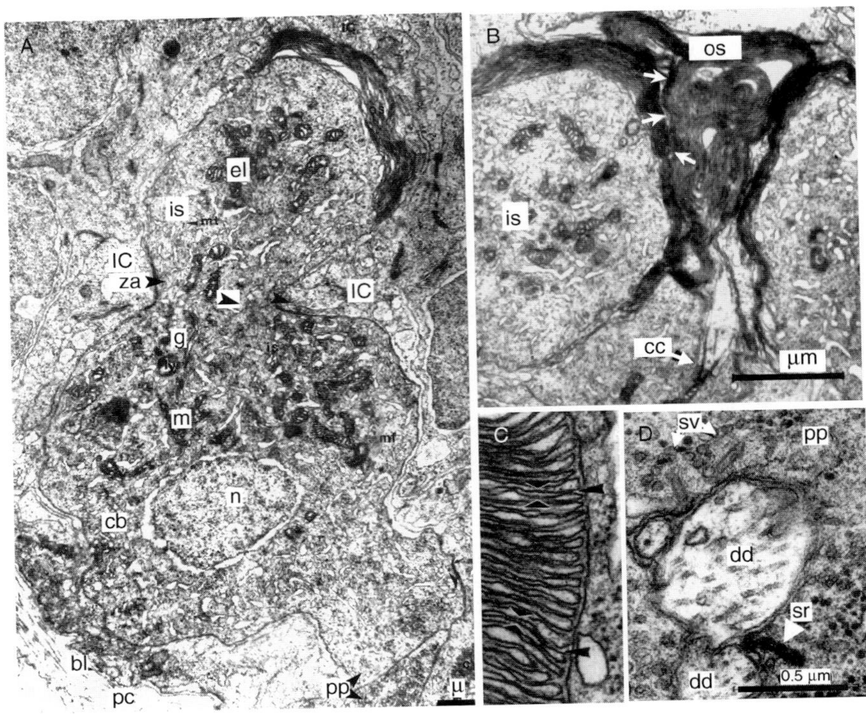

Fig. 6.3. Ultrastructure of the fish pineal photoreceptor cells. (A) In the apical part, the infoldings of the plasma membrane that make the outer segment cap the inner segment (is). The "is" contains numerous mitochondria that form the elipsoïd (el), and is divided into two parts by a neck showing a densification of the membrane (*zonula adherens*, za). The cell body (cb) contains the nucleus (n), Golgi apparatus (g), and endoplasmic reticulum. From the "cb" starts the photoreceptor pedicle (pp) filled with synaptic vesicles (sv). IC, intersticial cell; bl, basal lamina; pc, pericapillary spaces. (B) Detail of the "is" and outer segmment (os) from several photoreceptor cells in the pineal lumen; a $9 \times 2 + 0$ conective cilium (cc) rises from one "is." (C) Detail of the organization of an "os" membrane infoldings delimiting disks that are open to the extracellular spaces (black arrow heads). (D) A photoreceptor pedicle (pp) contains several synaptic ribbons (sr) surrounded by "sv"; the "sr" contact second-order neuron dendrites (dd).

2.2. The Cell Types of the Pineal

2.2.1. Photoreceptor Cells

The photoreceptor cells resemble structurally to the cone photoreceptors found in the retina. They display a polarized organization, with four distinct cell compartments: the outer and inner segments, cell body, and synaptic pedicle(s) (Figures 6.2B and 6.3A). The *outer segment*, which protrudes into the pineal lumen, develops from a $9 \times 2 + 0$ cilium[3] at the apical part of the cell (Figure 6.3A–C). Independent of the fish habitat, it is composed of flattened stacks (20–70) made by infoldings of the plasma membrane, which cover the apical part of the inner segment (Figure 6.3) (Ekström and Meissl, 1997; Falcón, 1999; Wagner and Mattheus, 2002; Bowmaker and Wagner, 2004). The *inner segment* is itself divided into two parts, separated by a neck at the level of the outer limiting membrane of the epithelium. The apical part of the inner segment protrudes into the pineal lumen; it contains the centrosome formed by a distal centriole (from which arises the $9 \times 2 + 0$ cilium) and a proximal centriole (of the $9 \times 3 + 0$ type). Mitochondria accumulate in the inner segment, forming the ellipsoid. The *cell body* contains the nucleus surrounded by the Golgi apparatus and a network of rough endoplasmic reticulum. Finally, one or several *synaptic pedicles* with synaptic buttons constitute the basal part. The synaptic pedicles contact dendrites of the so-called second-order neurons through ribbon type synapses (Figures 6.2B and 6.3D). They are filled with synaptic vesicles (350–500 Å) and few dense-core (secretory-like) vesicles (600–1000 Å). Bundles of microfilaments and microtubules, as well as numerous glycogen particles, are seen all over the photoreceptors.

There are variations around this general scheme. Thus, in a few fish species the outer segment may show a more or less disrupted aspect. In some cases, there is just a slight disorganization of the outer segment; in other cases the stacks are almost absent and only the bulbous extension with the $9 \times 2 + 0$ cilium remains (Falcón, 1999). At the other pole of these cells, the pedicle may not contact neurons anymore. In this case, the synaptic ribbons are facing interstitial cells or other synaptic ribbons of a neighboring photoreceptor, suggesting intense communication between these cell types. Pedicles reaching the basal lamina of the pineal epithelium may occasionally be observed. Intermediate situations are observed between the cone-type photoreceptors and the so-called "modified" photoreceptors within the same pineal gland. A regional distribution of the cone-type and modified

[3]A cilium is composed of nine pairs (9×2) of microtubules concentrically organized around a central pair (+1). In photoreceptor cells the central pair is lacking (+0).

photoreceptors has been observed within the pineal epithelium of the pike. Thus, modified photoreceptors are seen in areas where neurons are absent (Falcón, 1979b). This suggests that the light signal captured by these cells might not be transmitted to brain centers.

The diversity of photoreceptor types again reflects the diversity observed between fish and mammals. In mammals and snakes, the pinealocytes show no inner and outer segments, and no direct photosensitivity, while intermediate situations are observed in sauropsids (Collin and Oksche, 1981; Ekström and Meissl, 2003).

2.2.2. Second-Order Neurons

Wake (1973) first visualized the acetylcholinesterase activity of the goldfish pineal neurons. Since then, more species have been studied and other methods have been used to investigate these cells (Ekström and van Veen, 1984; Ekström *et al.*, 1987a,b; Ekström and Meissl, 1997). The neurons are usually located on the basal third of the pineal epithelium (Figure 6.4). Several types of neurons have been identified, based on their morphology, often displaying a regional distribution. In the pike for example, scattered large multipolar neurons are seen in the most distal part of the vesicle, while numerous but small bipolar or pseudounipolar neurons are more frequent in the pineal stalk (Falcón, 1979c; Falcón and Mocquard, 1979) (Figure 6.4). The ratio of photoreceptors/neurons varies from a species to another, but also within the epithelium in a same species (Falcón and Mocquard, 1979; McNulty, 1979). Most of these neurons send their axons to the brain; however, electrophysiological and (immuno)histochemical observations

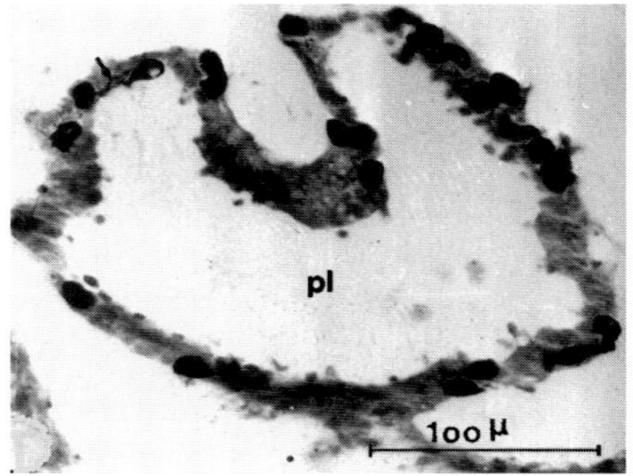

Fig. 6.4. The second-order neurons. They are here visualized by the acetylcholinesterase method in the pineal stalk of the pike (Falcón and Mocquard, 1979). pl, pineal lumen.

suggest that some interneurons may also be present (Ekström and Meissl, 1988, 1989).[4]

The axons of the neurons converge dorsally to the pineal stalk, to form a tract that enters the brain at the level of the subcommissural organ (Figures 6.1C–D and 6.2A). Neuronal tracing methods allowed identification of the projection areas (Hafeez and Zerihun, 1974; Ekström *et al.*, 1994; Jimenez *et al.*, 1995; Yañez and Añadon, 1996). In all species studied, the main projections are found in the habenula, and the pretectal, thalamic, preoptic, hypothalamic, and mesencephalic tegmental areas. It is of interest to note that most of these areas receive input also from the retina, and clear-cut overlapping has been demonstrated in the habenula, preoptic nuclei, dorsal thalamus, and pretectal areas (Ekström *et al.*, 1994).

In addition to this "pinealofugal" innervation, some fibers from central (and unknown) origin enter the pineal organ. This "pinealopetal" innervation is much less abundant and consists of neurons containing catecholamines, FMRF-amide, neuropeptide Y, growth hormone-releasing hormone (GHRH) or gonadotropin-releasing hormone (GnRH) (Ekström and Meissl, 1989; Rao *et al.*, 1996; Subhedar *et al.*, 1996).

2.2.3. Interstitial Cells

They are also called the glial or supporting cells. These cells occupy the whole epithelium height. They are in contact with the cerebrospinal fluid in their apical part, where they display numerous microvilli and some $9 \times 2 + 2$ cilia. The basal part, which contains the nucleus, is larger. It lays on the basal lamina in such a way that it isolates the photoreceptors and neurons from the perivascular spaces. These cells are thus well positioned to import compounds from the blood vessels. The exact function of the interstitial cells remains enigmatic. In addition to a supposed supportive role, there is ultrastructural indication that the cells are involved in an intense synthetic activity; the Golgi apparatus is well developed, and the rough and smooth endoplasmic reticulums form an abundant interconnected network (Falcón, 1979a).

The interstitial cells most probably provide nutrients and other products, such as growth factors, to other cell types of the epithelium, thus promoting survival and differentiation of the cellular constituents. It has been shown that dissociated fish pineal cells reconstruct, *in vitro*, the general architecture of a pineal epithelium (Bégay *et al.*, 1992; Bolliet *et al.*, 1997). The cells are organized concentrically around a lumen at one pole, and a network

[4]These interneurons are putative cholinergic, GABAergic, and substance P-containing neurons.

of collagen fibers is reconstructed at the other pole. The photoreceptors display the same structural features as observed *in vivo*, including an outer segment, ellipsoid, and interphotoreceptor gap and tight junctions (Bolliet *et al.*, 1997). However, when the photoreceptor cells are isolated and cultured without interstitial cells, no such reorganization is observed; they dedifferentiate and loose their polarized structure, and their metabolic activity is also affected (Falcón *et al.*, 1992a; Bolliet *et al.*, 1997). This is consistent with the observation that the retinal counter part of the pineal glia, the Müller cells, express growth factors, neurotransmitter transporters, and antioxidant agents, which mediate protective effects from various neurological insults (Garcia *et al.*, 2002; Garcia and Vecino, 2003; Harada *et al.*, 2003).

The pineal organ, unlike the retina, does not possess an organized pigment epithelium cell layer; this led to suggest interstitial cells might ensure some of the functions of the retinal pigment epithelium (RPE), such as phagocytosis of shed photoreceptor outer segments (Section 3.1.5). There is indication for the presence of an active phagocytotic activity (Falcón, 1979a). Also, they might be involved in the recycling of the retinal chromophore after the all-*cis*-retinal isomerized into all-*trans*-retinal (see later). In the retina, this process takes place in RPE cells for the rods, and in Müller cells for the cones (Mata *et al.*, 2002). Finally, interstitial cells of the trout pineal organ have been shown to take up and metabolize radioactive arachidonic acid, which incorporates preferentially into the phosphatidylinositol (PI) pool (Falcón and Henderson, 2001). The association of the labeling to the plasma membranes suggests the cells might be involved in the production of PI-derived second messengers, which would act locally.

2.2.4. Other Cell Types

Macrophages are abundant in the pineal lumen (Falcón, 1979a; Omura *et al.*, 1986; McNulty *et al.*, 1988). They exhibit a strong acid phosphatase activity, suggesting these cells play a role in the breakdown of phagocytized photoreceptor outer segments.

Another unidentified cell type, of unknown lineage and function, has also been described in the pineal organ of the pike, dogfish, and blind cave fish (Rüdeberg, 1969; Herwig, 1976; Falcón, 1979a). It has been suggested that these cells, which are also observed in the perivascular spaces, might be the precursors of the macrophages seen in the lumen (Rüdeberg, 1969; Herwig, 1976). The structural features of these macrophages clearly indicate that they are involved in the phagocytosis of photoreceptor outer segments and other constituents.

2.2.5. Intercellular Contacts

In addition to the above-mentioned synaptic ribbons, other specialized structures connect the cell types of the pineal epithelium. *Zonula adherens* junctions are seen between the different cell types at the level of the outer-limiting membrane. Desmosomes are seen between interstitial cells, while photoreceptor cells contact each other through *fascia adherens* and *macula occludens* (or gap) junctions (Falcón, 1979a,b).

2.3. Conclusions

The pineal organ of fish displays all the anatomical and cellular characteristics of a photoreceptive organ. It could be compared to a very simplified retina, containing only cones, no rods, and contacting directly the ganglion cells. The presence of some interneurons has been reported. However, the bipolar, amacrine, and horizontal cells network, as seen in the retina, is virtually absent. This complex retinal network allows the first step of integration in the visual process (Dowling, this volume). The absence of such a network in the fish pineal is consistent with the idea that the organ functions as a luminance detector.

3. THE FISH PINEAL ORGAN: A LIGHT SENSOR

The analogies between the retinal cones and the pineal photoreceptors are found not only at the structural level but also at the functional level. The mechanisms of phototransduction have been well studied in the rod and cone photoreceptors of the retina (Dowling, this volume; Korenbrot and Rebrik, 2002; Maeda *et al.*, 2003). In the retina, signal transduction is triggered when the photon reaches rhodopsin, a G-protein–coupled receptor. This activates the α-subunit of the G-protein transducin (Tα), which in turn activates the cyclic GMP (cGMP) phosphodiesterase. The subsequent decrease in cGMP results in the closure of a cationic cGMP-gated channel (CNG), which allows entry of $Na^+$and Ca^{2+}in the dark. Inactivation of rhodopsin is ensured by the successive action of rhodopsin kinase and arrestin. The resulting effect is that the photoreceptors are hyperpolarized on illumination and depolarized in the dark. Depolarization allows the release of glutamate, an excitatory neurotransmitter, at the synaptic terminals of the photoreceptors (Thoreson and Witkovsky, 1999; Vigh *et al.*, 1997). Sensitivity to light and response kinetics characterize the major differences between retinal rods and cones; and, the presence of different types of cones, with different spectral sensitivities, allows color vision (Collin *et al.*, 2003).

3.1. Components of the Phototransduction Cascade

3.1.1. Opsins

The photopigment molecules of vertebrate photoreceptor cells consist of an opsin protein coupled to a vitamin A–derived aldehyde (11-*cis*-retinal). Studies in rainbow trout indicated the retina and pineal organ both contains 11-*cis*-retinal and 11-*cis*-3-dehydroretinal. They are believed to be the chromophores of the pineal photopigments because they isomerize to all-*trans* forms on light stimulation (Tabata *et al.*, 1985). Immunocytochemical studies using antibodies directed against retinal opsins indicated opsin-like proteins are present in the outer segment of many fish pineal photoreceptor cells (Vigh-Teichmann *et al.*, 1980, 1982, 1983, 1992; van Veen *et al.*, 1984; Ekström *et al.*, 1987b; Östholm *et al.*, 1987; Fejer *et al.*, 1997; Vigh *et al.*, 1998; Forsell *et al.*, 2002). This was possible most probably because vertebrate photopigment opsins are remarkably similar. With the use of more specific antibodies directed against cone or rod retinal opsins, it appeared that some pineal photoreceptors express conelike opsin whereas others express rodlike opsin (Philp *et al.*, 2000a).

Cloning and *in situ* hybridization approaches allowed identification of various photoreceptor genes of the opsin family, displaying either retinal- or pineal-specific expression. Thus, in the zebrafish and salmon, the counterpart of the retinal rhodopsin (rho) genes, is the so-called extraretinal rhodopsin (ER), which displays ~75% homologies with retinal rhodopsin (Mano *et al.*, 1999; Philp *et al.*, 2000a). *ER* has been found in the eel and medaka genomes, suggesting this gene family is widely expressed among fish. Identified regulatory elements, located within the gene promoter, direct a specific gene expression in the pineal (Asaoka *et al.*, 2002). Such a dichotomy between the pineal and retinal genes results from gene duplication that occurred at the emergence of the fish lineage. Careful examination of the exon/intron structure of *rho* and *ER* genes indicated the latter is the true orthologue of the rho gene in other vertebrate classes (Bellingham *et al.*, 2003). In the halibut, pineal counterparts of retinal green and ultraviolet (UV) cone opsins, the so-called halibut pineal opsin 1 (HPO1) and HPO4, have been cloned. Probes directed against HPO1 and HPO4 detected transcripts in photoreceptor cells from the pineal organ and the retinal photoreceptors of this species (Forsell *et al.*, 2001, 2002). In the pineal organ of adults, the green and UV opsins were expressed in different photoreceptor cells. The halibut probe detected green opsin transcripts, with a similar pattern, in other fish species, including zebrafish, Atlantic salmon, cod, haddock, herring, turbot, and three cichlids. In contrast, a halibut UV-opsin probe hybridized only with transcripts in the retinal photoreceptors of this species. The absence of signal in the pineal organ of the other fish mentioned above indicates either they do not express UV-opsin or their UV-opsin genes display low homology with the UV-opsin gene of the halibut.

Another photoreceptor gene family isolated from fish is vertebrate ancient (VA) opsin, which has been isolated from carp, salmon, and smelt fish (Moutsaki *et al.*, 2000; Philp *et al.*, 2000b; Minamoto and Shimizu, 2002, 2003). Unlike ER, which shows specific pineal expression, VA opsin is expressed in the pineal photoreceptors as well as in the amacrine and horizontal cells of the retina, and cells of the *nuclei dorsomedialis* of the hypothalamus (Philp *et al.*, 2000b).

3.1.2. Transducin and Arrestin

Arrestin is important in the visual process. It caps rhodopsin after it has been phosphorylated by rhodopsin kinase, thus preventing contact with any transducin molecule (Maeda *et al.*, 2003). Immunocytochemistry allowed visualizing the presence of arrestin in the photoreceptor cells of the fish pineal organ as well as in those of the retina of the same species (Mirshahi *et al.*, 1984; Collin *et al.*, 1986; Ekström *et al.*, 1987b; Ekström and Meissl, 1990; Forsell *et al.*, 1997). It was shown that arrestin and serotonin (another photoreceptor-specific molecule; see later) colocalize in the same photoreceptor cells in cultures from trout pineal (Kroeber *et al.*, 1998). Similarly, transducin Tα was also evidenced in the fish pineal photoreceptor cells (van Veen *et al.*, 1986; Ekström *et al.*, 1987b; Philp *et al.*, 2000b).

3.1.3. Cyclic Nucleotide-Gated Channel

The CNG channels play a key role in the phototransduction cascade; their opening/closure is triggered by cGMP. They are composed of two subunits (α and β) displaying a tetrameric organization. The α, but not the β, subunits are able to form functional homomeric channels. The α-subunit of trout pineal CNG channel is the only one cloned to date in fish pineal organ (Decressac *et al.*, 2002). It displays strong homology with the CNG3 from chicken, bovine, and mouse cones. Transcripts of trout CNG3 are expressed only in photoreceptor cells. Patch-clamp studies indicated that the native and expressed channels have properties similar to those of cone CNG3 channels. They are activated by cGMP, insensitive to cyclic AMP (cAMP), and blocked by intracellular Mg^{2+} ions.

3.1.4. Intracellular Messengers

In trout pineal organs or dissociated trout pineal cells in culture, cGMP levels are modulated by light. A 15-min illumination reduces by 30–40% the levels measured in the dark (Falcón *et al.*, 1990, 1992a) in a way similar to that found in the frog retina (Blazynski and Cohen, 1986). Pertussis toxin (which uncouples transducin from rhodopsin) or phosphodiesterase inhibitors block the effects of light in the trout pineal photoreceptors. This is consistent with the idea that light activates a phosphodiesterase through a GTP-binding protein in these cells. In another teleost, the pike, cGMP levels measured *in vitro* every 3 h, remained rather constant along the LD cycle (Falcón and Gaildrat, 1997). In contrast, cGMP efflux was high during night and low during day, suggesting extrusion participates in the regulation of intracellular levels. However, under constant darkness (DD), both pineal cGMP content and extrusion varied on a circadian manner; levels were high during the subjective dark and low during the subjective light phases. This indicates that circadian clocks, expressed in the pineal photoreceptor cells of the pike pineal organ, control cGMP metabolism (see later). Among other consequences, one would be that light sensitivity is modulated by the circadian clocks in the pineal as is the case in the retina (Walker and Olton, 1979; Bassi and Powers, 1986; Lu *et al.*, 1995; Li and Dowling, 1998; Ren and Li, 2004). Altogether, these results point to a role of cGMP in the phototransduction process of the teleost pineal photoreceptor. Studies in the lamprey pineal also support this view (Uchida *et al.*, 2001).

3.1.5. Lipids

Although not being directly involved in the phototransduction process, polyunsaturated fatty acids (PUFA) are critical to photoreceptor structure, function, and development (Bell *et al.*, 1995). The retinal photoreceptor cell has less "fluid" lipid environment in the plasma membrane relative to disks membrane, and bleached rhodopsin regenerates faster in the latter than in the former (Boesze-Battaglia and Allen, 1998). It has been shown that the pineal organ and retina of fish display a similar and unique lipid composition, compared to other neuronal and nonneuronal structures (Henderson *et al.*, 1994). Among the PUFA of interest, docosahexaenoïc acid (DHA) appears as a major constituent of the outer segment membranes in the retinal photoreceptors (Bazan *et al.*, 1992; Chen and Anderson, 1993; Brown, 1994). In the pineal organ, DHA is the major constituent in all lipid classes, mainly found in the phosphatidylcholine and phosphatidylethanolamine glycerophospholipids pools (Henderson *et al.*, 1994). In culture, trout pineal organs take up and metabolize PUFA including ^{14}C-DHA. Radioautographic studies indicated that ^{3}H-DHA is taken up preferentially by photoreceptor cells (Falcón and Henderson, 2001). Electron microscopy indicated that the labeling was mainly associated with the outer segment membranes, Golgi apparatus, and mitochondria of the ellipsoid. The labeling in the outer segment is in keeping with the idea that DHA is involved in phototransduction rather than synaptic transmission, and DHA-containing phospholipids could be required for the proper localization of rhodopsin in the membrane (Weisinger *et al.*, 1999).

DHA is also known for being specifically required for photoreceptor survival and differentiation (Rotstein *et al.*, 2003), and this is particularly relevant in fish where the retina and the pineal organ generate new photoreceptors all life long. The accumulation of DHA in mitochondria is important to their integrity, which is essential to photoreceptor survival (Falcón and Henderson, 2001; Rotstein *et al.*, 2003).

Incorporation of ^{3}H-DHA may also be observed in interstitial cells of trout pineal organ, providing the end of the incubation is followed by 12–18 h of chase (Falcón and Henderson, 2001). This supports the idea that recycling of photoreceptor outer segment membranes operates in the interstitial cells.

3.2. Electrical Responses

The first direct evidence that the pineal organ of ectothermic vertebrates was a light sensor came from the pioneer work of Dodt and Heerd (1962) and Dodt (1963). Since then, many studies involving extra- or intracellular recordings have been performed, resulting in the collection of spike discharges from neurons, variations in membrane potential from photoreceptor cells, electropinealograms (EPG), and early receptor potentials (ERP).

3.2.1. Responses from Photoreceptor Cells

Three types of recordings can be obtained from the photoreceptor cells: ERP, EPG, and membrane potential.

The *ERP* results from charge redistribution in rhodopsin associated with protein conformational changes; and it is believed that ERP are generated mainly by cone photoreceptors in the retina (Cone and Pak, 1971; Sullivan and Shukla, 1999). They might therefore present potential interest for studies in the pineal organ, which contains only cone photoreceptors. In both, the retina and pineal organ, bright light stimuli induce ERP appearing with extremely short latency (100 μsec) (Falcón and Tanabe, 1983). Latency, recovery, and spectral sensitivity are similar in both organs. Spectral sensitivity curves obtained in light-adapted fish indicated a maximal absorption at 617 nm; however, after dark adaptation the maximal sensitivity was at 533 nm in the pineal organ. This would indicate the presence of at least two populations of cells with different photopigments.

The *EPG* is also a mass response of cells with, however, a much longer latency than the ERP. It is analogous to the electroretinogram (ERG)

(Meissl and Dodt, 1981). The EPG has been mainly investigated in frogs and lizards. The responses obtained in the four fish species studies display similar characteristics: pike (Falcón and Meissl, 1981); ayu, trout, and gigi (Hanyu *et al.*, 1969; Tabata, 1982a,b). Latency (±50 msec) and amplitude of the response depends on both the duration and intensity of the stimulus at a given wavelength. The response may remain long after the end of the stimulus. In the pike, the ERP, and EPG spectral sensitivity curves were similar. The other three teleost investigated displayed a single sensitivity peak at λ_{max} 525 nm. There are reasons to believe that the EPG is generated by the cone photoreceptors in the pineal organ, including the observation that EPG can be recorded from pieces of pike pineal epithelium containing no neurons (Section 2.2.2.; Falcón and Meissl, 1981). Also, the EPG is not affected in trout pineal treated with aspartate, which only affects the responses postsynaptic to the photoreceptors in the retina (Tabata, 1982a,b).

The *membrane potentials*, measured through intracellular recordings, have been studied in a very limited number of species (Figure 6.5) (trout, goldfish, European minnow; Ekström and Meissl, 1997). The photoreceptor cell is depolarized in the dark-adapted state, and the resting potential is at −20/−30 mV. In trout, flashes of bright light result in cell hyperpolarization by up to 30 mV (Meissl and Ekström, 1988a,b). But in some cases, potentials of −90 mV could be recorded (our unpublished data). The characteristics

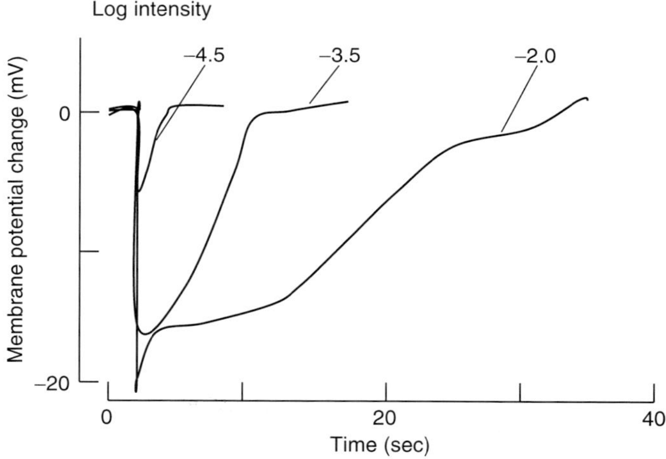

Fig. 6.5. Intracellular recordings from the pineal organ of the trout. Recordings were obtained from trout pineal photoreceptor cells. Light induces cell hyperpolarization. The responses decrease in latency and increase in duration as light intensity increases. (Unpublished data from Dr. H. Meissl, graph adapted from Falcón, 1999.)

of the trout pineal photoreceptor response, its similarities and differences with the retinal photoreceptor, have been extensively detailed elsewhere (Ekström and Meissl, 1997). Some differences are however important to recall. For example, the pineal photoreceptor cannot discriminate between rapidly changing light stimuli, and the latency and recovery of its response is longer than that of the retinal photoreceptor. Also, during prolonged stimulation the pineal photoreceptor maintains the same response amplitude. In other words, the membrane potential reflects the ambient level of illumination (Figure 6.5). This is an ideal situation for a luminance detector. Most of the pineal photoreceptors respond maximally to wavelength at 520–530 nm. Some also respond at a λ_{max} of 500 nm, supporting microspectrofluorimetric studies indicating the presence of two photopigments (Kusmic and Gualtieri, 2000).

3.2.2. Responses from Second-Order Neurons

The spike discharges are extracellular recordings from single neurons or from pineal tract (Hamasaki and Eder, 1977; Meissl and Dodt, 1981; Meissl *et al.*, 1986; Ekström and Meissl, 1997). The usual pattern of activity is a maintained discharge in the dark, which is inhibited upon illumination (Figure 6.6). There is a linear relationship between the logarithm of light intensity and the frequency of the discharges over a range of 5–6 log units (Falcón and Meissl, 1981). This provides a wide range of sensitivity because only 1–2 log units are absorbed by the skull (Meissl and Dodt, 1981). Two types of responses have been recorded:

1. The *chromatic response* consists of a long lasting inhibition upon exposure to UV light ($\lambda_{max} = \sim 355$ nm) and a stimulation by green ($\lambda_{max} = \sim 520$ nm) or red light ($\lambda_{max} = \sim 620$ nm) (Figure 6.6A and B). Such a response has been described in fish (pike, trout, and angelfish), lampreys, frogs, and lizards (Meissl and Dodt, 1981; Uchida and Morita, 1994). Studies in frog have shown that the two effects interact and depend also on light energy and previous adaptation state. Several mechanisms may account for the chromatic response. One hypothesis is based on the presence of functionally polarized interneurons that would transfer information from one cone system to another (Meissl and Dodt, 1981). Immunocytochemical, pharmacological, and electrophysiological studies do support the view that gamma-aminobutyric acid (GABA) interneurons are present in trout (Ekström *et al.*, 1987a; Ekström and Meissl, 1988, 1989; Meissl and Ekström, 1991; Brandstätter and Hermann, 1996). Another possible explanation would be that two photopigments are coexpressed in a single photoreceptor cell. One would drive cell depolarization, and

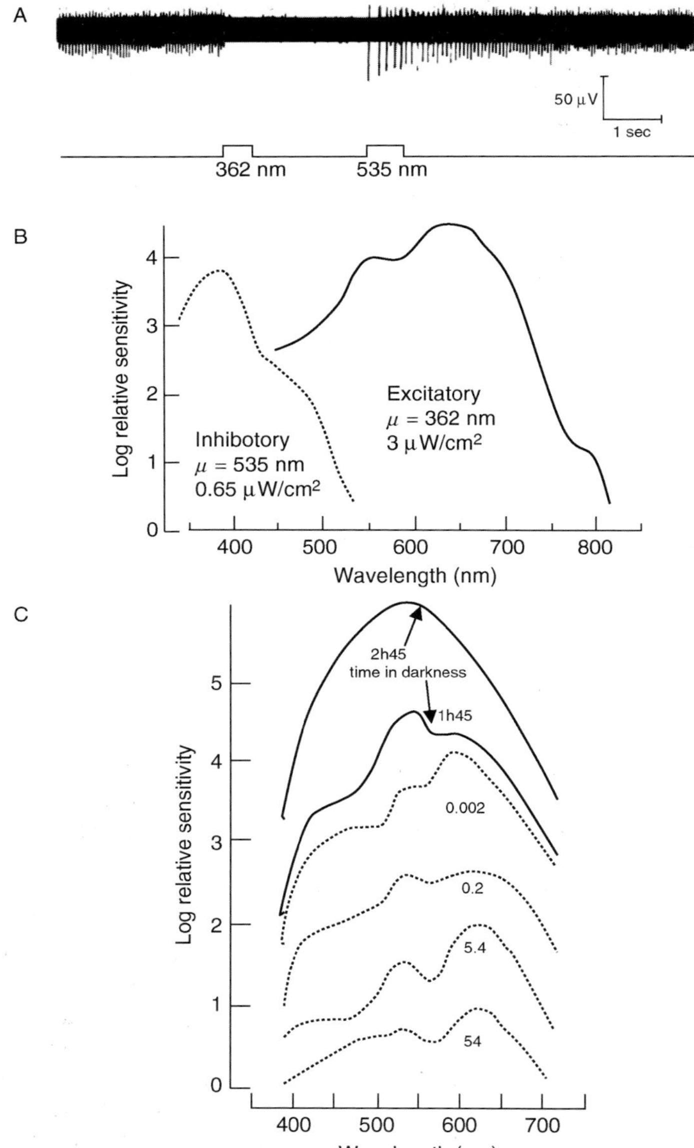

Fig. 6.6. Chromatic (A and B) and achromatic (C) responses recorded from pike pineal neurons. (A) Inhibition of the maintained neuronal discharges in response to UV (362 nm) light, and excitation in response to green (535 nm) light. (B) Spectral sensitivities of the inhibitory

the other cell hyperpolarization. This was suggested from studies in the lizard parietal eye (Solessio and Engbretson, 1993). The presence of two photopigments in a single cone photoreceptor has been demonstrated in the retina of several vertebrate species (Jacobs *et al.*, 2004).

2. The *achromatic (or luminance) response* is the most common and is characterized by an inhibitory effect of light at all wavelengths from UV to red light (Figure 6.6). The absolute light threshold of the chromatic response varies from $\sim 10^{-3}$ to $\sim 10^{-5}$ lm/m^2 (bright sunlight is 10^5 lm/m^2) in dark-adapted pineal organs, not considering the absorbance of the overlaying skull and skin (Meissl and Dodt, 1981). Spectral sensitivity curves show one or two peaks, usually in the blue/green range (490–530 nm) (Figure 6.6C). In all species investigated, there is a close parallelism between the pineal and the retinal spectral sensitivity curves. The pike is special because the λ_{max} is found at 620 nm in light-adapted pineal organs and at 530 nm in dark-adapted organs (Figure 6.6C) (Falcón and Meissl, 1981). This is a remarkable adaptation to the environment because a shift of sensitivity (Purkinje shift) operates at the light-to-dark transition, that is, when underwater absorbance becomes higher for red than for green wavelengths. In addition to increasing the sensitivity range during day, this might provide a mechanism preparative to phase transition.

3.3. Release of a Neurotransmitter

There is indication that glutamate is the neurotransmitter of the retinal photoreceptor cells (Ayoub and Dorst, 1998; Dowling, this volume; Kreitzer *et al.*, 2003). The neurotransmitter of the fish pineal photoreceptor is also most probably an excitatory neurotransmitter. Immunocytochemical investigations indicated that both glutamate and aspartate, colocalize in pineal photoreceptors of the ray, charr, and goldfish (Vigh *et al.*, 1995a). Immunoreactivity was detected in the perikarya, but more intense reactions were seen in the synaptic pedicles, where both aminoacids colocalize. A similar distribution pattern was seen in the retinal photoreceptors or the same species, as well as in the related photosensory pineal organ of lampreys, frogs, and lizards (Vigh *et al.*, 1995a,b, 1997; Debreceni *et al.*, 1997). Biochemical

(interrupted line) and excitatory (line) thresholds during steady exposure to weak green or UV background, respectively. (C) Spectral sensitivity curves measured in darkness and during continuous exposure to different levels of white light. (Adapted from Falcón and Meissl, 1981.)

methods also allowed identification of aspartate and glutamate in the fish pineal organ, (Meissl *et al.*, 1978; McNulty, 1988). In the goldfish, prolonged illumination induced an increase in both glutamate content and size of photoreceptor synaptic ribbons (McNulty, 1988). Similarly, dark adaptation induced a twofold increase in glutamate content of *Xenopus* retinal photoreceptors (Schmitz and Witkovsky, 1997). In the frog pineal organ, application of either one neurotransmitter mimics the action of dark (Meissl and George, 1984a,b); the excitatory neurotransmitters act probably through both *N*-methyl-D-aspartate (NMDA) and non-NMDA receptors (Jamieson, 1997).

Gamma-amino-butyric-acid is another neurotransmitter identified in the trout pineal organ (Meissl and Ekström, 1991). The GABA cells are located mainly in the rostral part of the organ. GABA was shown to alter the spiking activity of the second-order neurons by either suppressing or stimulating it, depending on the previous adaptation state. According to Meissl and Ekström (1991) these GABAergic cells belong to a network of interneurons that modulate light sensitivity during light or dark adaptation.

3.4. Conclusions

In brief, the photoreceptors of the fish pineal organ resemble, in many respects, the cone photoreceptors of the retina; and the phototransduction mechanisms, as summarized in Figure 6.7, are likely to be the same: the structural features of a pineal photoreceptor are those of a cone (outer segment, number of synaptic ribbons), they contain many of the proteins of the phototransduction cascade (Figure 6.7), and some of the cloned genes (e.g., CNG3) appear to belong to the cone family of genes. More generally, the pineal and the retinal photoreceptors share a unique and specific lipid composition, and their electrical responses are also very similar. There are however some differences in the time course and recovery of the responses, which are much longer in the pineal than in the retinal photoreceptors. In contrast, the intensity range of the response is wider for the pineal than for the retinal photoreceptors. The neuronal output, which conveys the information to brain centers, reflects the response of the photoreceptor cells to light stimuli. This information provides indication on the ambient illumination and, under certain circumstances, on the spectral composition. This seems particularly true at twilight. The pineal organ of fish is "a dosimeter" of ambient illumination and "a day-length indicator."

The output of the pineal photoreceptor cell is the excitatory neurotransmitter which is released into the synaptic cleft. However, other messengers are produced in response to the photoperiodic information, among which is the hormone melatonin.

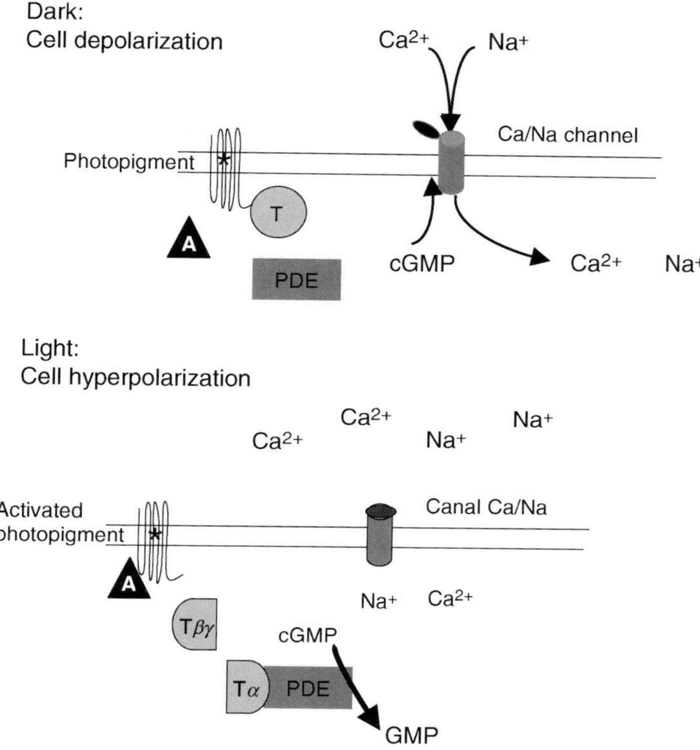

Fig. 6.7. Phototransduction molecules identified in the fish pineal photoreceptor cells, and possible mechanisms of the phototransduction. See text for details. A, arrestin; cGMP, cyclic guanosine monophosphate; PDE, phosphodiesterase; T, transducin (α and $\beta\gamma$ subunits).

4. THE FISH PINEAL ORGAN: A MELATONIN FACTORY

4.1. Melatonin is Synthesized Within the Photoreceptor Cells

Figure 6.8 details the pathway leading to melatonin synthesis. Melatonin is synthesized from tryptophan in four enzymatic steps (Klein *et al.*, 1981, 1997; Ganguly *et al.*, 2002). Tryptophan hydroxylase (TPOH) catalyzes the conversion of tryptophan into 5-hydroxytryptophan, which is then decarboxylated by the aromatic amino acid decarboxylase to produce serotonin. The arylalkylamine *N*-acetyltransferase (AANAT) converts serotonin to *N*-acetylserotonin, which is then *O*-methylated by the action of the hydroxyindole-*O*-methyltransferase (HIOMT) to produce melatonin. Other

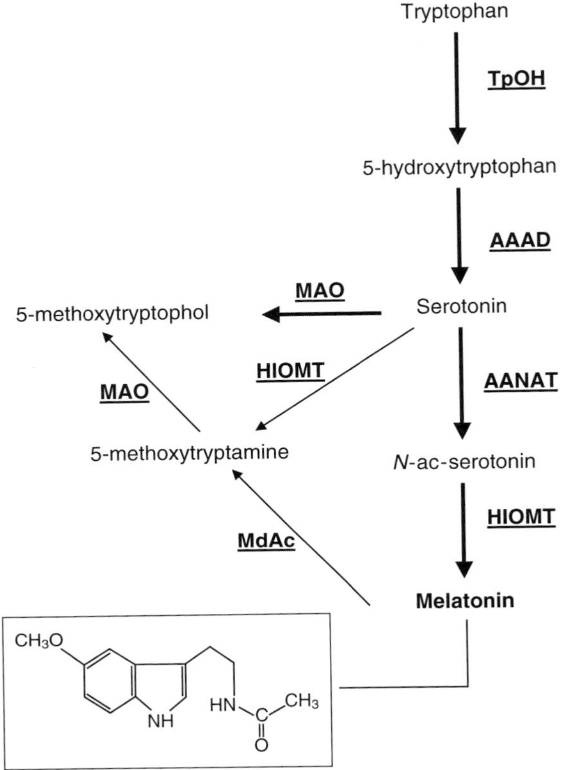

Fig. 6.8. Melatonin biosynthesis pathway. Thick arrows correspond to the main pathway. AAAD, aromatic amino acid decarboxylase; AANAT, arylalkylamine-*N*-acetyltransferase; *N*-ac-serotonin, *N*-acetylserotonin; HIOMT, hydroxyindole-*O*-methyltransferase; MdAcC, melatonin decarboxylase; MAO, monoamine oxydase (the pathways involving MAO are indirect); TpOH, tryptophan hydroxylase.

indole compounds—5-hydroxyindole acetic acid and 5-hydroxytryptophol—are produced by the fish pineal organ after oxidative deamination of serotonin; this is catalyzed by monoamine oxidase (MAO) (Figure 6.8). These compounds as well as serotonin are also substrates for HIOMT. In addition to being release in the blood or cerebrospinal fluid (CSF), melatonin can be deacetylated *in situ* to produce 5-methoxytryptamine and 5-methoxytryptophol (Falcón *et al.*, 1985; Yañez, 1996). Similar mechanisms operate in the retina (Cahill and Besharse, 1989, 1995). In the pineal organ, melatonin synthesis occurs in photoreceptor cells because melatonin and other indole compounds (serotonin, *N*-acetylserotonin) and enzymes of the indole metabolic pathway (MAO, HIOMT) have been located exclusively in photoreceptor cells using a variety of methods (histochemical, immunocytochemical, radioautographic). Moreover, single and isolated photoreceptors have been shown to release melatonin *in vitro* (Bolliet *et al.*, 1997). In the retina, cells located in the basal inner nuclear layer (bipolar and/or amacrine cells?) and the ganglion cell layer also expressed the genes encoding AANAT1 and HIOMT, albeit to a lesser degree than in the photoreceptors of the outer nuclear layer (Besseau *et al.*, 2005). This would suggest that melatonin synthesis is not specific to photoreceptor cells or that photoreception is not a specific property of the "classical" photoreceptor cells of the outer nuclear layer (Foster and Hankins, 2002; Foster *et al.*, 2003; Foster and Bellingham, 2004).

4.2. The LD Cycle Synchronizes a Rhythm in Melatonin Production

The alternation of light and darkness synchronizes the rhythmic production of melatonin both in the pineal organ and retina with, however, a 180° reversed pattern (Figure 6.9). Tools are now available to study this regulation: genes have been cloned, enzyme activities have been measured, and antibodies have been obtained, which allow detection or quantification of the corresponding antigens. There seems to be an inverse relationship between the levels of serotonin and those of *N*-acetylserotonin and melatonin. In the pineal organ of fish, as well as in most vertebrate species, *N*-acetylserotonin and melatonin content are high at night and low during day (Figure 6.9) (Falcón, 1999). No difference has been detected between deep-sea and demersal fish in the overall amounts and patterns of pineal melatonin content (Wagner and Mattheus, 2002). Melatonin is highly lipophylic, and thus crosses easily the cell membrane. Actually, only pineal melatonin, not retinal melatonin, is released into the blood stream and CSF as soon as it is synthesized (Figure 6.9). This has been inferred from *in vitro* studies in more than 30 fish species (Zachmann *et al.*, 1992a; Ekström and Meissl, 1997; Iigo *et al.*, 1997; Falcón, 1999) (Table 6.1). In superfusion systems, melatonin collected in the culture medium increases at the onset of darkness and decreases at the onset of light, reflecting the pattern observed *in vivo* in the plasma (Figures 6.9 and 6.10) (Falcón, 1999; Masuda *et al.*, 2003a; Bayarri *et al.*, 2004a,b). Light is clearly inhibitory; both *in vivo* and *in vitro*, unexpected light at night decreases pineal melatonin content and release it (Figure 6.10B); the amplitude of the response is proportional to the intensity of the light stimulus (Max and Menaker, 1992; Zachmann *et al.*, 1992b; Bolliet *et al.*, 1995; Bayarri *et al.*, 2002). Lunar light at night may also modulate the amplitude of the nocturnal melatonin surge, as shown in the rabbitfish (Rahman *et al.*, 2004). Spectral sensitivity curves indicate melatonin production is highly sensitive to blue and green wavelengths

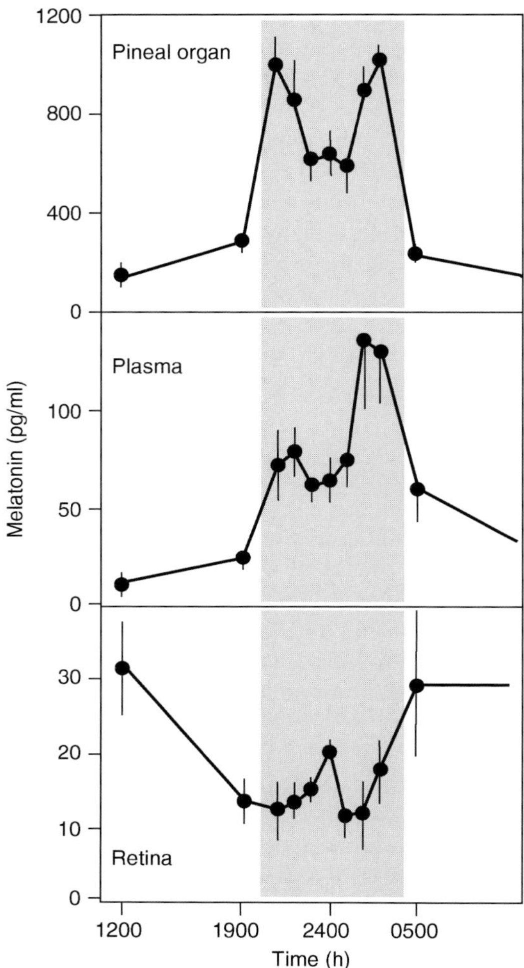

Fig. 6.9. *In vivo* LD variations in pineal organ, plasma, and retina melatonin content in the white sucker (*Catostomus commersoni*). Gray boxes correspond to scotophase. (Adapted from Zachmann *et al.*, 1992b.)

(Max and Menaker, 1992; Bayarri *et al.*, 2002). It appears that the light-dependent inhibition of the nervous and neurohormonal messages display the same spectral sensitivity curve.

Investigations by Klein and Weller (1970) in the rat demonstrated that the rhythm in melatonin biosynthesis results from the rhythmic activity of

Table 6.1
Photoperiodic Versus Circadian Control of Melatonin Release in Isolated Fish Pineal Organs in Culture

Species	Scientific name	Common name	Habitat	LD	DD	References
Agnatha						
Petromyzontiformes Petromyzontidae	*Petromyzon marinus*	Sea lamprey	Temperate euryhaline	+	+	Bolliet *et al.*, 1993
	Lethenteron camtschaticum[a]	Arctic lamprey	Polar euryhaline	+	+	Samejima *et al.*, 1997
Gnathostomata						
Chondrichthyes						
Elasmobranchii						
Carcharhiniformes Sphyrnidae	*Sphyrna lewini*	Scalloped hammerhead	Marine	+	–	Okimoto and Stetson, 1997
Squaliformes Squalidae	*Squalus acanthias*	Piked dogfish	Marine	+	–	Okimoto and Stetson, 1997
Actinopterygii						
Teleostei						
Anguilliformes Anguillidae	*Anguilla anguilla*	Eel	Temperate euryhaline	+	+	Bolliet *et al.*, 1996
Clupeiformes Clupeidae	*Alosa pseudoharengus*	Alewife	Temperate euryhaline	+	+	Bolliet *et al.*, 1996
Cypriniformes Cyprinidae	*Carassius auratus*	Goldfish	Temperate freshwater	+	+	Kezuka *et al.*, 1989
	Notemigonus crysoleucas	Golden shiner	Temperate freshwater	+	+	Bolliet *et al.*, 1996
	Semotilus atromaculatus	Creek chub	Temperate freshwater	+	+	Bolliet *et al.*, 1996
	Cyprinus carpio carpio	Common carp	Temperate freshwater	+	+	Bolliet *et al.*, 1996
	Danio rerio	Zebrafish	Tropical freshwater	+	+	Cahill, 1996

(*continued*)

Table 6.1 *(continued)*

Species	Scientific name	Common name	Habitat	LD	DD	References
Catostomidae	*Catostomus commersonii*	White sucker	Temperate freshwater	+	+	Zachmann *et al.*, 1992a
Siluriformes Ictaluridae	*Ictalurus punctatus*	Channel catfish	Temperate freshwater	+	+	Bolliet *et al.*, 1996
Salmoniformes Salmonidae	*Oncorhynchus mykiss*	Rainbow trout*	Temperate euryhaline	+	–	Gern and Greenhouse, 1988
	Oncorhynchus masou	Cherry salmon	Temperate euryhaline	+	–	Iigo *et al.*, 1997
Osmeriformes Plecoglossidae	*Plecoglossus altivelis*	Ayu	Temperate euryhaline	+	+	Iigo *et al.*, 2004
Esociformes Esocidae	*Esox lucius*	Pike*	Temperate freshwater	+	+	Falcón *et al.*, 1989
Gadiformes						
Gadidae	*Gadus morhua*	Atlantic cod	Marine	+	+	Bolliet *et al.*, 1996
Phycidae	*Urophycis tenuis*	White hake	Marine	+	+	Bolliet *et al.*, 1996
Cyprinodontiformes Fundulidae	*Fundulus heteroclitus*	Killifish	Temperate euryhaline	+	+	Roberts *et al.*, 2003
Poeciliidae	*Limia vittata*	Cuban molly	Tropical euryhaline	+	+	Okimoto and Stetson, 1997
	Poecilia sphenops	Molly	Tropical euryhaline	+	+	Okimoto and Stetson, 1997
	Poecilia reticulata	Guppy	Tropical euryhaline	+	+	Okimoto and Stetson, 1997
	Poecilia velifera	Sail-fin molly	Tropical euryhaline	+	+	Okimoto and Stetson, 1997
	Xiphophorus helleri	Green swordtail	Tropical freshwater	+	+	Okimoto and Stetson, 1997
	Xiphophorus maculatus	Southern platyfish	Tropical freshwater	+	+	Okimoto and Stetson, 1997

Scorpaeniformes Hemitrypteridae	*Hemitripterus americanus*	Sea raven	Marine	+	+	Bolliet *et al.*, 1996
Perciformes						
Moronidae	*Morone americana*	White perch	Temperate euryhaline	+	nt	Okimoto and Stetson, 1997
	Morone saxatilis	Striped bass	Temperate euryhaline	+	nt	Okimoto and Stetson, 1997
Mugilidae	*Mugil cephalus*	Mullet	Marine	+	nt	Okimoto and Stetson, 1997
Cichlidae	*Oreochromis mossambicus*	Mozambique tilapia	Tropical euryhaline	+	nt	Okimoto and Stetson, 1997
Centrarchidae	*Lepomis gibbosus*	Pumpkinseed	Temperate freshwater	+	+	Bolliet *et al.*, 1996
	Micropterus salmoides	Largemouth bass	Temperate freshwater	+	+	Bolliet *et al.*, 1996
Percidae	*Perca flavescens*	Yellow perch	Temperate freshwater	+	+	Bolliet *et al.*, 1996
Scombridae	*Scomber scombrus*	Atlantic mackerel	Marine	+	+	Bolliet *et al.*, 1996
Carangidae	*Caranx ignoblis*	White jack	Marine	+	nt	Okimoto and Stetson, 1997
Pleuronectiformes Pleuronectidae	*Pseudopleuronectes americanus*	White flounder	Marine	+	+	Bolliet *et al.*, 1996

[a]Former name *Lampreta japonica.*

Melatonin release by isolated pineal organs was measured under either alternating photoperiodic (LD column) or constant darkness (DD column). +, rhythmic release; –, no rhythmic release; nt, not tested. In two species (*) AANAT activity was also measured, giving similar results. Fish classification is based on Nelson (1994).

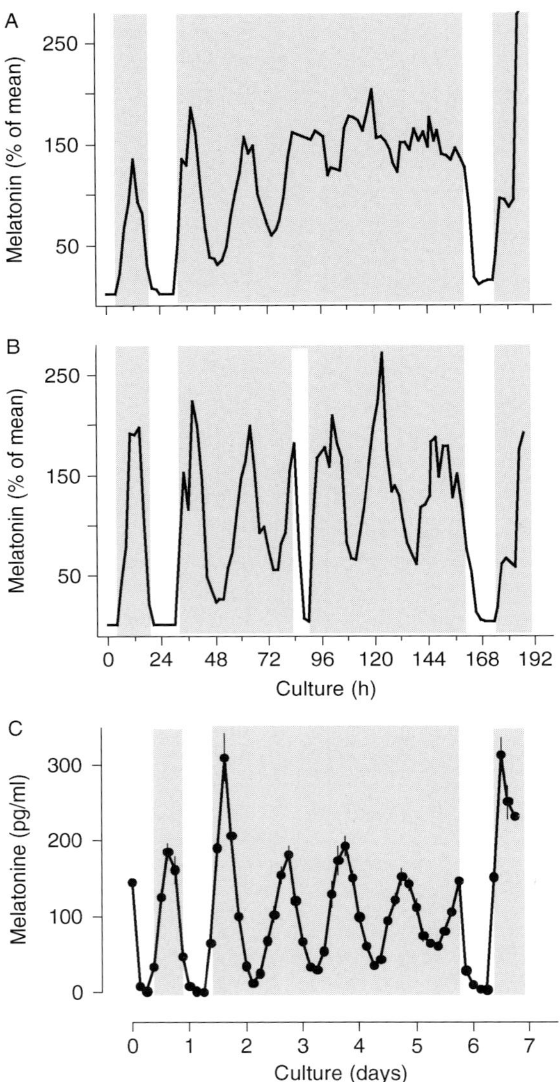

Fig. 6.10. *In vitro* secretion of melatonin content by cultured pike pineal organs (A and B) and cells (C). Under LD (first 24 h), melatonin secretion is high during the dark phase (gray boxes) and low during day time. The rhythms persist under constant darkness but tends to dampen (A and C). A 6-h pulse of light during the third cycle under DD induces a rebound of the rhythm and a phase delay (C). In all cases, LD after DD synchronizes the rhythm again to the prevailing photoperiod. (A and B) Superfusion (1 ml/h, 2 h collection interval). (C). Static culture (0.25×10^6 cells/well, 1 ml/dish, 2-h collection interval). [Adapted from Falcón, 1999 (A and B) and Bolliet *et al.*, 1997 (C).]

AANAT, which is increased at night (Klein *et al.*, 1981, 1997). Studies in fish led to similar conclusions (Figure 6.11); in this case, however, light acts directly on the photoreceptor cells to inhibit nocturnal AANAT activity (Figure 6.11C) (Falcón *et al.*, 1989). In contrast, HIOMT activity remains

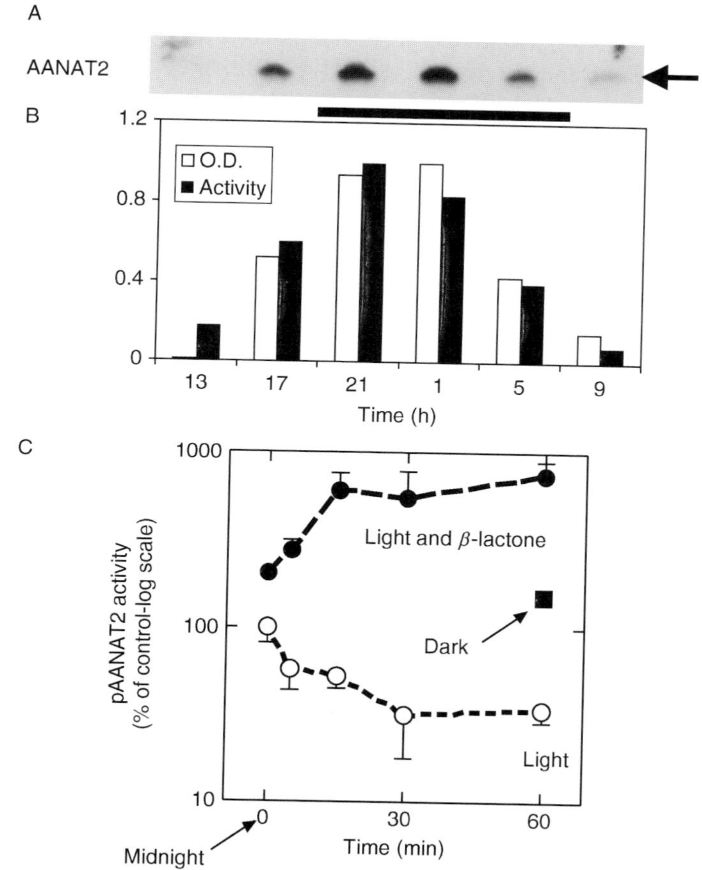

Fig. 6.11. AANAT2 protein amount and activity during the 24-h LD cycle in the pineal organ of the pike. (A) AANAT2 protein amount detected on Western blots from pineal extracts is higher at night (black bar) than during day. (B) There is a good correlation between the intensity of the spots density (OD) and AANAT2 activity, when both are measured from the same extracts. (C) Midnight AANAT2 activity is inhibited by light (more than 70% inhibition after 1 h). This is prevented in the presence of a proteasome inhibitor, which even induces a further increase of nocturnal AANAT2 activity versus controls. (A and B) *in vivo*; (C) *in vitro* (Falcón *et al.*, 2001).

rather constant throughout the LD cycle (Falcón *et al.*, 1987; Morton and Forbes, 1988). AANAT soon appeared as a key regulatory component in the rhythmic production of melatonin, which prompted efforts to clone the gene, measure activity, get species-specific antibodies and study the 3D structure of the protein (Klein *et al.*, 1997; Obsil *et al.*, 2001; Ganguly *et al.*, 2002). AANAT was first cloned in mammals (Borjigin *et al.*, 1995; Coon *et al.*, 1995) and then in fish, including pike, trout, zebrafish, and seabream (Bégay *et al.*, 1998; Mizusawa *et al.*, 1998; 2000; Coon *et al.*, 1999; Gothilf *et al.*, 1999; Zilberman-Peled *et al.*, 2004). Fish are unique among vertebrates because they possess at least two AANAT genes (AANAT1 and AANAT2),[5] whereas a single gene has been identified in other vertebrates. The deduced amino acid sequences of the two AANATs are 75% and 66% similar in pike and seabream respectively (Coon *et al.*, 1999; Zilberman-Peled *et al.*, 2004). AANAT2 is expressed in the pineal organ, and AANAT1 is expressed in the retina, and is more similar to the AANAT found in other vertebrates. This situation resembles that described for the genes encoding photopigment molecules (see later). In the case of AANAT2, two regions of the gene determine pineal specific expression in zebrafish; one is located in the 5′-flanking region, and the other 6 kb downstream the transcribed region (Appelbaum *et al.*, 2004). The encoded AANAT1 and AANAT2 proteins have distinct affinities for serotonin and differ in their relative affinities for indole-ethylamines versus phenylethylamines (Table 6.1) (Falcón and Bolliet, 1996; Coon *et al.*, 1999; Benyassi *et al.*, 2000; Zilberman-Peled *et al.*, 2004).

Regulation of pineal AANAT2 expression varies from a species to another. Transcripts abundance varies on a daily basis; they are more abundant at night than during day, in pike and zebrafish pineal organs, whereas levels remain constant throughout the LD cycle in trout (Figure 6.12) (Bégay *et al.*, 1998; Coon *et al.*, 1999; Gothilf *et al.*, 1999). Similarly, TPOH mRNA abundance varies in the pineal organ of pike and zebrafish, but not trout, suggesting a similar intracellular mechanism might regulate *Tpoh* and *Aanat2* expressions (Bégay *et al.*, 1998). In the case AANAT2 transcripts abundance is regulated, transcription inhibitors prevent the nocturnal increase in AANAT2 activity (Falcón *et al.*, 1998), while these are without effect when there is no variation in AANAT2 mRNA abundance (Falcón *et al.*, 1998). In this latter situation, melatonin release may however be inhibited suggesting another component of the melatonin biosynthesis pathway is transcriptionally regulated (Mizusawa *et al.*, 2001). In all species

[5]Analysis of the data banks (fugu and puffer fish genomes, and fish AANAT1 sequences available) indicates that genome duplication, at some point of teleost fish evolution, resulted in the appearance of two AANAT1 genes (Coon and Klein, 2006).

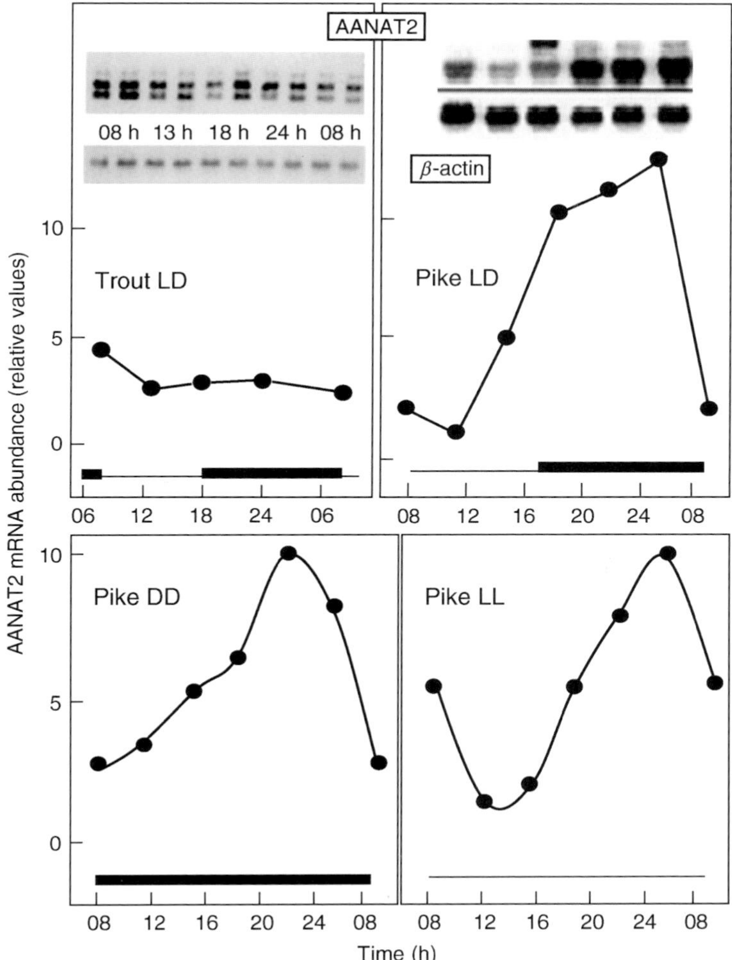

Fig. 6.12. LD variations in AANAT2 mRNA abundance in the pineal organs of the trout and pike. (Above) The Northern blots from pineal extracts were hybridized with probes directed against AANAT2 mRNA and the intensity of each band was normalized to the corresponding value obtained with a β-actin probe. Black bar corresponds to darkness. (Below) Variations in AANAT2 mRNA abundance are maintained under LL and DD only in pike. (Trout is adapted from Bégay *et al.*, 1998 and pike is adapted from Coon *et al.*, 1999.)

investigated AANAT2 activity and melatonin secretion are higher at night (Falcón *et al.*, 1987). Both rhythms have the same phase, and are shifted by 4–6 h compared to the rhythm of transcripts abundance in the case this varies (Figures 6.10–6.12). The nocturnal rise in AANAT2 activity involves *de novo* protein synthesis, because protein is elevated at night and reduced during day as shown on Western blots from pike pineal protein extracts and using specific antibodies directed against pike pineal AANAT2 (Figure 6.11A and B) (Falcón *et al.*, 2001). Conversely, the light-induced inhibition of AANAT2 activity results from protein degradation because illumination of dark adapted pineal organs decreases both AANAT2 activity and immuno-detected AANAT2 protein (Figure 6.11C) (Falcón *et al.*, 2001).

4.3. Conclusions

In addition to the nervous message, the photoreceptors of the fish pineal organ produce melatonin, a neuroendocrine message. The hormone is released into the blood and CSF in a rhythmic manner (Figure 6.13). The synthesis and release of melatonin is controlled by the environmental LD cycle in such a manner that it increases in the dark, and it decreases upon illumination. The retina also produces, but does not release, melatonin: pinealectomy, but not eye enucleation, reduces plasma melatonin levels (Iigo and Aida, 1995). Moreover, the melatonin peak occurs during day in the retina, and at night in the pineal organ and plasma (Figure 6.9). It seems that retinal melatonin exerts local autocrine and/or paracrine effects and is metabolized *in situ* (Falcón *et al.*, 2003). AANAT2 plays a crucial role in the regulation of pineal melatonin secretion from serotonin. The nocturnal melatonin rise results from an increase in AANAT2 protein amount and activity (Figures 6.11 and 6.13), whereas light induces opposite effects. In some but not all cases, the LD cycle also controls expression of the *Aanat2*

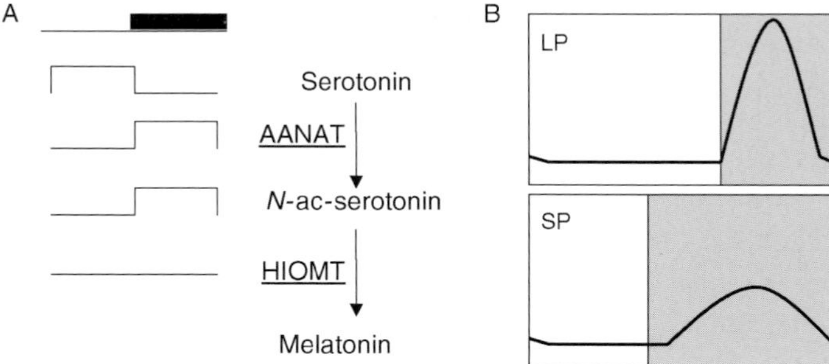

Fig. 6.13. Schematic presentation of the factors displaying rhythmicity in the melatonin biosynthesis pathway. Black bar in A and gray boxes in B correspond to the photophase. LP, long photoperiod; SP, short photoperiod.

gene. In temperate regions, the duration and amplitude of the pineal and plasma melatonin rhythms vary along with seasons. Usually, the rhythm is of high amplitude/short duration under long photoperiod, and of low amplitude/long duration under short photoperiod (Figure 6.13) (Kezuka *et al.*, 1988; Masuda *et al.*, 2003a). Thus, the melatonin signal reflects the prevailing photoperiod, and has the potential to provide the animal with accurate information on daily and calendar time, as is the case for the nervous signal sent to brain centers (see in an earlier section).

5. INTRACELLULAR REGULATION OF ARYLALKYLAMINE *N*-ACETYLTRANSFERASE 2

What mechanisms link the phototransduction process to the regulation of AANAT2 expression and/or activity? In other words, what intracellular mechanisms operate between the photoreceptive pole and the control of AANAT2 expression and activity? Two strategies have been developed among fish. The first strategy is common to all species investigated, and involves intracellular second messengers that regulate AANAT2 protein and AANAT2 activity; the second strategy, developed by many but not all fish, involves an additional control of *Aanat2* gene expression (Falcón, 1999). In this latter case, this appears to be mediated by intrapineal biological clocks.

5.1. Regulation of AANAT2 Protein

Early studies in mammals had shown that cAMP plays a crucial role in the control of AANAT activity (Klein *et al.*, 1981, 1997; Ganguly *et al.*, 2002). In the rat, norepinephrine (NE) released at night acts synergistically through α_1 and β_1 adrenergic receptors to increase AANAT activity. Binding to the β_1 receptors activates the adenylyl cyclase, which results in an increase in cAMP, AANAT activity, and melatonin release. Activation of the α_1 receptors pathway prompts the formation of diacyl glycerol (DAG) and accumulation of intracellular calcium ($[Ca^{2+}]_i$). Both also activate the adenylyl cyclase through a protein kinase C (PKC)-dependent mechanism, thus contributing to a further increase in cAMP accumulation. The increase in cAMP activates PKA, which phosphorylates two conserved sites in the AANAT of all vertebrates. In rat, this allows protection and activation of AANAT, which bounds 14-3-3 proteins (Klein *et al.*, 1997, 2002; Ganguly *et al.*, 2002).

Studies in fish are few and concern a very limited number of species. It is believed that both Ca^{2+} and cAMP contribute to the intracellular regulation of melatonin secretion by photoreceptor cells (Falcón, 1999; Falcón *et al.*, 2001). Analogs of cAMP or compounds that induce intracellular accumulation of cAMP have two effects: (1) they increase the amount of immunodetected AANAT2 protein, stimulate AANAT2 activity and thus melatonin release; (2) they prevent or reduce the inhibitory effects of light (Falcón *et al.*, 1992a, 2001; Thibault *et al.*, 1993a,b; Kroeber *et al.*, 2000). Light modulates intracellular cAMP levels in fish pineal organ. In cultured pike pineal organs peak cAMP levels are seen at the phase transitions (Falcón and Gaildrat, 1997). Light inhibits by ~30% intracellular cAMP levels of dark-adapted trout pineal cells and organs in culture (Falcón *et al.*, 1990, 1992b). Pharmacological studies led to suggest that light inhibits adenylyl cyclase and activates the cAMP phosphodiesterase, and that changes in cAMP turnover could regulate AANAT2 (Falcón *et al.*, 1992b). Some studies report no effect of light on cAMP accumulation, but attention must be paid to the temperature, which must be maintained as close as possible to the fish-preferred temperature. This is because cAMP metabolism is temperature sensitive in fish (see later) and an inappropriate temperature may mask the effects of light. Thus, one way through which light inhibits melatonin secretion would be a light-dependent decrease in cAMP. In the rat pineal gland, a decrease in cAMP induces AANAT dephosphorylation and dissociation from the 14-3-3, which eventually results in AANAT protein degradation through the proteasome (Klein *et al.*, 1997; Ganguly *et al.*, 2002). A similar situation probably operates in the fish pineal organ for two reasons. First, the phosphorylation sites are highly conserved between the fish and mammalian AANATs, and second, inhibitors of the proteasome prevent the light-induced degradation of AANAT2 protein and AANAT2 activity in trout, pike, and seabream (Falcón *et al.*, 2001; Figure 6.14).

How does light modulate cAMP levels in fish? As a second messenger of phototransduction, cGMP could be the mediator. However, there is no evidence for an involvement of cGMP in the control of cAMP levels or melatonin secretion (Falcón *et al.*, 1990; Thibault *et al.*, 1993a,b). Another candidate is calcium (Ca^{2+}) whose intracellular levels are controlled by light. In trout photoreceptor cells, two pathways allow Ca^{2+} to enter the cell in the dark. One is the cGMP-gated channel (Section 3.1.3) the other is an L-type voltage-gated channel (Bégay *et al.*, 1994a; Meissl *et al.*, 1996). Voltage-clamp studies showed the channel is open in the dark-adapted state, that is, when the photoreceptor cells are depolarized and the cyclic nucleotide-dependent channels are opened (Bégay *et al.*, 1994a) (Figure 6.14). Activation of this L-type channel at night induces rapid surges in $[Ca^{2+}]_i$ (Kroeber *et al.*, 2000). Conversely, illumination results in inactivation of the cGMP-gated channels, cell hyperpolarization, and eventually closure of the L-type voltage-gated channels (Figure 6.14). In trout pineal cells, the nocturnal

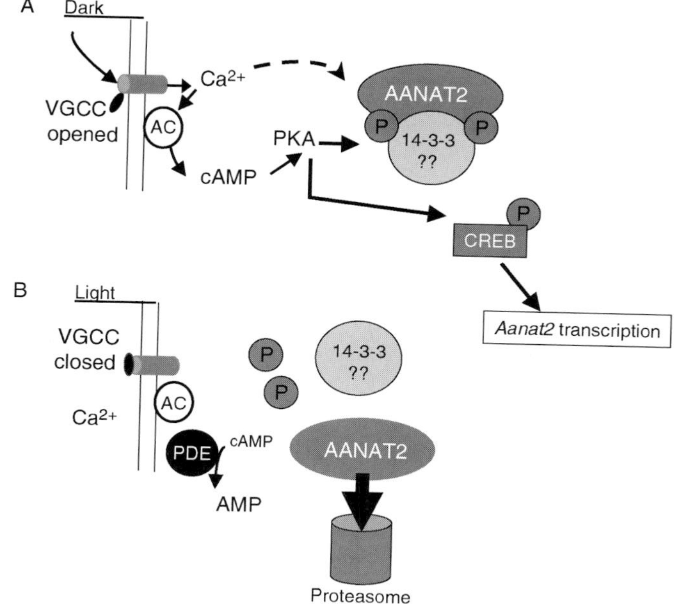

Fig. 6.14. Possible mechanisms involved in the light/dark control of melatonin production. In the dark, the photoreceptor cell is depolarized. This allows entry of Ca^{2+} through voltage-gated calcium channels (VGCC). Ca^{2+} acts through the adenylyl cyclase (AC) to increase cAMP. Phosphorylation of AANAT2, through activation of protein kinase A (PKA), allows association to 14-3-3 protein and activation. PKA might also act through the cAMP response element-binding protein (CREB) to induce AANAT2 transcription. In the light, the photoreceptor cell hyperpolarizes thus inducing closure of the VGCC. Intracellular Ca^{2+} and cAMP decrease eventually resulting in AANAT2 dephosphorylation and degradation after dissociation from 14-3-3. See text for further details and references.

entry of Ca^{2+} contributes to increase melatonin secretion, and this is mediated through Ca^{2+}-binding proteins (Bégay *et al.*, 1994a,b). Blockers of the Ca^{2+}-binding proteins inhibit both melatonin secretion and cAMP accumulation in the dark with, however, faster and more pronounced effects on the former than on the latter. This would indicate that $[Ca^{2+}]_i$ acts through two pathways to stimulate the nocturnal rise in melatonin secretion. One would be independent of cAMP and the other would involve cAMP, perhaps via activation of the PKC/adenylyl cyclase pathway, as reported in mammals (see in an earlier section). Another possible mechanism could be a calmodulin-mediated stimulation of Ca^{2+} on adenylyl cyclase. A Ca^{2+} calmodulin-dependent adenylyl cyclase is highly expressed in retinal photoreceptors (Xia *et al.*, 1993). The light-induced decrease in $[Ca^{2+}]_i$, following photoreceptor hyperpolarization, might contribute to lower AANAT2 activity through inhibition of cAMP synthesis (Falcón *et al.*, 1992b).

5.2. Regulation of *AANAT2* Gene

In addition to acting on the AANAT2 protein, cAMP might also act through the *Aanat2* gene, as reported in rodents (Ganguly *et al.*, 2002). In the rat pineal gland, the nocturnal increase in cAMP is followed by a large increase in AANAT transcripts abundance. The cAMP-dependent activation of PKA phosphorylates the cyclic AMP-response element (CRE)-binding protein (CREB) protein; in turn, pCREB binds to CRE of a number of genes, including *Aanat*. The degree of enhancement depends on the abundance of inducible cAMP early repressor (ICER), an inhibitory transcription factor. Dephosphorylation of CREB would contribute to turn off the system during the second half of the night. In trout, antibodies directed against CREB or pCREB detect one or two band of proteins in pineal extracts, respectively (Kroeber *et al.*, 2000). Phospho-CREB accumulates in the nuclei of dark-adapted photoreceptor cells in the presence of agents that induce cAMP accumulation. Under an LD cycle, a slight drop in the pCREB/CREB ratio is detected at the end of the L phase. However, tAANAT2 mRNA abundance remains constant throughout the LD cycle, and transcription inhibitors do not affect tAANAT2 activity (Bégay *et al.*, 1998; Falcón *et al.*, 1998). Thus, it is unlikely that the cAMP-induced accumulation of pCREB at night stimulates *Aanat2* transcription in trout, unless the rate of transcription is compensated for by a corresponding mRNA degradation. A different situation occurs in pineal organs showing a transcriptional regulation of *Aanat2*. In pike, transcription inhibitors prevent the nocturnal rise in AANAT2 activity, whether the cAMP pathway has been prestimulated or not. This, and the observation that CREB responsive elements are present in the pike and zebrafish *Aanat2* promoters, supports the view that the cAMP pathway is involved in the transcriptional regulation of the gene in fish (Figure 6.14).

5.3. Conclusions

The available data suggest that the dark-induced photoreceptor depolarization favors intracellular increases of Ca^{2+} and cAMP; and both participate in the nocturnal increase in melatonin secretion through mechanisms involving *de novo* AANAT2 protein synthesis (Figure 6.14). It is believed that PKA phosphorylation or AANAT2 protects the enzyme from degradation. Light-induced depolarization of the photoreceptor induces a reduction in intracellular Ca^{2+} and cAMP, and thus dephosphorylation of AANAT2; eventually, AANAT2 follows the proteasome pathway, resulting in the suppression of melatonin secretion (Figure 6.14). This "simple" scheme probably applies to all fish species investigated, and can be summarized as follows: AANAT2 increases in the dark because it is inhibited by light during day. This is the only mechanism operating in salmonids. But in a majority of species, other pathways involving intrinsic pineal pacemakers also contribute to regulating the rhythm in melatonin secretion.

6. PHOTOPERIODIC VERSUS CIRCADIAN CONTROL MELATONIN PRODUCTION

6.1. Clock or No Clock

In trout, salmon, and rabbitfish, melatonin responses to variations in environmental light are of an on-and-off type (Gern and Greenhouse, 1988; Thibault *et al.*, 1993a; Iigo *et al.*, 1997; Rahman *et al.*, 2004). Thus, *in vivo* (rabbitfish) or *in vitro* (salmonids), AANAT2 activity and/or melatonin release remain low under constant light (LL), and DD (Table 6.1). In addition, sudden changes in ambient illumination induce immediate changes in enzyme activity and hormone release in trout. This mechanism is facilitated because AANAT mRNA levels remain constant, making translation possible at any time of the LD cycle (Bégay *et al.*, 1998). The pineal organ of salmonids and rabbitfish are simple light sensors with regard to the production of melatonin, but they appear as exceptions among the fish species studies. In the great majority of fish, the response to the alternation of light and dark is more than passive (Falcón, 1999):

1. In contrast to the abrupt inhibitory effects of light at night, unexpected dark at midnight does not necessarily induce a rise in AANAT2 activity and melatonin release, as in the case in salmonids (Falcón *et al.*, 1987, 1989).
2. Under LD conditions, both AANAT2 activity and melatonin production often increase slightly before lights off, and the decreases anticipate the lights on.
3. Under DD, AANAT2 activity and melatonin production remain rhythmic for several 24-h cycles (Figure 6.10; Table 6.1). This is not the case under LL because light inhibits AANAT2 activity (Falcón *et al.*, 1987, 1989; Kezuka *et al.*, 1988; Iigo *et al.*, 1991, 2003, 2004; Bolliet *et al.*, 1994, 1996, 1997; Cahill, 1997; Okimoto and Stetson, 1999a,b; Roberts *et al.*, 2003; Bayarri *et al.*, 2004b).

4. Both, under DD and LL, *Tpoh* and/or *Aanat2* gene expressions remain rhythmic (Figure 6.12) (Bégay *et al.*, 1998; Coon *et al.*, 1998, 1999; Gothilf *et al.*, 1999).

Altogether it appears that *in vivo* and/or *in vitro* the rhythms of gene expression, enzyme activity, and melatonin release free run under constant conditions, with a period either longer or shorter than 24 h, depending on the species. They are driven by a circadian clock (or oscillator) located within the pineal organ (Falcón, 1999). Studies in the pike have shown that each photoreceptor cell contains a circadian mechanism, so that one pineal organ is made of multioscillatory cellular units (Bolliet *et al.*, 1997). Photoreceptor cells, obtained from dissociated pineal organs and cultured either alone or together, maintain a rhythmic secretion of melatonin under LD and DD for up to a week (Figure 6.10). Addition of interstitial cells to the culture improves the overall amount of melatonin secreted and the amplitude of the rhythm but does not change its period. The shape of the oscillations depends also very much on the seeding density in the wells, in such a way that the endogenous melatonin rhythm dampens faster at low or high cell densities, than at intermediate densities (Bolliet *et al.*, 1997). The damping observed at low densities could reflect uncoupling between the cellular oscillators because cell-to-cell communication is not favored under these conditions (see Section 2.2.3). At high cellular densities the damping occurs toward high melatonin values. In this case, cells aggregate, and cell-to-cell communication is favored; it may be that the over accumulation of some agent in the extracellular spaces uncouples the oscillators, or disrupts the oscillations.

6.2. The Circadian Clock and the Melatonin Output

What are the molecular mechanisms of the circadian clocks in fish? To date, most of our knowledge derives from studies in the suprachiasmatic nuclei (SCN) of the mammalian hypothalamus and the zebrafish. In mammals, the generation of the circadian oscillations results from a transcriptional autoregulated feedback loop involving a set of genes (Figure 6.15B) (Korf *et al.*, 2003; Sato *et al.*, 2004). The transcription factors BMAL1 and CLOCK form heterodimers that activate transcription of several genes through specific 'CTCGTG' E-box enhancers in their promoters. These include three *Period* (*Per1–3*), two *Cryptochrome* (*Cry1*, *2*) genes, two genes related to the retinoic acid orphan receptor family, *Rev*-Erb-α, and *Rorα*, and other clock-controlled genes. On the one hand, phosphorylation of the PER proteins by members of the casein kinase I (CKI) family allow formation of CKI–PER–CRY complexes that enter the cell nucleus and inhibit transcription—including their own—after binding to the BMAL1–CLOCK

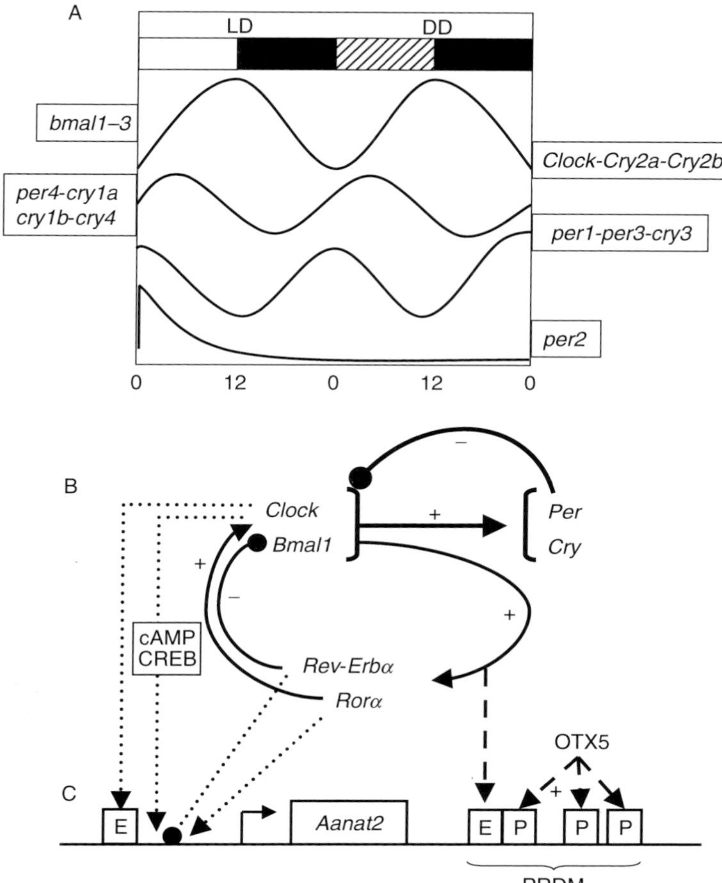

Fig. 6.15. Clock genes and *Aanat2* gene regulation in fish. (A) Phase relationship of clock genes variations in zebrafish under LD and DD; modified from Cahill, 2002. (B) The circadian loop (filled lines), for details see text (adapted from Sato *et al.*, 2004). (C) Induction of *Aanat2* transcription by the clock machinery (interrupted lines) results from the concomitant action of CLOCK/BMAL on one E-box (E), and of OTX5 on three photoreceptor-conserved elements (P), in the 3′ untranslated region of the *Aanat2* gene, the photoreceptor restrictive downstream modulator (PRDM) (adapted from Appelbaum *et al.*, 2005). Other possible pathways are indicated by the dotted lines (see text for details).

complexes. On the other hand, REV-ERBα and RORα compete for the same regulatory elements in the *Bmal1* promoter to inhibit or transactivate transcription, respectively. This mechanism operates in about 24 h. Four *Per*, six *Cry*, three *Clock*, and three *Bmal* genes have been cloned in zebrafish

(Cahill, 2002; Hirayama *et al.*, 2003; Vallone *et al.*, 2004) (Figure 6.15A). The clock genes are expressed throughout the animal and in an embryonic cell line (Vallone *et al.*, 2004). The available data indicate *Clock1*, *Bmal1*, and *Bmal2* mRNAs display peak abundance at the end of day and beginning of night, and these rhythms are maintained under DD (Whitmore *et al.*, 1998; Cermakian *et al.*, 2000). Apparently, all possible zCLOCKs–zBMALs combinations activate transcription from E-box containing promoters, but to a different extent (Hirayama *et al.*, 2003). The *zCry* genes constitute two groups. One group (*Cry1a*, *Cry1b*, *Cry2a*, *Cry2b*) has a high sequence similarity to mammalian *Cry* genes and inhibits CLOCK: BMAL1 mediated transcription; the other group (*Cry3*, *Cry4*) resembles the *Drosophila Cry* gene and does not carry transcription inhibitor activity (Kobayashi *et al.*, 2000). These *Cry* genes also display circadian rhythmicity with, however, different patterns. It has been shown that zCRY1a and zPER2 regulate the subcellular distribution zCLOCK–zBMAL complexes in opposite directions in NIH3T3 cells (Hirayama *et al.*, 2003); zCRY1a would inhibit transcription by promoting retention of, and interacting with, zCLOCK–zBMAL in the nucleus. In contrast, zPER2 would maintain the heterodimer in the cytoplasm, thereby preventing the zCLOCK–zBMAL activity. In the embryonic zebrafish cell line, light induces transient expression of *zPer2* (Figure 6.15) through activation of the mitogen-activated protein kinase (MAPK) pathway (Cermakian *et al.*, 2002), while it represses *Per4* (Vallone *et al.*, 2004). In brief, the clock mechanisms seem very similar in fish and mammals (Cahill, 2002). However, it is yet impossible to determine if all the clock genes operate in the same circadian cell unit or if there is a tissue-specific expression of the different homologs cloned (as is the case for the AANATs for example).

Two major questions arise: (1) how is the clock machinery linked to the circadian control of *Aanat2* expression? and (2) how does light reset the circadian clock and the circadian oscillations in melatonin secretion? Regarding the first question, there is indication suggesting that transcription factors of the circadian clock might control directly *Aanat2* gene expression, that is, the *Aanat2* gene is a direct output of the clock (Figure 6.15C). The *Aanat2* promoter contains several consensus-binding sites for the CLOCK-BMAL heterodimers, and these are known to enhance *Aanat* transcription in vertebrates through E-box responsive elements (Chen and Baler, 2000; Chong *et al.*, 2000; Appelbaum *et al.*, 2004). REV-ERBα/RORα responsive elements are also present in fish *Aanat2* promoter. However, their functional involvement remains to be demonstrated, and the *Rev*-Erbα and *Ror*α are in the process of being cloned. Only one *Rev*-Erbα sequence is available to date, and the gene is not expressed in adult zebrafish pineal organ (Delaunay *et al.*, 2000). Studies in the zebrafish suggest the synchronization of the *Aanat2* gene expression rhythm to the prevailing LD cycle most probably involves the morning induction of PER2 and the evening induction of the CLOCK/BMAL heterodimer. First, *zPer2* is the only clock gene whose expression remains constant under DD; light induces a transient *zPer2* expression in the zebrafish pineal organ at any time of the circadian cycle under DD; and morpholino-induced knockdown of *zPer2* prevents the nocturnal rise of *zAanat2* (Ziv *et al.*, 2005). Second, *zfAanat2* expression is enhanced as a result of the synergistic action of the rhythmic clock heterodimer BMAL/CLOCK and the photoreceptor-specific homeogene OTX5; the dimer binds one E-box element and the OTX5 binds three photoreceptor-conserved elements (PCE), all four located in the 3′-end of the *zAanat2*, the so-called photoreceptor restrictive downstream modulator (PRDM; Appelbaum *et al.*, 2005) (Figure 6.15C).

In this context, it may be suggested that the light-induced expression of *zPer2* determines the phase of the oscillations. In zebrafish embryos, light pulses of different durations induce the same phase shift in *zAanat2* expression rhythm if they are applied at the same circadian time (Vuilleumier *et al.*, unpublished data). Conversely, under LL, dark pulses of different durations induce different phase shifts, even if they are applied at the same circadian time. This indicates the phase of the rhythm is determined by the time at which the dark-to-light transition occurs, that is, when *zPer2* is induced. It remains to know why light pulses applied early or late during the subjective night induce phase delays or phase advances in the melatonin secretion rhythm, respectively, whereas light pulses applied early during the subjective day have no effect (Falcón, 1999; Iigo *et al.*, 2003; Masuda *et al.*, 2003b). The final effect most probably results from the "clock-genes context" at the moment the pulse is applied.

If there is strong indication that the *Aanat2* gene is a target for clock genes, there are other pathways through which the clock may modulate its expression. One of these might be the cAMP/PKA/CREB pathway (Section 5.2.) because in pike, pineal cAMP content and extrusion are under circadian control. This suggests the adenylyl cyclase and/or the cAMP phosphodiesterase are under circadian control. In chicken pineal photoreceptor cells, a causal link has been established between the circadian variations in cAMP and AANAT activity (Ivanova and Iuvone, 2003) (Figure 6.15). In view of the broad range of functions under cAMP control, the clock may also regulate other metabolic pathways in photoreceptor cells.

6.3. Conclusions

In most of the fish species investigated, circadian clocks control melatonin production. These clocks are located in the photoreceptor cells, which also contain the input to the clock—the photoreceptive unit—and one output of the clock—the melatonin-producing unit. The clock allows the rhythm in *Aanat2* gene expression to persist under constant conditions (DD or LL) with a period that approximates 24 h. Light acts on the clock to reset its phase, which results in resetting the rhythm in *Aanat2* expression. Second messengers such as Ca^{2+} and cAMP might also participate in the circadian control of AANAT2. There is indication that cGMP metabolism is also controlled by the circadian clock because under DD, pineal cGMP content and efflux are higher during the subjective night than during the subjective day in pike (Falcón and Gaildrat, 1997). This would suggest that the circadian clock modulates sensitivity to light along the 24-h cycle, as is the case in the retina (Ren and Li, 2004). However, a direct link between cGMP and the circadian control of melatonin secretion has never been established.

7. NONPHOTIC REGULATION OF PINEAL ORGAN OUTPUT SIGNALS

7.1. External Factors

The LD cycle is one of the many parameters that influence the physiology and behavior of vertebrates. Fish are ectotherm vertebrates and as such are directly influenced by the external temperature. Like the external lighting environment, temperature fluctuates on a daily and seasonal basis, and this is also reflected at the level of the melatonin signal produced by the pineal photoreceptors. The effects of temperature vary from a species to another. Temperature acts directly on the pineal organ to modulate melatonin secretion, and this seems to be mediated through AANAT2 activity (Thibault *et al.*, 1993a,b). In cultured pike and trout pineal organs, temperature response curves of AANAT2 activity are bell shaped, and there is a good correlation between the peak of AANAT2 response to temperature and the fish best physiological temperature (Figure 6.16) (12–15 °C in trout; 18–25 °C in pike; and 27 °C in seabream). The response to temperature seems to be an intrinsic property of the enzyme protein itself because the activity measured from pineal homogenates or from the recombinant enzymes display the same response curve as those obtained from cultured organs (Table 6.2) (Falcón *et al.*, 1996; Coon *et al.*, 1999; Benyassi *et al.*, 2000; Zilberman-Peled *et al.*, 2004).

The consequence is that the pineal organ integrates information from both temperature and photoperiod cycles (Zachmann *et al.*, 1992c; Falcón *et al.*, 1994). Under DD, temperature cycles alone are able to synchronize a rhythm in melatonin secretion *in vitro* (Figure 6.16). The amplitude of the

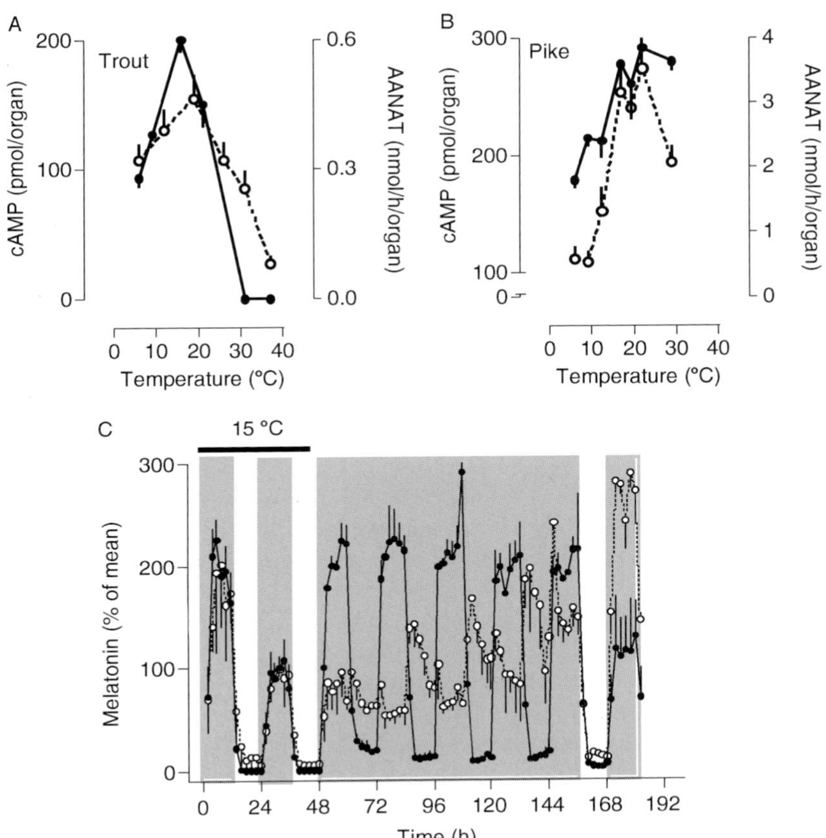

Fig. 6.16. Temperature, AANAT2 activity, and melatonin secretion. (A and B) Effects of temperature on cAMP accumulation and AANAT2 activity of pineal organs of trout and pike in culture. Peak values are the same for both parameters in each species, and differ from a species to another; see also Table 6.2. (C) Effects of temperature cycles on melatonin secretion by superfused pike pineal organs. After 48 h of culture under LD, organs were released under DD and a 10°C amplitude (10/20°C) temperature cycle. Filled circles/lines: the warm temperature was applied during the subjective dark phase resulting in the synchronization of a high-amplitude rhythm in melatonin secretion. Open circles/interrupted lines: the cold temperature was applied during the subjective dark phase; the rhythm first disappears and then tends to reappear, but peaks are during the subjective light. See text for more details. (Adapted from Bolliet *et al.*, 1996.)

oscillations will depend on the amplitude of the imposed thermocycles. However, in case a circadian oscillator drives the melatonin rhythm, temperature cycles are unable to entrain the circadian clocks, and temperature

Table 6.2
Comparison of the Kinetics of AANAT1 (Retina) and AANAT2 (Pineal Organ) in Pike (p) and Trout (t)

	Retina	Pineal organ
Linearity (prot. amount)	Yes	Yes
pH and molarity	6–0.2 mM	6–0.2 mM
Temperature (optimum)	37° (p), 25° (t)	20° (p), 12° (t)
Km AcCoA	0.1 mM	0.1 mM
Km tryptamine	0.02/0.08 mM	7/10 mM
Ki tryptamine	1 mM	Not inhibited
Km PEA	9.8 mM	Not acetylated

Activities were measured in organ homogenates or from the recombinant proteins. Data are from Bolliet *et al.*, 1996, Coon *et al.*, 1999, and Benyassi *et al.*, 2000. PEA: phenylethylamine.

pulses are unable to shift the phase of the clocks (Falcón *et al.*, 1994). Also, the period of the circadian clocks that drive the oscillations in melatonin secretion is insensitive to temperature—a phenomenon known as temperature compensation, but it may happen that the melatonin rhythm is no longer detectable at low temperatures (Bolliet *et al.*, 1994).

Altogether, it appears that the melatonin signal reflects both the prevailing photoperiod and environmental temperature. The latter affects mainly the amplitude of the rhythm. This has tremendous consequences on the shape of the melatonin oscillations along the seasons, more particularly in temperate and artic regions. It is believed that the seasonal variations in the amplitude of the plasma melatonin rhythm (Section 4.2) would result mainly from changes in temperature, while photoperiod changes would affect mainly the duration of the nocturnal surge. This may also be important during the daily cycle. The daily variations in temperature are not as dramatic as the annual variations are, but fish exhibit daily variations in depth/thermal preferences; and the pineal organ has been involved in this process (Zachmann *et al.*, 1992a,c).

7.2. Internal Factors

Although not extensive and exhaustive, some studies report that hormones and neurotransmitters may modulate pineal melatonin secretion.

7.2.1. Melatonin

2-iodo-melatonin inhibits melatonin secretion by trout pineal organs *in vitro*, indicating the hormone modulates its own production through a paracrine or autocrine feedback signal (Yañez and Meissl, 1995). This explains why melatonin production by trout and pike pineal photoreceptor cells in culture related with the frequency of medium renewal or seeding density in the dishes (Bégay *et al.*, 1992; Bolliet *et al.*, 1997). This might impact on the melatonin secretion rhythm. Indeed, in cultured pike pineal organs or cells, the shape—but not the period—of the circadian oscillations depends very much on the size of the organs or cell density (Bolliet *et al.*, 1997; Section 6.1). To date, melatonin receptors have not been identified in the fish pineal, and the effects might result from an interference with some metabolic pathway (e.g., inhibition of HIOMT activity). Melatonin has also been shown to directly inhibit the impulse frequency of the pineal neurons through yet unidentified mechanisms (Meissl *et al.*, 1990). It happens thus that one of the two output messages of the pineal organ can modulate the production of the two.

7.2.2. Neuromodulators

Adenosine is a ubiquitous neuromodulator that acts on an autocrine/paracrine mode. Interest in adenosine rose from the observation that it is produced locally by the fish pineal organ (Falcón *et al.*, 1988). There are extra- and intracellular potential sources: the first is the catabolism of extracellular ATP or extruded cAMP, via ecto-nucleotidases and 5′-nucleotidases; the second is the degradation of both cAMP and *S*-adenosyl-methionine (SAM); SAM is the cosubstrate of methyltransferases including HIOMT, the last enzyme in the melatonin pathway. Adenosine may cross the plasma membrane through specific transporters. In trout and pike pineal organ, there is indication that adenosine produced locally modulates AANAT2 activity and melatonin release through a cAMP-dependent mechanism involving cell surface adenosine receptors of high (A_1) and low (A_2) affinity (Falcón *et al.*, 1991, 1992b). Both are coupled to the adenylyl cyclase to respectively inhibit or activate it. The control by adenosine offers a fine-tuning of cAMP levels, depending on the balance between the intra- and extracellular concentrations of the nucleoside. This balance might vary during the LD cycle, providing the pineal with a continuous feedback modulation of melatonin secretion.

7.2.3. Neurotransmitters

As mentioned in Section 5.1, NE is the neurotransmitter that triggers the nocturnal rise in AANAT activity and melatonin release in the nonphotosensitive mammalian pineal gland (Klein *et al.*, 1981, 1997). The question rose to know whether this control already operates in fish. The answer is that this is species specific. In zebrafish and trout pineal organs, epinephrine and

NE have no effect on AANAT2 activity or melatonin release, whereas a slight inhibitory response to dopamine was reported in the former (Martin *et al.*, 1991; Martin and Meissl, 1992; Cahill, 1997). In pike, only NE has an effect. Low concentrations (10^{-8} M) further increase nocturnal AANAT2 activity, whereas high concentrations (10^{-6} M) are inhibitory (Falcón *et al.*, 1991). Pharmacological studies indicated that stimulation would be mediated by β-adrenergic receptors, and inhibition would be mediated by α-adrenergic receptors (α_1 and/or α_2). In the pike pineal organ, as in other fish tissues, the pharmacological tools available do not allow distinguishing between these two subtypes; and it has been suggested that the mammalian classification may simply not apply to fish (Johansson, 1984; Fabbri *et al.*, 1998). Fish are still waiting for a molecular characterization of these receptors. It is worthwhile mentioning that among vertebrates the α_2 receptor subtype is found in the avian pineal, whereas the β_1/α_1 receptor subtypes are found in the mammalian pineal. Fish would be located at a crossroads of the evolutionary process that has seen the progressive replacement of light by catecholamines in the control of melatonin secretion. In fish, however, very few nerve fibers innervate the pineal and catecholamines are thought to originate from the circulation (Fabbri *et al.*, 1998).

Although without effect on the melatonin biosynthesis, acetylcholine, dopamine, and NE all stimulate the firing rate of the second-order neurons in trout pineal organ (Martin and Meissl, 1992; Brandstätter *et al.*, 1995; Brandstätter and Hermann, 1996). The response increases with the applied light intensity, resulting in an attenuation of the ganglion cell response to light (Section 3.2.2). The effects of acetylcholine involve nicotinic and muscarinic receptors, and those of dopamine involve D_1 and D_2 receptors (Samejima *et al.*, 1994). Catecholamines might act, at least in part, via a cAMP-dependent mechanism (Martin *et al.*, 1991). The cooperative activation of the β-adrenergic and D_1 dopamine receptors on the one hand and D_2 dopamine receptors on the other, might underlay the increases in spike firing. Such a mechanism operates in *nucleus accumbens* neurons, where G-protein $\beta\gamma$ subunits released from the $G_{i/o}$-linked D_2 receptor act in combination with G_α subunits released from the D_1 receptor activation to stimulate specific adenylyl cyclases (Hopf *et al.*, 2003).

The last neurotransmitter that appears to modulate both the melatonin and the nervous signal is GABA (Ekström and Meissl, 1997). Two populations of intrapineal neurons produce GABA in trout (Ekström *et al.*, 1987a): one is located in the rostral pineal end-vesicle, and presumably constitutes a population of interneurons; the other is found in the pineal stalk, and their axons are sent to the brain. GABA is also present in glial cells in the pineal end-vesicle. Electrophysiological recordings indicated that GABA (and agonists) may inhibit or increase the spontaneous discharges of the pinealofugal neurons; bidirectional effects may also be observed in some cases, depending on the state of light or dark adaptation (Meissl and Ekström, 1991). These effects are mediated through $GABA_A$ receptors. GABA also reduces melatonin secretion through $GABA_A$ receptors (Meissl *et al.*, 1993). Benzodiazepines[6] have opposite effects. They increase melatonin production in the mesopic and partly in the photopic range of illumination, without showing clear effects in the dark-adapted organ (Meissl *et al.*, 1994).

7.2.4. Steroids

Sexual steroids and glucocorticoids modulate melatonin secretion by trout pineal photoreceptor cells (Bégay *et al.*, 1993, 1994c; Benyassi *et al.*, 2001). Specific probes detected mRNA corresponding to the 17-β-estradiol receptor in extracts from trout pineal organ and retina. Also, specific antibodies directed against trout glucocorticoid receptor labeled a single band of proteins in extracts from these organs. In isolated and cultured pineal cells, physiological concentrations of 17-β-estradiol inhibit the nocturnal rise in melatonin secretion in a dose-dependent manner (Bégay *et al.*, 1993, 1994b). Melatonin has been involved in the control of reproduction in fish (Zachmann *et al.*, 1992a; Boeuf and Falcón, 2002). These results would indicate the existence of a feedback loop mechanism involving sexual hormones. Over days of continuous treatment, the effects of 17-β-estradiol become stimulatory, and cells challenged with the steroid recover faster a high amplitude rhythm in melatonin release than the controls. This suggests that the steroid may have trophic effects, perhaps by promoting the synthesis of some growth factor by some pineal cells.

Glucocorticoids also inhibit melatonin secretion by trout pineal cells, whether they are treated for 6 h or up to 36 h in culture (Benyassi *et al.*, 2001). These receptor-mediated effects result in a 50% reduction in AANAT2 activity (HIOMT activity is not affected). Both the pineal organ and the corticoids have been involved in the control of reproduction, electrolyte and glucose balance, and the response to stress; the effects of glucocorticoids on melatonin secretion may be part of a loop in the regulation of these processes.

There is indication that the steroids might act directly on the *Aanat2* gene because its promoter contains several putative-binding sequences for both 17-β-estradiol and glucocorticoid receptors.

[6]Modulators of the $GABA_A$ receptors.

7.2.5. Peptides

Fibers from central origin have been shown to innervate the fish pineal organ. These include FMRF-amide, neuropeptide Y, substance P, GnRH, and GHRH, but there is no indication that these peptides are involved in the control of melatonin secretion (Ekström *et al.*, 1988; Ekström and Meissl, 1997; Miranda *et al.*, 2002). The atrial natriuretic peptide (ANP) modulates cyclic nucleotide levels in trout pineal organ (Falcón *et al.*, 1990). ANP induces large increases in cGMP and a faint increase in cAMP. The effects are however not correlated to melatonin secretion. The observation that the ANP-stimulated increase in cGMP accumulation is sensitive to light suggests the photoreceptor cells are the targets of the peptide. It would prove whether the ANP challenge affects the photoreceptor membrane potential and/or the neuronal firing rate.

7.3. Conclusions

It is obvious, from the above-mentioned data, that the photoreceptors of the fish pineal organ are multieffector cells, that is, they integrate information from different origin. The nervous output of the organ is modulated by other factors than the excitatory neurotransmitter released by the photoreceptor cells. The rhythm in melatonin secretion is synchronized by the alternation of light and dark, but other factors from external (temperature) or internal (neurotransmitters) origin may modulate the shape of the oscillations. It must be emphasized that only a very limited number of species have been investigated and more information is needed before a more exhaustive and clear-cut picture arises.

8. CONCLUSIONS AND PERSPECTIVES

Fish have two photoreceptive organs: the retina and the pineal organ, which share many common properties. The structural characteristics and functional properties of the pineal photoreceptors classify them into the cone family of photoreceptor cells, as found in the retina (Falcón, 1999; Klein, 2004). Many of the properties established for the retinal cones have been or are being extended to the pineal photoreceptors, as is the case for the phototransduction mechanism. The opposite also holds true, and the demonstration that the retinal photoreceptors make melatonin came after this was evidenced in the pineal (although it remains yet to determine which of the retinal rod and/or cones make melatonin). The pineal and retinal analogies have been largely emphasized in the literature (Collin and Oksche, 1981; O'Brien and Klein, 1986; Falcón, 1999; Klein, 2004). Actually, the more we know on the structural and functional genes express in each of these organs, the more it appears they are similar but distinct. This is the case for genes encoding proteins of the phototransduction and of the melatonin biosynthesis pathway. The genes encoding AANAT give a remarkable example. The presence of several AANATs in fish is a unique situation among vertebrates. It has been suggested that whole genome duplication occurred close to the origin of teleost fish (Amores *et al.*, 1998; Hoegg *et al.*, 2004; Jaillon *et al.*, 2004). The observation that other vertebrates possess only one AANAT (related to fish retinal AANAT1) would suggest that this whole genome duplication, which occurred ~230 million years ago, is responsible for the appearance of AANAT2 in teleost fish. The presence of two AANAT1s in some but not all teleost would result from another genome duplication, which occurred later during evolution of these fish (Coon and Klein, 2006). Alternatively, it might be that the appearance of two AANATs was concomitant to the appearance of the lateral eyes on the one hand and the pineal complex on the other. A more complete picture should arise after a search for AANAT genes will be performed in species at the base of chordates evolution, including amphioxus, lampreys, and chondrichtians. It is noteworthy that amphioxus and hagfish do not possess a pineal complex, whereas lampreys and other vertebrates do possess one. At least two whole genome duplications occurred between prechordates and early vertebrates (Taylor and Raes, 2004). It is believed that this led to key innovations such as acquisition of a new head and paired photosensory organs.

The search for AANAT in early chordates will also help determining when first appeared the melatonin signal. It has been suggested that AANAT and HIOMT genes might have been acquired through horizontal gene transfer from bacteria to chordates (Iyer *et al.*, 2004). The genome of the urochordate *Ciona* does not contain a gene for AANAT, indicating transfer occurred latter. According to Klein (2004), acquisition of AANAT provided considerable advantage by improving photosensitivity and protecting photoreceptors from light insults. Indeed, (aryl)alkylamine compounds, such as serotonin, can form adducts with retinaldehyde; acetylation of indole compounds by AANAT would represent a gain of function by preventing formation of such toxic adducts and make retinaldehyde more available for the phototransduction process.

The existence of two photoreceptive structures, the eye and the pineal organ, on the one hand and of two types of messages (nervous and hormonal) on the other raises the question of the redundancy of the outputs and beyond, of the need for two distinct photoreceptive structures. For example, the retinal and pineal nervous projections often overlap in the brain. Also, while providing information on rapid illumination changes, the pineal nervous message may also be an indicator of day length, which is what pineal melatonin does through the blood. This redundancy was solved during vertebrates' evolution, which led to an oversimplification, best evidenced by the dramatic morphological and functional changes undergone by the pineal cells. The mammalian pinealocyte has no phototransduction and circadian clock properties, and does not produce a nervous output. In other words, this cell is specialized in melatonin production. The melatonin surge is still nocturnal but under control by the eyes and hypothalamic circadian clocks located in the SCN (Collin and Oksche, 1981; Klein *et al.*, 1981; Korf *et al.*, 2003; Klein, 2004). Intermediate situations exist in lizards and birds. The possible mechanisms at the base of this evolutionary trend are not fully understood (for extensive discussion see Collin and Oksche, 1981; Ekström and Meissl, 2003).

The main function of the fish pineal organ is to integrate light information and elaborate messages that will impact on the animal's physiology. The photoreceptor cells occupy a key position because they are at the interface between the environment and the organism providing rapid (time scale of seconds) and less rapid (time scale of several hours) responses to environmental light. The downstream nervous information transmitted to brain centers via the ganglion cells provides the appropriate response to rapid changes. The hormonal melatonin response is slower and rhythmic, and synchronized by the 24-h LD cycle. Fish have developed several strategies to regulate the secretion of melatonin. In the pineal organ they all result in a nocturnal surge. In the general scheme, light acts in two ways: one is to inhibit the enzyme AANAT2 by promoting AANAT2 protein degradation; the other is to drive a rhythmic expression of the *Aanat2* gene. This is generally achieved through intracellular circadian clocks. A single photoreceptor cell thus appears as a full cellular circadian system, but in a few cases no clock is involved (Falcón, 1999).

Pineal photoreceptors are also multieffectors because they integrate other informative stimuli from external (temperature) and internal (hormones, neurotransmitters, and neuromodulators) origin. These factors do not affect the period of the circadian clocks that drive the melatonin rhythm, but they shape the amplitude of the melatonin oscillations. The pineal photoreceptor cannot be viewed as a simple light sensor anymore. Among the crucial questions to be answered is the identification of the pathways linking the phototransduction unit to the circadian clock unit, and these two units to the melatonin-producing machinery. Another interesting question regards the tissue specificity of gene expression. Studies in the zebrafish have shown that selective *Aanat2* pineal expression is determined by specific sequences located in the 5′- and 3′-flanking regions of the gene (Appelbaum *et al.*, 2004). This tissue-specific expression allows using the *Aanat* genes as markers of photoreceptor and circadian clock functions (Gothilf *et al.*, 1999).

This might also be of interest for studies related to photoreceptor differentiation and development. As was the case in the past, comparative studies between the retina and pineal will benefit from each other, keeping in mind that photoreceptor cells of the pineal organ are only of the cone type. The observation that melatonin secretion and AANAT1 activity are higher during day, not night, in the fish retina (Besseau *et al.*, 2005) further enhances the interest in such comparative studies.

ACKNOWLEDGMENTS

Many thanks to Michael Fuentès for his help in the iconography.

ABBREVIATIONS

AANAT, arylalkylamine (serotonin) *N*-acetyltransferase protein; ANP, atrial natriuretic peptide; BMAL, brain muscle ARNT-like protein; cAMP, adenosine cyclic 3′,5′-monophosphate; cGMP, guanosine cyclic 3′,5′-monophosphate; CLOCK, clock protein; CNG, cGMP-gated channel; CRE, cyclic AMP-response element; CREB, cyclic AMP-response element-(CRE)-binding protein; CKI, casein kinase I; CSF, cerebrospinal fluid; DAG, diacyl glycerol; DD, dark/dark (constant darkness); DHA, docosahexaenoïc acid; DNA, desoxyribonucleic acid; EPG, electropinealogram; ER, extraretinal rhodopsin; ERG, electroretinogram; ERP, early receptor potential; GABA, gamma aminobutyric acid; GnRH, gonadotropin releasing hormone; GTP, guanosine triphosphate; HIOMT, hydroxyindole-*O*-methyltransferase; ICER, inducible-cAMP-early-repressor protein; kinase A, cAMP-dependent kinase; LD, light/dark; LL: constant light; MAO, monoamine oxidase; mRNA, messenger RNA; NE, norepinephrine; *NMDA*, *N*-methyl-D-aspartate; PER, period protein; PKA, protein kinase A; PKC, protein kinase C;PUFA, polyunsaturated fatty acids; RPE, retinal pigment epithelium; RNA, ribonucleic acid; RPE, retinal pigment epithelium; SAM, S-adenosylmethionine; SCN, suprachiasmatic nuclei of the hypothalamus; Tα, α subunit of the G-protein transducin; TPOH, tryptophane hydroxylase.

REFERENCES

Amores, A., Force, A., Yan, Y. L., Joly, L., Amemiya, C., Fritz, A., Ho, R. K., Langeland, J., Prince, V., Wang, Y. L., Westerfield, M., Ekker, M., *et al.* (1998). Zebrafish hox clusters and vertebrate genome evolution. *Science* **282**(5394), 1711–1714.

Appelbaum, L., Toyama, R., Dawid, I. B., Klein, D. C., Baler, R., and Gothilf, Y. (2004). Zebrafish serotonin-*N*-acetyltransferase-2 gene regulation: Pineal-restrictive downstream module contains a functional E-box and three photoreceptor conserved elements. *Mol. Endocrinol.* **18,** 1210–1221.

Appelbaum, L., Anzulovich, A., Baler, R., and Gothilf, Y. (2005). Homeobox-clock protein interaction in zebrafish. A shared mechanism for pineal-specific and circadian gene expression. *J. Biol. Chem.* **280**(12), 11544–11551.

Asaoka, Y., Mano, H., Kojima, D., and Fukada, Y. (2002). Pineal expression-promoting element (PIPE), a *cis*-acting element, directs pineal-specific gene expression in zebrafish. *Proc. Natl. Acad. Sci. USA* **99,** 15456–15461.

Ayoub, G. S., and Dorst, K. (1998). Imaging of glutamate release from the goldfish retinal slice. *Vision Res.* **38,** 2909–2912.

Bassi, C. J., and Powers, M. K. (1986). Daily fluctuations in the detectability of dim lights by humans. *Physiol. Behav.* **38,** 871–877.

Baudelot, E. (1883). "Recherches sur le système nerveux des poissons." Masson, Paris.

Bayarri, M. J., Madrid, J. A., and Sanchez-Vazquez, F. J. (2002). Influence of light intensity, spectrum and orientation on sea bass plasma and ocular melatonin. *J. Pineal Res.* **32,** 34–40.

Bayarri, M. J., Munoz-Cueto, J. A., Lopez-Olmeda, J. F., Vera, L. M., Rol de Lama, M. A., Madrid, J. A., and Sanchez-Vazquez, F. J. (2004a). Daily locomotor activity and melatonin rhythms in Senegal sole (*Solea senegalensis*). *Physiol. Behav.* **81,** 577–583.

Bayarri, M. J., Rodriguez, L., Zanuy, S., Madrid, J. A., Sanchez-Vazquez, F. J., Kagawa, H., Okuzawa, K., and Carrillo, M. (2004b). Effect of photoperiod manipulation on the daily rhythms of melatonin and reproductive hormones in caged European sea bass (*Dicentrarchus labrax*). *Gen. Comp. Endocrinol.* **136,** 72–81.

Bazan, N. G., Gordon, W. C., and Rodriguez de Turco, E. B. (1992). Docosahexaenoic acid uptake and metabolism in photoreceptors: Retinal conservation by an efficient retinal pigment epithelial cell-mediated recycling process. *Adv. Exp. Med. Biol.* **318,** 295–306.

Bégay, V., Falcón, J., Thibault, C., Ravault, J. P., and Collin, J. P. (1992). Pineal photoreceptor cells: Photoperiodic control of melatonin production after cell dissociation and culture. *J. Neuroendocrinol.* **4,** 337–345.

Bégay, V., Valotaire, Y., Ravault, J. P., Collin, J. P., and Falcón, J. (1993). Photoreceptor cells of the pineal body in culture: Effect of 17 beta-estradiol on the production of melatonin. *C. R. Seances Soc. Biol. Fil.* **187,** 77–86.

Bégay, V., Bois, P., Collin, J. P., Lenfant, J., and Falcón, J. (1994a). Calcium and melatonin production in dissociated trout pineal photoreceptor cells in culture. *Cell Calcium* **16,** 37–46.

Bégay, V., Collin, J. P., and Falcón, J. (1994b). Calciproteins regulate cyclic AMP content and melatonin secretion in trout pineal photoreceptors. *Neuroreport* **5,** 2019–2022.

Bégay, V., Valotaire, Y., Ravault, J. P., Collin, J. P., and Falcón, J. (1994c). Detection of estrogen receptor mRNA in trout pineal and retina: Estradiol-17 beta modulates melatonin production by cultured pineal photoreceptor cells. *Gen. Comp. Endocrinol.* **93,** 61–69.

Bégay, V., Falcón, J., Cahill, G. M., Klein, D. C., and Coon, S. L. (1998). Transcripts encoding two melatonin synthesis enzymes in the teleost pineal organ: Circadian regulation in pike and zebrafish, but not in trout. *Endocrinology* **139,** 905–912.

Bell, M. V., Batty, R. S., Dick, J. R., Fretwell, K., Navarro, J. C., and Sargent, J. R. (1995). Dietary deficiency of docosahexaenoic acid impairs vision at low light intensities in juvenile herring (*Clupea harengus* L.). *Lipids* **30,** 443–449.

Bellingham, J., Tarttelin, E. E., Foster, R. G., and Wells, D. J. (2003). Structure and evolution of the teleost extraretinal rod-like opsin (errlo) and ocular rod opsin (rho) genes: Is teleost rho a retrogene? *J. Exp. Zool. Part B Mol. Dev. Evol.* **297,** 1–10.

Benyassi, A., Schwartz, C., Coon, S. L., Klein, D. C., and Falcón, J. (2000). Melatonin synthesis: Arylalkylamine *N*-acetyltransferases in trout retina and pineal organ are different. *Neuroreport* **11,** 255–258.

Benyassi, A., Schwartz, C., Ducouret, B., and Falcón, J. (2001). Glucocorticoid receptors and serotonin *N*-acetyltransferase activity in the fish pineal organ. *Neuroreport* **12,** 889–892.

Besseau, L., Benyassi, A., Møller, M., Coon, S. L., Weller, J. L., Boeuf, G., Klein, D. C., and Falcón, J. (2005). Melatonin pathway: Breaking the "High-at-Night" rule in trout retina. *Exp. Eye Res.* **82**(4), 620–627.

Blazynski, C., and Cohen, A. I. (1986). Rapid declines in cyclic GMP of rod outer segments of intact frog photoreceptors after illumination. *J. Biol. Chem.* **261,** 14142–14147.

Boesze-Battaglia, K., and Allen, C. (1998). Differential rhodopsin regeneration in photoreceptor membranes is correlated with variations in membrane properties. *Biosci. Rep.* **18,** 29–38.

Boeuf, G., and Falcón, J. (2002). Photoperiod and growth in fish. *Vie. Et. Milieu.* **51,** 237–246.

Bolliet, V., Ali, M. A., Anctil, M., and Zachmann, A. (1993). Melatonin secretion *in vitro* from the pineal complex of the lamprey *Petromyzon marinus*. *Gen. Comp. Endocrinol.* **89**(1), 101–106.

Bolliet, V., Bégay, V., Ravault, J. P., Ali, M. A., Collin, J. P., and Falcón, J. (1994). Multiple circadian oscillators in the photosensitive pike pineal gland: A study using organ and cell culture. *J. Pineal Res.* **16,** 77–84.

Bolliet, V., Falcón, J., and Ali, M. A. (1995). Regulation of melatonin secretion by light in the isolated pineal organ of the white sucker (*Catostomus commersoni*). *J. Neuroendocrinol.* **7,** 535–542.

Bolliet, V., Ali, M. A., Lapointe, F. J., and Falcón, J. (1996). Rhythmic melatonin secretion in different teleost species: An *in vitro* study. *J. Comp. Physiol. B.* **165,** 677–683.

Bolliet, V., Begay, V., Taragnat, C., Ravault, J. P., Collin, J. P., and Falcón, J. (1997). Photoreceptor cells of the pike pineal organ as cellular circadian oscillators. *Eur. J. Neurosci.* **9,** 643–653.

Borjigin, J., Wang, M. M., and Snyder, S. H. (1995). Diurnal variation in mRNA encoding serotonin *N*-acetyltransferase in pineal gland. *Nature* **378,** 783–785.

Bowmaker, J. K., and Wagner, H. J. (2004). Pineal organs of deep-sea fish: Photopigments and structure. *J. Exp. Biol.* **207,** 2379–2387.

Brandstätter, R., Fait, E., and Hermann, A. (1995). Acetylcholine modulates ganglion cell activity in the trout pineal organ. *Neuroreport* **6,** 1553–1556.

Brandstätter, R., and Hermann, A. (1996). Gamma-aminobutyric acid enhances the light response of ganglion cells in the trout pineal organ. *Neurosci. Lett.* **210,** 173–176.

Brown, M. F. (1994). Modulation of rhodopsin function by properties of the membrane bilayer. *Chem. Phys. Lipids* **73,** 159–180.

Cahill, G. M. (1996). Circadian regulation of melatonin production in cultured zebrafish pineal and retina. *Brain Res.* **708**(1–2), 177–181.

Cahill, G. M. (1997). Circadian melatonin rhythms in cultured zebrafish pineals are not affected by catecholamine receptor agonists. *Gen. Comp. Endocrinol.* **105,** 270–275.

Cahill, G. M. (2002). Clock mechanisms in zebrafish. *Cell Tissue Res.* **309,** 27–34.

Cahill, G. M., and Besharse, J. C. (1989). Retinal melatonin is metabolized within the eye of *Xenopus laevis*. *Proc. Natl. Acad. Sci. USA* **86,** 1098–1102.

Cahill, G. M., and Besharse, J. C. (1995). Circadian rhythmicity in vertebrate retinas: Regulation by photoreceptor oscillations. *Prog. Ret. Eye Res.* **14,** 267–291.

Cermakian, N., Whitmore, D., Foulkes, N. S., and Sassone-Corsi, P. (2000). Asynchronous oscillations of two zebrafish CLOCK partners reveal differential clock control and function. *Proc. Natl. Acad. Sci. USA* **97,** 4339–4344.

Cermakian, N., Pando, M. P., Thompson, C. L., Pinchak, A. B., Selby, C. P., Gutierrez, L., Wells, D. E., Cahill, G. M., Sancar, A., and Sassone-Corsi, P. (2002). Light induction of a vertebrate clock gene involves signaling through blue-light receptors and MAP kinases. *Curr. Biol.* **12,** 844–848.

Chen, H., and Anderson, R. E. (1993). Comparison of uptake and incorporation of docosahexaenoic and arachidonic acids by frog retinas. *Curr. Eye Res.* **12,** 851–860.

Chen, W., and Baler, R. (2000). The rat arylalkylamine *N*-acetyltransferase E-box: Differential use in a master *vs.* a slave oscillator. *Brain Res. Mol. Brain Res.* **81,** 43–50.

Chong, N. W., Bernard, M., and Klein, D. C. (2000). Characterization of the chicken serotonin *N*-acetyltransferase gene. Activation via clock gene heterodimer/E box interaction. *J. Biol. Chem.* **275,** 32991–32998.

Collin, J. P., and Oksche, A. (1981). Structural and functionnal relationships in the nonmammalian pineal organ. *In* "The Pineal Gland: Anatomy and Biochemistry" (Reiter, R. J., Ed.), Vol. I, pp. 27–67. CRC press, Boca Raton.

Collin, J. P., Mirshahi, M., Brisson, P., Falcón, J., Guerlotte, J., and Faure, J. P. (1986). Pineal-retinal molecular relationships: Distribution of "S-antigen" in the pineal complex. *Neuroscience* **19,** 657–666.

Collin, J. P., Voisin, P., Falcón, J., Faure, J. P., Brisson, P., and Defaye, J. R. (1989). Pineal transducers in the course of evolution: Molecular organisation, rhythmic metabolic activity and role. *Archiv. Histol. Cytol.* **52,** 441–449.

Collin, S. P., Knight, M. A., Davies, W. L., Potter, I. C., Hunt, D. M., and Trezise, A. E. (2003). Ancient colour vision: Multiple opsin genes in the ancestral vertebrates. *Curr. Biol.* **13,** 864–865.

Cone, R. A., and Pak, W. L. (1971). "Handbook of Sensory Physiology," p. 345. Springer-Verlag, Berlin-Heidelberg-New York.

Coon, S. L., and Klein, D. C. (2006). Evolution of arylalkylamine *N*-acetyltransferase: Emergence and divergence. *Mol. Cell. Endocrinol.* **252**(1–2), 1–10.

Coon, S. L., Roseboom, P. H., Baler, R., Weller, J. L., Namboodiri, M. A., Koonin, E. V., and Klein, D. C. (1995). Pineal serotonin *N*-acetyltransferase: Expression cloning and molecular analysis. *Science* **270,** 1681–1683.

Coon, S. L., Begay, V., Falcón, J., and Klein, D. C. (1998). Expression of melatonin synthesis genes is controlled by a circadian clock in the pike pineal organ but not in the trout. *Biol. Cell.* **90**(5), 399–405.

Coon, S. L., Begay, V., Deurloo, D., Falcón, J., and Klein, D. C. (1999). Two arylalkylamine *N*-acetyltransferase genes mediate melatonin synthesis in fish. *J. Biol. Chem.* **274,** 9076–9082.

Debreceni, K., Fejer, Z., Manzano e Silva, M. J., and Vigh, B. (1997). Immunoreactive glutamate in the pineal and parapineal organs of the lamprey (*Lampetra fluviatilis*). *Neurobiology* **5,** 53–56.

Decressac, S., Grechez-Cassiau, A., Lenfant, J., Falcón, J., and Bois, P. (2002). Cloning, localization and functional properties of a cGMP-gated channel in photoreceptor cells from fish pineal gland. *J. Pineal Res.* **33,** 225–233.

Delaunay, F., Thisse, C., Marchand, O., Laudet, V., and Thisse, B. (2000). An inherited functional circadian clock in zebrafish embryos. *Science* **289**(5477), 297–300.

Dodt, E. (1963). Photosensitivity of the pineal organ in the teleost, *Salmo Irideus* (Gibbons). *Experientia* **19,** 642–643.

Dodt, E., and Heerd, E. (1962). Mode of action of pineal nerve fibers in frogs. *J. Neurophysiol.* **25,** 405–429.

Ekström, P., and Meissl, H. (1988). Intracellular staining of physiologically identified photoreceptor cells and hyperpolarizing interneurons in the teleost pineal organ. *Neuroscience* **25,** 1061–1070.

Ekström, P., and Meissl, H. (1989). Signal processing in a simple vertebrate photoreceptor system: The teleost pineal organ. *Physiol. Bohemoslov.* **38,** 311–326.

Ekström, P., and Meissl, H. (1990). Electron microscopic analysis of S-antigen- and serotonin-immunoreactive neural and sensory elements in the photosensory pineal organ of the salmon. *J. Comp. Neurol.* **292,** 73–82.

Ekström, P., and Meissl, H. (1997). The pineal organ of teleost fishes. *Rev. Fish Biol. Fish* **7,** 284.

Ekström, P., and Meissl, H. (2003). Evolution of photosensory pineal organs in new light: The fate of neuroendocrine photoreceptors. *Philos. Trans. R. Soc. Lond. B Biol. Sci.* **358,** 1679–1700.

Ekström, P., and van Veen, T. (1984). Pineal neural connections with the brain in two teleost, the crucian carp and the European eel. *J. Pineal Res.* **1**(3), 245–261.

Ekström, P., van Veen, T., Bruun, A., and Ehinger, B. (1987a). GABA-immunoreactive neurons in the photosensory pineal organ of the rainbow trout: Two distinct neuronal populations. *Cell Tissue Res.* **250,** 87–92.

Ekström, P., Foster, R. G., Korf, H. W., and Schalken, J. J. (1987b). Antibodies against retinal photoreceptor-specific proteins reveal axonal projections from the photosensory pineal organ in teleosts. *J. Comp. Neurol.* **265,** 25–33.

Ekström, P., Honkanen, T., and Ebbesson, S. O. (1988). FMRFamide-like immunoreactive neurons of the nervus terminalis of teleosts innervate both retina and pineal organ. *Brain Res.* **460,** 68–75.

Ekström, P., Östholm, T., and Holmqvist, B. I. (1994). Primary visual projections and pineal neural connections in fishes, amphibians and reptiles. *Adv. Pineal Res.* **8,** 1–18.

Fabbri, E., Capuzzo, A., and Moon, T. W. (1998). The role of circulating catecholamines in the regulation of fish metabolism: An overview. *Comp. Biochem. Physiol. C. Pharmacol. Toxicol. Endocrinol.* **120,** 177–192.

Falcón, J. (1979a). L'organe pinéal du Brochet (*Esox lucius*, L.) I. Etude anatomique et cytologique. *Reprod. Nutr. Dev.* **19,** 445–465.

Falcón, J. (1979b). L'organe pinéal du Brochet (*Esox lucius*, L.) II. Etude en microscopie électronique de la différenciation et de la rudimentation des photorécepteurs; conséquences possibles sur l'élaboration des messages sensoriels. *Reprod. Nutr. Dev.* **19,** 661–688.

Falcón, J. (1979c). Unusual distribution of neurons in the pike pineal organ. *Prog. Brain Res.* **52,** 89–91.

Falcón, J. (1999). Cellular circadian clocks in the pineal. *Prog. Neurobiol.* **58,** 121–162.

Falcón, J., and Gaildrat, P. (1997). Variations in cyclic adenosine 3′,5′-monophosphate and cyclic guanosine 3′,5′-monophosphate content and efflux from the photosensitive pineal organ of the pike in culture. *Pflugers Arch.* **433,** 336–342.

Falcón, J., and Henderson, R. J. (2001). Incorporation, distribution, and metabolism of polyunsaturated fatty acids in the pineal gland of rainbow trout (*Oncorhynchus mykiss*) *in vitro*. *J. Pineal Res.* **31,** 127–137.

Falcón, J., and Meissl, H. (1981). The photosensotry function of the pineal organ of the pike (*Esox lucius*, L.). Correlation between structure and function. *J. Comp. Physiol.* **144,** 127–137.

Falcón, J., and Mocquard, J. P. (1979). L'organe pinéal du Brochet (*Esox lucius*, L.) III. Voies intrapinéales de conduction des messages photosensoriels. *Ann. Biol. Anim. Biochim. Biophys.* **19,** 1043–1061.

Falcón, J., and Tanabe, J. (1983). Early receptor potential of pineal organ and lateral eye of the pike. *Naturwissenschaften* **70,** 149–150.

Falcón, J., Balemans, M. G., van Benthem, J., and Collin, J. P. (1985). *In vitro* uptake and metabolism of [^{3}H]indole compounds in the pineal organ of the pike. I. A radiochromatographic study. *J. Pineal Res.* **2,** 341–356.

Falcón, J., Guerlotté, J., Voisin, P., and Collin, J. P. (1987). Rhythmic melatonin biosynthesis in a photoreceptive pineal organ: A study in the pike. *Neuroendocrinology* **45,** 479–486.

Falcón, J., Besse, C., Guerlotte, J., and Collin, J. P. (1988). 5′-Nucleotidase activity in the pineal organ of the pike. An electron-microscopic study. *Cell Tissue Res.* **251,** 495–502.

Falcón, J., Marmillon, J. B., Claustrat, B., and Collin, J. P. (1989). Regulation of melatonin secretion in a photoreceptive pineal organ: An *in vitro* study in the pike. *J. Neurosci.* **9,** 1943–1950.

Falcón, J., Thibault, C., Blazquez, J. L., Vaudry, H., Lin, N., and Collin, J. P. (1990). Atrial natriuretic factor increases cyclic GMP and cyclic AMP levels in a directly photosensitive fish pineal organ. *Pflügers Arch. Eur. J. Physiol.* **417,** 243–245.

Falcón, J., Thibault, C., Martin, C., Brun-Marmillon, J., Claustrat, B., and Collin, J. P. (1991). Regulation of melatonin production by catecholamines and adenosine in a photoreceptive pineal organ. An *in vitro* study in the pike and the trout. *J. Pineal Res.* **11,** 123–134.

Falcón, J., Bégay, V., Besse, C., Ravault, J. P., and Collin, J. P. (1992a). Pineal photoreceptor cells in culture: Fine structure and light control of cyclic nucleotide levels. *J. Neuroendocrinol.* **4,** 641–651.

Falcón, J., Thibault, C., Bégay, V., Zachmann, A., and Collin, J. P. (1992b). Regulation of the rhythmic melatonin secretion by the fish pineal photoreceptor cells. *In* "Rhythms in Fishes" (Ali, M. A. A., Ed.), pp. 162–198. Plenum Press, New York.

Falcón, J., Bolliet, V., Ravault, J. P., Chesneau, D., Ali, M. A., and Collin, J. P. (1994). Rhythmic secretion of melatonin by the superfused pike pineal organ: Thermo- and photoperiod interaction. *Neuroendocrinology* **60,** 535–543.

Falcón, J., Bolliet, V., and Collin, J. P. (1996). Partial characterization of serotonin *N*-acetyltransferases from northern pike (*Esox lucius*, L.) pineal organ and retina: Effects of temperature. *Pflugers Arch.* **432,** 386–393.

Falcón, J., Barraud, S., Thibault, C., and Begay, V. (1998). Inhibitors of messenger RNA and protein synthesis affect differently serotonin arylalkylamine *N*-acetyltransferase activity in clock-controlled and non-clock-controlled fish pineal. *Brain Res.* **797,** 109–117.

Falcón, J., Galarneau, K. M., Weller, J. L., Ron, B., Chen, G., Coon, S. L., and Klein, D. C. (2001). Regulation of arylalkylamine *N*-acetyltransferase-2 (AANAT2, EC 2.3.1.87) in the fish pineal organ: Evidence for a role of proteasomal proteolysis. *Endocrinology* **142**(5), 1804–1813.

Falcón, J., Besseau, L., Fazzari, D., Attia, J., Gaildrat, P., Beauchaud, M., and Boeuf, G. (2003a). Melatonin modulates secretion of growth hormone and prolactin by trout pituitary glands and cells in culture. *Endocrinology* **144,** 4648–4658.

Falcón, J., Gothilf, Y., Coon, S. L., Boeuf, G., and Klein, D. C. (2003b). Genetic, temporal and developmental differences between melatonin rhythm generating systems in the teleost fish pineal organ and retina. *J. Neuroendocrinol.* **15,** 378–382.

Fejer, Z., Szel, A., Rohlich, P., Gorcs, T., Manzano, E., Silva, M. J., and Vigh, B. (1997). Immunoreactive pinopsin in pineal and retinal photoreceptors of various vertebrates. *Acta Biol. Hung.* **48,** 463–471.

Forsell, J., Holmqvist, B., Helvik, J. V., and Ekström, P. (1997). Role of the pineal organ in the photoregulated hatching of the Atlantic halibut. *Int. J. Dev. Biol.* **41,** 591–595.

Forsell, J., Ekström, P., Flamarique, I. N., and Holmqvist, B. (2001). Expression of pineal ultraviolet- and green-like opsins in the pineal organ and retina of teleosts. *J. Exp. Biol.* **204,** 2517–2525.

Forsell, J., Holmqvist, B. I., and Ekström, P. (2002). Molecular identification and developmental expression of UV and green opsin mRNAs in the pineal organ of the Atlantic halibut. *Brain Res. Dev. Brain Res.* **136,** 51–62.

Foster, R. G., and Bellingham, J. (2004). Inner retinal photoreceptors (IRPs) in mammals and teleost fish. *Photochem. Photobiol. Sci.* **3**(6), 617–627.

Foster, R. G., and Hankins, M. W. (2002). Non-rod, non-cone photoreception in the vertebrates. *Prog. Retin. Eye Res.* **21,** 507–527.

Foster, R. G., Hankins, M., Lucas, R. J., Jenkins, A., Munoz, M., Thompson, S., Appleford, J. M., and Bellingham, J. (2003). Non-rod, non-cone photoreception in rodents and teleost fish. *Novartis Found. Symp.* **253,** 3–23.

Ganguly, S., Coon, S. L., and Klein, D. C. (2002). Control of melatonin synthesis in the mammalian pineal gland: The critical role of serotonin acetylation. *Cell Tissue Res.* **309,** 127–137.

Garcia, M., and Vecino, E. (2003). Role of Muller glia in neuroprotection and regeneration in the retina. *Histol. Histopathol.* **18,** 1205–1218.

Garcia, M., Forster, V., Hicks, D., and Vecino, E. (2002). Effects of muller glia on cell survival and neuritogenesis in adult porcine retina *in vitro*. *Invest. Ophthalmol. Vis. Sci.* **43,** 3735–3743.

Gern, W. A., and Greenhouse, S. S. (1988). Examination of *in vitro* melatonin secretion from superfused trout (*Salmo gairdneri*) pineal organs maintained under diel illumination or continuous darkness. *Gen. Comp. Endocrinol.* **71,** 163–174.

Gothilf, Y., Coon, S. L., Toyama, R., Chitnis, A., Namboodiri, M. A., and Klein, D. C. (1999). Zebrafish serotonin *N*-acetyltransferase-2: Marker for development of pineal photoreceptors and circadian clock function. *Endocrinology* **140,** 4895–4903.

Hafeez, M. A., and Zerihun, L. (1974). Studies on central projections of the pineal nerve tract in rainbow trout, Salmo gairdneri Richardson, using cobalt chloride iontophoresis. *Cell Tissue Res.* **154,** 485–510.

Hamasaki, D. I., and Eder, D. J. (1977). Adaptative radiation of the pineal system. *In* "Hadbook of Sensory Physiology: The Visual System in Vertebrates," (Crescitelli, F., Ed.), Vol. VII/5, pp. 497–548. Springer-Verlag, Berlin.

Hanyu, I., Niwa, H., and Tamura, T. (1969). A slow potential from the epiphysis cerebri of fishes. *Vision Res.* **9,** 621–623.

Harada, C., Harada, T., Quah, H. M., Maekawa, F., Yoshida, K., Ohno, S., Wada, K., Parada, L. F., and Tanaka, K. (2003). Potential role of glial cell line-derived neurotrophic factor receptors in Muller glial cells during light-induced retinal degeneration. *Neuroscience* **122,** 229–235.

Henderson, R. J., Bell, M. V., Park, M. T., Sargent, J. R., and Falcón, J. (1994). Lipid composition of the pineal organ from rainbow trout (*Oncorhynchus mykiss*). *Lipids* **29,** 311–317.

Herwig, H. J. (1976). Comparative ultrastructural investigations of the pineal organ of the blind cave fish, *Anoptichthys jordani*, and its ancestor, the eyed river fish, *Astyanax mexicanus*. *Cell Tissue Res.* **167,** 297–324.

Hirayama, J., Fukuda, I., Ishikawa, T., Kobayashi, Y., and Todo, T. (2003). New role of zCRY and zPER2 as regulators of sub-cellular distributions of zCLOCK and zBMAL proteins. *Nucleic Acids Res.* **31,** 935–943.

Hoegg, S., Brinkmann, H., Taylor, J. S., and Meyer, A. (2004). Phylogenetic timing of the fish-specific genome duplication correlates with the diversification of teleost fish. *J. Mol. Evol.* **59**(2), 190–203.

Hopf, F. W., Cascini, M. G., Gordon, A. S., Diamond, I., and Bonci, A. (2003). Cooperative activation of dopamine D1 and D2 receptors increases spike firing of nucleus accumbens neurons via G-protein betagamma subunits. *J. Neurosci.* **23,** 5079–5087.

Iigo, M., and Aida, K. (1995). Effects of season, temperature, and photoperiod on plasma melatonin rhythms in the goldfish, *Carassius auratus*. *J. Pineal Res.* **18,** 62–68.

Iigo, M., Kezuka, H., Aida, K., and Hanyu, I. (1991). Circadian rhythms of melatonin secretion from superfused goldfish (*Carassius auratus*) pineal glands *in vitro*. *Gen. Comp. Endocrinol.* **83,** 152–158.

Iigo, M., Hara, M., Ohtani-Kaneko, R., Hirata, K., Tabata, M., and Aida, K. (1997). Photic and circadian regulations of melatonin rhythms in fishes. *Biol. Signals* **6,** 225–232.

Iigo, M., Mizusawa, K., Yokosuka, M., Hara, M., Ohtani-Kaneko, R., Tabata, M., Aida, K., and Hirata, K. (2003). *In vitro* photic entrainment of the circadian rhythm in melatonin release from the pineal organ of a teleost, ayu (*Plecoglossus altivelis*) in flow-through culture. *Brain Res.* **982,** 131–135.

Iigo, M., Fujimoto, Y., Gunji-Suzuki, M., Yokosuka, M., Hara, M., Ohtani-Kaneko, R., Tabata, M., Aida, K., and Hirata, K. (2004). Circadian rhythm of melatonin release from the photoreceptive pineal organ of a teleost, ayu (*Plecoglossus altivelis*) in flow-thorough culture. *J. Neuroendocrinol.* **16,** 45–51.

Ivanova, T. N., and Iuvone, P. M. (2003). Circadian rhythm and photic control of cAMP level in chick retinal cell cultures: A mechanism for coupling the circadian oscillator to the melatonin-synthesizing enzyme, arylalkylamine *N*-acetyltransferase, in photoreceptor cells. *Brain Res.* **991,** 96–103.

Iyer, L. M., Aravind, L., Coon, S. L., Klein, D. C., and Koonin, E. V. (2004). Evolution of cell-cell signaling in animals: Did late horizontal gene transfer from bacteria have a role? *Trends Genet.* **20,** 292–299.

Jacobs, G. H., Williams, G. A., and Fenwick, J. A. (2004). Influence of cone pigment coexpression on spectral sensitivity and color vision in the mouse. *Vision Res.* **44,** 1615–1622.

Jaillon, O., Aury, J. M., Brunet, F., Petit, J. L., Stange-Thomann, N., Mauceli, E., Bouneau, L., Fischer, C., Ozouf-Costaz, C., Bernot, A., Nicaud, S., Jaffe, D., *et al.* (2004). Genome duplication in the teleost fish *Tetraodon nigroviridis* reveals the early vertebrate proto-karyotype. *Nature* **431,** 946–957.

Jamieson, D. (1997). Synaptic transmission in the pineal eye of young *Xenopus laevis* tadpoles: A role for NMDA and non-NMDA glutamate and non-glutaminergic receptors? *J. Comp. Physiol. A.* **181,** 177–186.

Jimenez, A. J., Fernandez-Llebrez, P., and Perez-Figares, J. M. (1995). Central projections from the goldfish pineal organ traced by HRP-immunocytochemistry. *Histol. Histopathol.* **10,** 847–852.

Johansson, P. (1984). Alpha-adrenoceptors: Recent development and some comparative aspects. *Comp. Biochem. Physiol. C.* **78,** 253–261.

Kezuka, H., Furukawa, K., Aida, K., and Hanyu, I. (1988). Daily cycles in plasma melatonin levels under long or short photoperiod in the common carp, *Cyprinus carpio*. *Gen. Comp. Endocrinol.* **72,** 296–302.

Klein, D. C. (2004). The 2004 Aschoff/Pittendrigh lecture: Theory of the origin of the pineal gland: A tale of conflict and resolution. *J. Biol. Rhythms* **19,** 264–279.

Klein, D. C., and Weller, J. L. (1970). Indole metabolism in the pineal gland: A circadian rhythm in *N*-acetyltransferase. *Science* **169,** 1093–1095.

Klein, D. C., Auerbach, D. A., Namboodiri, M. A., and Wheler, G. H. T. (1981). Indole metabolism in the mammalian pineal gland. *In* "The Pineal Gland" (Reiter, R. J., Ed.), pp. 199–227. CRC Press, Boca Raton.

Klein, D. C., Coon, S. L., Roseboom, P. H., Weller, J. L., Bernard, M., Gastel, J. A., Zatz, M., Iuvone, P. M., Rodriguez, I. R., Bégay, V., Falcón, J., Cahill, G. M., *et al.* (1997). The melatonin rhythm-generating enzyme: Molecular regulation of serotonin *N*-acetyltransferase in the pineal gland. *Recent Prog. Horm. Res.* **52,** 307–357.

Klein, D. C., Ganguly, S., Coon, S., Weller, J. L., Obsil, T., Hickman, A., and Dyda, F. (2002). 14-3-3 Proteins and photoneuroendocrine transduction: Role in controlling the daily rhythm in melatonin. *Biochem. Soc. Trans.* **30,** 365–373.

Kobayashi, Y., Ishikawa, T., Hirayama, J., Daiyasu, H., Kanai, S., Toh, H., Fukuda, I., Tsujimura, T., Terada, N., Kamei, Y., Yuba, S., Iwai, S., *et al.* (2000). Molecular analysis of zebrafish photolyase/cryptochrome family: Two types of cryptochromes present in zebrafish. *Genes Cells* **5,** 725–738.

Korenbrot, J. I., and Rebrik, T. I. (2002). Tuning outer segment Ca^{2+} homeostasis to phototransduction in rods and cones. *Adv. Exp. Med. Biol.* **514,** 179–203.

Korf, H. W., Von Gall, C., and Stehle, J. (2003). The circadian system and melatonin: Lessons from rats and mice. *Chronobiol. Int.* **20,** 697–710.

Kreitzer, M. A., Andersen, K. A., and Malchow, R. P. (2003). Glutamate modulation of GABA transport in retinal horizontal cells of the skate. *J. Physiol.* **546,** 717–731.

Kroeber, S., Schomerus, C., and Korf, H. W. (1998). A specific and sensitive double-immunofluorescence method for the demonstration of S-antigen and serotonin in trout and rat pinealocytes by means of primary antibodies from the same donor species. *Histochem. Cell Biol.* **109,** 309–317.

Kroeber, S., Meissl, H., Maronde, E., and Korf, H. W. (2000). Analyses of signal transduction cascades reveal an essential role of calcium ions for regulation of melatonin biosynthesis in the light-sensitive pineal organ of the rainbow trout (*Oncorhynchus mykiss*). *J. Neurochem.* **74,** 2478–2489.

Kusmic, C., and Gualtieri, P. (2000). Morphology and spectral sensitivities of retinal and extraretinal photoreceptors in freshwater teleosts. *Micron* **31,** 183–200.

Lerner, A. B., Case, J. D., Takahashi, Y., Lee, T. H., and Mori, W. (1958). Isolation of melatonin, the pineal factor that lightens melanocytes. *J. Am. Chem. Soc.* **80,** 2857–2858.

Li, L., and Dowling, J. E. (1998). Zebrafish visual sensitivity is regulated by a circadian clock. *Vis. Neurosci.* **15,** 851–857.

Lu, J., Zoran, M. J., and Cassone, V. M. (1995). Daily and circadian variation in the electroretinogram of the domestic fowl: Effects of melatonin. *J. Comp. Physiol. A* **177,** 299–306.

Maeda, T., Imanishi, Y., and Palczewski, K. (2003). Rhodopsin phosphorylation: 30 years later. *Prog. Retin. Eye Res.* **22,** 417–434.

Mano, H., Kojima, D., and Fukada, Y. (1999). Exo-rhodopsin: A novel rhodopsin expressed in the zebrafish pineal gland. *Brain Res. Mol. Brain Res.* **73,** 110–118.

Martin, C., and Meissl, H. (1992). Effects of dopaminergic and noradrenergic mechanisms on the neuronal activity of the isolated pineal organ of the trout, *Oncorhynchus mykiss*. *J. Neural. Transm.* **88,** 37–51.

Martin, C., Falcón, J., and Collin, J. P. (1991). Catecholamines regulate cAMP levels in the photosensitive trout pineal organ. *Adv. Pineal Res.* **5,** 137–140.

Masuda, T., Iigo, M., Mizusawa, K., Naruse, M., Oishi, T., Aida, K., and Tabata, M. (2003a). Variations in plasma melatonin levels of the rainbow trout (*Oncorhynchus mykiss*) under various light and temperature conditions. *Zool. Sci.* **20,** 1011–1016.

Masuda, T., Iigo, M., Mizusawa, K., and Aida, K. (2003b). Effects of macromolecule synthesis inhibitors on light-induced phase shift of the circadian rhythm in melatonin release from the cultured pineal organ of a teleost, ayu (*Plecoglossus altivelis*). *Zool. Sci.* **20,** 1405–1410.

Mata, N. L., Radu, R. A., Clemmons, R. C., and Travis, G. H. (2002). Isomerization and oxidation of vitamin a in cone-dominant retinas: A novel pathway for visual-pigment regeneration in daylight. *Neuron* **36,** 69–80.

Max, M., and Menaker, M. (1992). Regulation of melatonin production by light, darkness, and temperature in the trout pineal. *J. Comp. Physiol. A* **170,** 479–489.

McNulty, J. A. (1979). A comparative light and electron microscopic study of the pineal complex in the deep-sea fishes, *Cyclothone signata* and *C. acclinidens*. *J. Morphol.* **162**(1), 1–16.

McNulty, J. A., Rathbun, W. E., and Druse, M. J. (1988). Ultrastructural and biochemical responses of photoreceptor pinealocytes to light and dark *in vivo* and *in vitro*. *Life Sci.* **43,** 845–850.

Meissl, H., and Dodt, E. (1981). Comparative physiology of pineal photoreceptor organs. *In* "The Pineal Organ: Photobiology-Biochronometry–Endocrinology" (Oksche, A., and Pévet, P., Eds.), pp. 61–80. Elsevier, Amsterdam.

Meissl, H., and Ekström, P. (1988a). Dark and light adaptation of pineal photoreceptors. *Vision Res.* **28,** 49–56.

Meissl, H., and Ekström, P. (1988b). Photoreceptor responses to light in the isolated pineal organ of the trout, *Salmo gairdneri*. *Neuroscience* **25,** 1071–1076.

Meissl, H., and Ekström, P. (1991). Action of gamma-aminobutyric acid (GABA) in the isolated photosensory pineal organ. *Brain Res.* **562,** 71–78.

Meissl, H., and George, S. R. (1984a). Electrophysiological studies on neuronal transmission in the frog's photosensory pineal organ. The effect of amino acids and biogenic amines. *Vision Res.* **24,** 1727–1734.

Meissl, H., and George, S. R. (1984b). Photosensory properties of the pineal organ. Microiontophoretic application of excitatory amino acids onto pineal neurons. *Ophthalmic Res.* **16,** 114–118.

Meissl, H., Donley, C. S., and Wissler, J. H. (1978). Free amino acids and amines in the pineal organ of the rainbow trout (*Salmo gairdneri*): Influence of light and dark. *Comp. Biochem. Physiol. C* **61,** 401–405.

Meissl, H., Nakamura, T., and Thiele, G. (1986). Neural response mechanisms in the photoreceptive pineal organ of goldfish. *Comp. Biochem. Physiol. A* **84,** 467–473.

Meissl, H., Martin, C., and Tabata, M. (1990). Melatonin modulates the neural activity in photosensory pineal organ of the trout: Evidence for endocrine-neuronal interactions. *J. Comp. Physiol. A* **167,** 641–648.

Meissl, H., Anzelius, M., Östholm, T., and Ekström, P. (1993). Interaction of GABA, benzodiazepines and melatonin in the photosensory pineal organ of salmonid fish. *In* "Melatonin and the Pineal Gland: From Basic Science to Clinical Applications" (Touitou, Y., Arendt, J., and Pévet, P., Eds.), pp. 95–98. Elsevier, Amsterdam.

Meissl, H., Yanez, J., Ekström, P., and Grossmann, E. (1994). Benzodiazepines influence melatonin secretion of the pineal organ of the trout *in vitro*. *J. Pineal Res.* **17,** 69–78.

Meissl, H., Kroeber, S., Yanez, J., and Korf, H. W. (1996). Regulation of melatonin production and intracellular calcium concentrations in the trout pineal organ. *Cell Tissue Res.* **286,** 315–323.

Minamoto, T., and Shimizu, I. (2002). A novel isoform of vertebrate ancient opsin in a smelt fish, *Plecoglossus altivelis*. *Biochem. Biophys. Res. Commun.* **290,** 280–286.

Minamoto, T., and Shimizu, I. (2003). Molecular cloning and characterization of rhodopsin in a teleost (*Plecoglossus altivelis*, Osmeridae). *Comp. Biochem. Physiol. B* **134,** 559–570.

Miranda, L. A., Strobl-Mazzulla, P. H., and Somoza, G. M. (2002). Ontogenetic development and neuroanatomical localization of growth hormone-releasing hormone (GHRH) in the brain and pituitary gland of pejerrey fish *Odontesthes bonariensis*. *Int. J. Dev. Neurosci.* **20,** 503–510.

Mirshahi, M., Faure, J. P., Brisson, P., Falcón, J., Guerlotte, J., and Collin, J. (1984). S-antigen immunoreactivity in retinal rods and cones and pineal photosensitive cells. *Biol. Cell* **52,** 195–198.

Mizusawa, K., Iigo, M., Suetake, H., Yoshiura, Y., Gen, K., Kikuchi, K., Okano, T., Fukada, Y., and Aida, K. (1998). Molecular cloning and characterization of a cDNA encoding the retinal arylalkylamine *N*-acetyltransferase of the rainbow trout, *Oncorhynchus mykiss*. *Zool. Sci.* **15,** 345–351.

Mizusawa, K., Iigo, M., Masuda, T., and Aida, K. (2000). Photic regulation of arylalkylamine *N*-acetyltransferase-1 mRNA in trout retina. *Neuroreport* **11,** 3473–3477.

Mizusawa, K., Iigo, M., Masuda, T., and Aida, K. (2001). Inhibition of RNA synthesis differentially affects *in vitro* melatonin release from the pineal organs of ayu (*Plecoglossus altivelis*) and rainbow trout (*Oncorhynchus mykiss*). *Neurosci. Lett.* **309,** 72–76.

Morton, D. J., and Forbes, H. J. (1988). Pineal gland *N*-acetyltransferase and hydroxyindole-*O*-methyltransferase activity in the rainbow trout (*Salmo gairdneri*): Seasonal variation linked to photoperiod. *Neurosci. Lett.* **94,** 333–337.

Moutsaki, P., Bellingham, J., Soni, B. G., David-Gray, Z. K., and Foster, R. G. (2000). Sequence, genomic structure and tissue expression of carp (*Cyprinus carpio* L.) vertebrate ancient (VA) opsin. *FEBS Lett.* **473,** 316–322.

Nelson, J. S. (1994). "Fishes of the World," 3rd edn., John Wiley and Sons, New York, NY.

O'Brien, P. J., and Klein, D. C. (1986). "Pineal and Retinal Relationships." Academic Press, Orlando, FL.

Obsil, T., Ghirlando, R., Klein, D. C., Ganguly, S., and Dyda, F. (2001). Crystal structure of the 14-3-3zeta: Serotonin *N*-acetyltransferase complex. A role for scaffolding in enzyme regulation. *Cell* **105**(2), 257–267.

Okimoto, D. K., and Stetson, M. H. (1997). Photoregulation and melatonin release *in vitro* system from the pineal gland of cartilagineous and teleostean fishes: A comparative approach. *In* "Pineal Update, from Molecular Mechanisms to Clinical Implications" (Webb, S. M., Puig-Domingo, M., Moller, M., and Pevet, P., Eds.), pp. 121–126. PJD Publications Ltd, New York.

Okimoto, D. K., and Stetson, M. H. (1999a). Properties of the melatonim-generating system of the sailfin molly, *Poecilia velifera. Gen. Comp. Endocrinol.* **114,** 293–303.

Okimoto, D. K., and Stetson, M. H. (1999b). Presence of an intrapineal circadian oscillator in the teleostean family poeciliidae. *Gen. Comp. Endocrinol.* **114,** 304–312.

Omura, Y., and Oguri, M. (1969). Histological studies on the pineal organ of 15 species of teleost fishes. *Bull. Jap. Soc. Fish.* **35,** 991–1000.

Omura, Y., Korf, H. W., and Oksche, A. (1985). Vascular permeability (problem of the blood–brain barrier) in the pineal organ of the rainbow trout, *Salmo gairdneri. Cell Tissue Res.* **239** (3), 599–610.

Omura, Y., Ueno, S., and Ueck, M. (1986). Cytochemical demonstration of acid phosphatase activity in the pineal organ of the rainbow trout, *Salmo gairdneri. Cell Tissue Res.* **245,** 171–176.

Philp, A. R., Bellingham, J., Garcia-Fernandez, J., and Foster, R. G. (2000a). A novel rod-like opsin isolated from the extra-retinal photoreceptors of teleost fish. *FEBS Lett.* **468,** 181–188.

Philp, A. R., Garcia-Fernandez, J. M., Soni, B. G., Lucas, R. J., Bellingham, J., and Foster, R. G. (2000b). Vertebrate ancient (VA) opsin and extraretinal photoreception in the Atlantic salmon (*Salmo salar*). *J. Exp. Biol.* **203,** 1925–1936.

Rahman, M. S., Kim, B. H., Takemura, A., Park, C. B., and Lee, Y. D. (2004). Effects of moonlight exposure on plasma melatonin rhythms in the seagrass rabbitfish, *Siganus canaliculatus. J. Biol. Rhythms* **19,** 325–334.

Rao, S. D., Rao, P. D., and Peter, R. E. (1996). Growth hormone-releasing hormone immunoreactivity in the brain, pituitary, and pineal of the goldfish, *Carassius auratus. Gen. Comp. Endocrinol.* **102,** 210–220.

Ren, J. Q., and Li, L. (2004). A circadian clock regulates the process of ERG b- and d-wave dominance transition in dark-adapted zebrafish. *Vision Res.* **44,** 2147–2152.

Roberts, D., Okimoto, D. K., Parsons, C., Straume, M., and Stetson, M. H. (2003). Development of rhythmic melatonin secretion from the pineal gland of embryonic mummichog (*Fundulus heteroclitus*). *J. Exp. Zool. A* **296,** 56–62.

Rotstein, N. P., Politi, L. E., German, O. L., and Girotti, R. (2003). Protective effect of docosahexaenoic acid on oxidative stress-induced apoptosis of retina photoreceptors. *Invest. Ophthalmol. Vis. Sci.* **44,** 2252–2259.

Rüdeberg, C. (1969). Light and electron microscopic studies on the pineal organ of the dogfish, *Scyliorhinus canicula* L. *Z. Zellforsch. Mikrosk. Anat.* **96,** 548–581.

Samejima, M., Happe, H. K., Murrin, L. C., Pfeiffer, R. F., and Ebadi, M. (1994). Distribution of cholinergic and dopaminergic receptors in rainbow trout pineal gland. *J. Pineal Res.* **16,** 37–43.

Samejima, M., Tamotsu, S., Uchida, K., Moriguchi, Y., and Morita, Y. (1997). Melatonin excretion rhythms in the cultured pineal organ of the lamprey, *Lampetra japonica. Biol. Signals* **6**(4–6), 241–246.

Sato, T. K., Panda, S., Miraglia, L. J., Reyes, T. M., Rudic, R. D., McNamara, P., Naik, K. A., FitzGerald, G. A., Kay, S. A., and Hogenesch, J. B. (2004). A functional genomics strategy reveals Rora as a component of the mammalian circadian clock. *Neuron* **43,** 527–537.

Schmitz, Y., and Witkovsky, P. (1997). Dependence of photoreceptor glutamate release on a dihydropyridine-sensitive calcium channel. *Neuroscience* **78,** 1209–1216.

Solessio, E., and Engbretson, G. A. (1993). Antagonistic chromatic mechanisms in photoreceptors of the parietal eye of lizards. *Nature* **364,** 442–445.

Studnicka, F. K. (1905). Parietal organe. "Lehrbuch der vergleichenden mikroskopischen Anatomie der Wilbertiere. Teil V." Fischer, Jena.

Subhedar, N., Cerda, J., and Wallace, R. A. (1996). Neuropeptide Y in the forebrain and retina of the killifish, *Fundulus heteroclitus. Cell Tissue Res.* **283,** 313–323.

Sullivan, J. M., and Shukla, P. (1999). Time-resolved rhodopsin activation currents in a unicellular expression system. *Biophys. J.* **77,** 1333–1357.

Tabata, M. (1982a). The source cell of the pineal mass potential, electropinealogram (EPG) of rainbow trout. *Bull. Japan. Soc. Sci. Fish.* **48,** 477.

Tabata, M. (1982b). The electropinealogram (EPG) in teleost. *Bull. Japan. Soc. Sci. Fish* **48,** 151–155.

Tabata, M., Suzuki, T., and Niwa, H. (1985). Chromophores in the extraretinal photoreceptor (pineal organ) of teleosts. *Brain Res.* **338,** 173–176.

Taylor, J. S., and Raes, J. (2004). Duplication and divergence: The evolution of new genes and old ideas. *Annu. Rev. Genet.* **61,** 5–643.

Thibault, C., Collin, J. P., and Falcón, J. (1993a). Intrapineal circadian oscillator(s), cyclic nucleotides and melatonin production in pike pineal photoreceptor cells. *In* "Melatonin and the Pineal Gland: From Basic Science to Clinical Application" (Touitou, Y., Ed.), pp. 11–18. Elsevier, Amsterdam.

Thibault, C., Falcón, J., Greenhouse, S. S., Lowery, C. A., Gern, W. A., and Collin, J. P. (1993b). Regulation of melatonin production by pineal photoreceptor cells: Role of cyclic nucleotides in the trout (*Oncorhynchus mykiss*). *J. Neurochem.* **61,** 332–339.

Thoreson, W. B., and Witkovsky, P. (1999). Glutamate receptors and circuits in the vertebrate retina. *Prog. Retin. Eye Res.* **18,** 765–810.

Uchida, K., and Morita, Y. (1994). Spectral sensitivity and mechanism of interaction between inhibitory and excitatory responses of photosensory pineal neurons. *Pflugers Arch.* **427,** 373–377.

Uchida, K., Samejima, M., Kawata, M., Tamotsu, S., and Morita, Y. (2001). Effects of cGMP and cAMP on light responses of the photosensory pineal neurons in the lamprey, *Lampetra japonica. Biol. Signals Recept.* **10,** 389–398.

Underwood, H. (1989). The pineal and melatonin: Regulators of circadian function in lower vertebrates. *Experientia* **46,** 120–128.

Vallone, D., Gondi, S. B., Whitmore, D., and Foulkes, N. S. (2004). E-box function in a period gene repressed by light. *Proc. Natl. Acad. Sci. USA* **101,** 4106–4111.

van Veen, T., Ekström, P., Nyberg, L., Borg, B., Vigh-Teichmann, I., and Vigh, B. (1984). Serotonin and opsin immunoreactivities in the developing pineal organ of the three-spined stickleback, *Gasterosteus aculeatus* L. *Cell Tissue Res.* **237,** 559–564.

van Veen, T., Ostholm, T., Gierschik, P., Spiegel, A., Somers, R., Korf, H. W., and Klein, D. C. (1986). Alpha-transducin immunoreactivity in retinae and sensory pineal organs of adult vertebrates. *Proc. Natl. Acad. Sci. USA* **83,** 912–916.

Vigh, B., Debreceni, K., and Manzano e Silva, M. J. (1995a). Similar localization of immunoreactive glutamate and aspartate in the pineal organ and retina of various nonmammalian vertebrates. *Acta Biol. Hung.* **46,** 99–106.

Vigh, B., Vigh-Teichmann, I., Debreceni, K., and Takacs, J. (1995b). Similar fine structural localization of immunoreactive glutamate in the frog pineal complex and retina. *Arch. Histol. Cytol.* **58,** 37–44.

Vigh, B., Debreceni, K., Fejer, Z., and Vigh-Teichmann, I. (1997). Immunoreactive excitatory amino acids in the parietal eye of lizards, a comparison with the pineal organ and retina. *Cell Tissue Res.* **287,** 275–283.

Vigh, B., Rohlich, P., Gorcs, T., Manzano e Silva, M. J., Szel, A., Fejer, Z., and Vigh-Teichmann, I. (1998). The pineal organ as a folded retina: Immunocytochemical localization of opsins. *Biol. Cell* **90,** 653–659.

Vigh-Teichmann, I., Rohlich, P., Vigh, B., and Aros, B. (1980). Comparison of the pineal complex, retina and cerebrospinal fluid contacting neurons by immunocytochemical antirhodopsin reaction. *Z. Mikrosk. Anat. Forsch.* **94,** 623–640.

Vigh-Teichmann, I., Korf, H. W., Oksche, A., and Vigh, B. (1982). Opsin-immunoreactive outer segments and acetylcholinesterase-positive neurons in the pineal complex of *Phoxinus phoxinus* (Teleostei, Cyprinidae). *Cell Tissue Res.* **227,** 351–369.

Vigh-Teichmann, I., Korf, H. W., Nurnberger, F., Oksche, A., Vigh, B., and Olsson, R. (1983). Opsin-immunoreactive outer segments in the pineal and parapineal organs of the lamprey (*Lampetra fluviatilis*), the eel (*Anguilla anguilla*), and the rainbow trout (*Salmo gairdneri*). *Cell Tissue Res.* **230,** 289–307.

Vigh-Teichmann, I., Ali, M. A., and Vigh, B. (1992). Comparative ultrastructure and opsin immunocytochemistry of the retina and pineal organ in fish. *Prog. Brain Res.* **91,** 307–313.

Wagner, H. J., and Mattheus, U. (2002). Pineal organs in deep demersal fish. *Cell Tissue Res.* **307,** 115–127.

Wake, K. (1973). Acetylcholinesterase-containing nerve cells and their distribution in the pineal organ of the goldfish, *Carassius auratus*. *Z. Zellforsch. Mikrosk. Anat.* **145,** 287–298.

Walker, J. A., and Olton, D. S. (1979). Circadian rhythm of luminance detectability in the rat. *Physiol. Behav.* **23,** 11–15.

Weisinger, H. S., Vingrys, A. J., Bui, B. V., and Sinclair, A. J. (1999). Effects of dietary n-3 fatty acid deficiency and repletion in the guinea pig retina. *Invest. Ophthalmol. Vis. Sci.* **40,** 327–338.

Whitmore, D., Foulkes, N. S., Strahle, U., and Sassone-Corsi, P. (1998). Zebrafish Clock rhythmic expression reveals independent peripheral circadian oscillators. *Nat. Neurosci.* **1,** 701–707.

Xia, Z., Choi, E. J., Wang, F., Blazynski, C., and Storm, D. R. (1993). Type I calmodulin-sensitive adenylyl cyclase is neural specific. *J. Neurochem.* **60,** 305–311.

Yañez, J., and Añadon, R. (1996). Afferent and efferent connections of the habenula in the rainbow trout (*Oncorhynchus mykiss*): An indocarbocyanine dye (DiI) study. *J. Comp. Neurol.* **372,** 529–543.

Yañez, J., and Meissl, H. (1995). Secretion of methoxyindoles from trout pineal organs *in vitro*: Indication for a paracrine melatonin feedback. *Neurochem. Int.* **27,** 195–200.

Zachmann, A., Ali, M. A., and Falcón, J. (1992a). Melatonin and its effects in fishes: An overview. *In* "Rhythms in Fishes" (Ali, M. A., Ed.), pp. 149–165. Plenum Press, New York.

Zachmann, A., Knijff, S. C. M., Ali, M. A., and Anctil, M. (1992b). Effects of photoperiod and different intensities of light exposure on melatonin levels in the blood, pineal organ and retina of the brook trout (*Salvelinus frontinalis*, Mitchill). *Can. J. Zool.* **70,** 25–29.

Zachmann, A., Falcón, J., Knijff, S. C., Bolliet, V., and Ali, M. A. (1992c). Effects of photoperiod and temperature on rhythmic melatonin secretion from the pineal organ of the white sucker (*Catostomus commersoni*) *in vitro*. *Gen. Comp. Endocrinol.* **86,** 26–33.

Zilberman-Peled, B., Benhar, I., Coon, S. L., Ron, B., and Gothilf, Y. (2004). Duality of serotonin-*N*-acetyltransferase in the gilthead seabream (*Sparus aurata*): Molecular cloning and characterization of recombinant enzymes. *Gen. Comp. Endocrinol.* **138,** 139–147.

Ziv, L., Levkovitz, S., Toyama, R., Falcón, J., and Gothilf, Y. (2005). Functional development of the zebrafish pineal gland: Light-induced expression of period 2 is required for onset of the circadian clock. *J. Neuroendocrinol.* **17,** 314–320.

7

ELECTRORECEPTION: OBJECT RECOGNITION IN AFRICAN WEAKLY ELECTRIC FISH

GERHARD VON DER EMDE

1. Introduction
2. Fish Electrogenesis
3. Fish Electroreception
4. Passive Electrolocation
5. Production of Electric Signals
6. Active Electrolocation
   6.1. What Can Electric Fish Perceive About Their Environment During Active Electrolocation?
   6.2. How Can the Fish Extract Information About Objects During Active Electrolocation?
   6.3. Specializations of Certain Skin Regions for Active Electrolocation
   6.4. How Does the Brain Extract Information About Objects During Active Electrolocation?
   6.5. Higher Stages of the Electrosensory System

1. INTRODUCTION

Many groups of fishes perceive naturally occurring electric stimuli. They possess ampullary electroreceptor organs that respond to low-frequency electric fields. Two nocturnally active teleost groups produce weak electric signals (electric organ discharges, EOD), which they perceive with tuberous electroreceptor organs. In addition to "passive electrolocation," these weakly electric fish perform "active electrolocation," during which they detect alterations of their EOD caused by nearby objects. This enables them to perceive the complex impedance of objects and thus to identify animated objects. In addition, they can measure distance, size, shape, and other three-dimensional (3D) object properties.

When inspecting an object, the weakly electric fish *Gnathonemus petersii* employs two "electrical foveae" which are located on their moveable chin

Sensory Systems Neuroscience: Volume 25
FISH PHYSIOLOGY

DOI: 10.1016/S1546-5098(06)25007-6

appendage and in the nasal region. The former is used for object inspection, while the latter is used for object detection during foraging.

The brain of weakly electric fishes extracts information about objects by analyzing the input from the electroreceptor organs. The study of electroreceptive brain areas has revealed many general principals of neural processing of sensory and motor information.

2. FISH ELECTROGENESIS

Already the ancient Greeks and Egyptians knew that some fishes can produce electric signals because the discharges of strongly electric fishes (freshwater fishes, such as the electric eel from South America and the electric catfish from Africa; marine fishes, such as the electric ray) are clearly and painfully noticeable upon touching the animals. Before the discovery of electricity, the electric nature of these discharges was unidentified and they were attributed to some unknown and mysterious force. Strongly electric fish played an important role in the early history of the science of electricity and were debated by Galvani and Volta (Wu, 1984). Only after the nature of the EOD was discovered, it became apparent that strongly electric fishes use their discharges to stun or kill their prey and to defend themselves against predators (Moller, 1995).

Many years later, in 1951, Hans Lissmann at the University of Cambridge discovered that quite a number of fishes produce *weak* electric signals of only a few volts, which cannot be felt even when touching the fish (Lissmann, 1951). These weakly electric fishes live in tropical freshwaters of two continents: the African Mormyriformes consist of two families (the single member Gymnarchidae and the much larger Mormyridae), while in South America six families of the Gymnotiformes can be found. The purpose of weak EODs, which cannot be noticed by most aquatic organisms, was discovered 7 years later through a number of ingenious experiments by Lissmann and Machin (Lissmann, 1958; Lissmann and Machin, 1958). These experiments demonstrated that weakly electric fishes use their EODs for *active electrolocation*, that is, the detection and analysis of objects, and for *electrocommunication*.

3. FISH ELECTRORECEPTION

For electrolocation and electrocommunication to work, electric signals have to be sensed by the animals. After Lissmann's findings, researchers discovered that some fish use *ampullary electroreceptors* to sense weak,

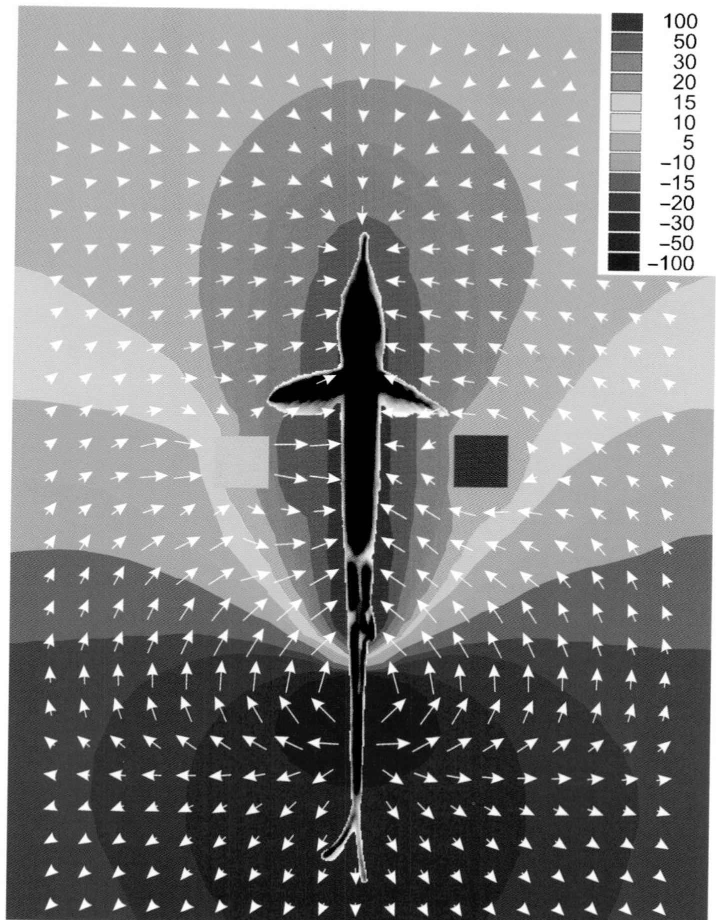

Chapter 7, Fig. 2.

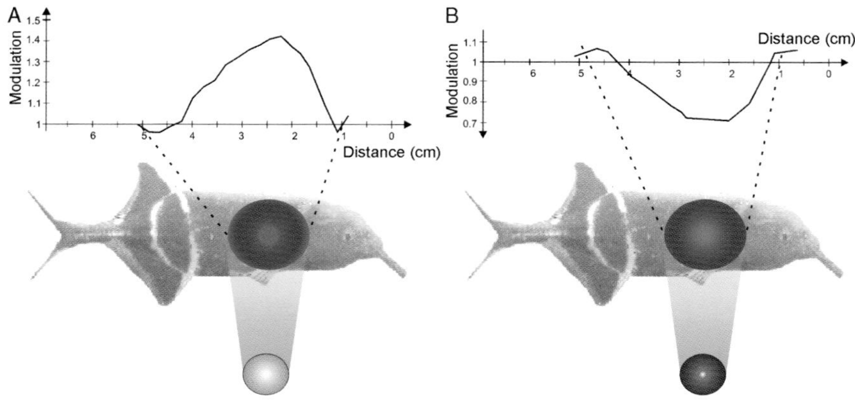

Chapter 7, Fig. 3.

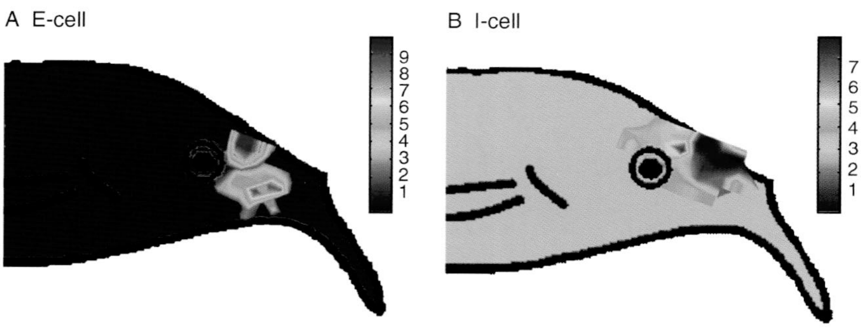

Chapter 7, Fig. 5.

low-frequency electric fields (Moller, 1995). All non-teleost fish (except for the Holosteans and Myxiniformes), the large and worldwide order Siluriformes, some members of the *Xenomystinae* (Bullock *et al.*, 1983), and all weakly electric fishes possess ampullary electroreceptor organs (Zakon, 1987; Northcutt, 1997). However, because ampullary receptors are sensitive to low-frequency electric signals up to about 50 Hz, they respond only weakly or not at all to the EODs produced by weakly electric fishes. For active electrolocation, weakly electric fish possess additional *tuberous* electroreceptors sensitive to higher frequencies and therefore responding optimally to their own discharges and those of conspecifics.

Each group of weakly electric fish has their own set of tuberous electroreceptor organs. Usually there are two types of organs: one type responds with a single action potential to the occurrence of an EOD or in wavefish (see later) to each cycle of the continuous signal. These are time coders, which inform the fish about the exact timing of the perceived EOD. The second type of tuberous organs constitutes amplitude coders. Depending on the amplitude of the received EOD they fire several action potentials, the number of which increases gradually with amplitude. Like ampullary receptor organs, tuberous organs are located in the skin of the animals. They have one or two chambers, which contain a variable number of electroreceptor cells (up to 50), and an electrical conductive canal that is loosely filled with cells and leads to the skin surface. The receptor cells are innervated by afferent nerve fibers, which gather in electroreceptive nerves and innervate the brain.

In this chapter, focus is on the African Mormyriformes. Excellent reviews on electrosensory processing and active electrolocation in Gymnotiformes can be found elsewhere (Budelli and Caputi, 2000; Castelló *et al.*, 2000; Zakon, 2003; Bodznick and Montgomery, 2005; Kawasaki, 2005; Nelson, 2005). Mormyriformes possess tuberous amplitude coding organs called *mormyromasts* playing a crucial role during active electrolocation, and time coders called *Knollenorgans* used exclusively for electrocommunication. Afferents of Knollenorgans respond with high sensitivity to EODs of other individuals by firing a single spike of a fixed latency. They thus "report" the occurrence of foreign EODs to the brain and therefore signal the presence of another electric fish up to a distance of several meters (Bennett, 1965; Knudsen, 1974). By analyzing the pattern of Knollenorgan responses on the electroreceptive skin surface, mormyrids can recognize the EOD-waveform of other fishes and thereby identify species- and sex-specific EODs during electrocommunication (Hopkins and Bass, 1981; Hopkins, 2005).

Mormyromast organs have two distinct types of sensory cells that are referred to as "A" and "B" (Szabo and Wersäll, 1970). These two types are separately innervated, and their primary afferents project to separate regions of the electrosensory lobe (ELL) in the medulla of the brain (Bell, 1990b). Physiologically, both types of fiber are silent in the absence of electrosensory stimulation, and both types respond to a brief EOD-like pulse of current with one or more spikes. Stimulation near threshold evokes a single spike at a latency of about 10 msec. When stimulus intensity increases, the latency of the first spike shows a smooth decrease to a minimum of about 2 msec and additional spikes are added. The smooth decrease in latency of the first spike suggested to Szabo and Hagiwara (1967) that mormyromast afferents may use spike latency as a code for stimulus intensity, and several physiological and behavioral observations support this hypothesis (Hall *et al.*, 1995).

4. PASSIVE ELECTROLOCATION

The process of attending to electric fields in the environment is called passive electrolocation (Kalmijn, 1966). Naturally occurring electric fields of biotic or abiotic origin are abundant in the environment of fishes. Biotic fields are produced by animals and plants when a more or less permeable membrane separates fluids of different ionic concentrations. All epithelia, especially those with thin boundaries, such as gills, the gut, or other body cavities, are electrical current sources because of leakage of charged particles (ions). As a result, animals (and sometimes plants) are surrounded by electrical potentials, which can additionally be modulated, for example, when fishes move their mouths or operculi. When animals are wounded, the electric currents increase considerably, making them even more detectable for electrosensory predators like, for example, sharks. Another, but usually weaker source of biotic voltages are contracting muscles. A moving animal thus constitutes a source of a slowly fluctuating but very variable electric field (Kalmijn, 1971; Peters and van Wijland, 1974).

In addition to the above-mentioned source of low-frequency electric fields, EODs of weakly electric fishes constitute important sources of high-frequency fields of biotic origin. They can be detected by other weakly electric fishes that have high-frequency tuberous electroreceptor organs, designed to detect the EOD of conspecifics. These fish are able to identify the sender identity (species and sex) by analyzing the waveform of their EOD, and in some cases by analyzing the temporal pattern of EOD production. In addition, they are capable of finding their way to the source of an EOD stimulus by following electric field lines, without actually knowing where exactly the stimulus is located (Davis and Hopkins, 1988; Hopkins, 2005).

Abiotic electric fields are also common in water habitats. They are caused by geochemical or seismographic processes and, in seawater, also by the flow of water through the Earth's magnetic field. Certain rock formations together with their surroundings can constitute sources of electric voltages and thus can serve as electrical "landmarks" for spatial orientation of electroreceptive fishes. All abiotic fields mentioned so far are DC or low-frequency sources, whose frequency often depend on the relative movement of the receiver, the electrosensory fish. High-frequency sources of abiotic fields are lightning (Hopkins, 1973) and also magnetic storms or seismic activity (Kalmijn, 2003).

Electrosensitive fish use ampullary electroreceptor organs for the detection of the low-frequency biotic or abiotic electric fields (Kalmijn, 1974). These receptor organs probably are the original type of electroreceptor organ, while tuberous organs might represent derived forms. Ampullary receptor organs in non-teleosts, that is, elasmobranch fishes, are called ampullae of Lorenzini (Lorenzini, 1678). Teleost electroreceptive species also possess ampullary organs, which were probably reinvented twice during evolution. All ampullary receptor organs consist of a chamber in the epidermis that can contain hundreds of electroreceptor cells. In marine species, a long canal filled with an electrically conducting substance with a length of up to 20 cm extends from the chamber to a pore in the skin connecting to the water. Because of physical reasons, canals are much shorter in freshwater species. The electroreceptor cells are embedded with most of their soma in the mass of supporting cells in the wall of the chamber. Like tuberous electroreceptor cells, they respond to the voltage difference between the opening of the pore at the skin and the subcutaneous tissue below. Even though non-teleost and teleost ampullary organs respond to opposite polarities, their basic physiological principles might be the same.

Lorenzini ampullae of marine species are extremely sensitive. They respond to voltage gradients of less than 0.01 μV/cm, which corresponds to a voltage of less than 1 V over 1000 km (Bennett and Clusin, 1979). This enables sharks and rays to sense voltages emanating from prey using passive electrolocation for foraging. In addition, elasmobranchs use their passive electric sense for orientation and navigation in their environment by sensing local oceanic electric fields such as those measured by (Pals *et al.*, 1982) on the ocean floor. Marine sharks and rays can employ their electric sense to indirectly measure the Earth's magnetic field and thereby acquire a compass sense (Kalmijn, 1974). Electric currents are induced across ocean currents and the bodies of the animals when they float or swim through the Earth's magnetic field. However, the exact mechanism for magnetoreception through the electric sense is still debated (Paulin, 1995; Kalmijn, 2003; Walker *et al.*, 2003).

Teleost ampullary receptor organs are less sensitive than the ampullae of Lorenzini. Nevertheless, teleost fish also use passive electrolocation for prey detection and orientation. The catfish *Ictalurus*, for example, was shown to be able to electrically detect a prey goldfish behind a layer of electrically transparent agar (Kalmijn, 1974). Other predatory catfishes (*Clarias* sp.) hunt for weakly electric mormyrids by detecting the low-frequency component of their EODs with ampullary electroreceptors (Hanika and Kramer, 1999). Weakly electric fishes that use active electrolocation (see later) can also passively electrolocate prey (von der Emde and Bleckmann, 1998) and electrical landmarks (Moller, 2002) by employing their low-frequency ampullary electroreceptor organs.

As mentioned earlier, passive electrolocation is a sense found in many aquatic vertebrates. One important functional role in many species is prey detection on a short range, augmented by vision, chemoreception, and mechanosensation (mechanosensory lateral line) (Collin and Whitehead, 2004). In addition to detecting prey, young and juvenile elasmobranchs also use electroreception to detect potential predators (Sisneros *et al.*, 1998). Weak bioelectric fields can also provide the stimulus for the localization of mates and conspecifics. In round stingrays, for example, buried females modulate the ionic potentials produced by their spiracles, mouth, and gill slits during ventilation in order to be located by searching males (Tricas *et al.*, 1995). Another important role passive electrolocation plays during orientation in the habitat, in marine species particularly during migration and geomagnetic orientation. Even though other senses are also involved in these behaviors, electroreception plays an essential part (Collin and Whitehead, 2004).

5. PRODUCTION OF ELECTRIC SIGNALS

Weakly electric fish produces two types of electrical signals: several families of the South American Gymnotiformes and the single member Gymnarchidae (Mormyriformes, Africa) emit wave-type EOD that consist of a continuous sinusoidal signal, which is only rarely interrupted. In contrast, all Mormyridae and the members of some gymnotiform families emit pulse-type EOD, which consist of brief pulses followed by a pause that is longer than the EOD pulse itself (Figure 7.1).

Each pulse of the EODs emitted by mormyrids is an all-or-nothing event and its waveform is species specific and often sex specific (Hopkins, 1988). EODs of different species have durations between 0.2 and several milliseconds and are produced with a variable temporal pattern. The basically biphasic EOD of *G. petersii*, for example, lasts for about 0.4 msec and the interval between EODs can vary from 8 msec during aggressive encounters, or 15 msec during periods of active electrolocation, to 500 msec or more during resting (Kramer, 1990). Variations in the temporal pattern of EOD

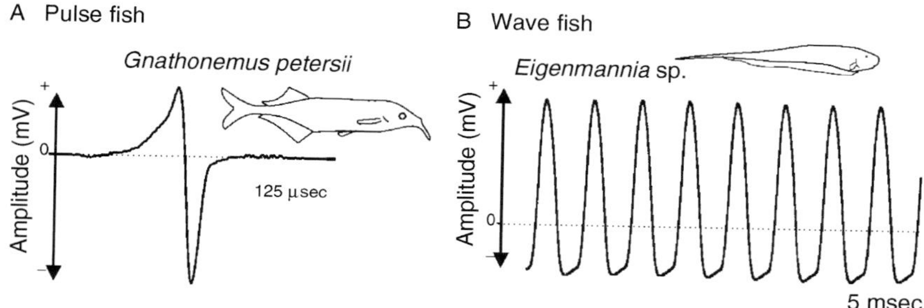

Fig. 7.1. Examples of a pulse-type EOD (A), produced by the African mormyrid *Gnathonemus petersii*, and a wave-type EOD (B), produced by the South American gymnotid *Eigenmannia virescens*.

production can correlate with certain behaviors. For example, fish emit a specific EOD rhythm at each stage of a trial when solving a conditioned electrolocation task. During active electrolocation of objects, EOD rate is very regular and fast (~70 Hz). In contrast, during swimming or during food search, EOD production can be extremely irregular and the average EOD rate is lower (von der Emde, 1992; Carlson, 2002).

Sensory stimulation can trigger the emission of EODs. A fish will respond with a so-called "novelty response," which consists of a transient increase of the EOD frequency, when it is stimulated by a novel, unexpected sensory signal of any modality (Post and von der Emde, 1999). Novelty responses evoked by low-intensity stimuli resemble "orienting responses" (Sokolov, 1963), reflexlike behaviors, which are found in all vertebrates and serve to orient the animal toward a novel stimulus. During active electrolocation, electric fish acquire information about the environment by analyzing the information provided by their self-produced EODs. Therefore, an increase in EOD rate during a novelty response will increase the flow of information and thus help the fish to analyze the source of the unexpected stimulus.

The temporal pattern of EOD production is regulated by the fish's brain. Each EOD is initiated by two action potentials running down the spinal cord to the electromotoneurons, whose synchronous response then leads to a single EOD. In mormyrids, the trigger for an EOD originates in the command nucleus (CN) of the brain. The synchronous discharge of its 20, multipolar neurons innervates the relay nucleus (RN) located right above it, whose axons run down the spinal cord toward the electromotoneurons. Together, the medullary RN and the CN constitute the central pattern generator which drives the electric organ.

Two important sources of afferent input to the CN are the precommand nucleus (PCN) located at the diencephalic/mesencephalic border, the thalamic dorsal posterior (DP) nucleus, and the ventroposterior (VP) nucleus of the torus semicircularis (Bell *et al.*, 1983; Carlson, 2003; von der Emde, 2004). Command neuron dendrites in the reticular formation also receive significant inputs from other sources. CN neurons do not show any intrinsic rhythmic activity of their own. If the connection between the midbrain and the CN is cut, no more EODs are initiated by the CN–RN complex. Thus, for the natural rhythm of EOD production and for electromotor responses to sensory stimuli, input from the midbrain is essential.

Stimulation of PCN, DP, and VP show that they are important sources of afferent input to the CN of the electromotor pathway. PCN and VP convey excitatory inputs to CN, while VP neurons are inhibitory, projecting also to PCN and DP. Probably, PCN and DP neurons integrate information of various origins and individually relay this to the CN in the medulla. PCN neurons may also have intrinsic, though normally nonsynchronized, pacemaker properties. Descending input from PCN, DP, and VP is integrated at the postsynaptic level in CN. When the result of this integration process is sufficient to activate CN neurons beyond their firing threshold, initiation of the electromotor command and the subsequent firing of the entire electromotor pathway follow. Thus, the integration of excitatory synaptic input is the most important factor contributing to the initiation of the firing of command neurons and thus the generation of the descending command signal (Grant *et al.*, 1999).

Collateral branches of the axons from CN to RN also project bilaterally to the bulbar command-associated nucleus and are the starting point of the so-called *corollary discharge pathway*, which runs from CN via several other command-associated nuclei to the electrosensory lateral line lobe (ELL) and other electrosensory areas of the mormyrid brain (Grant *et al.*, 1986; Bell and von der Emde, 1995; Bell *et al.*, 1995). The corollary discharge provides information about the timing of electromotor activity and has a strong effect on the processing of reafferent sensory input that is evoked by the EOD (see later).

6. ACTIVE ELECTROLOCATION

Even though electroreception is found in a number of aquatic animal groups (Bullock *et al.*, 1983; Heiligenberg, 1993; von der Emde, 1998b), which all use it as a passive electrolocation system, *active* electrolocation is only used by weakly electric fishes. The combination of electrogenic abilities (the production of weak electric signals with specialized electric organs) and epidermal electroreceptor organs can be found only in the South American Gymnotiformes and the African Mormyriformes. While an EOD is emitted, an electrical field builds up around the fish in the water (Figure 7.2). The field produced by the basically biphasic EOD of the mormyrid *G. petersii* is an asymmetric dipole field with one smaller pole at

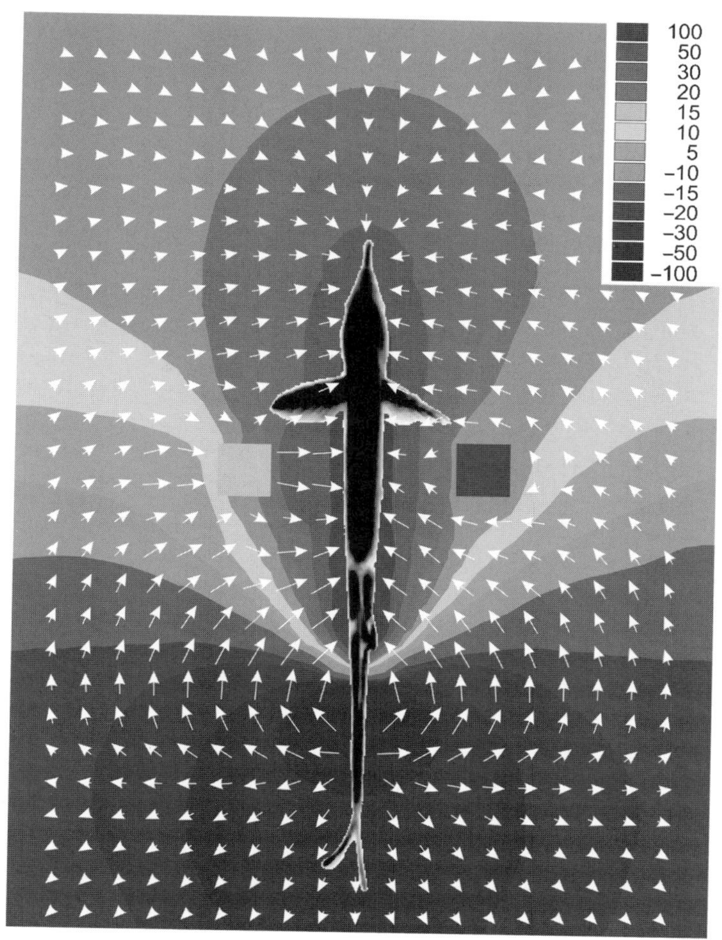

Fig. 7.2. The electric field existing during the peak of the first positive phase of the EOD of *G. petersii.* Equipotential areas are color coded (scale in upper right corner in mV/cm), while the direction of electrical field lines is indicated by white arrows. [Adapted from von der Emde and Schwarz (2001) with permission of Copyright 2002 Elsevier GmbH, Urban & Fisher Verlag.] (See Color Insert.)

the fish's tail and the other pole constituting the entire body of the fish anterior to the electric organ. Because water is a conducting medium, alternating electric current flows through the water and enters (or leaves) the fish's body mainly through the pores of the electroreceptor organs. The electroreceptor cells measure the electrical current flowing through them, which is proportional to the local electrical voltage between the inside and the outside of the fish. The electrical current density is highest at the tail and at the head of the fish (Caputi and Budelli, 2006). Because of the extremely constant EODs produced by the electric organ, the resulting electric field causes a specific spatial voltage pattern over the fish's skin surface that does not change as long as the electric field is not distorted by objects and the fish does not bend its body. (For a discussion of the effects of body movements on the electric field see Rother *et al.*, 2003; Caputi and Budelli, 2006.)

If the fish approaches an object with electric properties different from those of the surrounding water, the electric field is distorted. The 3D field distortions lead to a change in the voltage pattern within the "electric image" which the object casts onto the fish's skin surface (Rasnow and Bower, 1997; Caputi *et al.*, 1998; von der Emde *et al.*, 1998). Thus, the electric image is defined as the local modulation of the electric field at an area on the skin. In this skin area the voltage pattern is different in the presence of the object compared to the situation without the object.

A typical electric image has a center-surround (Mexican hat) spatial profile (Caputi *et al.*, 1998). For example, a good conductor produces an image with a large center region where the local EOD amplitude increases, surrounded by a small rim area where the amplitude decreases (Figure 7.3). The image of a nonconductor has an opposite appearance: in its center, local EOD amplitude decreases while it slightly increases in the surrounding rim area.

It is important to note that electric images are fundamentally different from optical images that are projected onto the retina of a vertebrate eye for vision. Because there are not any focusing mechanisms, electrical images are always blurred, or "out of focus." They would be in focus only if the distance between object and skin is zero. In addition, there is no one-to-one relationship between spatial object properties and image shape: electrical images are always strongly distorted compared to an optical projection of a 3D object onto a 2D surface. Optical images are determined by object shape, size, and other geometrical parameters. In contrast, electric images depend on several additional object properties, such as electrical properties, object depth, object distance, location along the fish's body, the angle relative to the longitudinal axis of the fish, the fish's skin and body properties and proportions, bending movements of the fish's body, the presence of additional objects, the background, and many more. All these parameters

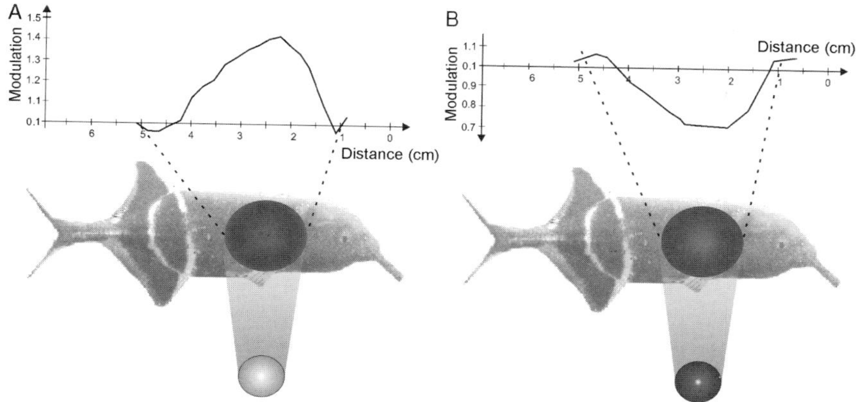

Fig. 7.3. Electric images of two spherical objects on the skin of *G. petersii.* In the lower diagrams the electric images are color coded on the fish's skin with a voltage increase shown in red and a decrease shown in blue. In the upper diagrams, the voltage modulation of a horizontal section through the image is shown. A modulation of 1 corresponds to no change in local EOD amplitude, while values larger or smaller than 1 indicate an increase or decrease of local voltages, respectively. (A) A typical image of a metal sphere is shown. It consists of a large center area in which the local voltage increases because of the presence of the conducting object, while in a smaller rim area local voltages decrease slightly. (B) An opposite voltage pattern is caused by the presence of a plastic sphere, which is an insulator. (See Color Insert.)

play a role in shaping the appearance of the electric image. However, despite these multiparametrical dependencies and for a human observer unpredictable appearance of an electric image, weakly electric fish are able to recognize a surprising variety of object properties even in complex natural scenes with multiple kinds of different objects (see later).

In addition to changing the amplitude of the locally occurring signal within the electric image, some objects also change the waveform of the local EOD (Heiligenberg, 1977; von der Emde, 1990). In mormyrids, the local waveform of the signal is only minimally altered by an object without capacitive properties, like most nonliving objects (stones, washed-out dead wood, and so on). However, living objects such as insect larvae (the main food item of many electric fish), other fishes, or living plants, possess complex impedances with a considerable capacitive component (Schwan, 1963; Heiligenberg, 1977; von der Emde and Bleckmann, 1992). These objects distort the local EOD waveform in addition to their effect on local EOD amplitude.

Waveform distortions are maximal for certain capacitive values. For example, with so-called "dipole objects" (von der Emde, 1990), maximal waveform distortions occur at a capacitive value of 1 nF. Larger or smaller capacitive values cause smaller distortions, and below 300 pF and above 300 nF no waveform distortions have been observed. Most living objects that can be found in the natural habitat of mormyrids have capacitive values that distort the EOD waveform (von der Emde and Schwarz, 2002). Like amplitude changes, local waveform changes depend on the distance of the object from the skin and on the size of the object (von der Emde and Ronacher, 1994).

6.1. What Can Electric Fish Perceive About Their Environment During Active Electrolocation?

The pioneers of electric fish research, Lissmann and Machin (1958), showed that electric fish can detect resistive properties of objects during active electrolocation. As noted earlier, the amplitude of the local EOD in the center of the electric image depends on the impedance of the object: low-impedance objects increase the EOD amplitude, while objects of high impedance decrease it (Figures 7.2 and 7.3). Lissmann assumed that the fish perceive a "black-and-white" electrical picture of their surroundings, that is, they only detect amplitude differences of the local EOD. However as pointed out earlier, animated objects also have capacitive properties. Heiligenberg (1973) suggested that in actively electrolocating fish, phase (timing-) modulations of the local EOD are caused by capacitive objects. However, because of the frequency composition of the EOD and the geometry of the electric field, the effects of capacitive objects on temporal properties of the EOD are rather small in mormyrids (maximally around 1 μsec with natural objects). Instead, relatively large effects occur in the waveform of the local EOD (von der Emde, 1990).

By conducting behavioral training experiments, we could show that mormyrids and also gymnotiforms can perceive capacitive object properties, a process that we have called "capacitance detection." We also showed that, in order to perform capacitance detection, the animals measure either EOD waveform distortions (mormyrids) or phase shifts (gymnotiforms) (von der Emde, 1990, 1998a; von der Emde and Ringer, 1992). Mormyrids can detect EOD waveform distortions independently of EOD amplitude changes, and thus can determine quantitatively the capacitance and the resistance of an object under investigation (von der Emde and Ronacher, 1994). It was shown that also some gymnotiforms can discriminate between different waveforms of their EODs during active electrolocation (Aguilera and Caputi, 2003).

Only a certain range of capacitive values can be detected by *G. petersii* and other mormyrids. Smaller fish species, most of which emit very short EOD pulses containing mostly high frequencies, can detect smaller capacitive values than fish with long-lasting (low frequency) EODs (von der Emde and Ringer, 1992). The detectable range of capacitances corresponds to the range of capacitive values of animated objects found in the natural habitat of the fish (von der Emde, 1990). Therefore, capacitance detection could be used to find living objects such as food items or other fishes. Because small fish can detect smaller capacitances, they might be able to detect smaller prey items, for example, smaller chironomid larvae.

How are capacitive object properties perceived, and how are they discriminated from purely resistive properties? Previous studies have shown that electric fish react with a novelty response to capacitive shunts in their vicinity (Harder *et al.*, 1967; Meyer, 1982). Because capacitances have impedance for alternating electrical currents, the perception of these stimuli could be based on the same sensory mechanisms as that for resistive objects: the receptors near the object detect a sudden change in peak intensity of the local EOD. Fish with high-frequency signals paying attention to amplitude modulations of their EOD should detect smaller capacitances than fish emitting long, low-frequency signals (Meyer, 1982) because the impedance of capacitances is lower for signals with higher frequencies. Further, should the fish merely detect amplitude changes of the signal, they would not be able to discriminate between a capacitive and resistive object causing the same change in amplitude. However, several studies have shown that several species of mormyrids (von der Emde, 1990; von der Emde and Ringer, 1992) and gymnotiforms (von der Emde, 1998a; Wagner, 1998) can do just this especially well.

For capacitance detection, *G. petersii* measures EOD waveform distortions. They do so by employing the two types of electroreceptor cells contained in each mormyromast organ, the A- and B-cells (von der Emde and Bleckmann, 1992). Especially the B-cells are extremely sensitive to even minute waveform distortions of the local EOD. Recording experiments from the afferent fibers innervating A- and B-cells revealed that it is the EOD waveform, and not the phase or the timing of the local EOD that is utilized by the fish to detect capacitive object properties. B-cells only responded to *natural* waveform distortions as they occur in the presence of capacitive objects. When stimuli were presented that were manipulated in their timing (or phase), B-cells did *not* respond with increased firing activity. In contrast, stimuli lacking timing cues but with a distorted waveform were quite effective (von der Emde and Bleckmann, 1997). This shows that B-receptor cells respond to the complex waveform changes during capacitance detection and do not utilize temporal cues. In contrast, gymnotiform wave species, such as *Eigenmannia*, might measure timing cues for capacitance detection. In these species, phase shifts caused by natural capacitive objects can be larger than 200 μsec, and thus are significantly more pronounced than in mormyrids (von der Emde, 1998a).

The detection of capacitive properties of objects through waveform distortions can be compared to the detection of color by the visual system, which utilizes the wavelength of the light reflected from an object. In vision, humans can determine color (frequency) and brightness (amplitude) of an object independently from one another (Garner, 1974), which is similar to the independent detection of an object's capacitance (waveform) and resistance (amplitude) during active electrolocation. Capacitance detection thus adds "color" to the otherwise "black-and-white" electrical picture of the environment and allows the fish to gather much more information (von der Emde, 1999). For example, during foraging the electrically "colored" food items, which have capacitive properties in contrast to the inanimate surroundings, "pop out" of an electrically "gray" background and are thus more easily detected and identified by the fish.

Besides finding out about an object's electrical properties, mormyrids can also localize objects in 3D space during active electrolocation. When presented with two objects at different distances, *G. petersii* can learn to choose that object which is located farther away than an alternative one. This discrimination is based only on distance, and is independent of the size or electrical properties of the objects (von der Emde *et al.*, 1998; Schwarz and von der Emde, 2001). Thus, the fish have a true sense of depth perception. Distance measurement is based on a novel sensory mechanism, by which *G. petersii* measures the *normalized maximal slope* of the electric image, which depends only on distance and not on size or other object properties. Fish can measure the slope of the electric image by determining the intensity change at the image's edges, or the "fuzziness" of the edges. For the *normalized* slope, they set this slope value in relation to the maximum intensity in the image's center. The slope/amplitude ratio is an unequivocal means to determine the distance of most types of objects. The farther away an object is, the more the electric image is "out of focus." Using the slope/amplitude ratio is a unique mechanism because it is very fast and requires only a single 2D receptive surface, which does not have to be moved (von der Emde *et al.*, 1998).

In experiments we found that, during active electrolocation, *G. petersii* can also determine the 3D *shape* of objects independently of its distance. Individual fish were trained in a two-alternative forced-choice procedure to discriminate between two differently shaped metal objects, for example, a cube and a cylinder. If they chose the cylinder by swimming through a nearby gate into another compartment of their tank, they received a food reward. In contrast, after swimming through a gate near the cube, they were chased back into their home compartment without any reward. Within about 2 weeks, they learned to avoid the negative object (S−) and only swam through the gate with the positive (reinforced) object (S+). Subsequent tests revealed that the fish had learned to remember both the shape of the negative object as an aversive stimulus and also the shape of the positive object as a positive stimulus (Davis and von der Emde, 2003).

To further investigate this issue, we conducted additional experiments by training fishes to discriminate between a certain object pair, each. A careful analysis of their choice behaviors during tests with new combinations of objects revealed that during discrimination the fish paid attention not only to object shape but in addition to many other object parameters. During training with a particular object combination, they had learned to remember not only the shapes of the positive and negative objects but also other object properties such as volume, height, material (metal or plastic), whether they possessed corners or rounded edges, how they were oriented in space, and many more. The fish assigned negative or positive attributes to each object parameter, depending on whether a particular feature was associated with the negative or the positive object during training. Some object properties, for example, large volumes or metal properties, were a priori assigned negative values, probably because of a genetic bias. The fish also had learned to weight the *relations* between the two objects: for example, they had learned to avoid the larger object and choose the smaller one. When during a test trial they had to choose between two novel objects, they decided on the basis of their previous training. They inspected both objects and compared their positive and negative properties. When this inspection resulted in more negative than positive characteristics of an object compared to the alternative one, they avoided the more negative and chose the positive object. The percentage of choice for the positive object depended on the difference between the two objects (S. Fetz and G. von der Emde, unpublished results).

Each individual fish had its own way of judging different object properties. Most animals placed much emphasis on the volume and material of an object. Object shape was also considered but played a secondary role. For example, when an S− object was large and made out of metal, fish learned quickly to avoid all large and low-resistive objects. When during a choice experiment a novel object combination with objects of identical volume and identical material were offered, fish decided between these objects according to object shape and some additional parameters such as height, possession of corners, and other local features (S. Fetz and G. von der Emde, unpublished results).

6.2. How Can the Fish Extract Information About Objects During Active Electrolocation?

In order to gain multidimensional information about an object during active electrolocation, electric fish have to analyze the electric images, which the object projects onto their skin. For different object parameters, different image parameters are useful. For example, a fish might want to know about the material of an object in order to decide whether the object is a potential food item. One way to achieve this would be to determine the object's complex impedance, that is, its capacitive and resistive values. If capacitive values fall into a certain range which is typical for an insect larva, the fish might decide to eat it. In order to determine the impedance of an object, the fish has to measure the waveform and the amplitude of the locally perceived signal within the electric image.

However, the amount of waveform distortions and the amount of amplitude change does not depend on object capacitance alone, but in addition on object size (including volume) and distance from the fish. The closer an object, the larger are the changes in waveform and amplitude. The size of the object has a similar effect: larger objects cause both larger waveform distortions and larger amplitude changes. It follows that a fish would be unable to determine the object's complex impedance if it only measures waveform distortions and/or only amplitudes of the locally occurring signals and does not know distance and size of the object. One might speculate that for measuring object size, the fish could use the width of the electric image that is projected onto the skin surface. However, even though object size and image width are correlated, image width also depends strongly on the distance of the object. Image width alone can therefore not be used as an independent indicator for object size.

If the fish were able to measure object distance, they could use this value to "calibrate" their waveform, amplitude, and image width measurements. As noted earlier, we have shown that *G. petersii* can accurately measure the distance of novel objects, and it is very likely that distance can be measured independently and in addition to waveform and amplitude of the signal (von der Emde *et al.*, 1998; Schwarz and von der Emde, 2001). There is good evidence that *G. petersii* measures an unknown object's distance by using the "slope/amplitude ratio." This would allow them to determine the size of the object by using the image's width. Finally, they can determine the object's complex impedance by measuring the waveform and amplitude of the local EODs, taking distance and size into account.

All the points just mentioned make it clear that for object evaluation fish have to perform several measurements simultaneously and integrate them all to get an accurate image of a 3D object. The fish have to measure at least four parameters of the electric image (peak amplitude, maximal slope, image width, and EOD waveform distortions) to determine an object's location, its distance, size, volume, and complex impedance. In order to determine the shape of an object, some additional parameters might be necessary such as dynamic parameters arising from swimming around the object and scanning it from different sides and angles.

It is important to note that things might be easier for very small, point-like targets, such as *Daphnia*. It appears to be possible that with such targets the gymnotiform *Apteronotus leptorhynchus* can infer object distance by measuring only relative image width (Rasnow, 1996; Assad *et al.*, 1999). This would simplify and expedite the identification and localization of small prey.

It appears that an accurate distance measurement is a crucial factor for determining all other object parameters. Therefore, it makes sense that *G. petersii* can measure object distance extremely fast, maybe even from a single electrical snapshot produced by one EOD. In many situations, fish might not have enough time to perform complicated, time-consuming processes (such as "electrosensory flow" and so on) for distance determination.

6.3. Specializations of Certain Skin Regions for Active Electrolocation

It was suggested for gymnotiforms (Castelló *et al.*, 2000; Caputi *et al.*, 2002) and for mormyrids (von der Emde and Schwarz, 2001) that some weakly electric fish possess an "electric fovea" in their skin. For gymnotiforms this suggestion was made for *Gymnotus carapo* (Castelló *et al.*, 2000; Aguilera *et al.*, 2001). In these fish, foveal and parafoveal regions were proposed to be located on the rostral pole of the fish's body, just below and above the snout. Here, the receptor organ density is exceptionally high and receptors from these regions project to a large area of the ELL in the brain (Castelló *et al.*, 1998), where they are overrepresented. In addition, the locally occurring EOD signal in these regions is different from that at other skin areas because of special electrical properties of the skin and different morphologies of the body and the electric organ. The authors proposed the term "prereceptor processing" for this effect, which occurs at the fovea and parafovea of *G. carapo* (Caputi *et al.*, 2002).

For the mormyrid *G. petersii*, former studies had shown that the distribution of electroreceptor organs varies at different skin areas (Harder, 1968; Quinet, 1971) which might lead to different spatial resolutions of the electric sense. We conducted behavioral, anatomical, and electrophysiological experiments in order to test whether *G. petersii* possesses a structure that can be called "electric fovea" and which is principally similar to the visual or acoustic fovea of other animals (Schwarz, 2000).

Our findings lead to the hypothesis that *G. petersii* has two foveal skin areas on its electroreceptive body surface. These foveae are used for active electrolocation and differ from all other skin areas with respect to their anatomy, behavioral use, and possibly physiology. One area is the electrical fovea is the "Schnauzenorgan," which is the long, movable chin appendix of *G. petersii*. A second region also specialized is the so-called "nasal region" located directly above the mouth and below the nostrils of *G. petersii*.

The densities of mormyromast electroreceptor organs at the Schnauzenorgan and to a lesser extent at the nasal region are considerably higher compared to all other skin areas of the fish (von der Emde and Schwarz, 2002; Hollmann and von der Emde, 2004). Mormyromast electroreceptor organs are the only electroreceptor organs involved in active electrolocation (Bell, 1990b). Our measurements indicate that mormyromast density is relatively low on the ventral and dorsal surfaces of the fish's trunk. On the lateral sides of the fish, the areas surrounding the lateral line canals are completely free of electroreceptor organs. Electroreceptor density is slightly higher on the head outside the nasal region, still higher on the nasal region, and the highest at the Schnauzenorgan. Here, receptor density increases dramatically toward the Schnauzenorgan's tip (Figure 7.4).

We found that mormyromast organs located at the Schnauzenorgan are morphologically different (e.g., they are smaller) from those at other skin

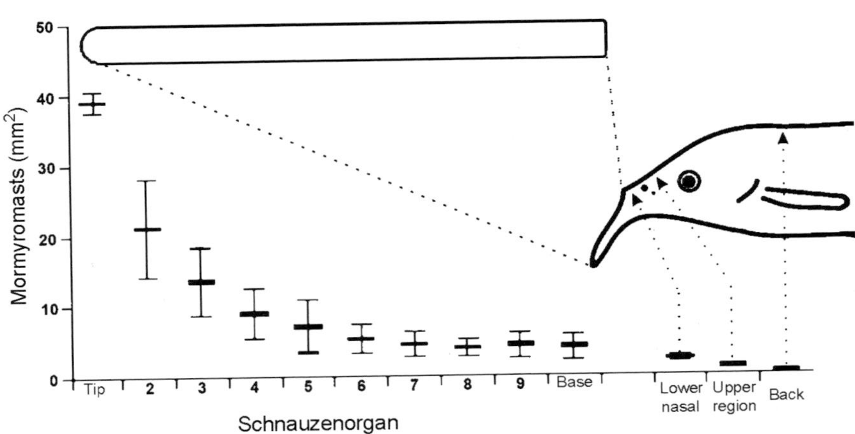

Fig. 7.4. Density (±standard deviation) of mormyromast electroreceptor organs on the body surface of *G. petersii*. The Schnauzenorgan is enlarged in the left of the diagram. It was divided into ten sectors. The lower nasal and upper nasal regions are the skin areas just below and above the nares. (This figure is adapted from unpublished data provided by M. Hollmann and G. von der Emde.)

areas. They also may be physiologically different, but this has to be proven by future experiments. Knollenorgan receptor organs, which are used exclusively for electrocommunication, and ampullary organs, which are used for passive electrolocation, decrease in their densities at the nasal region and on the Schnauzenorgan (von der Emde and Schwarz, 2002; Hollmann and von der Emde, 2004).

The skin at the nasal region of *G. petersii* differs in its electrical properties from that at other body regions (Caputi *et al.*, 1998; Schwarz, 2000). We found that the skin at the nasal region has the highest resistance of all skin areas, even slightly higher than that of the Schnauzenorgan. Capacitive skin properties are also highest at the nasal region and the Schnauzenorgan. This combination of high resistance and high capacitance may influence the electric field in a way to cause changes in field geometry and waveform distortions of the local EOD occurring both in the absence and the presence of objects. The complex local EOD properties caused by objects at the nasal region and the Schnauzenorgan may provide additional information for the fish about certain object properties (Assad *et al.*, 1999; Budelli and Caputi, 2000). Also the visual foveae of the eyes of many animals show structural specializations. In the mammalian retina, a foveal depression is found with several structural specializations ensuring that the fovea can serve its special functions.

Behavioral experiments have shown that both the nasal region and the Schnauzenorgan of *G. petersii* play important roles during active electrolocation of novel objects (Toerring and Moller, 1984; Schwarz and von der Emde, 2001). Objects are approached by the fish in such a way that they first influence the nasal region. If during this approach it turns out that an unknown object is large and potentially dangerous, the fish might turn around and inspect the object tail first (von der Emde, 1992). All other objects are approached head first, and at a closer distance, the Schnauzenorgan is moved over the object like a probe (T. Röver, G. Henscheid, and G. von der Emde, unpublished results). During foraging, *G. petersii* searches for prey insect larvae head first, using its nasal region like an electrical antenna. The fish swims with its head down at a constant angle of the body axis of $\sim 18°$, which lets the nasal region point forward and slightly upwards at an angle of $\sim 52°$. The tip of the Schnauzenorgan lightly touches the ground and is moved from left to right in a saccadelike manner while the fish moves forward. When a prey item is encountered, it is inspected by the Schnauzenorgan which is moved over it before the fish sucks in its prey (A. Padberg and G. von der Emde, unpublished data). This behavior is comparable to that of visually oriented animals with foveal eyes, which show special behaviors during visual object inspection. Mammals, for example, perform conjugate eye movements to fixate objects of attention on the fovea.

Saccadic eye movement keeps the fovea fixated on the object of interest as it, or the observer, moves.

We propose that the functions of the Schnauzenorgan and the nasal region of *G. petersii* are specialized for different tasks. *G. petersii* might possess two perceptual systems each associated with one electrical fovea: (1) a long-range guidance and detection system—nasal region; and (2) a shorter-range (food) identification system—Schnauzenorgan. During foraging at night, the forward and upward pointing nasal region may be used for detecting obstacles and other objects at a larger distance. Simultaneously, the Schnauzenorgan-saccades move the tip of the Schnauzenorgan over the ground where prey items are hidden. Thus, the Schnauzenorgan is used to detect and especially identify prey at close range. This hypothesis is supported by the presence of the anatomical specializations of the respective skin areas and especially by the behavior of the fish.

6.4. How Does the Brain Extract Information About Objects During Active Electrolocation?

Afferent fibers from mormyromast electroreceptors terminate in the ELL, a cerebellum-like cortical structure in the medulla. Fibers from A-type sensory cells terminate in the medial zone (MZ), fibers from B-type sensory cells terminate in the dorsolateral zone (DLZ), and fibers from ampullary receptors terminate in the ventrolateral zone (VLZ). The electroreceptive surface is somatotopically mapped within each of these zones.

Mormyromast afferent fibers terminate with mixed chemical–electrical synapses on small granular cells in the deeper layers of ELL. Axons of these granular cells terminate in turn on the basilar dendrites of larger cells located more superficially. Some of the granular cell effects on more superficial cells are excitatory and others inhibitory. The larger more superficial cells have apical dendrites as well as basilar dendrites. The apical dendrites extend throughout the molecular layer of ELL where they are contacted by parallel fibers that originate from an external granular cell mass known as the eminentia granularis posterior (EGp). The larger cells include two types of efferent cells, large ganglion (LG) and large fusiform (LF), that relay electrosensory information to higher levels of the system, as well as two types of Purkinje-like inhibitory interneurons known as medium ganglion cells (MG1 and MG2). MG1 cells appear to inhibit LF cells preferentially and MG2 cells appear to inhibit LG cells preferentially (Han *et al.*, 1999). LG cells are inhibited by electrosensory stimuli in the center of their receptive fields whereas LF cells are excited by such stimuli (Bell *et al.*, 1997), and findings suggest that MG1 cells are inhibited by electrosensory stimuli in the center of the receptive field whereas MG2 cells are excited (Mohr *et al.*, 2003).

The receptive fields of many ELL neurons have a center-surround organization (Bell *et al.*, 1997; Metzen and von der Emde, 2004). That means that they are excited (or inhibited) by a stimulus given in the center of their receptive fields and inhibited (or excited) by a stimulus presented within a small area surrounding the field's center (Figure 7.5). These surrounding areas are probably produced by input of neighboring interneurons which receive input from the surround area of the field and have an opposite effect on the cell than the direct or indirect input from afferent fibers bringing information from the field's center. The overall size of the receptive fields of most ELL neurons is less than 10 mm, which is not much larger than the receptive fields of single afferent fibers projecting from single mormyromast organs into the ELL (M. Metzen, J. Engelmann, and G. von der Emde, unpublished data)

These anatomical and physiological findings have been put together with the additional hypothesis of mutual inhibition between MG1 and MG2 cells. The circuit includes LF cells which are efferent "on cells" conveying increases in transcutaneous voltage to higher centers, and LG cells which are efferent "off cells" conveying decreases in transcutaneous voltage to higher centers. The hypothesized mutual inhibition between MG1 and

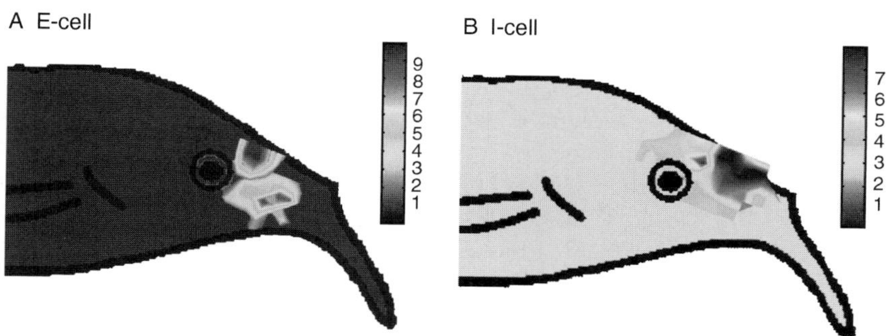

Fig. 7.5. Receptive fields of neurons recorded in the ELL of *G. petersii.* Most receptive fields have a center/surround organization, which means that neural responses increase when a stimulus is located in the field's center and decrease when the stimulus is in an area surrounding the center. Receptive fields are color coded with the number of spikes fired by the neuron in response to a local stimulus is coded according to the scale on the upper right. (A) Receptive fields of two E-cells are shown. These neurons are silent when the stimulus is given outside the receptive field (deep blue area) while they fire up to nine spikes at two regions near the eye. (B) The complex receptive field of a single I-cell is shown. Without stimulation, this cell fires ca. three spikes to the EOCD. A stimulus presented around the nares (deep blue region) inhibits this cell, while a stimulus given above the eye further excites it (yellow/red region). (This figure is adapted from unpublished data provided by M. Metzen and G. von der Emde.) (See Color Insert.)

MG2 cells could mediate contrast enhancement and other functions, as well as help shape the center/surround organization of receptive fields (Sawtell *et al.*, 2004).

Bidirectional topographically organized projections are present between the two mormyromast regions of ELL (Bell *et al.*, 1981). This suggests that the comparison between the responses of A- and B-type mormyromast afferents, necessary for distinguishing the capacitive and resistive properties of objects, could occur in ELL. No functional interaction between the two zones was observed. However, in a study directed at this question (von der Emde and Bell, 1994), cells in the MZ of ELL, for example, behaved just like the A-type afferents which terminate there and are unaffected by small waveform distortions, distortions that had marked effects in the DLZ where B-type afferents terminate. Thus, the comparison of information from A- and B-type seems to occur at a higher level of the system.

The mormyromast regions of ELL receive two different kinds of input with each EOD; EOD-evoked reafferent input from the periphery and corollary discharge signals associated with the EOD motor command. The electric organ corollary discharge (EOCD) signals are conveyed to ELL from various central structures. They can be examined in absence of sensory input by blocking the EOD with curare. The EOD motor command signal continues to be emitted under these conditions but without the normally consequent EOD. The effects of the EOCD can thus be examined in isolation and in combination with electrosensory stimuli controlled by the experimenter.

Interaction between EOD-evoked reafferent input and EOCD signals begins at the very first stage of central processing. Recordings of synaptic potentials show that the EOCD evokes a prominent excitatory postsynaptic potential (EPSP) in granular cells and arrives at these cells at the same time as EOD-evoked reafferent input. The prominent EOCD-driven EPSP appears to serve two different functions. The first is to selectively enhance transmission of the reafferent input that is evoked in mormyromast afferents by the fish's own EOD. Such reafferent input is the only input that can be used for active electrolocation. Input due to other voltage sources such as the discharges of nearby fishes are noise. The disruptive effects of such external noise stimuli are reduced by the EOCD enhancement of responses arriving just after the command to produce an EOD is send to the electric organ. This gating of reafferent input by the EOCD has been demonstrated behaviorally (Meyer and Bell, 1983; Hall *et al.*, 1995). The EOCD-driven enhancement of reafferent input from mormyromast afferents is one of the arguments for assigning these receptors their central role in active electrolocation, particularly since reafferent input in the other two classes of afferent fibers from the Knollenorgans and from ampullary organs is suppressed by the EOCD.

The second function of the EOCD-driven EPSP in granular cells is to provide a timing signal for deriving stimulus intensity from post-EOD latency of afferent spikes. The granular cell receives both EOCD-evoked and the afferent-evoked EPSPs. The sum of these two EPSPs within the granular cell grows larger as the latency of the afferent EPSP gets smaller, reaching a maximum when the peaks of the two EPSPs coincide at the minimal afferent latency. Thus, depolarization and activation of the granular cell by EOD-evoked reafferent responses depend on the latency of the spikes in the afferent fibers.

The EOCD EPSP in granular cells is evoked by a single EOCD-driven spike in fibers from the juxtalobar nucleus, one of the command-associated nuclei that is linked to the CN through a series of three other command-associated nuclei (Bell *et al.*, 1995; von der Emde and Bell, 1996). The timing of the EOD motor command is conveyed with great accuracy over this pathway that includes four synapses. A crude measurement showed that the time delay between the command signal in the CN and a spike in a juxtalobar cell varies by less than 50 μsec. This remarkable preservation of timing information argues in itself for the importance of accurate latency measurement in the mormyromast system.

Granular cells are strongly inhibited by electrosensory stimuli outside the excitatory centers of their receptive fields (Bell, 1990a). This inhibition appears to be mediated by a remarkable inhibitory interneuron in the deeper layers of ELL known as the large multipolar interneuron (LMI; Meek *et al.*, 1996). Both the dendrites and axons of these cells are myelinated and the cells appear to receive very little excitatory synaptic input. Both axonal and dendritic processes give rise to large inhibitory terminals that cover one-fourth to one-half of the surface area of granular cells. The lack of excitatory synaptic input to these cells as well as physiological findings (Han *et al.*, 2000), suggest that they are activated by a rapid nonsynaptic mechanism (e.g., ephaptically) through their large terminals on granular cells following primary afferent-induced depolarization of the granular cells. These anatomical and physiological specializations indicate a very rapidly acting and powerful lateral inhibition that could serve to enhance small differences in latency between neighboring groups of mormyromast afferents and thereby sharpen the electric image of an object (Meek *et al.*, 2001; Mohr *et al.*, 2003).

6.5. Higher Stages of the Electrosensory System

The ELL is only one of several major structures in the mormyrid brain that are concerned with the electrosensory system. At the top of the hierarchy is the valvula cerebelli, which is also known as the mormyrid gigantocerebellum because it is extraordinarily large and covers all the rest of the brain in most species (Nieuwenhuys and Nicholson, 1969). One-third to one-half of the valvula surface is devoted to the electrosensory system. The valvula receives input from and projects back to the lateral and preeminential nuclei. The lateral nucleus receives input from ELL and projects back to the preeminential nucleus as well as to other structures. The preeminential nucleus projects back to ELL both directly to the deep molecular layer and indirectly via EGp and the parallel fibers (Bell *et al.*, 1981; Merten *et al.*, 2000; Meek *et al.*, 2004; Sawtell *et al.*, 2004).

These anatomical findings demonstrate the presence of feedback from higher to lower stages of the mormyrid electrosensory system. Such feedback allows for the results of higher level processing, or for memories that may be stored at the higher levels, to be returned to lower levels for interaction with ascending information from the periphery. Extensive feedback from higher to lower centers is also a major feature of many other sensory systems, including the mammalian visual and auditory systems (Huffman and Henson, 1990; van Essen and Gallant, 1994), but the roles of such feedback are not completely understood.

Very little is known about the physiology of higher order electrosensory structures beyond ELL. Electrosensory responses have been recorded in the preeminential nucleus (von der Emde and Bell, 1995; Sawtell *et al.*, 2004), the lateral nucleus (Mohr and von der Emde, 1998; Goenechea, 2002), parts of the valvula cerebelli (Russell and Bell, 1978), and in the telencephalon (Prechtl *et al.*, 1998). Independent EOCD effects that cannot be explained by prior interaction in ELL have been observed in the preeminential nucleus and valvula cerebelli, and EOCD plasticity that appears to be different from the EOCD plasticity in ELL has been observed in preeminential nucleus (Goenechea and von der Emde, 2004; Sawtell *et al.*, 2004).

REFERENCES

Aguilera, P. A., and Caputi, A. A. (2003). Electroreception in *G. carapo*: Detection of changes in waveform of the electrosensory signals. *J. Exp. Biol.* **206,** 989–998.

Aguilera, P. A., Castelló, M. E., and Caputi, A. A. (2001). Electroreception in *Gymnotus carapo*: Differences between self-generated and conspecific-generated signal carriers. *J. Exp. Biol.* **204,** 185–198.

Assad, C., Rasnow, B., and Stoddard, P. K. (1999). Electric organ discharges and electric images during electrolocation. *J. Exp. Biol.* **202,** 1185–1193.

Bell, C. C. (1990a). Mormyromast electroreceptor organs and their afferent fibers in mormyrid fish. II. Intra-axonal recordings show initial stages of central processing. *J. Neurophysiol.* **63,** 303–318.

Bell, C. C. (1990b). Mormyromast electroreceptor organs and their afferent fibers in mormyrid fish. III. Physiological differences between two morphological types of fibers. *J. Neurophysiol.* **63,** 319–332.

Bell, C. C., and von der Emde, G. (1995). Electric organ corollary discharge pathways in mormyrid fish. II. The medial juxtalobar nucleus. *J. Comp. Physiol. A* **177,** 463–479.

Bell, C. C., Finger, T. E., and Russell, C. J. (1981). Central connections of the posterior lateral line lobe in mormyrid fish. *Exp. Brain Res.* **42,** 9–22.

Bell, C. C., Libouban, S., and Szabo, T. (1983). Pathways of the electric organ discharge command and its corollary discharges in mormyrid fish. *J. Comp. Neurol.* **216,** 327–338.

Bell, C. C., Dunn, K., Hall, C., and Caputi, A. (1995). Electric organ corollary discharge pathways in mormyrid fish. I. The mesencephalic command associated nucleus. *J. Comp. Physiol. A* **177,** 449–462.

Bell, C. C., Caputi, A., and Grant, K. (1997). Physiology and plasticity of morphologically identified cells in the mormyrid electrosensory lobe. *J. Neurosci.* **17,** 6409–6423.

Bennett, M. V. L. (1965). Electroreceptors in mormyrids. *Cold Spring Harb. Symp. Quant. Biol.* **30,** 245–262.

Bennett, M. V. L., and Clusin, W. T. (1979). Transduction at electroreceptors: Origin of sensitivity. *In* "Membrane Transduction Mechanisms" (Cone, R. A., and Dowling, J. E., Eds.), pp. 91–116. Raven Press, New York.

Bodznick, D., and Montgomery, J. C. (2005). The Physiology of low-frequency electrosensory systems. *In* "Electroreception" (Bullock, T. H., Hopkins, C. D., Popper, A. N., and Fay, R. R., Eds.), pp. 132–153. Springer, New York.

Budelli, R., and Caputi, A. A. (2000). The electric image in weakly electric fish: Perception of objects of complex impedance. *J. Exp. Biol.* **203,** 481–492.

Bullock, T. H., Bodznick, D. A., and Northcutt, R. G. (1983). The phylogenetic distribution of electroreception: Evidence for convergent evolution of a primitive vertebrate sense modality. *Brain Res. Rev.* **6,** 25–46.

Caputi, A. A., and Budelli, R. (2006). Peripheral electrosensory imaging by weakly electric fish. *J. Comp. Physiol. A* **192,** 587–600.

Caputi, A. A., Budelli, R., Grant, K., and Bell, C. C. (1998). The electric image in weakly electric fish: Physical images of resistive objects in *Gnathonemus petersii*. *J. Exp. Biol.* **201,** 2115–2128.

Caputi, A. A., Castelló, M. E., Aguilera, P. A., and Trujillo-Cenoz, O. (2002). Electrolocation and electrocommunication in pulse gymnotids: Signal carriers, pre-receptor mechanisms and the electrosensory mosaic. *J. Physiol. (Paris)* **96,** 493–505.

Carlson, B. A. (2002). Electric signalling behavior and the mechanisms of electric organ discharge production in mormyrid fish. *J. Physiol. (Paris)* **96,** 405–419.

Carlson, B. A. (2003). Single-unit activity patterns in nuclei that control the electromotor command nucleus during spontaneous electric signal production in the mormyrid *Brienomyrus brachyistius*. *J. Neurosci.* **23,** 10128–10136.

Castelló, M. E., Caputi, A. A., and Trujillo-Cenóz, O. (1998). Structural and functional aspects of the fast electrosensory pathway in the electrosensory lateral line lobe of the pulse fish *Gymnotus carapo*. *J. Comp. Neurol.* **401,** 549–563.

Castelló, M. E., Aguilera, P. A., Trujillo-Cenoz, O., and Caputi, A. A. (2000). Electroreception in *Gymnotus carapo*: Pre-receptional mechanisms and distribution of electroreceptor types. *J. Exp. Biol.* **203,** 3279–3287.

Collin, S. P., and Whitehead, D. L. (2004). The functional roles of passive electroreception in non-electric fishes. *Anim. Biol.* **54,** 1–15.

Davis, D., and von der Emde, G. (2003). Recognition of object shape during active electrolocation in electric fish. *In* "Proceedings of the 29th Göttingen Neurobiology Conference" (Zimmermann, H., and Elsner, N., Eds.), p. 1018. Georg Thieme Verlag, Göttingen.

Davis, E., and Hopkins, C. D. (1988). Behavioural analysis of electric signal localization in the electric fish; *Gymnotus carapo* (Gymnotiformes). *Anim. Behav.* **36,** 1658–1671.

Garner, W. R. (1974). "The Processing of Information and Structure." Wiley, London.

Goenechea, L. (2002). Antworten einzelner Neurone im elektrosensitiven Seitenlinienlobus und im Nukleus lateralis von *Gnathonemus petersii* bei einfachen und komplexen elektrosensorischen Reizen Institute for Zoology, p. 163. University of Bonn, Bonn.

Goenechea, L., and von der Emde, G. (2004). Responses of neurons in the electrosensory lateral line lobe (ELL) of the weakly electric fish, *Gnathonemus petersii,* to simple and complex electrosensory stimuli. *J. Comp. Physiol. A* **190,** 907–922.

Grant, K., Bell, C. C., Clausse, S., and Ravaille, M. (1986). Morphology and physiology of the brainstem nuclei controlling the electric organ discharge in mormyrid fish. *J. Comp. Neurol.* **245,** 514–530.

Grant, K., von der Emde, G., Gomes-Sena, L., and Mohr, C. (1999). Neural command of electromotor output in mormyrids. *J. Exp. Biol.* **202,** 1399–1407.

Hall, C., Bell, C., and Zelick, R. (1995). Behavioral evidence of a latency code for stimulus intensity in mormyrid electric fish. *J. Comp. Physiol. A* **177,** 29–39.

Han, V. Z., Bell, C. C., Grant, K., and Sugawara, Y. (1999). The mormyrid electrosensory lobe *in vitro*: Morphology of cells and circuits. *J. Comp. Neurol.* **404,** 359–374.

Han, V. Z., Grant, K., and Bell, C. C. (2000). Rapid activation of GABAergic interneurons and possible calcium independent GABA release in the mormyrid electrosensory lobe. *J. Neurophysiol.* **83,** 1592–1604.

Hanika, S., and Kramer, B. (1999). Electric organ discharges of mormyrid fish as a possible cue for predatory catfish. *Naturwissenschaften* **86,** 286–288.

Harder, W. (1968). Die Beziehungen zwischen Elektrorezeptoren, elektrischen Organen, Seitenlinienorganen und Nervensystem bei den Mormyridae (Teleostei, Pisces). *Z. Vergl. Physiol.* **59,** 272–318.

Harder, W., Schief, A., and Uhlemann, H. (1967). Zur Empfindlichkeit des schwachelektrischen Fisches *Gnathonemus petersii* (Mormyriformes; Teleostei) gegenüber elektrischen Feldern. *Z. Vergl. Physiol.* **54,** 89–108.

Heiligenberg, W. (1977). Principles of electrolocation and jamming avoidance in electric fish. A. neuroethological approach. *In* "Studies of Brain Function" (Braitenberg, V., Ed.), pp. 1–85. Springer-Verlag, Berlin/Heidelberg, New York.

Heiligenberg, W. (1993). Electrosensation. *In* "The Physiology of Fishes" (Evans, D. H., Ed.), pp. 137–160. CRC Press, Boca Raton.

Hollmann, M., and von der Emde, G. (2004). Electroreceptor organs in two "electrical foveae" of the weakly electric fish, *Gnathonemus petersii.* Proceedings of the 7th International Congress of Neuroethology, Nyborg/Denmark.

Hopkins, C. D. (1973). Lightning as background noise for communication among electric fish. *Nature* **242,** 268–269.

Hopkins, C. D. (1988). Neuroethology of electric communication. *Ann. Rev. Neurosci.* **11,** 497–535.

Hopkins, C. D. (2005). Passive electrolocation and the sensory guidance of oriented behavior. *In* "Electroreception" (Bullock, T. H., Hopkins, C. D., Popper, A. N., and Fay, R. R., Eds.), pp. 264–289. Springer, New York.

Hopkins, C. D., and Bass, A. H. (1981). Temporal coding of species recognition signals in an electric fish. *Science* **212,** 85–87.

Huffman, R. F., and Henson, O. W. (1990). The descending auditory pathway and acousticomotor systems: Connections with the inferior colliculus. *Brain Res. Rev.* **15,** 295–323.
Kalmijn, A. J. (1966). Electro-perception in sharks and rays. *Nature* **212,** 1232–1233.
Kalmijn, A. J. (1971). The electric sense of sharks and rays. *J. Exp. Biol.* **55,** 371–383.
Kalmijn, A. J. (1974). The detection of electric fields from inanimate and animate sources other than electric organs. *In* "Handbook of Sensory Physiology" (Fessard, A., Ed.), pp. 148–200. Springer-Verlag, Berlin.
Kalmijn, A. J. (2003). Physical principles of electric, magnetic, and near-field acoustic orientation in early aquatic. *In* "Sensory Processing in Aquatic Environments" (Collin, S. P., and Marshall, J., Eds.), pp. 77–91. Springer, New York.
Kawasaki, M. (2005). Physiology of electrosensory systems. *In* "Electroreception" (Bullock, T. H., Hopkins, C. D., Popper, A. N., and Fay, R. R., Eds.), pp. 154–194. Springer, New York.
Knudsen, E. I. (1974). Behavioral thresholds of electric signals in high frequency electric fish. *J. Comp. Physiol.* **91,** 333–353.
Kramer, B. (1990). Electrocommunication in teleost fishes: Behavior and experiments. Springer-Verlag, Berlin.
Lissmann, H. W. (1951). Continuous electric signals from the tail of a fish, *Gymnarchus niloticus* Cuv. *Nature* **167,** 201–202.
Lissmann, H. W. (1958). On the function and evolution of electric organs in fish. *J. Exp. Biol.* **35,** 156–191.
Lissmann, H. W., and Machin, K. E. (1958). The mechanism of object location in *Gymnarchus niloticus* and similar fish. *J. Exp. Biol.* **35,** 451–486.
Lorenzini, S. (1678). Osservazioni intorni alle Torpedini fatte de Stefano Lorenzini Fiorentino e dedicate al Serenissimo Fernando III, Pricipe di Toscana Onofri, Florence.
Meek, J., Grant, K., Sugawara, Y., Hafmans, T. G. M., Veron, M., and Denizot, J. P. (1996). Interneurons of the ganglionic layer in the mormyrid electrosensory lateral line lobe: Morphology, immunohistochemistry, and synaptology. *J. Comp. Neurol.* **375,** 43–65.
Meek, J., Hafmans, T. G. M., Han, V., Bell, C. C., and Grant, K. (2001). Myelinated dendrites in the mormyrid electrosensonry lobe. *J. Comp. Neurol.* **431,** 255–275.
Meek, J., Kirchberg, G., Grant, K., and von der Emde, G. (2004). Dye coupling without gap junctions suggests excitatory connections of GABAergic neurons. *J. Comp. Neurol.* **468,** 151–164.
Merten, U., Goenechea, L., and von der Emde, G. (2000). Responses to electric stimuli of single units in the midbrain of the electric fish, *Gnathonemus petersii. Zoology* **103,** 69.
Metzen, M., and von der Emde, G. (2004). Spatial dimensions of receptive fields and spike latencies of ELL neurons in *Gnathonemus petersii* (Teleostei). Proceedings of the 7th International Congress of Neuroethology, Nyborg/Denmark .
Meyer, J. H. (1982). Behavioral responses of weakly electric fish to complex impedances. *J. Comp. Physiol.* **145,** 459–470.
Meyer, J. H., and Bell, C. C. (1983). Sensory gating by a corollary discharge mechanism. *J. Comp. Physiol. A* **151,** 401–406.
Mohr, C., and von der Emde, G. (1998). Mapping of the nucleus lateralis (torus semicircularis) of the electrosensory system of *Gnathonemus petersii. In* "New Neuroethology on the Move Proceeding of the 26th Göttingen Neurobiology Conference 1998", (Elsner, N., and Wehner, R., Eds.), p. 54. Thieme, Stuttgart, New York.
Mohr, C., Roberts, P. D., and Bell, C. C. (2003). Cells of the mormyrid electrosensory lobe: I. Responses to the electric organ corollary discharge and to electrosensory stimuli. *J. Neurophys.* **90,** 1193–1210.
Moller, P. (1995). "Electric Fishes: History and Behavior." Chapman & Hall, London.

Moller, P. (2002). Multimodal sensory integration in weakly electric fish: A behavioral account. *J. Physiol. (Paris)* **96,** 547–556.

Nelson, M. E. (2005). Target detection, image analysis and modeling. *In* "Electroreception (Springer Handbook of Auditory Research)" (Bullock, T. H., Hopkins, C. D., Popper, A. N., and Fay, R. R., Eds.), pp. 290–317. Springer, New York.

Nieuwenhuys, R., and Nicholson, C. (1969). A survey of the general morphology, the fiber connections, the possible functional significance of the gigantocerebellum of mormyrid fish. *In* "Neurobiology of Cerebellar Evolution and Development" (Llinas, R., Ed.), pp. 107–134. American Medical Association, Chicago.

Northcutt, N. G. (1997). Evolution of Gnathostome lateral line ontogenies. *Brain Behav. Evol.* **50,** 25–37.

Pals, N., Peters, C., and Schoenhage, A. A. C. (1982). Local geo-electric fields at the bottom of the sea and their relevance for electrosensitive fish. *Netherlands J. Zool.* **32,** 479–494.

Paulin, M. G. (1995). Electroreception and the compass sense of sharks. *J. Theor. Biol.* **174,** 325–339.

Peters, R. C., and van Wijland, F. (1974). Electro-Orientation in the passive electric Catfish, *Ictalurus nebulosus* LeS. *J. Comp. Physiol.* **92,** 273–280.

Post, N., and von der Emde, G. (1999). The "novelty response" in an electric fish: Response properties and habituation. *Physiol. Behav.* **68,** 115–128.

Prechtl, J. C., von der Emde, G., Wolfart, J., Karamürsel, S., Akoev, G. N., Andrianov, Y. N., and Bullock, T. H. (1998). Sensory processing in the pallium of a teleost fish, *Gnathonemus petersii. J. Neurosci.* **18,** 7381–7393.

Quinet, P. (1971). Etude systematique des organes sensoriels de la peau des Mormyriformes (Pisces, Mormyriformes). *Ann. Mus. R. Afr. Cent. Tervuren (Belg.) Ser. 8* **190,** 1–97.

Rasnow, B. (1996). The effects of simple objects on the electric field of *Apteronotus. J. Comp. Physiol. A* **178,** 397–411.

Rasnow, B., and Bower, J. M. (1997). Imaging with electricity: How weakly electric fish might perceive objects. *In* "Computational Neuroscience Trends in Research 1997." Plenum Press, New York, London.

Rother, D., Migliaro, A., Canetti, R., Gomez, L., and Budelli, R. (2003). Electric images of two low resistance objects in weakly electric fish. *Biosystems* **71,** 171–179.

Russell, C. J., and Bell, C. C. (1978). Neuronal responses to electrosensory input in mormyrid valvula cerebelli. *J. Neurophysiol.* **41,** 1495–1510.

Sawtell, N. B., Mohr, C., and Bell, C. C. (2004). Recurrent feedback in the mormyrid electrosensory system: Cells of the preeminential and lateral toral nuclei. *J. Neurophysiol.* **93,** 2090–2103.

Schwan, H. P. (1963). Determination of biological impedances. *In* "Physical Techniques in Biological Research" (Nastuk, W. L., Ed.), pp. 323–407. Academic Press, New York.

Schwarz, S. (2000). *Gnathonemus petersii*: Three-dimensional object shape detection and the geometry of the self-produced electric field. Ph D Thesis, University of Bonn, Bonn.

Schwarz, S., and von der Emde, G. (2001). Distance discrimination during active electrolocation in the weakly electric fish *Gnathonemus petersii. J. Comp. Physiol. A* **186,** 1185–1197.

Sisneros, J. A., Tricas, T. C., and Luer, C. A. (1998). Response properties and biological function of the skate electrosensory system during ontogeny. *J. Comp. Physiol. A* **183,** 87–99.

Sokolov, E. N. (1963). "Perception and the Conditioned Reflex." Macmillan, New York.

Szabo, T., and Hagiwara, S. (1967). A latency change mechanism involved in sensory coding of electric fish (mormyrids). *Physiol. Behav.* **2,** 331–335.

Szabo, T., and Wersäll, J. (1970). Ultrastructure of an electroreceptor (Mormyromast) in a mormyrid fish, *Gnathonemus petersii*. II. *J. Ultrastruct. Res.* **30,** 473–490.

Toerring, M. J., and Moller, P. (1984). Locomotor and electric displays associated with electrolocation during exploratory behavior in Mormyrid fish. *Behav. Brain Res.* **12,** 291–306.

Tricas, T. C., Michael, S. W., and Sisneros, J. A. (1995). Electrosensory optimization to conspecific phasic signals for mating. *Neurosci. Lett.* **202,** 129–132.

van Essen, D. C., and Gallant, J. L. (1994). Neural mechanisms of form and motion processing in the primate visual system. *Neuron* **13,** 1–10.

von der Emde, G. (1990). Discrimination of objects through electrolocation in the weakly electric fish, *Gnathonemus petersii*. *J. Comp. Physiol. A* **167,** 413–421.

von der Emde, G. (1992). Electrolocation of capacitive objects in four species of pulse-type weakly electric fish. II. Electric signalling behavior. *Ethology* **92,** 177–192.

von der Emde, G. (1998a). Capacitance detection in the wave-type electric fish *Eigenmannia* during active electrolocation. *J. Comp. Physiol. A* **182,** 217–224.

von der Emde, G. (1998b). Electroreception. *In* "The Physiology of Fishes" (Evans, D. H., Ed.), pp. 313–343. CRC Press, Boca Raton, Florida.

von der Emde, G. (1999). Active electrolocation of objects in weakly electric fish. *J. Exp. Biol.* **202,** 1205–1215.

von der Emde, G. (2004). Distance and shape: Perception of the 3-dimensional world by weakly electric fish. *J. Physiol. (Paris)* **98,** 67–80.

von der Emde, G., and Bell, C. C. (1994). Responses of cells in the mormyrid electrosensory lobe to EODs with distorted waveforms: Implications for capacitance detection. *J. Comp. Physiol. A* **175,** 83–93.

von der Emde, G., and Bell, C. C. (1995).The nucleus prae-eminentialis of mormyrid electric fish: Field potentials, somatotopy and single unit activity 25th Annual Meeting, Society for Neuroscience, San Diego, CA, p. 184.

von der Emde, G., and Bell, C. C. (1996). Nucleus preeminentialis of mormyrid fish, a center for recurrent electrosensory feedback. I. Electrosensory and corollary discharge responses. *J. Neurophysiol.* **76,** 1581–1596.

von der Emde, G., and Bleckmann, H. (1992). Differential responses of two types of electroreceptive afferents to signal distortions may permit capacitance measurement in a weakly electric fish, *Gnathonemus petersii*. *J. Comp. Physiol. A* **171,** 683–694.

von der Emde, G., and Bleckmann, H. (1997). Waveform tuning of electroreceptor cells in the weakly electric fish, *Gnathonemus petersii*. *J. Comp. Physiol. A* **181,** 511–524.

von der Emde, G., and Bleckmann, H. (1998). Finding food: Senses involved in foraging for insect larvae in the electric fish *Gnathonemus petersii*. *J. Exp. Biol.* **201,** 969–980.

von der Emde, G., and Ringer, T. (1992). Electrolocation of capacitive objects in four species of pulse-type weakly electric fish. I. Discrimination performance. *Ethology* **91,** 326–338.

von der Emde, G., and Ronacher, B. (1994). Perception of electric properties of objects in electrolocating weakly electric fish: Two-dimensional similarity scaling reveals a City-Block metric. *J. Comp. Physiol. A* **175,** 801–812.

von der Emde, G., and Schwarz, S. (2001). How the electric fish brain controls the production and analysis of electric signals during active electrolocation. *Zoology* **103,** 112–124.

von der Emde, G., and Schwarz, S. (2002). Imaging of objects through active electrolocation in *Gnathonemus petersii*. *J. Physiol. (Paris)* **96,** 431–444.

von der Emde, G., Schwarz, S., Gomez, L., Budelli, R., and Grant, K. (1998). Electric fish measure distance in the dark. *Nature* **395,** 890–894.

Wagner, K. (1998). Kapazitätsdetektion bei zwei Spezies der südamerikanischen schwachelektrischen Fische *Brachyhypopomus pinnicaudatus* und *Gymnotus carapo,* Institut für Zoologie, Universität Bonn, Bonn, p. 93.

Walker, M. M., Diebel, C. E., and Kirschvink, J. L. (2003). Detection and use of the earth's magnetic field by aquatic vertebrates. *In* "Sensory Processing in Aquatic Environments" (Collin, S. P., and Marshall, J., Eds.), pp. 53–74. Springer, New York.

Wu, C. H. (1984). Electric fish and the discovery of animal electricity. *Am. Sci.* **72,** 598–607.

Zakon, H. H. (1987). The electroreceptors: Diversity in structure and function. *In* "Sensory Biology of Aquatic Animals" (Atema, J., Fay, R. R., Popper, A. N., and Tavolga, W. N., Eds.), pp. 813–850. Springer-Verlag, Berlin/Heidelberg, New York.

Zakon, H. H. (2003). Insight into the mechanisms of neuronal processing from electric fish. *Curr. Opin. Neurobiol.* **13,** 744–750.

8

MAGNETORECEPTION

MICHAEL M. WALKER
CAROL E. DIEBEL
JOSEPH L. KIRSCHVINK

1. Introduction
2. Introduction to Magnetic Field Stimuli
   2.1. The Earth's Magnetic Field
   2.2. Magnetic Fields as Experimental Stimuli in the Laboratory
3. How Can Magnetic Fields Be Detected?
   3.1. Magnetic Field Detection by Electrical Induction
   3.2. Magnetic Field Detection Based on Magnetite
4. Structure of Candidate Magnetite-Based Magnetoreceptors
5. Behavioral Responses to Magnetic Fields in the Laboratory
   5.1. Orientation Responses to Magnetic Field Direction
   5.2. Conditioned Responses to Spatial Variations in Magnetic Field Intensity
   5.3. Other Experiments Using Conditioned Responses
6. Neural Responses to Magnetic Fields in the Laboratory
   6.1. Induced Electrical Signals in Ampullary Electroreceptors
   6.2. Responses in the Trigeminal Nerve of Teleost Fish
7. Neuroanatomy
8. Use of the Magnetic Sense in Navigation
   8.1. Constraints on Theory and Experiment in the Study of Navigation by Fish
   8.2. Hypotheses on Magnetic Navigation Mechanisms
   8.3. Developing Experimental Approaches to Navigation
9. What is Known About the Navigational Abilities of Fish?
10. Concluding Remarks

1. INTRODUCTION

The opportunity to participate in both the discovery and the analysis of the structure, function, and use of a completely new sense is a rare event in the biological sciences (Bullock and Szabo, 1986). In the fishes, such an opportunity has occurred not once but twice in recent decades. The electric

Sensory Systems Neuroscience: Volume 25
FISH PHYSIOLOGY

DOI: 10.1016/S1546-5098(06)25008-8

sense was identified in the elasmobranch and teleost fishes during the 1960s (Murray, 1960; Dijkgraaf and Kalmijn, 1962; Kalmijn, 1966; von der Emde, this volume). Despite having been described nearly three centuries earlier, the function of the ampullae of Lorenzini was not conclusively identified until the work of Dijkgraaf and Kalmijn (1962), and Kalmijn (1966, 1982) in particular. Electrophysiological studies subsequently analyzed in detail the detection of electric fields via the electroreceptor cells located in the ampullae of Lorenzini of the elasmobranch fishes (Clusin and Bennett, 1979a,b) and in the tuberous electroreceptors of some freshwater teleost fishes (von der Emde, this volume).

In contrast with the electric sense, the existence of the magnetic sense was proposed a century before there was any experimental evidence for its existence (Wiltschko and Wiltschko, 1995). The original hypothesis that animals would have a magnetic sense was based on recognition that the Earth's magnetic field could be used by animals to navigate over long distances during homing and migration (Viguier, 1882). In contrast with the electric sense, the existence of the magnetic sense proved difficult to demonstrate experimentally and it is only in recent decades that the existence of the magnetic sense has been widely accepted. Thus, the first experimental evidence for the existence of the magnetic sense in the fishes was not reported until the work of Kalmijn (1978) and Quinn (1980), more than a decade after the electrical sense of elasmobranch fishes had been demonstrated at the behavioral and electrophysiological levels.

Vigorous debates continue, however, over the mechanisms for detection of magnetic fields and how animals might actually use the Earth's magnetic field to guide movement. In the fishes, there are at least two mechanisms proposed for magnetic field detection. Proponents of both hypotheses can clearly identify a detector mechanism located in specific receptor cells, afferent nerves that respond to magnetic field stimuli, and behavioral responses that permit psychophysical analysis of the capacities of the sense. Although many students of animal navigation also think the hypothesis that animals use the Earth's magnetic field in navigation is reasonable, there is as yet very little robust evidence in support of the hypothesis. This chapter begins by briefly characterizing the magnetic field as a stimulus in both the field and the laboratory. The central questions about the magnetic sense that are then addressed in the chapter are as follows:

- What is the evidence that animals, in this case fish, have a magnetic sense?
- What is the evidence that animals use a magnetic sense for navigation?
- What are the priorities for future research?

2. INTRODUCTION TO MAGNETIC FIELD STIMULI

2.1. The Earth's Magnetic Field

Along with providing the information about direction with which we are all familiar, the Earth's magnetic field provides two potential sources of information about location. These are as follows:

- Systematic variation in the intensity (or strength) and direction of the field generated in the Earth's core that might be translated by an animal's magnetoreceptor system into useful information for navigation over very large areas.
- Localized variation in intensity due to magnetic sources in the Earth's crust (e.g., magnetite in basalt, iron ore deposits) that could be used as magnetic landmarks or to identify specific locations such as seamounts.

At any point on the Earth's surface, the observed magnetic field can be described as a vector in three-dimensional space (Figure 8.1) (Skiles, 1985). The total field vector (TFV) is the sum of fields arising from a variety of sources, two of which are most relevant here. The primary source is the main field, which is produced in the core of the Earth and contains both dipole and nondipole components. Geophysicists have developed a periodically updated mathematical model [the International Geomagnetic Reference Field (IGRF)] that calculates the value of the main field at any point in space in and around the Earth. The dipole in the core dominates (generally >90%) the observed field and causes the magnitude (intensity) and direction of the vector to vary systematically between the magnetic equator and poles. Intensity of the main field varies from 25 to 65–70 microTesla (μT) or 2–5 nanoTesla/kilometer (nT/km) between the magnetic equator and poles, respectively. Similarly, the inclination of the Earth's magnetic field (the angle between the vector direction and the horizontal component of the field; Figure 8.1) varies from parallel to perpendicular to the Earth's surface between the magnetic equator and poles. The declination of the Earth's magnetic field is the angle between the directions of geographic and magnetic north and arises from the displacement of the magnetic poles relative to the geographic poles (Figure 8.1).

The second source of the Earth's magnetic field is the residual field (the field remaining after the IGRF has been subtracted from the observed field). The residual field (sometimes called magnetic anomalies) is produced by magnetized rocks in the crust of the Earth and causes slight variations in the observed TFV over a range of spatial scales. When mapped, these

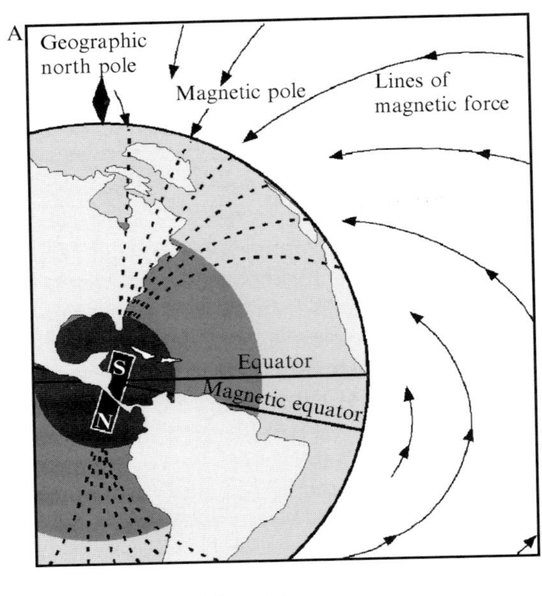

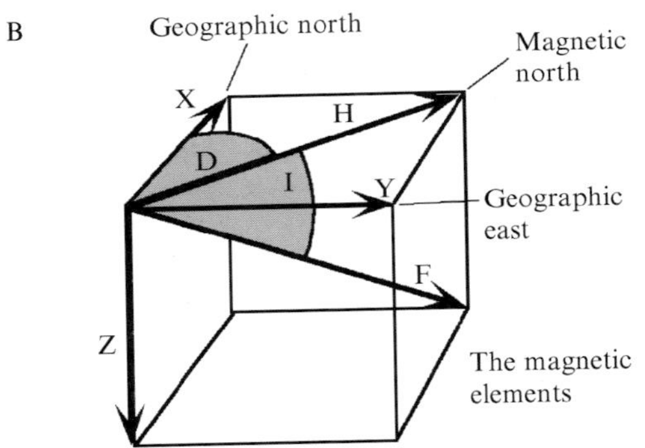

Fig. 8.1. The magnetic field of the Earth. [Adapted from Walker *et al.* (2002).] (A) The main field of the Earth (produced in the core) contains both dipole and nondipole components. The dipole component (represented by the bar magnet in the core of the Earth) is much greater than the nondipole component (not shown). The field due to the magnetic dipole in the core is represented by field lines, which show how the intensity (through increasing proximity of field lines) and inclination (through increasing angles of intersection of field lines with the surface of the Earth) of the field increase systematically between the magnetic equator and the magnetic poles. (B) The elements of the total magnetic field vector (labeled F) at the surface of the Earth. The total field vector can be resolved into components (arrows) in the X (north), Y (east), and Z (vertical) axes. The vector component (H) in the horizontal plane points in the direction of a handheld compass needle. The declination (D) is the angle between the H and X elements whereas the inclination (I) is the angle between the H and F elements of the field.

variations can be represented as a surface much like a physical topography. Of particular interest here are the new strips of seafloor produced by volcanic activity at mid-ocean spreading ridges. This seafloor becomes magnetized in the direction of the Earth's magnetic field at the time it cools to the temperature (the Curie temperature) below which magnetite in the magma spontaneously acquires a magnetic field. The magnetic field acquired by the rocks at the time of magnetization then adds to the field produced in the core but will subtract from the total field observed at the surface of the Earth when the polarity of the main field is reversed. Over geological time, successive reversals of the Earth's field result in distinct bands of positively and negatively magnetized seafloor (positive and negative magnetic anomalies) that are symmetrically arranged on opposite sides of spreading ridges (Vine, 1966). Cooling of magma through the Curie temperature is also responsible for the magnetic anomalies associated with the volcanoes that form seamount chains in the deep ocean.

As in space, the Earth's magnetic field varies over time scales ranging from seconds to decades. Short term variations in the TFV are produced by the interaction of the solar wind, a stream of charged particles flowing outwards from the sun with electric currents carried by charged particles flowing in the ionosphere (Skiles, 1985). Variations in the solar wind that cause changes in the magnitude of these currents can be caused by solar flares (magnetic storms), which can be powerful enough to disrupt radio transmissions and electric power systems. Similarly, at any point on the Earth, the observed field varies slightly with a 24-h period (the diurnal variation) because the interaction of the solar wind with electric currents in the ionosphere varies with the rotation of the Earth.

Over longer time periods, the values of the intensity, inclination, and declination of the Earth's field vary (the secular variation) as a consequence of the process of convection within the molten core of the Earth that generates the field. Heat-driven convection cells in the molten core spiral outwards through the molten core toward the boundary between the molten core and the solid mantle of the Earth (Glatzmaier and Roberts, 1995). This cyclonic convection is sufficient to cause the declination of the Earth's field to vary by as much as 30 min of arc per year whereas intensity and inclination of the field may vary by up to 120 nT and 20 min of arc per year, respectively. At much longer intervals (tens of thousands to hundreds of thousands of years), cyclonic convection can become disordered, resulting in substantial reduction of the intensity of the observed field, and can lead to reversal of the polarity of the field when convection becomes orderly again (Glatzmaier and Roberts, 1995). Estimates of the time taken for reversals to occur range from hundreds to a few thousand years and a mathematical model of cyclonic convection in the core simulated a reversal over about 1000 years (Glatzmaier and Roberts, 1995).

2.2. Magnetic Fields as Experimental Stimuli in the Laboratory

The first consideration in laboratory studies of the magnetic sense is the Earth's magnetic field and its interaction with built structures in the experimental situation. At the scale of a laboratory, the Earth's field is effectively invariant in space. That is, intensity, inclination, and declination of the field are constant apart from the changes (primarily in intensity) induced by temporal variations such as magnetic storms and the diurnal variation in the field. If required, the temporal variations can be controlled to some extent using feedback from a fluxgate magnetometer to programmable power supplies that drive electromagnetic coils which control the background field in an experimental space (see later).

The Earth's magnetic field in a laboratory space will be affected by the presence of iron and other materials in a building and by the constant and time-varying fields produced by laboratory equipment. Thus iron in girders and the bars used to reinforce concrete will result in spatial variations in both magnetic intensity and direction within the laboratory. Experimental spaces in which animals will be studied should therefore be isolated as much as possible from inhomogeneities due to iron in buildings so that the background field is uniform. This isolation can be achieved by locating the experimental space as close as possible to the center of a room or, even better, in a purpose-built structure that is free of magnetic materials such as iron. Once an experimental space has been established, the background field should be mapped in detail with a fluxgate magnetometer, as the values within the building will often be different from those that exist outside the building.

Control of magnetic fields used in laboratory experiments can be required at several levels and is usually achieved using electromagnetic coils. Because coils generate a variety of nonmagnetic artifacts that could interfere with experiments, it has become standard practice to generate experimental fields using double-wrapped coils (Kirschvink, 1992). Such coils have two sets of windings connected so that current flows in either parallel or antiparallel directions through the windings to generate an experimental field or no field, respectively. Under these conditions, the coils produce constant amounts of heat at constant current and so can have no influence on the outcomes of trials in which the experimental field produced by the coils is present or absent. Switching artifacts cannot be avoided and control tests (for example, with fields that are changed gradually rather than instantly) must be carried out to determine whether or not artifacts could be influencing the outcomes of experiments. Other nonmagnetic effects, for example, electrical fields and vibration, can be controlled relatively easily using appropriate shielding, isolation, or insulation.

Control of the background field throughout an experimental space, for example, in migratory orientation experiments, requires coil systems that produce a uniform field space. Although it has often been assumed that paired Helmholtz coils produce uniform fields, this is true only for a small space at their center. Substantially better designs for coils that use systems of three to five coils and generate larger uniform field volumes for a given coil size are now available (Kirschvink, 1992). In the simplest case, a single set of coils aligned with the inclination of the Earth's magnetic field can be used to reduce or increase the intensity of the field in an experimental space without changing the direction of the field significantly. To achieve independent control of the direction and intensity of the background field, a minimum of two, and preferably three, orthogonal coil systems are required. If two coil systems are used, the horizontal system must be aligned with the axis of the Earth's magnetic field and the second system must be oriented in the vertical plane. If three coil systems are used, control of the field in an experimental space will be easiest if one of the horizontal coil systems is aligned in the axis of the Earth's field in the laboratory and the other two systems are aligned in the remaining horizontal axis and the vertical axis, respectively.

A productive line of research has been the use of conditioning experiments to analyze the responses of animals to the presence and absence of localized magnetic anomalies superimposed on the uniform background field. The anomalies also introduce significant local variations in magnetic field direction into the experimental spaces so it is not certain which of these dimensions the animals are actually discriminating. The fields are produced by small magnetic coils associated with a response detector such as a microswitch that the animal activates by pressing on a key or paddle or through some other behavior that can be measured easily and objectively. The magnetic field produced by a coil in an experiment is typically a dipole that interacts with the background field to produce a localized dipole anomaly which increases and decreases the total field on opposite sides of the coil. The magnitude of the anomaly decays with distance from the center of the coil but has diffuse edges that make it difficult to determine the point at which the animal detects the stimulus. Kirschvink and Kobayashi-Kirschvink (1991) describe a coil system that uses two coplanar, concentric coils of equal dipole moment (area · current) but with antiparallel directions and different diameters (e.g., 2 and 10 cm). Close to the center of the coils, the small coil dominates and produces the anomaly. At larger distances, the two fields cancel and confine the anomaly to the area bounded by the larger of the two coils. This coil configuration has been used successfully to measure a threshold sensitivity to changes in the intensity of the anomaly at different frequencies (Walker and Bitterman, 1989; Kirschvink *et al.*, 1997).

There is now a significant opportunity to combine coil systems that produce uniform field spaces with the coplanar, concentric coil systems to determine the threshold sensitivity as a function of background field intensity. Such an experiment will provide the first test for the magnetic sense of Weber's law, which states that the ratio (the Weber fraction) of the threshold sensitivity to change in the intensity of a stimulus to the magnitude of the background intensity is a constant. The existence of some range of stimulus intensities over which the Weber fraction is constant is a widespread property of sensory systems. Demonstration of a constant value of the Weber fraction for the magnetic sense over a biologically relevant range of magnetic intensities would provide powerful psychophysical evidence that the magnetic sense shares a key sensory property with other senses.

Experiments in which the animal is restrained for electrophysiological recording or does not move from a constant position in behavioral experiments provide better conditions for control of the exposure of the animal to magnetic field stimuli. Such experiments can, however, present other challenges for control of stimulation. Thus, the ground plates used in electrophysiological recording experiments are often made of soft iron. Careful mapping with a fluxgate magnetometer of the magnetic field above the ground plate (at the level of the preparation) can be used to locate a point where the intensity of two of the field elements (typically the vertical and east-west components) approach zero, leaving just the horizontal component of the field. Aligning a coil with the remaining north-south component of the Earth's field makes it possible to change field intensity but not direction or to reverse field direction without changing intensity (Walker *et al.*, 1997). When the animal is restrained or does not move in experiments, a relatively small uniform field space is required simply to allow for slight variations in the position of the animal. Although success in obtaining reproducible behavioral responses to magnetic fields by animals that are not moving has been limited, the enhanced control of magnetic field stimulation under these experimental conditions potentially permits much more powerful psychophysical studies than has been possible with freely moving animals.

3. HOW CAN MAGNETIC FIELDS BE DETECTED?

3.1. Magnetic Field Detection by Electrical Induction

The fishes have been particularly important in the study of the magnetic sense because there are at least two plausible sensory mechanisms by which fish might detect magnetic fields. The first hypothesis proposed was that the marine elasmobranch fishes detect the electrical fields induced by their own movement and the movement of the saltwater medium through the Earth's magnetic field (Kalmijn, 1978, 1982). Because elasmobranch fishes are almost exclusively marine, they inhabit a medium that is electrically highly conductive. In the elasmobranch electroreceptor system, jelly-filled canals in the ampullae of Lorenzini are connected by a small pore to the external medium. As a consequence, an electrical circuit is formed between the fish and the saltwater medium. Because the jelly in the ampullary canals has a low electrical resistance, the voltage difference between the internal and external media is converted into a voltage drop across the electroreceptor cells located in the ampullae at the ends of the canals. The electroreceptor cells quickly accommodate to constant electrical fields, permitting the cells to operate at close to their threshold and detect electrical field stimuli in the 10–20 nV/cm range (Bodznick *et al.*, 2003). Variations in the membrane potential of the electroreceptor cells in response to variations in electrical field stimuli are then detected by neurons of the anterior lateral line nerve and transmitted to the brain. For a recent review of the structure and function of the electroreceptor system of the elasmobranchs, the reader is referred to Bodznick *et al.* (2003).

Because the fish and the saltwater form an electrical circuit, movement of the fish through the water will induce an electrical field that produces a voltage drop across the electroreceptor cells. Movement of saltwater through the Earth's magnetic field will also produce a voltage drop across the electroreceptor cells, even when the animal is passively moving with the water. The magnitude of the voltage drop depends on the intensity of the external magnetic field, the length of the canal, the orientation of the canal relative to the external magnetic field, and the relative conductivities of the internal and external media. The voltage drops generated by movement of both the fish and the water are sufficient to stimulate the electroreceptors and permit determination of magnetic field direction (Kalmijn, 1978, 1982), independent of whether the animal uses this information. Although the electroreceptors will not detect the constant electrical field signals produced by completely uniform movement of either the water or the fish, further theoretical analysis has shown that the low frequency movements of the body during swimming (side-to-side in sharks or up and down in rays) will still permit extraction of information about magnetic field direction (Paulin, 1995). Experimental evidence consistent with this hypothesis is considered in a later section.

3.2. Magnetic Field Detection Based on Magnetite

In contrast with the elasmobranch fishes, the teleost fishes have been important in the search for a magnetoreceptor system based on the magnetic mineral, magnetite. This hypothesis proposes that animals use the motion of or torque from chains of single-domain (SD) magnetite crystals to transduce magnetic field stimuli into mechanical signals that can be detected by the nervous system (Kirschvink and Gould, 1981; Kirschvink and Walker, 1985). Theoretical analyses demonstrate that magnetite-based magnetoreceptors should permit detection of both magnetic field direction and intensity, with the threshold sensitivity to changes in magnetic intensity potentially being as low as 10 nT (Kirschvink and Gould, 1981; Kirschvink and Walker, 1985). As discussed later, much of the experimental evidence that is consistent with the magnetite-based magnetoreception hypothesis has been obtained from the teleost fishes, and in particular from taxa that contain well-known migratory species. This evidence is considered in detail in the following sections.

The final hypothesis on the mechanism of detection of magnetic fields that is being investigated is based on the effects of Earth-strength magnetic fields on particular molecules in the retina. This hypothesis has not received significant attention in the fishes and is not considered further here.

4. STRUCTURE OF CANDIDATE MAGNETITE-BASED MAGNETORECEPTORS

The first evidence that animals produce magnetic particles suitable for use in magnetoreception came from studies of the rock magnetic properties of animal tissues. Following the discovery (Blakemore, 1975) of magnetotactic bacteria and their use of chains of SD magnetite for orientation, a search for magnetic material suitable for use in magnetoreception in animals identified magnetite in SD, multidomain (MD), and superparamagnetic (SPM) states in a variety of animals (Kirschvink *et al.*, 1985a). MD magnetite has a low magnetization that renders it unsuitable for use in magnetoreception because its interactions with the external magnetic field are weak (Kirschvink *et al.*, 2001).

In contrast, SPM magnetite has a high magnetization but the small size of the particles means that their moments are not fixed and align with an external magnetic field without physical movement of the particles. Broadly similar models for magnetic field detection using SPM magnetite have been proposed by Kirschvink and Gould (1981) and Davila *et al.* (2003) to explain responses to magnetic fields by honeybees (Gould *et al.*, 1978) and birds (Fleissner *et al.*, 2003). Both models depend on closely spaced SPM particles which are aligned such that their interactions in the external magnetic field exert forces of expansion or contraction within an elastic matrix. A further suggestion for the use of SPM magnetite in magnetoreception is through amplification of the flux density of the external magnetic field, which should enhance the frequency response and sensitivity of the SD receptors discussed later (Kirschvink *et al.*, 1997).

Models for the use of SD crystals in magnetic field detection are also based on the strength of their interaction with the external magnetic field. SD crystals have the maximum magnetization per unit volume for magnetite but must move if their moments are to align with the external magnetic field because the crystals are permanently magnetized (Kirschvink and Gould, 1981; Kirschvink and Walker, 1985). Thus, single domains have the strongest possible interaction with the relatively weak magnetic field of the Earth. Transduction of the magnetic field of the Earth into a signal that can be detected by the nervous system, however, requires the use of chains of SD magnetite such as those that are observed in the magnetotactic bacteria (Kirschvink and Walker, 1985) and fish (Mann *et al.*, 1988). The motion of these chains in response to the external magnetic field will then convert the magnetic field stimulus into a mechanical stimulus that could readily be detected using mechanically gated ion channels linked to the magnetite chain (Kirschvink, 1992). Thus MD particles are considered unlikely to play any role in magnetoreception whereas critical experimental tests have yet to resolve between models of magnetic field detection using SPM and SD particles.

The crystal and magnetic properties of SD magnetite have been used to identify chains of SD crystals within cells in the nose of the rainbow trout despite the small size (<50 nm) and extreme rarity (<5 ppb by volume) of the crystals. Reflections of laser light off crystal surfaces permitted detection of the chains of magnetite crystals in reflection mode confocal laser scanning microscopy (CLSM) (Figure 8.2A and B). Mapping the reflections in three dimensions then permitted imaging of single crystals in thin sections in the transmission electron microscope (Figure 8.2C and D) and unique identification of the crystals as magnetite using atomic and magnetic force microscopy (Figure 8.2E) (Walker *et al.*, 1997; Diebel *et al.*, 2000).

The cells containing the magnetite particles are 10–12 μm in length, have a distinctive multilobed shape, and are consistently located near the basal lamina of the olfactory epithelium (Figure 8.2A and B). The cells are relatively rare and were only found near the tips of the olfactory lamellae (distal to the cells of the olfactory sensory epithelium). The cells each have several processes that extend out to and are surrounded by tubular-shaped fibroblastic cells (with two processes) which help delineate the basal layer (Figure 8.2B). The chain of magnetite crystals in each cell is about 1-μm long (range 0.5–1.5 μm, $n = 4$; estimated from the CLSM; Diebel *et al.*, 2000) and the chains are estimated to have a magnetic to thermal energy ratio of about 4. The location of the chain of magnetite crystals within each cell suggests that

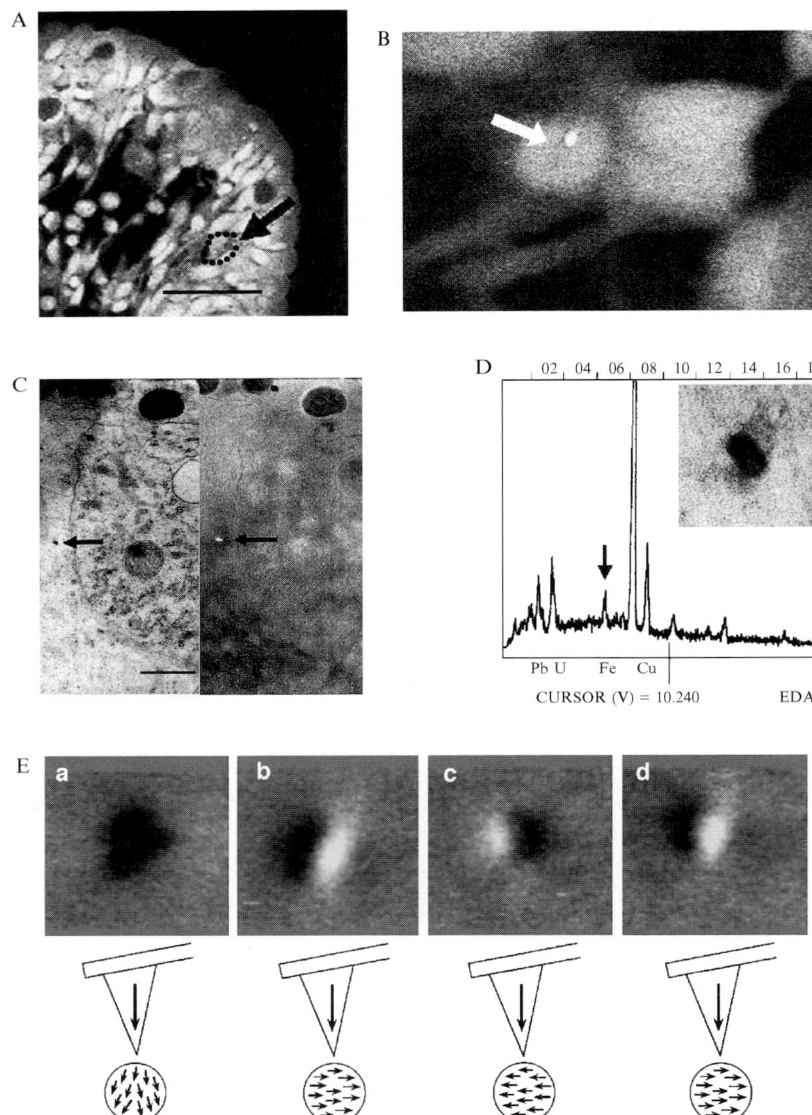

Fig. 8.2. Detection of intracellular magnetite. (A) Autofluorescence image of a magnetite-containing cell (dotted outline) viewed using a confocal laser scanning microscope in transmission mode. The slightly darker area at the lower right of the cell is where the reflection of the laser light by the magnetite has prevented light passing through the cell (scale bar 25 μm). [Adapted from Walker *et al.* (2000).] (B) CLSM optical slice (~30-μm wide) taken through a

a mechanical linkage of the chain to the cell could transduce the movement of the chain in response to the external magnetic field into changes in the membrane potential of the cell.

5. BEHAVIORAL RESPONSES TO MAGNETIC FIELDS IN THE LABORATORY

A variety of laboratory experiments have demonstrated that both teleost and elasmobranch fishes can detect magnetic fields. Fishes from at least two teleost orders have been shown to respond to magnetic field direction in orientation experiments whereas both teleost and elasmobranch fishes have been successfully conditioned to magnetic fields. In the paragraphs that follow, we examine key results from these experiments.

---

magnetite-containing cell within an olfactory lamella. The arrow points to the reflectance arising from particles contained within the cell. [Adapted from Diebel *et al.* (2000).] (C) Bright-field (left) and dark-field (right) transmission electron micrograph (TEM) of a crystal associated with the reflecting structure in a cell like those shown in (A) and (B). In bright-field TEM, both the crystal (arrow) and a much larger pigment granule (top center) are electron dense. In dark-field TEM, the crystal (arrow) reflects the electron beam strongly whereas the large pigment granule (upper right) does not (scale bar 1 μm). [Adapted from Walker *et al.* (1997).] (D) Energy dispersive analysis of X-ray emissions (EDAX) of the crystal in (C). Inset shows the crystal (length 50 nm) at higher magnification. The copper (Cu) peak is due to the copper grid used, and lead (Pb) and uranium (U) peaks are from TEM stains. The peak from iron (Fe) present in the crystal is indicated by an arrow. This peak was absent in control regions of the same section. [Adapted from Walker *et al.* (1997).] (E) MFM images that show the response of a single magnetic field source within a trout olfactory lamella in the presence of an applied field. The magnetic field applied in the plane of the sample was +1.4, +150, −150, and +130 mT for images **a–d**, respectively. MFM images (75-nm squares) are shown on top with a representation of the MFM tip and magnetization of the field source underneath. The MFM tip (inverted triangle) is permanently magnetized with a coercivity of +500 mT at right angles (arrow in inverted triangle) to the applied field. The small arrows within each circle under the tip represent the alignment of the individual magnetic dipole moments that might act as the field source. (**a**) Image shows a dark patch at the location of the magnetic field source. This dark patch indicates an attractive reaction between the tip and sample, consistent with the magnetic field from the MFM tip weakly magnetizing the field source and causing an attractive interaction. (**b–d**) MFM images show the nearly dipolar responses of the magnetic field source under a strong applied magnetic field. These are consistent with an MFM image of a field source comprising multiple single-domain particles of magnetite that are magnetized along the direction of the applied field. Note that the reversal of the field and dipolar response in **c** is consistent with the magnetization of the dipole moments in the field source flipping in the reversed applied field. In images **b–d**, the applied field was large enough to align completely the magnetic moments within the field, forming a dipole that interacted with the magnetized MFM tip to produce the light and dark patches in the images. [Adapted from Diebel *et al.* (2000).] Reversing the direction of magnetization of the dipole then resulted in the different locations of the light and dark patches in **b–d**.

5.1. Orientation Responses to Magnetic Field Direction

The critical assumption of orientation experiments is that the spontaneous directional choices made by animals placed in featureless orientation arenas match the directions they would choose in their normal environment (Emlen, 1975). Thus, diurnal birds that migrate at night will become spontaneously active at night as each migration season begins. When placed at night in a featureless orientation arena such as an Emlen funnel, the activity of the birds in the arena is consistent with the hypothesis that they are attempting to fly in the same direction as their flock mates are flying at the same time (Emlen, 1975).

Similar approaches have been used to study the migratory orientation of fish. Quinn (1980) captured sockeye salmon fry migrating from the gravel beds where they hatch to lakes where they spend their early life. The directions chosen by the fish when placed in an orientation arena were consistent with the hypothesis that they were orienting to the axis of the lake in which they would live until they reached the smolt stage and began their migration to the sea. In similar experiments, juvenile chinook salmon established a preferred orientation in which they faced into a current running from east to west in their home tank. When placed in a featureless arena where there was a radial current flow, the fish oriented preferentially in the east-west axis (Taylor, 1986, 1987) despite being able to choose any orientation direction. The orientation directions of the fish in the arenas used in both sets of experiments changed in response to variations in magnetic field direction produced using electromagnetic coils.

5.2. Conditioned Responses to Spatial Variations in Magnetic Field Intensity

Although the evidence to date is indirect, it appears that freely moving fish readily learn to discriminate changes in magnetic intensity in conditioning experiments. Two constraints on these experiments are that (1) the fields to be discriminated must be spatially distinctive and (2) the subjects must be required to move to produce the conditioned response. The simplest pair of spatially distinctive fields is the case where the animal discriminates the presence and absence of a magnetic intensity anomaly induced by an electromagnetic coil. Because the intensity of the Earth's magnetic field is constant within an experimental arena, the animal must therefore discriminate the presence and absence of intensity variations due to the coil. The animal must then move in order to gain exposure to the presence or absence of intensity variations in the experimental situation.

Yellowfin tuna have been trained to discriminate the presence and absence of a nonuniform magnetic field in experimental tanks (Walker, 1984).

Passing direct current through vertically oriented coils wound around the outside of the tank imposed localized, nonuniform fields of varying intensities on the uniform Earth's field within the tanks in which fish were trained (Figure 8.3B). Reversing the polarity of the current to the coils caused the nonuniform field to be added to or subtracted from the Earth's magnetic field in the tank (Figure 8.3D). Individually trained yellowfin tuna swam repeatedly through a hoop lowered into an experimental tank for a 30-sec trial period. At the end of each trial and depending on the presence or absence of the magnetic field produced by the coil, the fish were reinforced or unreinforced with food for swimming through the hoop. Discrimination learning was then detected as a change over time in the rates of response

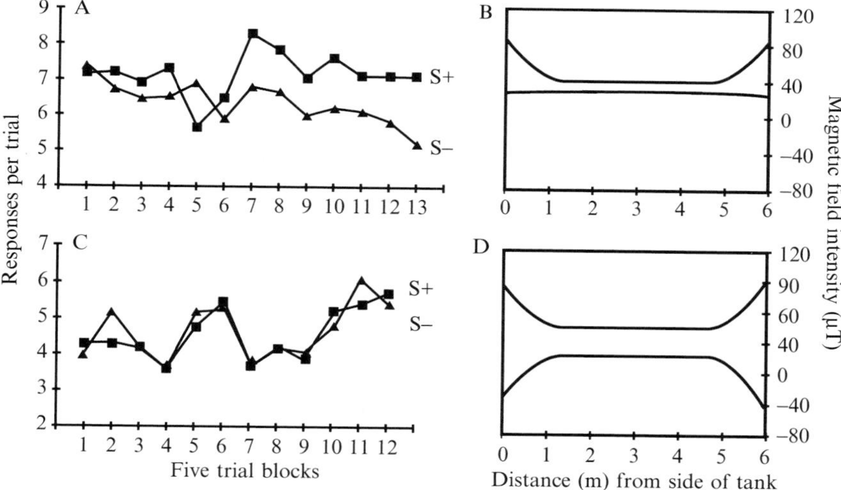

Fig. 8.3. Behavioral responses to magnetic fields by teleost fish. Magnetic discrimination learning in individually trained yellowfin tuna. [Adapted from Walker (1984).] In the training procedure, individual fish swam through a hoop during 30-sec trials. Depending on the magnetic field presented to the fish, response by the fish was reinforced with food (S+) given at the end of each trial or unreinforced (S−; triangles) no matter how often the fish responded. Each point is the mean of five S+ (squares) or S− (triangles) trials. (A) Fish ($n = 7$) were required to discriminate the presence and absence (indicated by varying and constant intensity values across the tank in B) of a localized magnetic field anomaly projected into the experimental tank by a coil through which direct current could be passed. To control for possible generalized effects of the experimental field on behavior, some of the fish were trained with the magnetic field of the Earth as S+ and the localized anomaly as S− whereas for the remaining fish the anomaly was S+ and the Earth's field was S−. (C) Fish ($n = 2$) were required to discriminate between two magnetic anomalies (indicated by the mirror image curves of varying intensity values across the tank in D) produced by reversing the polarity of the current passed through the coil.

during reinforced (S+) and unreinforced (S−) trials. In similar experiments, the pattern of discrimination learning by rainbow trout (*Oncorhynchus mykiss*; Walker *et al.*, 1997) was remarkably similar to the yellowfin tuna despite variations in the number of trials required for discrimination to appear.

Although the tuna learned to discriminate the presence and absence of the nonuniform field, they could not discriminate between the two nonuniform fields produced by reversing the polarity of the current to the coils. For fish tested with the presence and absence of the nonuniform field due to the coil, response rates during both S+ and S− trials remained similar over the first 6 five-trial blocks (Figure 8.3A and B). After 6 five-trial blocks, however, response rates were consistently higher in the presence of S+ than in the presence of S−. For fish trained with two nonuniform fields produced by reversing the polarity of the current to the coils, there was no separation of response rates to S+ or S− at any stage of the experiment (Figure 8.3C and D). The interpretation of these results was that the tuna distinguished the uniform and nonuniform fields based on the presence and absence of changes in intensity but failed to distinguish between the mirror image patterns of variation in intensity of the two nonuniform fields as the fish swam in the tank during trials (Walker, 1984).

Elasmobranch fishes have also been trained to discriminate the presence and absence of magnetic anomalies in experimental tanks. Hodson (2000) used the same approach as that used with the tuna to train short-tailed stingrays (*Dasyatis brevicaudata*) to discriminate the presence and absence of a magnetic intensity anomaly in an experimental tank. Meÿer *et al.* (2005) used activity conditioning to demonstrate that a mixed species group of sharks could be trained to search for food when a magnetic intensity anomaly was presented for 1 min at randomized intervals. The animals were required to swim to a target area in the center of the tank where food was only ever delivered if the magnetic anomaly was present. The animals quickly learned to recognize that the anomaly signaled food and entered the target area much more frequently when the anomaly was present than they did during control periods when the anomaly was absent (Figure 8.4). Because the magnetic fields were switched instantaneously in the above-mentioned experiments, there remains the possibility that the animals responded to the electrical transients associated with switching the magnetic field on and off. Reversible impairment of the responses by attachment of magnets adjacent to the olfactory epithelium in the short-tailed stingray suggests, however, that the electroreceptors do not play a significant role in magnetic field detection (Hodson, 2000; Kirschvink *et al.*, 2001).

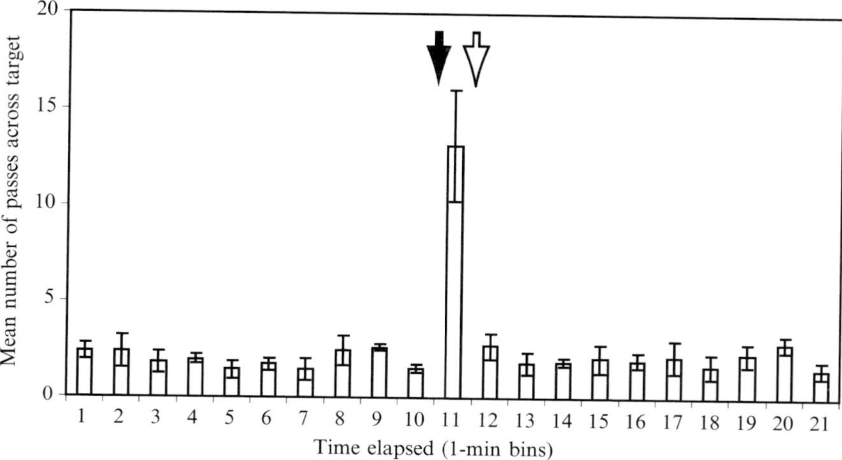

Fig. 8.4. Behavioral responses to magnetic fields by elasmobranch fish. Mean number of passes per minute by sharks across a 1.5 m^2 target during the 21-min duration of magnetoreception experiments. Means for each 1-min bin are derived from 11 trials. Error bars are 1 S.E.M. Shaded arrow, artificial magnetic field activated; unshaded arrow, artificial field turned off. [Adapted from Meÿer *et al.* (2005) with permission.]

5.3. Other Experiments Using Conditioned Responses

Classical conditioning to magnetic field stimuli has also been attempted with mixed results. Rommel and McCleave (1973) used cardiac conditioning to demonstrate sensitivity to electric field stimuli in American eels (*Anguilla rostrata*) and Atlantic salmon (*Salmo salar*) but obtained only equivocal results concerning the sensitivity of both species to changes in magnetic fields under the same conditions. Nishi and his colleagues (Nishi *et al.*, 2004, 2005; Nishi and Kawamura, 2005) reported cardiac conditioning of Japanese eels (*A. japonica*) to magnetic fields that differed in both magnetic field direction and intensity. Significant reductions in heartbeat rates (Figure 8.5B) occurred in 17 of 19 eels in both the marine and freshwater phases of the life cycle. Responses were recorded across changes in magnetic field intensity and direction ranging from 12,663 to 192,473 nT and from 21° to 80° easterly shift in the direction of the horizontal component of the magnetic field in the experimental tank (Figure 8.5A). The responses appear, however, to have been quite variable and did not exhibit any quantitative relationship between the magnitude of the stimulus and the cardiac response. This outcome could result from use of stimuli that were well above the threshold for detection,

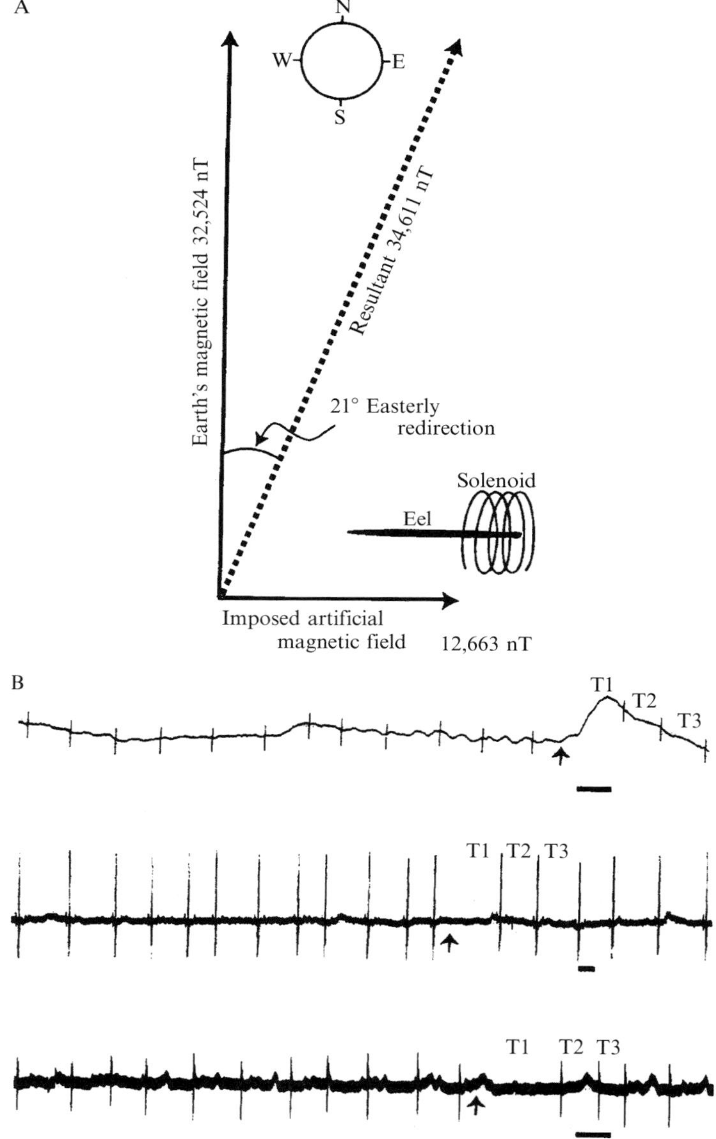

Fig. 8.5. Cardiac conditioning responses to magnetic fields in eels. (A) Experimental apparatus used in cardiac conditioning experiments with Japanese eels. The eel's head lay at the center of a solenoid (35-cm diameter) wrapped around a polyvinyl chloride (PVC) aquarium and oriented in the east-west axis. The solenoid in the apparatus produced a resultant field due to the combination of the geomagnetic field and the imposed magnetic field. The imposed field ranged between

suggesting that the eels discriminated all the magnetic field changes equally easily. In follow-up experiments, Nishi *et al.* (2005) showed that these responses did not occur when the eels had been made anosmic and argued that magnetoreception takes place in or around the nares of the eel. Taken together, these experimental results demonstrate that the magnetic sense can be analyzed using conditioning approaches in the same manner as better-known sensory systems.

6. NEURAL RESPONSES TO MAGNETIC FIELDS IN THE LABORATORY

6.1. Induced Electrical Signals in Ampullary Electroreceptors

As noted earlier, elasmobranch fishes are highly sensitive to electric fields and theoretically should be able to use their ampullary electroreceptors to detect electric current flows induced by their own or the water's movement through the Earth's magnetic field. Evidence consistent with this hypothesis has been obtained by recording responses to electrical and magnetic field stimuli in the Black Sea skate. Afferent nerves from ampullary electroreceptors in the wings (Brown and Ilyinsky, 1978) and primary area acousticolateralis in the dorsolateral region of the medulla oblongata of the brain (Andrianov *et al.*, 1974) responded to both electrical and magnetic field stimuli in the skate. A range of excitatory and inhibitory responses were detected in the primary afferent nerves that were tested under conditions of static and time-varying magnetic fields, water flow, and animal movement. Responses were recorded in the afferent nerves so long as the magnetic field was varying continuously and when either the fish or the water were moving, that is, when electrical fields were likely to be induced. The smallest field to which a response was detected was a field that was changing at a rate of 200 μT/sec. These results clearly confirmed that the electroreceptors do detect the electric current flows induced by magnetic field variations and by the movement of either the fish or saltwater through a static magnetic field. Because these experiments did not demonstrate that the induced electrical signals were either necessary or sufficient for use of magnetic fields

---

12,663 and 192,473 nT with the resultant intensity ranging from 35,611 to 187,298 nT. These intensity changes produced resultant fields whose directions ranged from 21° to 80° to the east of magnetic north. (B) Electrocardiograms showing changes in heartbeat rate in response to magnetic field changes in Japanese eels (scale bars = 1 sec). Arrows indicate the onset of the conditioning stimulus (magnetic field); T1–T3 indicate beats used to compare response to the magnetic field change with heartbeat rate during control tests in the absence of the experimental magnetic field. [A and B adapted from Nishi *et al.* (2004) with permission.]

by elasmobranch fishes, however, the mechanism of magnetic field detection in the elasmobranchs remains an open question.

6.2. Responses in the Trigeminal Nerve of Teleost Fish

The discovery of magnetite suitable for use in magnetoreception in front of the head in a variety of teleost fishes (Walker *et al.*, 1984; Kirschvink *et al.*, 1985b; Mann *et al.*, 1988; Diebel *et al.*, 2000) provided a focus for the search for the sensory nerve that might transmit magnetic field information to the brain. The olfactory (ON), trigeminal (TN), and anterior lateral line (ALLN) nerves are sensory nerves that innervate the front of the head and that could each potentially carry magnetic field information to the brain. The ON is the major source of afferent innervation for the olfactory mucosa. The TN is a mixed nerve that, *inter alia*, carries afferent signals from mechanoreceptor cells and is known to innervate the olfactory epithelium in birds and rats (Finger *et al.*, 1990; McKeegan *et al.*, 2005). The ALLN innervates the highly sensitive mechanoreceptors of the lateral line and, in the elasmobranchs, innervates the morphologically and developmentally related electroreceptors.

Responses to magnetic field stimuli were found to occur in the superficial ophthalmic (SO) branch of the TN of the trout (Walker *et al.*, 1997), the same branch of the TN system that responded to magnetic field stimuli in birds (Beason and Semm, 1987; Semm and Beason, 1990). The responsive units in the trout showed regular firing patterns except during phasic (transient) responses to a trebling of magnetic intensity presented as square waves at frequencies of 0.5 and 1 Hz (Figure 8.6A–C). Both excitatory and inhibitory responses were observed but each unit responded only to either the onset or the offset of a stimulus. The responses of the units could also be modulated by varying the presentation rate of the change in magnetic intensity. For the units shown in Figure 8.6B, poststimulus time histograms (PSTHs) showed that the latency (10–15 msec; the first point after the stimulus step and the period during which the firing rate was more than two standard deviations above the mean for each unit) and time course ($\sim$100 msec) of the responses by the two units exposed to both stimulation frequencies were similar. The peak amplitudes of the responses in the units in the upper and lower panels of Figure 8.6B decreased and increased, respectively, when the presentation rate of the step change in intensity increased from 0.5 to 1 Hz.

Surprisingly, no unit responded when magnetic field direction was reversed without a simultaneous change in intensity. As described in Section 2.2, the experimental apparatus permitted imposition of a field aligned either parallel or antiparallel to the background field in the experimental situation. Adding 50 μT parallel to the background field trebled intensity without changing magnetic field direction (upper panels of Figure 8.6C), whereas adding the same field antiparallel to the background field reversed the magnetic field direction without changing intensity (lower panels of Figure 8.6C). The unit presented in Figure 8.6C clearly did not respond to

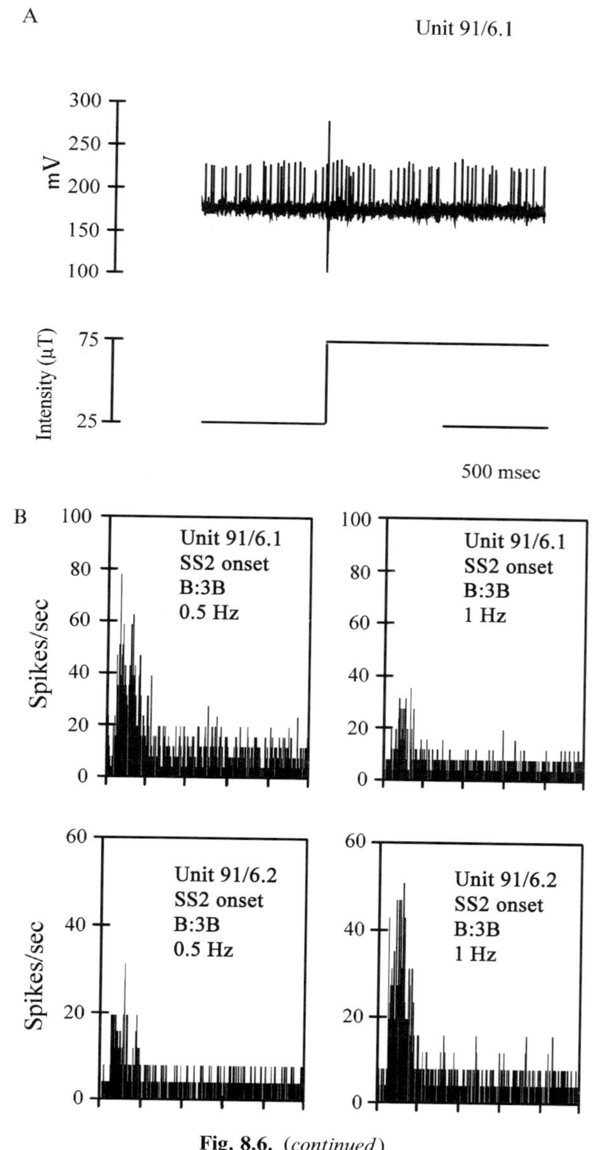

Fig. 8.6. (*continued*)

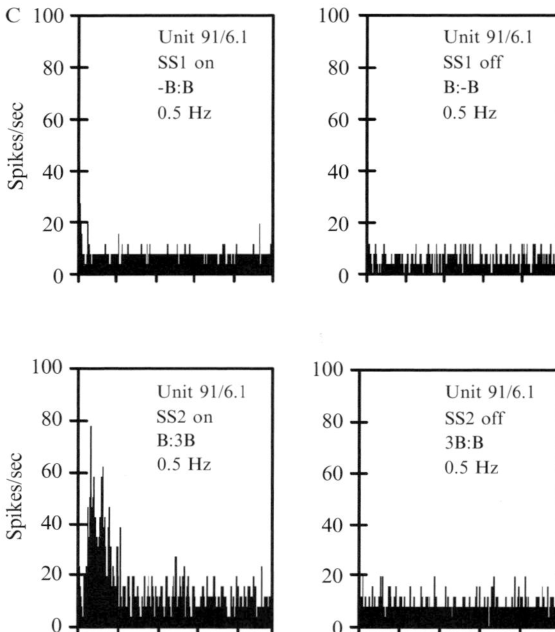

Fig. 8.6. Neural responses to magnetic field stimuli. (A) Spontaneous activity of a single unit (upper trace) in the TN of the rainbow trout in the background magnetic field followed by the activity for 1 sec following the onset of a stimulus (lower trace) that produced a step change in magnetic field intensity from 25 to 75 μT within the experimental tank. (B) Poststimulus time histograms of responses by two spontaneously active units to the same stimulus presented 128 times each at 0.5 and 1 Hz. Each plot begins at stimulus onset and is of 500-msec duration with the magnetic field remaining constant throughout the period shown in each panel. Sampling bin width is 2 msec in all panels and tick marks on the abscissa are at 100 msec intervals following stimulus onset. The top left panel in B is the same record as the bottom left panel in C. (C) Poststimulus time histograms of responses by one spontaneously active unit to the onsets and offsets of a trebling of magnetic intensity (lower two panels) or a reversal of magnetic field direction in the experimental tank (upper two panels). Each stimulus was presented 128 times at 0.5 Hz. Each plot begins at stimulus onset and is of 500 msec duration with the magnetic field remaining constant throughout the period shown in each panel. [All data adapted from Walker *et al.* (1997).]

reversal of the magnetic field direction and, as in the other units recorded, responded only to one of the two intensity changes (onset and offset of the experimental stimulus) presented to it. Of interest also are the differences in the firing rate of the unit during the last 300 msec of stimulation shown in Figure 8.6C. The mean and variance of the firing rate of the unit not only increased greatly during the phasic response to the change in intensity but also remained higher while the intensity was high (lower left panel in Figure 8.6C) and remained higher when the field switched back to the lower intensity (lower right panel in Figure 8.6C). In contrast, the mean and variance of the firing rate of the unit were uniformly low throughout the PSTHs when direction but not intensity was changing (upper panels in Figure 8.6C). This observation suggests that there may be sustained differences in firing rate that accommodate only slowly at different levels of magnetic intensity.

The pattern shown in Figure 8.6C is important also because it demonstrates that the units did not respond to the electrical artifacts induced by the sudden changes in magnetic fields in the experimental situation. The stimuli all changed the field by a constant 50 μT, which was sufficient either to reverse direction or to treble the intensity of the field in the experimental situation. As a consequence, the induced artifacts would have been the same in the top right and lower left panels and in the top left and lower right panels of Figure 8.6C. That is, the units appeared indifferent to the induced artifacts and only responded to one particular step change in the intensity of the magnetic field.

The neural responses to magnetic fields in the trout have neither been localized to any branch of the SO, shown to depend on magnetite such as that found in the cells in the nose, nor to underpin behavioral responses to magnetic fields by the trout. The responses to changes in magnetic intensity found in the TN are, however, consistent with detection of magnetic fields in front of the head of the trout and led to the search for detector cells associated with the TN discussed earlier.

7. NEUROANATOMY

In the first step toward testing the hypothesis that the magnetite-containing cells may be functionally linked to the TN, the SO branch of the TN was traced from the site where electrophysiological recordings of responses to magnetic field stimulation were made to the endings of the individual nerve cells as well as to the brain (Figure 8.7) (Walker *et al.*, 1997). Serial histological sections and DiI, a fluorescent lipophilic dye, placed on the cut ends of the TN were used to trace the nerve in both anterograde and retrograde directions. The dye migrated along both myelinated and unmyelinated fibers in the TN. Posterior to the orbit, the SO branch joined other branches of the TN and ended in cell bodies that make up part of the anterior ganglion (Figure 8.7C). From the ganglion, the labeled nerve tracts entered the anterior dorsal area of the medulla oblongata. Anterior to

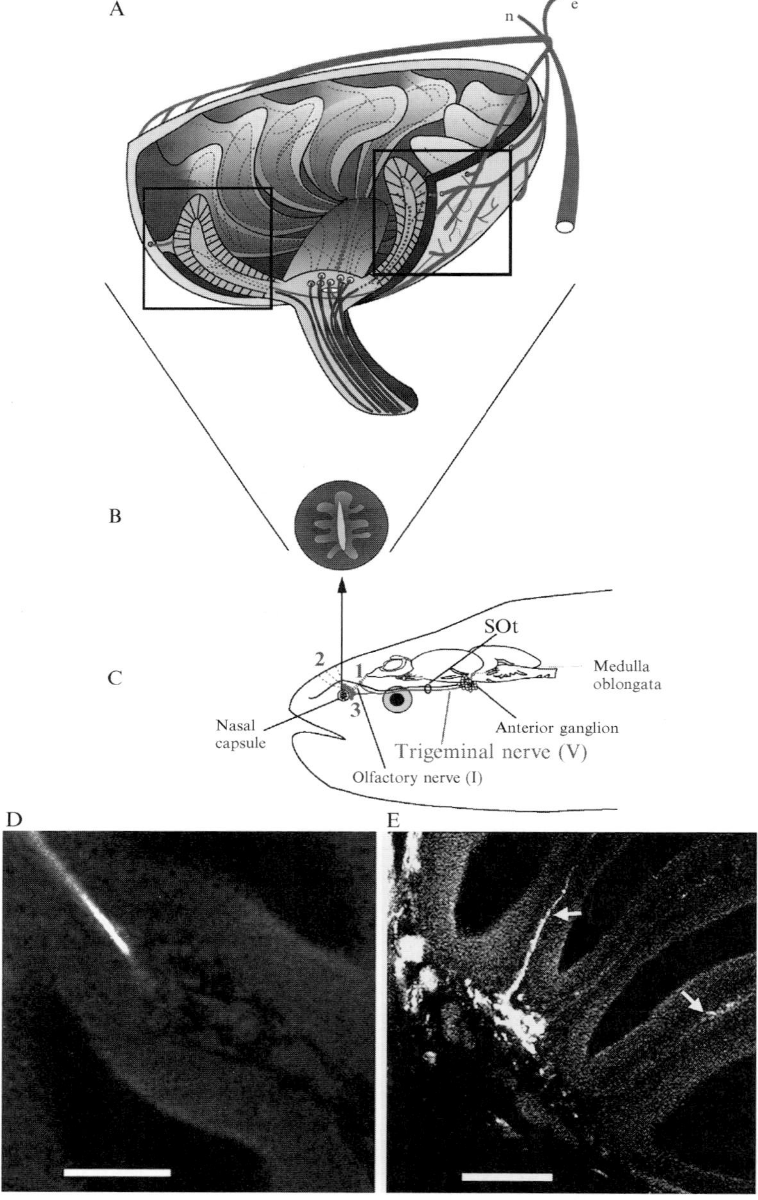

Fig. 8.7. Innervation of the head region and nasal capsule of the trout by the SO branch of the TN. (A–C) Schematic diagram of the innervation of the head region and nasal capsule of

the orbit, the SO branch has branches that innervate the skin, surround the olfactory nerve and olfactory capsule (processes 1–3 in Figure 8.7C), and also penetrate the olfactory lamellae within the olfactory capsule itself (Figure 8.7C). Fine branches of the SO penetrate the olfactory lamellae from both the top and from the base before terminating in finer processes within the olfactory lamellae, where the magnetite-containing cells are most often found (Figure 8.7A, D, and E).

Although it can be proposed that the candidate magnetoreceptor cells in the lamina propria of the olfactory lamellae are linked through the SO branch of the TN to the brain, afferent synaptic contacts between the nerve endings and the magnetoreceptor cells have not yet been identified. Detection of both magnetite and the endings of stained nerves in the confocal microscope has not yet been achieved due to the different media required for best detection of the magnetite and the labeled nerves. It has also been impossible to recognize the chains of magnetite crystals in the transmission electron microscope, at least in part because there is a very low probability that more than one crystal in a chain will fit within one thin section. A further difficulty in visualizing the magnetite in the electron microscope

---

the trout by the superficial ophthalmic (SO; shown in red in A–C) branch of the trigeminal nerve (TN) [A–C adapted from Walker *et al.* (1997).] (A) A three-dimensional diagram of the innervation by the SO into olfactory lamellae in the nasal capsule of the trout. One process innervates the nasal membrane and flap (n) and the other [top right (e)] is process 2 in C. Others form a network of nerves that surround the nasal capsule (right box). Within this network, the smaller branches have fine processes that pass through the nasal membrane, lining the nasal capsule, and innervate, at both the top and base, individual olfactory lamellae that form the olfactory rosette. The olfactory nerve (blue) is the combination of all axons of the olfactory sensory cells which are situated in the mucosa and send their axons to the olfactory bulb. The network of nerves surrounding the capsule generally lies in a fatty layer (light shading on outside surface of capsule; right box in A), which is typically found between the neurocranium (not shown) and the outer membrane (brown) that lines the nasal capsule. The pale area in the front two lamellae represents the folded layers of the olfactory epithelium that are separated internally by the lamina propria. New lamellae are formed in the area of the nasal capsule (not shown). The left box outlines the area shown in D. (B) Olfactory rosette within the trout nasal capsule (top view). The nasal flap that lies over the top of the olfactory rosette has been removed for clarity. (C) Innervation of the SO in the head region of the trout. The label SOt identifies the SO trunk where the SO branches of the trigeminal (V) and anterior lateral line (VIII) nerves pass together across the top of the eye before diverging to innervate the front of the head. (D) Optical slices showing two different branching patterns of DiI-labeled nerve processes entering trout olfactory lamellae. Optical slice through a single olfactory lamella (scale bar 100 μm). A labeled fine process from a branch of the SO branch can be seen entering the lamella through the top. [D and E adapted from Walker *et al.* (1997).] (E) Fine processes can also be seen entering the lamina propria of several lamellae (arrows) from their bases (scale bar 100 μm). These processes originate from a different branch of the SO than the one that innervates the area in D. (See Color Insert.)

is that the magnetite is very hard and, as a consequence, any crystal that is not completely contained within a thin section will catch the microtome knife and pull out of the section. There is thus only indirect evidence from magnetic impairment experiments in fish (Hodson, 2000) that the magnetoreceptor cells are functionally linked to the nervous system. Evidence that the magnetite will be linked to the SO branch of the TN in the trout has, however, come from conditioning experiments with pigeons, where bilateral sectioning of the SO branch abolished conditioned responses to magnetic fields by the birds (Mora *et al.*, 2004).

8. USE OF THE MAGNETIC SENSE IN NAVIGATION

8.1. Constraints on Theory and Experiment in the Study of Navigation by Fish

The only hypothesis ever proposed to explain the existence of the magnetic sense in animals is that the sense permits use of the Earth's magnetic field for navigation over long distances. Unfortunately, rigorous theoretical and methodological frameworks against which to assess hypotheses and evidence on how animals such as fish use the Earth's magnetic field for navigation have not yet been developed. Such frameworks are necessary because the experiments required to test such hypotheses present significant challenges in experimental design, control, and interpretation over and above the logistical difficulties of studying fish traveling in water. As a consequence, it has been difficult to achieve convincing evidence that fish do indeed use the Earth's magnetic field for navigation.

Developing a theoretical framework begins with the definition of navigation. For our purposes, navigation can be defined as the processes by which an animal first determines its current position relative to some goal that it cannot detect directly, and then sets a course to reach that goal. These two processes are consistent with the "map" and "compass" steps of true navigation originally proposed by Kramer (1953). The above definition excludes cases where fish can detect a goal at a substantial distance. Thus, salmon on their spawning migration that locate their natal river by following its chemical signature upstream are not navigating because they are in direct sensory contact with their goal.

This definition presents challenging problems in understanding how fish navigate. Fish that travel significant distances are likely to lose sensory contact with persistent features (such as visual, auditory, and olfactory landmarks) of the environment. Traveling fish must also overcome the effects of passive displacement during movement through water that may itself be moving with, against and across the directions taken by the animals as they travel (Figure 8.8). The risk of passive displacement during such movements generates strong selective pressure for the ability to monitor both position and direction, particularly when traveling over long distances to find small targets. These constraints require that external stimuli used in navigation provide consistent information which varies systematically such that locations can be identified uniquely over at least the portion of the biosphere used by a given species (Walker *et al.*, 2002).

As described earlier, the Earth's magnetic field provides consistent information about direction and potentially about location throughout the biosphere. Because this information varies systematically, locations can potentially be identified uniquely over large areas of the Earth (Walker *et al.*, 2002). The Earth's magnetic field will therefore permit animals such as fish to monitor position and direction while traveling at any depth and

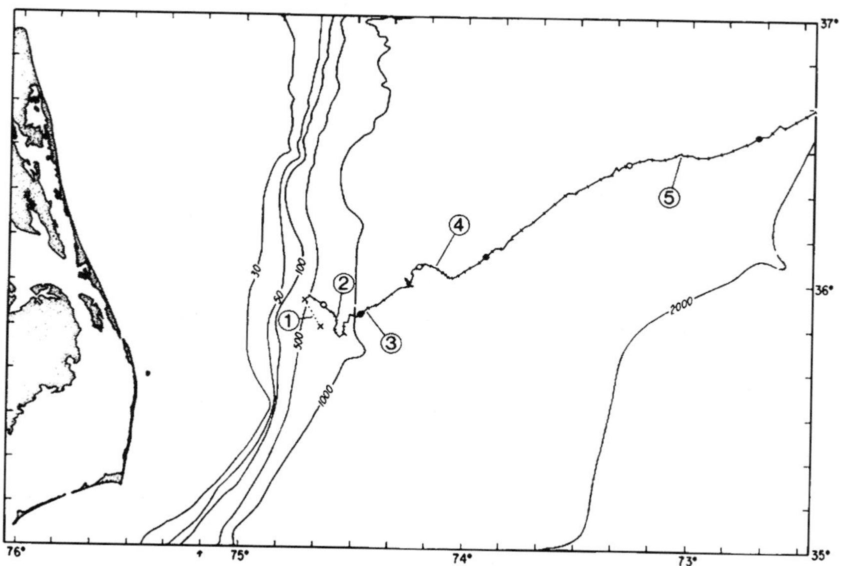

Fig. 8.8. Sustained directional swimming in a swordfish. Path of a swordfish tracked continuously for several days in the Atlantic Ocean near Cape Hatteras, North Carolina. Numbers at points along the track represent the following events: (1) The fish was caught on long-line gear. (2) The fish moved beneath a layer of cold surface water. (3) The fish emerged from beneath the cold surface layer 10 h later. (4) Nine hours later the fish was in the Gulf Stream. (5) By mid-afternoon on the following day, the fish had left the Gulf Stream and entered the Sargasso Sea. Filled circles represent sunset; open circles represent sunrise. [Adapted from Carey and Robison (1981).]

independent of any passive displacement due to water currents. Obtaining such information also depends on the sensory system(s) detecting variations in the stimuli to determine position with sufficient resolution to meet the navigational needs of the animal (Walker *et al.*, 2002).

The physics of proposed magnetoreceptor mechanisms do not exclude the possibility that the Earth's magnetic field is used for navigation, although the psychophysical experiments to determine whether or not fish have the necessary sensitivity have not yet been carried out. The results of sensory studies and field experiments with other vertebrates suggest, however, that the magnetic sense is widespread and that fish are likely to have the sensitivity required for use of the Earth's magnetic field in navigation (Klimley, 1993; Walker *et al.*, 1997; Diebel *et al.*, 2000). Thus, a magnetic sense based on the electroreceptors in the ampullae of Lorenzini could be useful in determining direction when setting a course toward a goal (Kalmijn, 1978; Paulin, 1995). Similarly, a magnetic sense based on SD magnetite would permit determination of direction and provide sufficient sensitivity to determine position with a resolution of a few kilometers (Kirschvink and Gould, 1981).

8.2. Hypotheses on Magnetic Navigation Mechanisms

As noted earlier, navigating animals must first determine their current location relative to a goal and then set and maintain a course for the goal (Kramer, 1953). The magnetic compasses of fish and other animals are well understood from laboratory studies. There is evidence that teleost and elasmobranch fishes respond to both magnetic intensity and direction in laboratory experiments (Wiltschko and Wiltschko, 1995) but almost no evidence that they respond to either of these dimensions of the Earth's magnetic field in nature. The use of magnetic field direction as a reference direction for setting and maintaining courses during long-distance navigation is relatively easy to understand. In contrast, there is no widely accepted model for how the Earth's magnetic field can be used to determine location. Because at least three models for magnetic position determination have been proposed (see later), caution will be required to ensure that the outcomes of experimental tests of predictions derived from any hypothesis on position determination are not ambiguous.

The simplest mechanism for determining location on the surface of the Earth is a system of paired coordinates in which two dimensions of one or more environmental stimuli vary systematically relative to each other. Models of bicoordinate position determination using the Earth's magnetic field fall into two classes: one derived from properties of the Earth's magnetic field and the other in which inferences about how the Earth's magnetic field might be

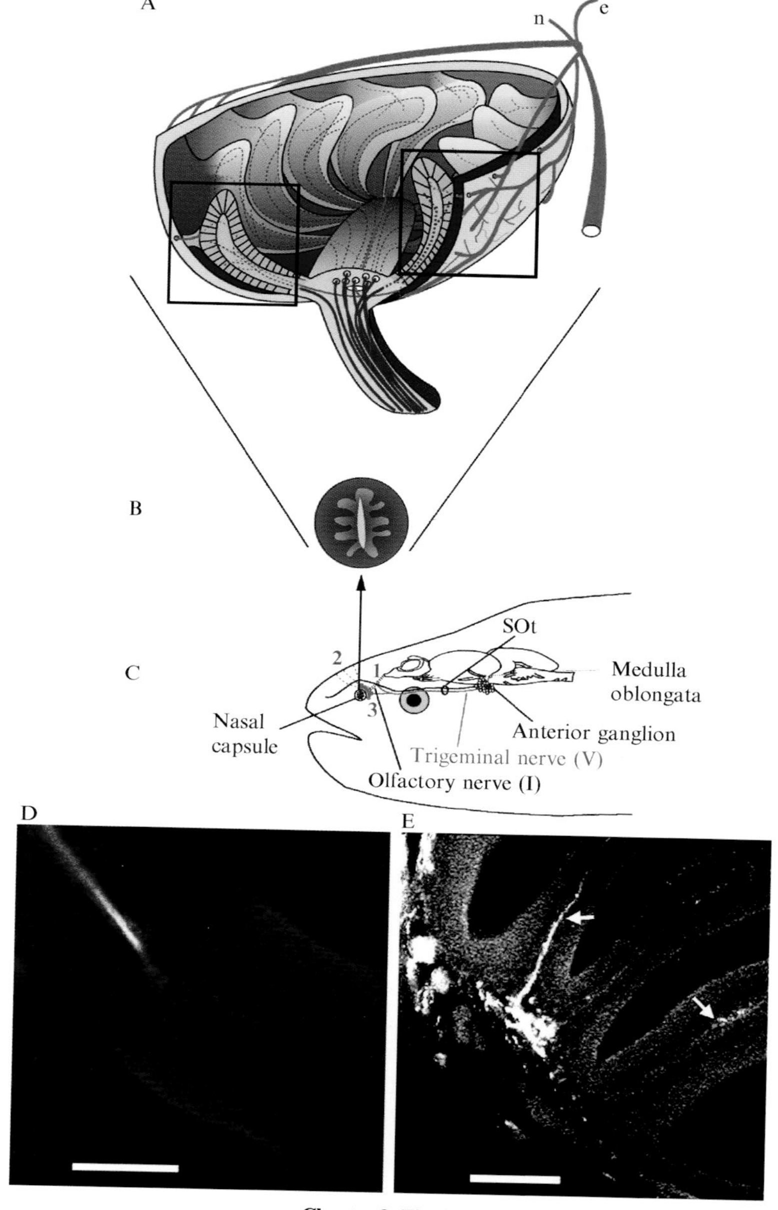

Chapter 8, Fig. 7.

used are made from the results of behavioral experiments. Ideally, such models should be able not only to explain existing observations but also to make testable predictions about navigation behavior. Although several models of position determination using the Earth's magnetic field have been suggested, none have yet found experimental support in field studies.

One model of position determination that is based on properties of the Earth's magnetic field exploits a regular pattern of magnetic anomalies originating from rocks in the deep ocean crust. The process of seafloor spreading in the deep ocean results in linear magnetic anomalies (magnetic lineations) that are symmetrically arranged on opposite sides of spreading ridges. A second pattern of magnetic anomalies, aligned at high angles to the axes of the magnetic lineations, arise from magnetization of cooling magma produced at fracture zones across the new crust produced by the spreading ridges (Vine, 1966). These intersecting anomaly patterns distributed over the whole of the deep ocean are stable over very long periods and remain present during reversals of the dipole field (Kirschvink *et al.*, 1986). The anomaly patterns could be used to guide movement over long distances and would require only that the magnetic sense be able to detect small changes in the total intensity of the Earth's magnetic field. If animals use the anomalies as proposed, the model predicts that animals should preferentially follow linear features that are consistent with their migration direction.

Two further models of navigation under discussion used an alternative approach of deriving models of position determination based on data from behavioral experiments. First, Lohmann and his colleagues (Lohmann and Lohmann, 1996) developed a model of position determination based on responses by hatchling sea turtles to magnetic inclination and intensity in laboratory orientation experiments. This model was challenged on the grounds that inclination and intensity of the Earth's magnetic field are normally highly correlated and are only aligned at high angles to each other over areas that are both comparatively small and ephemeral over evolutionary time (Courtillot *et al.*, 1997).

Although the nature of the second coordinate has yet to be resolved, subsequent laboratory orientation experiments have demonstrated that sea turtles could use magnetic intensity, inclination, or both to determine magnetic "latitude." In experiments, juvenile sea turtles have been exposed to magnetic fields that simulated displacement over several hundred kilometers to the north or south of the location where they had been captured. Under these conditions, the animals oriented in directions that would lead them to return to the location from where they were captured (Lohmann *et al.*, 2004). This result suggests that the turtles could, in effect, use the value of the magnetic field at a single point (the area of a small orientation arena) to determine latitude.

The second model of magnetic position determination based on existing behavioral data arises from systematic variations in the behavior of homing pigeons returning to a loft from release sites distributed across a region (Walker, 1998). The variations in the behavior of the pigeons at release sites were correlated with the systematic variations in the intensity and the direction in which the slope in the intensity of the Earth's main field is oriented (Walcott, 1978). The model proposed that the intensity of the main field, which varies systematically in the magnetic north-south direction, could be treated as a latitude. In contrast, the second coordinate (direction of the slope of the main field) varies at right angles (roughly east-west) to the intensity latitude and so could be treated as a "longitude."

Challenges to this model have been based on physical arguments (Reilly, 2002) and the sensitivity to changes in magnetic fields required for the model to work (Wallraff, 1999). The model does, however, make predictions about the position determination behavior of animals that could be tested experimentally by displacing animals away from their familiar areas in much the same way as occurs in homing pigeon experiments. The model predicts that the behavior of displaced animals should be dominated by the residual magnetic field of the Earth at the release site (Walker, 1998). Testing the prediction will require high-resolution magnetic field information to which the behavior of the animals can be correlated, and higher resolution (more fixes per unit time) data than have been acquired to date in the tracking of the animals.

8.3. Developing Experimental Approaches to Navigation

Efforts to develop a methodological framework for the experimental study of navigation by fish can benefit from the experience of studies with birds. Homing pigeons have been the experimental model of choice for the study of long-distance navigation because they will navigate on demand, returning directly to their loft after displacement to distant, unfamiliar locations. Homing pigeons differ from wild birds (e.g., bank swallows; Keeton, 1973) and the rock dove, the ancestral species of the homing pigeon, in that pigeons have been bred for their homing ability, and also feed, rest, and breed exclusively at the loft. It therefore seems reasonable to hypothesize that the motivation of the pigeons to return directly to their loft is higher than in wild birds which, depending on their needs, may travel to feed in any one of a number of locations as well as returning to their nest after being captured, displaced, and released.

The variable that has been used almost exclusively in studies of pigeon navigation has been the vanishing bearing, the compass bearing from a release site at which an individual bird disappears from view in binoculars. Because pigeons normally take up to several minutes to disappear from view, the vanishing bearing represents the outcome of the processes by which the bird determines its current location and then sets a course to its loft. It has therefore been very difficult to interpret the effects of experimental treatments on navigation behavior except where the effects can be interpreted as simple effects on the sun compass or the magnetic compass (Schmidt-Koenig, 1958; Keeton, 1971; Walcott and Green, 1974). In the last 5 years or so, high resolution tracks have been obtained from pigeons that have carried devices (Steiner *et al.*, 2000) which record positions taken from the satellites in the global positioning system (GPS). The use of geographical information systems combined with computer-intensive analytical techniques promise rapid advances in the ability to detect and analyze any behavior emitted as the animals seek to determine their current location and set a course for their loft.

The implication of the above-mentioned fact for experimental studies of navigation in fishes is that it will be important to achieve as much experimental control as possible over the behavior of the animals and, in particular, over their motivation to travel directly to a particular location that can be specified in advance. It will also be necessary in future experimental studies of navigation by fishes to develop methods for detecting navigational responses against the background of other behaviors that may be exhibited by free-living wild animals. Although it is not possible to obtain track records using the GPS, acoustic transmitters have long been used to track fish continuously. Maximizing the density and accuracy of position fixes on fish being tracked should therefore assist detection of navigational responses and also the effects of sensory treatments on behavior of fish that have been experimentally displaced.

9. WHAT IS KNOWN ABOUT THE NAVIGATIONAL ABILITIES OF FISH?

Descriptive evidence for the navigational abilities of fishes has come from mark-recapture and tracking studies that described point-to-point movements and continuous tracks recorded over periods from a few days up to years. Early studies demonstrated that representatives of different fish taxa travel over distances from tens to thousands of kilometers in both the shallow and deep water marine environments (Block *et al.*, 2001; Boustany *et al.*, 2002; Bonfil *et al.*, 2005; Weng *et al.*, 2005). During many of these movements, the fish clearly traveled independently of the flow directions of either local or major currents and could stay on the same course for long periods (Figure 8.8) (Carey and Robison, 1981).

Tracking of individual fish using ultrasonic transmitters has also provided clear evidence for the ability to return repeatedly to the same locations (Yuen, 1970; Carey and Robison, 1981; Klimley and Nelson, 1984; Klimley, 1993) and to travel in the same direction for extended periods (Carey and Robison, 1981). The earliest examples of the ability of fish to monitor position during travel include those of Yuen (1970), Carey and Robison (1981), and Klimley and Nelson (1984) who found that skipjack tuna, billfish, and scalloped hammerhead sharks made repeated daily movements to and from feeding or refuge areas. A skipjack tuna tracked over 5 days by Yuen (1970) fed at the Ka'ula Bank in the Hawai'ian Islands by day and made night journeys of 25–206 km away from the bank, returning to the bank at about the same time each morning. Scalloped hammerhead sharks and swordfish showed the same pattern, although the sharks left shallow water each night to feed and returned to shallow water by day (Figure 8.9) (Carey and Robison, 1981; Klimley and Nelson, 1984). The precise timing of these movements and the regular returns to the same locations are consistent with the hypothesis that both teleost and elasmobranch fishes can monitor their position accurately.

More direct evidence that fish can both determine and monitor their position has come from the homing movements made by displacement experiments with juvenile lemon sharks (Sundström *et al.*, 2001). Sharks were transported away from their previously determined activity space to 18 randomly chosen release sites 4–15 km offshore from Bimini Islands. After release, the sharks swam in an oscillating zigzag pattern along a consistent mean direction for 2–10 min before selecting a heading along a new axis (Figure 8.10). After repeating this pattern in several directions, the fish chose a "homeward" direction, much as homing pigeons appear to do (Dennis *et al.*, 2005). The fish then returned rapidly to shallow water areas around the Bimini Islands but did not simply retrace the path over which they were taken during the displacement. These observations are consistent with the ability of fish to navigate but do not tell us how navigation is achieved.

Evidence that fish use the Earth's magnetic field for navigation is, to our knowledge, limited to one study in which scalloped hammerhead sharks were tracked around seamounts in Baja, California (Klimley, 1993). As in the studies with tuna, the tracked fish made repeated round trip journeys each night, traveling away from El Bajo Espiritu Santu seamount into the pelagic area and returning each morning. Monte Carlo simulations demonstrated that the fish responded to features of the residual magnetic field but not to bathymetric features that were associated with the seamount. Klimley (1993) described the pattern of movement by the sharks as a magnetotopotaxis and suggested that the fish used the Earth's magnetic field in navigation. This conclusion is consistent with the prediction that

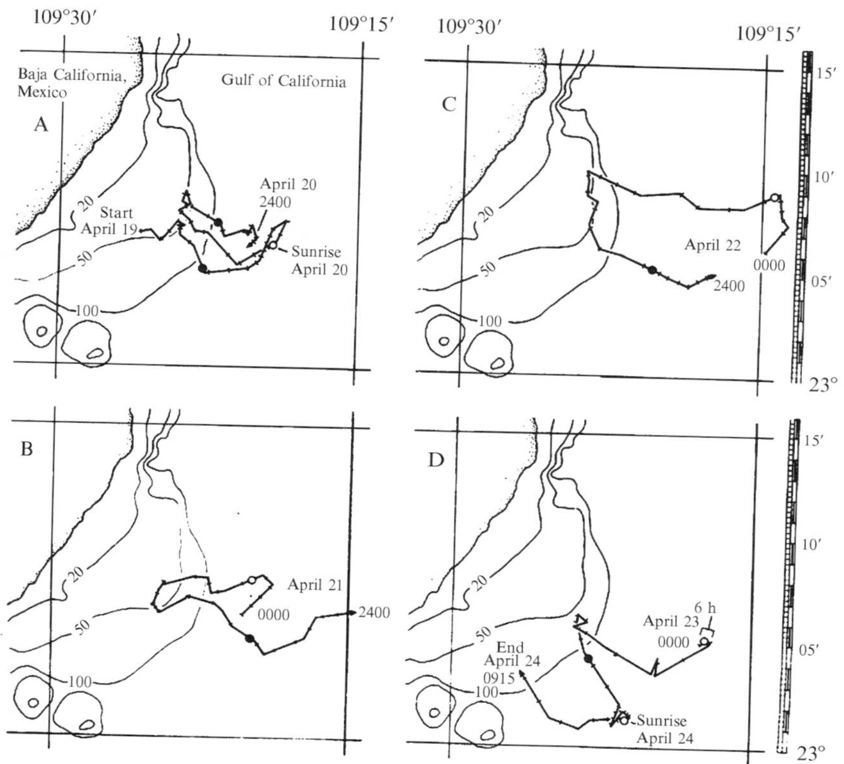

Fig. 8.9. Diurnal movement patterns in swordfish. Track of a swordfish showing repeated return movements over 4 days. Each day the fish moved onshore to the 50 fathom (91 m) depth contour on a bank and then moved out over deep water at night. The move toward shore began about 1 h before dawn and the move offshore began several hours before sunset. Filled circles represent sunset and open circles represent sunrise. [Adapted from Carey and Robison (1981).]

navigating animals might use the topography of the residual magnetic field to construct a familiar area map.

Attempts to manipulate the magnetic sense experimentally during navigation have been inconclusive. Yano *et al.* (1997) tracked salmon fitted with magnetic coils that reversed both the horizontal direction and the inclination of the magnetic field and increased the total intensity 3.5-fold within the heads of the fish. The coils were switched on and off at 11.25-min intervals and the behavior of the animals monitored in the presence and absence of the reversed field produced by the coils. Although the swimming speed of the fish was observed to decrease prior to any changes in swimming

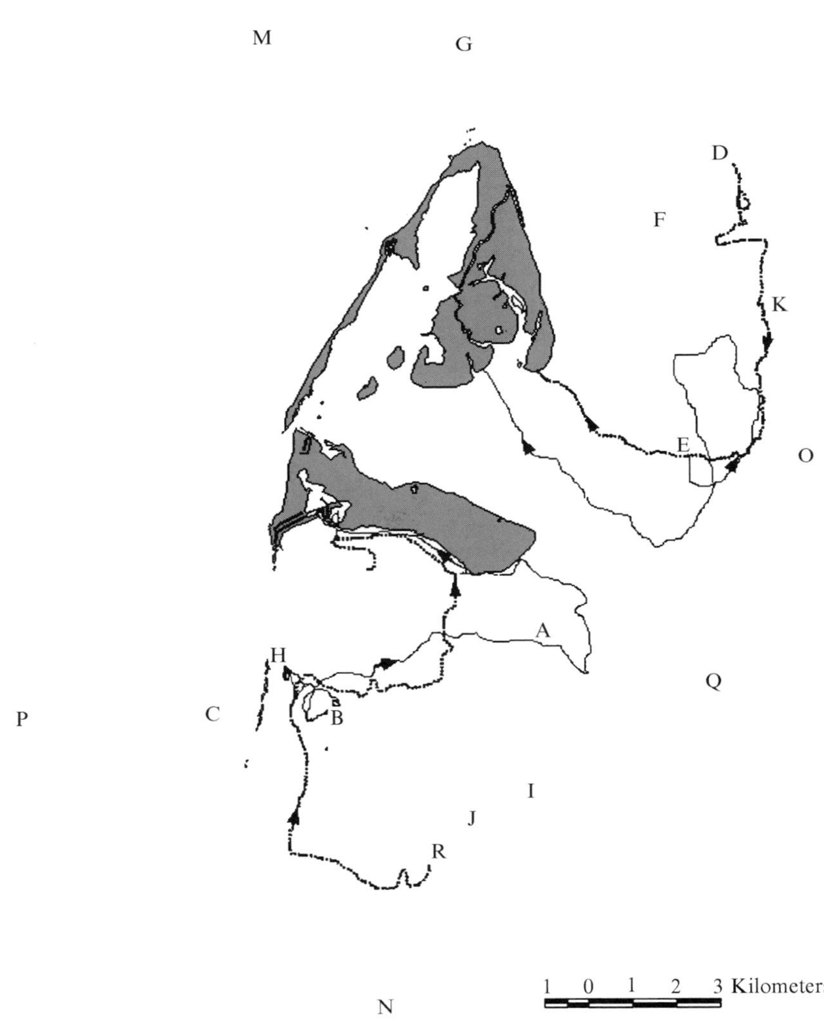

Fig. 8.10. Homing behavior of juvenile lemon sharks. Tracks from 4 (of 41) juvenile lemon sharks, *Negaprion brevirostris*, displaced to release points (A–R) located 4–15 km offshore from Bimini Islands, Bahamas. On release (at points B, D, E, and R), the sharks swam to the bottom and moved in a pattern of oscillating directions for 2–10 min before setting off on sustained, relatively straight courses that quickly brought them back to the islands. [Adapted from Sundström *et al.* (2001) with permission.]

direction, at no stage during the tracks was it possible to detect any effect on behavior of switching the magnetic coils on or off. This result illustrates the need to seek greater experimental control over the behavior of tracked animals, for example, by using animals that will return to a specified location after displacement.

10. CONCLUDING REMARKS

Fish have played a central role in studies of the structure and function of the magnetic sense. More is known about the crystal properties of biogenic magnetite, the candidate magnetite-based magnetoreceptor cells, and the magnetic field detection pathway in the fishes than in any other metazoan group. We are, however, only just beginning the systematic study of the magnetic sense in animals and there is much more yet to be discovered than has been learned to date. Thus, for a magnetite-based magnetoreceptor system, descriptions of the internal structure of candidate magnetoreceptor cells, their synaptic links to afferent nerves, and central projections in the brain are priorities for structural study of the magnetic sense in all animals, including the fishes, which are well suited for such work. Similarly, functional studies of these magnetoreceptor cells, the coding of magnetic field information for transmission to the brain, analysis of the psychophysical properties, and demonstration of dependence of sensory capacity on the structure of the sense are all required if we have to understand the magnetic sense in detail.

Despite no new data having been reported for some time, it is important to resolve the issue of whether or not the elasmobranch fishes detect magnetic fields using their electroreceptor system. Recordings from both peripheral and central neurons demonstrate that the electroreceptor system of elasmobranchs detects the electrical currents induced by fish and water movement through the Earth's magnetic field (Andrianov *et al.*, 1974; Brown and Ilyinsky, 1978). What the recordings do not show is whether the electrical information about the external magnetic field is distinguished from other electrical field information and whether use is made of the magnetic information obtained by the electroreceptor system. It is also not clear how sensitive the electroreceptor system will be to magnetic fields as the threshold sensitivity to magnetic field changes estimated from the recordings was 200 μT/sec (Andrianov *et al.*, 1974), equivalent to traveling from the magnetic equator to the magnetic pole in 0.25 sec.

Distinguishing between the electrical induction- and magnetite-based magnetoreception mechanisms will require further behavioral experiments like those in which magnets were attached over the predicted location of magnetite-based magnetoreceptors in the short-tailed stingray (Hodson, 2000; Kirschvink *et al.*, 2001). A further approach to distinguishing these mechanisms is bilateral sectioning of the SO branch of the TN, which abolished magnetic field discrimination in homing pigeons (Mora *et al.*, 2004). Failure to discriminate magnetic fields after bilateral sectioning of the SO of the TN would suggest the conclusion that their electroreceptors provide the elasmobranchs with no useful sensitivity to magnetic field stimuli even though the electroreceptors clearly detect the electrical signals that will frequently be induced in them.

The discovery that juvenile lemon sharks will home rapidly after being experimentally displaced (Sundström *et al.*, 2001) opens the way for more detailed studies of the role of the magnetic sense in navigation. Homing pigeons have been the only organisms that would home reliably after experimental displacement. In contrast with the pigeons, the likely motivation for the lemon sharks to home as rapidly as they did was to escape predation because "numerous large sharks that feed on smaller sharks inhabit the water around the Bimini Islands" (Sundström *et al.*, 2001). There are perhaps other fish species in which juvenile animals inhabit nurseries to which they will return after experimental displacement. Thus, the opportunity exists to begin comparative study of the navigation behavior of fish and homing pigeons. Although there is little powerful experimental evidence for the use of the Earth's magnetic field in navigation in any major taxonomic group, we suggest that there will be much to learn from such comparative studies just as from comparative studies of the magnetic sense itself in fish and birds. We look forward to exciting advances in our understanding of the properties and use of this most enigmatic of animal senses in the years to come.

REFERENCES

Andrianov, G. N., Brown, H. R., and Ilyinsky, O. B. (1974). Responses of central neurons to electrical and magnetic stimuli of the Ampullae of Lorenzini in the Black Sea Skate. *J. Comp. Physiol. A* **93,** 287–299.

Beason, R. C., and Semm, P. (1987). Magnetic responses of the trigeminal nerve system of the bobolink (*Dolichonyx oryzivorus*). *Neurosci. Lett.* **80,** 229–234.

Blakemore, R. P. (1975). Magnetotactic bacteria. *Science* **190,** 377–379.

Block, B. A., Dewar, H., Blackwell, S. B., Williams, T. D., Prince, E. D., Farwell, C. J., Boustany, A., Teo, S. L. H., Seitz, A., Walli, A., and Fudge, D. (2001). Migratory movements, depth preferences, and thermal biology of Atlantic Bluefin Tuna. *Science* **293,** 1310–1314.

Bodznick, D., Montgomery, J., and Tricas, T. C. (2003). Electroreception: Extracting behaviorally important signals from noise. *In* "Sensory Processing in Aquatic Environments" (Collin, S. P., and Marshall, N. J., Eds.), pp. 389–403. Springer-Verlag, NY.

Bonfil, R., Meÿer, M., Scholl, M. C., Johnson, R., O'Brien, S., Oosthuizen, H., Swanson, S., Kotze, D., and Paterson, M. (2005). Transoceanic migration, spatial dynamics, and population linkages of white sharks. *Science* **310,** 101–103.

Boustany, A. M., Davis, S. F., Pyle, P., Anderson, S. D., Le Boeuf, B. J., and Block, B. A. (2002). Satellite tagging: Expanded niche for white sharks. *Nature* **415,** 35–36.

Brown, H. R., and Ilyinsky, O. B. (1978). The ampullae of Lorenzini in the magnetic field. *J. Comp. Physiol. A* **125,** 333–341.

Bullock, T. H., and Szabo, T. (1986). Introduction. *In* "Electroreception" (Bullock, T. H., and Heiligenberg, W., Eds.), pp. 1–12. Wiley, New York.

Carey, F. G., and Robison, B. H. (1981). Daily patterns in the activities of swordfish, *Xiphias gladius*, observed by acoustic telemetry. *Fish Bull.* **79,** 277–292.

Clusin, W. T., and Bennett, M. V. L. (1979a). The oscillatory responses of skate electroreceptors to small voltage stimuli. *J. Gen. Physiol.* **73,** 685–702.

Clusin, W. T., and Bennett, M. V. L. (1979b). The ionic basis of oscillatory responses of skate electroreceptors. *J. Gen. Physiol.* **73,** 703–723.

Courtillot, V., Hulot, G., Alexandrescu, M., le Mouel, J.-L., and Kirschvink, J. L. (1997). Sensitivity and evolution of sea-turtle magnetoreception: Observations, modelling and constraints from geomagnetic secular variation. *Terra Nova* **9,** 203–207.

Davila, A. F., Fleissner, G., Winklhofer, M., and Petersen, N. (2003). A new model for a magnetoreceptor in homing pigeons based on interacting clusters of superparamagnetic magnetite. *Phys. Chem. Earth* **28,** 647–652.

Dennis, T. E., Rayner, M. J., and Walker, M. M. (2005). Spatially explicit behavioural response to geomagnetic intensity during homing in pigeons (Abstract). *In* "Fifth International Conference on Orientation and Navigation: Birds, Humans, and Other Animals," paper 18. Royal Institute of Navigation, University of Reading, London.

Diebel, C. E., Proksch, R., Green, C. R., Neilson, P., and Walker, M. M. (2000). Magnetite defines a magnetoreceptor. *Nature* **406,** 299–302.

Dijkgraaf, S., and Kalmijn, A. J. (1962). Verhaltensversuche zur Funktion der Lorenzischen Ampullen. *Naturwissenschaften* **49,** 400.

Emlen, S. T. (1975). Migration: Orientation and navigation. *In* "Avian Biology" (Farner, D. S., and King, J. R., Eds.), Vol. 5, pp. 129–219. Academic Press, New York.

Finger, T. E., St. Jeor, V. L., Kinnamon, J. C., and Silver, W. L. (1990). Ultrastructure of substance P- and CGRP-immunoreactive nerve fibers in the nasal epithelium of rodents. *J. Comp. Neurol.* **294,** 293–305.

Fleissner, G., Holtkamp-Rötzler, E., Hanzlik, M., Winklhofer, M., Fleissner, G., Petersen, N., and Wiltschko, W. (2003). Ultrastructural analysis of a putative magnetoreceptor in the beak of homing pigeons. *J. Comp. Neurol.* **458,** 350–360.

Glatzmaier, G. A., and Roberts, P. H. (1995). A three-dimensional self-consistent computer simulation of a geomagnetic field reversal. *Nature* **377,** 203–209.

Gould, J. L., Kirschvink, J. L., and Deffeyes, K. S. (1978). Bees have magnetic remanence. *Science* **201,** 1026–1028.

Hodson, R. B. (2000). Magnetoreception in the short-tailed stingray, *Dasyatis brevicaudata*. University of Auckland, New Zealand. Unpublished Masters Thesis.

Kalmijn, A. J. (1966). Electro-perception in sharks and rays. *Nature* **212,** 1232–1233.

Kalmijn, A. J. (1978). Experimental evidence of geomagnetic orientation in elasmobranch fishes. *In* "Animal Migration, Navigation and Homing" (Schmidt-Koenig, K., and Keeton, W. T., Eds.), pp. 347–353. Springer-Verlag, Berlin, New York.

Kalmijn, A. J. (1982). Electric and magnetic field detection in elasmobranch fishes. *Science* **218,** 916–918.

Keeton, W. T. (1971). Magnets interfere with pigeon homing. *Proc. Natl. Acad. Sci. USA* **68,** 102–106.

Keeton, W. T. (1973). Release-site bias as a possible guide to the "map" component in pigeon homing. *J. Comp. Physiol. A* **86,** 1–16.

Kirschvink, J. L. (1992). Uniform magnetic fields and double-wrapped coil systems: Improved techniques for the design of bioelectromagnetic experiments. *Bioelectromagnetics* **13,** 401–412.

Kirschvink, J. L., and Gould, J. L. (1981). Biogenic magnetite as a basis for magnetic field detection in animals. *Biosystems* **13,** 181–201.

Kirschvink, J. L., and Kobayashi-Kirschvink, A. (1991). Is geomagnetic sensitivity real? Replication of the Walker-Bitterman magnetic conditioning experiment in honey bees. *Am. Zool.* **31,** 169–185.

Kirschvink, J. L., and Walker, M. M. (1985). Particle-size considerations for magnetite-based magnetoreceptors. *In* "Magnetite Biomineralization and Magnetoreception by Living Organisms: A New Biomagnetism" (Kirschvink, J. L., Jones, D. S., and MacFadden, B. J., Eds.), pp. 243–254. Plenum Publishing Corporation, New York.

Kirschvink, J. L., Jones, D. S., and MacFadden, B. J. (Eds.). (1985a). "Magnetite Biomineralization and Magnetoreception by Living Organisms: A New Biomagnetism," Vol. xxi, p. 682. Plenum Publishing Corporation, New York.

Kirschvink, J. L., Walker, M. M., Chang, S.-B., Dizon, A. E., and Peterson, K. A. (1985b). Chains of single-domain magnetite particles in the chinook salmon, *Oncorhynchus tshawytscha. J. Comp. Physiol. A* **157,** 375–381.

Kirschvink, J. L., Dizon, A. E., and Westphal, J. A. (1986). Evidence from strandings for geomagnetic sensitivity in cetaceans. *J. Exp. Biol.* **120,** 1–24.

Kirschvink, J. L., Padmanabha, S., Boyce, C. K., and Oglesby, J. (1997). Measurement of the threshold sensitivity of honeybees to weak, extremely low frequency magnetic fields. *J. Exp. Biol.* **200,** 1363–1368.

Kirschvink, J. L., Walker, M. M., and Diebel, C. E. (2001). Magnetite-based magnetoreception. *Curr. Opin. Neurobiol.* **11,** 462–467.

Klimley, A. P. (1993). Highly directional swimming by scalloped hammerhead sharks, *Sphyrna lewini,* and substrate irradiance, temperature, bathymetry and geomagnetic field. *Marine Biol.* **117,** 1–22.

Klimley, A. P., and Nelson, D. R. (1984). Diel movement patterns of the scalloped shark (*Sphyrna lewini*) in relation to El Bajo Espiritu Santo: A refuging central position social system. *Behav. Ecol. Sociobiol.* **15,** 45–54.

Kramer, G. (1953). Wird die Sonnenhöhe bei der Heimfindeorientierung verwertet? *J. für. Ornithologie* **94,** 201–219.

Lohmann, K. J., and Lohmann, C. M. F. (1996). Detection of magnetic field intensity by sea turtles. *Nature* **380,** 59–61.

Lohmann, K. J., Lohmann, C. M. F., Ehrhart, L. M., Bagley, D. A., and Swing, T. (2004). Geomagnetic map used in sea turtle navigation. *Nature* **428,** 909–910.

Mann, S., Sparks, N. H. C., Walker, M. M., and Kirschvink, J. L. (1988). Ultrastructure, morphology and organization of biogenic magnetite from sockeye salmon, *Oncorhynchus nerka*: Implications for magnetoreception. *J. Exp. Biol.* **140,** 35–49.

McKeegan, D. E. F., Smith, F. S., Demmers, T. G. M., Wathes, C. M., and Jones, R. B. (2005). Behavioral correlates of olfactory and trigeminal gaseous stimulation in chickens, *Gallus domesticus. Physiol. Behav.* **84,** 761–768.

Meÿer, C. G., Holland, K. N., and Papastamatiou, Y. P. (2005). Sharks can detect changes in the geomagnetic field. *J. R. Soc. Interface* **2,** 129–130.

Mora, C. V., Davison, M. C., Wild, N. M., and Walker, M. M. (2004). Magnetoreception and its trigeminal mediation in the homing pigeon. *Nature* **432,** 508–511.

Murray, R. W. (1960). Electrical sensitivity of the ampullae of Lorenzini. *Nature* **187,** 957.

Nishi, T., and Kawamura, G. (2005). *Anguilla japonica* is already magnetosensitive at the glass eel phase. *Fish. Sci.* **67,** 1213–1224.
Nishi, T., Kawamura, G., and Matsumoto, K. (2004). Magnetic sense in the Japanese eel, *Anguilla japonica*, as determined by conditioning and electrocardiography. *J. Exp. Biol.* **207,** 2965–2970.
Nishi, T., Kawamura, G., and Sannomiya, S. (2005). Anosmic Japanese eel *Anguilla japonica* can no longer detect magnetic fields. *Fish. Sci.* **71,** 101–106.
Paulin, M. G. (1995). Electroreception and the compass sense of sharks. *J. Theor. Biol.* **174,** 325–339.
Quinn, T. P. (1980). Evidence for celestial and magnetic compass orientation in lake-migrating sockeye salmon fry. *J. Comp. Physiol. A* **137,** 243–248.
Reilly, W. I. (2002). Magnetic position determination by homing pigeons? *J. Theor. Biol.* **218,** 47–54.
Rommel, S. A., and McCleave, J. D. (1973). Sensitivity of American eels (*Anguilla rostrata*) and Atlantic salmon (*Salmo salar*) to weak electric and magnetic fields. *J. Fish. Res. Board Canada* **30,** 657–663.
Schmidt-Koenig, K. (1958). Experimentelle Einfluβnahme auf die 24-Stunden-Periodik bei Brieftauben und deren Auswirkungen unter besonderer Berücksichtigung des Heimfindevermögens. *Zietschrift für Tierpsychologie* **15,** 301–331.
Semm, P., and Beason, R. C. (1990). Responses to small magnetic field variations by the trigeminal system of the bobolink. *Brain Res. Bull.* **25,** 735–740.
Skiles, D. D. (1985). The geomagnetic field: Its nature, history and biological relevance. *In* "Magnetite Biomineralization and Magnetoreception by Living Organisms: A New Biomagnetism" (Kirschvink, J. L., Jones, D. S., and MacFadden, B. J., Eds.), pp. 43–102. Plenum Publishing Corporation, New York.
Steiner, I., Bürgi, C., Werffeli, S., Dell'Omo, G., Valenti, P., Tröster, G., Wolfer, D. P., and Lipp, H.-P. (2000). A GPS logger and software for analysis of homing in pigeons and small mammals. *Physiol. Behav.* **71,** 589–596.
Sundström, L. F., Gruber, S. H., Clermont, S. M., Correia, J. P. S., de Marignac, J. R. C., Morrissey, J. F., Lowrance, C. R., Thomassen, L., and Oliveira, M. T. (2001). Review of elasmobranch behavioural studies using ultrasonic telemetry with special reference to the lemon shark, *Negaprion brevirostris*, around Bimini Islands, Bahamas. *Env. Biol. Fishes* **60,** 225–250.
Taylor, P. B. (1986). Experimental evidence for geomagnetic orientation in juvenile salmon, *Oncorhynchus tshawytscha* Walbaum. *J. Fish Biol.* **28,** 607–623.
Taylor, P. B. (1987). Experimental evidence for juvenile chinook salmon, *Oncorhynchus tshawytscha* Walbaum, orientation at night and in sunlight after a 7° change in latitude. *J. Fish Biol.* **31,** 89–111.
Viguier, C. (1882). Le sens d'orientation et ses organes chez les animaux et chez l'homme. *Revue Philosophique de la France et de l'Étranger* **14,** 1–36.
Vine, F. J. (1966). Spreading of the ocean floor; new evidence. *Science* **154,** 1405–1415.
Walcott, C. (1978). Anomalies in the Earth's magnetic field increase the scatter of pigeons' vanishing bearings. *In* "Animal Migration, Navigation, and Homing" (Schmidt-Koenig, K., and Keeton, W. T., Eds.), pp. 143–151. Springer-Verlag, Berlin.
Walcott, C., and Green, R. P. (1974). Orientation of homing pigeons altered by a change in the direction of an applied magnetic field. *Science* **184,** 180–182.
Walker, M. M. (1984). Learned magnetic field discrimination in the yellowfin tuna, *Thunnus albacares*. *J. Comp. Physiol. A* **155,** 673–679.
Walker, M. M. (1998). On a wing and a vector: A model for magnetic navigation by homing pigeons. *J. Theor. Biol.* **192,** 341–349.
Walker, M. M., and Bitterman, M. E. (1989). Honeybees can be trained to respond to very small changes in geomagnetic field intensity. *J. Exp. Biol.* **145,** 489–494.

Walker, M. M., Kirschvink, J. L., Chang, S.-B. R., and Dizon, A. E. (1984). A candidate magnetic sense organ in the yellowfin tuna, *Thunnus albacares*. *Science* **224,** 751–753.

Walker, M. M., Diebel, C. E., Haugh, C. V., Pankhurst, P. M., Montgomery, J. C., and Green, C. R. (1997). Structure and function of the vertebrate magnetic sense. *Nature* **390,** 371–376.

Walker, M. M., Diebel, C. E., and Green, C. R. (2000). Structure, function and use of the magnetic sense in vertebrates. *J. Appl. Physics* **87,** 4653–4658.

Walker, M. M., Dennis, T. E., and Kirschvink, J. L. (2002). The magnetic sense and its use in long-distance navigation by animals. *Curr. Opin. Neurobiol.* **12,** 735–744.

Wallraff, H. G. (1999). The magnetic map of homing pigeons: An evergreen phantom. *J. Theor. Biol.* **197,** 265–269.

Weng, K. C., Castilho, P. C., Morrissette, J. M., Landeira-Fernandez, A. M., Holts, D. B., Schallert, R. J., Goldman, K. J., and Block, B. A. (2005). Satellite tagging and cardiac physiology reveal niche expansion in salmon sharks. *Science* **310,** 104–106.

Wiltschko, R., and Wiltschko, W. (1995). "Magnetic Orientation in Animals," Vol. xvii, p. 297. Springer, Berlin, New York.

Yano, A., Ogura, M., Sato, A., Sakaki, Y., Shimizu, Y., Baba, N., and Nagasawa, K. (1997). Effect of modified magnetic field on the ocean migration of maturing chum salmon. *Marine Biol.* **129,** 523–530.

Yuen, H. S. H. (1970). Behaviour of skipjack tuna, *Katsuwonus pelamis*, as determined by tracking with ultrasonic devices. *J. Fish. Res. Board Canada* **27,** 2071–2079.

9

NEURAL AND BEHAVIORAL MECHANISMS OF AUDITION

ANDREW H. BASS
ZHONGMIN LU

1. Introduction
2. Behavioral Studies of Audition
3. Peripheral and Central Auditory Pathways
   3.1. Auditory System
   3.2. Vocal Motor Inputs to Auditory System
4. Neurophysiological Mechanisms of Audition
   4.1. Encoding of Vocal Signals
   4.2. Directional Hearing
5. Auditory Lateral Line Integration
6. Vocal Modulation of Inner Ear and Lateral Line
7. Steroid Hormones and Seasonal Changes in Hearing
8. Future Directions

1. INTRODUCTION

The auditory system of fish, like that of other vertebrates, is adapted to the detection of sound, a mechanical disturbance that leads to the vibration of molecules in any medium. The past decade has witnessed major advances in our understanding of the behavioral and neural mechanisms of audition among teleost fish, the largest group of living fishes. Since there are several reviews of auditory mechanisms among teleosts (Bass and Ladich, in press; Bass and McKibben, 2003; Ladich and Bass, 2003a,b; Lu, 2004; Bass *et al.*, 2005), we will highlight the major points of those summaries, while providing more complete coverage of the recent findings in this field of study. The focus will be on two main topics, mechanisms of vocal/acoustic communication among sound producing/sonic fish and mechanisms of directional hearing among sonic and nonsonic species. Since much of this

Sensory Systems Neuroscience: Volume 25
FISH PHYSIOLOGY

DOI: 10.1016/S1546-5098(06)25009-X

study has been completed in two families of teleosts, the Batrachoididae (order Batrachoidiformes) and Eleotridae (order Perciformes), they are the primary focus of the discussions [for a more complete taxonomic overview of teleosts see Nelson (1994) and Pough *et al.* (2002)].

2. BEHAVIORAL STUDIES OF AUDITION

Batrachoidids include several genera commonly referred to as midshipman fish and toadfish that have long been a focus of underwater playback experiments with freely swimming animals in either their natural habitats or in large enclosures (Fine *et al.*, 1977; Bass and McKibben, 2003). Studies of the behavioral mechanisms of audition among midshipman fish have investigated the recognition of the spectral and temporal features of acoustic signals that mediate social interactions in reproductive and agonistic contexts. Nesting male midshipman produce a multiharmonic advertisement call known as a "hum" (after Hubbs, 1920; Ibara *et al.*, 1983) with a fundamental frequency (F0) near 100 Hz in its natural habitat (Figure 9.1A). Observations of the midshipman's nocturnal spawning behavior show that gravid females, whose ovaries contain mature eggs, orient toward the nest of a humming male (Brantley and Bass, 1994). The essential role of male acoustic courtship

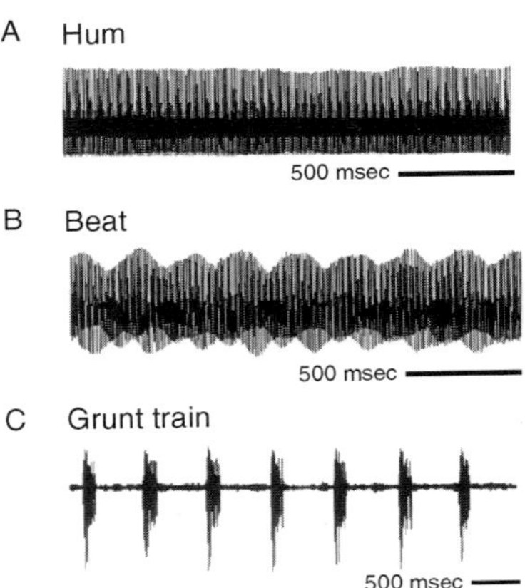

Fig. 9.1. Vocal signals of teleosts. (A) Type I male midshipman fish produce advertisement calls known as hums. (B) The hums of two neighboring males are often concurrent and produce an acoustic beat with a modulation rate determined by the difference in the fundamental frequencies between the two hums. (C) A series of brief grunts (grunt train) are generated in agonistic contexts. [Reprinted with permission from McKibben and Bass (2001a), Copyright 2001, American Institute of Physics.]

in mate attraction is shown by underwater playbacks. Gravid females, but not spent females that have released their eggs, swim directly head-on and often touch their head against an underwater speaker that broadcasts pure tones at frequencies like the F0s of natural hums (McKibben and Bass, 1998, 2001b). Two choice playback experiments, to our knowledge the first in teleosts, show that females make preferential choices based on a tone's frequency, as well as its duration and intensity. Increases in either the duration or intensity of playback tones make the signals more attractive to females, while frequency preference varies with ambient temperature. The F0 of hums increases with rising temperature at about 5 Hz/°C (Brantley and Bass, 1994; McKibben and Bass, 1998). At any one temperature, females prefer tone frequencies that are most similar to the F0 of natural hums (McKibben and Bass, 1998). For example, a female will choose a 90-Hz tone over an 80-Hz tone at 14.5 °C; 90 Hz is closer to the F0 of a male's hum at that temperature (McKibben and Bass, 1998). There is apparently an intense selection pressure to maintain a close coupling between female frequency selectivity and hum F0. One can only speculate as to why a particular F0 is produced at any one temperature. Perhaps it is shaped by the physiological adaptations of the sonic muscle to produce long duration hum signals (minutes to >1 h) at the low ambient temperatures (12–16 °C) found in their nesting habitat at night [Bass *et al.*, 1999; M. Marchaterre and A. Bass, unpublished observations; see Young and Rome (2001) for general discussion of sonic muscle physiology]. To date, recordings from natural populations show no correlation between F0 and male body size so that this coupling mechanism appears to mainly function in locating humming males in their nests. By contrast, larger males appear to make louder sounds (M. Marchaterre and A. Bass, unpublished observations), suggesting that intensity may be an important cue that contributes to female choice.

Midshipman males that build nests and acoustically court females are known as type I males. An alternative male reproductive morph known as a type II male does not exhibit these behavioral tactics but rather sneak or satellite spawns to essentially steal fertilizations away from a type I male (Brantley and Bass, 1994). The extreme divergence in reproductive tactics between type I and II males is paralleled by a divergence in a large suite of somatic, gonadal, neural, and endocrinological traits (Bass, 1996). Type I and II males will also approach tone playbacks, although their responses are not as consistent as those of gravid females (McKibben and Bass, 1998). What is especially interesting, however, is that each male morph approaches an underwater playback speaker in a morph-specific manner, both of which differ from the female's response behavior. Thus, type I males approach the speaker as if they were establishing a nest site by backing into the plastic frame holding the speaker (unlike the head-on approach of females;

see earlier) and sometimes showing digging-like motions resembling the nest-building behavior of type I males (Brantley and Bass, 1994). Type II males adopt a satellite-like posture after approaching a speaker by lining up alongside the plastic frame holding the speaker and sometimes placing their tail under the speaker frame as they do along the outer perimeter of a nest (Brantley and Bass, 1994). The positive phonotactic responses of type I and II males would be consistent with their use of the hum as a beacon to locate the nests of other males either for the purposes of building their own nest in a suitable area (type I males) or to sneak and satellite spawn (type II males).

The envelope shape of an acoustic waveform varies with the extent of amplitude and frequency modulation. A midshipman male's advertisement hum shows almost no modulation in its envelope shape (Figure 9.1A), except at signal onset and offset (Bass *et al.*, 1999). Modulation is introduced into the hum waveform when the hums of neighboring males overlap and generate acoustic beats that have a rate of amplitude modulation determined by the difference (dF) between the F0s of the two hums (Figure 9.1B). "Grunt trains," a series of brief (~100 msec), broadband signals produced by type I males when defending their nest against potential intruders, show dramatic fluctuations in amplitude with their sudden onset and offset (Figure 9.1C) (Brantley and Bass, 1994; Bass *et al.*, 1999).

McKibben and Bass (1998) tested the hypothesis that acoustic beats are more attractive signals than tones to gravid females because beats are indicative of a cluster of nesting males that females could choose from when spawning. Underwater playbacks show that females do not prefer a tone played from a single speaker to concurrent tones played from a pair of adjacent speakers that mimic the overlapping hums of two neighboring males (the nests of males are often close to one another) (Bass, 1996). Acoustic beats might however be used by individual females (or males) to locate the nest of a single humming male. The depth of modulation of a natural beat will vary with the relative contribution of each tone or hum from two separate sources (either underwater speakers or type I males). In the midshipman's auditory world, depth of modulation will vary with proximity to one of two nests (or underwater speakers) so that an individual's sensitivity to changes in depth of modulation could contribute to the spatial localization task of finding the nest of a humming male.

Additional playback experiments compared the responsiveness of gravid females to a tone stimulus played through one speaker to an acoustic beat signal played through a second speaker (rather than arising from concurrent playback of individual tones from adjacent speakers, see earlier). The results show that females prefer the single-source tone stimulus, suggesting that as long as a beat signal is resolved into two separate tones at their sources (as with two speakers), females do not reject the signals (McKibben and Bass, 1998).

The rejection of a single-source beat also suggests that the presence or absence of envelope modulation in an individual signal contributes to the recognition of hums from "nonhum" signals. The rejection of a single-source beat might arise from their perceptual similarity to amplitude modulated (AM)-like grunt trains (Figure 9.1). In support of this hypothesis, gravid females are not attracted to either natural or computer-synthesized grunt trains (McKibben and Bass, 1998). Other studies show that a single source beat becomes more attractive as its envelope shape becomes more flattened and humlike with decreases in either dF or depth of modulation (McKibben and Bass, 2001a). Similarly, pulse trains that vary in the duration of either the pulse or the silent gap between successive pulses become more attractive as they become more humlike (McKibben and Bass, 2001a). Thus, females are more responsive to playbacks with either increasing pulse duration or decreasing gap duration. Together, the results of the underwater playback studies support the hypothesis that envelope modulation contributes to mechanisms of signal recognition in midshipman fish.

Playback studies with the Gulf toadfish, *Opsanus beta*, investigate the relationship between vocal-acoustic behaviors and endocrinological state. The advertisement calls of toadfish are known as "boatwhistles." These signals are more complex than the hums of midshipman because they include an introductory gruntlike signal followed by a multiharmonic component known as a "hoot" [the hoot is the analogue of the hum; see Bass and McKibben (2003) for more extensive discussion]. Early studies with both *O. beta* and the oyster toadfish *O. tau* show that nesting males respond to playbacks of natural advertisement calls and pure tones (Fish, 1972; Winn, 1972). Like midshipman, pure tones can be used to mimic advertisement calls and gravid female toadfish approach advertisement call playbacks and tones. One advantage of experiments with toadfish is that vocal responses are easily evoked from nesting males using underwater playbacks; this has always been a difficult task with midshipman fish (A. Bass, unpublished observations). The vocal responses of toadfish males are influenced by a variety of spectral and temporal parameters of playback tones including frequency, duration, and repetition rate. In general, male responsiveness increases as the synthetic call becomes closer in appearance to natural calls, reminiscent of the increased attraction of gravid female midshipman fish to synthetic calls as they become more humlike. Remage-Healey and Bass (2005) have used this experimental paradigm to identify concurrent changes in circulating hormone profiles and vocal behavior. As per prior studies (Fish, 1972; Winn, 1972), males show increased call rate and call duration in response to playbacks of tones that resemble the duration and F0 of the boatwhistles of conspecific males. Increased calling is also accompanied by increased levels of 11-ketotestosterone (11 kT), the principal circulating androgen in many teleosts including batrachoidids [Sisneros *et al.* (2004a) and references therein]. Playbacks of either gruntlike signals or noise do not elicit increases in either calling behavior or 11 kT levels. The increases in both 11 kT levels and calling behavior occur within 20 min, consistent with the rapid effects of 11 kT on the vocal pattern generator positioned at the hindbrain–spinal cord junction (Remage-Healey and Bass, 2004; also see Section 3.2). Potentially, this experimental paradigm could be used to probe an individual's perceptual recognition of gradual changes in signal characteristics in the absence of a vocal response.

Behavioral studies of several species of weakly electric mormyrid fish that are also sonic have used a conditioning paradigm to assess their behavioral sensitivity to the temporal features of playback signals. Using the animal's own electric organ discharge rate as a metric for sound recognition, these studies have shown that mormyrids are able to discriminate interclick intervals consistent with the natural variation of pulse repetition rates in their calls (Marvit and Crawford, 2000; Fletcher and Crawford, 2001). Other studies of temporal discrimination in mormyrids and other sonic species have measured auditory sensitivity using the auditory brainstem response which is a compound potential recorded by electrodes placed on the skull that represents the summed auditory-related activity in the inner ear and brainstem (Kenyon *et al.*, 1998; Wysocki and Ladich, 2002, 2003). Studies of the ontogeny of sound discrimination abilities in sonic fish have also just begun (Wysocki and Ladich, 2001; Sisneros and Bass, 2005).

3. PERIPHERAL AND CENTRAL AUDITORY PATHWAYS

Hair cell-based sensory systems among fish include the auditory, vestibular, and lateral line modalities; in some species, the lateral line includes both mechanosensory and electrosensory divisions (Bullock and Heiligenberg, 1986). These three systems share a common embryonic origin from placodal tissues (Gibbs, 2004) and all, excepting the electrosensory system, are responsive to vibratory stimuli in the aquatic environment (Coombs and Janssen, 1988). Batrachoidids are perhaps the best-studied group of teleosts for central and peripheral mechanisms of all three hair cell-based systems (Bass and McKibben, 2003).

3.1. Auditory System

Like other vertebrates, the inner ear of teleosts is innervated by the eighth cranial nerve (VIIIth) and includes both otolithic and nonotolithic end organs, each with its own sensory epithelium (Figure 9.2). The saccule,

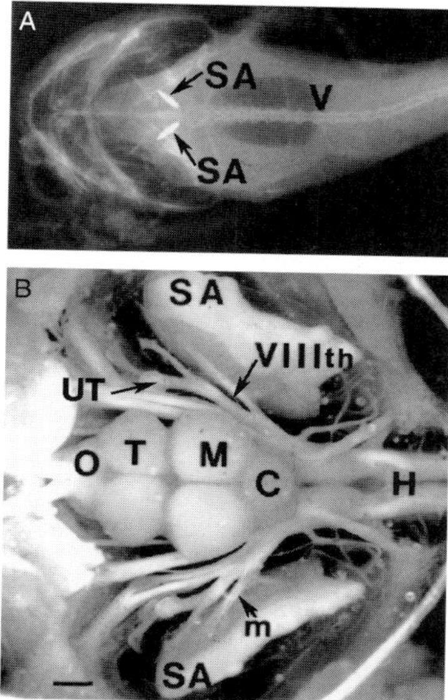

Fig. 9.2. Inner ear of teleosts. (A) X-ray image of a ventral view of a type I male midshipman fish highlighting the positions of the two saccular otoliths (SA) relative to the vertebral column (V). The saccule is the largest end organ of the inner ear and main auditory organ in this and many other species of teleosts. (B) Dorsal view of the brain and inner ear of a female midshipman fish. The sensory epithelium (macula, m) apposing the saccular otolith is indicated along with the branch of the eighth cranial nerve (VIIIth) that innervates saccular hair cells. C, cerebellum; H, hindbrain; M, midbrain; O, olfactory bulb; T, telencephalon; UT, utricle. Scale bar = 1 mm. [Adapted from Bass *et al.* (1999) with permission of MIT Press.]

utricle, and lagena are the otolithic organs, while the three semicircular canals are the nonotolithic ones. Some species have another nonotolithic sensory epithelium known as the macula neglecta that may also serve an auditory function as it does in elasmobranches (Corwin, 1981). Although each otolithic organ has been suggested to function in either vestibular and/or auditory sensation, the saccule appears to be the main organ of hearing in most species (Popper and Fay, 1999).

The central auditory pathways are similar across all major groups of teleosts. The brainstem circuitry that includes those pathways from the inner ear to the hindbrain and from the hindbrain to the midbrain are also very similar to other groups of fishes and tetrapods (McCormick, 1999; Bass *et al.*, 2005). There is insufficient evidence to evaluate the degree of similarity in the organization of forebrain auditory systems [but see McCormick (1999) for discussion]. Here, we summarize the available data for batrachoidids

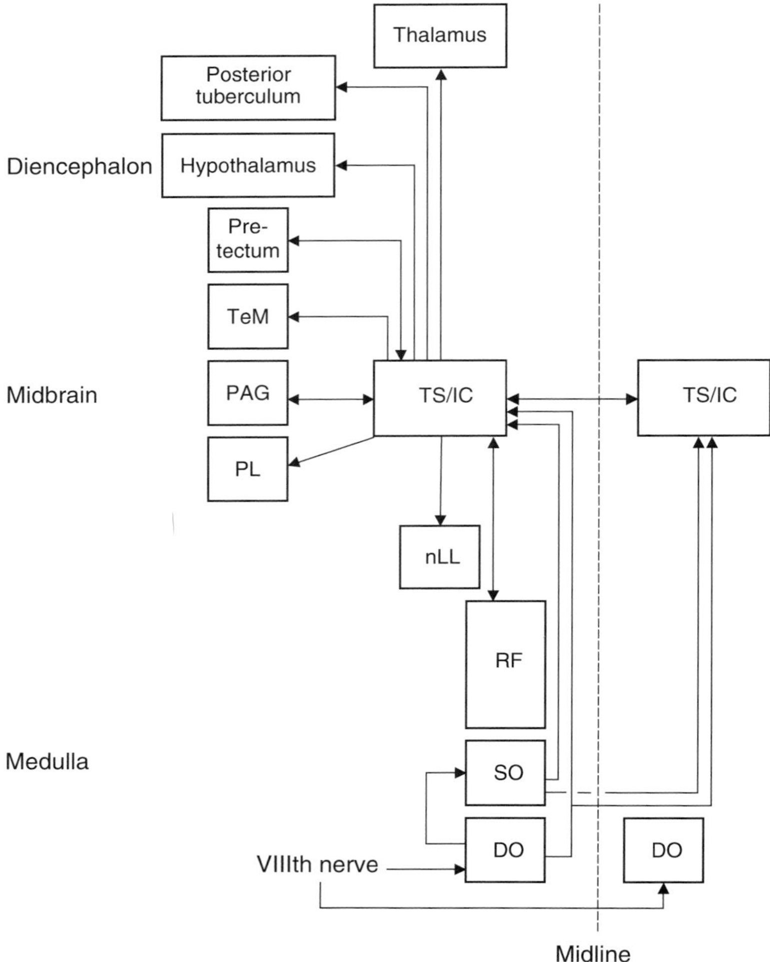

Fig. 9.3. Central auditory system of teleosts. Schematic overview of the ascending auditory system of midshipman fish and toadfish delineated by neurobiotin labeling of either a single eighth (VIIIth) cranial nerve or nucleus centralis, the main auditory region in the midbrain's torus semicircularis (homologue of inferior colliculus of birds and mammals; the abbreviation TS/IC implies this homology). The VIIIth nerve projects to the descending octaval nucleus (DO) in the medulla; this projection is mainly ipsilateral with some evidence for a lighter contralateral projection also (Bass *et al.*, 1994, 2000; McCormick, 1999; Kozloski and Crawford, 1998). The TS/IC has both lateral line and auditory subdivisions [see Weeg and Bass (2000) for lateral line pathways]. The auditory TS/IC receives inputs from DO, an adjacent secondary octaval nucleus (SO) that receives DO input, and the reticular formation (RF). TS/IC projects to several sites

based on tract-tracing studies using biotin-like tracers injected into the physiologically identified midbrain auditory center in the torus semicircularis (TS) that is known as nucleus centralis [Bass *et al.*, 2000, 2001; see Kozloski and Crawford (1998) and Prechtl *et al.* (1998) for sonic mormyrids and Tomchik and Lu (2004) for gobies]. The TS is considered the homologue of the inferior colliculus (IC) of birds and mammals; hence, the designation TS/IC in Figure 9.3 [see McCormick (1999) and Bass *et al.* (2005) for further discussion]. A descending octaval nucleus (DO) is the main hindbrain target of saccular afferents (Figure 9.3). The DO nucleus and a secondary octaval nucleus (SO) that is contiguous with and receives input from DO are the principal sources of medullary input to the TS. Other inputs to TS arise from the midline reticular formation in the hindbrain (RF), the contralateral TS, the midbrain periaqueductal gray (PAG), and the pretectum (a medial pretoral nucleus, see Bass *et al.*, 2000). As shown in Figure 9.3, the RF, PAG, and pretectum have reciprocal connections with TS; there are also reciprocal connections between the two DOs in gobies (not shown; S. Tomchik and Z. Lu, unpublished observations). The TS also projects to a nucleus (nLL) adjacent to the lateral lemniscus (the main fiber bundle carrying auditory-related axons from the medulla to TS), the midbrain tectum (TeM), the hypothalamus, the posterior tuberculum, and the dorsal thalamus that projects, in turn, to the telencephalon (Goodson and Bass, 2002).

3.2. Vocal Motor Inputs to Auditory System

While the central auditory pathways of sonic teleosts are similar to those of nonsonic fish (McCormick, 1999), sonic fish show additional integration sites between the vocal and auditory systems (Bass *et al.*, 1994, 2000; Goodson and Bass, 2002). Sound production, or vocalization, has independently evolved among several groups of teleosts. In many cases, the peripheral sound producing apparatus and the motor neurons that innervate sonic muscles have been identified (Bass and Ladich, in press). The most extensive studies of central vocal pathways have been in

---

including the contralateral TS/IC, a nucleus adjacent to the lateral lemniscus (nLL), which is the main fiber bundle carrying axons from the medulla to the TS/IC, vocal-related regions of the periaqueductal gray (PAG) and the paralemniscal tegmentum (PL), and the midbrain tectum (TeM). Forebrain targets of the TS/IC include the pretectum, the hypothalamus, the posterior tuberculum, and the dorsal thalamus. Reciprocal connections are indicated by a solid line connecting two arrowheads. [Adapted from Bass *et al.* (2005) with permission of Springer Science and Media.]

batrachoidids [see Bass and McKibben (2003) for extensive review]. Briefly, Pappas and Bennett (1966) first identified a central sonic motor system in a teleost fish in their studies of electrotonic coupling in the oyster toadfish (*O. tau*). Since that time, neurophysiological and anatomical studies have revealed an expansive pacemaker-motor neuron network positioned at the junction of the caudal medulla and rostral spinal cord (Bass and Baker, 1990; Bass *et al.*, 1994). A midline sonic motor nucleus (SMN) innervates the ipsilateral sonic muscle attached to the lateral wall of the swim bladder. A column of pacemaker-like neurons (PN) along the ventrolateral margin of each SMN provides the sole afferent input to each SMN and apparently establishes the rhythmic firing frequency of motor neurons. Each PN innervates motor neurons bilaterally, thereby providing a basis for the extensive coupling among sonic motor neurons that leads to the simultaneous contraction of both sonic muscles. Finally, anatomical evidence in midshipman shows a ventral medullary nucleus (VM) positioned rostral to the PN-SMN circuit that is extensively coupled across the midline by a ventral commissural pathway; VM appears to provide the sole input to the PN-SMN circuit (Bass *et al.*, 1994; Goodson and Bass, 2002; Kittelberger *et al.*, 2006). The rhythmic motor volley of the pacemaker-motor circuit is easily monitored by intracranial recordings from occipital nerve roots that carry the axons of sonic motor neurons that form the sonic nerve innervating each sonic muscle. The sonic motor volley is referred to as a "fictive vocalization" because it mimics the basic features of natural vocalizations, namely F0 and duration.

The anatomical studies have delineated a descending vocal motor pathway that interfaces with auditory-recipient nuclei at forebrain, midbrain, and hindbrain levels (Bass *et al.*, 1994, 2000; Goodson and Bass, 2002). Each of the vocal-acoustic centers (VAC) includes several interconnected nuclei that receive auditory input either from the inner ear via the VIIIth cranial nerve (Bass *et al.*, 1994, 2000, 2001) or from auditory-recipient nuclei in the TS (Figure 9.4) and dorsal thalamus (Th, Figure 9.4). A hindbrain region (hVAC, Figure 9.4) includes a division of DO that provides auditory input to the midbrain (TS, Figure 9.3). At midbrain levels (mVAC, Figure 9.4), the periaqueductal gray is a vocally active site that receives input from the TS (Figure 9.3) and projects to hindbrain vocal sites. Vocal-auditory integration sites in the forebrain (fVAC, Figure 9.4) include hypothalamic and telencephalic nuclei that receive TS input either directly or indirectly via the thalamus (Goodson and Bass, 2002). The hVAC also includes an efferent nucleus that sends axons directly to the inner ear; the physiological significance of vocal-auditory integration is just beginning to be investigated at the level of these VIIIth nerve efferents as discussed in section 6.

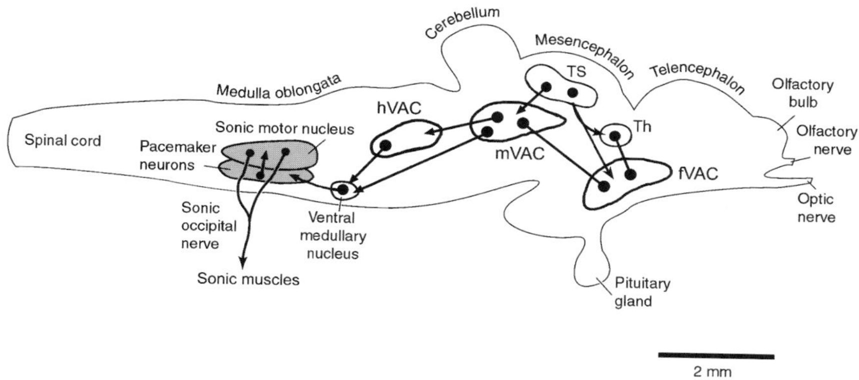

Fig. 9.4. Descending vocal motor system in sonic teleosts. A line drawing of a saggital view of the descending vocal control system in midshipman and toadfish is shown here. Each major level of the central nervous system has several nuclei that integrate both vocal and acoustic information; for simplicity, the nuclei at each level are clustered into forebrain (fVAC), midbrain (mVAC), and hindbrain (hVAC) vocal-acoustic integration centers (Goodson and Bass, 2002). Also shown here is the position of a vocal pattern generator that extends across the caudal hindbrain and rostral spinal cord and includes pacemaker neurons that innervate sonic motor neurons that innervate, in turn, sonic muscles attached to the walls of the swimbladder (Bass and Baker, 1990; Bass *et al.*, 1994). The pacemaker-motor neuron circuit establishes the contraction rate of the sonic muscles that determines the pulse repetition rate of broadband grunts and the fundamental frequency of multiharmonic hums. A ventral medullary nucleus links the hVAC and mVAC (which are also linked) to the vocal pattern generator. The hVAC includes the hindbrain DO nucleus that receives direct input from the inner ear via the VIIIth nerve (Figure 9.3). The mVAC and fVAC are reciprocally connected (dots connected by solid line) and receive inputs from the auditory midbrain (TS) and thalamus (Th). [Adapted from Bass and McKibben (2003) with permission from Elsevier.]

4. NEUROPHYSIOLOGICAL MECHANISMS OF AUDITION

This section includes two parts. The first will consider auditory mechanisms for the encoding of species-specific calls and the second will review mechanisms of directional hearing. There is a bias toward a review of directional sensitivity since other reviews have considered auditory encoding of vocalizations in some detail (Bass and McKibben, 2003; Bass *et al.*, 2005), and the most recent studies since those reviews focus on mechanisms of directional sensitivity.

4.1. Encoding of Vocal Signals

Auditory neuroscientists want to identify the spike train parameters that provide adequate information for a receiver to identify and distinguish the physical attributes of one acoustic signal from another signal. Extensive studies in nonsonic fish, especially goldfish and trout, have demonstrated many similarities in the encoding properties of auditory neurons between teleosts and tetrapods (Fay and Simmons, 1999; Feng and Schellart, 1999; Lu, 2004). Studies of sonic batrachoidids and mormyrids have further revealed that both peripheral and central auditory neurons encode the spectral and temporal properties of their vocal signals (Bass and McKibben, 2003; Lu, 2004). Studies in midshipman fish have identified neurons within the auditory midbrain that are broadly tuned to the principal frequency components of their calls [Bodnar and Bass, 1997; see Crawford (1993) for sonic momyrids]. These studies have also discussed the encoding of the dF of acoustic beats, in part, to provide insight into mechanisms that could lead to the behavioral discrimination of hums from nonhums (Section 2). The results demonstrate a peripheral to central transformation in the encoding of beat dF. Peripheral saccular neurons show robust encoding of tone frequency as measured by the degree of phase locking of spikes to each cycle of a tone stimulus that is referred to as the vector strength of synchronization or VS (Goldberg and Brown, 1969) (Figure 9.5) (McKibben and Bass, 1999). While VS values are high for each tone frequency that contributes to a synthetic beat, they are relatively low for dF, the modulation rate of the beat waveform (McKibben and Bass, 2001a). Midbrain neurons show the inverse, that is, poor encoding of a beat's component frequencies but robust encoding of dF (Bodnar and Bass, 1997, 1999). Importantly, midbrain neurons are tuned to dFs that overlap the natural range of dFs, that is, $\leq$10 Hz. Most of the midbrain neurons show the same dF sensitivity irrespective of the component tones that make up the beat stimulus, although some exhibit different patterns of tuning depending on the spectral composition of the beats. Spectral composition may be additionally encoded by other spike train parameters (Bodnar and Bass, 1999, 2001a). Midbrain auditory neurons also show sensitivity to the intensity, depth of modulation, and duration of beats (Bodnar and Bass, 1999, 2001b), all of which are behaviorally relevant parameters for signal recognition (Section 2). The majority of midbrain auditory neurons encode beat and AM-like stimuli in a similar manner, although some show differential encoding that could contribute to the distinction between multisource beats generated by two neighboring males and single source grunt trains from one male (Figure 9.6) (Bodnar and Bass, 1997). Peripheral to central transformations in the temporal encoding of acoustic signals has also been shown for mormyrids where the behaviorally relevant variable is pulse repetition rate (Kozloski and Crawford, 2000; Suzuki *et al.*, 2002). Together, the studies in batrachoididids and mormyrids reveal mechanisms of temporal coding shared with terrestrial vertebrates (Bass *et al.*, 2005).

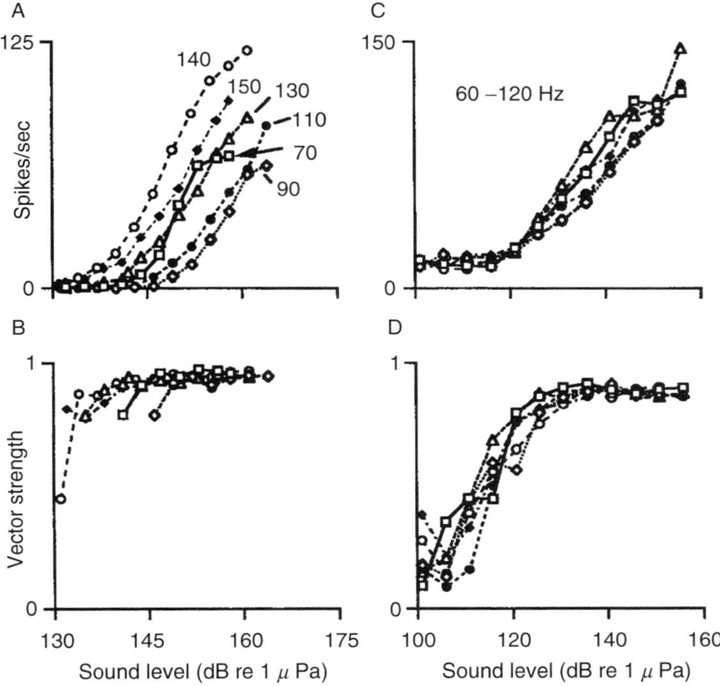

Fig. 9.5. Frequency encoding by VIIIth nerve afferents. Neurophysiological recordings in response to tone stimuli are shown here for two afferent fibers (A/B and C/D) within the saccular branch of the VIIIth nerve in midshipman fish. Responses are analyzed for changes in spike rate (A, C) and degree of phase locking (B, D; vector strength of synchronization) with increasing amplitude across a range of tone frequencies (70–150 Hz in A and B as indicated by symbols in A and at 10-Hz increments from 60 to 120 Hz in C and D). Vector strength measures the accuracy of phase locking to a periodic stimulus like the tone stimuli used here; values range from 0 for a uniform or random distribution to 1 for perfect synchronization (Goldberg and Brown, 1969). Vector strength provides the most robust code of frequency even at the lowest amplitudes tested. [Adapted from McKibben and Bass (1999) with permission of Springer Science and Media.]

4.2. Directional Hearing

Behavioral studies on directional hearing in fishes and hypotheses regarding sound localization have been reviewed (Lu, 2004). Here we emphasize experimental apparatuses that are used to generate well-calibrated directional stimuli and neurophysiological findings of coding of acoustic particle motion by peripheral and central auditory neurons.

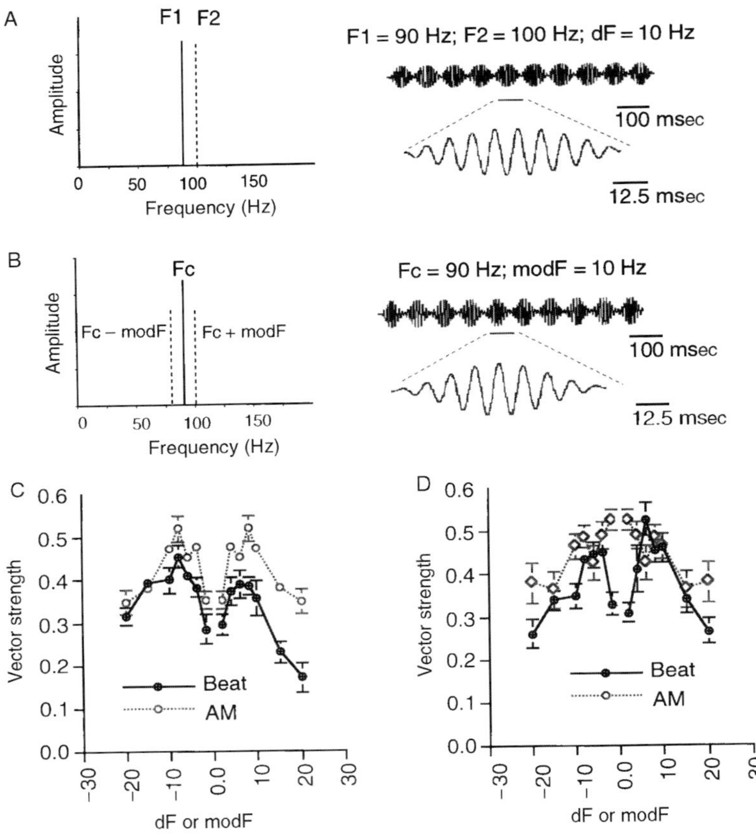

Fig. 9.6. Midbrain encoding of amplitude modulated sounds. (A and B) Examples of the power spectrum (left) and waveform (right) of representative stimuli used in neurophysiological studies of midbrain auditory encoding (Bodnar and Bass, 1997; Bass *et al.*, 2001). (A) Two tones with frequencies of F1 and F2 (left) interfere to produce a beat waveform with a rate of amplitude modulation determined by the difference between F1 and F2 (dF) that is 10 Hz in the example provided (right). (B) AM signals are produced by the modulation of a single tone stimulus at a particular rate or modulation frequency (modF). The power spectrum (left) has a carrier frequency (Fc) and two sidebands with frequencies of either Fc + modF or Fc − modF. The sidebands will vary as the modulation frequency changes. In this example, Fc is 90 Hz and modF is 10 Hz. The acoustic waveform (right) has an envelope shape resembling the beat waveform shown in "A." (C and D) Extracellular recordings from single neurons in the midbrain auditory nucleus (NC) of midshipman fish in response to either beat (filled circles) or AM (open circles) stimuli. F1 and Fc equal 90 Hz; positive and negative modF values are plotted to allow comparison with the full range of negative and positive dF stimuli. Plots show the vector strength of synchronization (±SE) versus either dF or modF. The majority of midbrain neurons (65%) show similar encoding patterns for both dF and modF stimuli (C) with the remainder (D) showing differences [Bodnar and Bass (1997), and text for further explanation]. [Adapted from Bodnar and Bass (1997); copyright by the Society for Neuroscience.]

4.2.1. Experimental Setups

A sound wave contains particle motion and pressure components that are thought to stimulate the fish ear through direct and indirect pathways, respectively [see Bass and Clark (2003) for a review of underwater acoustics]. A remaining need in studies of fish directional hearing is to independently manipulate particle motion and pressure components under laboratory conditions. Three types of devices have been successfully used to provide linear accelerations to mimic underwater acoustic particle motion (Sand, 1974; Fay, 1984; Schellart *et al.*, 1995). In particular, the shaker apparatus invented by Richard R. Fay provides accurate directional stimuli in three-dimensional space, which enhances our understanding of neural coding of acoustic particle motion by otolithic organ afferents and central auditory neurons (Fay and Edds-Walton, 1997a,b; Lu *et al.*, 1998, 2003, 2004; Edds-Walton *et al.*, 1999; Lu and Popper, 2001; Ma and Fay, 2002; Weeg *et al.*, 2002; Edds-Walton and Fay, 2003). It is composed of a water-filled, aluminum experimental dish with two orthogonal pairs of horizontal Brüel and Kjær mini shakers that are attached to the sides of the dish along the front-to-back and side-to-side axes, and a vertical Brüel and Kjær shaker that is connected to the center of the bottom of the dish (Lu *et al.*, 1996; Fay and Edds-Walton, 1997a). A fish is stabilized by a head holder that is firmly attached to the experimental dish. The apparatus can provide linear whole-body acceleration at the submicron level along any axis in three-dimensional space. Accelerations are calibrated using three PCB piezoelectronic accelerometers attached to the surface of the dish along the three orthogonal axes. The shaker apparatus has been upgraded with Tucker-Davis Technologies (TDT) digital-to-analog and analog-to-digital modules (system III) and custom-written, windows-based calibration and data acquisition programs in C++ (Lu *et al.*, 2003, 2004).

4.2.2. Detection of Acoustic Particle Motion

What are the behavioral capabilities of fish to detect acoustic particle motion? Using a cardiac conditioning method, Lu *et al.* (1996) measured behavioral detection thresholds of a hearing nonspecialist, the oscar (*Astronotus ocellatus*), when the fish is linearly accelerated at 100 Hz along different azimuthal axes and the vertical axis. The fish has behavioral detection thresholds ranging from −58 to −56 dB re: 1 μm, which corresponds to 1.2–1.6 nm in displacement (root mean square).

The saccule is often the largest among the three otolithic organs (Figure 9.2) and considered as the major auditory organ in non-clupeid fishes, including the midshipman, sleeper goby, and toadfish (Bass *et al.*, 1994; Fay and Edds-Walton, 1997a; Lu and Popper, 1998). What is the overall role of an otolithic organ in hearing sensitivity? Using an auditory brainstem recording technique, Lu and Xu (2002) investigated how the saccule in the sleeper goby contributes to the overall auditory sensitivity by comparing auditory thresholds before and after unilateral and bilateral removal of the saccular otoliths. For normal sleeper gobies in response to 100-Hz linear accelerations, directional thresholds range from 1.9 to 3.1 nm in the horizontal plane and from 1.2 to 2.2 nm in the midsagittal plane (Figure 9.7). Unilateral removal of a saccular otolith results in selective reductions of auditory sensitivity by 3–7 dB in the azimuthal axes (i.e., 30° and 60° in Figure 9.7A) that are close to the longitudinal axis (indicated by the arrow in Figure 9.7A) of the damaged saccule, but does not change the hearing threshold in the midsagittal plane (Figure 9.7B). At 100 Hz, bilateral removal of the saccular otoliths causes robust hearing losses of 27–35 dB along all the axes in both the horizontal and midsagittal planes (Figure 9.7A and B). These results reveal that the role of the saccule in directional hearing is consistent with its orientation. The saccule in the sleeper goby is positioned vertically, with the longitudinal axis of each saccule deviating about 40° off the midsagittal plane [see Figure 9.2 for a midshipman that resembles a sleeper goby; also see Lu *et al.* (1998)]. Since the two saccular epithelia in the sleeper goby (and midshipman, Figure 9.2) are approximately perpendicular to each other, the intact saccule cannot compensate for the auditory sensitivity reduction along the horizontal stimulus axes near the longitudinal

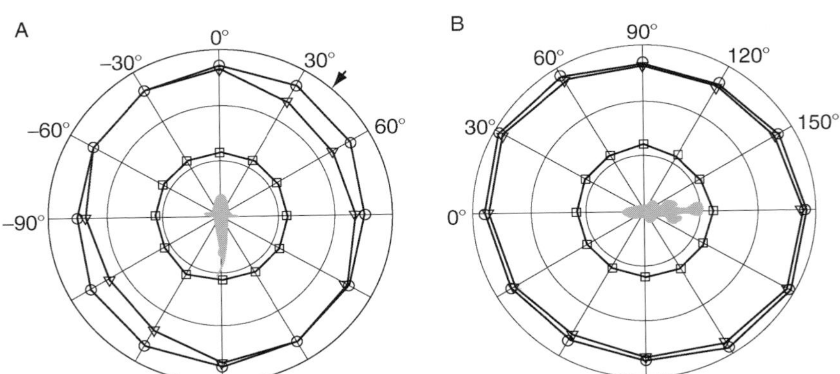

Fig. 9.7. Average auditory thresholds in the horizontal (A) and midsagittal (B) planes for sleeper gobies ($n = 9$) in response to 100-Hz linear accelerations. Thresholds were determined from auditory brainstem responses using a correlation method. Circles, intact saccules; triangles, removal of the right saccular otolith; squares, removal of both saccular otoliths. The *arrow* in A represents the longitudinal axis of the right saccule whose otolith is surgically removed. For both polar plots, the outer, middle, and inner circles represent displacement scales of 1, 10, and 100 nm.

axis of the damaged saccule. The intact saccule, however, can compensate for the functional loss of the damaged saccule in the midsagittal plane, which is consistent with the same vertical orientation of both saccules. The studies on the sleeper goby also demonstrate that among the three otolithic organs, the saccule plays the dominant role in directional hearing because the robust hearing loss (up to 35 dB) from bilateral damage of the saccules cannot be compensated by the other otolithic organs (lagenas and utricles).

4.2.3. Peripheral Coding of Acoustic Particle Motion

Studies in goldfish (*Carassius auratus*) and cod (*Gadus morhua* L.) were the first to investigate the responses of individual saccular afferents to directional stimuli (linear vibrations) (Fay and Olsho, 1979; Hawkins and Horner, 1981). The most sensitive saccular, lagenar, and utricular afferents in the goldfish can respond to 140-Hz linear displacements as small as 0.1 nm, which is equivalent to 0.077 mm/sec^2 (Fay, 1984). But threshold ranges for the three otolithic organ afferents have not been reported, and statistical comparisons cannot be made. Systematic studies report directional response properties of single saccular afferents of the toadfish (*O. tau*), sleeper goby, and midshipman to 100-Hz linear accelerations. They show similar ranges of auditory thresholds with the lowest sensitivity of 0.1 nm and the highest sensitivity greater than 100 nm (Fay and Edds-Walton, 1997a; Lu *et al.*, 1998; Weeg *et al.*, 2002). The lowest threshold of these hearing nonspecialists is the same as that of hearing specialists such as the goldfish, suggesting that saccular afferents of both hearing specialists and nonspecialists have similar capacities to encode acoustic particle motion. This displacement threshold of fish is equivalent to the vibrational sensitivity of bullfrog saccular and lagenar afferents (Koyama *et al.*, 1982) and corresponds to the lowest threshold in displacement of the basilar membrane of the guinea pig (Allen, 1997).

Studies on directional responses have been conducted on lagenar and utricular afferents in the sleeper goby using the same shaker apparatus and methods as those for saccular afferents in the same species (Lu *et al.*, 2003, 2004). Results show that lagenar and utricular afferents are about 30 dB less sensitive than saccular afferents, corresponding to the size differences among the three otolithic organs and the results from the auditory brainstem recordings involving saccular otolith removal (see Section 4.2.2). Threshold variations among afferents innervating different otolithic organs in the sleeper goby are perhaps present in other hearing nonspecialists such as the toadfish and midshipman.

Saccular afferents in fishes, including the cod, goldfish, midshipman, sleeper goby, and toadfish, encode particle motion in a directional manner (Hawkins and Horner, 1981; Fay, 1984; Fay and Edds-Walton, 1997a;

Lu *et al.*, 1998; Weeg *et al.*, 2002). Extracellular single-unit studies have shown that individual saccular afferents have specific best response axes and the range of afferents' best response axes is generally consistent with the orientation of the saccule. Response directionality of saccular afferents is thought to derive from morphological and physiological polarity of sensory hair cells in the saccule (Popper, 1976). Flock (1964) shows that each sensory hair cell in fish otolithic organs has a ciliary bundle containing an eccentrically placed kinocilium and a cluster of stereocilia whose length decreases with increasing distance from the kinocilium. Hudspeth and Corey (1977) demonstrate that *in vitro* hair cells in the frog saccule are directionally sensitive to stimulation by bending the ciliary bundle of hair cells. A hair cell has the most excitatory response when its ciliary bundle is bent from the stereocilia toward the kinocilium, the direction of the morphological polarity of the hair cell. The excitatory response of the hair cell gradually decreases and become inhibitory as the stimulus direction deviates away from the morphological polarity up to 180°. Does the response directionality of hair cells faithfully replay onto afferents innervating the hair cells? Using whole cell recording, neuronal tracer injection, and confocal imaging techniques, Lu and Popper (2001) demonstrate that the response directionality of each single saccular afferent strongly correlates with the average morphological polarization of the hair cells that the afferent innervates (Figure 9.8), indicating that the directionality of saccular afferents results directly from that of the saccular hair cells.

Two studies have investigated the morphological pattern of afferent innervation of hair cells in the saccule following intracellular labeling of afferents with a low-molecular-weight neuronal tracer, neurobiotin. For saccular afferents of the toadfish (*O. tau*), Edds-Walton *et al.* (1999) show maximum arbor widths between 84 and 543 μm (median = 251) and the number of dendritic terminals between 8 and 111 (median = 39). Lu and Popper (2001) report for saccular afferents of the sleeper goby that dendritic arbor innervation areas range from 893 to 21,393 μm^2 (median = 2617 μm^2) and the number of dendritic endings from 10 to 54 (median = 22). In general, each saccular afferent in either the toadfish or sleeper goby innervates a small portion of the saccular macula with hair cells having similar morphological polarizations. Some saccular afferents in both species innervate hair cells with opposing or opposite morphological polarizations. It is clear that single saccular afferents receive inputs from multiple hair cells. It is further likely that a hair cell could be innervated by more than one saccular afferent, but convincing evidence is needed. Confocal microscopic and intracellular labeling results have demonstrated that a saccular ganglion cell can make more than one synapse on a hair cell (Figure 9.8C) (Lu and Popper, 2001).

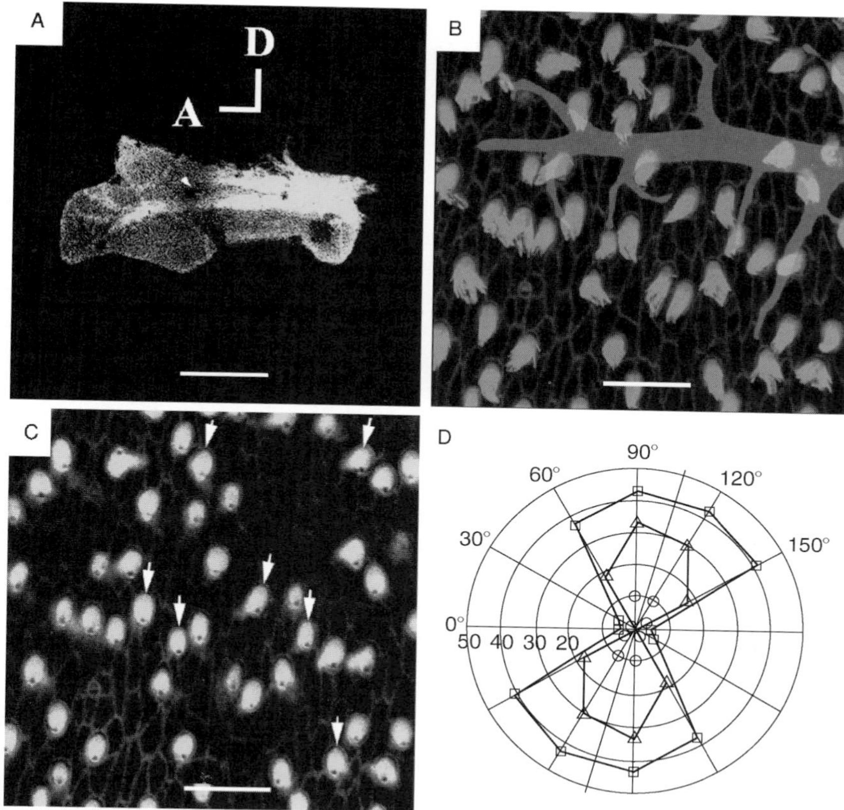

Fig. 9.8. Morphophysiology of a saccular afferent of the sleeper goby. (A) Saccular epithelium whole mount with dendritic terminals of a saccular ganglion cell indicated by the arrowhead. The saccular ganglion cell is intracellularly injected with a neuronal tracer, neurobiotin. The epithelium is processed in a solution of Texas red 595 avidin to mark dendritic terminals of the saccular ganglion cell, and then in an Oregon green phalloidin solution to label ciliary bundles of hair cells. Scale bar = 500 μm. (B) Merged image of 3D structures of dendritic terminals of the saccular ganglion cell and ciliary bundles of hair cells in the saccule. Scale bar = 15 μm. (C) Single optical section of the hair cell ciliary bundles shown in B. Note that phalloidin labels stereocilia of ciliary bundles but not kinocilia (the eccentrically positioned *dark circles* of the ciliary bundles). The arrows indicate morphological polarizations of the hair cells innervated by the saccular ganglion cell. Scale bar = 15 μm. (D) Z-axis functions of the saccular afferent in response to 100-Hz linear accelerations at three stimulus levels (5.6, 10, and 17.8 nm). The sagittal best sensitivity of the saccular ganglion cell is 3.2 nm. $Z = N \times R^2$, where N is the number of spikes and R is the strength of phase locking. The solid line shows the sagittal best response axis (i.e., 106°) of the saccular afferent, which is close to the average morphological polarization (93°) of the hair cells that the afferent innervates. Six stimulus axes are from 0° to 150° with 30° steps. [Adapted from Lu and Popper (2001) with permission of Springer Science and Business Media.]

Directional responses of saccular afferents are stimulus level dependent (Lu *et al.*, 1998; Weeg *et al.*, 2002). To respond to linear accelerations, most saccular afferents have an auditory dynamic range from 15 to 30 dB (Fay and Edds-Walton, 1997a; Lu *et al.*, 1998). The degree of directional tuning of saccular afferents is high at stimulus levels just above threshold, showing narrow directional response profiles. Directional response profiles of saccular afferents become circular at stimulus levels near the midpoint between threshold and saturation. Saccular responses become omnidirectional at high stimulus levels where responses at different stimulus axes are saturated.

Studies have shown that lagenar and utricular afferents of the sleeper goby are also directionally sensitive to linear accelerations (Lu *et al.*, 2003, 2004). Lagenar afferents have the same directional selectivity as saccular afferents. However, utricular afferents have higher directional selectivity than saccular and lagenar afferents. The high directional selectivity of utricular afferents cannot be explained by the cosine function of responses of hair cells in the utricle, suggesting that some utricular hair cells may be more directionally selective than others. Like saccular afferents, the response directionality of lagenar and utricular afferents is also stimulus-level dependent.

In summary, all three otolithic organs in the sleeper goby and probably in other fishes contribute to directional hearing. In the sleeper goby, since saccular afferents are more sensitive to acoustic particle motion than lagenar and utricular afferents, the saccule plays the most important role in directional hearing. Perhaps the same is the case for sonic batrachoidids given the similarly large size of the saccule (Figure 9.2) and the neurophysiological recordings from saccular afferents (Fay and Edds-Walton, 1997a,b; McKibben and Bass, 1999, 2001a; Weeg *et al.*, 2002; Sisneros and Bass, 2003, 2005; Sisneros *et al.*, 2004). It appears that the orientation of each otolithic organ determines its specific role in directional hearing, that is, the horizontally oriented utricle mainly serves for particle motion detection in azimuth, while the vertically positioned saccule and lagena primarily function in the detection of elevation (Figure 9.9). Three pairs of nonparallel otolithic organs in the fish inner ears form a three-dimensional, biological accelerometer to detect acoustic particle motion. The response dynamic ranges of saccular afferents overlap very little with those of lagenar and utricular afferents. Therefore, saccular, lagenar, and utricular afferents together broaden the response dynamic range of directional hearing.

4.2.4. Central Coding of Acoustic Particle Motion

The encoding of acoustic particle motion has been studied for neurons in the hindbrain's DO nucleus and in the nucleus centralis, the main auditory nucleus of the midbrain's TS (Figure 9.3). Wubbels and Schellart (1997)

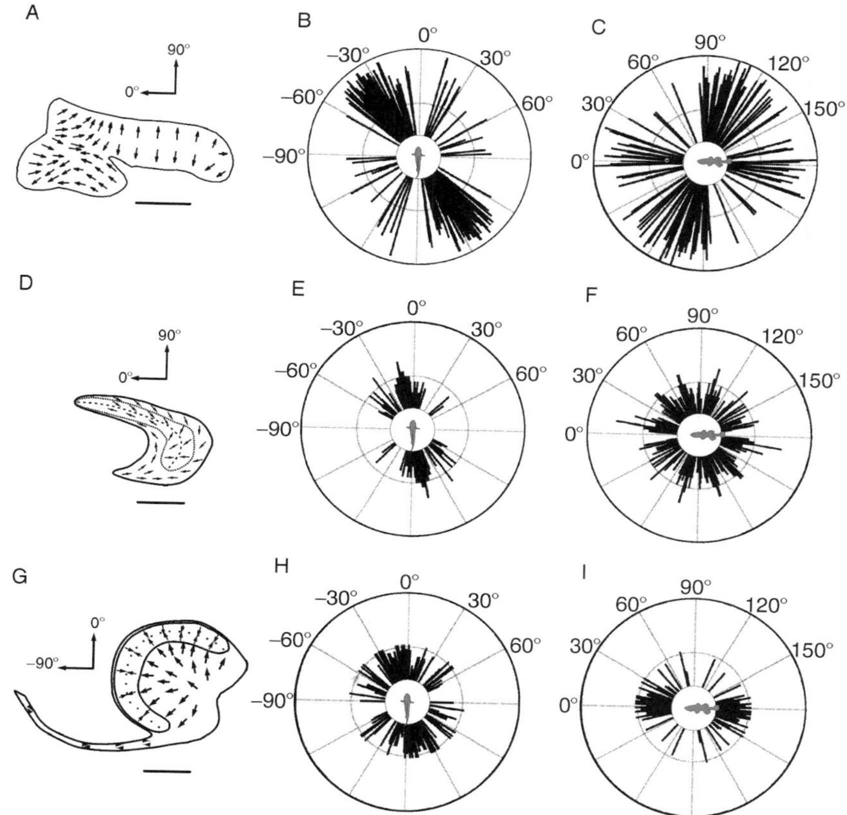

Fig. 9.9. Morphological polarizations of hair cells in otolithic organ maculae (A, saccular; D, lagenar; G, utricular) and auditory thresholds in the horizontal and midsagittal planes for saccular (B and C), lagenar (E and F), and utricular (H and I) afferents of sleeper gobies. Arrows in A, D, and G indicate hair cell morphological polarizations, and scale bars in A, D, and G are 500, 200, and 200 μm. For the middle and right columns, each dark line is a horizontal or sagittal best response axis, and the length of each line represents the directional threshold of an afferent (the longer the line, the more sensitive the afferent). For each polar plot, the outer, middle, and inner circles indicate displacement scales of 0.1, 30, and 1000 nm.

studied neural encoding of sound direction in the TS of trout (*Oncorhynchus mykiss*) in the horizontal plane. About 45% of the TS neurons are directionally sensitive, while 75% of these directionally selective neurons are phase locked. Some variations in directional selectivity are observed in different regions of the TS. Neurons in the medial TS show uniform directional tuning to the rostrocaudal axis; neurons in the lateral part had various best response axes. Therefore, a crude topographical representation of directional selectivity is present in the TS of the trout, which is apparently not as well organized as those in terrestrial vertebrates (Knudsen and Konishi, 1978).

Investigations of directional responses have also been conducted on DO and TS neurons in toadfish and TS neurons in goldfish (Fay and Edds-Walton, 1999; Ma and Fay, 2002; Edds-Walton and Fay, 2003). A major finding of these studies is that DO and TS neurons are more directionally selective than saccular afferents. It is hypothesized that inhibition accounts for the increased sharpness of directional selectivity, that is, a central neuron receives simultaneous directional excitatory and inhibitory inputs with different best excitatory/inhibitory response axes, and the inhibitory input sharpens the excitatory input (Figure 9.10). The simple model shown in Figure 9.10 can explain the sharpened directional response profiles of auditory medullar and TS neurons. Although inhibition is present in TS (Lu and Fay, 1993, 1996), the inhibitory neural circuits underlying the increased sharpness of directional tuning remain to be investigated. A study on the sleeper goby suggests that the DO receives inputs from all three otolithic organs, suggesting that a single DO neuron could receive afferent inputs from multiple otolithic organs (Tomchik and Lu, 2004). Thus, excitatory

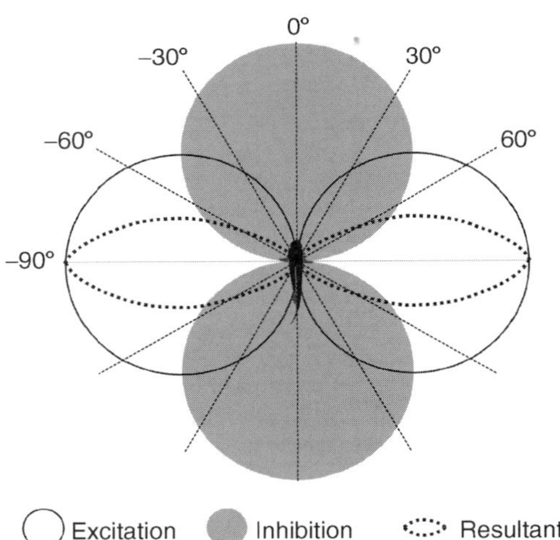

Fig. 9.10. Model of the increased sharpness of directional selectivity in the brain. It is a polar plot of response versus stimulus axis for an auditory neuron in the DO or TS that is hypothesized to receive both excitatory (open circles) and inhibitory (gray circles) cosine inputs. The neuron's directional response profile, indicated by dotted lines, is the resultant of addition of the excitatory (positive) and inhibitory (negative) inputs. Resultant responses that are less than 0 are plotted as 0. [Based on the hypothesis by Fay and Edds-Walton (1999).]

and inhibitory inputs that converge on single DO neurons could come from either different or the same otolithic organ(s).

5. AUDITORY LATERAL LINE INTEGRATION

An investigator studying the auditory system of teleosts also tries to understand the organization of the lateral line system. Peripheral lateral line information is relayed to the brain by as many as six distinct pairs of cranial nerves (Northcutt *et al.*, 2000). Although the peripheral lateral line system has distinct targets in the medulla and midbrain, neuroanatomical and neurophysiological evidence demonstrates integration of the two modalities in both hindbrain and midbrain nuclei (Weeg and Bass, 2000, 2002; Edds-Walton and Fay, 2003). Weeg and Bass (2002) provide evidence from single unit recordings of primary, lateral line afferents for high levels of phase locking over a range of 20–100 Hz to a stimulus waveform generated by a vibrating sphere equalized for velocity across frequencies; this range of frequency encoding overlaps the low frequency range of the amplitude spectrum of midshipman calls. Edds-Walton and Fay (2003) report bimodal, auditory, and lateral line units in the midbrain of the closely related toadfish (*O. tau*); this integration may occur in deep layers of the TS (Weeg and Bass, 2000; Edds-Walton and Fay, 2003). Bimodal units are also known for nonvocal species and are perhaps a common trait of the midbrain of all teleosts (Feng and Schellart, 1999).

6. VOCAL MODULATION OF INNER EAR AND LATERAL LINE

Given that one source of acoustic input is often self-generated, that is, an animal's own vocal signal interactions between vocal and acoustic pathways are perhaps important in the processing of extrinsic acoustic signals. The mapping in midshipman fish and toadfish of a central vocal-auditory link via neurons that directly innervate the inner ear (Bass *et al.*, 1994) provides a simple vertebrate preparation for assessing the role of central vocal pathways in the modulation of peripheral sensitivity during self-vocalization. Studies have now provided the evidence for such a mechanism in midshipman fish (Weeg *et al.*, 2005). Either electrical or neurochemical (glutamate) stimulation of midbrain vocal sites evokes a rhythmic sonic motor volley, the fictive vocalization (Section 3.2), whose temporal features parallel the temporal modulation of the firing pattern of hindbrain efferent neurons that innervate the inner ear. Batrachoidids are a particularly advantageous preparation for these studies because the axons of efferent neurons to different divisions of the inner ear are present in identifiable bundles that are apposed to the afferent branches of the VIIIth cranial nerve that innervate each end organ (Highstein and Baker, 1986). One bundle of caudal efferents innervates multiple end organs (lagena, posterior lateral line, posterior semicircular canal); a second bundle only innervates the saccule, the main auditory end organ (Sections 3.1 and 4.2). Saccular efferents show features especially well adapted for adjusting peripheral sensitivity to conspecific signaling during self-vocalization. All of the recordings from neurons that contribute to the saccular bundle show an increase in the firing rate of efferents that is temporally correlated with the onset and offset of the entire fictive vocalization with the onset and offset of each fictive sound pulse. As pointed out by Lin and Faber (1988) in their study of goldfish, the activation of inhibitory efferents to the saccular epithelium might participate in maintaining the sensitivity of hair cells to external sound sources since the hair cell-afferent nerve synapse habituates with repetitive stimulation (Furukawa and Matsura, 1978; Furukawa, 1981; Winslow and Sachs, 1987; Lin and Faber, 1988). Thus, the vocal motor system essentially decreases the sensitivity of the saccule (and other parts of the inner ear and lateral line system) to reafferent signals generated by one's own vocalizations. Given the extensive similarities in auditory mechanisms between teleosts and other vertebrate groups, including birds and mammals (Fay and Popper, 1999), the presence of efferents to all parts of the inner ear among vertebrates (Roberts and Meredith, 1992; Manley, 2000) and that the most fundamental mechanisms of vertebrate audition originated among fish, the vocal-auditory interaction discovered in midshipman fish likely represents a neural-based mechanism that can adjust peripheral auditory sensitivity among all sonic vertebrates. The significance of vocal motor modulation of lateral line activity perhaps also relates to their sensitivity to vocal signals (Section 5). The influence of lateral line efferents on encoding mechanisms has also been described in the context of self-stimulatory locomotor acts, namely swimming (Tricas and Highstein, 1991).

7. STEROID HORMONES AND SEASONAL CHANGES IN HEARING

Studies of midshipman fish have shown that steroid hormones can modulate the temporal properties of the vocal pattern generator in the hindbrain spinal cord [Remage-Healey and Bass, 2004; also see Goodson and Bass (2000a,b) for studies on neuropeptides]. Parallel studies now show steroid-dependent seasonal shifts in the frequency encoding properties of the saccular epithelium. During the spring and summer, midshipman fish migrate from offshore sites more than 100 m in depth to build nests in the shallow intertidal zone (Bass, 1996; Sisneros *et al.*, 2004a). As discussed in Section 2, observations of the nocturnal spawning behavior of midshipman fish plus underwater playback experiments show that gravid females use the midshipman's advertisement hum to detect and locate males in nests; spent females that have released their eggs do not show a positive phonotactic response (McKibben and Bass, 1998). These observations led to the hypothesis that seasonal variation in the reproductive state of female midshipman fish can influence the response properties of their peripheral auditory system. A subsequent study investigated seasonal plasticity in frequency encoding by the primary afferents from the saccule, the main auditory end organ (Section 3.1). Recordings from individual saccular afferents of adult female midshipman collected from wild populations during both the summer (from nests) and winter (from offshore trawls) show a dramatic improvement during the summer months in the degree of phase locking to the upper harmonic frequencies that are prominent in the type I male's courtship hum (Figure 9.11) (Sisneros and Bass, 2003). This enhancement is not related to either temperature or the effects of captivity. Although a hum's F0 is highly stable and prominent, the second harmonic that is close to 200 Hz often has the most energy compared with either F0 or the other harmonics that range up to 400–600 Hz.

What might be the behavioral advantage for an increased encoding of the upper harmonics of the advertisement call? Many teleosts, including midshipman, use sound communication signals in shallow water environments, that is, less than 5 m in depth. Sound transmission in these environments will vary with depth and substrate composition. The cutoff frequency (f_0) below which sound transmission is negligible is estimated by the following equation:

$$f_0 = \frac{(c_\mathrm{w}/4h)}{\sqrt{1 - c_\mathrm{w}^2/c_\mathrm{s}^2}}$$

where c_w, the speed of sound in water; c_s, speed of sound in the bottom substrate; f_0 is in Hz; and h, water depth [from Rogers and Cox (1988); also see Jensen *et al.* (1994) and Premus and Spiesberger (1997)].

Water depth is considered to have the most profound effect on transmission distance at depths <100 m (Jensen *et al.*, 1994). Fine and Lenhardt (1983) show for the toadfish *O. tau* that the upper harmonics will have a greater transmission distance than F0 in the very shallow water environments like those inhabited by nesting male toadfish and midshipman during the breeding season (Bass and Clark, 2003). During the breeding season,

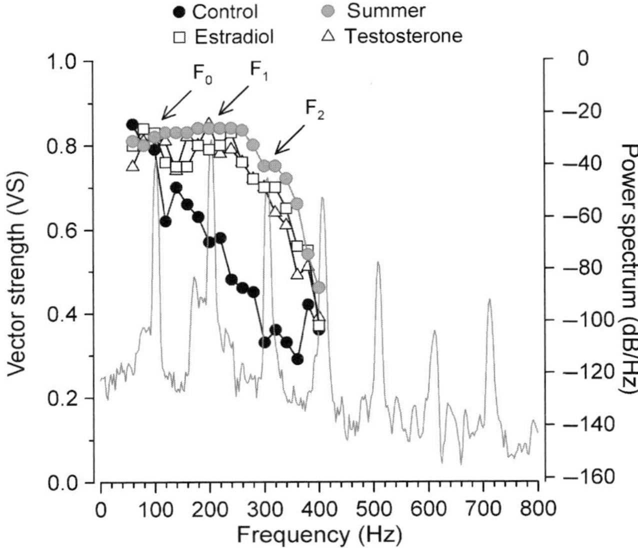

Fig. 9.11. Vocal-auditory coupling in midshipman fish. The Y-axis to the left shows vector strength values (VS, see Figure 9.5 for an explanation of VS) for the responses to tonal stimuli obtained from single unit recordings from the afferents that innervate the saccular epithelium in a female midshipman fish. The Y-axis to the right shows relative amplitude values (dB/Hz) for the power spectrum of the advertisement hum of a type I male (see Figure 9.1A for acoustic waveform). Frequency is plotted along the X-axis for both measures. Shown here are median VS values obtained from the saccular afferents of nonreproductive females that are either untreated controls (dark-shaded circles), treated with testosterone (triangles), or treated with 17β-estradiol (squares). Nonreproductive females show high VS values for frequencies close to the hum's fundamental frequency (F0), while steroid-treated females show a wider range of similar robust encoding that extends up to the hum's second and third harmonics (F_1 and F_2). The encoding observed for steroid-treated females overlaps that observed for wild-caught reproductive females from the summer breeding season (light-shaded circles). [Adapted from Sisneros *et al.* (2004b).]

midshipman males may be calling in water depths of up to 5 m (A. Bass and M. Marchaterre, unpublished observations). At these depths, the cutoff frequency will be close to 100 Hz, which is similar to the F0 of hums. Hence, an enhanced sensitivity to the upper harmonics perhaps aids fish to locate humming males during the approach from deeper offshore sites (type I males also nest in a rocky-gravel substrate which should help shift the cutoff frequency to a lower value, away from the hum's F0) (Rogers and Cox, 1988; Bass and Clark, 2003). Earlier saccular nerve recordings in nonreproductive type I males also show that the addition of a second harmonic to a stimulus can enhance F0 encoding (McKibben and Bass, 2001a). Thus, as an individual gets close to a male's nest, the enhanced encoding of F0 will further improve the ability to find a type I male. Together, the data support the hypothesis that increased sensitivity to the upper harmonics during the breeding season contributes to signal detection especially in shallow water and noisy environments like those where type I male midshipman fish nest and acoustically court females. Males may also show a seasonal shift in peripheral frequency encoding that could aid them in locating suitable nesting areas where other males are already calling; the peripheral encoding properties of nonreproductive males resemble those of nonreproductive females (McKibben and Bass, 1999).

A subsequent study has shown that steroid hormones, either testosterone or 17β-estradiol, can induce changes in the frequency-encoding properties of the peripheral auditory system of winter nonreproductive females so that they come to resemble those of summer reproductive females (Figure 9.11) (Sisneros *et al.*, 2004b). Females naturally exhibit increases in circulating levels of both testosterone and 17β-estradiol during the spring months prior to their migration into nests in the intertidal zone during the summer; these increases accompany gonadal recrudescence (Sisneros *et al.*, 2004a). Estrogen receptors are present in the cochlea of humans and rodents (Stenberg *et al.*, 1999, 2001), suggesting that estrogen may influence the shifts in frequency sensitivity at differing stages of the human menstrual cycle (McFadden, 1998). Estrogen receptor alpha (ERα) has also been identified in the saccular epithelium of midshipman fish after cloning a partial cDNA for the *ER*α gene (Sisneros *et al.*, 2004b; Forlano *et al.*, 2005), suggesting that steroid binding to inner ear end organs may be a trait shared by many vertebrate groups. Moreover, the ERα studies together with the neurophysiological studies in midshipman fish may be indicative of more widespread influences of estrogen and perhaps other steroids on a range of peripheral auditory mechanisms (McFadden, 1998). The studies in midshipman have used *in situ* hybridization methods to show that ERα mRNA transcripts in the saccule are present over regions of the nerve that lies close to the hair cells (Forlano *et al.*, 2005). The latter study also shows that aromatase, the enzyme that converts testosterone to estradiol, is present within a subpopulation of saccular nerve ganglion cells [see Forlano *et al.* (2001) for studies of aromatase expression in midshipman fish]. The presence of aromatase in auditory ganglion cells and of ERα mRNA over the saccular nerve apposed to the hair cell epithelium suggests that estrogen, derived either from the gonads or from the aromatization of testosterone in ganglion cells (and in the brain), can contribute to changes in the encoding of acoustic stimuli by the peripheral auditory system.

8. FUTURE DIRECTIONS

There is a continuing need for the coupling of neurophysiological and behavioral studies to fully understand just why and how mechanisms of signal recognition and directional hearing have evolved among teleosts. Such a combinatorial, neuroethological approach will help distinguish between hearing mechanisms that are specifically adaptations to hearing in an underwater environment from those that are shared by all vertebrates regardless of the transmission medium.

The resurgence of underwater playback studies in batrachoidids (Section 2) provides the opportunity for establishing the behavioral relevance of neural mechanisms considered essential for recognition and localization tasks (Sections 4.1 and 4.2) involved in, for example, finding a male calling from his nest. There remains an essential need to independently manipulate particle motion and pressure components of acoustic stimuli to completely understand their separate and overlapping representations in both the peripheral and central auditory systems (Section 4.2). Similarly, there is a necessity to accurately measure both components of natural acoustic stimuli (e.g., vocalizations) for the design of appropriate stimuli in neurophysiological and behavioral playback experiments and, more generally, for a more complete understanding of how the auditory system of fish is adapted to the detection of sound fields in underwater habitats. Moreover, there need to be neurophysiological studies within a single species that distinguish the contribution of any one end organ of the inner ear to the performance of either auditory and/or vestibular tasks. Are there distinct neuronal populations for the separate processing of auditory and vestibular stimuli as observed among terrestrial vertebrates? Might there also be multimodal integration sites for auditory and vestibular stimuli as is now suggested for the auditory and lateral line systems (Section 5)? Further study of the organization of central pathways (Section 3.1) together with the use of behaviorally relevant auditory (Section 4) and vestibular (Aksay *et al.*, 2000) stimuli will help resolve these issues.

The discovery of central pathways for vocal motor modulation of efferents to the inner ear and lateral line system (Sections 3.2 and 6) provides an opportunity to understand central mechanisms of sensorimotor integration that lead to changes in the sensitivity of the auditory and lateral line systems to external stimuli. Given the numerous examples of convergence in anatomical (McCormick, 1999; Bass *et al.*, 2005) and neurophysiological (Fay and Popper, 1999; Lu, 2004; Bass *et al.*, 2005) mechanisms of audition between teleosts and other vertebrate groups, the peripheral mechanism that midshipman fish have apparently evolved for maintaining sensitivity to conspecific signals during self-vocalization will perhaps be common to all sonic vertebrates.

Finally, the discovery of steroid hormone influences on hearing (Section 7) opens the door for further integrative studies at the intersection of endocrinological and auditory mechanisms. These studies can range from the biophysical (e.g., potential steroid influences on the electrical resonance properties of hair cells) to the behavioral (e.g., steroid influences on mate choice) levels of analysis.

ACKNOWLEDGMENTS

Thanks to M. Marchaterre and T. Natoli for help with the figures and references. The study is supported in part by NIH-NIDCD grants to Andrew H. Bass (DC00092) and Zhongmin Lu (DC03275).

REFERENCES

Aksay, E., Baker, R., Seung, H. S., and Tank, D. W. (2000). Anatomy and discharge properties of pre-motor neurons in the goldfish medulla that have eye-position signals during fixations. *J. Neurophysiol.* **84,** 1035–1049.

Allen, J. B. (1997). OHCs shift the excitation pattern via BM tension. *In* "Diversity in Auditory Mechanics" (Lewis, E. R., Long, G. R., Lyon, R. F., Narins, P. M., Steele, C. R., and Hetcht-Poinar, E., Eds.), pp. 167–175. World Scientific Publishers, Singapore.

Bass, A. H. (1996). Shaping brain sexuality. *Am. Sci.* **84,** 352–363.

Bass, A. H., and Baker, R. (1990). Sexual dimorphisms in the vocal control system of a teleost fish: Morphology of physiologically identified neurons. *J. Neurobiol.* **21,** 1155–1168.

Bass, A. H., and Clark, C. W. (2003). The physical acoustics of underwater sound communication. *In* "Springer Handbook of Auditory Research, Acoustic Communication" (Simmons, A. M., Fay, R. R., and Popper, A., Eds.), Vol. 16, pp. 15–64. Springer, New York.

Bass, A. H., and Ladich, F. (in press). Recent studies of vocal-acoustic communication: From behavior to neurons. *In* "Fish Bioacoustics" (Popper, A., Fay, R., and Webb, J., Eds.). Springer, New York.

Bass, A. H., and McKibben, J. R. (2003). Neural mechanisms and behaviors for acoustic communication in teleost fish. *Prog. Neurobiol.* **69,** 1–26.

Bass, A. H., Marchaterre, M. A., and Baker, R. (1994). Vocal-acoustic pathways in a teleost fish. *J. Neurosci.* **14,** 4025–4039.

Bass, A. H., Bodnar, D. A., and Marchaterre, M. A. (1999). Complementary explanations for existing phenotypes in an acoustic communication system. *In* "Neural Mechanisms of Communication" (Hauser, M., and Konishi, M., Eds.), pp. 493–514. MIT Press, Cambridge.

Bass, A. H., Bodnar, D. A., and Marchaterre, M. A. (2000). Midbrain acoustic circuitry in a vocalizing fish. *J. Comp. Neurol.* **419,** 505–531.

Bass, A. H., Bodnar, D. A., and Marchateere, M. A. (2001). Acoustic nuclei in the medulla and midbrain of the vocalizing gulf toadfish. *Opsanus beta. Brain Behav. Evol.* **57,** 63–79.

Bass, A. H., Rose, G. J., and Pritz, M. B. (2005). Auditory midbrain of fish, amphibians and reptiles: Models systems for understanding auditory function. *In* "The Inferior Colliculus" (Winer, J. A., and Schreiner, C. E., Eds.), pp. 459–492. Springer, New York.

Bodnar, D. A., and Bass, A. H. (1997). Temporal coding of concurrent acoustic signals in auditory midbrain. *J. Neurosci.* **17,** 7553–7564.

Bodnar, D. A., and Bass, A. H. (1999). A midbrain combinatorial code for temporal and spectral information in concurrent acoustic signals. *J. Neurophysiol.* **81,** 552–563.

Bodnar, D. A., and Bass, A. H. (2001a). Coding of concurrent signals by the auditory midbrain: Effects of duration. *J. Comp. Physiol. A* **187,** 381–391.

Bodnar, D. A., and Bass, A. H. (2001b). The coding of concurrent vocal signals by the auditory midbrain: The effects of stimulus level and depth of modulation. *J. Acoust. Soc. Am.* **109,** 809–825.

Brantley, R. K., and Bass, A. H. (1994). Alternative male spawning tactics and acoustic signals in the plainfin midshipman fish, *Porichthys notatus* (Teleostei, Batrachoididae). *Ethology* **96,** 213–232.

Bullock, T. H., and Heiligenberg, W. (1986). "Electroreception." John Wiley & Sons, New York.

Coombs, S., and Janssen, J. (1988). Water flow detection by the mechanosensory lateral line. *In* "Comparative Perception" (Stebbins, W. C., and Berkeley, M. A., Eds.), Vol. 2, pp. 89–123. John Wiley & Sons, New York.

Corwin, J. T. (1981). Peripheral auditory physiology in the lemon shark: Evidence of parallel otolithic and non-otolithic sound detection. *J. Comp. Physiol. A* **142,** 379–390.

Crawford, J. D. (1993). Central auditory neurophysiolology of a sound producing fish: The mesencephalon of *Pollimyrus isidori* (Mormyridae). *J. Comp. Physiol. A* **172,** 139–152.

Edds-Walton, P. L., and Fay, R. R. (2003). Directional selectivity and frequency tuning of midbrain cells in the oyster toadfish, *Opsanus tau. J. Comp. Physiol. A* **189,** 527–543.

Edds-Walton, P. L., Highstein, S. M., and Fay, R. R. (1999). Dendritic arbors and central projections of physiologically characterized auditory fibers from the saccule of the toadfish, *Opsanus tau. J. Comp. Physiol. A* **411,** 212–238.

Fay, R. R. (1984). The goldfish codes the axis of acoustic particle motion in three dimensions. *Science* **225,** 951–954.

Fay, R. R., and Edds-Walton, P. L. (1997a). Directional response properties of saccular afferents of the toadfish, *Opsanus tau. Hear. Res.* **111,** 1–21.

Fay, R. R., and Edds-Walton, P. L. (1997b). Diversity in frequency response properties of saccular afferents of the toadfish, *Opsanus tau. Hear. Res.* **113,** 235–246.

Fay, R. R., and Edds-Walton, P. L. (1999). Sharpening of directional auditory input in the descending octaval nucleus of the toadfish, *Opsanus tau. Biol. Bull.* **197,** 240–241.

Fay, R. R., and Olsho, L. W. (1979). Discharge patterns of lagenar and saccular neurons of the goldfish eighth nerve: Displacement sensitivity and directional characteristics. *Comp. Biochem. Physiol. A* **62,** 377–386.

Fay, R. R., and Popper, A. N. (1999). "Springer Handbook of Auditory Research, Volume 11: Comparative Hearing: Fish and Amphibians." Springer, New York.

Fay, R. R., and Simmons, A. M. (1999). The sense of hearing in fishes and amphibians. *In* "Comparative Hearing: Fishes and Amphibians" (Fay, R. R., and Popper, A. N., Eds.), pp. 269–318. Springer, New York.

Feng, A. S., and Schellart, N. A. M. (1999). Central auditory processing in fish and amphibians. *In* "Springer Handbook of Auditory Research, Vol. 11: Comparative Hearing: Fish and Amphibians" (Fay, R. R., and Popper, A. N., Eds.), pp. 218–268. Springer, New York.

Fine, M. L., and Lenhardt, M. L. (1983). Shallow-water propagation of the toadfish mating call. *Comp. Biochem. Physiol. A* **76,** 225–231.

Fine, M. L., Winn, H. E., and Olla, B. L. (1977). Communication in fishes. *In* "How Animals Communicate" (Sebeok, T. A., Ed.), pp. 472–518. Indiana University Press, Bloomington.

Fish, J. F. (1972). The effect of sound playback on the toadfish. *In* "Behavior of Marine Animals" (Winn, H. E., and Olla, B. L., Eds.), pp. 386–434. Plenum, New York.

Fletcher, L. B., and Crawford, J. D. (2001). Acoustic detection by sound-producing fishes (Mormyridae): The role of gas-filled tympanic bladders. *J. Exp. Biol.* **204,** 175–183.

Flock, Å. (1964). Structure of the macula utriculi with special reference to directional interplay of sensory responses as revealed by morphological polarization. *J. Cell Biol.* **22,** 413–431.

Forlano, P. M., Deitcher, D. L., Myers, D. A., and Bass, A. H. (2001). Neuroanatomical basis for high aromatase levels in teleost fish: Aromatase enzyme and mRNA expression identify glia as source. *J. Neurosci.* **21,** 8943–8955.

Forlano, P. M., Deitcher, D. L., and Bass, A. H. (2005). Distribution of estrogen receptor alpha mRNA in the brain and inner ear of a vocal fish with comparisons to sites of aromatase expression. *J. Comp. Neurol.* **483,** 91–113.

Furukawa, T. (1981). Effects of efferent stimulation on the saccule of goldfish. *J. Physiol.* **315,** 203–215.

Furukawa, T., and Matsura, S. (1978). Adaptive rundown of excitatory post-synatpic potentials at synapses between hair cells and eighth nerve fibers in the goldfish. *J. Physiol.* **276,** 193–209.

Gibbs, M. A. (2004). Lateral line receptors: Where do they come from developmentally and where is our research going? *Brain Behav. Evol.* **64,** 163–181.

Goldberg, J. M., and Brown, P. B. (1969). Response of binaural neurons of dog superior olivary complex to dichotic tone stimuli: Some physiological mechanisms of sound localization. *J. Neurophysiol.* **32,** 613–636.

Goodson, J. L., and Bass, A. H. (2000a). Forebrain peptide modulation of sexually polymorphic vocal motor circuitry. *Nature* **403,** 769–772.

Goodson, J. L., and Bass, A. H. (2000b). Vasotocin innervation and modulation of vocal-acoustic circuitry in the teleost *Porichthys notatus. J. Comp. Neurol.* **422,** 363–379.

Goodson, J. L., and Bass, A. H. (2002). Vocal-acoustic circuitry and descending vocal motor pathways in teleost fish: Convergence with terrestrial vertebrates reveals conserved traits. *J. Comp. Neurol.* **448,** 298–321.

Hawkins, A. D., and Horner, K. (1981). Directional characteristics of primary auditory neurons from the cod ear. *In* "Hearing and Sound Communication in Fishes" (Tavolga, W. N., Popper, A. N., and Fay, R. R., Eds.), pp. 311–327. Springer, New York.

Highstein, S. M., and Baker, R. (1986). Organization of the efferent vestibular nuclei and nerves of the toadfish, *Opsanus tau. J. Comp. Neurol.* **243,** 309–325.

Hubbs, C. L. (1920). The bionomics of *Porichthys Notatus* Girard. *Am. Nat.* **54,** 380–384.

Hudspeth, A. J., and Corey, D. P. (1977). Sensitivity, polarity, and conductance change in the response of vertebrate hair cells to controlled mechanical stimuli. *Proc. Natl. Acad. Sci. USA* **74,** 2407–2411.

Ibara, R. M., Penny, L. T., Ebeling, A. W., van Dykhuizen, G., and Cailliet, G. (1983). The mating call of the plainfin midshipman fish, *Porichthys notatus. In* "Predators and Prey in Fishes" (Noakes, D. G. L., Lundquist, D. G., Helfman, G. S., and Ward, J. A., Eds.), pp. 205–212. Dr. W. Junk Publishers, The Hague, Netherlands.

Jensen, F. B., Kuperman, W. A., Porter, M. B., and Schmidt, H. (1994). Computational ocean acoustics. *Am. Inst. Physics.* New York.

Kenyon, T. N., Ladich, F., and Yan, H. Y. (1998). A comparative study of hearing ability in fishes: The auditory brainstem response approach. *J. Comp. Physiol. A* **182,** 307–318.

Kittelberger, J. M., Land, B. R., and Bass, A. H. (2006). Midbrain peraqueductal gray and vocal patterning in a teleost fish. *J. Neurophysiol.* **96,** 71–85.

Knudsen, E. I., and Konishi, M. (1978). A neural map of auditory space in the owl. *Science* **200,** 795–797.

Koyama, H., Lewis, E. R., Leverenz, E. L., and Baird, R. A. (1982). Acute seismic sensitivity of the bullfrog ear. *Brain Res.* **250,** 168–172.

Kozloski, J., and Crawford, J. D. (1998). Functional neuroanatomy of auditory pathways in the sound-producing fish *Pollimyrus. J. Comp. Neurol.* **401,** 227–252.

Kozloski, J., and Crawford, J. D. (2000). Transformations of an auditory temporal code in the medulla of a sound-producing fish. *J. Neurosci.* **20,** 2400–2408.

Ladich, F., and Bass, A. H. (2003a). Underwater sound generation and acoustic reception in fishes with some notes on frogs. *In* "Sensory Processing in Aquatic Environments" (Colin, S., and Marshall, N. J., Eds.), pp. 173–193. Springer, New York.

Ladich, F., and Bass, A. H. (2003b). Audition. *In* "Catfishes" (Arratia, G., Kapoor, B. G., Chardon, M., and Diogo, R., Eds.), pp. 701–730. Science Publishers, Inc., Enfield, NH.

Lin, J. W., and Faber, D. S. (1988). Synaptic transmission mediated by single club endings on the goldfish mauthner cell. I. Characteristics of electrotonic and chemical postsynaptic potentials. *J. Neurosci.* **8,** 1302–1312.

Lu, Z. (2004). Neural mechanisms of hearing in fishes. *In* "The Senses of Fish: Adaptations for the Reception of Natural Stimuli" (von der Emde, G., Mogdans, J., and Kapoor, B. G., Eds.), pp. 147–172. Narosa Publishing House, New Delhi.

Lu, Z., and Fay, R. R. (1993). Acoustic response properties of single units in the torus semicircularis of the goldfish, *Carassius auratus*. *J. Comp. Physiol. A* **173,** 33–48.

Lu, Z., and Fay, R. R. (1996). Two-tone interaction in primary afferents and midbrain neurons of the goldfish, *Carassius auratus*. *Aud. Neurosci.* **2,** 257–273.

Lu, Z., and Popper, A. N. (1998). Morphological polarizations of sensory hair cells in the three otolithic organs of a teleost fish: Fluorescnet labeling of ciliary bundles. *Hear. Res.* **126,** 47–57.

Lu, Z., and Popper, A. N. (2001). Neural response directionality correlates of hair cell orientation in a teleost fish. *J. Comp. Physiol. A* **187,** 453–465.

Lu, Z., and Xu, Z. (2002). Effects of saccular otolith removal on hearing sensitivity of the sleeper goby (*Dormitator latifrons*). *J. Comp. Physiol. A* **188,** 595–602.

Lu, Z., Popper, A. N., and Fay, R. R. (1996). Behavioral detection of acoustic particle motion by a teleost fish (*Astronotus ocellatus*): Sensitivity and directionality. *J. Comp. Physiol. A* **179,** 227–233.

Lu, Z., Song, J., and Popper, A. N. (1998). Encoding of acoustic directional information by saccular afferents of the sleeper goby, *Dormitator latifrons*. *J. Comp. Physiol. A* **182,** 805–815.

Lu, Z., Xu, Z., and Buchser, W. J. (2003). Acoustic response properties of lagenar nerve fibers in the sleeper goby, *Dormitator latifrons*. *J. Comp. Physiol. A* **189,** 889–905.

Lu, Z., Xu, Z., and Buchser, W. J. (2004). Coding of acoustic particle motion by utricular fibers of the sleeper goby, *Dormitator latifrons*. *J. Comp. Physiol. A* **190,** 923–938.

Ma, L. W., and Fay, R. R. (2002). Neural representations of the axis of acoustic particle motion in nucleus centralis of the torus semicircularis of the goldfish, *Carassius auratus*. *J. Comp. Physiol. A* **188,** 301–313.

Manley, G. A. (2000). Cochlear mechanisms from a phylogenetic viewpoint. *Proc. Natl. Acad. Sci. USA* **97,** 11736–11743.

Marvit, P., and Crawford, J. D. (2000). Auditory discrimination in a sound-producing electric fish (*Pollimyrus*): Tone frequency and click-rate difference detection. *J. Acoust. Soc. Am.* **108,** 1819–1825.

McCormick, C. A. (1999). Anatomy of the central auditory pathways of fish and amphibians. *In* "Springer Handbook of Auditory Research, Comparative Hearing: Fish and Amphibians" (Popper, A. N., and Fay, R. R., Eds.), Vol. 11, pp. 155–217. Springer, New York.

McFadden, D. (1998). Sex differences in the auditory system. *Dev. Neuropsychol.* **14,** 261–298.

McKibben, J. R., and Bass, A. H. (1998). Behavioral assessment of acoustic parameters relevant to signal recognition and preference in a vocal fish. *J. Acoust. Soc. Am.* **104,** 3520–3533.

McKibben, J. R., and Bass, A. H. (1999). Peripheral encoding of behaviorally relevant acoustic signals in a vocal fish: Single tones. *J. Comp. Physiol. A* **184,** 563–576.

McKibben, J. R., and Bass, A. H. (2001a). Peripheral encoding of behaviorally relevant acoustic signals in a vocal fish: Harmonic and beat stimuli. *J. Comp. Physiol. A* **187,** 271–285.

McKibben, J. R., and Bass, A. H. (2001b). Effects of temporal envelope modulation on acoustic signal recognition in a vocal fish. *J. Acoust. Soc. Am.* **109,** 2934–2943.

Nelson, J. S. (1994). "Fishes of the World." John Wiley & Sons, New York.

Northcutt, R. G., Holmes, P. H., and Albert, J. S. (2000). Distribution and innervation of lateral line organs in the channel catfish. *J. Comp. Neurol.* **421,** 570–592.

Pappas, G. D., and Bennett, M. V. L. (1966). Specialized junctions involved in electrical transmission between neurons. *Ann. NY Acad. Sci.* **137,** 495–508.

Popper, A. N. (1976). Ultrastructure of the auditory regions in the inner ear of the lake whitefish. *Science* **192,** 1020–1023.

Popper, A. N., and Fay, R. R. (1999). The auditory periphery in fishes. *In* "Springer Handbook of Auditory Research, Comparative Hearing: Fish and Amphibians" (Popper, A. N., and Fay, R. R., Eds.), Vol. 11, pp. 43–100. Springer, New York.

Pough, F. H., Janis, C. M., and Heiser, J. B. (2002). "Vertebrate Life." Prentice-Hall, Upper Saddle River, New Jersey.

Prechtl, J. C., von der Emde, G., Wolfart, J., Karamürsel, S., Akoev, G. N., Andrianov, Y. N., and Bullock, T. H. (1998). Sensory processing in the pallium of a mormyrid fish. *J. Neurosci.* **18,** 7381–7393.

Premus, V., and Spiesberger, J. L. (1997). Can acoustic multipath explain finback (*B. physalus*) 20-Hz doublets in shallow water? *J. Acoust. Soc. Am.* **101,** 1127–1138.

Remage-Healey, L., and Bass, A. H. (2004). Rapid, hierarchical modulation of vocal patterning by steroid hormones. *J. Neurosci.* **24,** 5892–5900.

Remage-Healey, L., and Bass, A. H. (2005). Simultaneous, rapid, elevations in steroid hormones and vocal signaling during playback challenge: A field experiment in Gulf toadfish. *Horm. Behav.* **47,** 297–305.

Roberts, B. L., and Meredith, G. E. (1992). The efferent innervation of the ear: Variations on an enigma. *In* "The Evolutionary Biology of Hearing" (Webster, D. B., Fay, R. R., and Popper, A. N., Eds.), pp. 185–210. Springer, New York.

Rogers, P. H., and Cox, M. (1988). Underwater sound as a biological stimulus. *In* "Sensory Biology of Aquatic Animals" (Atema, J., Fay, R. R., Popper, A. N., and Tavolga, W. N., Eds.), pp. 130–149. Springer, New York.

Sand, O. (1974). Directional sensitivity of microphonic potentials from the perch. *J. Exp. Biol.* **60,** 881–899.

Schellart, N. A. M., Wubbels, R. J., Schreurs, W., Faber, A., and Goossens, J. H. L. M. (1995). Two-dimensional vibrating platform in nm range. *Med. Biol. Eng. Comput.* **33,** 217–220.

Sisneros, J. A., and Bass, A. H. (2003). Seasonal plasticity of peripheral auditory frequency sensitivity. *J. Neurosci.* **23,** 1049–1058.

Sisneros, J. A., and Bass, A. H. (2005). Ontogenetic changes in the response properties of individual, primary auditory afferents in the vocal plainfin midshipman fish *Porichthys notatus* Girard. *J. Exp. Biol.* **208,** 3121–3131.

Sisneros, J. A., Forlano, P. M., Knapp, R., and Bass, A. H. (2004a). Seasonal variation of steroid hormone levels in an intertidal-nesting fish, the vocal plainfin midshipman. *Gen. Comp. Endocrinol.* **136,** 101–116.

Sisneros, J. A., Forlano, P. M., Deitcher, D. L., and Bass, A. H. (2004b). Steroid-dependent auditory plasticity leads to adaptive coupling of sender and receiver. *Science* **305,** 404–407.

Stenberg, A. E., Wang, H., Sahlin, L., and Hultcrantz, M. (1999). Mapping of estrogen receptors α and β in the inner ear of mouse and rat. *Hear. Res.* **136,** 29–34.

Stenberg, A. E., Wang, H., Fish, J., III, Schrott-Fischer, A., Sahlin, L., and Hultcranz, M. (2001). Estrogen receptors in the normal adult and developing human ear and in Turner's syndrome. *Hear. Res.* **157,** 87–92.

Suzuki, A., Kozloski, J., and Crawford, J. D. (2002). Temporal encoding for auditory computation: Physiology of primary afferent neurons in sound-producing fish. *J. Neurosci.* **22,** 6290–6301.

Tomchik, S. M., and Lu, Z. (2004). Octavolateral projections and organization in the medulla of a teleost fish, the sleeper goby (*Dormitator latifrons*). *J. Comp. Neurol.* **481,** 96–117.

Tricas, T. C., and Highstein, S. M. (1991). Action of the octavolateralis efferent system upon the lateral line of free-swimming toadfish, *Opsanus tau*. *J. Comp. Physiol. A* **169,** 25–38.

Weeg, M. S., and Bass, A. H. (2000). Midbrain lateral line circuitry in a vocalizing fish. *J. Comp. Neurol.* **418,** 841–864.

Weeg, M. S., and Bass, A. H. (2002). Frequency response properties of lateral line superficial neuromasts in a vocal fish, with evidence for acoustic sensitivity. *J. Neurophysiol.* **88,** 1252–1262.

Weeg, M. S., Fay, R. R., and Bass, A. H. (2002). Tuning and directional responses of saccular afferents of the midshipman. *J. Comp. Physiol. A* **188,** 631–641.

Weeg, M. S., Land, B. R., and Bass, A. H. (2005). Temporal modulation of efferents to the inner ear and lateral line by central vocal pathways. *J. Neurosci.* **25,** 5967–5974.

Winn, H. E. (1972). Acoustic discrimination by the toadfish with comments on signal systems. *In* "Behavior of Marine Animals" (Winn, H. E., and Olla, B. L., Eds.), pp. 361–385. Plenum, New York.

Winslow, R. L., and Sachs, M. B. (1987). Effect of electrical stimulation of the crossed olivocochlear bundle on auditory nerve response to tones in noise. *J. Neurophysiol.* **57,** 1002–1021.

Wubbels, R. J., and Schellart, N. A. M. (1997). Neural coding of sound direction in the auditory midbrain of the rainbow trout. *J. Neurophysiol.* **77,** 3060–3074.

Wysocki, L. E., and Ladich, F. (2001). The ontogenetic development of auditory sensitivity, vocalization and acoustic communication in the labyrinth fish, *Trichopsis vittata*. *J. Comp. Physiol. A* **187,** 177–187.

Wysocki, L. E., and Ladich, F. (2002). Can fishes resolve temporal characteristics of sounds? New insights using auditory brainstem responses. *Hear. Res.* **169,** 36–46.

Wysocki, L. E., and Ladich, F. (2003). The representation of conspecific sounds in the auditory brainstem of teleost fishes. *J. Exp. Biol.* **206,** 2229–2240.

Young, I. S., and Rome, L. C. (2001). Mutually exclusive muscle designs: The power output of the locomotory and sonic muscles of the oyster toadfish (*Opsanus tau*). *Proc. R. Soc. Lond. B. Biol. Sci.* **268,** 1965–1970.

10

THE LATERAL LINE SYSTEM OF FISH

HORST BLECKMANN

1. Introduction
2. The Lateral Line Periphery
    2.1. Morphology
    2.2. Natural Lateral Line Stimuli
    2.3. Behavior
    2.4. Multimodal Guidance of Behavior
    2.5. Physiology
    2.6. Responses to a Moving Vibrating Sphere
    2.7. Responses to Moving Objects
    2.8. Running Water
    2.9. Response Masking in Running Water
    2.10. Central Pathways for Lateral Line Information Processing
    2.11. Descending Recurrent Pathways
3. Central Physiology
    3.1. Ongoing Activity
    3.2. Latency
    3.3. Dynamic Response Properties
    3.4. Thresholds
    3.5. Phasic Versus Tonic Responses
    3.6. Frequency Encoding
    3.7. Dynamic Amplitude Range
    3.8. Lateral Line Maps
    3.9. Receptive Field Organization
    3.10. Moving Object Stimuli
    3.11. Running Water Conditions
    3.12. Multimodality
4. Conclusions

1. INTRODUCTION

Whenever an aquatic animal moves it inevitably causes water displacements and pressure fluctuations, that is, hydrodynamic stimuli (Kalmijn, 1988a). Consequently hydrodynamic stimuli are constantly present in

Sensory Systems Neuroscience: Volume 25
FISH PHYSIOLOGY

DOI: 10.1016/S1546-5098(06)25010-6

natural waters where they provide a wealth of sensory information. This probably is the reason why most, if not all, aquatic animals have developed a sensory system for the detection of water movements, pressure fluctuations, or both (Bleckmann, 1994). In cartilaginous and bony fishes and in aquatic amphibians this system is called the mechanosensory lateral line. In most fish species parts of the lateral line are visible externally as a row of small pores found on the trunk and head. According to Parker (1904), these pores were first described by Stennon in 1664.

Living fishes constitute about 25,000 species, that is, fish comprise about 50% of all vertebrates. No wonder that fish live in a wide range of aquatic habitats like the deep sea, tide pools, fast running streams, sweet water ponds, and large lakes. As diverse as these habitats are the sensory systems of fish. Fish may or may not have a highly developed visual and/or acoustic system. They may have a keen sense of smell and taste, and they may be able to sense animate and inanimate electric fields. Irrespective of whether these sensory systems are present all cartilaginous and bony fish possess a mechanosensory lateral line (Northcutt, 1989). This suggests that the lateral line is one of the most important sensory systems of fish.

This chapter will briefly review the morphology and behavioral relevance of the lateral line system and then focus in greater detail on the peripheral and central processing of lateral line information. (For further general information on the fish lateral line refer to Coombs and Janssen, 1989; Bleckmann, 1993, 1994, 2004; Montgomery *et al.*, 2000; Janssen, 2004; Mogdans *et al.*, 2004).

2. THE LATERAL LINE PERIPHERY

2.1. Morphology

The smallest functional unit of the lateral line system is the neuromast, a sensory structure that occurs freestanding on the skin (superficial neuromasts or SN) or in fluid-filled canals (canal neuromasts or CN) that usually open to the environment through a series of pores (Figure 10.1). In many fish SN occur in pits or on pedestals raised above the skin. Not only the design, alignment, and number of SN but also the design and number of lateral line canals can be very different in different fish species (Figure 10.2), even if closely related (Coombs *et al.*, 1988; Webb, 1989a). SN are usually smaller than CN (Münz, 1989). Fish with a reduced canal system often show a concomitant increase in the number of SN (Coombs *et al.*, 1988). The variability of lateral line canals includes their number, placement and branching pattern, canal width, canal compartmentalization, and the number, size, and

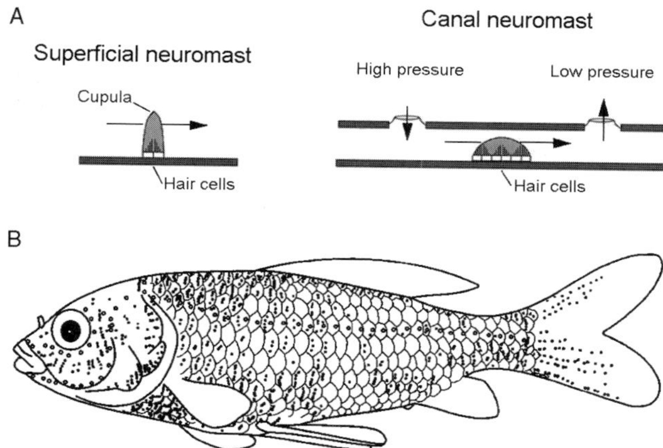

Fig. 10.1. (A and B) Morphology of the lateral line periphery. (A) Diagrammatical representation of a SN and a lateral line canal with a CN (redrawn from Coombs and Montgomery, 1999). Arrows indicate direction of water flow. (B) Distribution of SNs (dots) in the goldfish, *Carassius auratus*. Circles indicate canal pores. The drawing of the goldfish was gratefully provided by Grotefeld.

placement of canal pores (Coombs *et al.*, 1988; Webb, 1989b). These morphological variations most likely have functional implications that we have only begun to explore. In many fish species there are even enormous differences between different parts of the peripheral lateral line (e.g., between the head and trunk lateral lines of some surface-feeding fish and between the dorsal and ventral lateral lines of rays). These differences suggest that different parts of the lateral line system on the same animal may have different functions (Denton and Gray, 1988). With the exception of a few major taxa (e.g., Notopteroidei and Clupeomorpha), most fish taxa can not be classified by their lateral line characteristics (Coombs *et al.*, 1988; Webb, 1989b).

Lateral line neuromasts are composed of sensory hair cells, supporting cells, and mantle cells, which separate a neuromast from the surrounding tissue. Lateral line hair cells are similar in morphology to the hair cells in the auditory and vestibular organs of vertebrates (Flock and Wersäll, 1962). They carry a hair bundle at their apical surface that is composed of up to 150 stereovilli and a single true kinocilium (Figure 10.3A). The stereovilli grow longer from one edge of the hair bundle to the other. The kinicilium always occurs eccentrically at the tall edge of the bundle, thus hair cells have a morphological polarization. The hair cells of a lateral line neuromast are oriented into two opposing directions parallel to the major (usually long) axis of the neuromast (Flock and Wersäll, 1962) (Figure 10.3B). The hair bundles protrude into a cupula (Figure 10.3B), which connects the bundles

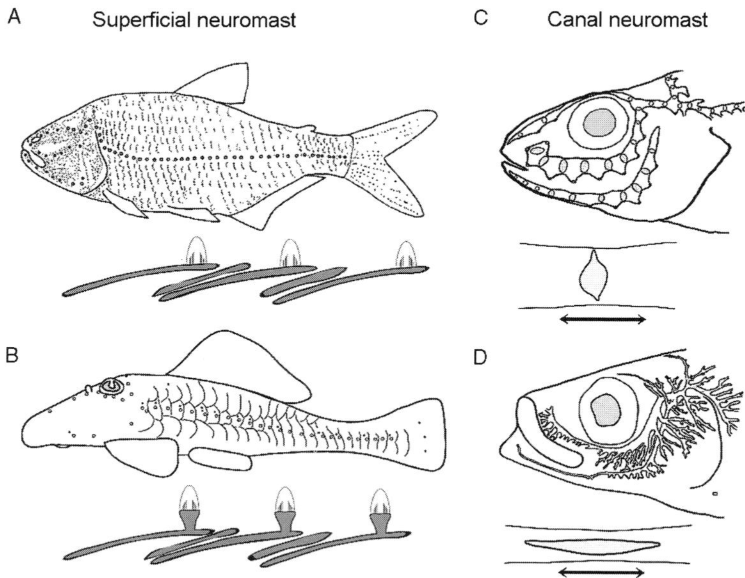

Fig. 10.2. (A–D) Examples of the morphological diversity of the fish lateral line. The peripheral lateral line of the blind cave fish, *Astyanax hubbsi* (redrawn from Schemmel, 1967) (A) and the running water fish, *Ancistrus* sp. (Grotefeld, unpublished data) (B). Dots indicate SNs, circles represent canal pores. Note that *Astyanax* has much more SNs than *Ancistrus*. (C) The head of the freshwater fish, *Percarina demidoffi*. Note the widened head lateral line canals (redrawn from Jakubowski, 1967). (D) Head of a herring *Sprattus sprattus*, a fish that lives in the open ocean. The black lines indicate the course of the lateral line canals (redrawn from Blaxter *et al.*, 1983).

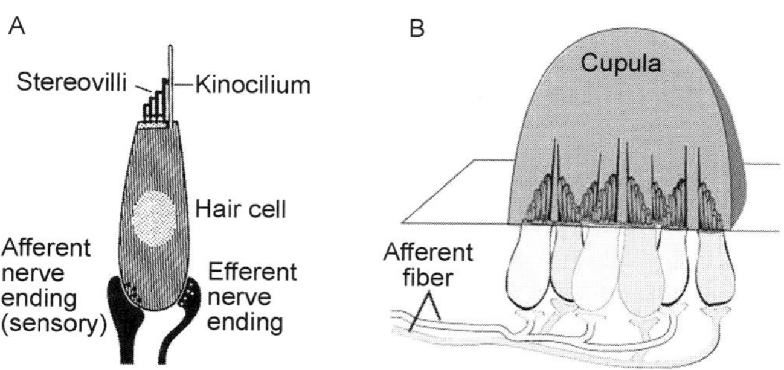

Fig. 10.3. (A) A lateral line hair cell with a single true kinocilium and several stereovilli (Flock, 1971b). (B) A lateral line neuromast with antagonistically aligned hair cells, a cupula, and two afferent nerve fibers. Efferent innervation is not shown.

with the water surrounding the fish or with canal fluid. Within a given animal, but also across species, cupula length and shape may vary substantially (Teyke, 1990). Fish belonging to different species may have less than 10 or more than 1000 SN on each body side (Puzdrowski, 1989). Lateral line neuromasts are innervated by afferent nerve fibers. In addition, lateral line neuromasts may receive an efferent innervation (Münz, 1989). In general, CN are innervated by more nerve fibers than SN. While efferent terminals are present in all CN so far investigated, it is still an open question whether all SN receive an efferent innervation (Münz, 1989). In cichlid fish a single afferent fiber may contact more than one neuromast. If so, it innervates two or several neighboring SN but never an SN and a CN (Münz, 1989). Whether this is a general feature in fish is not known. A single afferent nerve fiber innervates only hair cells with the same directional orientation, even if that fiber contacts more than one neuromast (Münz, 1985). The morphological data already indicate that there are at least two parallel pathways that carry lateral line information to the brain: One pathway that encodes SN and one that encodes CN information.

2.2. Natural Lateral Line Stimuli

In natural habitats hydrodynamic stimuli are caused by inanimate and animate sources. Water motions caused by animals consist of short-lasting transients, long-lasting frequency- and amplitude-modulated water motions, or a mixture of both (Bleckmann, 1994). In terms of displacement, the main spectral amplitude of hydrodynamic stimuli generated by the swimming movements of animals often is below 10 Hz (Enger *et al.*, 1989; Bleckmann *et al.*, 1991a). However, higher-frequency water oscillations may also occur. For instance, the base frequency of the swimming appendages of smaller zooplankton shows a range of 10 Hz up to about 45 Hz (Newbury, 1972). During steady locomotion a subundulatory swimming fish (e.g., a trout) generates a wake (Figure 10.4). The flow in the wake consists of a chain of slightly deformed vortex rings. The rings of the ladderlike arrangement are composed of start–stop vortices that are generated during propulsion by the caudal fin. The almost circular vortex rings result from a similarity of the distance covered during a half fin beat and the fin height (Blickhan *et al.*, 1992). The wake behind a swimming fish (*Carassius auratus*) can show a clear vortex structure for at least 30 sec, particle velocities significantly higher than background noise could still be detected 3 min after a fish (body length 19 cm) had intercepted the measuring plane (Hanke *et al.*, 2000). Fish wakes can contain frequencies up to about 100 Hz (Bleckmann *et al.*, 1991a). They provide some information about the size, swimming style, and swimming speed of the wake generator (Hanke and

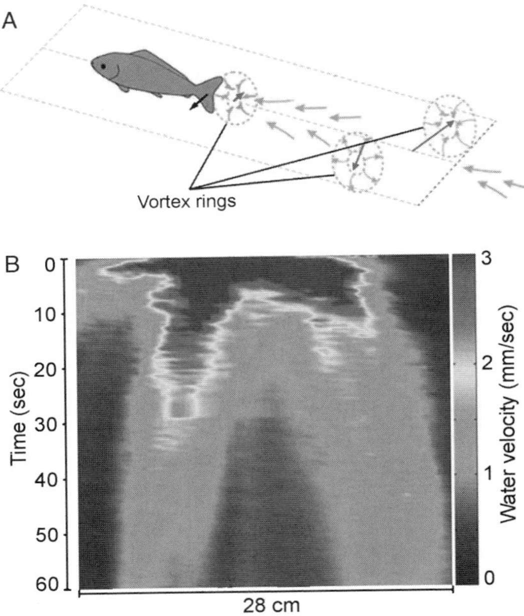

Fig. 10.4. Color coded water velocity behind a swimming fish (*Carassius auratus*) averaged over the columns of the camera field. Dark red indicates water velocities >3 mm/sec and dark blue <0.1 mm/sec. The x-axis shows the width of the camera field, and the y-axis the time that has passed since the fish entered the camera field (after Hanke *et al.*, 2000). (See Color Insert.)

Bleckmann, 2004). The complicated flow patterns and the turbulences in fish wakes provide useful information for prey detection and prey localization (Dehnhardt *et al.*, 2001; Pohlmann *et al.*, 2004).

Some fish use self-generated water motions for lateral line perception. For instance, during swimming or gliding, blind cave fish generate low-frequency water motions and pressure alterations (Campenhausen *et al.*, 1981; Weissert and Campenhausen, 1981). If a blind cave fish passes or approaches an object the self-produced water motions and pressure fluctuations get altered in a predictable way (Hassan, 1992). By analyzing these alterations, cave fish obtain lateral line information about the size, shape, spacing, and distance of nearby objects (Campenhausen *et al.*, 1981; Weissert and Campenhausen, 1981).

Some fish use capillary water surface waves for prey detection. One source of capillary surface waves are terrestrial insects fallen into the water. Surface waves caused by terrestrial insects are usually long lasting, have displacement amplitudes below 100 μm, and are irregular in time course

(Lang, 1980; Bleckmann, 1985). Aquatic and semiaquatic invertebrates and vertebrates that contact the water–air interface from below to breathe or feed incidentally also generate surface waves. These waves are usually more regular in time course and are shorter than insect generated waves (Bleckmann, 1985). The spectra of fish and frog generated surface waves are relatively narrowband with an upper frequency limit of 20–40 Hz. In contrast the frequency content of surface waves generated by terrestrial (prey) insects fallen into the water in addition have suprathreshold components up to at least 60 Hz (Lang, 1980; Bleckmann, 1985).

It should be stressed that the lateral line system of many fish is constantly exposed to background water motions caused—for instance—by wind, temperature or salinity gradients, and by gravity. As long as it does not move, a fish that lives in a pond, a lake, or in the deep ocean may face little hydrodynamic noise. In contrast fish that live along the ocean shoreline, in tide pools, or in rapidly running rivers often face turbulent water conditions which may interfere with lateral line perception (Bleckmann, 1994).

2.3. Behavior

Besides visual, acoustic, and olfactory cues, lateral line information provides the basis for many behavioral decisions (Bleckmann, 1993). Although many casual observations and carefully designed experiments point in this direction, still little is known about the full behavioral capabilities of the lateral line, especially in fish that are constantly exposed to background water motions. Hofer (1908) was the first to demonstrate that pike (*Esox lucius*) with an intact lateral line can sense weak water motions. It is now clear that many fish use this ability for prey detection, predator avoidance, intraspecific communication, schooling behavior, object discrimination, and object entrainment and rheotaxis (Bleckmann, 1993, 1994; Montgomery *et al.*, 1997).

The mottled sculpin, *Cottus bairdi*, is a nocturnal feeder that uses the lateral line to detect prey. Mottled sculpins inhabit not only huge lakes (e.g., Lake Michigan) but also occupy fast water areas in streams. The peripheral lateral line of *Cottus* consists of SN and of CN (Janssen *et al.*, 1987). *Cottus* exhibits an orienting response to a water jet (Janssen *et al.*, 1990) or to a vibrating sphere (Coombs and Janssen, 1990) that—in the absence of visual cues—relies on lateral line input (Coombs, 1994). Blinded sculpins determine the direction and the distance of a dipole source as long as the lateral line is intact on the side of the fish facing the dipole (Coombs and Conley, 1995; Janssen and Corcoran, 1998). The ability to estimate source distance is restricted to short ranges (about one fish body length).

Surface-feeding fish like *Aplocheilus lineatus* and *Pantodon bucholzi* detect capillary water surface waves with their cephalic lateral line) (Schwartz, 1970; Hoin-Radkovski *et al.*, 1984). At 100 Hz behavioral thresholds are as low as 0.01 μm peak-to-peak water displacement amplitude (Bleckmann, 1980). Surface-feeding fish not only determine the direction but also the distance to a surface wave source. For distance determination they use the damping and dispersion properties of water surface waves thus indicating that they exploit the physical properties of the water surface for prey localization (Bleckmann *et al.*, 1989a). Surface-feeding fish not only localize a wave source but in addition can discriminate surface waves with a frequency difference limen of about 15%, that is, they can distinguish a 20-Hz from a 23-Hz wave train (Bleckmann *et al.*, 1981; Vogel and Bleckmann, 1997). Surface-feeding fish (*A. lineatus*) also discriminate pure sine wave stimuli from sine waves which show abrupt frequency changes (e.g., from 20 to 25 Hz or from 60 to 63 Hz) but fail to discriminate surface waves (carrier frequency 20, 40, and 60 Hz) that differ only in amplitude modulation (modulation frequency 10 and 20 Hz, modulation depth up to 80%) (Vogel and Bleckmann, 1997). Thus the ability to analyze the frequency content of a surface wave stimulus appears to be much more important than the ability to analyze its amplitude content.

The midwater fish *C. auratus* discriminates the hydrodynamic stimuli caused by moving objects, that is, this fish uses lateral line input to discriminate the direction of object motion, object speed, size, and shape (Vogel and Bleckmann, 2000). The blind cavefish *Astyanax mexicanus* (Characidae) readily passes through a barrier of rods without touching them. *Astyanax* makes use of the fact that any nearby stationary object interacts hydrodynamically with its body and thus alters the stimulus pattern of the lateral line in a predictable way (Campenhausen *et al.*, 1981). *Astyanax* even uses hydrodynamic information to develop an inner map of its aquatic environment (Teyke, 1989; Burt de Perera, 2004).

The lateral line is usually viewed as a system for close range detection. However, some predatory fish can sense and even track the hydrodynamic trail caused by prey fish (Pohlmann *et al.*, 2001, 2004). Assuming that prey fish like herring swim with a speed of 1 m/sec (Videler and Wardle, 1991), theoretically they are still detectable for a predator intercepting the wake after the fish has covered a distance of about 60 m. This detection range may be even larger for a predator intercepting the water motions caused by a large fish or by a school of fish which, for instance, can easily consist of 10^5 or 10^7 individuals (Pitcher and Parrish, 1993).

Most behavioral experiments designed to uncover lateral line function have been done in still water, that is, under conditions that are quite unnatural for many fish. River fish may use lateral line input to ease their life in rapid water flow with much turbulence created by submerged rocks, branches, or roots. Trout prefer specific locations from which they only leave to seize pieces of drifting debris. Trout (*Salvelinus fontinalis*, *Salmo trutta*) may even use hydrodynamic information to maintain position and to capture energy from vortices while moving through turbulent flow (Sutterlin and Waddy, 1975; Liao *et al.*, 2003; Beal *et al.*, 2006).

2.4. Multimodal Guidance of Behavior

The integration of sensory information from different modalities into a coherent "gestalt" is one of the accomplishments of the mammalian cortex (Kandel *et al.*, 2000) but may also be realized to a certain degree in the brains of lower vertebrates. Consequently some researchers have studied the roles played by different sensory systems in organizing complex behaviors of fishes. In the largemouth bath, *Micropterus salmoides*, and the pike, *Esox masquinongy*, vision is crucial to the initial detection of, and orientation to, prey. Lateral line and vision together determine the optimum distance and angular deviation for the initiation of a strike toward the prey. Blinded fish strike accurately at prey at close ranges and small angular deviations, indicating that lateral line information is only sufficient for the final phases of a strike (New, 2002). Modeling results suggest that in weakly electric fish (*Apteronotus albifrons*) electrosensory and mechanosensory information are integrated for prey detection and prey localization (Nelson *et al.*, 2002). Prey movements (i.e., hydrodynamic stimuli) shorten search times in mormyrid fish (*Gnathonemus petersii*) only when active electrolocation and vision are not possible (von der Emde and Bleckmann, 1998). This indicates that the mechanosensory lateral line is important for the detection of prey if other sensory modalities are eliminated. Both lateral line and chemosensory informations play an important role in the prey capture of trout (Montgomery *et al.*, 2002). Catfish (*Siluris glanis*) can track prey fish on the basis of the hydrodynamic trails left behind by their swimming motion. Most likely they do so by integrating hydrodynamic, chemosensory, tactile, and visual information (Pohlmann *et al.*, 2001, 2004). In general, many of the above-mentioned behavioral experiments demonstrate a hierarchy of the senses involved in prey capture with different modalities playing critical roles in succeeding phases.

2.5. Physiology

Hair cells are displacement detectors. Displacement of the ciliary bundle toward the kinocilium causes a depolarization, displacement in the opposite direction a hyperpolarization of the hair cell (Kroese and van Netten, 1989).

Receptor potentials follow the deflection of the cilia (cupula) within a fraction of a microsecond (Kroese and van Netten, 1989). Consequently primary lateral line afferents respond with latencies in the ms range to a water wave stimulus. In general, the response amplitude of a hair cell depends on both the angle of ciliary bundle deflection and the stimulus amplitude (Figure 10.5). The membrane potential of a hair cell therefore does not provide unequivocal information (Flock, 1965).

Under natural conditions, water motions (SN) and canal fluid motions (CN) cause lateral line cupulae to move, which in turn deflect the ciliary bundles of the underlying hair cells. The relationship between cupula movement and the amplitude of the receptor potential of a neuromast is linear but finally reaches saturation (Figure 10.5B). Lateral line neuromasts may already show a physiological response if a homogenous water current (measured outside the boundary layer) surpasses a velocity of 0.1 mm/sec. Response saturation may occur around a velocity of 1.5 mm/sec (Görner, 1962). Primary lateral line afferents show ongoing activity from less than 1 up to about 80 impulses/sec; ongoing activity may be regular, irregular, or bimodal (Bleckmann and Topp, 1981; Münz, 1985; Görner and Mohr, 1989; Mohr and Bleckmann, 1998). Depending on the alignment of the hair cells innervated by a primary afferent nerve fiber (see in an earlier section), a water jet directed against the cupula causes either a decrease or an increase

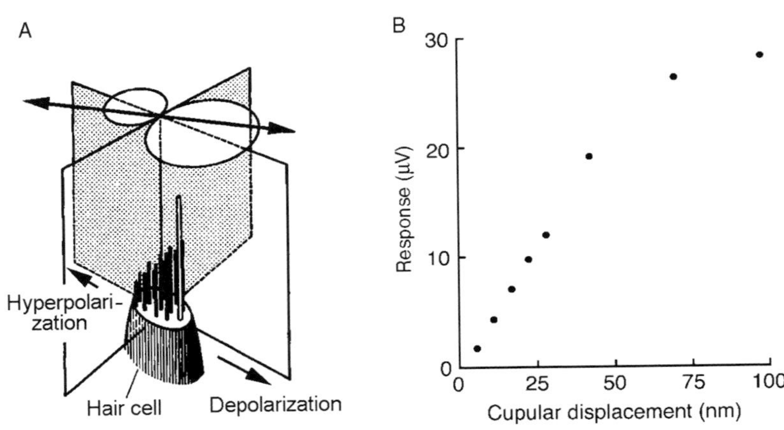

Fig. 10.5. (A) The directional sensitivity of a lateral line hair cell. The membrane potential varies with the cosine of the angle between the direction of maximum sensitivity and the direction of the applied displacement stimulus. Top arrow marks the long axis of the neuromast (redrawn from Flock, 1971a). (B) Relationship between the cupula displacement and the electrical response (amplitude of microphonic potential) of a lateral line neuromast stimulated with sinusoidal (125 Hz) fluid motions (redrawn from Kroese and van Netten, 1989).

in ongoing discharge rate (Görner, 1963). Since the work of Harris and Bergeijk (1962) researchers have stimulated the lateral line with constant volume vibrating spheres. At stimulus levels just above threshold, units start to phase lock to sinusoidal water motions. With increasing stimulus amplitude, the degree of phase locking increases and finally reaches a plateau at stimulus levels about 20 dB (one order of magnitude) above threshold. Due to the directional sensitivity of hair cells primary lateral line afferents phase lock only to one half of a full cycle of a sinusoidal wave stimulus (Figure 10.6C). On average, discharge rates start to increase above ongoing

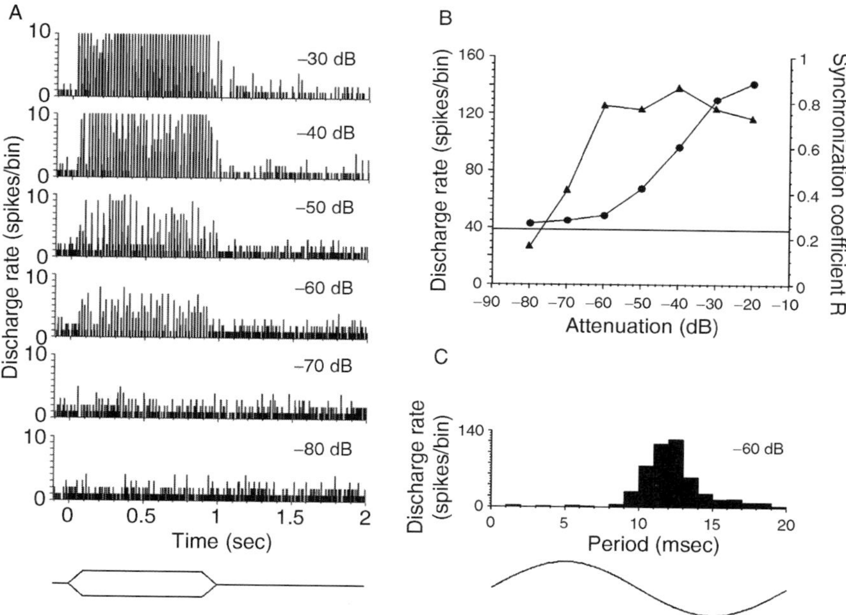

Fig. 10.6. (A–C) Characterization of the responses of a goldfish posterior lateral line nerves (PLLN) fiber to different levels of a 50 Hz constant-amplitude sine wave stimulus generated by a vibrating sphere of 8 mm diameter. (A) PST histograms (bin width 2 msec) of the responses to ten repetitions of the stimulus. The bottom trace represents the envelope of the stimulus. Vibration amplitudes were 144, 38, 15, and 4 μm for –30 to –60 dB attenuation, respectively, and <4 μm for attenuations greater than –60 dB. At levels of –40 dB and stronger, the unit responded with at least one highly phase-locked spike per cycle. (B) Input–output function of the fiber shown in A. Discharge rates (line connecting circles, left-hand axis) and synchronization coefficients R (lines connecting triangles, right-hand axis) are plotted as function of stimulus level (rel. dB). An attenuation of –20 dB corresponds to displacement amplitude of 425 μm. (C) Raster diagram of the distribution of spikes within each cycle of the 50-Hz stimulus and the corresponding period histogram. Graph was derived from the data shown in A (–60 dB attenuation) (after Bleckmann *et al.*, 2001).

discharge rate at stimulus levels about 10 dB higher than those that cause phase locking (Figure 10.6). The dynamic amplitude range of primary afferents is between 11 and 50 dB (Mogdans and Bleckmann, 1999) but may reach 90 dB (Elepfandt and Wiedemer, 1987). In the most sensitive preparations, a stimulus amplitude of 0.02 μm (stimulus frequency 50–100 Hz) already causes a neural response (Bleckmann and Topp, 1981; Kroese and Schellart, 1992). Due to inertial and frictional forces, afferents that innervate SN respond within their operating range (0 to about 70 Hz) approximately in proportion to the velocity of the water surrounding the cupula. Within lateral line canals, fluid flow depends on the pressure differences between neighboring canal pores. As a consequence, afferents that innervate CN respond (water motion parallel to the surface of the fish) within their operating range (>0 Hz to about 150 Hz) approximately in proportion to outside water acceleration (Kalmijn, 1988b). In consequence two types of afferents can be distinguished: type I afferents function as velocity detectors, type II afferents are more sensitive to water acceleration (Münz, 1985). It should be mentioned that even fibers that innervate only SN can differ in their reponse properties. Such fibers may be low-pass, band-pass, broadly tuned, or complex (Weeg and Bass, 2002).

It is important to note that CN situated in narrow canals can also respond to a homogeneous water jet, that is, to a DC stimulus, provided the jet is oriented perpendicular to the surface of the fish and hits a single canal pore (Fest and Hofmann, unpublished data).

Both, type I and type II afferents encode the amplitude of a sinusoidal wave stimulus by the degree of phase coupling and—at higher stimulus amplitudes—by firing rate. Thus the central analysis of water motions can be based on a phase-locking code and—if stimulus amplitudes are large—on a spike rate code (Elepfandt and Wiedemer, 1987). Primary lateral line afferents also phase lock to the amplitude modulation frequency of a sinusoidal wave stimulus (Mogdans and Bleckmann, 1999). When tested near threshold with a vibrating sphere, the receptive fields (RF) of primary afferents are small (Figure 10.7, top). This agrees well with the morphological data (see in an earlier section) which show that a single lateral line afferent innervates only one (either an SN or a CN) or only a few neighboring SN. If stimulated with a vibrating object the responses of primary afferents quickly diminish with source distance (Caird, 1978b). This is due to the fact that the flow generated by a dipole source attenuates at a rate of $1/\text{distance}^3$. For physical reasons (cf. Kalmijn, 1988b) a fish not firmly attached to the substrate can only detect the spatial derivative of the imposed local flow which—close to a dipole source—attenuates at a rate close to $1/\text{distance}^4$ (Denton and Gray, 1988). Thus in case of dipole stimulation lateral line function is restricted to a few centimeters. Despite the variations

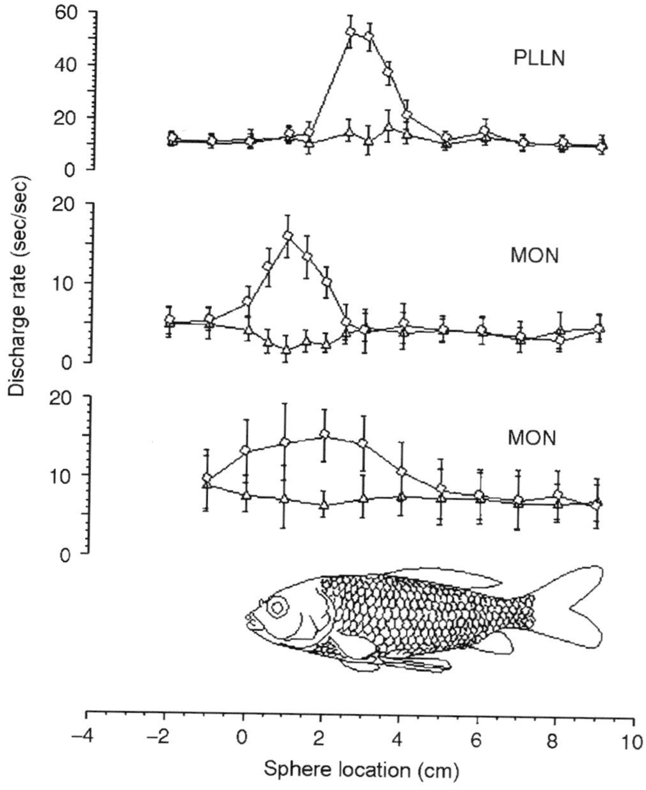

Fig. 10.7. The RF of a goldfish PLLN fiber (top) and of two MON units (center and bottom). Circles represent average unit responses to 20 repetitions of a 50-Hz sine wave stimulus (duration 1 sec) presented at different locations along the side of the fish. Triangles represent average ongoing activity during a period of 1 sec before stimulus presentation. Displacement amplitudes were 80 μm (PLLN) and 100 μm (MON). Note the differences in discharge rate between PLLN and MON units (different scaling of y axis). From top to bottom the width of the RF measured at 50% of the maximum evoked rate were 1.6, 1.7, and 3.9 cm. The fish drawing is scaled to match the size of the fish in the experiment in which data for the top and center graph were recorded (tip of snout at 0 cm, caudal peduncle at 7.4 cm). The fish used in the experiment in which the bottom trace was obtained was 4-mm smaller (after Bleckmann *et al.*, 2001).

in lateral line morphology (see in an earlier section), the response properties of primary lateral line afferents, if stimulated in still water with a vibrating sphere, are quite similar in different fish species (Münz, 1985; Bleckmann and Münz, 1990; Coombs and Montgomery, 1992;). Whether this is still true if we investigate the various lateral line systems under natural stimulus and noise conditions remains to be shown.

2.6. Responses to a Moving Vibrating Sphere

If a vibrating sphere is moved slowly along the side of a fish primary lateral line afferents that innervate trunk CN show spike rates that correspond to changes in pressure gradient amplitudes. 180° phase shifts in the responses correspond to reversals in the direction of the pressure gradient, and changes in sphere distance alter the location of the side peaks and the 180° phase shifts as expected from theoretical calculations (Coombs *et al.*, 1996). Calculations of lateral line excitation patterns for a linear array of CN at different distances from a dipolar source indicate that the information about source azimuth is contained in the location of the maximum pressure-difference amplitude, whereas information about source distance is contained in the spread of excitation. This distance cue is robust and unambiguous; that is, if size or source vibration amplitude is increased at a given distance, the level but not the spread of lateral line excitation increases. Thus even though peak excitation levels may be identical for a distant, high-amplitude (or large) source and a nearby low-amplitude or small source, there is sufficient information in the spread of excitation to distinguish between the two. Thus the spatial representation of source distance along a two-dimensional array of sensors may provide fish with a mechanism of depth perception (Hassan, 1989). Unlike visual images, which get smaller as the source moves further away, hydrodynamic images get larger, as do electrosensory images (von der Emde *et al.*, 1998).

2.7. Responses to Moving Objects

Animate sources of natural hydrodynamic stimuli rarely vibrate with constant amplitude and frequency. Instead they move around while searching for mates, prey, or shelter. To create more complex lateral line stimuli, researchers have stimulated the lateral line of fishes with small moving objects (Bleckmann and Zelick, 1993; Müller and Bleckmann, 1993; Mogdans and Bleckmann, 1998; Engelmann and Bleckmann, 2004). Moving objects cause low-frequency transient water motions that are followed by an ill-defined long-lasting wake (Mogdans and Bleckmann, 1998). Water motions are always associated with changes in hydrodynamic pressure and vice versa. However, while the wake caused by a moving object may contain several high-velocity peaks the pressure changes caused by that object are prominent only during the initial transient but are small in the object's wake (Mogdans and Bleckmann, 1998).

Type I afferents, that is, afferents that most likely innervate SN, respond to a moving object with a well-defined single peak of excitation followed by a decrease in neural activity or vice versa. This response pattern inverses when object motion direction is reversed (Figure 10.8, top). Type I afferents, in addition, discharge numerous unpredictable bursts of spikes after the initial, well-defined response, that is, after the object has passed the fish (Mogdans and Bleckmann, 1998). Type II afferents, that is, afferents that most likely receive input from CN, also respond to a moving object with a single peak of excitation followed by inhibition or vice versa. Like in type I afferents this temporal response pattern inverses when object motion direction is reversed. However, in contrast to type I afferents the neural activity of type II afferents is barely affected by the wake of a moving object (Figure 10.8, bottom) (Mogdans and Bleckmann, 1998).

2.8. Running Water

In the natural habitat, fish rarely face still water conditions because either the water moves, the fish moves, or both move. To fully comprehend lateral line perception, researchers have started to investigate how water flow affects the lateral line system of fishes. If exposed to running water (test range up to 15 cm/sec), type I afferents respond with a burstlike increase in ongoing activity (Voigt *et al.*, 2000; Engelmann *et al.*, 2002b). In contrast, ongoing activity of type II afferents, that is, of afferents that most likely receive input from CN, barely changes if fish are exposed to water flow. According to the directional sensitivity of lateral line hair cells and the innervation pattern of lateral line neuromasts (see in an earlier section) about 50% of flow-sensitive primary lateral line afferents should respond with an increase and 50% with a decrease in neural activity. This has, however, not been found since nearly all flow-sensitive lateral line afferents increased their discharge rate if the fish was exposed to unidirectional water flow (Engelmann *et al.*, 2000, 2002a; Voigt *et al.*, 2000). Moreover, if the direction of water flow is reversed primary afferents continue to respond with an increase in neural activity (Figure 10.9). One possible explanation for this finding is the scales of fish, which, if exposed to water flow, create microturbulences that may stimulate lateral line neuromasts irrespective of flow direction (Chagnaud *et al.*, unpublished data; Engelmann *et al.*, 2002a).

2.9. Response Masking in Running Water

In running water (10 cm/sec) the responses of type I afferents to a vibrating sphere stimulus are masked (Figure 10.10, left), both, in terms of discharge rate and in terms of phase locking. Obviously gross water flow drives SN into saturation and therefore renders them useless for the detection of water motions generated by a vibrating sphere. In contrast, vibrating sphere evoked discharge rates and phase locking of type II afferents are barely affected by running water (Figure 10.10, right) (Engelmann *et al.*, 2000, 2002a).

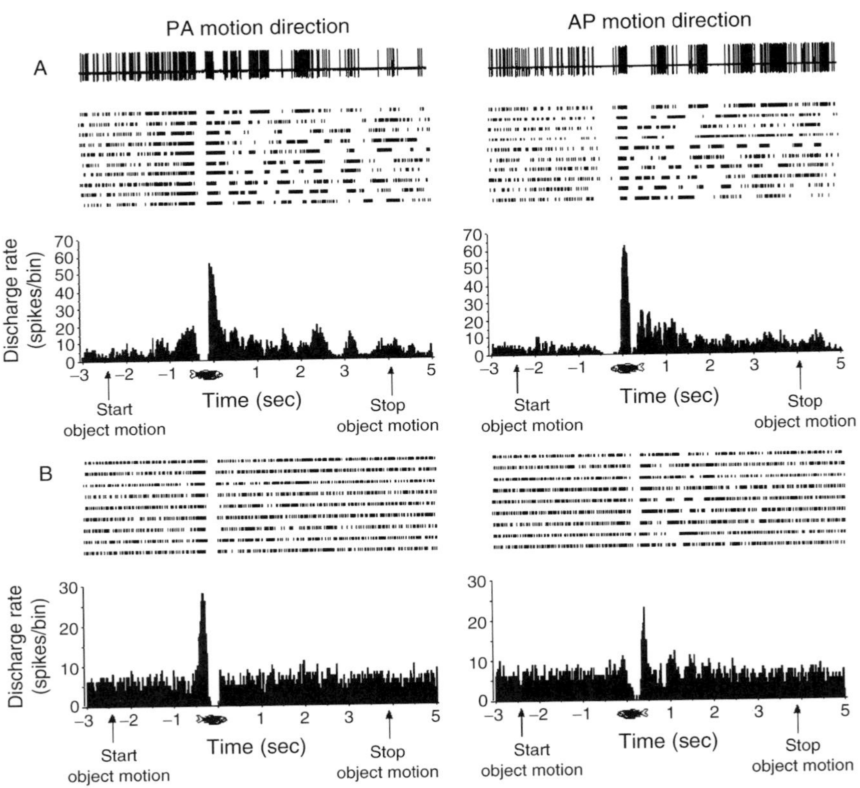

Fig. 10.8. (A and B) Responses of goldfish PLLN fibers to an object passing the fish laterally with a speed of 15 cm/sec. Raster diagrams and PSTHs (bin width 20 msec) of the responses to 10 stimulus presentations are shown. Left: motion direction was from posterior to anterior. Right: motion direction was from posterior to anterior. The fish symbol indicates the size, location, and orientation of the fish relative to the path of the moving object. At zero time, the lateral distance between object and fish was minimal. Negative times refer to the object's approach of the fish. Top: example of a unit that responded with a triphasic main response pattern that consisted of excitation followed by inhibition and again by excitation in the PA direction. When object motion was in the opposite direction, the main response pattern inversed, that is, the unit responded with inhibition followed by excitation and again inhibition. The unit continued to fire unpredictable bursts of spikes for a long time after the object had passed the fish. Bottom: Example of a unit that responded with a biphasic discharge pattern which consisted of excitation followed by inhibition in the PA direction. When object motion was in the opposite direction, the main response pattern inversed, that is, the unit responded with inhibition followed by excitation. Compared to the unit shown in A this unit barely responded after the object had passed the fish (after Mogdans and Bleckmann, 1998).

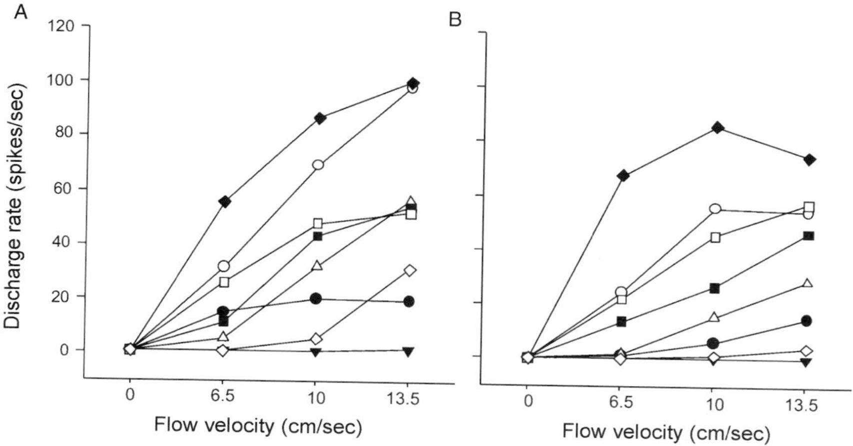

Fig. 10.9. (A and B) Responses of anterior lateral line nerves (ALLN) fibers to unidirectional water flow. Flow was either from anterior to posterior (A) or from posterior to anterior (B). Identical symbols refer to recordings made from the same primary afferent. Note that all flow-sensitive afferents responded with an increase in neural activity to unidirectional water flow, irrespective of flow direction (Chagnaud and Bleckmann, unpublished data).

In goldfish unidirectional water flow (10 cm/sec) also masks the responses of type I afferents to an object that passes the fish laterally. However, the degree of masking depends on object motion direction. When the object passes the fish from anterior to posterior, that is, when the object moves with the flow, responses of type I fibers are masked. When object motion direction is opposite to flow direction, the responses of type I fibers may or may not be masked (Kröther *et al.*, 2002). In general, the responses of type II afferents are again less affected by running water than the responses of type I afferents. Nevertheless, in the few type II afferents recorded, responses to anterior–posterior object motion were weaker in running water than in still water. In contrast the responses to posterior–anterior object motion were comparable in still and running water (Kröther *et al.*, 2002). Trout (*Onchorhynchus mykiss*) inhabit lakes and running water. In general the responses of trout afferents to a moving object were weaker than those observed in goldfish. According to Kröther *et al.* (2002), the effects of running water on lateral line responses to a moving object are largely due to hydrodynamic effects. In still water the changes in water velocity and pressure caused by a moving object are independent of object

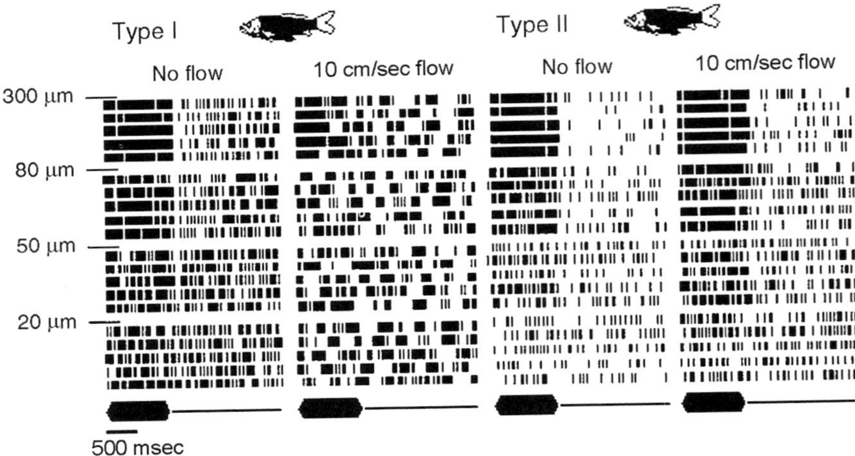

Fig. 10.10. Raster diagrams of the responses of type I and type II fibers in the goldfish PLLN to 10 repetitions of a 50-Hz sine wave stimulus generated by a vibrating sphere of 8-mm diameter. Distance between fish and sphere was 6 mm. Left: responses to the sphere without background water flow. Right: responses to the sphere in the presence of a 10 cm/sec background flow. Flow direction was from anterior to posterior. Type I fibers are putatively innervating SN. Numbers on the far left give peak-to-peak displacement amplitudes of sphere vibration. Note that background flow masked the response to the vibrating sphere. Type II fibers are putatively innervating CN. The responses of type II fibers to the vibrating sphere were not masked by background flow (after Engelmann *et al.*, 2000).

motion direction. In running water these changes are small if the object moves with the flow. When the object moves against the flow, only a transient change in water velocity is discernable from the increased background flow whereas the change in pressure is even greater than the pressure fluctuations measured in still water.

2.10. Central Pathways for Lateral Line Information Processing

Information of the peripheral lateral line enters the medulla through at least three pairs of lateral line nerves. These are the two anterior lateral line nerves (ALLN) that carry information of cephalic neuromasts, and the two posterior lateral line nerves (PLLN) that carry information from trunk neuromasts (Northcutt, 1989; Reiner and Northcutt, 1992). A pair of middle lateral line nerves innervates some SN and CN just behind the eye (Norris, 1925; Allis, 1889; Puzdrowski, 1989; Song, 1989; Bleckmann *et al.*, 1991b; Song and Northcutt, 1991a). Roots of the various lateral line nerves enter

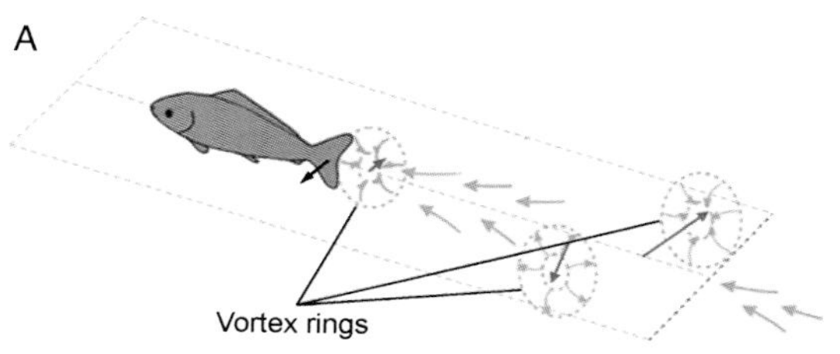

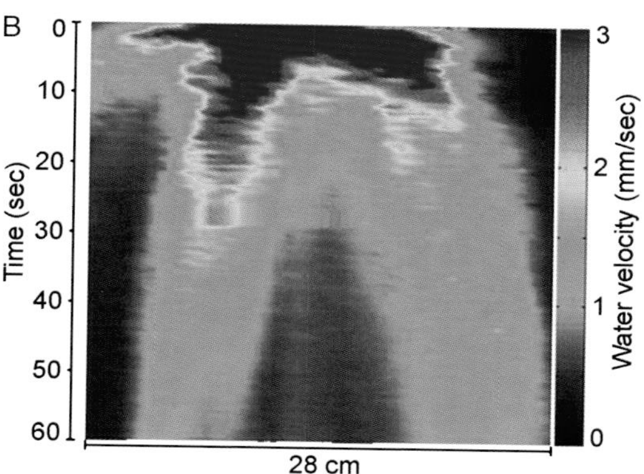

Chapter 10, Fig. 4.

the ipsilateral brainstem and bifurcate into ascending and descending branches which terminate in the medial octavolateralis nucleus (MON) (Claas and Münz, 1981; Finger and Tong, 1984; New and Bodznick, 1985; DeRosa and Fine, 1988; Bleckmann *et al.*, 1991b; Song and Northcutt, 1991b; Puzdrowski and Leonard, 1993; Wullimann, 1998). The terminal pattern of lateral line afferents implies that the activity in any single fiber reaches many parts of the MON. This appears to be ideal for the comparison and cross-correlation of the input that reaches the brain from different head and body areas since the temporal and spatial pattern of lateral line excitation can be used by a fish to locate prey (Bleckmann and Schwartz, 1982; Denton and Gray, 1983; Käse and Bleckmann, 1987; Coombs and Conley, 1997; Franosch *et al.*, 2003). It should be mentioned that some primary lateral line projections always reach the ipsilateral cerebellar granular eminence (Claas *et al.*, 1981; McCormick, 1989; Puzdrowski, 1989; Bleckmann *et al.*, 1991b; Song and Northcutt, 1991b; Wullimann *et al.*, 1991) and, in a few species, the corpus and valvula cerebelli (Claas and Münz, 1981; Wullimann *et al.*, 1991). The functional significance of these projections is not known but they may be involved in the establishment of a central representation of expected sensory input (see later).

The second-order projections from the MON ascend in the lateral longitudinal fascicle and terminate bilaterally, albeit with a stronger contralateral component, in the lateral portion of the torus semicircularis (Callens *et al.*, 1967; O'Bennar, 1976; Altman and Dawes, 1981; Lowe, 1986; Bleckmann *et al.*, 1987, 1989b; Weeg and Bass, 2000) and in the deep layers of the optic tectum (Boord and Montgomery, 1989; McCormick, 1989). The final ascending pathway for the mechanosensory lateral line involves the relay of information from midbrain to various diencephalic nuclei (Finger and Bullock, 1982; Bleckmann *et al.*, 1987; Wullimann, 1998; Weeg and Bass, 2000). Although the forebrain circuitry of the lateral line is not well understood, some telencephalic areas also receive lateral line information. Known areas include the medial pallium in batoid elasmobranches (Bleckmann *et al.*, 1987, 1989b) and the dorsal part of the telencephalon in teleosts (Echteler, 1985a; Striedter, 1991; Prechtl *et al.*, 1998; Wullimann, 1998) (Figure 10.11).

2.11. Descending Recurrent Pathways

The MON receives descending input from parts of the cerebellum (Montgomery *et al.*, 1995).The MON also receives descending input from the nucleus preeminentialis, which gets input from the torus semicircularis (Wullimann, 1998; Weeg and Bass, 2000). In some teleosts the MON receives additional input from the ipsilateral sensory trigeminal nucleus

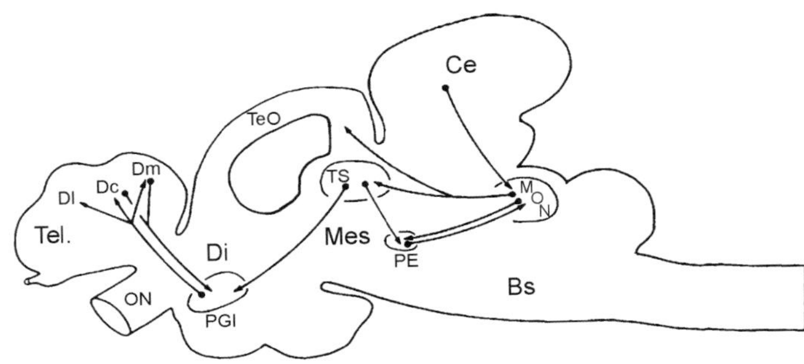

Fig. 10.11. Summary of the lateral line pathways in the brain of teleost fish (cyprinids). Bs, brain stem; Ce, cerebellum; Dl, Dc, and Dm, lateral, central, and medial dorsal telencephalic areas; Di, diencephalons; MON, medial octavolateralis nucleus; ON, optic nerve; PE, preeminential nucleus; PGl, lateral preglomerular nucleus; TeO, tectum opticum; Tel., telencephalon; TS, torus semicircularis. [Redrawn and slightly altered from Wullimann (1998).]

(McCormick and Hernandez, 1996). Discussed functions of descending recurrent inputs include gain control and the establishment of central representations of expected sensory input (Montgomery and Bodznick, 1994).

Lateral line neuromasts are innervated by efferent nerve fibers. The somata of these fibers are located in a single rhombencephalic nucleus, the octavolateralis efferent nucleus, which may exhibit rostral and caudal subdivisions (Puzdrowski, 1989; Song and Northcutt, 1991b; Schellart *et al.*, 1992; Wagner and Schwartz, 1992; Tomchik and Lu, 2005). In certain euteleosts, a minor efferent lateral line supply originates in the diencephalon (Roberts and Meredith, 1989). In fish, octavolateralis efferents respond to visual, somatosensory, vestibular, and lateral line stimuli (Roberts and Russell, 1972; Tricas and Highstein, 1990, 1991). The efferent system modulates the overall sensitivity of the mechanosensory lateral line (Roberts and Meredith, 1989; Tricas and Highstein, 1990, 1991; Tomchik and Lu, 2006).

3. CENTRAL PHYSIOLOGY

3.1. Ongoing Activity

In contrast to primary lateral line afferents, most central lateral line neurons show only a weak or even no ongoing activity. If ongoing activity occurs the discharge rates are usually much lower (<2 Hz) than in primary afferents (Alnaes, 1973; Bleckmann *et al.*, 1987; Schellart and Kroese, 1989; Müller and Bleckmann, 1993; Wojtenek *et al.*, 1998). In general, ongoing activity decreases along the lateral line pathway, that is, in midbrain and forebrain lateral line centers ongoing activity is even weaker than in the hindbrain MON (Bleckmann *et al.*, 1989b).

3.2. Latency

Electric lateral line nerve shocks and vibrating spheres placed close to the fish have been intensively used to stimulate central lateral line units. In the thornback guitarfish, *Platyrhinoidis triseriata*, the response latencies of central units to a lateral line nerve shock or to a vibrating sphere stimulus usually exceed 5 msec. In general response latencies are longer in the midbrain than in the MON. In the midbrain of *Platyrhinoidis* the longest response latencies (24–80 msec) occurred in the anterior nucleus. Diencephalic and telencephalic lateral line responses have onset latencies of more than 17 msec (Bleckmann *et al.*, 1987). In general, response latencies decrease with decreasing stimulus rise time (Bleckmann *et al.*, 1989b).

3.3. Dynamic Response Properties

One of the most striking differences between primary lateral line afferents and central lateral line neurons are their responses to repetitive stimuli. Primary lateral line afferents in general show no or only a weak response decrement to a repetitive stimulus regime (e.g., one stimulus train per second). In contrast, higher-order lateral line neurons display a pronounced response decrement if the lateral line nerve is stimulated with an electric shock or if the lateral line is stimulated with a vibrating sphere in quick succession. In the midbrain of *Platyrhinoidis* a complete recovery of lateral line responses may require an interstimulus interval of up to 5 sec (Bleckmann *et al.*, 1987, 1989b). Units and/or evoked potentials recorded from diencephalic and telencephalic lateral line areas may already show a response decrement at rates higher than one stimulus every 10 sec (Bleckmann *et al.*, 1987, 1989b), and in one diencephalic lateral line area (posterior lateral tuberal nucleus) of *Platyrhinoidis* a response decrement was already manifested if the interstimulus interval was less than 55 sec (Bleckmann *et al.*, 1987). These findings suggest that patterns of noise are filtered out at higher brain levels while a stimulus caused by a passing organism still may catch the fish's interest.

3.4. Thresholds

Another striking difference between primary lateral line afferents and central lateral line units is sensitivity. While nearly all primary afferents are extremely sensitive to sinusoidal water motions (see in an earlier section), many central lateral line units do not respond to such stimuli even if the sphere vibration amplitude is 1000 μm (Mogdans and Kröther, 2001; Plachta *et al.*, 2003). Lateral line units that are insensitive to a vibrating sphere stimulus may require a certain temporal and/or spatial pattern of water motions in order to respond (see also later). However, many central lateral line units do respond to the water motions caused by a stationary vibrating sphere. Minimal displacement thresholds of these units are usually in the frequency range 75–150 Hz. In this range the most sensitive medullary and midbrain lateral line units already respond if the peak-to-peak water vibration amplitude (stimulus frequency 100 Hz) is 0.01 μm (Bleckmann *et al.*, 1989b). In *Platyrhinoidis* the displacement thresholds of diencephalic and telenecephalic lateral line regions to a single-frequency vibrating sphere stimulus are two to three orders of magnitude higher (e.g., diencephalon, 0.2 μm; telencephalon, 1.9 μm) than the displacement thresholds of the most sensitive midbrain and hindbrain lateral line units (Bleckmann *et al.*, 1989b).

3.5. Phasic Versus Tonic Responses

Like primary lateral line afferents, some MON cells respond with a sustained discharge to long-lasting single-frequency wave stimuli (Mogdans and Goenechea, 1999; Engelmann and Bleckmann, 2004). These cells often exhibit a high degree of phase coupling whereas cells with responses unlike those of primary afferents often only show a weak or no phase coupling at all (Coombs *et al.*, 1998). If phase coupling occurs, it usually increases with increasing stimulus amplitude (Bleckmann *et al.*, 1989b). Most central lateral line units that phase couple respond only to one-half of a full cycle of a sinusoidal wave stimulus (Bleckmann *et al.*, 1989b; Bleckmann and Zelick, 1993) thus indicating that lateral line afferents with opposing directional sensitivity (e.g., headward or tailward) do not converge onto such interneurons. Most central lateral line units show only a phasic increase (or decrease) of neural activity to single-frequency wave stimuli, irrespective of stimulus duration (Bleckmann *et al.*, 1989b; Müller and Bleckmann, 1993; Plachta *et al.*, 1999; Engelmann and Bleckmann, 2004). However, in these units an additional tonic and/or off-response component may occur at high stimulus intensities. (Bleckmann *et al.*, 1989b; Müller and Bleckmann, 1993). In midbrain units an inhibition may change to tonic excitation if the vibrating sphere is moved to another rostrocaudal position (Engelmann and Bleckmann, 2004). The phasic responses of central lateral line units to single-frequency wave stimuli may indicate that stimulus frequency is less preserved in the temporal discharge pattern of lateral line neurons at higher brain levels. However, it may also be that many of these cells—like cortical acoustic neurons in monkeys (Wang *et al.*, 2005)—would discharge in a sustained manner if driven by their respective yet unknown preferred stimulus. For example, whether phasic central lateral line units respond more strongly to more complex stimuli than to single-frequency water motions has never been tested. It has also never been tested whether some central lateral line units need a certain spatial stimulus pattern to be driven optimally. Central lateral line units that respond with periods of intermittent excitation and inhibition or with a short burst of activity after stimulus offset have also been found. In other cells, responses take considerable time to reach a maximal level of discharge.

3.6. Frequency Encoding

Some units in the MON, midbrain, and forebrain of fish are highly sensitive to a vibrating sphere stimulus, often with minimal displacement thresholds in the frequency range 75–150 Hz (Bleckmann *et al.*, 1987; Bleckmann and Münz, 1990). One means by which frequency information can be encoded is by spatial segregation of neurons within a given brain area according to their preferred frequency. Such a tonotopy is commonly observed in the acoustic system of terrestrial vertebrates (Merzenich and Reid, 1974; Scheich *et al.*, 1979; Pettigrew *et al.*, 1981; Hattori and Suga, 1997) and may be present within the acoustic part of the torus semicircularis of teleosts (Echteler, 1985b). Primary lateral line afferents have a uniform frequency response, and fibers with continuously distributed characteristic frequencies do not exist. This already suggests that a tonotopic organization is unlikely in the central lateral line. In *Platyrhinoidis*, for instance, different functional lateral line subsystems based on frequency were not found (Bleckmann *et al.*, 1987). In general medullary, midbrain, and forebrain lateral line units are more likely to be wideband, that is, vibrating sphere stimuli cause neural responses in the frequency range from below 10 Hz up to about 200 Hz (Bleckmann and Bullock, 1989). Midbrain lateral line units of trout yield two peaks of increased sensitivity that coincide with the best frequencies (in terms of displacement) of SN and CN (Schellart and Kroese, 1989). If stimulated with isodisplacement stimuli (test range 33–200 Hz), toral lateral line units of goldfish are either low frequency (33 Hz), mid frequency (50 to 100 Hz), or high frequency (>200 Hz) (Plachta *et al.*, 1999). In the catfish *Ancistrus*, some toral lateral line units respond with a sustained firing rate to certain stimulus frequencies. If the stimulus frequency is raised by as little as 2 Hz only the phasic response component remains (Müller and Bleckmann, 1993). From the auditory system of both invertebrates and vertebrates the suppression of a response to an excitatory frequency by simultaneous presentation of an inhibitory frequency is well known (Evans, 1975). Such a suppression may increase the selectivity of a neuron to a behaviorally relevant frequency range (Boyd *et al.*, 1984). Although some frequency selectivity does occur in the central lateral line, all physiological studies indicate that the lateral line system of fishes is less concerned with the analysis of frequency. Nevertheless fish have the behavioral ability to discriminate water wave frequencies (Bleckmann *et al.*, 1981; Frühbeis, 1984). Whether this frequency discrimination is solely based on lateral line information or on additional auditory input is not known.

3.7. Dynamic Amplitude Range

In several fish species the dynamic amplitude range of central lateral line areas has been determined with vibrating sphere stimuli. Although in many cases the highest stimulus intensities used were not sufficient to saturate the neural responses, dynamic amplitude ranges of up to 90 dB have been found in higher-order lateral line units. In *Platyrhinoidis* some medullary units encode a stimulus amplitude of up to 150 μm, while other units already show saturation at a peak-to-peak displacement of 6 μm (Bleckmann *et al.*, 1989b). Thus in terms of the upper stimulus amplitude that can be encoded there is a range fractioning. In some central lateral line units saturation is followed by a high-intensity decrease of neural responses (Bleckmann *et al.*, 1989b; Schellart and Kroese, 1989).

Similar data have been obtained with moving object stimuli. Spike counts usually increase monotonically with object velocity. Although the slopes of the velocity curves may be fairly similar, there is again a range fractioning. Some units may not respond at object speeds <10 cm/sec while other units already reach saturation at an object speed of only 5 cm/sec (Wojtenek *et al.*, 1998). In many units spike rates increase with object size, in other units peak spike rates and number of spikes are not affected in a systematic fashion by object size (Wojtenek *et al.*, 1998). Since spike rates depend on object speed and object size, unit responses are ambiguous. In some units discharge rates are suppressed if an object passes the fish laterally (Wojtenek *et al.*, 1998).

3.8. Lateral Line Maps

Sculpin can position their snout near where a hydrodynamic stimulus occurred along their trunk (Hoekstra and Janssen, 1986). This already indicates that the central lateral line preserves some spatial information. Several anatomical and physiological studies have demonstrated central lateral line maps. In all bony fishes primary lateral line afferents distribute such in the MON that fibers from the ALLN are represented medially or ventromedially and fibers from the PLLN laterally or dorsolaterally (McCormick, 1989; Bleckmann *et al.*, 1991b; Song and Northcutt, 1991b). The projections from the anterior and posterior lines run parallel but do not mix. Moreover, the anterior lateral line is represented in the MON in which the projection of anterior neuromasts lies ventrolateral to the projections of more posterior neuromasts. The same applies to the projections of posterior lateral line neuromasts (Alexandre and Ghysen, 1999). Thus the primary projections of the lateral line of fishes is doubly somatotopic suggesting that at least the spatial information along the rostrocaudal axis is preserved in the CNS to some degree. There may also be a crude somatotopy with the dorsolateral and ventrolateral surfaces of the trunk represented ventrally and dorsally, respectively, in the MON of some fishes (Song and Northcutt, 1991b; Ivry and Baldo, 1992). There are no indications that horizontally and vertically oriented SN map separately (Song and Northcutt, 1991b; Ivry and Baldo, 1992). Whether the two hair cell populations in a neuromast (see in an earlier section) map differentially is also not known (Fritzsch, 1989). An anatomical study indicates that lateral line fibers innervating CN terminate in a single well-defined field in the MON (New *et al.*, 2000). If this finding can be extended to other species, this could mean that the CN system maintains a high spatial resolution throughout the ascending lateral line pathway and thus can be used for object localization. This idea is in line with the observation that flow insensitive toral lateral line neurons, which are believed to receive input from flow insensitive MON cells, encode the rostrocaudal location of a moving object (Plachta *et al.*, 2003; Engelmann and Bleckmann, 2004) and that these cells mediate orientation to prey (Coombs *et al.*, 2001). Anatomical studies point in the same direction, that is, there is a rough topography of projections from the MON to the nucleus ventrolateralis of the torus semicircularis (Weeg and Bass, 2000). Physiological studies (Zittlau *et al.*, 1986; Bartels *et al.*, 1990) have revealed computed lateral line maps in the tectum of the clawed frog *Xenopus laevis* and the Axolotl *Ambystoma mexicanum* that represent the direction of surface wave propagation. Whether computed lateral line maps do also exist in the tectum of surface-feeding fish is not known. Presumably more lateral line maps will be uncovered in the near future.

3.9. Receptive Field Organization

The excitatory RF of MON and toral lateral line units range from small single or double peaked to large and multipeaked, covering most or even the whole surface of the fish (Figure 10.7, middle and bottom) (Alnaes, 1973; Caird, 1978; Wubbels *et al.*, 1993; Coombs *et al.*, 1998; Mogdans and Goenechea, 1999; Plachta *et al.*, 1999; Engelmann and Bleckmann, 2004;

Kröther *et al.*, 2004). This indicates that some central cells receive input from a restricted portion of the lateral line periphery, whereas other cells receive input from neuromasts that are widely distributed across the head and trunk of the fish. Given the small RF of primary afferents (Figure 10.7, top), the high convergence of lateral line input on many central cells is surprising. However, as in other modalities, the central units with a small RF again indicate that lateral line subsystems exist at least up to the level of the midbrain that preserve high spatial resolution along the neuraxis. The excitatory RF of some MON and toral lateral line cells include regions in which dipole stimulation leads to inhibition of ongoing activity (Engelmann and Bleckmann, 2004). Whether dipole stimulation can also inhibit evoked activity is not known. Anatomical (New *et al.*, 1996) and modeling data (Coombs *et al.*, 1996; Montgomery and Coombs, 1998) suggest that primarylike RFs in central lateral line neurons are sharpened by neural mechanisms based on lateral inhibition. However, physiological evidence that confirms this is missing.

Lateral line units in the torus semicircularis of teleost fish receive input from a large but restricted portion of the contralateral body surface. The rostrocaudal position of the recording site corresponds to the rostrocaudal position of the RF (Knudsen, 1977; Bleckmann and Zelick, 1993; Plachta *et al.*, 2003; Engelmann and Bleckmann, 2004). A similar somatotopic organization exists in the mesencephalic nuclear complex of *Platyrhinoidis* (Bleckmann *et al.*, 1989b). In the corpus and valvula cerebelli of *Carassius*, units that respond to ipsilateral or contralateral stimulation of the lateral line in the tail or body region are found in a more posterior part than those that respond to stimulation of the head region. RFs are usually large, diffuse, and overlapped with one another, and no pattern of specific topographic projection exists (Kotchabhakdi, 1976). The lateral line units recorded in the eminentia granularis of the catfish *Ictalurus* primarily respond to ipsilateral stimulation (Lee and Bullock, 1984). Again RFs are large and difficult to delimit. They overlap with one another and do not fall into any recognizable pattern. Units in the valvula, which may (or may not) be multimodal, have their RFs confined to the head region (Lee and Bullock, 1984). Mapping of lateral line (mechanosensory) cerebellar RFs in *Platyrhinoidis* again reveals a complex somatotopy. The RFs are large and include ipsilateral and contralateral body areas, such as the tail and anal fins, are mostly represented in a relatively small region in the caudal tip of the posterior lobe of the cerebellar corpus; more rostral body areas (pectoral fins, trunk, and head area) are represented in the rest of the posterior lobe of the cerebellar corpus and the caudal part of the posterior lobe of the cerebellar corpus (Fiebig, 1988). Almost no information is available on diencephalic lateral line RFs. Bleckmann *et al.* (1987) recorded units in the lateral tuberal nucleus of *Platyrhinoidis*. In one case in which the RF was mapped it was large and complex, restricted to the anterior half of the body, and followed the course of the infraorbital and body lateral line canals. Telencephalic RFs, determined only with evoked potentials following a vibrating sphere stimulus, are again large and may differ with respect to the number, polarity, and latencies of the peaks (Bleckmann *et al.*, 1989b).

3.10. Moving Object Stimuli

In goldfish up to 30% of all central (MON and torus semicircularis) lateral line units do not respond to a stationary vibrating sphere, even if sphere vibration amplitudes of 800 μm are applied (Wojtenek *et al.*, 1998). This amplitude is substantially higher than the amplitudes that cause rate saturation in primary lateral line afferents (Mohr and Bleckmann, 1998). Nevertheless many MON and toral cells insensitive to vibrating sphere stimuli readily respond with excitation or inhibition if a sphere or another small object passes the fish laterally (Bleckmann and Zelick, 1993; Müller and Bleckmann, 1993; Müller *et al.*, 1996; Mogdans *et al.*, 1997; Wojtenek *et al.*, 1998; Plachta *et al.*, 1999). Thus, up to the level of the midbrain, there are at least two functionally separated lateral line pathways, one of which processes the well-defined and spatially restricted water motions caused by a stationary vibrating sphere, and the other processes the gross water motions caused by an object that passes the fish laterally. Although the responses of MON and toral cells to moving object stimuli are highly variable, two consistent response types can be distinguished. Units may show a response while the object passes the fish and to the wake of the passing object. Units with these response properties most likely receive input from SN. Other central units also respond to a passing object with excitation, however, these units do not respond to the wake caused by the object. Units of this type probably receive input from CN. Unlike primary afferents, many central lateral line units are neither in terms of spike rate nor in terms of their temporal response pattern directionally sensitive (Figure 10.12A). This indicates that the two populations of hair cells in a neuromast finally converge onto these units. In terms of peak spike rate or total number of spikes, many central lateral line neurons are directionally sensitive. In these neurons object motion in one direction causes a decrease in ongoing discharge rate while object motion in the opposite direction causes excitation (Figure 10.12B) (Wojtenek *et al.*, 1998). Although these neurons are rare, some medullary neurons of both *Carassius* (Mogdans *et al.*, 1997) and the catfish *Ancistrus* (Müller *et al.*, 1996) respond only if the object moves from either anterior to posterior or posterior to anterior. Neurons of this type are more common in the midbrain than in the hindbrain (Figure 10.12C). Moreover, most directionally sensitive midbrain

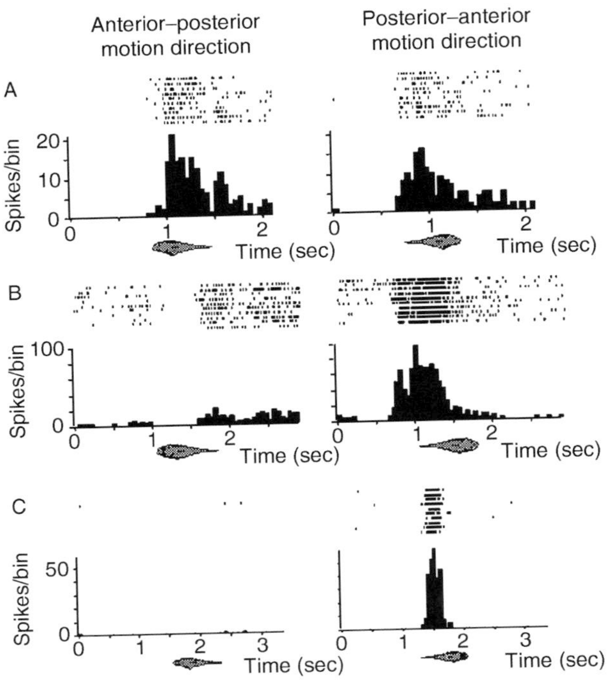

Fig. 10.12. (A–C) Effects of the direction of object motion on the temporal discharge patterns of lateral line units in the midbrain of goldfish. (A) Example of a unit in which the temporal pattern of the response barely changed when the direction of object motion was reversed. (B) Data from a unit for which the temporal response pattern clearly inversed when object motion direction was reversed. Note that the responses to PA object motion direction is characterized by a broad excitatory peak. In contrast, the response to AP object motion direction shows a broad zone of suppressed neural activity that is followed by a period of increased neural activity. (C) Data from a unit that responded with a sharp peak of excitation to PA object motion but did not respond to AP object motion. Data in A–C were recorded from units in different animals but in response to almost identical stimuli, that is, object speed was between 15 and 20 cm/sec and minimal object distance was 1 cm (after Wojtenek *et al.*, 1998).

lateral line units of *Carassius*, like those of the catfish *Ancistrus* (Müller *et al.*, 1996), prefer object movements from caudal to rostral.

3.11. Running Water Conditions

If MON units are stimulated with a stationary vibrating sphere in the presence of background flow, at least four types of units can be distinguished. Type MI units respond to running water with either an increase or a decrease in neural activity. If stimulated with a vibrating sphere, the response rates and/or the degree of phase coupling decrease under running water conditions. Type MI units most likely receive excitatory or inhibitory input from type I afferents, that is, from SN. Ongoing discharge rates of type MII units are hardly altered by running water. Moreover, these units do not change their responses to dipole stimuli in the presence of water flow. Type MII units most likely receive input from CN. Like type MII units, the ongoing discharge rates of type MIII units are not altered in running water. However, the responses of these units to dipole stimuli presented in background flow are significantly masked (Figure 10.13) (Kröther *et al.*, 2004). Most likely excitatory input to type MIII units, mediated via type II primary afferents, is inhibited by input from type I afferents. Like type MI units, type MIV units are flow sensitive. However, the responses of these units to a vibrating sphere

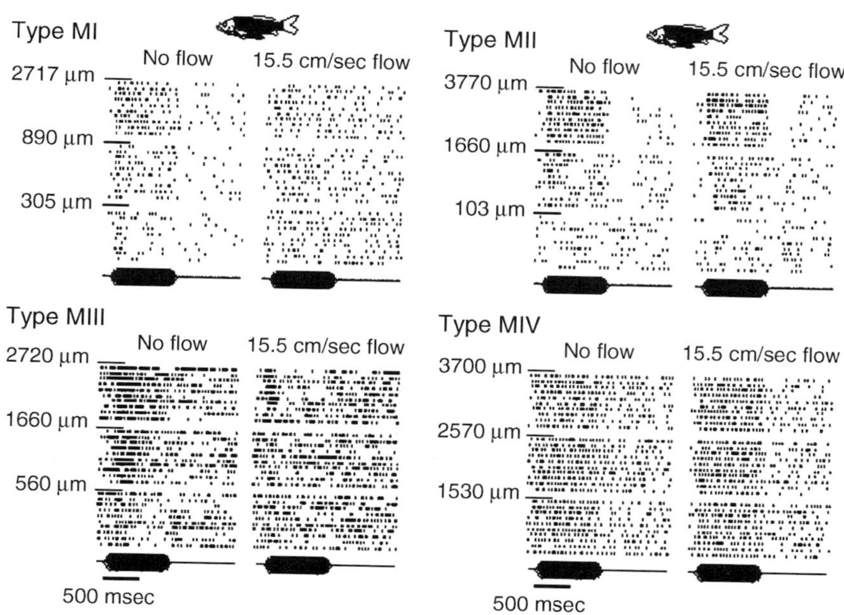

Fig. 10.13. Responses of goldfish type MI, MII, MIII, and MIV cells in the medial octavolateralis nucleus to a 50-Hz sine wave stimulus generated by a vibrating sphere of 8-mm diameter. Raster diagrams of the responses to 10 stimulus repetitions are shown for three peak-to-peak displacement amplitudes. Data were recorded in still water (no flow) and in running water (velocity 15.5 cm/sec). Flow direction was from anterior to posterior. The background flow masked the responses of type MI and MIII cells, but not the responses of type MII and MIV cells (after Kröther *et al.*, 2002).

stimulus are not affected by running water. This suggests that type MIV units also receive input from both SN and CN (Kröther *et al.*, 2004). In goldfish, toral lateral line units are again either flow sensitive or flow insensitive (flow speed 10 cm/sec), that is, toral units receive either input from flow-sensitive or from flow-insensitive MON units. Ninety percent of all flow-insensitive toral lateral line units respond with a short burst if an object passes the fish, that is, these units probably receive input from the canal system. Like MON units, flow-sensitive toral units show sustained increases or decreases in ongoing discharge rate if exposed to water flow. In still water, units of both types may respond to a vibrating sphere and/or to a sphere passing the fish laterally. Units that respond to a vibrating sphere are localized more ventrally than units that respond only to a moving sphere (Engelmann and Bleckmann, 2004). In running water, responses of flow-sensitive units to the vibrating sphere are masked (Engelmann and Bleckmann, 2004). The responses of 12 out of 14 flow-insensitive units were also masked by laminar water flow (Engelmann and Bleckmann, 2004). Most likely units of this type receive excitatory input from type MIII units. Thus like in the MON a complete separation of SN and CN input does not exist at the level of the torus semicircularis.

The effects of running water on the responses of MON and toral lateral line units to a moving object are diverse and depend on object motion direction. In running water, the responses of nearly all flow sensitive units that respond to a moving object in still water are masked when the object passes the fish laterally. Some MON flow-sensitive units respond to a moving object under running water conditions even though they do not respond to the moving object in still water. Most of the flow-sensitive central units studied so far did not respond to anterior–posterior object motion direction in running water. However, all of these units responded to posterior–anterior object motion direction in both still and running water. Most likely this can be explained, like in primary afferents (see in an earlier section), by peripheral hydrodynamic effects (Engelmann and Bleckmann, 2004; Kröther *et al.*, 2004).

3.12. Multimodality

Most physiological studies of fish sensory biology have been restricted to a single modality. However, fish may use hydrodynamic, visual, acoustic, tactile, electrical, and chemical cues to detect and identify a stimulus source and to separate this source from other stimulus (noise) sources (Kalmijn, 1988b; von der Emde and Bleckmann, 1992; Hara, 1993; Hawkins, 1993; Braun *et al.*, 2002). To ease this task, inputs from different sensory modalities should converge in higher brain centers.

At least up to the level of the midbrain the lateral line system of fishes appears to maintain a largely separated central pathway. With possible exceptions (Andrianov and Ilyinsky, 1973) second-order lateral line units of the medulla are unimodal (Bleckmann and Bullock, 1989). However, midbrain and especially forebrain lateral line units may be bi- or trimodal, that is, higher-order lateral line units may also respond to either visual, somatosensory, acoustic, vibratory, ampullary, or tuberous input (Bleckmann and Bullock, 1989; Bleckmann and Zelick, 1993; Kirsch *et al.*, 2002). Multisensory integration can be reached by at least two mechanisms: algebraic interaction such as dendritic summation, and multiplicative interaction involving facilitation. In *Platyrhinoidis* some diencephalic lateral line units respond only when the animal receives both lateral line and corresponding visual information (Bleckmann *et al.*, 1987). Lateral line evoked discharges may be suppressed by ampullary (Bleckmann *et al.*, 1989b) or visual (Tricas and Highstein, 1990) input. In other cases ampullary input facilitates lateral line responses (Müller and Bleckmann, 1993).

4. CONCLUSIONS

Lateral line stimuli and background noise. One problem in lateral line research is our lack of knowledge about the temporal and spatial characteristics of natural hydrodynamic stimuli, including self-generated and external background noise. One of the reasons for this lack of knowledge is the difficulty to measure and quantify the temporal and spatial characteristics of natural water motions. Since many fish species live in strikingly different habitats (e.g., lakes, ponds, or fast running rivers) they are probably exposed to very different hydrodynamic conditions. This may be one of the reasons why the peripheral lateral line system shows such a high morphological diversity (Coombs *et al.*, 1988; Webb, 1989a). Despite this convincing hypothesis recordings from primary afferents while stimulating the fish with a stationary vibrating sphere repeatedly failed to demonstrate clear form-function relationships (Bleckmann and Münz, 1990; Coombs and Montgomery, 1992; Montgomery *et al.*, 1994). This has raised the question whether such relationships do exist at all (Northcutt, 1988; Webb, 1989a). However, researchers failed to demonstrate a clear form-function relationships may not be too surprising if we take into account that most, if not all, sensory systems are adapted to both natural stimuli and natural background noise. The same probably holds true for the mechanosensory lateral line. Therefore we may uncover form-function relationships only if we test the various types of lateral lines under the hydrodynamic conditions they are designed for. To meet these conditions we need more studies about the spatial and temporal distributions of fish in relation to their natural hydrodynamic environment.

Behavior. We still have no idea about the full behavioral capabilities of the fish lateral line. Although we know that many fish use lateral line input to detect, discriminate, and localize stationary or moving sources (see in an earlier section), nearly all studies have been done in still water and with still water fish. We have no idea how the lateral lines of still water and running water fish perform under laminar and turbulent flow conditions. We do know that fish can detect vortex streets but we do not know the kind of information fish can extract from the wakes caused by animate or inanimate sources. Besides asking more and more refined questions about the function of the lateral line of standard still water fish, we should also investigate fish that are highly specialized with respect to this sensory modality, for example, fish with multiple lateral lines like *Xiphister atropurpureus*, or fish that have a close peripheral contact between the lateralis and the otic system, like some catfish (Bleckmann *et al*., 1991b) and mormyrids (Stipetic, 1939).

Central physiology. Although researchers have traced the central lateral line pathway fairly well (McCormick and Braford, 1988), we still have little knowledge about the exact circuit diagrams and the functions of the various lateral line nuclei. For instance, on entering the brain individual primary lateral line nerve fibers bifurcate into an ascending and a descending branch which terminate throughout the entire MON (Claas and Münz, 1981; Bleckmann *et al*., 1991b; Alexandre and Ghysen, 1999). The reason for this massive arborization is still not known. The second-order cells in the MON also receive direct and indirect input from parallel fibers that have their origin in the cerebellum and metencephalon (New *et al*., 1996). One function of this circuitry may be the cancellation of sensory stimulation related to the fish's own body movements (Montgomery and Bodznick, 1994) but to make this a firm conclusion we need more data. In terms of lateral line processing we still do not know the function of the torus semicircularis, the cerebellum, the tectum, and the forebrain. From amphibian studies we know that tectal lateral line units are involved in object localization (Claas *et al*., 1989; Bartels *et al*., 1990). There are some indications that forebrain lateral line centers of fish are involved in the building of spatial cognitive maps (Teyke, 1989; Rodriguez *et al*., 1994; Salas *et al*., 1996a,b). To learn more about the functions of the various lateral line areas we should change to more natural stimuli, and we should apply these stimuli under quasinatural noise conditions. The study done by Engelmann *et al*. (2002a) shows nicely that background flow (noise) in combination with local dipole stimuli is a powerful tool to uncover lateral line functions.

Multimodal integration. In general very little is known about multimodal information processing in the brain of fishes. We clearly need more experiments in which we determine the number of stimulus modalities to which higher-order lateral line neurons respond. Unfortunately there are serious difficulties with this type of experiment since higher-order neurons of fish often require large (>30 sec) stimulus repetition rates (Bleckmann *et al.*, 1987; Kirsch *et al.*, 2002); or even worse, higher-order lateral line neurons may not respond at all to unimodal simple (biologically not meaningful) stimuli. Researchers studying the response properties of central lateral line neurons face another problem because both lateral line and pressure- and/or motion-sensitive inner ear receptors may be stimulated by mechanically disturbances of the surrounding water (Kalmijn, 1988b). In addition spinal and trigeminal receptors in the skin need to be assessed for their relative contributions, at least with stronger stimuli (for a thorough discussion of multimodal information processing in the fish brain see Braun *et al.*, 2002). It should be mentioned that in some fish air-filled cavities are directly linked to the lateral line (Denton *et al.*, 1979; Webb and Smith, 2000). While all this makes lateral line research even more difficult it opens the possibility to study the multimodal integration of sensory information processing.

ACKNOWLEDGMENTS

I thank J. Mogdans and two anonymous reviewers for their helpful comments on an earlier draft of this chapter. The original research of the author was generously supported by the DFG, the BMBF, DARPA, BfG, DAAD, and the EU.

REFERENCES

Alexandre, D., and Ghysen, A. (1999). Somatotopy of the lateral line projection in larval zebrafish. *Proc. Natl. Acad. Sci. USA* **13,** 7758–7762.

Allis, E. P. (1889). The anatomy and development of the lateral line system in *Amia calva*. *J. Morphol.* **2,** 463–540.

Alnaes, E. (1973). Unit activity of ganglionic and medullary second order neurones in the eel lateral line system. *Acta Physiol. Scand.* **88,** 160–174.

Altman, J. S., and Dawes, E. A. (1981). Mapping of lateral line and auditory input to the brain of *Xenopus laevis*. *J. Physiol.* **317,** 78–79.

Andrianov, G. N., and Ilyinsky, O. B. (1973). Some functional properties of central neurons connected with the lateral-line organs of the catfish (*Ictalurus nebulosus*). *J. Comp. Physiol. A* **86,** 65–376.

Bartels, M., Münz, H., and Claas, B. (1990). Representation of lateral line and electrosensory systems in the midbrain of the axolotl, *Ambystoma mexicanum*. *J. Comp. Physiol. A* **167,** 47–356.

Beal, D. N., Hover, F. S., Triantafyllou, M. S., Liao, J. C., and Lauder, G. V. (2006). Passive propulsion in vortex wakes. *J. Fluid. Mech.* **549,** 385–402.

Blaxter, J. H. S., Gray, J. A. B., and Best, A. B. C. (1983). Structure and development of the free neuromasts and the lateral line system of the herring. *J. Mar. Biol. Ass. UK* **63,** 247–260.

Bleckmann, H. (1980). Reaction time and stimulus frequency in prey localization in the surface-feeding fish *Aplocheilus lineatus*. *J. Comp. Physiol. A* **140,** 163–172.

Bleckmann, H. (1985). Discrimination between prey and non-prey wave signals in the fishing spider *Dolomedes triton* (Pisauridae). *In* "Acoustic and Vibrational Communication in Insects" (Kalmring, K., and Elsner, N., Eds.), pp. 215–222. Paul Parey, Berlin.

Bleckmann, H. (1993). Role of the lateral line and fish behavior. *In* "Behaviour of Teleost Fishes" (Pitcher, T. J., Ed.), 2nd edn., pp. 201–246. Chapman and Hall, London, New York, Tokyo.

Bleckmann, H. (1994). Reception of hydrodynamic stimuli in aquatic and semiaquatic animals. *In* "Progress in Zoology" (Rathmayer, W., Ed.), Vol. 41, pp. 1–115. Gustav Fischer, Stuttgart, Jena, New York.

Bleckmann, H. (2004). 3-D-orientation with the octavolateralis system. *J. Physiol. Paris* **98,** 53–63.

Bleckmann, H., and Bullock, T. H. (1989). Central nervous physiology of the lateral line system, with special reference to cartilaginous fishes. *In* "The Mechanosensory Lateral Line. Neurobiology and Evolution" (Coombs, S., Görner, P., and Münz, H., Eds.), pp. 387–408. Springer, New York.

Bleckmann, H., and Münz, H. (1990). Physiology of lateral-line mechanoreceptors in a teleost with highly branched, multiple lateral lines. *Brain Behav. Evol.* **35,** 240–250.

Bleckmann, H., and Schwartz, E. (1982). The functional significance of frequency modulation within a wave train for prey localization in the surface-feeding fish *Aplocheilus lineatus*. *J. Comp. Physiol. A* **145,** 331–339.

Bleckmann, H., and Topp, G. (1981). Surface wave sensitivity of the lateral line organs of the topminnow *Aplocheilus lineatus*. *Naturwissenschaften* **68,** 624–625.

Bleckmann, H., and Zelick, R. (1993). The responses of peripheral and central mechanosensory lateral line units of weakly electric fish to moving objects. *J. Comp. Physiol. A* **172,** 115–128.

Bleckmann, H., Waldner, I., and Schwartz, E. (1981). Frequency discrimination of the surface-feeding fish *Aplocheilus lineatus*—a prerequisite for prey localization? *J. Comp. Physiol. A* **143,** 485–490.

Bleckmann, H., Bullock, T. H., and Jorgensen, J. (1987). The lateral line mechanoreceptive mesencephalic, diencephalic, and telencephalic regions in the thornback ray, *Platyrhinoidis triseriata* (Elasmobranchii). *J. Comp. Physiol. A* **161,** 67–84.

Bleckmann, H., Tittel, G., and Blübaum-Gronau, E. (1989a). The lateral line system of surface-feeding fish: Anatomy, physiology, and Behavior. *In* "The Mechanosensory Lateral Line. Neurobiology and Evolution" (Coombs, S., Görner, P., and Münz, H., Eds.), pp. 501–526. Springer, New York.

Bleckmann, H., Weiss, O., and Bullock, T. H. (1989b). Physiology of lateral line mechanoreceptive regions in the elasmobranch brain. *J. Comp. Physiol. A* **164,** 459–474.

Bleckmann, H., Breithaupt, T., Blickhan, R., and Tautz, J. (1991a). The time course and frequency content of hydrodynamic events caused by moving fish, frogs, and crustaceans. *J. Comp. Physiol. A* **168,** 749–757.

Bleckmann, H., Niemann, U., and Fritzsch, B. (1991b). Peripheral and central aspects of the acoustic and lateral line system of a bottom dwelling catfish, *Ancistrus* spec. *J. Comp. Neurol.* **314,** 452–466.

Bleckmann, H., Mogdans, J., and Dehnhardt, G. (2001). Lateral line research: The importance of using natural stimuli in studies of sensory systems. *In* "Ecology of Sensing" (Barth, F. G., and Schmid, A., Eds.), pp. 149–167. Springer-Verlag, Berlin/Heidelberg, New York.

Blickhan, R., Krick, C., Breithaupt, T., Zehren, D., and Nachtigall, W. (1992). Generation of a vortex-chain in the wake of a subundulatory swimmer. *Naturwissenschaften* **79,** 220–221.

Boord, R. L., and Montgomery, J. C. (1989). Central mechanosensory lateral line centers and pathways among the elasmobranchs. *In* "The Mechanosensory Lateral Line. Neurobiology and Evolution" (Coombs, S., Görner, P., and Münz, H., Eds.), pp. 323–340. Springer, New York.

Boyd, P., Kühne, R., Silver, S., and Lewis, B. (1984). Two-tone suppression and song coding by ascending neurones in the cricket *Gryllus campestris* L. *J. Comp. Physiol. A* **154,** 423–430.

Braun, C. B., Coombs, S., and Fay, R. R. (2002). What is the nature of multisensory interaction between octavolateralis sub-systems? *Brain Behav. Evol.* **59,** 162–176.

Burt de Perera, T. (2004). Spatial parameters encoded in the spatial map of the blind Mexican cave fish, *Astyanax fasciatus. Anim. Behav.* **68,** 291–295.

Caird, D. M. (1978). A simple cerebellar system: The lateral line lobe of the goldfish. *J. Comp. Physiol. A* **127,** 61–74.

Callens, M., Vandenbussche, E., and Greenway, P. H. (1967). Convergence of retinal and lateral line stimulation on tectum opticum and cerebellar neurons. *Arch. Int. Physiol. Biochem.* **75,** 148–150.

Campenhausen, C. V., Riess, I., and Weissert, R. (1981). Detection of stationary objects in the blind cave fish *Anoptichthys jordani* (Characidae). *J. Comp. Physiol. A* **143,** 369–374.

Claas, B., and Münz, H. (1981). Projection of lateral line afferents in a teleost brain. *Neurosci. Lett.* **23,** 287–290.

Claas, B., Fritzsch, B., and Münz, H. (1981). Common efferents to lateral line and labyrinthine hair cells in aquatic vertebrates. *Neurosci. Lett.* **27,** 231–235.

Claas, B., Münz, H., and Zittlau, K. E. (1989). Direction coding in central parts of the lateral line system. *In* "The Mechanosensory Lateral Line. Neurobiology and Evolution" (Coombs, S., Görner, P., and Münz, H., Eds.), pp. 409–419. Springer, New York.

Coombs, S. (1994). Nearfield detection of dipole sources by the goldfish (*Carassius auratus*) and the mottled sculpin (*Cottus bairdi*). *J. Exp. Biol.* **190,** 109–129.

Coombs, S., and Conley, R. A. (1995). Source distance determination by the mottled sculpin lateral line. *In* "Nervous Systems and Behaviour" (Burrows, M., Mattheson, T., Newland, P. L., and Schuppe, H., Eds.), p. 349. Thieme Verlag, Stuttgart.

Coombs, S., and Conley, R. A. (1997). Dipole source localization by mottled sculpin II. The role of lateral line excitation patterns. *J. Comp. Physiol. A* **180,** 401–416.

Coombs, S., and Janssen, J. (1989). Water flow detection by the mechanosensory lateral line. *In* "Comparative Perception" (Stebbins, W. C., and Berkley, M., Eds.), pp. 89–123. John Wiley, New York.

Coombs, S., and Janssen, J. (1990). Behavioral and neurophysiological assessment of lateral line sensitivity in the mottled sculpin, *Cottus bairdi. J. Comp. Physiol. A* **167,** 557–567.

Coombs, S., and Montgomery, J. C. (1992). Fibers innervating different parts of the lateral line system of an Antarctic Notothenioid, *Trematomus bernachii*, have similar frequency responses despite large variation in the peripheral morphology. *Brain Behav. Evol.* **40,** 217–233.

Coombs, S., and Montgomery, J. C. (1999). The enigmatic lateral line system. *In* "Comparative Hearing: Fish and Amphibians" (Fay, R. R., and Popper, A. N., Eds.), Vol. 11, pp. 319–362. Springer, New York.

Coombs, S., Janssen, J., and Webb, J. F. (1988). Diversity of lateral line systems: Evolutionary and functional considerations. *In* "Sensory Biology of Aquatic Animals" (Atema, J., Fay, R. R., Popper, A. N., and Tavolga, W. N., Eds.), pp. 553–593. Springer, New York.

Coombs, S., Hastings, M., and Finneran, J. (1996). Modeling and measuring lateral line excitation patterns to changing dipole source locations. *J. Comp. Physiol. A* **178,** 359–371.

Coombs, S., Mogdans, J., Halstead, M., and Montgomery, J. (1998). Transformation of peripheral inputs by the first-order lateral line brainstem nucleus. *J. Comp. Physiol. A* **182,** 609–626.

Coombs, S., Braun, C. B., and Donovan, B. (2001). The orienting response of Lake Michigan mottled sculpin is mediated by canal neuromasts. *J. Exp. Biol.* **204,** 33–348.

Dehnhardt, G., Mauck, B., Hanke, W., and Bleckmann, H. (2001). Hydrodynamic trail-following in harbor seals (*Phoca vitulina*). *Science* **293,** 102–104.

Denton, E. J., and Gray, J. A. B. (1983). Mechanical factors in the excitation of clupeid lateral lines. *Proc. Royal. Soc. Lond. B* **218,** 1–26.

Denton, E. J., and Gray, J. A. B. (1988). Mechanical factors in the excitation of the lateral lines of fishes. *In* "Sensory Biology of Aquatic Animals" (Atema, J., Fay, R. R., Popper, A. N., and Tavolga, W. N., Eds.), pp. 595–617. Springer, New York.

Denton, E. J., Gray, J. A.B, and Blaxter, J. H. S. (1979). The mechanics of the clupeid acoustico-lateralis system: Frequency responses. *J. Mar. Biol. Ass. UK* **59,** 27–47.

DeRosa, F., and Fine, M. L. (1988). Primary connections of the anterior and posterior lateral line nerves in the oyster toadfish. *Brain Behav. Evol.* **31,** 312–317.

Echteler, S. M. (1985a). Organization of central auditory pathways in a teleost fish, *Cyprinus carpio*. *J. Comp. Physiol. A* **156,** 267–280.

Echteler, S. M. (1985b). Tonotopic organization in the midbrain of a teleost fish. *Brain Res.* **338,** 387–391.

Elepfandt, A., and Wiedemer, L. (1987). Lateral-line responses to water surface waves in the clawed frog, *Xenopus laevis*. *J. Comp. Physiol. A* **160,** 667–682.

Engelmann, J., and Bleckmann, H. (2004). Coding of lateral line stimuli in the goldfish midbrain in still- and running water. *Zoology* **107,** 135–151.

Engelmann, J., Hanke, W., Mogdans, J., and Bleckmann, H. (2000). Hydrodynamic stimuli and the fish lateral line. *Nature* **408,** 51–52.

Engelmann, J., Hanke, W., and Bleckmann, H. (2002a). Lateral line reception in still- and running water. *J. Comp. Physiol. A* **188,** 513–526.

Engelmann, J., Kröther, S., Mogdans, J., and Bleckmann, H. (2002b). Responses of primary and secondary lateral line units to dipole stimuli applied under still and running water conditions. *Bioacoustics* **12,** 58–160.

Enger, P. S., Kalmijn, A. J., and Sand, O. (1989). Behavioral investigations on the functions of the lateral line and inner ear in predation. *In* "The Mechanosensory Lateral Line. Neurobiology and Evolution" (Coombs, S., Görner, P., and Münz, H., Eds.), pp. 575–587. Springer, New York.

Evans, E. F. (1975). Cochlear nerve and cochlear nucleus. *In* "Handbook of Sensory Physiology" (Keidel, W. D., and Neff, W. D., Eds.), Vol. 5/2, pp. 1–108. Springer, Berlin/Heidelberg, New York.

Fiebig, E. (1988). Connections of the corpus cerebelli in the thornback guitarfish, *Platyrhinoidis triseriata* (Elasmobranchii): A study with WGA-HRP and extracellular granule cell recording. *J. Comp. Neurol.* **268,** 567–583.

Finger, T. E., and Bullock, T. H. (1982). Thalamic center for the lateral line system in the catfish *Ictalurus nebulosus*: Evoked potential evidence. *J. Neurobiol.* **13,** 39–47.

Finger, T. E., and Tong, S. L. (1984). Central organization of eighth nerve and mechanosensory lateral line systems in the brainstem of Ictalurid catfish. *J. Comp. Neurol.* **229,** 129–151.

Flock, A. (1965). Electronmicroscopic and electrophysiological studies on the lateral line canal *organ. Acta Otolaryngol.* **199,** 1–90.

Flock, A. (1971a). Sensory transduction in hair cells. I. Principles of receptor physiology. *In* "Handbook of Sensory Physiology" (Loewenstein, W. R., Ed.), pp. 396–441. Springer, New York.

Flock, A. (1971b). The lateral line organ mechanoreceptors. *In* "Fish Physiology" (Hoar, W. S., and Randall, D. J., Eds.), Vol. 5, pp. 241–263. Academic Press, New York.

Flock, A., and Wersäll, J. (1962). A study of the orientation of sensory hairs of the receptor cells in the lateral line organ of a fish with special reference to the function of the receptors. *J. Cell Biol.* **15,** 19–27.

Franosch, J. M. P., Sobotka, M. C., Elepfandt, A., and van Hemmen, J. L. (2003). Minimal model of prey localization through the lateral-line system. *Phys. Rev. Lett.* **91,** 158101-1–158101-4.

Fritzsch, B. (1989). Diversity and regression in the amphibian lateral line and electrosensory system. *In* "The Mechanosensory Lateral Line. Neurobiology and Evolution" (Coombs, S., Görner, P., and Münz, H., Eds.), pp. 99–114. Springer, New York.

Frühbeis, B. (1984). Verhaltensphysiologische Untersuchungen zur Frequenzunterscheidung und Empfindlichkeit durch das Seitenlinienorgan des blinden Höhlenfisches *Anoptichthys jordani.* Dissertation, Universität Mainz, pp. 1–73.

Görner, P. (1962). Beitrag zum Bau und zur Arbeitsweise des Seitenorgans von *Xenopus laevis. In* "Verhandlungen der Deutschen Zoologischen Gesellschaft" (Herre, W., and Prell, H., Eds.), pp. 193–198. Gustav Fischer, Stuttgart.

Görner, P. (1963). Untersuchungen zur Morphologie und Elektrophysiologie des Seitenlinienorgans vom Krallenfrosch (*Xenopus laevis* Daudin). *J. Comp. Physiol. A* **47,** 316–338.

Görner, P., and Mohr, C. (1989). Stimulus localization in *Xenopus*: Role of directional sensitivity of lateral line stitches. *In* "The Mechanosensory Lateral Line. Neurobiology and Evolution" (Coombs, S., Görner, P., and Münz, H., Eds.), pp. 543–560. Springer, New York.

Hanke, W., and Bleckmann, H. (2004). The hydrodynamic trails of *Lepomis gibbosus* (Centrarchidae), *Colomesus psittacus* (Tetraodontidae) and *Thysochromis ansorgii* (Cichlidae) investigated with scanning particle image velocimetry. *J. Exp. Biol.* **207,** 1585–1596.

Hanke, W., Brücker, C., and Bleckmann, H. (2000). The ageing of the low frequency water disturbances caused by swimming goldfish and its possible relevance to prey detection. *J. Exp. Biol.* **203,** 1193–1200.

Hara, O. (1993). Role of olfactation in fish behavior. *In* "Behaviour of Teleost Fishes" (Pitcher, T. J., Ed.), 2nd edn., pp. 171–199. Chapman and Hall, London, New York.

Harris, G. G., and van Bergeijk, W. A. (1962). Evidence that the lateral line organ responds to near-field displacements of sound sources in water. *J. Acoust. Soc. Am.* **34,** 1831–1841.

Hassan, E. S. (1989). Hydrodynamic imaging of the surroundings by the lateral line of the blind cave fish *Anoptichthys jordani. In* "The Mechanosensory Lateral Line. Neurobiology and Evolution" (Coombs, S., Görner, P., and Münz, H., Eds.), pp. 217–228. Springer, New York.

Hassan, E. S. (1992). Mathematical description of the stimuli to the lateral line system for fish, derived from a three-dimensional flow field analysis. I. The case of moving in open water and of gliding towards a plane surface. *Biol. Cybern.* **66,** 443–452.

Hattori, T., and Suga, N. (1997). The inferior colliculus of the mustached bat has the frequency-vs-latency coordinates. *J. Comp. Physiol. A* **180,** 271–284.

Hawkins, A. D. (1993). Underwater sound and fish behaviour. *In* "Behaviour of Teleost Fishes" (Pitcher, T. J., Ed.), 2nd edn., pp. 129–169. Chapman and Hall, London, New York.

Hoekstra, D., and Janssen, J. (1986). Lateral line receptive field in the mottled sculpin. *Copeia* **1,** 91–96.

Hofer, B. (1908). Studien über die Hautsinnesorgane der Fische I. Die Funktion der Seitenorgane bei den Fischen. *Ber. Kgl. Bayer. Biol. Versuchsstation München* **1,** 115–168.

Hoin-Radkovski, I., Bleckmann, H., and Schwartz, E. (1984). Determination of source distance in the surface-feeding fish *Pantodon buchholzi* (Pantodontidae). *Anim. Behav.* **32,** 840–851.

Ivry, R. B., and Baldo, J. V. (1992). Is the cerebellum involved in learning and cognition? *Curr. Opin. Neurobiol.* **2,** 212–216.

Jakubowski, M. (1967). Cutaneous sense organs of fishes. VIII. The structure of the system of lateral-line canal organs in the Percidae. *Acta Biol. Cracov. Ser. Zool.* **10,** 69–81.

Janssen, J. (2004). Lateral line sensory ecology. *In* "The Senses of Fish. Adaptations for the Reception of Natural Stimuli" (von der Emde, G., Mogdans, J., and Kapoor, B. G., Eds.), pp. 231–264. Narosa Publishing House, New Delhi.

Janssen, J., and Corcoran, J. (1998). Distance determination via the lateral line in the mottled sculpin. *Copeia* **1998**(3), 657–661.

Janssen, J., Coombs, S., Hoekstra, D., and Platt, C. (1987). Anatomy and differential growth of the lateral line system of the mottled sculpin, *Cottus bairdi* (Scorpaeniformes: Cottidae). *Brain Behav. Evol.* **30,** 210–229.

Janssen, J., Coombs, S., and Pride, S. (1990). Feeding and orientation responses of mottled sculpin, *Cottus bairdi*, to water jets. *Envir. Biol. Fish* **29,** 43–50.

Kalmijn, A. J. (1988a). Detection of weak electric fields. *In* "Sensory Biology of Aquatic Animals" (Atema, J., Fay, R. R., Popper, A. N., and Tavolga, W. N., Eds.), pp. 151–186. Springer, New York.

Kalmijn, A. J. (1988b). Hydrodynamic and acoustic field detection. *In* "Sensory Biology of Aquatic Animals" (Atema, J., Fay, R. R., Popper, A. N., and Tavolga, W. N., Eds.), pp. 83–130. Springer, New York.

Kandel, E. R., Schwartz, J. H., and Jessell, T. M. (2000). "Principles of Neural Science," 3rd edn. McGraw-Hill, New York.

Käse, R., and Bleckmann, H. (1987). Prey localization by surface wave-ray tracing—fish trag bugs like oceanographers track storm. *Experientia* **43,** 290–293.

Kirsch, J. A., Hofman, M. A., Mogdans, J., and Bleckmann, H. (2002). Responses of diencephalic neurons to sensory stimulation in the goldfish, *Carassius auratus*. *Brain Res. Bull.* **57,** 419–421.

Knudsen, E. I. (1977). Distinct auditory and lateral line nuclei in the midbrain of catfishes. *J. Comp. Neurol.* **173,** 417–432.

Kotchabhakdi, N. (1976). Functional organization of the goldfish cerebellum. Information processing of input from peripheral sense organs. *J. Comp. Physiol. A* **112,** 75–93.

Kroese, A. B. A., and Schellart, N. A. M. (1992). Velocity- and acceleration-sensitive units in the trunk lateral line of the trout. *J. Neurophysiol.* **68,** 2212–2221.

Kroese, A. B. A., and van Netten, S. M. (1989). Sensory transduction in lateral line hair cells. *In* "The Mechanosensory Lateral Line. Neurobiology and Evolution" (Coombs, S., Görner, P., and Münz, H., Eds.), pp. 265–284. Springer, New York.

Kröther, S., Mogdans, J., and Bleckmann, H. (2002). Brainstem lateral line responses to sinusoidal wave stimuli in still- and running water. *J. Exp. Biol.* **205,** 1471–1484.

Kröther, S., Bleckmann, H., and Mogdans, J. (2004). Effects of running water on brainstem lateral line responses in trout, *Oncorhynchus mykiss*, to sinusoidal wave stimuli. *J. Comp. Physiol. A* **190,** 437–448.

Lang, H. H. (1980). Surface wave discrimination between prey and nonprey by the back swimmer *Notonecta glauca* L. (Hemiptera, Heteroptera). *Behav. Ecol. Sociobiol.* **6,** 233–246.

Lee, L. T., and Bullock, T. H. (1984). Sensory representation in the cerebellum of the catfish. *J. Comp. Physiol. A* **13,** 157–169.

Liao, J. C., Beal, D. N., Lauder, G. V., and Triantafyllou, M. S. (2003). The Kármán gait: Novel body kinematics of rainbow trout swimming in a vortex street. *J. Exp. Biol.* **206,** 1059–1073.

Lowe, D. A. (1986). Organisation of lateral line and auditory areas in the midbrain of *Xenopus laevis*. *J. Comp. Neurol.* **245,** 498–513.

McCormick, C. A. (1989). Central lateral line mechanosensory pathways in bony fish. *In* "The Mechanosensory Lateral Line. Neurobiology and Evolution" (Coombs, S., Görner, P., and Münz, H., Eds.), pp. 341–364. Springer, New York.

McCormick, C. A., and Braford, M. R. J. (1988). Central connections of the octavolateralis system: Evolutionary considerations. *In* "Sensory Biology of Aquatic Animals" (Atema, J., Fay, R. R., Popper, A. N., and Tavolga, W. N., Eds.), pp. 733–756. Springer, New York.

McCormick, C. A., and Hernandez, D. V. (1996). Connections of the octaval and lateral line nuclei of the medulla in the goldfish, including the cytoarchitecture of the secondary octaval population in goldfish and catfish. *Brain Behav. Evol.* **47,** 113–138.

Merzenich, M. M., and Reid, M. D. (1974). Representation of the cochlea within the inferior colliculus of the cat. *Brain Res.* **77,** 397–415.

Mogdans, J., and Bleckmann, H. (1998). Responses of the goldfish trunk lateral line to moving object. *J. Comp. Physiol. A* **182,** 659–676.

Mogdans, J., and Bleckmann, H. (1999). Peripheral lateral line responses to amplitude modulated hydrodynamic stimuli.. *J. Comp. Physiol. A* **185,** 173–180.

Mogdans, J., and Goenechea, L. (1999). Responses of medullary lateral line units in the goldfish, *Carassius auratus,* to sinusoidal and complex wave stimuli. *Zoology* **102,** 227–237.

Mogdans, J., and Kröther, S. (2001). Brainstem lateral line responses to sinusoidal wave stimuli in the goldfish, *Carassius auratus. Zoology* **104,** 153–166.

Mogdans, J., Bleckmann, H., and Menger, N. (1997). Sensitivity of central units in the goldfish, *Carassius auratus,* to transient hydrodynamic stimuli. *Brain Behav. Evol.* **50,** 261–283.

Mogdans, J., Kröther, S., and Engelmann, J. (2004). Neurobiology of the fish lateral line: Adaptations for the detection of hydrodynamic stimuli in running water. *In* "The Senses of Fish. Adaptations for the Reception of Natural Stimuli" (von der Emde, G., Mogdans, J., and Kapoor, G. B., Eds.), pp. 265–287. Narosa Publishing House, New Delhi.

Mohr, C., and Bleckmann, H. (1998). Electrophysiology of the cephalic lateral line of the surface-feeding fish *Aplocheilus lineatus. Comp. Bioch. Physiol. A* **119,** 807–815.

Montgomery, J. C., and Bodznick, D. (1994). An adaptive filter that cancels self-induced noise in the electrosensory and lateral line mechanosensory systems of fish. *Neurosci. Lett.* **174,** 145–148.

Montgomery, J. C., and Coombs, S. (1998). Peripheral encoding of moving sources by the lateral line system of a sit-and-wait predator. *J. Exp. Biol.* **201,** 91–102.

Montgomery, J. C., Coombs, S., and Janssen, J. (1994). Form and function relationships in the lateral line systems: Comparative data from six species of Antarctic notothenioid fish. *Brain Behav. Evol.* **44,** 299–306.

Montgomery, J. C., Coombs, S., Conley, R. A., and Bodznick, D. (1995). Hindbrain sensory processing in lateral line, electrosensory, and auditory systems: A comparative overview of anatomical and functional similarities. *Auditory Neurosci.* **1,** 207–231.

Montgomery, J. C., Baker, C. F., and Carton, A. G. (1997). The lateral line can mediate rheotaxis in fish. *Nature* **389,** 960–963.

Montgomery, J. C., Garton, G., Voigt, R., Baker, C., and Diebel, C. (2000). Sensory processing of water currents by fishes. *Phil. Trans. R. Soc. Lond. B* **355,** 1325–1327.

Montgomery, J. C., Macdonald, F., Baker, C. F., and Carton, A. G. (2002). Hydrodynamic contributions to multimodal guidance of prey capture behavior in fish. *Brain Behav. Evol.* **59,** 190–198.

Müller, H. M., and Bleckmann, H. (1993). The responses of octavolateralis cells to moving sources. *J. Comp. Physiol.* **179,** 455–471.

Münz, H. (1985). Single unit activity in the peripheral lateral line system of the cichlid fish *Sarotherodon niloticus* L. *J. Comp. Physiol. A* **157,** 555–568.

Münz, H. (1989). Functional organization of the lateral line periphery. *In* "The Mechanosensory Lateral Line. Neurobiology and Evolution" (Coombs, S., Görner, P., and Münz, H., Eds.), pp. 285–298. Springer, New York.

Müller, H. M., Fleck, A., and Bleckmann, H. (1996). The responses of central octavolateralis cells to moving sources. *J. Comp. Physiol. A* **179,** 455–471.

Nelson, M. E., MacIver, M. A., and Coombs, S. (2002). Modeling electrosensory and mechanosensory images during the predatory behavior of weakly electric fish. *Brain Behav. Evol.* **59,** 199–210.

New, J., Braun, C. B., and Walter, K. (2000). Central projections of nerve fibers innervating individual canal neuromast organs in the muskellunge, *Esox masquinongy. Soc. Neurosc.* **26,** 146.

New, J. G. (2002). Multimodal integration in the feeding behaviors of predatory teleost fishes. *Brain Behav. Evol.* **59,** 177–189.

New, J. G., and Bodznick, D. (1985). Segregation of electroreceptive and mechanoreceptive lateral line afferents in the hindbrain of chondrostean fishes. *Brain Res.* **336,** 89–98.

New, J. G., Coombs, S., McCormick, C. A., and Oshel, P. E. (1996). Cytoarchitecture of the medial octavolateralis nucleus in the goldfish, *Carassius auratus. J. Comp. Neurol.* **366,** 534–546.

Newbury, T. K. (1972). Vibration perception by chaetognaths. *Nature* **236,** 459–460.

Norris, H. W. (1925). Observation upon the peripheral distribution of the cranial nerves of certain ganoid fishes (*Amia*, *Lepidosteus*, *Polyodon*, *Scaphirhynchus*, and *Acipenser*). *J. Comp. Neurol.* **39,** 345–432.

Northcutt, G. R. (1988). Sensory and other neural traits and the adaptationist program: Mackarels of San Marco? *In* "Sensory Biology of Aquatic Animals" (Atema, J., Fay, R. R., Popper, A. N., and Tavolga, W. N., Eds.), pp. 869–879. Springer, New York/Heidelberg, Berlin.

Northcutt, R. G. (1989). The phylogenetic distribution and innervation of craniate mechanoreceptive lateral lines. *In* "The Mechanosensory Lateral Line. Neurobiology and Evolution" (Coombs, S., Görner, P., and Münz, H., Eds.), pp. 17–78. Springer, New York.

O'Bennar, J. D. (1976). Electrophysiology of neural units in the goldfish optic tectum. *Brain Res. Bull.* **1,** 529–541.

Parker, G. H. (1904). The function of the lateral-line organs in fishes. *Bull. US Bur. Fish* **24,** 185–207.

Pettigrew, A. G., Anson, M., and Chung, S. H. (1981). Hearing in the frog: A neurophysiological study of the auditory response in the midbrain. *Proc. Royal Soc. Lond. B* **212,** 433–457.

Pitcher, T. J., and Parrish, J. K. (1993). Functions of shoaling behaviour in teleosts. *In* "Behaviour of Teleost Fishes" (Pitcher, T. J., Ed.), 2nd edn., pp. 363–439. Chapman and Hall, London.

Plachta, D., Mogdans, J., and Bleckmann, H. (1999). Responses of midbrain lateral line units of the goldfish, *Carassius auratus*, to constant-amplitude and amplitude modulated water wave stimuli. *J. Comp. Physiol. A* **185,** 405–417.

Plachta, D., Hanke, W., and Bleckmann, H. (2003). A hydrodynamic topographic map and two hydrodynamic subsystems in a vertebrate brain. *J. Exp. Biol.* **206,** 3479–3486.

Pohlmann, K., Grasso, F. W., and Breithaupt, T. (2001). Tracking wakes: The nocturnal predatory strategy of piscivorous catfish. *Proc. Natl. Acad. Sci. USA* **98,** 7371–7374.

Pohlmann, K., Atema, J., and Breithaupt, T. (2004). The importance of the lateral line in nocturnal predation of piscivorous catfish. *J. Exp. Biol.* **207,** 2971–2978.

Prechtl, J. C., von der Emde, G., Wolfart, J., Karamürsel, S., Akoev, G. N., Andrianov, Y. N., and Bullock, T. H. (1998). Sensory processing in the pallium of a mormyrid fish. *J. Neurosci.* **18,** 7381–7393.

Puzdrowski, R. L. (1989). Peripheral distribution and central projections of the lateral-line nerves in goldfish, *Carassius auratus. Brain Behav. Evol.* **34,** 110–131.

Puzdrowski, R. L., and Leonard, R. B. (1993). The octavolateral systems in the stingray, *Dasyatis sabina*. I. Primary projections of the octaval and lateral line nerves. *J. Comp. Neurol.* **332,** 21–37.

Reiner, A., and Northcutt, R. G. (1992). An immunohistochemical study of the telencephalon of the Senegal bichir (*Polypterus senegalus*). *J. Comp. Neurol.* **319,** 359–386.

Roberts, B. L., and Meredith, G. E. (1989). The efferent system. *In* "The Mechanosensory Lateral Line. Neurobiology and Evolution" (Coombs, S., Görner, P., and Münz, H., Eds.), pp. 445–459. Springer, New York.

Roberts, B. L., and Russell, I. J. (1972). The activity of lateral-line efferent neurones in stationary and swimming dogfish. *J. Exp. Biol.* **57,** 435–448.

Rodriguez, F., Duran, E., Vargas, J. P., Torres, B., and Salas, C. (1994). Performance of goldfish trained in allocentric and egocentric maze producers suggests the presence of a cognitive mapping system in fishes. *Anim. Learn. Behav.* **22,** 409–420.

Salas, C., Broglio, C., Rodríguez, F., López, J. C., Portavella, M., and Torres, B. (1996a). Telencephalic ablation in goldfish impairs performance in a 'spatial constancy' problem but not in a cued one. *Behav. Brain Res.* **79,** 193–200.

Salas, C., Rodríguez, F., Vargas, J. P., Durán, E., and Torres, B. (1996b). Spatial learning and memory deficits after telencephalic ablation in goldfish trained in place and turn maze procedures. *Behav. Neurosci.* **110,** 965–980.

Scheich, H., Bonke, B. A., and Langner, G. (1979). Functional organization of some auditory nuclei in the Guinea fowl demonstrated by the 2-deoxyglucose technique. *Cell Tissue Res.* **204,** 17–27.

Schellart, N. A. M., and Kroese, A. B. A. (1989). Interrelationship of acousticolateral and visual systems in the teleost midbrain. *In* "The Mechanosensory Lateral Line. Neurobiology and Evolution" (Coombs, S., Görner, P., and Münz, H., Eds.), pp. 421–443. Springer, New York.

Schellart, N. A. M., Prins, M., and Kroese, A. B. A. (1992). The pattern of trunk lateral line afferents and efferents in the rainbow trout (*Salmo gairdneri*). *Brain Behav. Evol.* **39,** 371–380.

Schemmel, C. (1967). Vergleichende Untersuchungen an den Hautsinnesorganen ober- und unterirdisch lebender *Astyanax*-Formen. *Z. Morphol. Tiere* **61,** 255–316.

Schwartz, E. (1970). Ferntastsinnesorgane von Oberflächenfischen. *Z. Morphol. Tiere* **67,** 40–57.

Song, J. (1989). The lateral line system in the florida gar, *Lepisosteus platyrhincus* Dekay. Dissertation, pp. 1–160. University of Michigan.

Song, J., and Northcutt, R. G. (1991a). Morphology, distribution and innervation of the lateral-line receptors of the Florida gar, *Lepisosteus platyrhincus*. *Brain Behav. Evol.* **37,** 10–37.

Song, J., and Northcutt, R. G. (1991b). The primary projections of the lateral-line nerves of the Florida gar, *Lepisosteus platyrhincus*. *Brain Behav. Evol.* **37,** 38–63.

Stipetic, E. (1939). Über das Gehörorgan der Mormyriden. *Z. vergl. Physiol.* **26,** 740–752.

Striedter, G. F. (1991). Auditory, electrosensory, and mechanosensory lateral line pathways through the diencephalon and telencephalon of channel catfish. *J. Comp. Neurol.* **312,** 311–331.

Sutterlin, A. M., and Waddy, S. (1975). Possible role of the posterior lateral line in obstacle entrainment by brook trout (*Salvelinus fontinalis*). *J. Fish Res. Bd. Canada* **32,** 2441–2446.

Teyke, T. (1989). Learning and remembering the environment in the blind cave fish *Anoptichthys jordani*. *J. Comp. Physiol. A* **164,** 655–662.

Teyke, T. (1990). Morphological differences in neuromasts of the blind cave fish *Astyanax hubbsi* and the sighted river fish *Astyanax mexicanus*. *Brain Behav. Evol.* **35,** 23–30.

Tomchik, S. M., and Lu, Z. (2005). Octavolateral projections and organization in the medulla of a teleost fish, the sleeper goby (*Dormitator latifrons*). *J. Comp. Neurol.* **481,** 96–117.

Tomchik, S. M., and Lu, Z. (2006). Auditory physiology and anatomy of octavolateral efferent neurons in a teleost fish. *J. Comp. Physiol. A* **192,** 51–68.

Tricas, T. C., and Highstein, S. M. (1990). Visually mediated inhibition of lateral line primary afferent activity by the octavolateralis efferent system during predation in the free-swimming toadfish, *Opsanus tau. Exp. Brain Res.* **83,** 233–236.

Tricas, T. C., and Highstein, S. M. (1991). Action of the octavolateralis efferent system upon the lateral line of free-swimming toadfish, *Opsanus tau. J. Comp. Physiol. A* **169,** 25–37.

Videler, J. J., and Wardle, C. S. (1991). Fish swimming stride by stride: Speed limits and endurance. *Rev. Fish Biol. Fish.* **1,** 23–40.

Vogel, D., and Bleckmann, H. (1997). Surface wave discrimination in the topminnow *Aplocheilus lineatus. J. Comp. Physiol. A* **180,** 671–681.

Vogel, D., and Bleckmann, H. (2000). Behavioral discrimination of water motions caused by moving objects. *J. Comp. Physiol. A* **186,** 1107–1117.

Voigt, R., Carton, A. G., and Montgomery, J. C. (2000). Responses of anterior lateral line afferent neurons to water flow. *J. Exp. Biol.* **203,** 2495–2502.

von der Emde, G., and Bleckmann, H. (1992). Differential responses of two types of electroreceptive afferents to signal distortions may permit capacitance measurement in a weakly electric fish, *Gnathonemus petersii. J. Comp. Physiol. A* **171,** 683–694.

von der Emde, G., and Bleckmann, H. (1998). Finding food: Senses involved in foraging for insect larvae in the electric fish *Gnathonemus petersii. J. Exp. Biol.* **201,** 969–980.

von der Emde, G., Schwarz, S., Gomez, L., Budelli, R., and Grant, K. (1998). Electric fish measure distance in the dark. *Nature* **395,** 890–894.

Wagner, T., and Schwartz, E. (1992). Efferent lateral-line neurons of teleosts and their projection to lateral-line segments. *In* "Rhythmogenesis in Neurons and Networks. Proceedings of the 20th Göttingen Neurobiology Conference" (Elsner, N., and Richter, W. D., Eds.), p. 251. Georg Thieme Verlag, Stuttgart.

Wang, X., Lu, T., Snider, R. K., and Liang, L. (2005). Sustained firing in auditory cortex evoked by preferred stimuli. *Nature* **435,** 341–345.

Webb, J. F. (1989a). Developmental constraints and evolution of the lateral line system in teleost fishes. In "The Mechanosensory Lateral Line. Neurobiology and Evolution," pp. 79–98. Springer, New York.

Webb, J. F. (1989b). Gross morphology and evolution of the mechanoreceptive lateral-line system in teleost fishes. *Brain Behav. Evol.* **33,** 34–53.

Webb, J. F., and Smith, W. L. (2000). The laterophysic connection in chaetodontid butterfly fish: Morphological variation and speculations on sensory function. *Phil. Trans. R. Soc. Lond. B* **355,** 1125–1129.

Weeg, M. S., and Bass, A. H. (2000). Central lateral line pathways in a vocalizing fish. *J. Comp. Neurol.* **418,** 41–64.

Weeg, M. S., and Bass, A. H. (2002). Frequency response properties of lateral line superficial neuromasts in a vocal fish, with evidence for acoustic sensitivity. *J. Neurophysiol.* **88,** 1252–1262.

Weissert, R., and von Campenhausen, C. (1981). Discrimination between stationary objects by the blind cave fish *Anoptichthys jordani. J. Comp. Physiol. A* **143,** 75–382.

Wojtenek, W., Mogdans, J., and Bleckmann, H. (1998). The responses of midbrain lateral line units of the goldfish *Carassius auratus* to moving objects. *Zoology* **101,** 69–82.

Wubbels, R. J., Kroese, A. B. A., and Schellart, N. A. M. (1993). Response properties of lateral line and auditory units in the medulla oblongata of the rainbow trout (*Oncorhynchus mykiss*). *J. Exp. Biol.* **179,** 77–92.

Wullimann, M. F. (1998). The central nervous system. *In* "The Physiology of Fishes" (Evans, D. H., Ed.), pp. 245–282. CRC Press, New York.

Wullimann, M. F., Hofmann, M. H., and Meyer, D. L. (1991). The valvula cerebelli of the spiny eel, *Macrognathus aculeatus*, receives primary lateral-line afferents from the rostrum of the upper jaw. *Cell Tissue Res.* **266,** 285–293.

Zittlau, K. E., Claas, B., and Münz, H. (1986). Directional sensitivity of lateral line units in the clawed toad *Xenopus laevis* Daudin. *J. Comp. Physiol. A* **158,** 469–474.

11

NEUROMODULATORY FUNCTIONS OF TERMINAL NERVE-GnRH NEURONS

HIDEKI ABE
YOSHITAKA OKA

1. Introduction
2. Electrophysiological and Morphological Features of Single TN-GnRH Neurons Revealed by Intracellular Recording and Labeling
3. GnRH Release Demonstrated by RIA
4. Pacemaker Mechanism of TN-GnRH Neurons
5. Modulation of Pacemaker Frequencies of TN-GnRH Neurons by GnRH
6. Autocrine/Paracrine Control of TN-GnRH Neuron Pacemaker Frequencies by GnRH
7. Cellular Mechanisms of Modulation of Pacemaker Frequencies by GnRH
   7.1. Early Phase: Transient Decrease of Pacemaker Activity
   7.2. Late Phase: Subsequent Increase of Firing Activity
   7.3. Significance of Cell-to-Cell Electrical Interactions in the Cluster of TN-GnRH Neurons
   7.4. Physiological Significance of the Pacemaker Activity and Its Modulation
8. Multimodal Sensory Inputs to TN-GnRH System
   8.1. Neuroanatomical Evidence for Multimodal Sensory Inputs to TN-GnRH System
   8.2. Glutamatergic Excitatory Inputs to TN-GnRH Neurons
   8.3. Inhibitory Inputs to TN-GnRH Neurons
   8.4. Fine Structural Evidence for Synaptic Inputs to TN-GnRH Neurons
9. Neuromodulatory Action of GnRH Peptides
10. Distribution of GnRH Receptors in the Brain
11. Nonsynaptic Release of GnRH
12. Modulation of Neural Functions by GnRH
13. Behavioral Functions of TN-GnRH System
14. Working Hypothesis

1. INTRODUCTION

Animals flexibly adjust their physiological and behavioral actions in response to a changing environment. The peptidergic nervous system is believed to play an important role in this process. In this chapter, we will

Sensory Systems Neuroscience: Volume 25
FISH PHYSIOLOGY

DOI: 10.1016/S1546-5098(06)25011-8

review recent literature on an extrahypothalamic GnRH (*go*nadotropin-*r*eleasing *h*ormone) peptidergic neuron system, the terminal nerve (TN) GnRH system, which we suggest is neuromodulatory and is involved in the control of the motivational state of the animal. Fish brains have been extensively used for the study of GnRH systems because the GnRH systems are highly developed in fish. Studies in fish are revealing the versatile physiological and behavioral functions of the TN-GnRH systems.

The TN was first anatomically described as the last macroscopically identifiable cranial nerve number zero in elasmobranches by Fritsch (Demski and Schwanzel-Fukuda, 1987). Although the TN had subsequently been identified in various other vertebrate species including human embryos, teleosts, urodele amphibians, and so on, the nature of this nerve was enigmatic, and very little attention had been paid to it until the immunohistochemical report of Schwanzel-Fukuda and Silverman (1980). They first described the presence of extensive GnRH[1] immunoreactive cells and fibers in a structure that has been anatomically defined as the TN. This finding raised a new and exciting hypothesis for the function of this enigmatic cranial nerve; it was speculated that the GnRH-immunoreactive TN (TN-GnRH) system might be involved in the neural mechanisms of sexual behavior because GnRH was considered the key molecule in the control of reproduction. Shortly after this report, Demski and Northcutt (1983) and Springer (1983) reported that the TN neurons of the goldfish, which are located at the rostral base of the olfactory bulb, project to the retina, supracommissural part of the area ventralis of the telencephalon, and the preoptic area. Since the latter two brain areas were implied in the control of sexual behaviors in fish, Demski and Northcutt (1983) proposed a hypothesis that the TN may be a new chemosensory system that is involved in the control of sexual behavior. Although the chemosensory nature of the TN was denied by Fujita *et al.* (1991) (see later), the involvement of TN-GnRH system in the fine-tuning of behavioral motivation, including that for sexual behavior, should be an attractive hypothesis in a different context. This is going to be the main topic of the present chapter.

[1]GnRH was originally called LHRH (luteinizing-hormone releasing hormone) since its discovery in the hypothalamus of mammalian brains. GnRH was first identified as a hypophysiotropic decapeptide hormone that is produced in the hypothalamus and facilitates gonadotropin release from the pituitary gonadotropes. Recent anatomical studies have shown that GnRH neurons and their fibers are found not only in the hypothalamus but also in the "extrahypothalamic" areas. Such "extrahypothalamic" GnRH systems have been found in the TN and midbrain (mammals: Schwanzel-Fukuda and Silverman, 1980; Schwanzel-Fukuda *et al.*, 1985, 1987; Somoza *et al.*, 2002; teleosts: Amano *et al.*, 1991; Kim *et al.*, 1995; Munz and Stumpf, 1981; Parhar *et al.*, 1994; Yamamoto *et al.*, 1995). Although the function of the hypothalamic GnRH system has been well studied, the functional significance of "extrahypothalamic" GnRH system is elusive and has only poorly been studied yet.

Many immunohistochemical studies have demonstrated the presence of GnRH-immunoreactive neurons and fibers in wide areas of the brain, including the TN system, in almost all vertebrate species reported to date (Demski and Northcutt, 1983; Somoza *et al.*, 2002). Among vertebrates, it is well known that the extrahypothalamic as well as hypothalamic GnRH systems are most developed and thus are most intensively studied in teleosts; both hypothalamic and extrahypothalamic GnRH neurons and fibers are extensively distributed, and the neuronal cell bodies tend to be grouped as clusters. There is now a general agreement on the presence of three functionally as well as anatomically different GnRH systems in teleost fishes (Oka, 1997), and a similar principle applies to most vertebrate species in general (Somoza *et al.*, 2002). We took advantage of the brain of a teleost fish, the dwarf gourami (*Colisa lalia*, a tropical freshwater fish), and have clearly identified the following three anatomically as well as functionally distinctive GnRH neuronal systems, using combined immunohistochemistry, brain lesioning, high-performance liquid chromatography (HPLC) and radioimmunoassay (RIA): (1) conventional hypophysiotropic POA-GnRH system, (2) TN-GnRH system, and (3) midbrain GnRH system (Yamamoto *et al.*, 1995). Among these GnRH systems, we have been interested in the biological significance of the main extrahypothalamic GnRH system, the TN-GnRH system. Although we still do not have a global picture of the functional significance of the TN-GnRH system, we now have several lines of evidence to show that the TN-GnRH system serves as a "neuromodulatory system" that controls the motivational or arousal state of the animal. Furthermore, we hypothesize that peptidergic neuromodulatory systems, including the TN-GnRH system, play pivotal roles during biological adaptations to environmental changes; the animals have neurobiological mechanisms for adapting to the environmental changes. In this chapter, we will try to review results of research on the physiology of TN-GnRH neurons to supply building blocks for substantiating our hypothesis.

2. ELECTROPHYSIOLOGICAL AND MORPHOLOGICAL FEATURES OF SINGLE TN-GnRH NEURONS REVEALED BY INTRACELLULAR RECORDING AND LABELING

Although GnRH-producing neurons have attracted the attention of many scientists in the field of reproductive biology, it has been extremely difficult to study the electrophysiological and morphological features of single GnRH neurons. This is because in most vertebrate species the cell bodies of hypothalamic as well as extrahypothalamic GnRH neurons are rather small (usually about 10 μm or so in diameter) and diffusely distributed along the basal regions of the olfactory forebrain. They form a diffuse continuum along the medial part of the olfactory nerve (and TN), basal olfactory bulbs, basal telencephalon, preoptic area, and the basal hypothalamus (Parhar, 2002). The distribution of some TN-GnRH neurons often overlaps with that of hypothalamic GnRH neurons. Although a limited number of studies tried to identify GnRH neurons by *post hoc* immunohistochemistry of intracellularly recorded and labeled neurons (Kelly *et al.*, 1984), technical difficulties have prevented the electrophysiological study of GnRH neurons.

We took advantage of the brain of a tropical fish, the dwarf gourami (*C. lalia*), and started detailed electrophysiological and morphological studies of GnRH neurons, especially the extrahypothalamic TN-GnRH neurons (Oka, 1992). We start by briefly summarizing results of these studies. For a detailed review, see Oka (1997).

The TN-GnRH neurons of the dwarf gourami have the following advantages over GnRH neurons of other vertebrate species in that: (1) the cell bodies are large (about 20–40 μm in diameter), (2) they form a tight cell cluster without intercalating glial cells in the ventral-most part of the brain beneath the meningeal membrane, and (3) the whole brain of this fish can be maintained *in vitro* for a long period without oxygenation (Oka and Matsushima, 1993). We have developed a preparation in which the whole brain is dissected out from the skull and the ventral connective tissue is removed. The brain is pinned down ventral side up in a chamber, and the cluster of TN-GnRH neurons can be clearly seen under a stereomicroscope. Thus, we can easily record electrical activities with microelectrodes or patch pipettes from the visually identified TN-GnRH neurons in a semi-intact whole brain *in vitro* preparation. This preparation has a great advantage over most other vertebrate brains, where recording from the identified GnRH neurons is extremely difficult because they are small and scattered. Furthermore, in the dwarf gourami, we have also found that the TN-GnRH neurons show exocytotic peptide release activity even from the cell bodies (Oka and Ichikawa, 1991, 1992; Ishizaki *et al.*, 2004). This is also advantageous because the peptides are usually released from small nerve terminals or varicosities, which are very difficult to access for direct measurements of their electrical or release activities. Thus, we believe that the whole brain *in vitro* preparation of the dwarf gourami is ideal for studying the molecular and cellular mechanism of GnRH release from TN-GnRH neurons in addition to their physiological characteristics. Furthermore, we have developed a thick brain slice preparation in which the exposed TN-GnRH neurons can be more precisely observed and accessed under an upright microscope with infrared differential interference contrast (IR-DIC) optics for electrophysiological analyses (Haneda and Oka, 2004). After the ventral meningeal membrane of the forebrain is carefully removed, thick brain slices of about

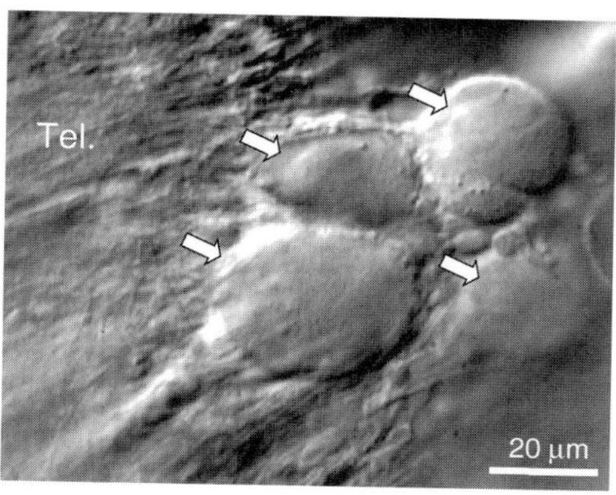

Fig. 11.1. Infrared differential interference contrast microscopy (IR-DIC) image of the cluster of TN-GnRH neurons in a sagittal slice of the dwarf gourami brain (rostral is to the upper right). Large arrowheads indicate TN-GnRH neurons. Tel., telencephalon.

1 mm containing TN-GnRH neurons are manually cut out with razor blades in a low-Na^+ Ringer solution. The slice containing GnRH neurons is allowed to recover for at least 1 h before the experiments in a standard Ringer solution. Under the upright microscope with IR-DIC optics, the exposed TN-GnRH neurons are easily identified visually in the brain slice (Figure 11.1). The cell bodies of TN-GnRH neurons are large and make a tight cell cluster in the ventral surface of the olfactory bulb-telencephalon border. Whole-cell patch-clamp recordings are usually made from spherical TN-GnRH neurons that are fully exposed to the surface (not buried deep in the cluster) so as to maximize the space-clamp quality during the voltage clamp.

Electrophysiological recording from single TN-GnRH neurons revealed that the majority of these neurons have quite regular pacemaker activities (Figure 11.2) (Oka, 1992, 1997; Oka and Matsushima, 1993). The intrinsic nature of this activity was demonstrated by the voltage dependence of the pacemaker frequency, rhythm resetting, and persistence of rhythmicity after synaptic isolation (Oka and Matsushima, 1993). In addition, irregular or burst firing patterns were also encountered, although the occurrence was much less frequent. Progress in promoter-driven transgenic techniques has made it possible to record from mammalian GnRH neurons. With this technique GnRH neurons can be GFP-tagged and identified in slices or dissociated cell preparations (in mice, Skynner *et al.*, 1999; Spergel *et al.*, 1999; Suter *et al.*, 2000a,b; DeFazio and Moenter, 2002; DeFazio *et al.*, 2002; Moenter *et al.*, 2003; Nunemaker *et al.*, 2002, 2003a,b; or in rats, Kato *et al.*, 2003). Although regular pacemaker activity like that of gourami TN-GnRH

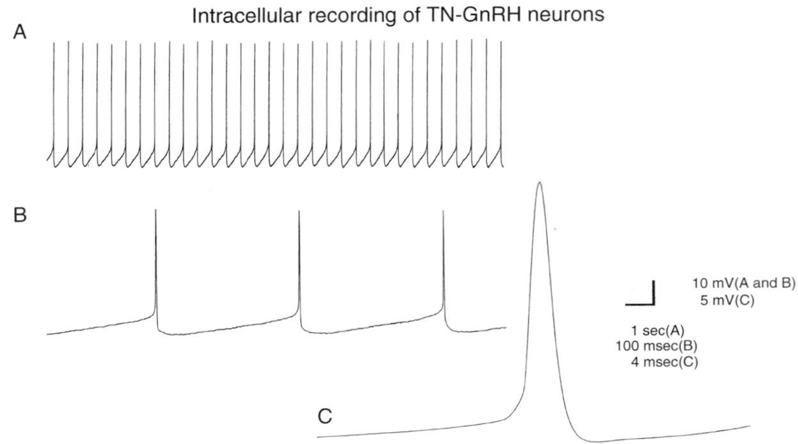

Fig. 11.2. (A–C) Intracellular recordings of spontaneous regular pacemaker activity of TN-GnRH cells in a whole brain *in vitro* preparation. (A) Typical example of regular beating discharge mode. (B) and (C) show recordings from the same cell on different time scales, and (C) shows a spike averaged from 30 traces. (Modified from Oka and Matsushima, 1993.)

neurons has not been recorded from the mammalian GnRH neurons, irregular or burst firing patterns have been recorded from GFP-labeled hypothalamic GnRH neurons in brain slices (Nunemaker *et al.*, 2002, 2003b) as well as from GT1–1 cells (cell lines that produce GnRH; Charles and Hales, 1995) and cultured embryonic GnRH neurons (Abe and Terasawa, 2005). It has been suggested that burst firing facilitates peptide secretion by enhancing the calcium influx required for release (Dutton and Dyball, 1979; Cazalis *et al.*, 1985). Here, the different firing patterns may reflect a change in membrane excitability either due to intrinsic cellular properties and/or due to synaptic inputs. Thus, it may be a future problem to examine whether the GnRH neurons switch between these different firing patterns according to different physiological conditions, and to determine the physiological significance and underlying mechanisms of these changes.

With regard to the frequency of pacemaker activity of TN-GnRH neurons, the following observation may be relevant to their functional significance. We recorded from two different GnRH neurons in the brain of a fish and compared the frequencies of the two neurons. We found a highly significant correlation between the frequencies of these two neurons. This means that the frequencies of the pacemaker potentials are similar among different neurons recorded from the brain of an individual fish. The pacemaker frequencies recorded from TN-GnRH neurons varied among different fish and ranged from 0.9 to 6.7 Hz. Thus, we suggested that the pacemaker frequencies of TN-GnRH neurons might be dependent on the physiological conditions of the fish and therefore vary from fish to fish under different conditions. However, there was no discernible sexual difference in the frequency or pattern of pacemaker activities.

Anatomical observation of intracellularly labeled cells (Figure 11.11, lower left) showed that, regardless of discharge patterns, all labeled TN-GnRH neurons have profusely branched axons that project to a wide range of brain areas, which have been shown to receive a dense network of GnRH-immunoreactive fibers (Oka and Ichikawa, 1990; Yamamoto *et al.*, 1995). These areas include ventral and dorsal telencephalic areas, preoptic areas, thalamus, inferior lobe of hypothalamus, midbrain tegmentum, torus semicircularis, and optic tectum. Rostrocaudally, the projection areas innervated by a single GnRH neuron ranged from the olfactory bulb to the medulla, and in some cases, as far as the rostral spinal cord. This explains how a small number of TN-GnRH neurons (~10–20 on each side) can be the source of widespread GnRH-immunoreactive fibers throughout the brain. Multiple axonal projections of TN-GnRH neurons enable simultaneous regulation of excitability and/or transmitter release from target neurons in widespread areas of the brain (Oka and Matsushima, 1993). In contrast to these wide projections in the brain, projection of TN-GnRH neurons to the pituitary was never observed, which clearly indicates that they do not serve as a hypophysiotropic hormone system but rather act as a neuromodulatory system in the brain.

In summary, regular intrinsic pacemaker activity and widespread projections in the brain are characteristics of TN-GnRH neurons. There are a few studies that report on the electrical activity of TN neurons, although neurons were not proven to be GnRH-immunoreactive (Fujita *et al.*, 1985, 1991; White and Meredith, 1987). All neurons showed pacemaker-like spontaneous activity, which suggests that the pacemaker activity is an important physiological characteristic common to all TN neurons. Furthermore, it is intriguing to find that other candidate modulatory neurons demonstrate physiological and morphological characteristics that are similar to those of TN-GnRH neurons. These neural groups include serotonergic neurons in the dorsal raphe (Aghajanian, 1990), noradrenergic neurons in the locus coeruleus (Williams *et al.*, 1984), dopaminergic neurons in the substantia nigra and ventral tegmental area (Grace, 1988, 1991), and histaminergic neurons in the hypothalamus (Haas and Reiner, 1988; Wada *et al.*, 1991). All of these neurons show spontaneous pacemaker-like activities, which probably arise from intrinsic ion channel properties. Endogenous pacemaker activity is presumably one of the essential features common to all peptidergic and monoaminergic modulatory neurons. Similarities between these modulatory neurons and the TN-GnRH system are evident in their patterns of projections as well. Although there have been no intracellular staining or tracing study of axonal projections comparable to ours (Oka and Matsushima, 1993), it has been demonstrated by immunocytochemical methods that serotonergic, noradrenergic, dopaminergic, and histaminergic projections are widely distributed throughout the brain. These substances are synthesized by a rather small number of neurons whose soma are restricted to discrete loci (Wada *et al.*, 1991). Thus, the morphological and physiological characteristics of TN-GnRH cells, as demonstrated in our study, are not specific to this system but are features that are shared with other modulator neurons, including monoaminergic and peptidergic neurons.

3. GnRH RELEASE DEMONSTRATED BY RIA

If the TN-GnRH neurons function as a peptidergic neuromodulatory system, they should be able to release GnRH peptides in the brain. In order to examine whether GnRH peptides are actually released in the brain from the GnRH neurons, we measured the release of GnRH peptides into the medium from the brain-pituitary slices by using an RIA (Ishizaki *et al.*, 2004). To measure GnRH release from different GnRH systems, and to examine whether there are differences between them, we conducted a static incubation of brain-pituitary slices under various conditions and measured GnRH released into the incubation medium by RIA. The slices were divided into two parts, a wedge of tissue containing GnRH neurons in the preoptic area and axon terminals in the attached pituitary (defined as "POA-GnRH" slices), and the other slice of brain containing the cell bodies and fibers of TN-GnRH neurons in the rostral forebrain and midbrain tegmentum-GnRH neurons in the midbrain (defined as "TN-GnRH" slices). We demonstrated that GnRH release was evoked by a high $[K^+]_o$ depolarizing stimulus (in both slices) via Ca^{2+} influx through voltage-gated Ca^{2+} channels. Our experiments using specific Ca^{2+} channel blockers suggest that GnRH release from POA-GnRH slices induced by depolarization is mainly dependent on the Ca^{2+} influx through ω-conotoxin-sensitive N-type Ca^{2+} channels. Depolarization-induced GnRH release from TN-GnRH slices is dependent on both nifedipine-sensitive L- and ω-conotoxin-sensitive N-type Ca^{2+} channels. In either slice, ω-agatoxin-sensitive P/Q-type Ca^{2+} channels were not involved. The most prominent difference between the GnRH release from the POA-GnRH and TN-GnRH slices, however, was the presence of a sexual difference in the GnRH release. Males demonstrated greater release from the POA-GnRH slices than females. There was no difference between the sexes in release of GnRH from the TN-GnRH slice. In TN-GnRH slices, our evidence suggested that the store-operated Ca^{2+} influx might be involved in the basal (spontaneous and unstimulated) GnRH release.

Finally, we found that glutamate (Glu) application significantly increased the GnRH release in the TN-GnRH but not in the POA-GnRH slices in a dose-dependent manner. In accordance with these results, Glu applications increased the frequency of pacemaker potentials of TN-GnRH cells in a dose-dependent manner. Furthermore, we now have evidence for the existence of ionotropic as well as metabotropic Glu receptors (GluRs) and their depolarizing effects in TN-GnRH neurons (Kiya and Oka, 2003; see in a later section). Thus, our data have shown that Glu increases the pacemaker frequency, which can potentially lead to the increased release of GnRH from the TN-GnRH neurons.

4. PACEMAKER MECHANISM OF TN-GnRH NEURONS

The pacemaker potential and its frequency or pattern seem to be important for TN-GnRH neurons functioning as neuromodulators. Therefore, we focused on the mechanisms of pacemaker frequency generation (Oka, 1995, 1996; Abe and Oka, 1999).

We first found a novel tetrodotoxin (TTX)-resistant persistent Na^+ current, $I_{Na(slow)}$, which supplies the persistent depolarizing drive and plays an important role in the generation of pacemaker potentials in TN-GnRH neurons (Oka, 1995). This current was discovered because the pacemaker potentials were rather resistant to TTX but were readily blocked by substituting Na^+-impermeant ions (tetramethylammonium or choline) for Na^+ in the perfusing solution. With Na^+-impermeant ions extracellullarly, the resting membrane potential became more hyperpolarized than the control level. The $I_{Na(slow)}$ current was further characterized using voltage-clamp recording (Oka, 1996). The $I_{Na(slow)}$ currents could be isolated pharmacologically by blocking K^+ currents, Ca^{2+} currents, and the conventional fast Na^+ current. The current was characterized by its resistance to TTX, dependence on external Na^+, slow activation, very slow and little inactivation, and wide overlap of steady-state activation and inactivation curves (window currents) near the resting potential. These characteristics are distinct from those of conventional fast Na^+ currents and are relevant for the generation of persistent inward currents necessary for the pacemaker activity of TN-GnRH neurons.

Next we searched for the candidate outward currents that should counteract the persistent depolarizing drive supplied by $I_{Na(slow)}$. We demonstrated, by using the whole-cell voltage-clamp recording, that the TN-GnRH cells have at least four different types of voltage-dependent K^+ currents.

Among these, a tetraethylammonium (TEA)-sensitive K^+ current that is slowly activating, long lasting, and has lower threshold for activation appeared to be the most likely candidate that contributes to the repolarizing phase of the pacemaker potentials of TN-GnRH cells. This current is only partially inactivated at a steady state of ~ -60 to -40 mV, which is equivalent to the resting membrane potential of the TN-GnRH neurons. Other currents were almost inactivated at such voltages. Furthermore, in current-clamp recordings bath application of TEA together with TTX reversibly blocked the pacemaker potentials. The characteristics of the TEA-sensitive K^+ current, $I_{K(v)}$, are similar to those of the delayed rectifier K^+ current described in the GnRH secretory GT1 cell line (Bosma, 1993), in embryonic GnRH neurons (Kusano *et al.*, 1995), and in many other excitable cells (Rudy, 1988). These currents activate slowly, do not inactivate during the test pulse, are sensitive to extracellular TEA, and are insensitive to 4-aminopyridine (4AP), a chemical that blocks the A-type voltage sensitive potassium channel. The steady-state inactivation of these currents occurs in a relatively positive membrane potential range. However, the voltage activation thresholds of these currents are different; TEA-sensitive K^+ currents of the TN-GnRH neurons have relatively low activation thresholds (around -40 mV), but the delayed rectifier K^+ currents of GT1 cells and embryonic GnRH neurons have relatively high activation thresholds (around -20 mV).

From these results, we concluded that the basic pattern of pacemaker activities of the TN-GnRH neurons, especially their subthreshold component, is generated as follows. When the TN-GnRH cells are at the resting potential, a considerable amount of $I_{Na(slow)}$ is deinactivated (relieved from inactivated status) which supplies the persistent depolarizing drive, and the membrane potentials show gradual depolarization. When the membrane potential reaches the activation threshold for the $I_{K(V)}$, outward current develops, and the net flux of current reverses to outward. Then, the membrane potential hyperpolarizes and deactivates the K^+ current and the next cycle begins. The action potential phase of the pacemaker activity is obviously composed of TTX-sensitive Na^+ currents and delayed rectifier K^+ currents, and several type(s) of Ca^{2+} currents (see later).

5. MODULATION OF PACEMAKER FREQUENCIES OF TN-GnRH NEURONS BY GnRH

As mentioned earlier, the firing frequency or mode of the pacemaker activity of TN-GnRH neurons appears to change according to the physiological condition of the animal (e.g., arousal, motivational status, hormonal milieu, and so on) which may be influenced by environmental factors. The firing frequency and/or mode of pacemaker activity is reported to affect the efficacy of exocytotic release of the GnRH peptide from GnRH neurons (Peng and Horn, 1991; Nunemaker *et al.*, 2001, 2003b; Moenter *et al.*, 2003). The released GnRH peptide then in turn influences the excitability of target neurons in various brain regions. Such peptide-induced alteration in the excitability of neurons may be the neural basis for long-lasting changes in animal behavior (Oka, 1997, 2002). Therefore, the study of the nature and the mechanisms of modulation of pacemaker activities of TN-GnRH neurons by hormones or transmitters should give us invaluable information about the control mechanisms of the neuromodulatory TN-GnRH system. During the survey of possible candidates that modulate the pacemaker frequency of TN-GnRH neurons, we found that salmon GnRH, the molecular species of GnRH produced by TN-GnRH neurons, affects the pacemaker activity (Abe and Oka, 2000; Figure 11.3). In Ringer solution, TN-GnRH neurons showed a slow regular beating discharge (Figure 11.3Ba). During the bath application of sGnRH, the firing frequency of pacemaker activity was transiently decreased (early phase; Figure 11.3Bb) and subsequently increased (late phase; Figure 11.3Bc). This biphasic modulation is clearly seen in an electrophysiological recording trace on a slower time scale (Figure 11.3A) or in a frequency plot (Figure 11.3C). The latency of the transient decrease in firing frequency was about 20–90 sec, but this varied from neuron to neuron even at the same concentration of sGnRH. The duration of the transient decrease in firing frequency increased steadily with decreasing sGnRH concentration. Furthermore, when the sGnRH concentration was 0.2 nM, this decrease in firing frequency became persistent and the late-phase frequency increase was not observed. This suggests that the onset of the late-phase increase of the pacemaker potential may be dependent on the sGnRH concentration and the early-phase decrease less dependent on it, although it is difficult to assess quantitatively. Xu *et al.* (2004) reported a similar dose-dependent switch in the response of mouse GFP-labeled GnRH neurons to GnRH. Such biphasic modulation of pacemaker activity was also shown by bath application of mammalian GnRH. In addition, we found that a pre- or coperfused GnRH antagonist inhibited the modulation of pacemaker activity by bath application of sGnRH. The results strongly suggest that modulation by GnRH peptide of pacemaker activity of TN-GnRH neurons is caused by GnRH receptor (GnRH-R) activation, although there does not seem to be a selectivity of different molecular species of GnRH for the receptor activation. However, this does not mean that GnRH released from the GnRH neurons that belong to the other GnRH systems (other molecular species of GnRH) activates the TN-GnRH neurons under physiological conditions because immunoreactive fibers of the other

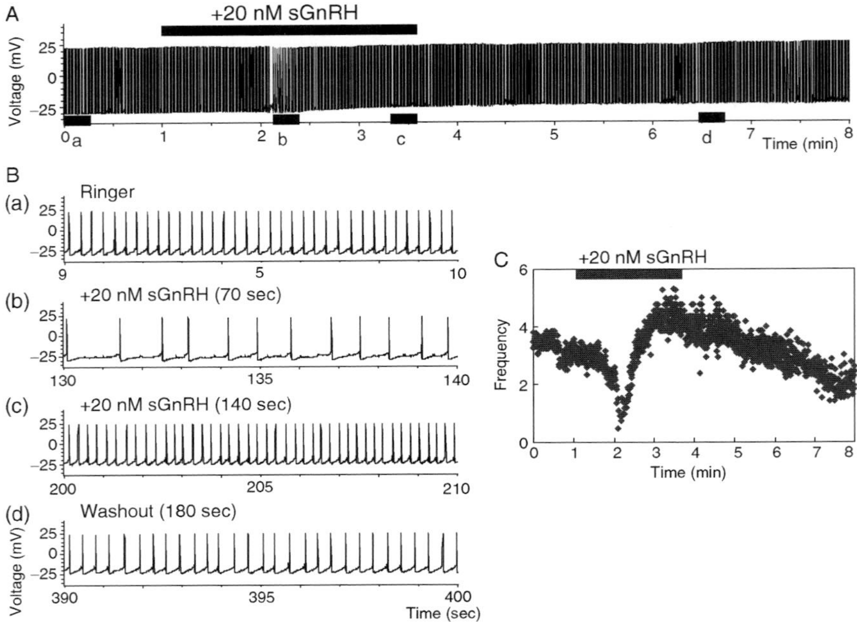

Fig. 11.3. Modulation of pacemaker frequency by sGnRH. (A) In a current-clamp whole-cell recording from a TN-GnRH neuron, bath application of sGnRH, the same molecular species of GnRH peptide produced by TN-GnRH neurons themselves, biphasically modulated their pacemaker activity. (B) Bath application of sGnRH transiently decreased (b) and subsequently increased the frequency of pacemaker activity (c). Following washout, the firing frequency of pacemaker activity recovered (d). (C) Frequency of pacemaker activity plotted against the time course. (Modified from Abe and Oka, 2000.)

GnRH systems are not distributed near the TN-GnRH neuronal cell bodies (Yamamoto *et al.*, 1995).

6. AUTOCRINE/PARACRINE CONTROL OF TN-GnRH NEURON PACEMAKER FREQUENCIES BY GnRH

There are alternative possible mechanisms to explain the biphasic modulations of the frequency of pacemaker activity of TN-GnRH neurons mentioned earlier. First, GnRH-Rs may exist on the cell surface of TN-GnRH neurons, and the pacemaker activity of TN-GnRH neurons may be directly modulated by the downstream cell signaling pathways. Second, GnRH-Rs may exist on the cell surface of non-GnRH neurons, and the pacemaker activity of TN-GnRH neurons may be indirectly modulated by these neurons. To test these alternative possibilities, we dialyzed the patch-clamped GnRH neuron with guanosine 5′-[β-thio] diphosphate (GDP-β-S), a GDP derivative that is a competitive inhibitor of many G-protein–mediated processes, by including it in the patch pipette solution. This test procedure is possible because GnRH-Rs are members of the G-protein–coupled receptor family (Stojilkovic *et al.*, 1994b). After the diffusion of GDP-β-S into the TN-GnRH neuron, bath application of sGnRH failed to evoke any modulation of the firing frequency of pacemaker activity. From these results, we suggested that the G-protein–coupled process mediates this biphasic modulation of the pacemaker activity by sGnRH in TN-GnRH neurons. It has been reported that GT1–7 cells express GnRH-Rs (Krsmanovic *et al.*, 1993; Stojilkovic *et al.*, 1994a,b). Moreover, coexpression of the GnRH peptide and the GnRH-R has been shown by double immunolabeling by reverse transcription-polymerase chain reaction (RT-PCR) of the cytoplasm of laser-captured mouse GnRH neurons (Krsmanovic *et al.*, 1999; Martinez-Fuentes *et al.*, 2004) or by single cell RT-PCR of GFP-labeled mouse hypothalamic GnRH neurons (Xu *et al.*, 2004). We have also demonstrated the presence and the molecular nature of the GnRH-Rs expressed on the cell surface of TN-GnRH neurons in the dwarf gourami brain by *in situ* hybridization and single cell patch-RT-PCR methods (Hajdú *et al.*, 2006). It is highly probable that GnRH-Rs exist on the cell surface of TN-GnRH neurons and play an important role in modulating the ion channel(s) underlying the pacemaker activity via G-protein–mediated signaling pathways.

Another line of evidence to support the presence of autocrine/paracrine control of TN-GnRH neuron pacemaker activities comes from the fine structural and electrochemical demonstration of somatodendritic release of GnRH. Because the TN-GnRH neurons of the dwarf gourami are large and form a tight cell cluster at a characteristic location (boundary between the olfactory bulb and the telencephalon), we can easily identify them at the electron microscopic level (Oka and Ichikawa, 1991). In electron micrographs, the TN-GnRH neuronal cell bodies are closely apposed with each other without intervening glial cells. Within the cytoplasm of TN-GnRH neurons, there are many dense-cored vesicles (DCVs) budding from highly stacked Golgi apparatus. Immunoelectron microscopy revealed that these DCVs in the cytoplasm contain mature GnRH peptides (Oka and Ichikawa, 1992). Furthermore, coated pits or vesicles, which suggest membrane retrieval after an active exocytosis, are frequently distributed beneath the plasma membrane of the cell body and small somatic processes of TN-GnRH neurons (Oka and Ichikawa, 1991). All of the fine structural evidence confirms our hypothesis of exocytotic GnRH release from the somatodendritic area. Somatodendritic release of neurohypophysial peptides by exocytosis from the hypothalamic magnocellular neurons has also been reported (Pow and Morris, 1989), and, thus, the somatodendritic release may be a phenomenon common to various kinds of peptidergic neurons.

Somatodendritic release of GnRH from the TN-GnRH neurons was also demonstrated by using our newly developed electrochemical method for the real-time measurement of GnRH release (Figure 11.4) (Ishizaki and Oka, 2001). This is an electrochemical recording method using carbon fiber electrodes (CFE), which was developed for the real-time measurement of electroactive substances such as monoamines. Since the sGnRH molecule contains two Trp and one Tyr residues, which are electroactive (readily oxidizable) amino acids, GnRH can also be measured electrochemically. The CFE was gently pressed against the surface of TN-GnRH cell bodies while the

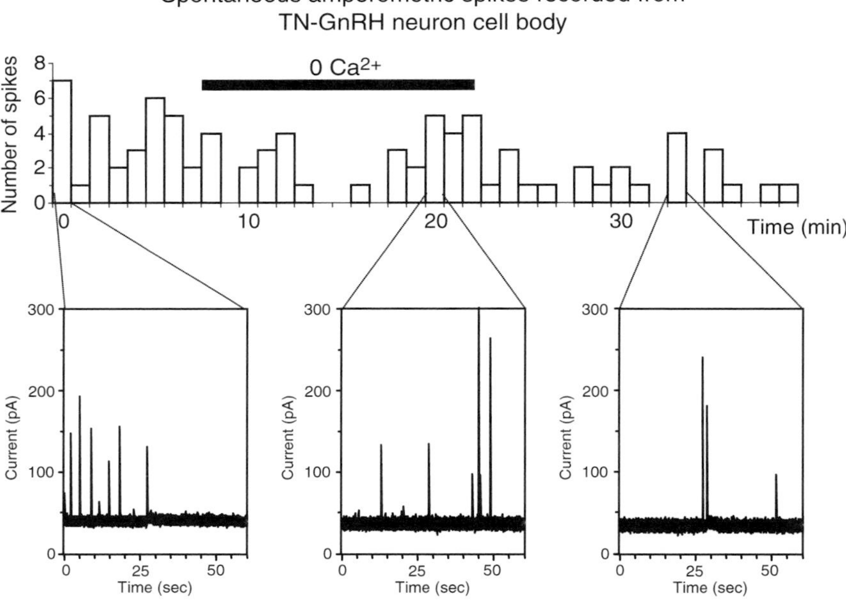

Fig. 11.4. Real-time measurement of GnRH release from the surface of the TN-GnRH neuronal cell body. A carbon fiber microelectrode (CFE) was gently pressed against the exposed surface of a cell body in a whole brain *in vitro* preparation. The voltage of the CFE was held at +900 mV, the redox potential for GnRH peptides. Spontaneous amperometric spikes arising from the released and oxidized GnRH were recorded. The spontaneous GnRH release was independent of the presence or absence (indicated by the bar labeled "0 Ca^{2+}") of extracellular Ca^{2+} in the perfusing medium (Ishizaki and Oka, unpublished observation).

voltage was held at 900 mV, the redox potential for GnRH. Spontaneous amperometric spikes were recorded, which indicates a somatic release of GnRH peptides. The spontaneous GnRH release was independent of the presence of extracellular Ca^{2+} in the perfusing medium.

The morphological, molecular biological, and electrochemical evidence suggest the presence of autocrine/paracrine regulation of pacemaker activity of TN-GnRH neurons by GnRH peptides. It is noteworthy that such autocrine/paracrine control of GnRH neurons by GnRH ligands is not unique to TN-GnRH neuronal activity but has also been shown to underlie the control of hormone release from cultured hypothalamic neurons (Krsmanovic *et al.*, 1999), as well as the control of electrical activity, intracellular Ca^{2+} dynamics, and pulsatile GnRH release of the embryonic GnRH neurons (Martinez-Fuentes *et al.*, 2004). Furthermore, the autocrine/paracrine action of GnRH has been suggested to play an important role in promoting differentiation and migration of GnRH neurons from the olfactory placodes (Romanelli *et al.*, 2004). As emphasized in this chapter, the TN-GnRH neurons of the dwarf gourami brain provide opportunities to study mechanisms of autocrine/paracrine actions of GnRH at the cellular and molecular levels. We can directly measure GnRH release from the exposed somatodendritic area of GnRH neurons in thick brain slice preparation. This gives more precise GnRH release data than attempting to measure GnRH release from small synaptic terminals or varicosities in the brain. In addition, our preparation enables multidisciplinary studies of visually identified GnRH neurons without the production of GFP transgenic animals. Most importantly, the dwarf gourami preparation provides a model in which electrophysiological, morphological, and behavioral information has been and is easily obtained and analyzed to give a more thorough description of the unique characteristics of the TN-GnRH system.

7. CELLULAR MECHANISMS OF MODULATION OF PACEMAKER FREQUENCIES BY GnRH

7.1. Early Phase: Transient Decrease of Pacemaker Activity

Let us first discuss the mechanisms responsible for the early-phase decrease of pacemaker potentials by GnRH. Calcium is a very common intracellular signaling molecule. Calcium can enter the cell from the extracellular space or it can be released from internal stores. The two receptors involved in calcium release from internal stores are the IP_3-stimulated IP_3 receptor/channel and the caffeine/calcium-stimulated ryanodine receptor/channel, both of which are located in the endoplasmic reticulum of certain neurons.

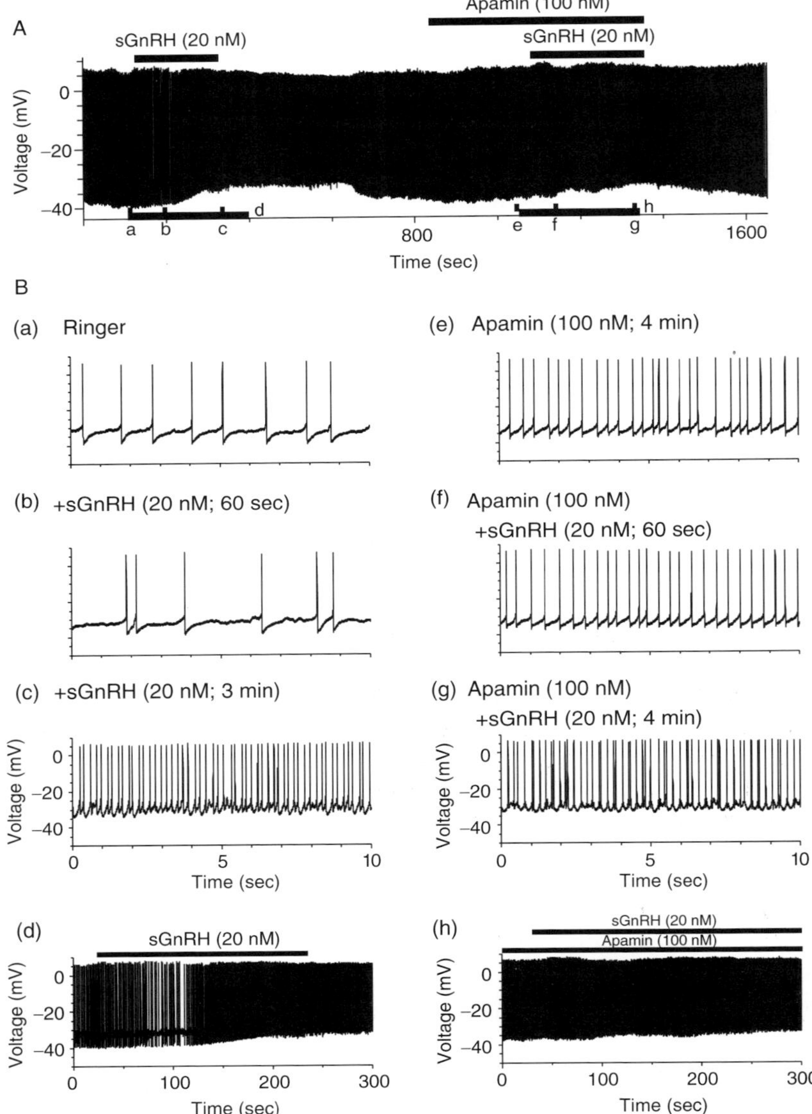

Fig. 11.5. Inhibition of the SK-type Ca^{2+}-activated K^{+} channel disrupts the transient decrease of pacemaker frequency after GnRH application. (A) Continuous recording of the pacemaker activity. (B) In Ringer solution, bath application of sGnRH (20 nM) transiently decreased (b) and subsequently increased (c) the frequency of pacemaker activity (the overall time course of these changes is shown in d). By bath application of apamin (100 nM), which inhibits the

In our experiments, we tested the hypothesis that intracellularly applied inhibitors of IP_3 and ryanodine receptors cause disruption of GnRH-induced changes in pacemaker potentials. The ryanodine receptor antagonist, Ruthenium red, and the IP_3 receptor antagonist, heparin (applied intracellularly via the recording pipette), disrupted the early-phase decrease of pacemaker potentials but left the late-phase increase intact (Abe and Oka, 2002). This early-phase decrease in pacemaker potential is hypothesized to occur through activation of Ca^{2+}-dependent K^+ currents by Ca^{2+} released from the intracellular stores. This was confirmed by current-clamp recording in which apamin, a specific blocker of the small conductance Ca^{2+}-dependent K^+ channel (SK channel), inhibited the early-phase decrease of pacemaker potentials (Figure 11.5; Abe and Oka, 2002). Van Goor *et al.* (1999a) also reported on the presence of SK-type Ca^{2+}-sensitive K^+ currents and their possible involvement in GnRH-induced activation in GT1–7 cells. Furthermore, in voltage-clamp recordings, we found a transient potassium current that is dependent on the presence of external Ca^{2+} ions. This current is kinetically different from the 4AP-sensitive potassium A-current that we previously found in these neurons (Abe and Oka, 1999). We obtained evidence to suggest that this tentative $I_{K(Ca)}$ is facilitated by sGnRH and contributes to the early-phase transient decrease of frequency of pacemaker potentials.

7.2. Late Phase: Subsequent Increase of Firing Activity

Concerning the late-phase increase of pacemaker frequency by sGnRH, we expected that sGnRH may modulate $I_{Na(slow)}$ or $I_{K(V)}$ or both, since the basic rhythm of the pacemaker potential was found to be generated by the interplay of $I_{Na(slow)}$ and $I_{K(V)}$ currents (see in an earlier section). However, our preliminary experiments showed that the bath application of sGnRH failed to evoke any noticeable modulation of the $I_{Na(slow)}$ or $I_{K(V)}$.

Therefore, we examined whether Ca^{2+} currents, typically thought to be involved in the generation of pacemaker activities, were modulated by GnRH (Abe and Oka, 2002). In a rather preliminary study, we used relatively wide-range blockers of voltage-dependent Ca^{2+} currents, such as multivalent cations (Ni^{2+}, La^{3+}, and so on). We recorded the whole-cell currents of TN-GnRH neurons under voltage clamp and examined the effects of sGnRH

SK-type Ca^{2+}-activated K^+ channel, the frequency of pacemaker activity was increased (e). Furthermore, bath application of apamin blocked the transient decrease (f) but not the subsequent increase in the frequency of pacemaker activities by sGnRH (g) (the overall time course of these changes is shown in h). Following washout by Ringer solution, the firing frequency decreased again (Abe and Oka, 2002).

in perfusing solution. We found that the persistent component of the Ca^{2+} current appeared to be upregulated by sGnRH.

We carried out a detailed voltage-clamp study using specific voltage-dependent Ca^{2+} channel blockers (Haneda and Oka, 2004) and found the involvement of voltage-dependent T-, L-, N-, and R-type Ca^{2+} channels in TN-GnRH neurons (Figure 11.6). These L-, N-, and R-type currents constituted 30.7 ± 3.1%, 41.0 ± 3.9%, and 23.6 ± 1.6%, of high-voltage activated (HVA) currents, respectively. These currents were recorded after inactivation of low-voltage activated (LVA) (T-type) current by holding the cells at −60 mV (Figure 11.7). Considering the activation kinetics, the L-, N-, and R-type Ca^{2+} currents, together with the TTX-sensitive conventional sodium current, $I_{Na(fast)}$, may be involved in the action potential phase (suprathreshold phase) of the pacemaker activities, Ishizaki *et al.* (2004) measured GnRH release from brain slices containing TN- and midbrain GnRH neurons by radioimmunoassay and showed that depolarization-induced GnRH release was significantly reduced by the application of nifedipine (blocker of L-type Ca^{2+} currents) and ω-conotoxin GVIA (blocker of N-type Ca^{2+} currents), but not by ω-agatoxin TK (blocker of P-type Ca^{2+} currents). These results further support the hypothesis that TN-GnRH neurons express L-, N-, R-, and T-type voltage-dependent Ca^{2+} currents. It has been demonstrated that embryonic GnRH neurons (Kusano *et al.*, 1995)

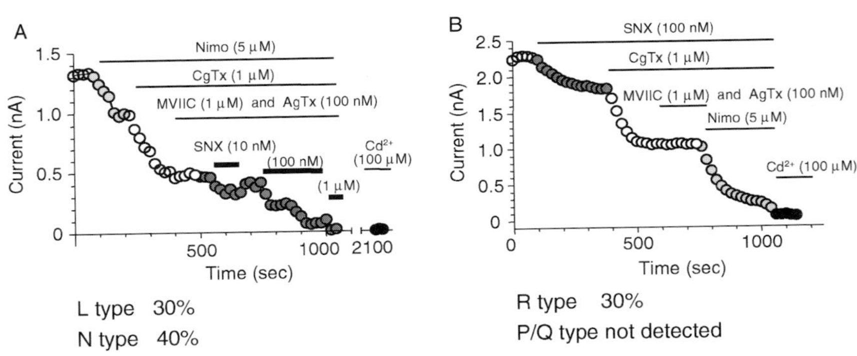

Fig. 11.6. Pharmacological characterization of the HVA Ca^{2+} currents in TN-GnRH neurons. (A) Time course of peak inward currents (absolute values) during sequential addition of specific blockers of HVA channels. Blockers were present during the periods indicated by the bars above the plots. Nimo, Nimodipine; CgTx, ω-conotoxin GVIA; MVIIC, ω-conotoxin MVIIC; AgTx, ω-agatoxin TK. The current component that was insensitive to all these blockers was blocked by SNX-482 (SNX) in a dose-dependent manner. (B) The effects of SNX were tested without prior addition of the other HVA channel blockers. It should be noted that SNX does not reduce the component sensitive to CgTx or Nimo. (Modified from Haneda and Oka, 2004.)

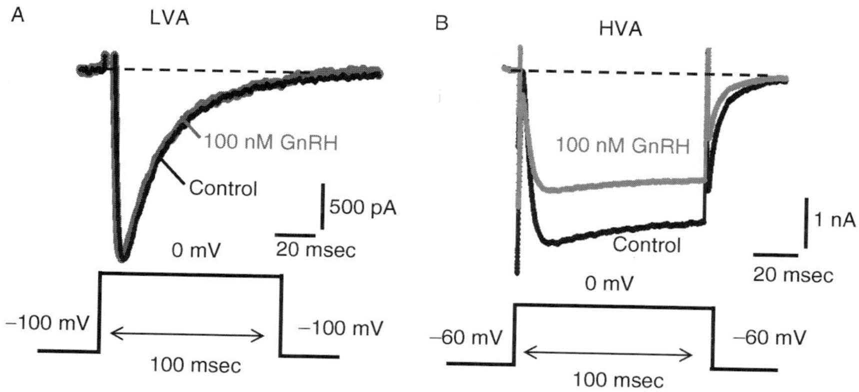

Fig. 11.7. HVA- but not LVA-Ca^{2+} currents are suppressed by GnRH HVA currents in TN-GnRH neurons. The current trace elicited by a depolarizing command pulse to 0 mV from a holding potential of −100 mV (A) or from that of −60 mV (B) show low-voltage-activated (LVA) and high-voltage-activated (HVA) Ca^{2+} currents, respectively. Note that only HVA- but not LVA-Ca^{2+} currents are suppressed by 100 nM GnRH (modified from Haneda and Oka, 2004). (See Color Insert.)

and GT1 cells (Bosma, 1993; Hales *et al.*, 1994; Javors *et al.*, 1995; Costantin and Charles, 1999) express several types of voltage-dependent Ca^{2+} currents. Kusano *et al.* (1995) reported that embryonic mouse GnRH neurons show T- and L-type Ca^{2+} currents, and GT1–7 cells show T-, N-, and L-type Ca^{2+} currents. Watanabe *et al.* (2004) reported on the expression profiles of voltage-dependent Ca^{2+} channels in GT1–7 cells, and Kato *et al.* (2003) and Nunemaker *et al.* (2003a) reported the existence of those in GFP-labeled hypothalamic GnRH neuron of rats and mice. GFP-labeled hypothalamic GnRH neurons showed T-, N-, L-, P/Q-, and R-type, whereas GT1–7 cells showed T-, N-, L-, and R-type Ca^{2+} currents. R-type Ca^{2+} current appears to be a major Ca^{2+} current component in GT1–7 cells, GFP-labeled hypothalamic GnRH neurons, and TN-GnRH neurons. However, some electrophysiological properties appear to be different due to possible channel heterogeneity (Tottene *et al.*, 1996) or differential expression of Ca^{2+} channel β subunits (Arikkath and Campbell, 2003). It has been suggested that tightly coupled HVA Ca^{2+} currents in active zones are involved in triggering neurotransmitter release in a very fast and localized manner (Rios and Stern, 1997; Neher, 1998). P/Q-type currents are reported to be most efficiently coupled to secretory machinery in neuromuscular junctions (Urbano *et al.*, 2003), and R-type current is more tightly coupled to the control of rapid secretory responses than N- or P/Q-type currents in chromaffin cells (Albillos *et al.*, 2000; Aldea *et al.*, 2002). The study of the differential distribution of different Ca^{2+} channel types and their coupling to GnRH release machinery in TN-GnRH neuron is an important problem.

Surprisingly, in a study (Haneda and Oka, 2004), HVA currents, representing up to 40% of the total current, were inhibited by the application of sGnRH, whereas LVA currents were not inhibited (Figure 11.7). The inhibition of HVA currents was dose dependent (EC_{50} was 11.5 nM) and channel type-specific among different HVA currents. The N- and R-type currents were preferentially inhibited, but L-type currents had lower sensitivity (Figure 11.8). Seemingly, there is a discrepancy of results concerning sGnRH-induced modulation of voltage-dependent Ca^{2+} currents compared with our former preliminary study (Abe and Oka, 2002). Earlier literature on the modulation of Ca^{2+} currents have shown many examples of voltage-dependent Ca^{2+} currents that are either inhibited or stimulated by the activation of G-protein–coupled receptors (Dolphin, 1998; Meir *et al.*, 1999). There are examples of modulation of HVA currents via the voltage-dependent inhibitory mechanisms, which could be relieved by a strong

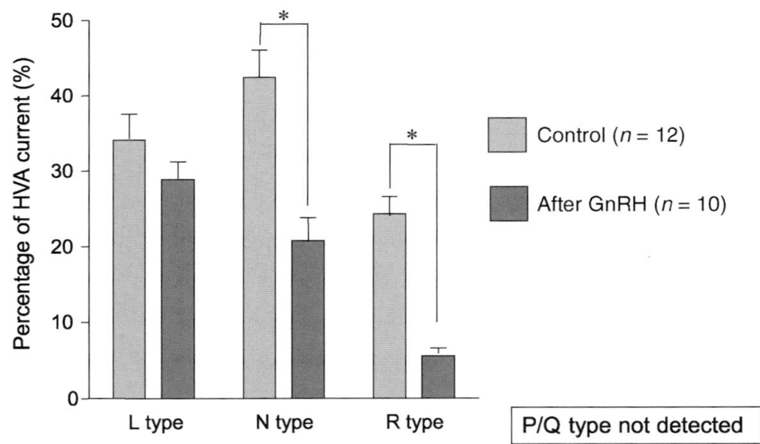

Fig. 11.8. Each HVA Ca^{2+} current subtype has different sensitivity for the GnRH-induced inhibition. Peak inward currents elicited by step depolarizations from −60 to −10 mV were plotted as a function of time, while HVA antagonists were sequentially applied (as in Figure 11.6). GnRH-insensitive current was categorized as L- (Nimo-sensitive) or N- (CgTX-sensitive) or R-type (residual, Cd^{2+}-sensitive) current. The graph shows the percentage of each GnRH-insensitive HVA current component compared with that obtained without GnRH treatment. In the control as well as the GnRH-treated groups, the percentage was defined as the ratio of the current amplitude of each subtype to the total HVA current amplitude before the GnRH treatment. There were statistically significant reductions in N- and R-type (*, $P < 0.05$) but not in L-type currents by GnRH treatment. (Modified from Haneda and Oka, 2004.)

depolarizing prepulse (Bean, 1989; Elmslie *et al.*, 1990). In voltage-dependent inhibition, the activation of the currents was modified to become slower, which has been referred to as "kinetic slowing" (Ikeda, 1996). This voltage-dependent inhibition was generally caused by the activation of G_i/G_o- or G_Z- or G_S-coupled receptors. The activation of a $G_{q/11}$-coupled receptor causes a voltage-independent inhibition, which is not accompanied by kinetic slowing (Hille, 1994). Since our results indicated a lack of prominent kinetic slowing or voltage-dependent relief of inactivation, the channel inhibition by GnRH observed here may belong to the voltage-independent inhibition mediated through the activation of a $G_{q/11}$-coupled receptor. On the other hand, HVA currents have been known to be positively modulated via the protein kinase C (PKC) pathway (Bourinet *et al.*, 1992; Yang and Tsien, 1993; Stea *et al.*, 1995; Zamponi and Snutch, 1998; Meir *et al.*, 1999). For example, Bosma and Hille (1992) reported that immortalized gonadotropes (αT3–1 cells), which also have GnRH-Rs, express HVA Ca^{2+} currents, and these currents were augmented by application of GnRH or phorbol esters. Bannister *et al.* (2004) reported, using a heterologous expression system, that muscarinic acetylcholine receptors evoke both inhibition and stimulation of $Ca_V2.3$, which are expressed in neuronal and neuroendocrine cells where they are believed to form native R-type Ca^{2+} channels. Muscarinic inhibition of $Ca_V2.3$ is mediated by $G_{\beta\gamma}$ subunits, whereas stimulation is mediated by $G_{\alpha q/11}$, diacylglycerol, and a Ca^{2+}-independent PKC. Similar phenomena have been reported for N- and P/Q-type Ca^{2+} currents (Swartz, 1993; Zhu and Ikeda, 1994; Zamponi *et al.*, 1997; Hamid *et al.*, 1999), and L-type currents have been known to be upregulated via both $G_{\alpha S}$/cAMP/PKA and $G_{\alpha q}$/PKC pathway (Kamp and Hell, 2000). Heterologous expression study using COS-7 cell line revealed that the GnRH-R is equally coupled to $G_{\alpha q}$ and $G_{\alpha 11}$ subunits (Grosse *et al.*, 2000). Krsmanovic *et al.* (2001) reported that GnRH-Rs in mouse hypothalamic neurons and GT1–7 cells are also coupled to $G_{\alpha i/o}$ subunits. Krsmanovic *et al.* (2003) further reported that G-protein coupling of the GnRH-R showed an agonist concentration-dependent switch from $G_{\alpha S}$ at low concentrations to $G_{\alpha i}$ at high concentrations. Based on this new information of ligand concentration-dependent selection of G proteins, a plausible mechanism for bidirectional modulation of TN-GnRH neuronal currents might be the use of multiple G-protein pathways. With this mechanism, the activity of TN-GnRH neurons would depend on extracellular concentration of GnRH and on the delicate intracellular milieu (e.g., stimulation via G_s- and $G_{q/11}$-coupled pathways at low sGnRH concentrations and inhibition via $G_{i/o}$- and $G_{\beta\gamma}$-coupled pathways at high sGnRH concentrations).

Alternatively, a possible contribution of other yet unidentified currents in the late-phase increase of pacemaker activity by sGnRH still remains to be determined.

For example, GnRH neurons of the rat express cyclic nucleotide gated (CNG) channels (El-Majdoubi and Weiner, 2002). GT1-1 cells also express CNG channels (Vitalis *et al.*, 2000), and the gating of these channels by cAMP has been shown (Charles *et al.*, 2001). In GT1-1 cells, cAMP analogs increased the frequency of calcium oscillations that have been associated with action potential bursts. In addition, manipulations of cAMP levels in rat hypothalamic GnRH neurons altered the pulsatile luteinizing hormone release pattern that in turn reflected the release pattern of GnRH (Paruthiyil *et al.*, 2002). Calcium influx through voltage-independent channels that are activated by depletion of Ca^{2+} stores has also been shown to be involved in late-phase increased pacemaker frequency in GT-1 cells (Van Goor *et al.*, 1999b). As reported in adrenal chromaffin cells, this kind of Ca^{2+} influx may further be involved in the release of GnRH peptides (Fomina and Nowycky, 1999). LeBeau *et al.* (2000) showed in a model study that both CNG channel-mediated current and store-operated Ca^{2+} currents contribute in the late-phase increase in GnRH-induced firing frequency of GT1-7 cells. Similar mechanisms may exist in TN-GnRH neurons. We have found evidence to suggest that simultaneous upregulation of Na^+-selective conductances and downregulation of K^+-selective conductances may underlie the GnRH-induced inward currents (our unpublished observations). Future studies should be designed to identify the nature of the conductances that are modulated by GnRH-induced signaling pathways. The $[Ca^{2+}]_i$ dynamics and subsequent modulation of GnRH release activity induced by GnRH application also needs to be carefully analyzed. The use of pharmacological interference with or genetic manipulations of the intracellular signaling pathways will be necessary to understand the mechanisms of GnRH-receptor signaling and subsequent effects on the various ionic currents and intracellular Ca^{2+} dynamics that are involved with the modulation of GnRH neuron pacemaker frequencies.

7.3. Significance of Cell-to-Cell Electrical Interactions in the Cluster of TN-GnRH Neurons

In many vertebrate species, the TN-GnRH neurons tend to form cell clusters (Schwanzel-Fukuda and Silverman, 1980; Stell *et al.*, 1984; Oka and Ichikawa, 1990; Steven *et al.*, 2003). This cell clustering is especially prominent in the dwarf gourami where TN-GnRH neurons are gathered into one tight cell cluster without intervening glial cells (Oka and Ichikawa, 1991; Oka, 1997). We have found that the TN-GnRH neurons in the dwarf gourami are electrically coupled with each other in the cluster (Haneda and Oka, unpublished). They seem to have synchronized pacemaker activities via these electrical couplings. Thus, the activity of the TN-GnRH neurons is synchronized not only by GnRH autocrine/paracrine control but also by their electrical interactions. The TN-GnRH neurons form not only morphological cluster but also a physiologically coordinated functional unit. In fact this neural unit provides positive feedback autocontrol in that the effect of GnRH on TN-GnRH neurons is the facilitation of increased pacemaker frequency (Abe and Oka, 2000), and this upregulation of pacemaker frequency leads to an increased release of GnRH (Ishizaki *et al.*, 2004). The effects of this positive autocontrol will be kind of a positive feedback upregulation of pacemaker activity. Furthermore, the electrical coupling among TN-GnRH neurons will synchronize the pacemaker activity of TN-GnRH neurons and keep them firing in unison. Our previous morphological study using single cell intracellular labeling (Oka and Matsushima, 1993) showed that the projection areas of each TN-GnRH neuron do not overlap completely. Although the axonal projection of each TN-GnRH neuron covers a very large area of brain, each neuron covers a distinct region not innervated completely by any other TN-GnRH neuron; for example, one neuron may preferentially project densely in the optic tectum and midbrain tegmentum, while another may project mainly to the telencephalic areas, and so on. The coordinated activity of TN-GnRH neurons as a cluster enables them to act as a functional syncytium. Because it has been suggested that GnRH peptides are released from the varicosities distributed along TN-GnRH axons and these peptides stimulate GnRH-Rs in many brain regions (Oka, 2002), the physiologically coordinated activity of TN-GnRH neuronal clusters would be advantageous for the simultaneous neuromodulation of the entire brain.

7.4. Physiological Significance of the Pacemaker Activity and Its Modulation

Here we discuss the physiological significance of pacemaker activity and its modulation by sGnRH in the TN-GnRH neurons. As we have stated earlier, the release of GnRH peptides is suggested to be related to the pacemaker frequency of TN-GnRH neurons. For example, in our studies, application of Glu increased the pacemaker frequency of TN-GnRH neurons and increased GnRH release from brain slices. Therefore, stimuli that lead to the modulation of TN-GnRH pacemaker activity will affect the release of GnRH from the varicose axons of TN-GnRH neurons thereby significantly influencing the neuromodulatory capability of TN-GnRH system in the brain.

Likewise, it has been well established that the release of GnRH from the preoptic/hypothalamic GnRH neurons is pulsatile, and this pulsatile release of GnRH is essential for the control of the hypothalamo-pituitary-gonadal axis and hence reproductive function in mammals (Terasawa, 1998, 2003;

Terasawa *et al.*, 1999). Although the TN- and preoptic/hypothalamic-GnRH systems consist of anatomically and functionally different cell populations in the forebrain, synthesize distinct molecular forms of GnRH, and have different embryonic origins (Parhar, 2002; Whitlock *et al.*, 2003), electrical recordings from hypothalamic GnRH neurons visualized in transgenic mice and rats have revealed electrophysiological characteristics that share some characteristics common to those already demonstrated in fish TN-GnRH neurons (Oka and Matsushima, 1993; Oka, 1997). Studies have revealed that GnRH modulates the firing frequency and/or pulsatile release of GnRH in GT1–7 cells (Zheng *et al.*, 1997; Van Goor *et al.*, 1999a,b, 2000) and in mouse hypothalamic GnRH neurons (Krsmanovic *et al.*, 1999; Martinez-Fuentes *et al.*, 2004; Xu *et al.*, 2004). This autocrine/paracrine control of peptide release is seen in other peptidergic systems. For example, the release of oxytocin from a single oxytocin neuron into the brain environment stimulates its own activity and thus further oxytocin release (Freud-Mercier and Richard, 1984; Moos *et al.*, 1984). The autocrine/paracrine positive feedback mechanism is probably a general feature of peptidergic neurosecretory neurons and secretory cells whose autoinduced synchronized activity leads to the facilitation of peptide release.

8. MULTIMODAL SENSORY INPUTS TO TN-GnRH SYSTEM

8.1. Neuroanatomical Evidence for Multimodal Sensory Inputs to TN-GnRH System

In our model of the neuromodulatory functions of the TN-GnRH neurons, we hypothesized that the pacemaker activity, which is closely related to the neuromodulatory function of these neurons, is under the control of various environmental and physiological conditions through hormonal or synaptic inputs (Oka, 1997, 2002). The modulation of TN-GnRH neuron pacemaker activity influences release of GnRH from the extensively projecting axonal branches, which simultaneously modulate neural activities in wide brain areas (see in an earlier section). In line with this hypothesis, Yamamoto and Ito (2000) demonstrated multimodal sensory inputs to the TN-GnRH neurons anatomically. The main afferent nucleus to the TN-GnRH system from the midbrain, the nucleus tegmento-terminalis (NTT[2]), receives somatosensory and visual information, and

[2]Yamamoto and Ito (2000) originally used the term "nucleus tegmento-olfactorius," but later renamed it as "nucleus tegmento-terminalis" to indicate its heavy projection to the TN-GnRH neurons, not the olfactory bulb itself.

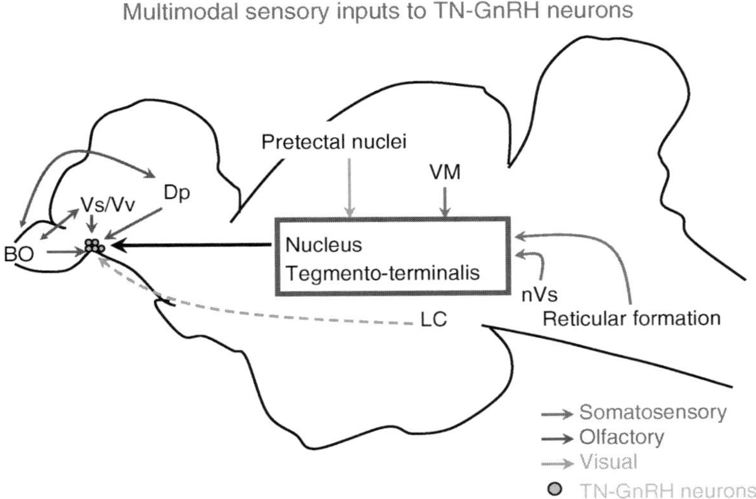

Fig. 11.9. Schematic diagram of the multimodal sensory inputs to the TN-GnRH neurons. The main afferent nucleus, termed the nucleus tegmento-terminalis (NTT), receives somatosensory and visual information, and the afferent sources in the telencephalic regions receive olfactory information. BO, bulbus olfactorius; Dp, area dorsalis telencephali pars posterior; LC, locus coeruleus; nVs, nucleus sensorius nervi trigemini; VM, nucleus ventromedialis thalami; Vs, area ventralis telencephali pars supracommissuralis; Vv, area ventralis telencephali pars ventralis (modified from Yamamoto and Ito, 2000). (See Color Insert.)

the afferent sources to the TN-GnRH system from telencephalon receive olfactory information. All of these nuclei send heavy axonal projections to the TN-GnRH neurons. These results suggest that such projections represent the neuronal pathways that convey sensory inputs from environmental conditions to the TN-GnRH system to change the activity of the TN-GnRH neurons (Figure 11.9; modified from Yamamoto and Ito 2000). In order to understand the nature of synaptic inputs and their physiological functions, we should first analyze, by using electrophysiological techniques, the neuronal pathways suggested by these anatomical studies.

8.2. Glutamatergic Excitatory Inputs to TN-GnRH Neurons

Glutamate, the major neurotransmitter of the vertebrate brain, is also known to be an important regulator of neuroendocrine secretion in the GnRH system (van den Pol *et al.*, 1990). Thus, we hypothesized that the TN-GnRH neurons receive glutamatergic multimodal sensory inputs, which synaptically regulate the release of GnRH. When we applied Glu to slices of brain *in vitro*, the frequency of pacemaker potentials of TN-GnRH neurons were increased in a dose-dependent manner within a certain concentration range (cf. Kiya and Oka, 2003). Furthermore, incubation of TN-GnRH brain slices in a Ringer solution containing Glu (0.5–5 mM) significantly increased the GnRH release in a dose-dependent manner (Ishizaki and Oka, unpublished data; cf. Ishizaki *et al.*, 2004). Thus, GluR activation can potentially increase the frequency of pacemaker activity and, consequently, the release of GnRH from TN-GnRH neurons. To test this hypothesis we analyzed electrophysiologically the type of GluRs expressed by TN-GnRH neurons using various specific GluR agonists and antagonists.

First, we pharmacologically identified the type of GluRs in the TN-GnRH neurons (Figure 11.10). To isolate the membrane potential responses, experiments were performed in the presence of TTX, a blocker of conventional fast sodium channels. Bath application of kainic acid (KA), a specific agonist of non-*N*-methyl-D-aspartate (non-NMDA) type of GluRs, depolarized the TN-GnRH neurons (Figure 11.10Aa). However, application of KA in the presence of 6-cyano-7-nitroquinoxaline-2,3-dione (CNQX), a specific antagonist of non-NMDA type of GluRs, cells showed no depolarizing response (Figure 11.10Ab). Similarly, bath application of NMDA, a specific agonist of NMDA type of GluRs, depolarized the TN-GnRH neurons in an Mg^{2+}-free Ringer solution (Figure 11.10Ac). In addition, application of NMDA in the presence of D-2-amino-5-phosphonopentanoate (D-AP5), a specific antagonist of NMDA type of GluRs, leads to only subtle depolarizing responses (Figure 11.10Ad). In contrast, bath application of 3,5-dihydroxyphenylglycine (3,5-DHPG), a specific agonist of group 1 metabotropic GluRs (mGluRs), or (2*R*, 4*R*)-4-aminopyrrolidine-2,4-dicarboxylate [(2*R*, 4*R*)-APDC], a specific agonist of group 2 mGluRs, had no effect (Figure 11.10Ba and Bb). Furthermore, bath application of (1*S*, 3*R*)-1-aminocyclopentane-1,3-dicarboxylic acid [(1*S*, 3*R*)-ACPD], a nonspecific agonist of group 1 and 2 mGluRs, had no effect. Unexpectedly, L-(+)-2-amino-4-phosphonobutyric acid (L-AP4) (100 μM), a specific agonist of group 3 mGluRs, hyperpolarized the membrane (Figure 11.10Bc). When L-AP4 was applied in the presence of (*RS*)-α-cyclopropyl-4-phosphonophenylglycine (CPPG), a specific antagonist of group 3 mGluRs, cells showed almost no hyperpolarizing response.

To confirm that we have identified all types of GluRs in the TN-GnRH neurons, we applied Glu in the presence of a cocktail of various blockers. In spite of the combined presence of TTX, CNQX, and D-AP5 in the Ringer solution, and GDP-β-S, a general blocker of G-protein–coupled signaling pathways in the pipette solution, a large depolarizing response to Glu remained (Figure 11.10D). Response to Glu was also examined in the combined presence of TTX, CNQX, D-AP5, and *N*-ethyl maleimide, a membrane-permeable

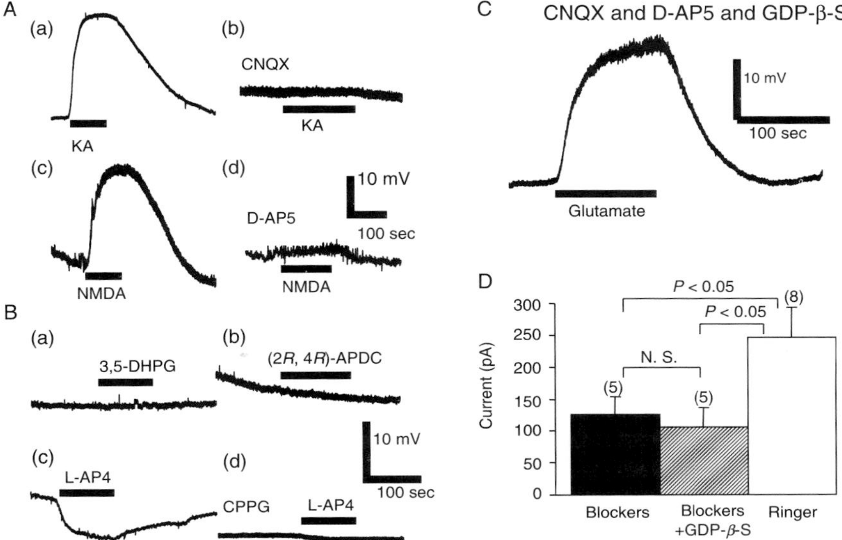

Fig. 11.10. TN-GnRH neurons have both iGluRs and mGluRs. The effects of specific agonists and antagonists of several types of GluRs were examined in the presence of TTX (0.75 μM). Each record is from a different cell. (A) iGluRs were examined. Kainic acid (KA) depolarized the cell (a). In the presence of CNQX (AMPA receptor blocker), cells showed no depolarizing response (b). In the Mg^{2+}-free Ringer solution, NMDA depolarized the cell (c). In the presence of D-AP5 (NMDA receptor blocker), cells only showed a small depolarizing response (d). (B) mGluRs were examined. 3,5-DHPG or (2*R*, 4*R*)-APDC had no effect on the cells (a and b). L-AP4 hyperpolarized the cells (c). In the presence of CPPG, cells showed almost no hyperpolarizing response (d). (C) A large voltage response to Glu (1 mM) remained in the combined presence of CNQX (10 μM) and D-AP5 (25 μM) in the Ringer bath solution and GDP-β-S (1 mM) in the pipette solution ($n = 9$). (D) In the presence of blockers (TTX 0.75 μM, CNQX 10 μM, D-AP5 25 μM, and CPPG 10 μM), Glu-induced current (246 ± 47 pA, $n = 8$) was significantly blocked (126 ± 29 pA, $n = 5$)($P < 0.05$). However, further addition of intracellular GDP-β-S (1 mM) made no significant difference (106 ± 32 pA, $n = 5$). (Modified from Kiya and Oka, 2003.)

inhibitor of pertussis-toxin (PTX)-sensitive G-proteins, in a Ringer solution. However, the response to Glu still remained. To quantitatively analyze this residual Glu-induced response, we measured the Glu-induced current in the voltage-clamp mode (Figure 11.10D). The cells were voltage clamped at −60 mV. In the presence of TTX, bath application of Glu evoked an inward current. This current was significantly blocked in the presence of blocker cocktail (TTX, CNQX, D-AP5, and CPPG). However, further addition of intracellular GDP-β-S made no significant difference. Since the concentration of each blocker in the cocktail was high enough to block the voltage response caused by each ionotropic GluR (iGluR; Figure 11.10D), it is possible that a pharmacologically novel type of iGluR is present, in addition to the conventional iGluRs and mGluRs, in the TN-GnRH neurons.

Thus, we could demonstrate that Glu increases the pacemaker frequencies of the TN-GnRH neurons, and that non-NMDA and NMDA types of iGluRs and group 3 mGluRs are present in the TN-GnRH neurons (Kiya and Oka, 2003). As the anatomical study of Yamamoto and Ito (2000) indicated the possible multimodal sensory inputs from the NTT to the TN-GnRH neurons as described earlier, may be the source of glutamatergic inputs to the TN-GnRH neurons and may change the pacemaker activity of the TN-GnRH neurons. It is well known that activation of iGluRs, especially the NMDA type of GluRs, facilitates the influx of extracellular Ca^{2+}, which plays a crucial role in the exocytosis. Thus, glutamatergic inputs to the TN-GnRH neurons may increase not only the pacemaker frequencies but also the release of GnRH peptides from the axonal branches.

It is noteworthy that Kiya and Oka (2003) also demonstrated the possible presence of pharmacologically novel type of iGluRs in addition to three other conventional types of GluRs. In the experiments using specific agonists and antagonists, each antagonist could clearly block the response of each type of conventional receptor. In spite of this, about half of all types of Glu responses in the TN-GnRH neurons remained in the presence of blocker cocktail that should block all the conventional iGluRs as well as mGluRs. This is obviously too large to be regarded as an unblocked conventional GluR current component. Although the possibility remains that this current is derived from differences in the pharmacology between teleost GluRs and mammalian ones or from contribution of Glu transporters, it is possible that a pharmacologically novel type of iGluR is present in the TN-GnRH neurons. GluRs in the central nervous system (CNS) have been intensively investigated and have been largely categorized into five major groups: non-NMDA and NMDA type of iGluRs, and group 1–3 mGluRs. However, the type of iGluR suggested in the present study has not been reported to date. From the quantitative analysis with voltage-clamp method, about half of all types of Glu responses in the TN-GnRH neurons remained. Although we can only speculate at present that the novel type of iGluRs is dominantly expressed and has important functions in the TN-GnRH neurons, further studies will elucidate this challenging problem.

8.3. Inhibitory Inputs to TN-GnRH Neurons

In addition to the excitatory synaptic inputs to the TN-GnRH neurons, the existence of inhibitory synaptic inputs is also probable. Fujita *et al.* (1985) first reported on the electrophysiological characteristics of the TN neurons in the Asian carp (*Cyprinus carpio*). In their pioneering study, they recorded from TN neurons in the olfactory nerve layer of the Asian carp by using intracellular recoding techniques, although they could not identify whether the neurons were GnRH-immunoreactive or not. Their recording also revealed pacemaker-like activities (Fujita *et al.*, 1985), although they were not as regular as those recorded from the TN-GnRH neurons of the dwarf gourami. They also described a long-lasting slow inhibitory post-synaptic potential (IPSP)-like synaptic potential after stimulation of the medial olfactory tract (Fujita, 1987), which suppressed spontaneous activity. Later, Fujita *et al.* (1991) studied the effects of various sensory stimuli on the spontaneous pacemaker-like activity of TN neurons recorded extracellularly in the goldfish with the hope to confirm the TN chemosensory hypothesis of Demski and Northcutt (1983). The only sensory input to the TN neurons in the goldfish that affected their activity was nociceptive in nature (pressure and tail pinch stimuli), and it inhibited the spontaneous pacemaker-like activity. Thus, it is probable that synaptic inputs that underlie the IPSP-like potentials found in the carp contribute to this inhibition of spontaneous pacemaker-like activity. However, the origin of this IPSP-like potential still remains to be determined.

White and Meredith (1987) also recorded intracellularly from the TN ganglion cells *in vitro* in elasmobranchs, the bonnethead shark and stingray, and found spontaneous pacemaker-like activities as well as IPSP-like and excitatory postsynaptic potential (EPSP)-like responses after stimulation of their central and peripheral axons[3]. Although these electrophysiological studies are preliminary and need further clarification, they recorded IPSP- and EPSP-like potentials in response to nerve stimulations; this proves that the TN "ganglion" is not like the sympathetic or dorsal root ganglia that lack synapses. Although the afferent sources of these IPSP-like responses are not known in the carp, goldfish, or elasmobranchs, they may be one of the list of candidate brain areas identified by anatomical tract tracing study of Yamamoto and Ito in two teleost species (Yamamoto and Ito, 2000).

As discussed earlier, the excitatory and inhibitory synaptic inputs to TN-GnRH neurons are suggested to relay the multimodal sensory inputs from the environment to the neuromodulatory TN-GnRH neurons and thus affect the activity of TN-GnRH neurons. Along this line, it is an important future problem to identify the electrophysiological and pharmacological nature and the sources of inhibitory and excitatory synaptic inputs, and to study how the natural sensory inputs, such as visual, olfactory, or somato-sensory, affect the activity of TN-GnRH neurons. Such experiments will require conducting intracellular or patch-clamp recordings from semi-intact

[3]The elasmobranch TN neurons are separate from the brain and form clusters as terminal nerve ganglion, which has both central and peripheral nerve components.

TN-GnRH neurons *in vivo* while applying natural sensory stimuli to the sensory organs.

8.4. Fine Structural Evidence for Synaptic Inputs to TN-GnRH Neurons

There are only a few reports on the fine structural evidence for synaptic inputs to the TN-GnRH neurons (Wedekind, 1979; Matsutani and Uchiyama, 1986; Oka and Ichikawa, 1991). Matsutani and Uchiyama (1986) reported the presence of three morphologically different types of synaptic terminals in the "nucleus olfactoretinalis" (equivalent to the structure including TN-GnRH and non-GnRH neurons) of black scraper. They noted that S (synaptic terminals with spherical synaptic vesicles) and F2 synapses [those with flattened synaptic vesicles and small dense-cored vesicle (DCV)] terminate on large cells (equivalent to TN-GnRH neurons), while S, F1 (synaptic terminals with flattened vesicles), and F2 synapses terminate on medium-sized cells (non-GnRH TN neurons). By combining brain lesioning and electron microscopy (EM) they suggested a telencephalic nucleus Vs (supracommissural part of the area ventralis of the telencephalon) as the possible origin of S terminals to the large TN-GnRH neurons. The telencephalic origin of afferents to the TN-GnRH neurons is in good agreement with those reported by Yamamoto and Ito (2000).

Wedekind (1979) also noted synaptic terminals in deep invaginations of the cell body as well as on the dendrites of TN-GnRH neurons in a labyrinth fish. They described the fine structure of synaptic terminals as having small and elongated synaptic vesicles and possible glycogen granules. They did not know the origins of these synaptic inputs, and it is rather difficult to discuss the homology of synaptic inputs among different fish species.

In the dwarf gourami, Oka and Ichikawa (1991) studied the TN-GnRH neurons and TN non-GnRH neurons ultrastructurally. TN-GnRH neurons bear very few thick primary dendrites, but numerous thin processes, "somatic processes," arise directly from the cell body. These somatic processes are thin (usually less than 0.5 μm in diameter) and have intricate branching patterns. The somatic processes are always intermixed with other similar processes, astrocytic processes with glial filaments, small myelinated and unmyelinated axons, and other small, unidentified profiles. They coined the term the "glomeruloid complex" for this complex of structures. Synaptic inputs that impinge directly upon TN-GnRH cell bodies are extremely rare. On the other hand, many synaptic terminals are found in the glomeruloid complex. Thus, the glomeruloid complex can be regarded as a major target of synaptic inputs, as well as a site of special cell-to-cell contact among TN-GnRH neurons. Most of the synaptic terminals found in the glomeruloid complex had clear spherical synaptic vesicles and asymmetrical postsynaptic membrane thickenings (typical of excitatory synapses), while a smaller number of terminals had several DCVs in addition to the spherical vesicles. The DCVs found in such synaptic terminals are much smaller than the DCVs distributed in the cytoplasm of TN-GnRH cells, which suggests that they contain catecholamines. It is conceivable that the synaptic terminals observed in the dwarf gourami, which have asymmetrical postsynaptic thickenings and spherical synaptic vesicles, correspond to the S terminals of Matsutani and Uchiyama (1986), whereas those with several DCVs compare to their F2 terminals. Thus, the two EM studies in different fish species suggest at least two different types of synaptic inputs to the TN-GnRH neurons.

By combining the morphological and electrophysiological techniques, we will better understand the synaptic control mechanisms underlying the multimodal sensory input to the TN-GnRH neurons. Of course, it will be necessary in the next step to study the effects of natural sensory stimuli on the physiology of TN-GnRH neurons as well as the electrophysiological and morphological studies of these synaptic inputs to them.

9. NEUROMODULATORY ACTION OF GnRH PEPTIDES

Experimental evidence for a neuromodulatory action of GnRH (causing late slow EPSP and increase in the neuronal excitability) has originally been suggested in the study of peripheral ganglionic cells in the sympathetic ganglia of bullfrog as early as 1970s (Jan and Jan, 1983). It was reported that GnRH is coreleased with acetylcholine from the preganglionic sympathetic cells and evokes late slow EPSP in the postganglionic cells. This late slow EPSP occurs in the time course of tens of seconds to minutes, and so differs considerably from the conventional fast synaptic transmission mediated by ionotropic receptors. It is now known that the slow response is mediated by metabotropic receptors, G-protein–coupled GnRH-Rs. Brown *et al.* (1981) showed that this extremely slow and long-lasting potential change results from the closing of noninactivating "M-type" K^+ current that is active at the resting membrane potential range and causes an increase in the excitability of the postganglionic neuronal membranes. For example, the most striking effect of M-current inhibition is a reduction in spike frequency accommodation as follows (Jones, 1987). In the bullfrog sympathetic neurons, during maintained depolarizing current injections, the cell normally fires only once or with a brief burst of action potentials. This is due to slow activation of the M-current upon depolarization. When the M-current is inhibited, however, the cell remains relatively depolarized and can fire action potentials repeatedly. This kind of increase in the excitability may also function in the brain and may be one of the main physiological functions of neuromodulation by GnRH.

Despite intensive efforts of researchers in many groups, signaling elements that couple GnRH-R activation to M-current modulation have remained enigmatic. However, it has been shown that activation of phospholipase C initiates M-current modulation and that recovery requires ATP and phosphoinositide 4-kinase (Suh and Hille, 2002). This study suggested that breakdown of phosphatidylinositol 4,5-bisphosphate (PIP_2) is a crucial determinant of M-channel modulation. In addition, the identity of gene products comprising M channels was clarified by Wang *et al.* (1998) who demonstrated that K^+ channels of the KCNQ subtype comprise M channels. This brought new excitement to the field since polymorphisms in human KCNQ genes are linked to diseases such as epilepsy, arrhythmias, and deafness, thus underscoring the importance of these channels in mediating neuronal excitability (Jentsch, 2000).

Importantly, GnRH peptide is also involved in the neuromodulation of Ca^{2+} currents, which are involved in various physiological events in neural as well as nonneural cells. Elmslie *et al.* (1990) reported that GnRH modulate N- and P/Q-type Ca^{2+} currents, and the signal transduction mechanisms of the modulation have been well studied (e.g., Dolphin, 1998). Furthermore, Ford *et al.* (2003) reported in bullfrog sympathetic ganglia that GnRH exerts long-term neurotrophic regulation of the N-type Ca^{2+} channel expression in addition to generation of late slow EPSP by inhibition of the M-type K^+ current. We characterized Ca^{2+} channels expressed in the TN-GnRH neuronal cell bodies and found a selective modulation of channels by autocrine/paracrine action of GnRH peptides released from the TN-GnRH neurons (Haneda and Oka, 2004; see Section 7).

On the one hand, it has been suggested that HVA Ca^{2+} currents tightly coupled to exocytotic machinery in the active zones are involved in triggering neurotransmitter release in a very fast and localized manner (Rios and Stern, 1997; Neher, 1998). Since the N- and P/Q-type Ca^{2+} channels are localized in presynaptic active zones and are involved in the Ca^{2+} influx necessary for the transmitter release, it is suggested that GnRH peptide modulates the transmitter release via modulation of these Ca^{2+} currents. However, this has not yet been demonstrated experimentally. Complex meshlike presynaptic axon terminals that form huge glomerulus-like structures in a thalamic nucleus of certain teleost fish brain have been discovered quite recently (Tsutsui, personal communication). These may be large enough to be amenable to the simultaneous study of intracellular Ca^{2+} dynamics and electrical activities, especially of Ca^{2+} channels, in the presynaptic terminals, which has been difficult in the presynaptic terminals of most vertebrate brains. Furthermore, this thalamic nucleus seems to be one of the targets of the TN-GnRH neuromodulatory system (distribution of GnRH fibers as well as GnRH-Rs; Oka *et al.*, unpublished observations). Synaptic transmission in this thalamic nucleus has already been well studied by Tsutsui *et al.* (Tsutsui *et al.*, 2001; Tsutsui and Oka, 2002). It should therefore be a very exciting future problem to study the modulation of calcium dynamics and the resultant synaptic transmissions in these synaptic terminals by the neuromodulatory action of TN-GnRH system.

On the other hand, there is only little information on the possible neuromodulatory function of GnRH in the CNS. In the rat hippocampus, GnRH induces a long-lasting depolarization associated with increased input resistance, decrease of the afterhyperpolarization following a train of action potentials, and reduction of accommodation of repetitive cell discharge (Wong *et al.*, 1990). Yang *et al.* (1999) reported that GnRH treatment induced a long-term enhancement of excitatory synaptic transmission that was mediated by iGluRs. They also showed that this GnRH-R–mediated synaptic potentiation was associated with stimulation of PKC. However, the mechanisms have not been studied further, probably because of technical difficulties in experiments using the brain instead of the peripheral ganglionic neurons.

10. DISTRIBUTION OF GnRH RECEPTORS IN THE BRAIN

After the finding of extensive extrahypothalamic GnRH neuronal systems (wide distribution of GnRH-immunoreactive fibers; see in an earlier section) in the vertebrate brain, the progress of research on the anatomical distribution of GnRH "receptors" in the brain has been and still is rather slow. The earlier literature reported on the anatomical distribution of GnRH binding sites by using the radiolabeled GnRH ligands. These studies reported results suggestive of GnRH "action at a distance" (Jan and Jan, 1983); the distribution of binding sites for the radiolabeled ligands was rather different from the distribution of GnRH-immunoreactive fibers in the extrahypothalamic areas. The situation where GnRH ligands and receptors are located at a distance is similar to that first proposed in the neuromodulatory action of GnRH after diffusion in the sympathetic ganglia.

After the cloning of genes encoding GnRH-Rs (Millar *et al.*, 2004), some authors have reported the anatomical distribution of GnRH-R mRNA-expressing neurons by using *in situ* hybridization techniques or immunohistochemistry in teleost brains (goldfish, Peter *et al.*, 2003; rainbow trout, Madigou *et al.*, 2000; Tilapia, Soga *et al.*, 2005). Since several different subtypes of GnRH-R genes have been cloned to date, we have to pay special attention to the selection of specific gene probes that are specific to a certain type of GnRH-R. Although the number of literature on the specific anatomical distribution of different subtypes of GnRH-R genes is sparce, the available literature seems to suggest differential distribution of different subtypes of GnRH-Rs in different areas of the brain (Peter *et al.*, 2003; Soga *et al.*, 2005). However, because of the inherent technical limitations in each experimental method, immunohistochemistry, *in situ* hybridization, and autoradiography, combinations of these methods may be required to determine the anatomical and functional relationships between the GnRH release sites and the receptors. Such experiments are needed in order to prove the possibility of "GnRH action at a distance" in the CNS as originally suggested in the autonomic ganglia (Jan and Jan, 1983).

11. NONSYNAPTIC RELEASE OF GnRH

As already described earlier, the axons of the TN-GnRH neurons are distributed widely throughout the brain from the olfactory bulb to the spinal cord. Therefore, GnRH peptides can potentially modulate the excitabilities and/or the transmitter release of the target neurons simultaneously via the widely projecting axonal branches when GnRH peptides are released from these axons. Hydrophilic macromolecules, such as peptides, are generally released from the cells via exocytosis, and this applies to the release of GnRH from the TN-GnRH neurons. The ultrastructural studies of TN-GnRH neurons (Matsutani and Uchiyama, 1986; Demski and Fields, 1988; Oka and Ichikawa, 1991) have shown that their cell bodies contain numerous membrane-bound DCVs. As shown in Section 6 , there is both morphological and physiological evidence for the release of GnRH peptides in the somatodendritic area. In the GnRH fiber varicosities, GnRH immunoreactivities have also been demonstrated in DCVs (Oka and Ichikawa, 1992), but none of these structures showed any evidence of the presence of synaptic active zones, that is, postsynaptic densities, widened synaptic clefts, or accumulation of synaptic vesicles in the presynaptic membrane. Thus, we hypothesize that TN-GnRH cells release GnRH nonsynaptically from DCV-containing fiber varicosities as well as somatodendritic areas and that it exerts its modulatory action on GnRH-Rs located on nearby as well as distant target neurons. There is morphological evidence favoring the idea of nonsynaptic release of neuropeptides (Thureson-Klein and Klein, 1990). Similar mechanisms have also been suggested for GnRH release in the midbrain central gray of the rat (Buma, 1989). Such mode of peptidergic action may be relevant for long-lasting and widespread neuromodulation. In this respect, it will be an important future problem to study the anatomical relationship between the GnRH release sites and the GnRH-Rs at both light and electron microscopic levels.

12. MODULATION OF NEURAL FUNCTIONS BY GnRH

We now turn to the discussion of the modulation by GnRH peptides of neural function as a whole. As we mentioned in the Introduction, teleost retinas are characterized by a dense projection of "retinopetal" fibers from the TN-GnRH neurons. These fibers are known to synapse on dopaminergic interplexiform cells (Zucker and Dowling, 1987), and GnRH and other neuropeptides (e.g., FMRFamide) affect the activity of retinal ganglion cells (Stell *et al.*, 1984; Walker and Stell, 1986). Umino and Dowling (1991) reported that when the retina was superfused with Ringer solution containing GnRH, horizontal cells depolarized and their response to small spots increased, whereas their responses to full-field lights decreased. Their results suggested that GnRH acts by stimulating the release of dopamine from interplexiform cells. Furthermore, Behrens *et al.* (1993) reported that GnRH elicits light adaptive formation of horizontal cell spinules *in vitro* by stimulating the dopaminergic interplexiform cells. Li and Dowling (2000) reported on visual system-defective *night blindness b* mutants of zebrafish. These mutant fish were characterized by degeneration of the centrifugal projection from the TN-GnRH neurons to the retina and the simultaneous degeneration of retinal dopaminergic interplexiform cells (known to be the target of retinopetal projection; Zucker and Dowling, 1987) and disappearance of the effect of dopamine on the outward potassium current of bipolar cells. Maaswinkel and Li (2003) found that olfactory stimulation with amino acids increases behavioral visual sensitivity, but this effect was eliminated after disruption of the TN-GnRH projections to the retina or after the destruction of dopaminergic interplexiform cells. This visual defect was similar to that of the *night blindness b* mutants. Based on the behavioral study, they suggested that the TN system and dopaminergic interplexiform cells are the main candidates for olfactory modulation of visual sensitivity. Although the functional relationship between the TN-GnRH and olfactory systems are not known, the existence of anatomical projection from the olfactory bulb to the TN-GnRH neurons (Yamamoto and Ito, 2000) may support their suggestions.

A differential usage of GnRH-R subtypes has also been suggested in the retina (Grens *et al.*, 2005). On the basis of anatomical study on the differential localization of GnRH-R subtypes, Grens *et al.* (2005) suggested that GnRH derived from the retinopetal axons of TN-GnRH system could broadly influence processing of retinal signals in both lateral processing circuits (represented by amacrine cells) through one type of GnRH-R (GnRH-R1) and in the vertical throughput pathway (represented by retinal ganglion cells) through another subtype (GnRH-R2). Thus, the TN-GnRH fibers projecting to the retina seem to have a definitive physiological function.

The visual system is not the only system modulated by TN-GnRH neurons. There is evidence that the olfactory system is also modulated by the TN-GnRH peptidergic system. It has been found in some species that the axons of TN-GnRH neurons are present in the olfactory nerve and can be traced to the lamina propria of the olfactory epithelium (mudpuppy, Eisthen *et al.*, 2000; dwarf gourami, Wirsig-Wiechmann and Oka, 2002; reviewed by Wirsig-Wiechmann *et al.*, 2002). This suggests an intriguing possibility that GnRH peptides derived from the TN-GnRH system may have access to the olfactory receptor neurons and may modulate responses to odorants including sex pheromones (Wirsig-Wiechmann *et al.*, 2002). In this context, Eisthen *et al.* (2000) reported a result showing that GnRH modulates the sensitivity of olfactory receptor neurons by modulating their Na^+ and K^+ channel properties. They used the olfactory receptor neurons from mudpuppies (*Necturus maculosus*) and did voltage-clamped whole-cell recordings to examine the effects of GnRH on voltage-activated currents in olfactory receptor neurons from olfactory mucosal slices. They found that GnRH slowly but reversibly increases the magnitude, but does not alter the kinetics, of a TTX-sensitive inward current (most probably conventional Na^+ current involved in the generation of action potentials) and a certain outward current. This effect appeared to be seasonal with more animals responding to GnRH during the courtship and mating season. They suggest that GnRH increases the excitability of olfactory receptor neurons and that the TN-GnRH neurons function to modulate the odorant sensitivity of olfactory receptor neurons. GnRH modulation of odorant responses in the olfactory epithelium has also been reported in axolotls (Park and Eisthen, 2003). Although this hypothesis needs to be further tested by examining the olfactory responses to the odorants and the currents need to be isolated for more rigorous electrophysiological analysis, it provides very attractive working hypotheses to be tested in future studies.

Thus, the study of the neuromodulatory function of GnRH on the visual and olfactory systems and of the possible bidirectional interactions among the visual, olfactory, and the TN-GnRH neuromodulatory systems should be one of the most exciting topics to be studied in the near future. Since the projection of GnRH fibers arising from the TN-GnRH system and the distribution of GnRH-Rs is extensive throughout the brain, we may expect various kinds of behavioral effects that result from the neuromodulatory action of GnRH. It will be an interesting and challenging future topic to study the neuromodulatory action of GnRH on GnRH target neurons in the brain (such as olfactory bulb, ventral telencephalon, or optic tectum). The whole brain *in vitro* preparation of dwarf gourami will allow the relationship between TN-GnRH neuron spontaneous discharge mode and pacemaker frequency and brain function to be examined.

13. BEHAVIORAL FUNCTIONS OF TN-GnRH SYSTEM

Concerning the function of TN, Demski and Northcutt (1983) hypothesized that the TN was a chemosensory nerve that sent pheromonal information to the preoptic area. They reported in the goldfish that the TN neurons, which are located at the rostral base of the olfactory bulb, project to the retina, supracommissural part of the area ventralis of the telencephalon (Vs) and the preoptic area. They also reported that electrical stimulation of the optic nerve induced sperm release. They supposed that electrical stimulation of the optic nerve should antidromically stimulate the retinopetal TN fibers and lead to the collateral activation of the preoptic area and Vs. It had been previously suggested in teleosts that the preoptic area and Vs are important for the facilitation of sexual behaviors (Koyama *et al.*, 1984; Satou *et al.*, 1984). Demski and Northcutt (1983) hypothesized that the TN is a new chemosensory system in vertebrates, and that pheromonal stimulation facilitates sexual behavior via its projection to the preoptic area. However, this possibility was later refuted by Fujita *et al.* (1991). They stimulated the olfactory epithelium of the goldfish with identified sex pheromones (17α20β-dihydroxy-4-pregnen-3-one and prostaglandin F2α) and recorded the electrical activity of TN neurons and mitral cells (the principal neuron that receives synaptic input from olfactory receptor cells and projects to the secondary olfactory center). Whereas the mitral cells responded to the sex pheromones, the TN neurons did not. Thus, the experimental results do not support Demski's hypothesis that the TN-GnRH system is chemosensory. We have also shown behaviorally that the total lesion of bilateral TN-GnRH neurons in the male dwarf gourami does not interfere with the overall performance of the sexual behaviors, although it affected the motivational or arousal state of the fish (Yamamoto *et al.*, 1997; see later).

Yamamoto *et al.* (1997) analyzed the behavioral changes in the dwarf gourami after bilateral electrolytic lesions of the TN-GnRH neurons. More than 2 weeks after the lesion, the GnRH fibers in the brain that arise from the TN-GnRH cells degenerated. However, the sexual behavior of these TN-defective male remained almost unchanged except for a characteristic impairment in one of the repertoires of male sexual behavior and nest building. The occurrence of mating trials in which males showed no nest building at all was increased. The fish seemed to require stronger stimuli for the initiation of nest building, but the overall frequency of nest building was not significantly affected once the behavior was triggered. From these results, Yamamoto *et al.* (1997) suggested that the TN-GnRH system is involved in the control of the threshold for nest-building behavior initiation. In other words, the motivational level of that behavior was affected. The only available data concerning the behavioral function of TN-GnRH system in vertebrates other than teleosts is from a study by Wirsig and Leonard (1987). They examined the effects of bilateral TN transections on the sexual behavior of the male hamsters (the hamster TN axons form bundles and can be surgically ablated). They found that TN lesions produced a decrease in mating frequency or an increase in the number of intromissions required to reach ejaculation. The latter effect indicates the increased threshold for the ejaculation, which may be regarded as the decreased level of motivation for the behavior after the TN lesion.

More detailed behavioral experiments are necessary to understand the behavioral functions of the neuromodulatory TN-GnRH system. For example, the performance of sexual behavior repertoires should be quantitatively analyzed after application of agonists or antagonists that are specific to the GnRH and/or GnRH-Rs exclusively expressed in the TN-GnRH system (cf. Kauffman *et al.*, 2005). Since it has been reported that GnRH passes the blood–brain barrier (Barrera *et al.*, 1991), we can use either intraventricular or systemic injections of drugs. This kind of behavioral experiment, however, requires knowledge of the forms and specificity of GnRH/GnRH-Rs in the three different GnRH neural systems. Another possibility is to use the molecular biological tools such as small-interfering RNAs (siRNAs) for specific knockdown of GnRH-Rs, targeted ablation using toxin-expressing transgenic techniques, mutant fish techniques, and so on. Another future direction for the behavioral study may be to record electrical activities of TN-GnRH neurons from freely moving fish in various behavioral contexts.

14. WORKING HYPOTHESIS

Our present working hypothesis concerning the neuromodulatory function of TN-GnRH neurons is illustrated in Figure 11.11. The TN-GnRH neurons show regular pacemaker activities whose basic rhythm is dependent on the interplay between the TTX-resistant persistent sodium current, $I_{Na(slow)}$, and a TEA-sensitive voltage-dependent potassium current, $I_{K(V)}$. In addition, a nifedipine-sensitive L-type Ca^{2+} current, an ω-conotoxin-sensitive N-type Ca^{2+} current, and an SNX-482-sensitive R-type Ca^{2+}

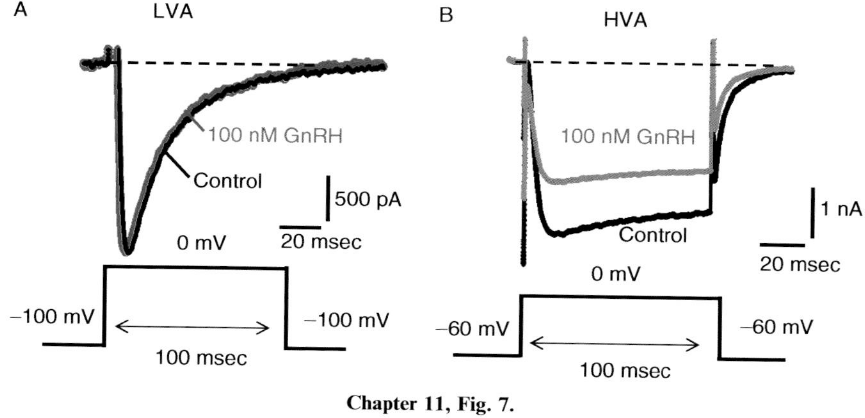

Chapter 11, Fig. 7.

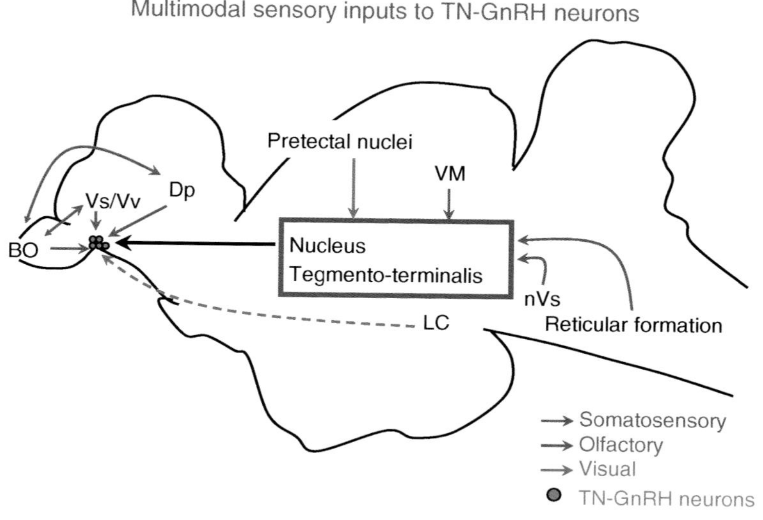

Chapter 11, Fig. 9.

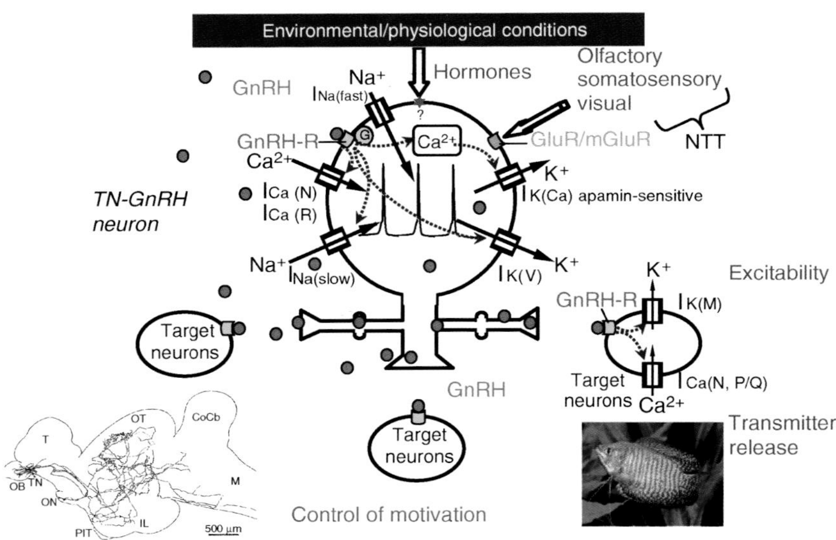

Chapter 11, Fig. 11.

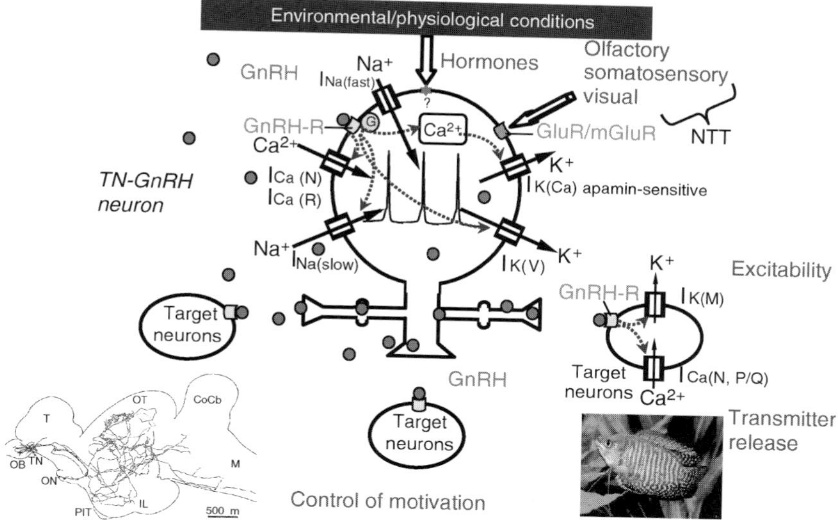

Fig. 11.11. Diagram illustrating the neuromodulator functions of TN-GnRH neurons, including some hypotheses. See text for details. (See Color Insert.)

current are observed in the TN-GnRH cell bodies and are probably activated during the action potential phase of the pacemaker activity. The TN-GnRH neurons release GnRH not only from the varicosities and axon terminals but also from the somatodendritic areas. The GnRH peptides released from the somatodendritic areas of the GnRH neurons facilitate their own activity (autocrine) and/or the activity of neighboring GnRH neurons (paracrine) and may cause synchronized positive feedback facilitation of multiple GnRH neurons. During this kind of modulation, the released GnRH peptide binds to the G-protein–coupled GnRH-Rs in the cell membrane of TN-GnRH neurons and may function in the following manner: (1) GnRH-R activation facilitates Ca^{2+} release from intracellular Ca^{2+} stores. The increased intracellular Ca^{2+} activates apamin-sensitive Ca^{2+}-dependent K^+ current(s) and decreases the frequency of pacemaker potentials transiently. The increased intracellular Ca^{2+} may further be involved in the somatodendritic release of GnRH peptides. (2) The downstream signaling pathway increases a yet unidentified slow inward cationic current(s) and decreases a potassium current(s) simultaneously, and the resultant slow depolarization increases the frequency of pacemaker potential (Haneda and Oka, unpublished data). In the GnRH-producing GT1 cell line, it has been suggested that Ca^{2+} influx through voltage-independent channels [those activated by depletion of Ca^{2+} stores, store-operated channel (SOC)] and/or by a cyclic nucleotide-gated current may occur through GnRH-R activation and is involved in the increase of pacemaker frequencies (Van Goor *et al.*, 1999b). This kind of calcium influx may further be involved in the release of GnRH peptides (Fomina and Nowycky, 1999). Considering the similarities in the electrophysiological and other properties among various types of GnRH and GnRH-related neurons, these mechanisms may also apply to the TN-GnRH neurons.

Changes in the environmental factors (such as ambient temperature or photoperiod) and physiological conditions influence GnRH neurons via transmitter receptors such as GluR and hormone receptors. The properties of the ion channels in TN-GnRH neurons that underlie pacemaker activities may be modified by downstream signal transduction mechanisms similar to the one described earlier. In fact, there are several lines of experimental evidence to support this idea. In the dwarf gourami and Tilapia, Yamamoto and Ito (2000) reported anatomical evidence for the somatosensory and visual inputs to the TN-GnRH neurons from the NTT in the midbrain tegmentum, and olfactory inputs from the olfactory bulb and the primary olfactory projection areas in the telencephalon. On the other hand, the TN-GnRH neurons have various kinds of hormone receptors besides GnRH-Rs. It has been reported in Tilapia (*Oreochromis niloticus*) that TN-GnRH neurons express receptors for thyroid hormone, testosterone, and cortisol, and these hormones are involved in the regulation of GnRH genes in TN-GnRH neurons (Soga *et al.*, 1998; Parhar *et al.*, 2000).

The neural and hormonal modulation of pacemaker activity will lead to an altered release of GnRH from various parts of TN-GnRH neurons. Since GnRH is known to modulate ion channels that control neuronal excitability or transmitter release, GnRH released from TN-GnRH neurons may modulate the excitability of target neurons or presynaptic release of transmitters in widespread brain regions simultaneously via extensive, multiple axonal branches. Finally, these cellular events may lead to finely tuned control of the motivational or arousal state underlying various animal behaviors.

ACKNOWLEDGMENTS

We would like to extend our gratitude to all of our colleagues and friends for help and discussion. Special thanks are due to Drs. Yamamoto, Ishizaki, and Mrs. Kiya and Haneda who contributed to lots of data presented here. This research was supported by grants-in-aid for Scientific Research from the Ministry of Education, Culture, Sports, Science and Technology (MEXT) of Japan to Yoshitaka Oka (#15370032 and #17023016) and the Sasakawa Scientific Research Grant from the Japan Science Society to Hideki Abe (#10–179).

REFERENCES

Abe, H., and Oka, Y. (1999). Characterization of K^+ currents underlying pacemaker potentials of fish gonadotropin-releasing hormone cells. *J. Neurophysiol.* **81,** 643–653.

Abe, H., and Oka, Y. (2000). Modulation of pacemaker activity by salmon gonadotropin-releasing hormone (sGnRH) in terminal nerve (TN)-GnRH neurons. *J. Neurophysiol.* **83,** 3196–3200.

Abe, H., and Oka, Y. (2002). Mechanisms of the modulation of pacemaker activity by GnRH peptides in the terminal nerve-GnRH neurons. *Zool. Sci.* **19,** 111–128.

Abe, H., and Terasawa, E. (2005). Firing pattern and rapid modulation of activity by estrogen in primate luteinizing hormone releasing hormone-1 neurons. *Endocrinology* **146,** 4312–4320.

Aghajanian, G. K. (1990). Use of brain slices in the study of serotonergic pacemaker neurons of the brainstem raphe nuclei. *In* "Preparations of Vertebrate Central Nervous System *In Vitro*" (Jahnsen, H., Ed.), pp. 25–47. Wiley, Chichester.

Albillos, A., Neher, E., and Moser, T. (2000). R-type Ca^{2+} channels are coupled to the rapid component of secretion in mouse adrenal slice chromaffin cells. *J. Neurosci.* **20,** 8323–8330.

Aldea, M., Jun, K., Shin, H. S., Andres-Mateos, E., Solis-Garrido, L. M., Montiel, C., Garcia, A. G., and Albillos, A. (2002). A perforated patch-clamp study of calcium currents and exocytosis in chromaffin cells of wild-type and alpha(1a) knockout mice. *J. Neurochem.* **81,** 911–921.

Amano, M., Oka, Y., Aida, K., Okumoto, N., Kawashima, S., and Hasegawa, Y. (1991). Immunocytochemical demonstration of salmon GnRH and chicken GnRH-II in the brain of masu salmon, *Oncorhynchus masou. J. Comp. Neurol.* **314,** 587–597.

Arikkath, J., and Campbell, K. P. (2003). Auxiliary subunits: Essential components of the voltage-gated calcium channel complex. *Curr. Opin. Neurobiol.* **13,** 298–307.

Bannister, R. A., Melliti, K., and Adams, B. A. (2004). Differential modulation of $Ca_V2.3$ Ca^{2+} channels by $G_{\alpha q/11}$-coupled muscarinic receptors. *Mol. Pharmacol.* **65,** 381–388.

Barrera, C. M., Kastin, A. J., Fasold, M. B., and Banks, W. A. (1991). Bidirectional saturable transport of LHRH across the blood-brain barrier. *Am. J. Physiol.* **261,** E312–E318.

Bean, B. P. (1989). Neurotransmitter inhibition of neuronal calcium currents by changes in channel voltage dependence. *Nature* **340,** 153–156.

Behrens, U. D., Douglas, R. H., and Wagner, H. J. (1993). Gonadotropin-releasing hormone, a neuropeptide of efferent projections to the teleost retina induces light-adaptive spinule formation on horizontal cell dendrites in dark-adapted preparations kept *in vitro*. *Neurosci. Lett.* **164,** 59–62.

Bosma, M. M. (1993). Ion channel properties and episodic activity in isolated immortalized gonadotropin-releasing hormone (GnRH) neurons. *J. Membr. Biol.* **136,** 85–96.

Bosma, M. M., and Hille, B. (1992). Electrophysiological properties of a cell line of the gonadotrope lineage. *Endocrinology* **130,** 3411–3420.

Bourinet, E., Fournier, J., Nargeot, J., and Charnet, P. (1992). Endogenous *Xenopus*-oocyte Ca-channels are regulated by protein kinases a and c. *FEBS Lett.* **299,** 5–9.

Brown, D. A., Constanti, A., and Adams, P. R. (1981). Slow cholinergic and peptidergic transmission in sympathetic ganglia. *Fed. Proc.* **40,** 2625–2630.

Buma, P. (1989). Characterization of luteinizing hormone-releasing hormone fibres in the mesencephalic central grey substance of the rat. *Neuroendocrinology* **49,** 623–630.

Cazalis, M., Dayanithi, G., and Nordmann, J. J. (1985). The role of patterned burst and interburst interval on the excitation-coupling mechanism in the isolated rat neural lobe. *J. Physiol.* **369,** 45–60.

Charles, A., Weiner, R., and Costantin, J. (2001). cAMP modulates the excitability of immortalized hypothalamic (GT1) neurons via a cyclic nucleotide gated channel. *Mol. Endocrinol.* **15,** 997–1009.

Charles, A. C., and Hales, T. G. (1995). Mechanisms of spontaneous calcium oscillations and action potentials in immortalized hypothalamic (GT1–7) neurons. *J. Neurophysiol.* **73,** 56–64.

Costantin, J. L., and Charles, A. C. (1999). Spontaneous action potentials initiate rhythmic intercellular calcium waves in immortalized hypothalamic (GT1–1) neurons. *J. Neurophysiol.* **82,** 429–435.

DeFazio, R. A., and Moenter, S. M. (2002). Estradiol feedback alters potassium currents and firing properties of gonadotropin-releasing hormone neurons. *Mol. Endocrinol.* **16,** 2255–2265.

DeFazio, R. A., Heger, S., Ojeda, S. R., and Moenter, S. M. (2002). Activation of A-type gamma-aminobutyric acid receptors excites gonadotropin-releasing hormone neurons. *Mol. Endocrinol.* **16,** 2872–2891.

Demski, L. S., and Fields, R. D. (1988). Dense-cored vesicle-containing components of the terminal nerve of sharks and rays. *J. Comp. Neurol.* **278,** 604–614.

Demski, L. S., and Northcutt, R. G. (1983). The terminal nerve: A new chemosensory system in vertebrates? *Science* **220,** 435–437.

Demski, L. S., and Schwanzel-Fukuda, M. (1987). The terminal nerve (nervus terminalis): Structure, function, and evolution. Introduction. *Ann. NY Acad. Sci.* **519,** ix–xi.

Dolphin, A. C. (1998). Mechanisms of modulation of voltage-dependent calcium channels by G proteins. *J. Physiol.* **506,** 3–11.

Dutton, A., and Dyball, R. E. (1979). Phasic firing enhances vasopressin release from the rat neurohypophysis. *J. Physiol.* **290,** 433–440.

Eisthen, H. L., Delay, R. J., Wirsig-Wiechmann, C. R., and Dionne, V. E. (2000). Neuromodulatory effects of gonadotropin releasing hormone on olfactory receptor neurons. *J. Neurosci.* **20,** 3947–3955.

El-Majdoubi, M., and Weiner, R. I. (2002). Localization of olfactory cyclic nucleotide-gated channels in rat gonadotropin-releasing hormone neurons. *Endocrinology* **143,** 2441–2444.

Elmslie, K. S., Zhou, W., and Jones, S. W. (1990). LhRH and GTP-γ-S modify calcium current activation in bullfrog sympathetic neurons. *Neuron* **5,** 75–80.

Fomina, A. F., and Nowycky, M. C. (1999). A current activated on depletion of intracellular Ca^{2+} stores can regulate exocytosis in adrenal chromaffin cells. *J. Neurosci.* **19,** 3711–3722.

Ford, C. P., Dryden, W. F., and Smith, P. A. (2003). Neurotrophic regulation of calcium channels by the peptide neurotransmitter luteinizing hormone releasing hormone. *J. Neurosci.* **23,** 7169–7175.

Freud-Mercier, M. J., and Richard, P. (1984). Electrophysiological evidence for facilitatory control of oxytocin neurons by oxytocin during suckling in the rats. *J. Physiol.* **352,** 447–466.

Fujita, I. (1987). Electrophysiology of the terminal nerve in cyprinids. *Ann. NY Acad. Sci.* **519,** 69–79.

Fujita, I., Satou, M., and Ueda, K. (1985). Ganglion cells of the terminal nerve: Morphology and electrophysiology. *Brain Res.* **335,** 148–152.

Fujita, I., Sorensen, P. W., Stacey, N. E., and Hara, T. J. (1991). The olfactory system, not the terminal nerve, functions as the primary chemosensory pathway mediating responses to sex pheromones in male goldfish. *Brain Behav. Evol.* **38,** 313–321.

Grace, A. A. (1988). *In vivo* and *in vitro* intracellular recording from rat midbrain dopamine neurons. *Ann. NY Acad. Sci.* **537,** 51–76.

Grace, A. A. (1991). Regulation of spontaneous activity and oscillatory spike firing in rat midbrain dopamine neurons recorded *in vitro*. *Synapse* **7,** 221–234.

Grens, K. E., Greenwood, A. K., and Fernald, R. D. (2005). Two visual processing pathways are targeted by gonadotropin-releasing hormone in the retina. *Brain Behav. Evol.* **66,** 1–9.

Grosse, R., Schmid, A., Schoneberg, T., Herrlich, A., Muhn, P., Schultz, G., and Gudermann, T. (2000). Gonadotropin-releasing hormone receptor initiates multiple signaling pathways by exclusively coupling to $G_{q/11}$ proteins. *J. Biol. Chem.* **275,** 9193–9200.

Haas, H. L., and Reiner, P. B. (1988). Membrane properties of histaminergic tuberomammilary neurones of the rat hypothalamus *in vitro*. *J. Physiol.* **399,** 633–646.

Hajdú, P., Ikemoto, T., Akazome, Y., Park, M. K., and Oka, Y. (2006). Terminal nerve GnRH neurons express multiple GnRH receptors in a teleost, dwarf gourami (*Colisa lalia*) (in submission).

Hales, T. G., Sanderson, M. J., and Charles, A. C. (1994). GABA has excitatory actions on GnRH-secreting immortalized hypothalamic (GT1–7) neurons. *Neuroendocrinology* **59,** 297–308.

Hamid, J., Nelson, D., Spaetgens, R., Dubel, S. J., Snutch, T. P., and Zamponi, G. W. (1999). Identification of an integration center for cross-talk between protein kinase c and G protein modulation of N-type calcium channels. *J. Biol. Chem.* **274,** 6195–6202.

Haneda, K., and Oka, Y. (2004). Selective modulation of voltage-gated calcium channels in the terminal nerve gonadotropin-releasing hormone (TN-GnRH) neurons of a teleost, the dwarf gourami (*Colisa lalia*). *Endocrinology* **145,** 4489–4499.

Hille, B. (1994). Modulation of ion-channel function by G-protein-coupled receptor. *Trends Neurosci.* **17,** 531–536.

Ikeda, S. R. (1996). Voltage-dependent modulation of N-type calcium channels by G-protein beta gamma subunits. *Nature* **380,** 255–258.

Ishizaki, M., and Oka, Y. (2001). Amperometric recording of gonadotropin-releasing hormone release activity in the pituitary of the dwarf gourami (teleost) brain-pituitary slices. *Neurosci. Lett.* **299,** 121–124.

Ishizaki, M., Iigo, M., Yamamoto, N., and Oka, Y. (2004). Different modes of gonadotropin-releasing hormone (GnRH) release from multiple GnRH systems as revealed by radioimmunoassay using brain slices of a teleost, the dwarf gourami (*Colisa lalia*). *Endocrinology* **145,** 2092–2103.

Jan, Y. N., and Jan, L. Y. (1983). A LHRH-like peptidergic neurotransmitter capable of "action at a distance" in autonomic ganglia. *Trends Neurosci.* **6,** 320–325.

Javors, M. A., King, T. S., Chang, X., Klein, N. A., and Schenken, R. S. (1995). Partial characterization of K^+-induced increase in $[Ca^{2+}]_{cyt}$ and GnRH release in GT1–7 neurons. *Brain Res.* **694,** 49–54.

Jentsch, T. J. (2000). Neuronal KCNQ potassium channels: Physiology and role in disease. *Nat. Rev. Neurosci.* **1,** 21–30.

Jones, S. W. (1987). Luteinizing hormone-releasing hormone as a neurotransmitter in bullfrog sympathetic ganglia. *Ann. NY Acad. Sci.* **519,** 310–322.

Kamp, T. J., and Hell, J. W. (2000). Regulation of cardiac L-type calcium channels by protein kinase a and protein kinase c. *Circ. Res.* **87,** 1095–1102.

Kato, M., Ui-Tei, K., Watanabe, M., and Sakuma, Y. (2003). Characterization of voltage-gated calcium currents in gonadotropin-releasing hormone neurons tagged with green fluorescent protein in rats. *Endocrinology* **144,** 5118–5125.

Kauffman, A. S., Wills, A., Millar, R. P., and Rissman, E. F. (2005). Evidence that the type-2 gonadotrophin-releasing hormone (GnRH) receptor mediates the behavioral effects of GnRH-II on feeding and reproduction in musk shrews. *J. Neuroendocrinol.* **17,** 489–497.

Kelly, M. J., Ronnekleiv, O. K., and Eskay, R. L. (1984). Identification of estrogen-responsive LHRH neurons in the guinea pig hypothalamus. *Brain Res. Bull.* **12,** 399–407.

Kim, M. H., Oka, Y., Amano, M., Kobayashi, M., Okuzawa, K., Hasegawa, Y., Kawashima, S., Suzuki, Y., and Aida, K. (1995). Immunocytochemical localization of sGnRH and cGnRH-II in the brain of goldfish, *Carassius auratus*. *J. Comp. Neurol.* **356,** 72–82.

Kiya, T., and Oka, Y. (2003). Glutamate receptors in the terminal nerve gonadotropin-releasing hormone neurons of the dwarf gourami (teleost). *Neurosci. Lett.* **345,** 113–116.

Koyama, Y., Satou, M., Oka, Y., and Ueda, K. (1984). Involvement of the telencephalic hemispheres and the preoptic area in sexual behavior of the male goldfish, *Carassius auratus*: A brain-lesion study. *Behav. Neural. Biol.* **40,** 70–86.

Krsmanovic, L. Z., Stojilkovic, S. S., Mertz, L. M., Tomic, M., and Catt, K. J. (1993). Expression of gonadotropin-releasing hormone receptors and autocrine regulation of neuropeptide release in immortalized hypothalamic neurons. *Proc. Natl. Acad. Sci. USA* **90,** 3908–3912.

Krsmanovic, L. Z., Martinez-Fuentes, A. J., Arora, K. K., Mores, N., Navarro, C. E., Chen, H. C., Stojilkovic, S. S., and Catt, K. J. (1999). Autocrine regulation of gonadotropin-releasing hormone secretion in cultured hypothalamic neurons. *Endocrinology* **140,** 1423–1431.

Krsmanovic, L. Z., Mores, N., Navarro, C. E., Tomic, M., and Catt, K. J. (2001). Regulation of Ca^{2+}-sensitive adenylyl cyclase in gonadotropin-releasing hormone neurons. *Mol. Endocrinol.* **15,** 429–440.

Krsmanovic, L. Z., Mores, N., Navarro, C. E., Arora, K. K., and Catt, K. J. (2003). An agonist-induced switch in G protein coupling of the gonadotropin-releasing hormone receptor regulates pulsatile neuropeptide secretion. *Proc. Natl. Acad. Sci. USA* **100,** 2969–2974.

Kusano, K., Fueshko, S., Gainer, H., and Wray, S. (1995). Electrical and synaptic properties of embryonic luteinizing hormone-releasing hormone neurons in explant cultures. *Proc. Natl. Acad. Sci. USA* **92,** 3918–3922.

LeBeau, A. P., Van Goor, F., Stojilkovic, S. S., and Sherman, A. (2000). Modeling of membrane excitability in gonadotropin-releasing hormone-secreting hypothalamic neurons regulated by Ca^{2+}-mobilizing and adenylyl cyclase-coupled receptors. *J. Neurosci.* **20,** 9290–9297.

Li, L., and Dowling, J. E. (2000). Disruption of the olfactoretinal centrifugal pathway may relate to the visual system defect in *night blindness b* mutant zebrafish. *J. Neurosci.* **20,** 1883–1892.

Maaswinkel, H., and Li, L. (2003). Olfactory input increases visual sensitivity in zebrafish: A possible function for the terminal nerve and dopaminergic interplexiform cells. *J. Exp. Biol.* **206,** 2201–2209.

Madigou, T., Mananos-Sanchez, E., Hulshof, S., Anglade, I., Zanuy, S., and Kah, O. (2000). Cloning, tissue distribution, and central expression of the gonadotropin-releasing hormone receptor in the rainbow trout (*Oncorhynchus mykiss*). *Biol. Reprod.* **63,** 1857–1866.

Martinez-Fuentes, A. J., Hu, L., Krsmanovic, L. Z., and Catt, K. J. (2004). Gonadotropin-releasing hormone (GnRH) receptor expression and membrane signaling in early embryonic GnRH neurons: Role in pulsatile neurosecretion. *Mol. Endocrinol.* **18,** 1808–1817.

Matsutani, S., and Uchiyama, H. I. (1986). Cytoarchitecture, synaptic organization and fiber connections of the nucleus olfactoretinalis in a teleost (*Navodon modestus*). *Brain Res.* **373,** 126–138.

Meir, A., Ginsburg, S., Butkevich, A., Kachalsky, S. G., Kaiserman, I., Ahdut, R., Demirgoren, S., and Rahamimoff, R. (1999). Ion channels in presynaptic nerve terminals and control of transmitter release. *Physiol. Rev.* **79**(3), 1019–1088.

Millar, R. P., Lu, Z. L., Pawson, A. J., Flanagan, C. A., Morgan, K., and Maudsley, S. R. (2004). Gonadotropin-releasing hormone receptors. *Endocr. Rev.* **25,** 235–275.

Moenter, S. M., DeFazio, R. A., Pitts, G. R., and Nunemaker, C. S. (2003). Mechanisms underlying episodic gonadotropin-releasing hormone secretion. *Front. Neuroendocrinol.* **24,** 79–93.

Moos, F., Freund-Mercier, M. J., Guerne, J. M., Stoeckel, M. E., and Richard, P. (1984). Release of oxytocin and vasopressin by magnocellular nuclei *in vitro*: Specific facilitatory effect of oxytocin on its own release. *J. Endocrinol.* **102,** 63–72.

Munz, H., and Stumpf, W. E. J. (1981). LHRH systems in the brain of platyfish. *Brain Res.* **221,** 1–13.

Neher, E. (1998). Vesicle pools and Ca^{2+} microdomains: New tools for understanding their roles in neurotransmitter release. *Neuron* **20,** 389–399.

Nunemaker, C. S., DeFazio, R. A., Geusz, M. E., Herzog, E. D., Pitts, G. R., and Moenter, S. M. (2001). Long-term recordings of networks of immortalized GnRH neurons reveal episodic patterns of electrical activity. *J. Neurophysiol.* **86,** 86–93.

Nunemaker, C. S., DeFazio, R. A., and Moenter, S. M. (2002). Estradiol-sensitive afferents modulate long-term episodic firing patterns of GnRH neurons. *Endocrinology* **143,** 2284–2292.

Nunemaker, C. S., DeFazio, R. A., and Moenter, S. M. (2003a). Calcium current subtypes in gonadotropin-releasing hormone neurons. *Biol. Reprod.* **69,** 1914–1922.

Nunemaker, C. S., Straume, M., DeFazio, R. A., and Moenter, S. M. (2003b). Gonadotropin-releasing hormone neurons generate interacting rhythms in multiple time domains. *Endocrinology* **144,** 823–831.

Oka, Y. (1992). Gonadotropin-releasing hormone (GnRH) cells of the terminal nerve as a model neuromodulator system. *Neurosci. Lett.* **142,** 119–122.

Oka, Y. (1995). Tetrodotoxin-resistant persistent Na^+ current underlying pacemaker potentials of fish gonadotrophin-releasing hormone neurons. *J. Physiol.* **482,** 1–6.

Oka, Y. (1996). Characterization of TTX-resistant persistent Na^+ current underlying pacemaker potentials of fish gonadotropin-releasing hormone (GnRH) neurons. *J. Neurophysiol.* **75,** 2397–2404.

Oka, Y. (1997). The gonadotropin-releasing hormone (GnRH) neuronal system of fish brain as a model system for the study of peptidergic neuromodulation. *In* "GnRH Neurons: Genes to Behavior" (Parhar, I. S., and Sakuma, Y., Eds.), pp. 245–276. Brain shuppan, Tokyo.

Oka, Y. (2002). Physiology and release activity of GnRH neurons. *Prog. Brain Res.* **141,** 259–281.

Oka, Y., and Ichikawa, M. (1990). Gonadotropin-releasing hormone (GnRH) immunoreactive system in the brain of the dwarf gourami (*Colisa lalia*) as revealed by light microscopic immunocytochemistry using a monoclonal antibody to common amino acid sequence of GnRH. *J. Comp. Neurol.* **300,** 511–522.

Oka, Y., and Ichikawa, M. (1991). Ultrastructure of the ganglion cells of the terminal nerve in the dwarf gourami (*Colisa lalia*). *J. Comp. Neurol.* **304,** 161–171.

Oka, Y., and Ichikawa, M. (1992). Ultrastructural characterization of gonadotropin-releasing hormone (GnRH)-immunoreactive terminal nerve cells in the dwarf gourami. *Neurosci. Lett.* **140,** 200–202.

Oka, Y., and Matsushima, T. (1993). Gonadotropin-releasing hormone (GnRH)-immunoreactive terminal nerve cells have intrinsic rhythmicity and project widely in the brain. *J. Neurosci.* **13,** 2161–2176.

Parhar, I. S. (2002). Cell migration and evolutionary significance of GnRH subtypes. *Prog. Brain Res.* **141,** 3–17.

Parhar, I. S., Koibuchi, N., Sakai, M., Iwata, M., and Yamaoka, S. (1994). Gonadotropin-releasing hormone (GnRH): Expression during salmon migration. *Neurosci. Lett.* **172,** 15–18.

Parhar, I. S., Soga, T., and Sakuma, Y. (2000). Thyroid hormone and estrogen regulate brain region-specific messenger ribonucleic acids encoding three gonadotropin-releasing hormone genes in sexually immature male fish, *Oreochromis niloticus*. *Endocrinology* **141,** 1618–1626.

Park, D., and Eisthen, H. L. (2003). Gonadotropin releasing hormone (GnRH) modulates odorant responses in the peripheral olfactory system of axolotls. *J. Neurophysiol.* **90,** 731–738.

Paruthiyil, S., El Majdoubi, M., Conti, M., and Weiner, R. I. (2002). Phosphodiesterase expression targeted to gonadotropin-releasing hormone neurons inhibits luteinizing hormone pulses in transgenic rats. *Proc. Natl. Acad. Sci. USA* **99,** 17191–17196.

Peng, Y. Y., and Horn, J. P. (1991). Continuous repetitive stimuli are more effective than bursts for evoking LHRH release in bullfrog sympathetic ganglia. *J. Neurosci.* **11,** 85–95.

Peter, R. E., Prasada, R., Baby, S. M., Illing, N., and Millar, R. P. (2003). Differential brain distribution of gonadotropin-releasing hormone receptors in the goldfish. *Gen. Comp. Endocrinol.* **132,** 399–408.

Pow, D. V., and Morris, J. F. (1989). Dendrites of hypothalamic magnocellular neurons release neurohypophysial peptides by exocytosis. *Neuroscience* **32,** 435–439.

Rios, E., and Stern, M. D. (1997). Calcium in close quarters: Microdomain feedback in excitation-contraction coupling and other cell biological phenomena. *Annu. Rev. Biophys. Biomol. Struct.* **26,** 47–82.

Romanelli, R. G., Barni, T., Maggi, M., Luconi, M., Failli, P., Pezzatini, A., *et al.* (2004). Expression and function of gonadotropin-releasing hormone (GnRH) receptor in human olfactory GnRH-secreting neurons: An autocrine GnRH loop underlies neuronal migration. *J. Biol. Chem.* **279,** 117–126.

Rudy, B. (1988). Diversity and ubiquity of K channels. *Neuroscience* **25,** 729–749.

Satou, M., Oka, Y., Kusunoki, M., Matsushima, T., Kato, M., Fujita, I., and Ueda, K. (1984). Telencephalic and preoptic areas integrate sexual behavior in hime salmon (landlocked red salmon, *Oncorhynchus nerka*): Results of electrical brain stimulation experiments. *Physiol. Behav.* **33,** 441–447.

Schwanzel-Fukuda, M., and Silverman, A. J. (1980). The nervus terminalis of the guinea pig: A new luteinizing hormone-releasing hormone (LHRH) neuronal system. *J. Comp. Neurol.* **191,** 213–225.

Schwanzel-Fukuda, M., Morrel, J. I., and Pfaff, D. W. (1985). Ontogenesis of neurons producing luteinizing hormone-releasing hormone (LHRH) in the nervus terminalis of the rat. *J. Comp. Neurol.* **238,** 348–364.

Schwanzel-Fukuda, M., Garcia, M. S., and Morrell, J. I. P. (1987). Distribution of luteinizing hormone-releasing hormone in the nervus terminalis and brain of the mouse detected by immunocytochemistry. *J. Comp. Neurol.* **255,** 231–244.

Skynner, M. J., Slater, R., Sim, J. A., Allen, N. D., and Herbison, A. E. (1999). Promoter transgenics reveal multiple gonadotropin-releasing hormone-I-expressing cell populations of different embryological origin in mouse brain. *J. Neurosci.* **19,** 5955–5966.

Soga, T., Sakuma, Y., and Parhar, I. S. (1998). Testosterone differentially regulates expression of GnRH messenger RNAs in the terminal nerve, preoptic and midbrain of male tilapia. *Brain Res. Mol. Brain Res.* **60,** 13–20.

Soga, T., Ogawa, S., Millar, R. P., Sakuma, Y., and Parhar, I. S. (2005). Localization of the three GnRH types and GnRH receptors in the brain of a cichlid fish: Insights into their neuroendocrine and neuromodulator functions. *J. Comp. Neurol.* **487,** 28–41.

Somoza, G. M., Miranda, L. A., Strobl-Mazzulla, P., and Guilgur, L. G. (2002). Gonadotropin-releasing hormone (GnRH): From fish to mammalian brains. *Cell. Mol. Neurobiol.* **22,** 589–609.

Spergel, D. J., Kruth, U., Hanley, D. F., Sprengel, R., and Seeburg, P. H. (1999). GABA- and glutamate-activated channels in green fluorescent protein-tagged gonadotropin-releasing hormone neurons in transgenic mice. *J. Neurosci.* **19,** 2037–2050.

Springer, A. D. (1983). Centrifugal innervation of goldfish retina from ganglion cells of the nervus terminalis. *J. Comp. Neurol.* **214,** 404–415.

Stea, A., Soong, T. W., and Snutch, T. P. (1995). Determinants of PKC-dependent modulation of a family of neuronal calcium channels. *Neuron* **15,** 929–940.

Stell, W. K., Walker, S. E., Chohan, K. S., and Ball, A. K. (1984). The goldfish nervus terminalis: A luteinizing hormone-releasing hormone and molluscan cardioexcitatory peptide immunoreactive olfactoretinal pathway. *Proc. Natl. Acad. Sci. USA* **81,** 940–944.

Steven, C., Lehnen, N., Kight, K., Ijiri, S., Klenke, U., Harris, W. A., *et al.* (2003). Molecular characterization of the GnRH system in zebrafish (*Danio rerio*): Cloning of chicken GnRH-II, adult brain expression patterns and pituitary content of salmon GnRH and chicken GnRH-II. *Gen. Comp. Endocrinol.* **133,** 27–37.

Stojilkovic, S. S., Krsmanovic, L. Z., Spergel, D. J., and Catt, K. J. (1994a). Gonadotropin-releasing hormone neurons: Intrinsic pulsatility and receptor-mediated regulation. *Trends Endocrinol. Metab.* **5,** 201–209.

Stojilkovic, S. S., Reinhart, J., and Catt, K. J. (1994b). Gonadotropin-releasing hormone receptors: Structure and signal transduction pathways. *Endocrinol. Rev.* **15,** 462–499.

Suh, B. C., and Hille, B. (2002). Recovery from muscarinic modulation of M current channels requires phosphatidylinositol 4,5-bisphosphate synthesis. *Neuron* **35,** 507–520.

Suter, K. J., Song, W. J., Sampson, T. L., Wuarin, J. P., Saunders, J. T., Dudek, F. E., and Moenter, S. M. (2000a). Genetic targeting of green fluorescent protein to gonadotropin-releasing hormone neurons: Characterization of whole-cell electrophysiological properties and morphology. *Endocrinology* **141,** 412–419.

Suter, K. J., Wuarin, J. P., Smith, B. N., Dudek, F. E., and Moenter, S. M. (2000b). Whole-cell recordings from preoptic/hypothalamic slices reveal burst firing in gonadotropin-releasing hormone neurons identified with green fluorescent protein in transgenic mice. *Endocrinology* **141,** 3731–3736.

Swartz, K. J. (1993). Modulation of Ca^{2+} channels by protein kinase C in rat central and peripheral neurons: Disruption of G protein-mediated inhibition. *Neuron* **11,** 305–320.

Terasawa, E. (1998). Cellular mechanism of pulsatile LHRH release. *Gen. Comp. Endocrinol.* **112,** 283–295.

Terasawa, E. (2003). Pulse generation in LHRH neurons. *In* "Neuroplasticity, Development, and Steroid Hormone Action" (Handa, R., Hayashi, S., Terasawa, E., and Kawata, M., Eds.), pp. 153–168. CRC Press, Boca Raton.

Terasawa, E., Schanhofer, W. K., Keen, K. L., and Luchansky, L. (1999). Intracellular Ca^{2+} oscillations in luteinizing hormone-releasing hormone neurons derived from the embryonic olfactory placode of the rhesus monkey. *J. Neurosci.* **19,** 5898–5909.

Thureson-Klein, A. K., and Klein, R. L. (1990). Exocytosis from neuronal large dense-cored vesicles. *Int. Rev. Cytol.* **121,** 67–126.

Tottene, A., Moretti, A., and Pietrobon, D. (1996). Functional diversity of P-type and R-type calcium channels in rat cerebellar neurons. *J. Neurosci.* **16,** 6353–6363.

Tsutsui, H., and Oka, Y. (2002). Slow removal of Na^{+} channel inactivation underlies the temporal filtering property in the teleost thalamic neurons. *J. Physiol.* **539,** 743–753.

Tsutsui, H., Yamamoto, N., Ito, H., and Oka, Y. (2001). Encoding of different aspects of afferent activities by two types of cells in the corpus glomerulosum of a teleost brain. *J. Neurophysiol.* **85,** 1167–1177.

Umino, O., and Dowling, J. E. (1991). Dopamine release from interplexiform cells in the retina: Effects of GnRH, FMRFamide, bicuculline, and enkephalin on horizontal cell activity. *J. Neurosci.* **11,** 3034–3046.

Urbano, F. J., Piedras-Renteria, E. S., Jun, K., Shin, H. S., Uchitel, O. D., and Tsien, R. W. (2003). Altered properties of quantal neurotransmitter release at endplates of mice lacking P/Q-type Ca^{2+} channels. *Proc. Natl. Acad. Sci. USA* **100,** 3491–3496.

van den Pol, A. N., Wuarin, J. P., and Dudek, F. E. (1990). Glutamate, the dominant excitatory transmitter in neuroendocrine regulation. *Science* **250,** 1276–1278.

Van Goor, F., Krsmanovic, L. Z., Catt, K. J., and Stojilkovic, S. S. (1999a). Control of action potential-driven calcium influx in GT1 neurons by the activation status of sodium and calcium channels. *Mol. Endocrinol.* **13,** 587–603.

Van Goor, F., Krsmanovic, L. Z., Catt, K. J., and Stojilkovic, S. S. (1999b). Coordinate regulation of gonadotropin-releasing hormone neuronal firing patterns by cytosolic calcium and store depletion. *Proc. Natl. Acad. Sci. USA* **96,** 4101–4106.

Van Goor, F., Krsmanovic, L. Z., Catt, K. J., and Stojilkovic, S. S. (2000). Autocrine regulation of calcium influx and gonadotropin-releasing hormone secretion in hypothalamic neurons. *Biochem. Cell Biol.* **78,** 359–370.

Vitalis, E. A., Costantin, J. L., Tsai, P. S., Sakakibara, H., Paruthiyil, S., Iiri, T., Martini, J. F., Taga, M., Choi, A. L., Charles, A. C., and Weiner, R. I. (2000). Role of the cAMP signaling pathway in the regulation of gonadotropin-releasing hormone secretion in GT1 cells. *Proc. Natl. Acad. Sci. USA* **97,** 1861–1866.

Wada, H., Inagaki, N., Yamatodani, A., and Watanabe, T. (1991). Is the histaminergic neuron system a regulatory center for whole-brain activity? *Trends Neurosci.* **14,** 415–418.

Walker, S. E., and Stell, W. K. (1986). Gonadotropin-releasing hormone (GnRH), molluscan cardioexcitatory peptide (FMRFamide), enkephalin and related neuropeptides affect goldfish retinal ganglion cell activity. *Brain Res.* **384,** 262–273.

Wang, H. S., Pan, Z., Shi, W., Brown, B. S., Wymore, R. S., Cohen, I. S., Dixon, J. E., and McKinnon, D. (1998). KCNQ2 and KCNQ3 potassium channel subunits: Molecular correlates of the M-channel. *Science* **282,** 1890–1893.

Watanabe, M., Sakuma, Y., and Kato, M. (2004). High expression of the R-type voltage-gated Ca^{2+} channel and its involvement in Ca^{2+}-dependent gonadotropin-releasing hormone release in GT1–7 cells. *Endocrinology* **145,** 2375–2383.

Wedekind, H. P. (1979). Electron microscopic study of the giant cells in the olfactory bulb of labyrinth fish (belontiidae, perciformes). *Cell Tissue Res.* **199,** 509–517.

White, J., and Meredith, M. (1987). Synaptic interactions in the nervus terminalis ganglion of elasmobranchs. *Ann. NY Acad. Sci.* **519,** 33–49.

Whitlock, K. E., Wolf, C. D., and Boyce, M. L. (2003). Gonadotropin-releasing hormone (GnRH) cells arise from cranial neural crest and adenohypophyseal regions of the neural plate in the zebrafish, *Danio rerio*. *Dev. Biol.* **257,** 140–152.

Williams, J. T., North, R. A., Shefner, S. A., Nishi, S., and Egan, T. M. (1984). Membrane properties of rat locus coeruleus neurones. *Neuroscience* **13,** 137–156.

Wirsig, C. R., and Leonard, C. M. (1987). Terminal nerve damage impairs the mating behavior of the male hamster. *Brain Res.* **417,** 293–303.

Wirsig-Wiechmann, C. R., and Oka, Y. (2002). The terminal nerve ganglion cells project to the olfactory mucosa in the dwarf gourami. *Neurosci. Res.* **44,** 337–341.

Wirsig-Wiechmann, C. R., Wiechmann, A. F., and Eisthen, H. L. (2002). What defines the nervus terminalis? Neurochemical, developmental, and anatomical criteria. *Prog. Brain Res.* **141,** 45–58.

Wong, M., Eaton, M. J., and Moss, R. L. (1990). Electrophysiological actions of luteinizing hormone-releasing hormone: Intracellular studies in the rat hippocampal slice preparation. *Synapse* **5,** 65–70.

Xu, C., Xu, X. Z., Nunemaker, C. S., and Moenter, S. M. (2004). Dose-dependent switch in response of gonadotropin-releasing hormone (GnRH) neurons to GnRH mediated through the type I GnRH receptor. *Endocrinology* **145,** 728–735.

Yamamoto, N., and Ito, H. (2000). Afferent sources to the ganglion of the terminal nerve in teleosts. *J. Comp. Neurol.* **428,** 355–375.

Yamamoto, N., Oka, Y., Amano, M., Aida, K., Hasegawa, Y., and Kawashima, S. (1995). Multiple gonadotropin-releasing hormone (GnRH)-immunoreactive systems in the brain of the dwarf gourami, *Colisa lalia*: Immunohistochemistry and radioimmunoassay. *J. Comp. Neurol.* **355,** 354–368.

Yamamoto, N., Oka, Y., and Kawashima, S. (1997). Lesions of gonadotropin-releasing hormone-immunoreactive terminal nerve cells: Effects on the reproductive behavior of male dwarf gouramis. *Neuroendocrinology* **65,** 403–412.

Yang, J., and Tsien, R. W. (1993). Enhancement of N- and L- type calcium channel currents by protein kinase c in frog sympathetic neurons. *Neuron* **10,** 127–136.

Yang, S. N., Lu, F., Wu, J. N., Liu, D. D., and Hsieh, W. Y. (1999). Activation of gonadotropin-releasing hormone receptors induces a long-term enhancement of excitatory postsynaptic currents mediated by ionotropic glutamate receptors in the rat hippocampus. *Neurosci. Lett.* **260,** 33–36.

Zamponi, G. W., and Snutch, T. P. (1998). Modulation of voltage-dependent calcium channels by G proteins. *Curr. Opin. Neurobiol.* **8,** 351–356.

Zamponi, G. W., Bourinet, E., Neison, D., Nargeot, J., and Snutch, T. P. (1997). Cross talk between G proteins and protein kinase c mediated by the calcium channel $\alpha 1$ subunit. *Nature* **385,** 442–446.

Zheng, L., Krsmanovic, L. Z., Vergara, L. A., Catt, K. J., and Stojilkovic, S. S. (1997). Dependence of intracellular signaling and neurosecretion on phospholipase d activation in immortalized gonadotropin-releasing hormone neurons. *Proc. Natl. Acad. Sci. USA* **94,** 1573–1578.

Zhu, Y., and Ikeda, S. R. (1994). Modulation of Ca^{2+}-channel currents by protein kinase c in adult at sympathetic neurons. *J. Neurophysiol.* **72,** 1549–1560.

Zucker, C. L., and Dowling, J. E. (1987). Centrifugal fibres synapse on dopaminergic interplexiform cells in the teleost retina. *Nature* **330,** 166–168.

INDEX

A

AANAT2 gene, 271, 275
  expression, 282
  regulation of, 278
AANAT2 protein, 271
Absolute visual sensitivity, 188–95
Acanthopterygii, olfactory organ in, 12–13
Accessory optic system (AOS), 217
  in teleost, 187–8
Acetylcholine and chemotransduction mechanisms, 131–2
N-acetylserotonin, 265
Acipenser baeri, *see* Siberian sturgeon
Acipenser oxyrhynchus, *see* Sturgeon
Active electrolocation, 307
A-delta fibers, 154–7
  composition, 155
A-Delta nociceptors in fish, electrophysiological properties of, 159
Adenosine 5′-triphosphate (ATP), 54
β-Adrenergic receptors, 288
Aequidens pulcher, *see* Blue acara
Aetobatus narinari, *see* Spotted eagle ray
African lungfish (*Protopterus annectens*), 168
  ventilatory responses in, 106
African Mormyriformes, 308, *see also* Mormyriformes
ALLN, *see* Anterior lateral line nerves
Allodynia, GABA in, 166–7
Amacrine cells, 223
Amago salmon (*Oncorhynchus rhodurus*)
  taste buds in, 47
American eel (*Anguilla rostrata*), 107–8
Amia calva, *see* Bowfin
Amino acids, 206, 209
  feeding behavior and, 77–82
  for gustation, gustatory responses to, 63–8
  for gustation, receptors, 71–2
    L-Arg receptor, 71–2
    T1R, 72
    T2R, 72
L-Amino acids, 4–5
DL-2-Amino-4-phosphonobutyric acid (APB), 192
Ampullary electroreceptor organs, fishes, 308–9
Ampullary electroreceptors
  induced electrical signals, 355–6
  organs, 311
Amygdala, 205
Anableps anableps, *see* Four-eyed fish
Analgesia in fish, 171
Angelfish (*Pterophyllum scalare*), 206
Anguilla rostrata, *see* American eel
Animal responses
  avoidance learning, 170
  in vivo observations, 170–2
ANP, *see* Atrial natriuretic peptide
Anterior lateral line nerves (ALLN), 356, 428, 434
AOS, *see* Accessory optic system
Apomorphine, 219
Apteronotus leptorhynchus, *see* Ghost fish
L-Arg receptor, 71–2
Arrestin, 255
Arylalkylamine *N*-acetyltransferase 2 (AANAT2), intracellular regulation of
  regulation of *AANAT2* gene, 278
  regulation of AANAT2 protein, 275–8
Atlantic cod (*Gadus morhua*)
  cardiac stroke volume in, 100, 102

Atlantic salmon (*Salmo salar*), 107–8
feeding behavior in, 82
Atrial natriuretic peptide (ANP), 290
Audition system in fishes, 377–405
acoustic particle motion, 391–9
central coding of, 396–9
coding of, 393–7
detection of, 391–3
peripheral coding of, 393–7
central coding of acoustic particle motion, 396–9
coding of acoustic particle motion, 393–7
detection of acoustic particle motion, 391–3
directional hearing, 389–99
acoustic particle motion, 391–9
experimental setups, 391
lateral line integration, 399
neural and behavioral mechanisms, 377–405
neurophysiological mechanisms, 387–99
acoustic particle motion, 391–9
directional hearing, 389–99
peripheral and central auditory pathways, 382–7
vocal motor inputs to, 384–7
peripheral coding of acoustic particle motion, 393–7
steroid hormones and seasonal hearing changes, 400–3, 405
vocal modulation of inner ear and lateral line, 399–400, 404
vocal motor inputs to, 384–7
vocal signals encoding, 386–90
Australian lungfish (*Neoceratodus forsteri*)
ventilatory responses in, 106
Aversive behavior, 83–4

B

Barred snapper (*Hoplopagrus guentheri*)
nasal cavity in, 6
Basses (*Serranidae*), 181
Batrachoidids
behavioral mechanisms of audition, 378–82, 388
peripheral and central auditory pathways, 382–7
vocal modulation of inner ear and lateral line, 399
Benzodiazepines, 289
Bet-receptor, of gustation, 64–7
Bile acids, 4–5
gustation and, 68–9
Billfish, 368
Bipolar cells, 222–3
Black-tip shark (*Carcharhinus melanopterus*), 156
Blennies (*Blenniidae*), 181
Blenniidae, *see* Blennies
Blood–brain barrier, 248
Blue acara (*Aequidens pulcher*), 208
Bmal1, 281
Bmal2, 282
Bony rayed fishes
olfactory chamber in, 13–4
Bowfin (*Amia calva*)
ventilatory responses in, 106
Bradycardia, 99
hypoxia-induced, 101, 103–4
Branchial chemoreceptors, 97, *see also* Chemoreceptors
Branchial neuroepithelial cells, *see* Neuroepithelial cells
Branchial vasoconstriction, 105
Brown bullhead, 107–8
Brown trout (*Salmo trutta*)
feeding behavior in, 82
Butterflyfish (*Chaetodon rainfordi*), 181

C

Candidate magnetite-based magnetoreceptors, 346–9
Capacitance detection, 318–20
Carassius auratus, *see* Goldfish
Carassius carassius, *see* Crucian carp
Carbonic anhydrase, treatment of trout with, 121
Carcharhinus melanopterus, *see* Black-tip shark
Cardiac output management, 102–3
Cardiorespiratory responses, chemoreceptor-driven, 98
cardiovascular responses, 99–105
endocrine responses, 110
ventilatory responses, 105–9

Cardiorespiratory system, branchial chemoreceptors regulation of, 138–9
cardiorespiratory responses, chemoreceptor-driven, 98
cardiovascular responses, 99–105
endocrine responses, 110
ventilatory responses, 105–9
central integration and efferent pathways, 133–8
chemoreceptors, 110–1
chemosensory cells, 125–9
chemotransduction mechanisms, 129–33
location and orientation of, 111–25
Cardiorespiratory system, in fish, 97
Cardiovascular responses, chemoreceptor-driven, 99–103
to acute aquatic hypercapnia, 99, 101
to acute aquatic hypoxia, 99–101
blood flow redistribution, 105
bradycardia, 99, 101, 103–4
hypertension and, 103–4
other cardiovascular adjustments, 105
Carp (*Cyprinus carpio*), 155, 165, 169–70
Cascade model, 221–2
Casein kinase I (CKI), 280
Catecholamines, and endocrine responses, 110
Catfish (*Clarias batrachus*), 168–9
Cationic cGMP-gated channel, 253
Catostomus commersoni, *see* White sucker
Central nervous system
brain structure, 160–5
pathways to brain, 165–6
Central pathways for lateral line information processing, fishes, 428–30
Central physiology, of fishes, 423, 430–43
dynamic amplitude range, 434
dynamic response properties, 431
frequency encoding, 433–4
latencies, 431
lateral line maps, 434–5
moving object stimuli, 430, 438
multimodality, 440–3
ongoing activity, 430–1
phasic *vs.* tonic responses, 432–3
receptive field organization, 423, 435–7
running water conditions, 438–40
thresholds, 431–2
Cerebral hemispheres, telencephalon, 160
Cerebrospinal fluid (CSF), 264
C fiber, 155, 157
CGZ, *see* Circumferential germinal zone
Chaetodon rainfordi, *see* Butterflyfish
Channel catfish (*Ictalurus punctatus*), 109, 166
feeding behavior in, 78
OSN in, 15
trigeminal fibers of, 62
Chemicals, miscellaneous, gustatory responses
bile acids, 68–9
carboxylic acids, 71
CO_2 and pH, 70–1
quinine, strychnine, and tetrodotoxin, 69–70
Chemoreceptors, for cardiorespiratory system regulation, 110–1
cardiorespiratory responses by
cardiovascular responses, 99–105
endocrine responses, 110
ventilatory responses, 105–9
chemosensory cells morphology, 125–9
chemotransduction mechanisms, 129–33
development, their plasticity, and consequences on cardiorespiratory responses, 128–9
location and orientation of, 111–25
chemoreceptors sensitivity to CO_2 and O_2, 123–5
CO_2/pH sensitive chemoreceptors, 117–23
oxygen-sensitive chemoreceptors, 111–7
Chemosensory cells, 125–9
Chemotransduction mechanisms, in chemoreceptors, 129–33
acetylcholine and nicotine in, 131–2
dopamine in, 133
intracellular Ca^{2+} concentration and increase in, 129
K^+ channel inhibition and, 130
Chimeras, 160
Chorda tympani nerve, 56
Chromatic processing, 221
Chromophores in fish, 183
Cichlid (*Haplochromis burtoni*), 181, 201
Ciliary marginal zone (CMZ), 186
Ciliated OSNs, 3, 5, 14–7
odorant responses and, 21, 22
Ciona, 291
Circadian organization of fish, 245
Circadian pacemakers, 245
Circadian rhythm, 196, 200

Circumferential germinal zone (CGZ), 184
Clarias batrachus, *see* Catfish
Clock genes and *Aanat2* gene regulation in fish, 281
CMZ, *see* Ciliary marginal zone
CNG channel, *see* Cyclicnucleotide-gated channel
Cod, directional responses, 393
CO_2/H^+-driven behavior, 84
Collagen fibers, 252
Color channel, 220
Command nucleus (CN), 313–4
Cone bipolar cells, 181
Cone functions and flicker ERG, 192
Contrast visual sensitivity
  concepts, 214
    in brain areas, 216–7
    as determined by conditioned response, 215
    determined by OKR, 215
    determined by OMR, 216
    dual channel hypothesis of, 217–8
    roles of dopamine, 218–9
  development and dysfunction of, 219–20
CO_2 receptors, gustation and, 70–1
Coregonus clupeaformis, *see* Whitefish
Cornea, 181
Cortex, 160–2, 165, 169, 216–7, 419
Cortical development from fishes to mammals, 162
Cortisol and endocrine responses, 110
Corythoichtyes paxtoni, *see* Pipefish
CO_2 sensitive chemoreceptors, 110–1, 117–23
  branchial location of, 117–8
Cownose ray (*Rhinoptera bonasus*), 156
Cranial nerve I, 2
(CRE)-binding protein, 278
Crucian carp (*Carassius carassius*), 22, 167, 221
  odorant responses, 21–2
Crypt cells, 3, 4, 14–7
Cryptochrome (*Cry1*, *2*) genes, 280
CSF, *see* Cerebrospinal fluid
Cutaneous buds, *see* Taste buds
Cyanide, 111–2
Cyclicnucleotide-gated (CNG) channel, 255
Cyprinus carpio, *see* Carp

D

DAG, *see* Diacyl glycerol
Damselfish (*Dascyllus* sp.), 183
Danio aequipinnatus, *see* Giant danio
Danio rerio, *see* Zebrafish
Dark-adapted goldfish, 189
Dasyatis sabina, *see* Stingray
Descending nucleus, 165
Desmosomes, 253
DHA, *see* Docosahexaenoïc acid
Diacyl glycerol (DAG), 273
Diameter, of receptive fields, 158
Diencephalon, 163, 244
  substance P, 167
Dissociated trout pineal cells, 255
DLR, *see* Dorsal light reflex
Docosahexaenoïc acid (DHA), 256–7
  in mitochondria, 257
Dogfish (*Scyliorhinus canicula*), 165–7
  cardiac stroke volume in, 100, 102
Dopamine, 197–9, 208
Dopaminergic interplexiform cells (DA-IPC), 197, 200, 205
Dorsal light reflex (DLR), 184–5
Dorsal posterior (DP) nucleus, 314
Dorsal thalamus, 251
Dorsolateral zone (DLZ), 326, 328
Dorsomedial telencephalon, 188
Drosophila Cry gene, 282
Dual channel hypothesis, 220
Dwarf gourami fishes
  GnRH systems in, 457–8
  TN-GnRH neurons of, 458–9, 467

E

Early receptor potential (ERP), 257
Earth's magnetic field
  animals use in navigation, 338, 362–4
  information about location, 339–42, 363–4
Eels
  cardiac conditioning responses in magnetic field, 352–4, 356
Elasmobranch fishes, 156, 165–6, 311, 338, 345, 371
  electroreceptor system, 345

reticular formation, 165
spinal cord of, 165
Elasmobranch species, 156
Electrical foveae, 307
Electrical induction, magnetic field detection by, 344–5
Electrical signals, 312–4
Electric fields, as shark repellents, 156
Electric fovea, 323–4
Electric image, 316–7
Electric organ corollary discharge (EOCD) signals, 328–9
Electric organ discharges (EOD), 307, 310–9, 322–3, 325, 328–9
pulse-type, 312–3
response to, 313
wave-type, 312–3
Electrogenesis, 308
Electroolfactogram (EOG) technique, 4–5
Electrophysiological recordings (ERG), 191–3
Electropinealograms (EPG)
in frogs and snakes, 258
Electroreception, 308–10
Electroreceptive brain areas, 308
Electroretinogram (ERG), 185, 257–8
Electrosensitive fish, 311
Electrosensory flow process, 323
Electrosensory lateral line lobe (ELL), 310, 314, 326–30
Electrosensory system, stages in, 329–30
ELL, *see* Electrosensory lateral line lobe
Eminentia granularis posterior (EGp), 326
End buds, *see* Taste buds
Endocrine responses, chemoreceptors-driven, 110
Endogenous opioids, 154, 168–9
Enkephalins, 168–9
EOD, *see* Electric organ discharges
EOD rhythm, 313
Epaulette shark (*Hemiscyllium ocellatum*)
cardaic responses in, 105
EPG, *see* Electropinealograms
Eptatretus stouti, *see* Pacific hagfish
ER, *see* Extraretinal rhodopsin
ERG, *see* Electrophysiological recordings; Electroretinogram
ERP, *see* Early receptor potential
Escape response assay, 190
Esox lucius, see *Pike*
Excitatory postsynaptic potential (EPSP), 328–9
Extrahypothalamic GnRH (gonadotropin releasing hormone) peptidergic neuron system, *see* GnRH neurons
Extrahypothalamic TN-GnRH neurons, 458
Extraretinal rhodopsin (ER), 254

F

Fascia adherens, 253
Feeding behavior, 75–82
feeding stimulants, 76–7
gustatory, tactile, or olfactory, 75–6
patterns by food extract and chemical stimulation, 79
triggered by amino acids, 77–82
Fenestrated blood vessels, 247–8
FERG, *see* Flicker ERG
FFF, *see* Flicker fusion frequency
Filefish (*Stephanolepsis cirrhifer*), 168
Fish lens, 181
Fish nociceptors, 158
Fish pineal organ
in culture, photoperiodic *versus* circadian control of melatonin release in isolated, 267–9
melatonin factory
LD cycle synchronizes rhythm in melatonin production, 265
within photoreceptor cells, 263–5
Fish pineal organ, light sensor, 253
components of phototransduction cascade
cyclicnucleotide-gated channel (CNG), 255
intracellular messengers, 255–6
lipids, 256–7
opsins, 254–5
transducin and arrestin, 255
electrical responses
from photoreceptor cells, 257–9
release of a neurotransmitter, 261–2
from second-order neurons, 259–61
Fish pineal photoreceptor cells, 254
ultrastructure of, 248
Fish retina, 196
Flicker ERG, 192

Flicker fusion frequency (FFF), 219
FMRF-amide, 207, 251
Four-eyed fish (*Anableps anableps*), 180–1
Fovea, 181
Fungiform gustatory cells, 54

G

GABA, 166–7
Gadus morhua, *see* Atlantic cod
Ganglion cells, 326
Gar (*Lepisosteus*)
  ventilatory responses in, 106
Gene expression, nociception, 169
Ghost fish (*Apteronotus leptorhynchus*), 167–8
GHRH, *see* Growth hormone-releasing hormone
Giant danio (*Danio aequipinnatus*), 183
Ginglymostoma cirratum, *see* Nurse shark
Glial cells, *see* Interstitial cells
Glucocorticoids, 289
mGluR6-receptor, 219
Glutamate, 204, 220
cGMP efflux, 256
cGMP-gated channel, 276
Gnathonemus petersii, 307
GnRH, *see* Gonadotropin releasing hormone
GnRH-immunoreactiveTN (TN-GnRH) system, *see* TN-GnRH neurons/system
GnRH neurons
  autocrine/paracrine control of TN-GnRH neurons pacemaker frequencies by, 466–9, 468
  cellular mechanisms of modulation of pacemaker frequencies of TN-GnRH neurons by, 469–78
    subsequent increase of firing activity, 471–6
    transient decrease of pacemaker activity, 469–70, 471
  electrophysiological and morphological studies of, 458
  infrared differential interference contrast microscopy (IR-DIC) image of, 459
  modulation of neural functions by, 489–91, 494
  modulation of odorant responses by, 490
  molecular and cellular mechanism of, 458
  neuromodulation of Ca^{2+} currents, 485–7
  neuromodulatory action of, 485–7
  nonsynaptic release of, 488–9
  pacemaker frequencies of TN-GnRH neurons
    autocrine/paracrine control, 466–9, 486
    cellular mechanisms of modulation of, 469–78
    modulation of, 464–6
  receptors distribution in brain, 487–8
  release demonstrated by RIA, 462–3
  release from TN-GnRH neurons, 458, 462–3, 488, 493
  TN-GnRH neurons advantages over, 458
Goldfish (*Carassius auratus*), 155, 163, 169, 180, 205, 218
  directional responses, 393, 398
  electrophysiological activity, 164
  TN neurons of, 456
  vocal modulation of inner ear and lateral line, 400
  voltage of electric shock, 171
Goldfish pineal neurons, acetylcholinesterase activity of, 250
Gonadotropin releasing hormone (GnRH), 205, 251
  peptidergic neuron system, 456
G-protein $\beta\gamma$ subunits, 288
G-protein-coupled receptor proteins, 2, 16–7
Granule cells, 23–4
Green sunfish (*Lepomis cyanellus*), 181
Growth hormone-releasing hormone (GHRH), 251
Guppy (*Poecilia reticulata*), 183
Gustation, 85–6, *see also* Gustatory system, structural organization
  amino acid receptors, 71–2
    L-Arg receptor, 71–2
    T1R, 72
    T2R, 72
  definition, 45
  functional properties
    responses to chemical stimuli, 63–73
    responses to mechanical/tactile stimuli, 73–4

gustatory behaviors
aversive behavior, 83–4
feeding behavior, 75–82
transduction sequences for, 72–3
Gustatory amino acids, *see* Amino acids, for gustation
Gustatory behaviors
aversive behavior, 83
CO_2/H^+-driven behavior, 84
feeding behavior, 75–82
feeding stimulants, 76–7
gustatory, tactile, or olfactory, 75–6
triggered by amino acids, 77–82
Gustatory cells
heterogeneity of, within taste buds, 51–2
in invertebrates, 46
sensory epithelial layers in, 54
types, 49
Gustatory cortex, 60
Gustatory receptors, 64, *see also* Amino acids
Gustatory responses
to chemical stimuli, 63
amino acid receptors, 71–2
to amino acids, 63–8
to miscellaneous chemicals, *see* Chemicals, miscellaneous
transduction mechanisms, 72–3
to mechanical/tactile stimuli, 73–4
Gustatory system, structural organization
central gustatory nuclei and pathways
alliance with trigeminal nerve, 61–2
descending projections to facial and vagal lobes, 61
primary gustatory nuclei, 56–8
secondary and tertiary gustatory nuclei, 58–60
telencephalic gustatory areas, 60–1
cranial nerves in, 56
facial nerve in, 56, 58, 61
glossopharyngeal nerve in, 56–7
taste buds
comparison with higher vertebrate taste buds, 53–4
development, 52–3
distribution, 46–7
heterogeneity of gustatory cells within taste bud, 51–2
SCCs and common chemical sense, 55–6
structure, 47–51
vagal nerve in, 56–8, 61
Gymnarchidae, 308, 312
Gymnotiformes, 308–9, 315, 318
kollenorgans receptor organs, 309, 325
mormyromasts electroreceptor organs, 309, 325

H

Habenula, 251
Hagfishes, telencephalic structure in, 160
Halibut pineal opsin 1 (HPO1), 254
Haplochromis burtonii, *see* Cichlid
Hawaiian goatfish (*Parupeneus porphyreus*)
feeding behavior in, 75
Hemiscyllium ocellatum, *see* Epaulette shark
Hepatic glucose levels, 245
Heterodontus francisci, *see* Horn shark
Himantura sp., *see* Longtailed ray
Holosteans, 309
Homing pigeons behavior, to magnetic inclination and intensity, 366, 372
Hoplopagrus guentheri, *see* Barred snapper
Horn shark (*Heterodontus francisci*), 165
Horseradish peroxidase, 164–5
Humans, spinothalamic tract in, 161
6-Hydroxydopamine (6-OHDA), 197–8
Hypercapnia
cardiovascular responses and, 99, 101, 103–4
chemotransduction mechanism, 129–31
effects of, on heart rate, arterial blood pressure, and cardiac output in teleosts, 102
hyperventilation during, 109
and ventilatory responses, chemoreceptors-driven, 106–9
Hyperventilation
during aquatic hypercapnia, 109
during hypoxia, 106
O_2 content and, 112
Hypophysiotropic POA-GnRH system, 457
Hypothalamic monoamine content, 245
Hypoxemia, 112
Hypoxia
cardiovascular responses, chemoreceptors-driven and, 99, 101, 103–4
chemotransduction mechanism, 129–31

Hypoxia (*continued*)
hyperventilation during, 106
ventilatory responses, chemoreceptors-driven and, 106

I

IASP, *see* International Association for the Study of Pain
Ictalurus, 311
Ictalurus punctatus, *see* Channel catfish
IGRF, *see* International Geomagnetic Reference Field
Induced electrical signals
ampullary electroreceptors, 355–6
INL, *see* Inner nuclei layer
Inner nuclei layer (INL), 184–5
International Association for the Study of Pain (IASP), 153–4
International Geomagnetic Reference Field (IGRF), 339
Interstitial cells, 251–2
Intracellular messengers, 255–6

J

Japanese goatfish (*Upeneoides bendsasi*), feeding behavior in, 75
Juvenile lemon sharks, homing behavior of, 370, 372

K

Killifish (*Cyprinodontiformes*)
olfactory organ in, 12
Kollenorgans receptor organs, 309, 325

L

Lampreys, telencephalic structure in, 160
Large fusiform (LF), 326–7
Large ganglion (LG), 326–7
Lateral line periphery of fishes
behavioral capabilities of, 417–9, 442
multimodal guidance, 419
central pathways for information processing, 428–30
descending recurrent pathways, 429–30
directional sensitivity of, 420
morphology, 412–5
natural lateral line stimuli, 415–7
physiology, 419–23, 420–1, 423
responses
to masking running water, 425, 427–8
to moving objects, 424–6
to moving vibrating sphere, 424
to running water, 425, 427–8
Lateral line system, 412–30, 441–2
central physiology, 423, 430–43
periphery, 412–30, 441–2
Lateral olfactory tract (LOT), 27, 31
L-cone opsin mRNA expression, 201–2
LD cycle, *see* Light-dark (LD) cycle
Lepidosiren paradoxa, *see* South American lungfish
Lepisosteus, *see* Gar
Lepomis cyanellus, *see* Green sunfish
Lesions of pallia, 162
Leu-receptor, of gustation, 64–7
Light/dark control of melatonin production, 277
Light/dark (LD) cycle, 195, 244
Lingual nerves, in humans, 62
Lipids, 256–7
Locomotors activity, variations of, 245
Longtailed ray (*Himantura* sp.), 156
Lorenzini ampullae, 311, 338, 345, 364
LOT, *see* Lateral olfactory tract
Luminance detector, 253

M

Macula occludens, 253
Magnetic field
behavioral responses in laboratory, 349–55
in American eels, 353
conditioned responses to spatial changes in magnetic field intensity, 350–3
by elasmobranch fish, 352–3

in Japanese eels, 353
orientation responses to direction, 350
by teleost fish, 351
cardiac conditioning responses in eels, 353–4, 356
detection by electrical induction, 344–5
as experimental stimuli in laboratory, 342–4
intensity conditioned responses to spatial changes in, 350–3
in laboratory
behavioral responses, 349–55
as experimental stimuli, 342–4
neural responses, 355–9
orientation responses to direction, 350
neural responses in laboratory, 355–9
induced electrical signals in ampullary electroreceptors, 355–6
in trigeminal nerve of teleost fish, 356–9
orientation responses to direction of, 350
stimuli, 339–44
Earth's magnetic field, 339–42
as experimental stimuli in laboratory, 342–4
Magnetic navigation mechanisms, 364–6
Magnetic sense in fishes, 338
Magnetoreception, 311, 371
Magnetoreceptors
candidate magnetite-based, 346–9, 371–2
Magnetotopotaxis, 368
MAO, *see* Monoamine oxidase
Mechanoreception, gustation and, 73–4
Medial olfactory tract (MOT), 29, 31–2
Medial zone (MZ), 326–8
Melanosomes, 182
Melatonin, 279
administration, 245
biosynthesis pathways, 264
factory, fish pineal organ
LD cycle synchronizes rhythm in melatonin production, 265
within photoreceptor cells, 263–5
synthesis, 264
2-iodo-Melatonin, 286
Mesencephalic nucleus, 165
Mesencephalic tectum, 165
L-Methionine, 206
Microvillous OSNs, 3–4, 14–7
odorant responses and, 15, 29
Midbrain GnRH system, 457
Midshipman
auditory organ, 383, 391–2
behavioral mechanisms of audition, 378–82, 378
encoding of acoustic signals, 388
steroid hormones and seasonal hearing changes, 400–3, 402
vocal modulation of inner ear and lateral line, 399–400
Midwater fish
behavioral capabilities of lateral line system in, 418
Mitogen-activated protein kinase (MAPK), 282
Mitral cells, 24, 30
Modified photoreceptors, 249
Moleculer markers of nociception
GABA, 166–7
global gene expression, 169
NMDA, 167–8
opioids, endogenous opioids, and enkephalins, 168–9
substance P and preprotachykinins, 167
Monoamine oxidase (MAO), 264
Mormyridae, 308, 312–3, 318
auditory sensitivity, 382, 388
Mormyriformes, 309, 315
Mormyromast electroreceptor organs, 324
Morone americana, *see* White perch
Morphology, lateral line system in fishes, 412–5
MOT, *see* Medial olfactory tract
Müller cells as scaffolds, 184–5
Multidomain (MD) magnetite crystals, 347–8
Myxiniformes, 309

N

Nasal cavity, and olfaction, 6–13
in agnathans, 6
gross morphology of, in teleost fishes, 7–10
Nasal region, 324
Nasal sac ventilation
in round goby, 12
Navigation
definition, 362
experimental approaches to, 366–7
by fish, 362–4, 367–71

Navigation (*continued*)
magnetic navigation mechanisms, 364–6
magnetic sense use in, 362–71
Navigational abilities of fishes, 362–4, 367–71
Nbb zebrafish, 212
NE, *see* Norepinephrine
Neoceratodus forsteri, *see* Australian lungfish
Neumeyer's interpretation, 218
Neural activity in cortical area, 163
Neural and behavioral mechanisms of audition, 377–405
Neural apparatus
nociceptor anatomy, 154–6
nociceptor electrophysiology, 156–60
Neuroanatomy, 359–62, 360
Neuroepithelial cells, of gill epithelium, 125–8
immunoreactivity of, 127
Neurokinin (NK1) receptors, 167
Neuropeptide Y, 166, 251, 290
Neurotransmitter(s), 130, 220
glutamate as, 133–4
release of, in fish pineal organ, 261–2
Nicotine and chemotransduction mechanisms, 131–2
Night blindness mutations, 211
NMDA, *see* *N*-methyl-D-aspartate
N-methyl-D-aspartate (NMDA), 154, 167–8
Nociception
central nervous system pathways and, 154
and International Association for the Study of Pain (IASP), 153–4
moleculer markers of
GABA, 166–7
global gene expression, 169
NMDA, 167–8
opioids, endogenous opioids, and enkephalins, 168–9
substance P and preprotachykinins, 167
substance P in, 167
Nociceptor(s), 154, 156–7
electrophysiology, 156–60
Non-Fourier motion detection, 216
Norepinephrine (NE), 273
Novelty responses to EOD, 314
Noxious chemicals, as shark repellents, 156
Noxious stimuli, 153–4
Nucleic acids, odorant responses and, 29
Nuclei dorsomedialis, 255
Nucleotides, 4–5
Nucleus accumbens neurons, 288
Nucleus tegmento-terminalis (NTT), 478–9
glutamatergic excitatory inputs from, 479–82, 481
Nucleus ventromedialis, 168
Nurse shark (*Ginglymostoma cirratum*), 165

O

Odorant receptors (OR), 19–21
genes, 19
odorant binding onto, 16
Odorous molecules, 4
OKR, *see* Optokinetic response
Olfaction, *see* Olfactory system
Olfactory bulb(s), 22
centrifugal fibers in, 23–4
concentric layers in, 23
granule cells in, 23–4
information flow, 24–5, 29–31
mitral cells in, 22–4
morphological arrangement of, in lampreys, zebrafish, salmonids, and round goby, 5
neural composition, 23–4
ruffed cells in, 23, 24
telencephalon, 160
Olfactory chamber, 12–3
in bony rayed fishes, 13
in round goby, 23
Olfactory epithelial responses, 4–5
Olfactory epithelium, 6, 205
in Acanthopterygii, 12–3
gross morphology of, in teleost fishes, 7–10
Olfactory glomeruli, 14, 23
Olfactory nerve, *see* Cranial nerve I
Olfactory (ON) nerve, 356
Olfactory organ, *see* Nasal cavity; Olfactory epithelium
Olfactory pathway, primary, in teleost fish, 3
Olfactory responses, 45
Olfactory sensory neurons (OSN), 2
cell types
ciliated OSN, 3, 5 14–7
crypt cells, 3, 4, 14–7
microvillous OSNs, 3–4, 14–7
morphological arrangement of, in lampreys, zebrafish, salmonids, and round goby, 5

morphology and central projections, 13–5
odorant receptors, 19–21
odorant responses, 21–2
transduction of olfactory signals, 15–9
Olfactory signals
central processing of, 31–2
transduction of, 17–9
nitric oxide role, 19
Olfactory system, 2–4
nasal cavity in, 6–13
olfactory bulb and, 24–30
olfactory signals, processing of, 31
OSN and, 13–25
in tetrapods, 2
Ommatidia, 184
OMR, *see* Optomotor response
Oncorhynchus gorbuscha, *see* Pink salmon
Oncorhynchus mykiss, *see* Rainbow trout
ONL, *see* Outer nuclei layer
Opioids, 154, 168–9
receptors, 168–9
Opsins, 183, 254–5
mRNA expression, circadian regulation of, 201–4
Optic tectum (OT), 186–7, 205
Optokinetic nystagmus, 188, 215, 219
Optokinetic response (OKR), 189
Optomotor response (OMR), 214
OR, *see* Odorant receptors
Orbitofrontal cortex, gustatory neurons in, 85
Orienting responses to EOD, 314
OR-type genes, 19
O_2-sensitive K^+ channel, detection in zebrafish, 116
OSN, *see* Olfactory sensory neurons
OT, *see* Optic tectum
Outer nuclei layer (ONL), 184–5
Oxygen-sensitive chemoreceptors, 110–7
association of bradycardia and hyperventilation with, 113
NaCN application and, 114
stimulant, *see* Cyanide

P

Pacific hagfish (*Eptatretus stouti*), 164
Pacific sanddab, 102, 107–8
Pallium, 160
afferent nuclei to, 161
efferent projections to, 161
lessions, 162
Parabrachial nucleus (PBN), 60
Parkinson's disease, 219
Parupeneus porphyreus, *see* Hawaiian goatfish
Passive electrolocation, 307, 310–2
PBN, *see* Parabrachial nucleus
PCN, *see* Precommand nucleus
Pertussis toxin, 256
Petromyzon marinus, 157, 165
Petromyzonol sulfate, 6
PGN, *see* Primary gustatory nuclei
Phagocytosis of shed photoreceptor, 252
Phosphatidylcholine, 256
Phosphatidylethanolamine glycerophospholipids pools, 256
Phosphatidylinositol (PI) pool, 252
Phosphodiesterase inhibitors, 256
Photoperiodic *versus* circadian control melatonin production
circadian clock and melatonin output, 280–3
clock or no clock, 279–80
Photopigments, 182–3, 221
spectral tuning of, 224
Photoreceptor cells, 248
types of recordings, 257
Photoreceptor gene, 255
Phototransduction process, 256
pH sensitive chemoreceptors, 110–1, 117–23
Pigeons, *in vivo* microdialysis, 199
Pike (*Esox lucius*), 170
Pike pineal organs, secretion of melatonin content by cultured, 270
Pineal, functional organization
anatomy, 246–8
cell types
intercellular contacts, 253
interstitial cells, 251–2
others, 252
photoreceptor cells, 249–50
second-order neurons, 250–1
Pineal epithelium, 250
Pineal gland of fish, 254
Pineal hormone, 245
Pineal lumen, 252
Pinealocytes, 250
Pinealofugal intervation, 251

Pinealopetal intervation, 251
Pineal organ(s), 252
epithelium and photoreceptor cells from, 247
gross anatomy of, 246
LD variations in AANAT2 mRNA abundance in, 273
of teleost, 245
of trout, 258
Pineal organ output signals, nonphotic regulation of
external factors, 284–6
internal factors
melatonin, 286–7
neuromodulators, 287
neurotransmitters, 287–9
peptides, 290
steroids, 289
Pink salmon (*Oncorhynchus gorbuscha*), 224
Pipefish (*Corythoichtyes paxtoni*), 181
PKC/adenylyl cyclase, 277
Plasma ion levels, 245
Plasticity, developmental, of respiratory control in zebrafish, 128–9
Platyrhinoidis triseriata, see Thornback guitarfish
PLLN, *see* Posterior lateral line nerves
POA-GnRH slices, 462–3
Poecilia reticulata, see Guppy
Polymodal nociceptors, 157–8
Polyunsaturated fatty acids (PUFA), 256
Posterior lateral line nerves (PLLN), 428, 435
Precommand nucleus (PCN), 314
Preoptic nuclei, 251
Preprotachykinins, 167
Pretectal areas, 251
Pretectum in teleosts, 187–8
Primary gustatory nuclei (PGN), 56–8
Prionotus carolinus, see Sea robin
Pro-receptor, of gustation, 64–7
Prostaglandins, 4–5, 30
Protein kinase C (PKC) activity in rats, 198
Protopterus aethiopicus
ventilatory responses in, 106
Protopterus annectens, see also African lungfish
ventilatory responses in, 106
Protopterus dolloi
ventilatory responses in, 106–7
Pterophyllum scalare, see Angelfish
PUFA, *see* Polyunsaturated fatty acids
Puffer toxin, *see* Tetrodotoxin
Purinergic receptors (P2X), 54
Purkinje shift, 193, 197

Q

Quinidine, 127–8
Quinine
aversive behavior and, 83
gustation and, 69–70
Quinpirole, 208

R

Rabbitfish, 279
Rainbow trout (*Oncorhynchus mykiss*), 154–7, 163, 167, 169, 183
carbonic anhydrase with, treatment of, 121
cardaic responses in, 104–5
cardiac stroke volume in, 102
effectiveness of amino acids tested on palatine nerve responses in, 67
electrophysiological activity, 164
quinine hydrochloride in, 69
taste buds in, 47
taurocholic acid (TCA), 68
Rana catesbeiana, 205
Rana pipiens, 205
Receptors, for gustation, *see* Amino acids
Red drum (*Sciaenops ocellatus*), 166
Relay nucleus (RN), 313–4
Retina, 205, 251
histological sections of, 211
photoreceptors in, 258
signal transduction, 253
TN projection, 205–7
Retinal ganglion cells (RGC), 181
Retinal Müller cells, 244
Retinal pigment epithelium (RPE), 252
cells, 182
Retinitis pigmentosa, mammalian models of, 213
RFamide-related peptides, 169
Rhinobatus battilium, see Shovelnose ray
Rhinoptera bonasus, see Cownose ray
Rhodopsin, 183

Rhodopsin/porhyropsin ratio, 224–5
Rhombencephalon, 163
  substance P in, 167
Roach (*Rutilus rutilus*), 183
Rod bipolar cells, depolarization of, 198
Rod-cone dominance
  shifts, 204
  transition, 196–7
Rodopsin mRNA, 201
Round goby (*Neogobius melanostomus*)
  nasal sac ventilation in, 12
  olfactory chamber in, 18
RPE, *see* Retinal pigment epithelium
Ruffed cells, 23
Rutilus rutilus, *see* Roach

S

Saccade, 215
Salmon, 279
Salmonids, 201
Salmo salar, *see* Atlantic salmon
Sauropsids, 250
Saxitoxin (STX), gustation and, 69–70
Scalloped hammerhead sharks, 368
SCC, *see* Solitary chemosensory cells
Schnauzenorgan, 324–6
Sciaenops ocellatus, *see* Red drum
Scyliorhinus canicula, *see* Dogfish
Sea catfish (*Plotosus lineatus*)
  taste buds in, 47
Sea robin (*Prionotus carolinus*), 166
Sea turtles hatching
  magnetic inclination and intensity, 365
Secondary gustatory nuclei (SGN), 58–60
Second-order neurons, 244, 250–1
Sensory nerves, 156
Septum, 205
Serotonin (5-HT), 54
Serranidae, *see* Basses
Sexual steroids, 289
SGN, *see* Secondary gustatory nuclei
Shark repellents, 156
Shovelnose ray (*Rhinobatus battilium*), 156
Siberian sturgeon (*Acipenser baeri*), 221
Signal-to-noise ratio in cones, 182
Siluriformes, 309
Single-domain (SD) magnetite crystals, 346–8
Skipjack tuna, 368
Sleeper goby
  auditory organ, 391–2
  directional responses, 393, 396
  morphophysiology of a saccular afferent of, 395, 397
Slope/amplitude ratio, 322
Solitary chemosensory cells (SCC), 55–6
Somatosensory system, 153
South American Gymnotiformes, 312
South American lungfish (*Lepidosiren paradoxa*)
  pulmonary ventilation in, 122
  ventilatory responses in, 106
Spectral tuning of photopigments, 224
Spectral visual sensitivity
  spectral coding
    amacrine, RGCs, and visual brain areas, 223
    bipolar cells, 222–3
    horizontal cells, 221–2
  tuning, 224–6
Spinothalamic tract in humans, 161
Spiny dogfish, 107–8
Spotted eagle ray (*Aetobatus narinari*), 156
Stephanolepsis cirrhifer, *see* Filefish
Steroids, 4–5
Stickleback (*Gasterosteus aculeatus*), 166
Stimulants, feeding
  Bet and amino acids mixture extract, 76–7
  Gly and Ala, 77
  short-necked clam extract, 76
Stingray (*Dasyatis sabina*), 156
Stress hormones, and endocrine responses, 110
Striate cortex V1, 217
Striatum, 160
Strychnine, gustation and, 69–70
Sturgeon (*Acipenser oxyrhynchus*), 165
STX, *see* Saxitoxin
Subpallium, 160
Substance P, 154, 167
α-subunit of G-protein transducin, 253
Superparamagnetic (SPM) magnetite crystals, 347–8
Surface-feeding fish
  behavioral capabilities of lateral line system in, 418
Swordfish, 368–9
  directional swimming, 363
Synaptic pedicles, 249

T

Tactile stimulation, gustation and, 73–4
Tambaqui (*Colossoma macropomum*), 109
Taste bud(s)
  cells, *see* Gustatory cells
  comparison with higher vertebrate taste buds, 53–4
  development, 52–3
  distribution, 46–7
  epithelial cells in, 53
  heterogeneity of gustatory cells within taste bud, 51–2
  primordia, 53
  SCCs and common chemical sense, 55–6
  structure, 47–51
  in teleosts, distribution of, 47
Taste disks, 54
Taste perception, 46
Tectal neurons, 187
Telencephalic structure
  in hagfishes, 160
  in lampreys, 160
Telencephalon, 160, 166
  medium, 160
  substance P in, 167
Teleost fishes, 159, 165–6, 338, 345
  auditory system of, 399
  central auditory pathways, 383–4
  electroreceptive species, 311
  GnRH systems in, 457
  inner ear of, 383
  pineal organ, 245
  skin, 154
  steroid hormones and seasonal hearing changes, 401
  trigeminal nerve
    neural responses to magnetic field in, 356–9, 357–8
  trigeminal nuclei in, 164–5
  vocal signals of, 378, 379
Teleost retina, 185
  spectral opponency in, 223
Teleosts, absolute visual sensitivity, 188–9, 194–5
  measurement
    behavioral tests, 189–91
    dark adaptation, 193–4
    electrophysiological recordings (ERG), 191–3
Teleosts, chemosensory modulation of visual sensitivity
  olfactory stimulation, 205–7
  retinal projection of TN, 208–9
  terminal nerve (TN), 204–5
  visual sensitivity modulation by TN, 207–8
Teleosts, circadian regulation of visual sensitivity
  modulation of rod and cone sensitivity, 195–6
  modulation of rod–cone dominance
    effects of dopamine, 197–8
    evidence of rod-cone dominance transitions in outer retina, 196–7
  regulation of dopamine and melatonin release
    assays to retinal dopamine determination, 198–9
    light entrains circadian rhythm of dopamine and melatonin release, 200–1
    retinal dopamine and melatonin release follow
      circadian rhythm, 199–200
  regulation of opsin mRNA expression, 201–4
Teleosts, inherited and acquired impairments of visual sensitivity
  acquired impairments, 212–3
  night blindness mutations, 210–2
Teleosts, visual system in, 180–1, 188
  brain areas involved in vision, 186–8
  structure of eye
    arrangement of photoreceptors, 184
    optics, 181–2
    photopigments, 182–3
    significance of stem, progenitor, and precursor cells in adult retina, 184–6
Terminal buds, *see* Taste buds
Terminal nerve (TN), 204–5, 209
  GnRH neuron system, *see* TN-GnRH neurons/system
Terrestrial vertebrates, 156
Tertiary gustatory nuclei (pTGN), 58–60

Tetrodotoxin (TTX)
aversive behavior, 83–4
gustation and, 69–70
TGN, *see* Tertiary gustatory nuclei
Thalamic gustatory nucleus, *see* Tertiary gustatory nuclei
Thalamus, 161
Thornback guitarfish (*Platyrhinoidis triseriata*), 165
TN-GnRH neurons/system
behavioral functions of, 491–2
bilateral electrolytic lesions of, 491–2
characteristics of, 461–2
of dwarf gourami, 458–9
electrophysiological and morphological features of, 457–62
intracellular recording and labeling, 457–62
multimodal sensory inputs to, 478–85
fine structural evidence for synaptic inputs, 484–5
glutamatergic excitatory inputs from NTT, 479–82
inhibitory inputs to, 482–4
neuroanatomical evidence for, 478–9
neuromodulatory function hypothesis, 492–4
pacemaker frequency, 460–1
autocrine/paracrine control by GnRH, 466–9, 486
cell-to-cell electrical interactions significance, 476–7
cellular mechanisms of modulation by GnRH, 469–78
mechanisms of, 463–4
modulation by GnRH, 464–6
physiological significance of modulation, 477–8
subsequent increase of firing activity, 471–6
transient decrease of pacemaker activity, 469–71
pacemaker mechanisms of, 463–4
pharmacological characterization of HVA Ca^{2+} currents in, 472
physiological significance of pacemaker activity and its modulation, 477–8
release of GnRH from, 458, 462–3, 488
retinopetal fibers from, 489–90
spontaneous amperometric spikes from, 468
Toadfish
auditory organ, 391
directional responses, 393, 398
steroid hormones and seasonal hearing changes, 401
vocal-acoustic behaviors and endocrinological state relationship, 381
vocal modulation of inner ear and lateral line, 399
Torus semicircularis (TS), 186–7
Total lipid content, 245
Tpoh, 271
Transducin, 255
T1R genes, gustutory responses and, 72
Trigeminal (TN) nerve, 155, 356
and gustation, 61–2
adaptations of, in vertebrates, 62
of teleost fish
neural responses to magnetic field in, 356–9, 357–8
Trout pineal organs, 255
TRP2 channels, 17
T1R receptors, 52, 72
T2R receptors, 52, 72
TS, *see* Torus semicircularis
Tuberous electroreceptors, 309

U

Unmyelinated C fibers, 154, 156
Upeneoides bendsasi, *see* Japanese goatfish

V

Vagal motor output, 135
Vascular resistance, 103–5
Vasoconstriction, branchial, 105
Ventilatory responses, chemoreceptors-driven, 105
effect of hypercapnia, 106–9
effect of hypoxia, 106
other ventilatory responses, 109
Ventrolateral zone (VLZ), 326
Ventroposterior (VP) nucleus, 314
Vertebrate ancient (VA) opsin, 255

Visual sensitivity, 189
  circadian rhythm of, 195
Visual streaks, 181
Vitamin A in fish, 183
Vitreal dopamine accumulation, changes in, 202
VLZ, *see* Ventrolateral zone
Vole, 205
Vomeronasal system, in tetrapods, 2
V2R-type receptor genes, 20–1

W

Wagon wheel effect, 216
Weak electric signals, *see* Electric organ discharges
Weakly electric fishes
  active electrolocation, 309, 314–30
    brain extract information about objects, 326–9
    electrosensory system, 329–30
    extract information about objects, 322–3
    perceive about environment, 318–21
    specializations of skin regions, 323–6
Weber's law, 344
Whitefish (*Coregonus clupeaformis*), 14
  odorant responses, 24
White perch (*Morone americana*), 201
White sucker (*Catostomus commersoni*), 266

X

Xenomystinae, 309
Xenopus, 200
  exogenous dopamine, 197
  retinal photoreceptors, 262
Xenopus laevis, 205

Y

Yellowfin tuna (*Thunnus albacares*), 103
Yohimbine, 141

Z

zAanat2 expression, 283
Zebrafish (*Danio rerio*), 155, 160–1, 180, 183, 193, 195
  brain, 163
  changes in behavioral threshold in, 207
  L-cone opsin mRNA in LD, circadian expression of, 203
  developmental plasticity of respiratory control in, 128–9
  retinal sensitivity in, 194
  larvae, 212, 216
  models, 213
zfAanat2 expression, 283
Zonula adherens junctions, 253
zPer2, 283

OTHER VOLUMES IN THE FISH PHYSIOLOGY SERIES

VOLUME 1 Excretion, Ionic Regulation, and Metabolism
Edited by W. S. Hoar and D. J. Randall

VOLUME 2 The Endocrine System
Edited by W. S. Hoar and D. J. Randall

VOLUME 3 Reproduction and Growth: Bioluminescence, Pigments, and Poisons
Edited by W. S. Hoar and D. J. Randall

VOLUME 4 The Nervous System, Circulation, and Respiration
Edited by W. S. Hoar and D. J. Randall

VOLUME 5 Sensory Systems and Electric Organs
Edited by W. S. Hoar and D. J. Randall

VOLUME 6 Environmental Relations and Behavior
Edited by W. S. Hoar and D. J. Randall

VOLUME 7 Locomotion
Edited by W. S. Hoar and D. J. Randall

VOLUME 8 Bioenergetics and Growth
Edited by W. S. Hoar, D. J. Randall, and J. R. Brett

VOLUME 9A Reproduction: Endocrine Tissues and Hormones
Edited by W. S. Hoar, D. J. Randall, and E. M. Donaldson

VOLUME 9B Reproduction: Behavior and Fertility Control
Edited by W. S. Hoar, D. J. Randall, and E. M. Donaldson

VOLUME 10A Gills: Anatomy, Gas Transfer, and Acid-Base Regulation
Edited by W. S. Hoar and D. J. Randall

VOLUME 10B Gills: Ion and Water Transfer
Edited by W. S. Hoar and D. J. Randall

VOLUME 11A The Physiology of Developing Fish: Eggs and Larvae
Edited by W. S. Hoar and D. J. Randall

VOLUME 11B The Physiology of Developing Fish: Viviparity and Posthatching Juveniles
Edited by W. S. Hoar and D. J. Randall

VOLUME 12A The Cardiovascular System
Edited by W. S. Hoar, D. J. Randall, and A. P. Farrell

VOLUME 12B The Cardiovascular System
Edited by W. S. Hoar, D. J. Randall, and A. P. Farrell

VOLUME 13 Molecular Endocrinology of Fish
Edited by N. M. Sherwood and C. L. Hew

VOLUME 14 Cellular and Molecular Approaches to Fish Ionic Regulation
Edited by Chris M. Wood and Trevor J. Shuttleworth

VOLUME 15 The Fish Immune System: Organism, Pathogen, and Environment
Edited by George Iwama and Teruyuki Nakanishi

VOLUME 16 Deep Sea Fishes
Edited by D. J. Randall and A. P. Farrell

VOLUME 17 Fish Respiration
Edited by Steve F. Perry and Bruce Tufts

VOLUME 18 Muscle Growth and Development
Edited by Ian A. Johnson

VOLUME 19 Tuna: Physiology, Ecology, and Evolution
Edited by Barbara A. Block and E. Donald Stevens

VOLUME 20 Nitrogen Excretion
Edited by Patricia A. Wright and Paul M. Anderson

VOLUME 21 The Physiology of Tropical Fishes
Edited by Adalberto L. Val, Vera Maria F. De Almeida-Val, and David J. Randall

VOLUME 22 The Physiology of Polar Fishes
Edited by Anthony P. Farrell and John F. Steffensen

VOLUME 23 Fish Biomechanics
Edited by Robert E. Shadwick and George V. Lauder

VOLUME 24 Behaviour and Physiology of Fish
Edited by Katherine A. Sloman, Rod W. Wilson, and Sigal Balshine

VOLUME 25 Sensory Systems Neuroscience
Edited by Toshiaki J. Hara and Barbara S. Zielinski